Advances in Fungal Biotechnology

Advances in Fungal Biotechnology

Editor

MAHENDRA RAI

Prof. & Head
Dept. of Biotechnology
S.G.B. Amravati University
Maharashtra, India

I.K. International Publishing House Pvt. Ltd.

NEW DELHI • BANGALORE

Published by

I.K. International Publishing House Pvt. Ltd.
S-25, Green Park Extension, Uphaar Cinema Market
New Delhi – 110 016 (India)
E-mail: info@ikinternational.com
website: www.ikbooks.com

ISBN 978-81-89866-53-2

Reprint 2013

Published by Krishan Makhijani for I.K. International Publishing House Pvt. Ltd., S-25, Green Park Extension, Uphaar Cinema Market, New Delhi – 110 016 and Printed by Rekha Printers Pvt. Ltd., Okhla Industrial Area, Phase II, New Delhi – 110 020.

Preface

There has been a paradigm shift in Mycology from traditional to modern. The mushrooms, which were once treated as poisons are now utilized as panacea. These fungi produce industrially useful enzymes. The fungi are used as biocontrol agents and beneficial endophytes. Their potential as weedicides is gaining attention. There are some mycoviruses which are being used for the control of diseases caused by fungi. On the otherhand, some phytopathogens are now behaving as human pathogens due to immunocompromising capacity of the hosts. Fungi present in indoor environment are also causing several fatal diseases.

There are potential applications of fungi in the field of enzyme production, for example, xylanases produced by fungi are commercially very useful.

Some fungi are phyto-friendly representing microscopic forms (arbuscular mycorrhizal fungi) to macroscopic (ectomycorrhizal fungi). They are used in sustainable plant production. Such fungi show multifunctionality as biofertilizers, bioprotectors and bioremediators.

The fungi produce a wide variety of secondary metabolites and represent a colourful world of useful pigments and antibiotics. These pigments are used as potential dyes and also in pharmaceuticals.

In the past, the concept of identification of fungi was based on the type of host on which they occur, and morphological and cultural characters of fungi. But the method involved has been time-consuming and tedious. Now, molecular markers are used for the identification and differentiation of fungi. These methods are rapid, user-friendly and more authentic.

The information technology has played a pivotal role in fungal biotechnology. Internet has become an important part of mycology. A separate chapter on this aspect has been included in this book.

The present book gives an idea of how the traditional mycology has shifted to modern mycology. In the past two decades, fungal biotechnology has progressed with fast pace. Therefore, the recent advances in the field has been incorporated in the book. This book would be immensely useful for mycologists, biotechnologists, microbiologists and botanists.

I wish to thank all the contributors for submission of valuable manuscripts. I would like to thank Mr. Aniket Gade for help in edition. I wish to thank my students: Alka Karwa, Ravindra Ade, Avinash Ingle, Dnyaneshwar Rathod and Chandan Vora, who helped in editing of manuscripts.

Finally, I would like to thank Shivangi, Shivani and Aditya for their unconditional love, support and patience during editing of the book.

Editor

List of Contributors

Waldemar Zangaro, Universidade Estadual de Londrina - UEL. CCB/Dept.de Biologia Animal e Vegetal. UEL. CCB/Dept.de Microbiologia, Laboratório de Ecologia Microbiana. Caixa Postal 6001 86051-990, Londrina PR Brazil

Marco A. Nogueira, UEL. CCB/Dept.de Microbiologia, Laboratório de Ecologia Microbiana. Caixa Postal 6001 86051-990, Londrina PR Brazil

Galdino Andrade, UEL. CCB/Dept.de Microbiologia, Laboratório de Ecologia Microbiana. Caixa Postal 6001 86051-990, Londrina PR Brazil; E-mail: andradeg@uel.br

K.L. Bailey, Agriculture and Agri-Food Canada, 107 Science Place, Saskatoon, Saskatchewan, Canada, S7N 0X2. Tel: (306) 956-7200, Fax: (306) 956-7247, Email: BaileyK@agr.gc.ca

S.M. Boyetchko, Agriculture and Agri-Food Canada, 107 Science Place, Saskatoon, Saskatchewan, Canada, S7N 0X2. Tel: (306) 956-7200, Fax: (306) 956-7247

G. Peng, Agriculture and Agri-Food Canada, 107 Science Place, Saskatoon, Saskatchewan, Canada, S7N 0X2. Tel: (306) 956-7200, Fax: (306) 956-7247

R.K. Hynes, Agriculture and Agri-Food Canada, 107 Science Place, Saskatoon, Saskatchewan, Canada, S7N 0X2. Tel: (306) 956-7200, Fax: (306) 956-7247

W.G. Taylor, Agriculture and Agri-Food Canada, 107 Science Place, Saskatoon, Saskatchewan, Canada, S7N 0X2. Tel: (306) 956-7200, Fax: (306) 956-7247

W.M. Pitt, Agriculture and Agri-Food Canada, 107 Science Place, Saskatoon, Saskatchewan, Canada, S7N 0X2. Tel: (306) 956-7200, Fax: (306) 956-7247

Petr Baldrian, Laboratory of Biochemistry of Wood-Rotting Fungi, Institute of Microbiology ASCR, Vídeòská 1083, 14220 Praha 4, Czech Republic,. Tel.: +420241062315, Fax: +420241062384, E-mail: baldrian@biomed.cas.cz

Carla C.C.R. de Carvalho, Centro de Engenharia Biológica e Química, Instituto Superior Técnico, Av. Rovisco Pais, 1049-001 Lisboa, Portugal, Centro de Engenharia Biológica e Química, Instituto Superior Técnico, Tel: +351-21-8419594; Fax: +351-21-8419062; E-mail: ccarvalho@ist.utl.pt

Deepak K. Rahi, Department of Microbiology, Panjab University, Chandigarh, India deepakrahi10@rediffmail.com / dkrio@pu.ac.in

Sonu Rahi, Department of Botany and Microbiology, M. H. College of Home, Science and Science for Women, Jabalpur, M.P., India

Akhilesh Pandey, Department of Bioscience, R.D. University, Jabalpur, M.P., India

Ramcharan Rajak, Seedling Acadamy of Science, Design technology & Management, Jaipur, Rajasthan

Ebere C. Anyanwu, Cahers Neurosciences Research, 41 MCH, 190 College Road, Birmingham B8 3TG, UK; E-mail: ebereanyanwu@msn.com

Ijeoma Kanu, Department of Microbiology, Abia State University, Uturu, Nigeria; E-mail: Ijay2001kay@yahoo.com

Hanna Dahm, Department of Microbiology, Institute of General and Molecular Biology, Nicolaus Copernicus University, Gagarina 9, 87-100 Toruñ, Poland; Phone: +48 56 611 4384; Fax: +48 56 611 4772; E-mail dahm@biol.uni.torun.pl

Agnieszka Ciesielska, Department of Microbiology, Institute of General and Molecular Biology, Nicolaus Copernicus University, Gagarina 9, 87-100 Toruñ, Poland; Phone: +48 56 611 4384; Fax: +48 56 611 4772; E-mail dahm@biol.uni.torun.pl

László Radócz, Department of Plant Protection, University of Debrecen, 138. Böszörményi str. H-403, Telephone: + 36 52 508659, Fax: +36 52 413385, E-mail radocz@agr.unideb.hu

Jixun Zhan, Southwest Center for Natural Products Research and Commercialization, Office of Arid Lands Studies, College of Agriculture and Life Sciences, The University of Arizona, 250 East Valencia Road, Tucson, Arizona 85706-6800, U.S.A, Tel: 1-520-741-1691. Fax: 1-520-741-0113;

A. A. Leslie Gunatilaka, Southwest Center for Natural Products Research and Commercialization, Office of Arid Lands Studies, College of Agriculture and Life Sciences, The University of Arizona, 250 East Valencia Road, Tucson, Arizona 85706-6800, U.S.A, Tel: 1-520-741-1691. Fax: 1-520-741-0113; E-mail: leslieg@ag.arizona.edu.

Marta M. Rubio-Texeira, Whitehead Institute, Massachusetts Institute of Technology, 9 Cambridge Center -653, Cambridge, MA 02142, Tel: 00-1-617 258-5190; Fax: 00-1-617-258-7226; Email: rubio@wi.mit.edu

Miguel Marin, Department of Physiology, Genetics and Microbiology, University of Alicante, Carretera San Vicente del Raspeig s/n, E-03690 San Vicente del Raspeig (Alicante), Spain; Telephone: +34 965 90 94 94; Fax: +34 965 90 95 69; E-mail: mmarinz@ua.es

Neetu Dahiya, Department of Biotechnology, Panjab University, Chandigarh, PIN-160014, India

Nelson Durán, Núcleo de Ciências Ambientais, Universidade de Mogi das Cruzes-UMC, Mogi Das Cruzes, S.P., Brazil

Roseli De Conti, Instituto de Química, Universidade Estadual de Campinas-UNICAMP, C.P. 6154, Campinas, CEP 13083-970, S.P., Brazil

Maria Francisca S. Teixeira, Mycology Department, Faculty of Science, Universidade Federal de Amazonias-UFAM, Manaus, Brazil

Maria Papagianni, Department of Hygiene and Technology of Food of Animal Origin, School of Veterinary Medicine, Aristotle University of Thessaloniki, Greece; Tel: +30-2310-999804. Fax: +30-2310-999829; E-mail: mp2000@vet.auth.gr

Maria de Lourdes T. M. Polizeli, Departamento de Biologia, Faculdade de Filosofia, Ciências e Letras de Ribeirão, Preto – Universidade de São Paulo, Av. Bandeirantes, 3900, Bairro Monte Alegre, 14040-901 Ribeirão Preto – SP, Brazi; Telephone + 55 16 36024680; Fax +55 16-36331758 E -mail: polizeli@ffclrp.usp.br

Sonali Wankhede, Department of Biotechnology, SGB Amravati University, Amravati-444 602, Maharashtra, India

Mahendra Rai, Department of Biotechnology, SGB Amravati University, Amravati-444 602, Maharashtra, India; E-mail: mkrai123@rediffmail.com

A. Arif Khan, Department of Microbiology, College of Life Sciences, Cancer Hospital and Research Institute, Gwalior (M.P.) 474009, India

Preeti Bhatnagar, Department of Microbiology, College of Life Sciences, Cancer Hospital and Research Institute, Gwalior (M.P.) 474009, India

Sudhir K. Jain, Department of Microbiology, College of Life Sciences, Cancer Hospital and Research Institute, Gwalior (M.P.) 474009, India

S.K. Deshmukh, Department of Natural Products, Nichoals Piramal Research Centre, Goregaon, Mumbai (Maharashtra) India

Tara Dubey, Manager, Microbiology Laboratory, Forensic Analytical, Hayward, CA 94544, USA, E:mail: tdubey@hotmail.com and tdubey@aerotechP&k.com

Thomas J. Volk, Dept of Biology, 3024 Cowley Hall, University of Wisconsin-La Crosse, La Crosse WI 54601, E-mail:volk.thom@uwlax.edu, web page [http://TomVolkFungi.net]

Contents

1

Developing Weed Control Technologies with Fungi

K.L. Bailey, S.M. Boyetchko, G. Peng, R.K. Hynes, W.G. Taylor and W.M. Pitt

Agriculture and Agri-Food Canada, 107 Science Place, Saskatoon, Saskatchewan, Canada, S7N 0X2. Tel: (306) 956-7200, Fax: (306) 956-7247; E-mail: BaileyK@agr.gc.ca
Saskatoon Research Centre Contribution Number 1709.

INTRODUCTION

Weeds are the primary pest in agriculture, forestry, and urban environments throughout the world. The annual loss in world food production due to weeds is estimated at 10–15% (Riches, 2001). For centuries, people have been relatively dependent on the chemicals for weed control. Salts, oils, ashes, and smelter sludge were used for the control of weeds until the late 19th century, when Bordeaux mixture was found to have both fungicidal and herbicidal properties and this led to the use of copper sulphate and arsenic acids as herbicides (CropLife Canada, 2002a). Weed control was revolutionized in the 1940s with the development of the first synthetic herbicides known as 2, 4-D and MCPA (CropLife Canada, 2002a). The advent of inexpensive, broad-spectrum herbicides that were less toxic than the prior alternatives improved both crop yield and quality, thereby reducing the need for mechanical tillage and hand weeding (Delaplane, 2000).

In today's global market, herbicides account for 37% of all pesticide active ingredients with a value over $14 billion USD in 2001 (Fig. 1.1, Kiely *et al.*, 2004). Proportionately, the United States has the largest share of the world's herbicide market ($6.4 billion), with about 78% used in agriculture, 12% by industry, commercial or government agencies, and 10% by the home and garden sector. However, the trends from 1982–2001 showed an increase in the amount of active ingredient applied by the home and garden sector and a decline by the agricultural, industrial, commercial, and government sectors (Fig. 1.2). While Canada has a smaller market share valued at approximately $1.0 billion USD per year. Similar patterns were observed until 2001 when herbicide sales for non-agricultural uses in the forestry, golf, and turf and landscape markets declined by 11.6% (CropLife Canada, 2001, 2002b, 2003). This reduction followed the release of a government policy paper that examined the use of pesticides in Canada, their

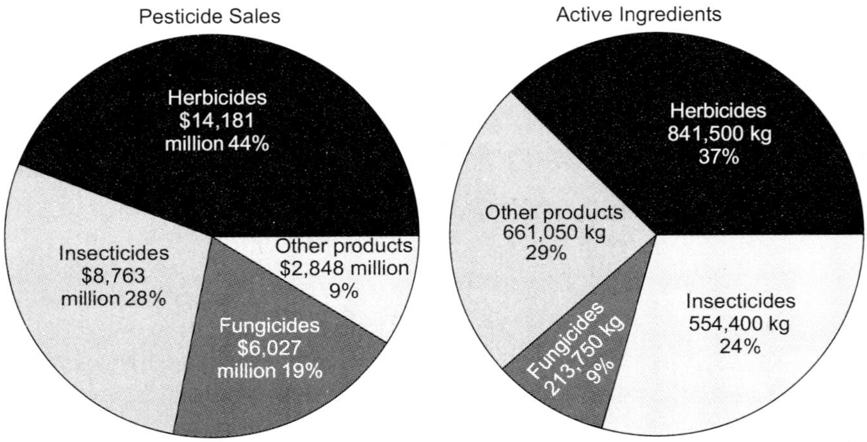

Figure 1.1 World pesticide expenditures ($USD) at the user level (left) and the world amount of active ingredient used by pesticide type (right) in 2001. Other products include nematicides, fumigants, rodenticides, and other miscellaneous product types. Data sourced from Kiely *et al.* (2004).

impact on health and the environment, and the economic implications of their use and regulation (Caccia, 2000). The report stimulated legislative amendments to Canada's *Pest Control Products Acts* (Anonymous, 2002), which strengthened human health and environmental protection considerations by supporting greater pesticide risk reduction and expediting the registration of new lower-risk products to be used as part of integrated pest management (IPM) programmes. Biopesticides, which use naturally occurring bacteria, viruses, or fungi as the active ingredient, are recognized as one of the new reduced-risk products stimulating resurgence in research to develop and commercialize such products.

BIOHERBICIDE PROTOTYPES

Bioherbicides are defined as plant pathogens which when applied inundatively in a fashion similar to synthetic herbicides, reduce the competitive ability of the target weeds below a desired threshold (Crump *et al.*, 1999). The earliest ideas for bioherbicides were conceived in the late 1970s when researchers from the United States described a new root and stem rot of milkweed vine caused by *Phytophthora palmivora* Butler (Burnett *et al.*, 1973) and endemic anthracnose of northern jointvetch caused by *Colletotrichum gloeosporioides* (Penz) Sacc. f.sp. *aeschynomene* (Daniel *et al.*, 1973). Research resulted in the registration and subsequent commercialization of these two pathogens as DeVine® in 1980 and Collego® in 1981, respectively (Bowers, 1986; Kenney, 1986; Templeton *et al.*, 1989; Charudattan, 2000, 2005). A fortuitous observation of seedling blight caused by *Colletotrichum gloeosporioides* f.sp. *malvae* (Cgm) on round-leaved mallow (Mortensen, 1988; Makowski and Mortensen, 1992) then resulted in the registration of a third bioherbicide BioMal® in 1992 (Mortensen, 1998; Charudattan, 2000; Mortensen and Bailey, 2002; Boyetchko *et al.*, 2006).

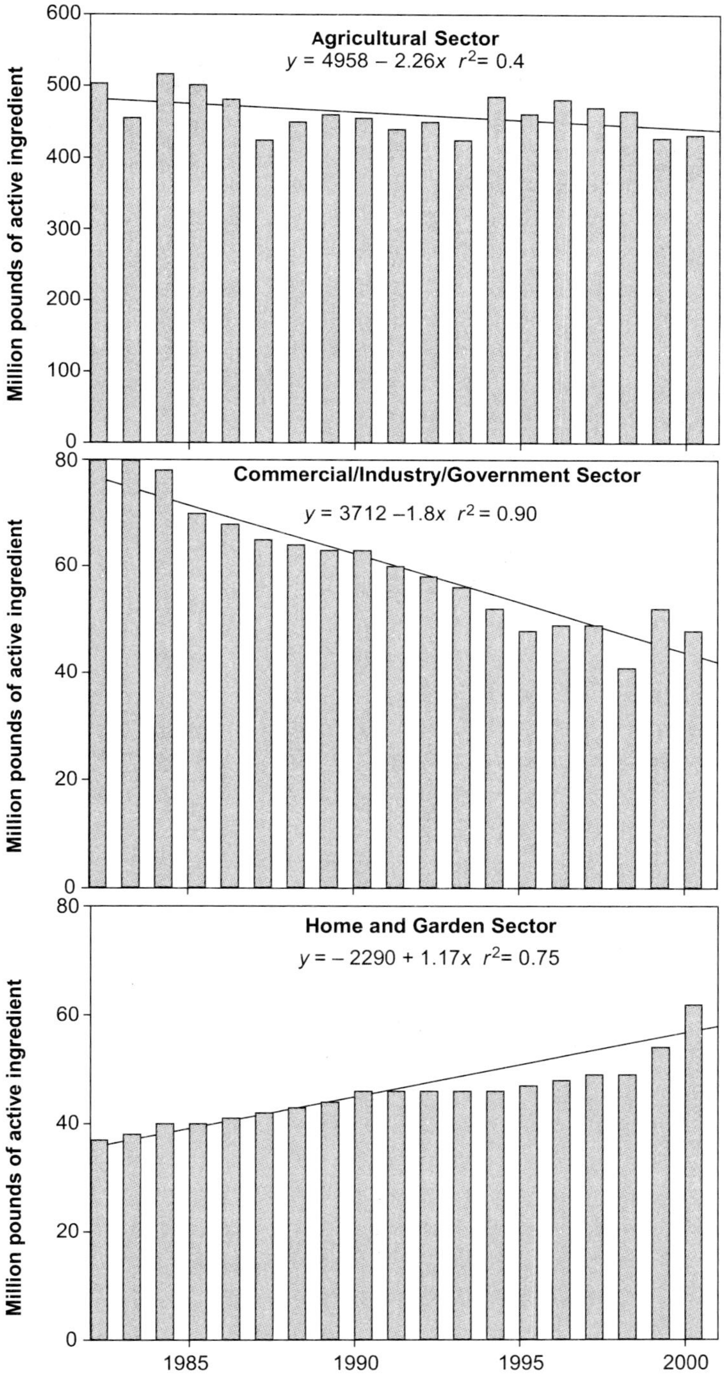

Figure 1.2 Annual amount of herbicide active ingredient used in the United States from 1982-2001 by market sector. Data sourced from Kiely *et al*. (2004).

Devine®, Collego®, and BioMal® may be considered prototype bioherbicides in North America and were developed using the same strategic approach as chemical herbicides, albeit with a few exceptions unique to living organisms; namely the bioproducts must: (i) comprise a virulent strain to give satisfactory and consistent field performance, (ii) possess a rapid mode of action, (iii) facilitate the application of large quantities of propagules with dose related activity, (iv) produce infective propagules by submerged fermentation that can be processed and stabilized efficiently, (v) be rapidly dispersed in water-based formulations for easy application, (vi) facilitate post-emergent application to foliage via spray technology with the ease of chemical equivalents, (vii) constraints of the system must be resolved by the tripartite interaction of the disease triangle, and (viii) products must adhere to the concept of narrow host-range specificity (i.e., one pathogen strain: one weed controlled).

The success of these prototype bioherbicides created optimism among researchers that more products would soon follow. Unfortunately, there has been less than ten bioherbicides registered around the globe over the last two decades, and only five were considered to be "verifiable successes" (Charudattan, 2000, 2001, 2005; Boyetchko and Peng, 2004; Boyetchko, 2005). While naïve from its inception, researchers had neither adequately considered the constraints of a chemical paradigm on a biological system (Meneley, 2000), nor they had fully conceived the biological, environmental, technological, commercial, and regulatory complexities that would emerge during the development (Cross and Polonenko, 1996). These constraints and challenges have now been more adequately addressed, and so we remain optimistic (Auld and Morin, 1995; Mortensen, 1998; Charudattan, 2000; Boyetchko and Peng, 2004; Hallett, 2005; Bailey and Mupondwa, 2006).

Critical reflections on the experiences with the prototypes suggested different approaches that should be taken for evaluating the scientific and the commercial potential of candidate microorganisms for biological control (Templeton *et al.*, 1989; Zorner *et al.*, 1993; Cross and Polonenko, 1996). All recognized one important factor—when commercialization is the ultimate goal, then the science must consider factors deemed important to the industry. Learning from past experiences and our research team's collective experiences in developing BioMal® and other forthcoming bioherbicides, we have developed a strategic framework to resolve some of the challenges with this biotechnology. This article focuses on our strategy for bioherbicide development and how we integrate multiple disciplines to advance the scientific and the technological innovations towards commercial application (Fig. 1.3), while also addressing the considerations of industry and regulatory agencies that ultimately shape the final product.

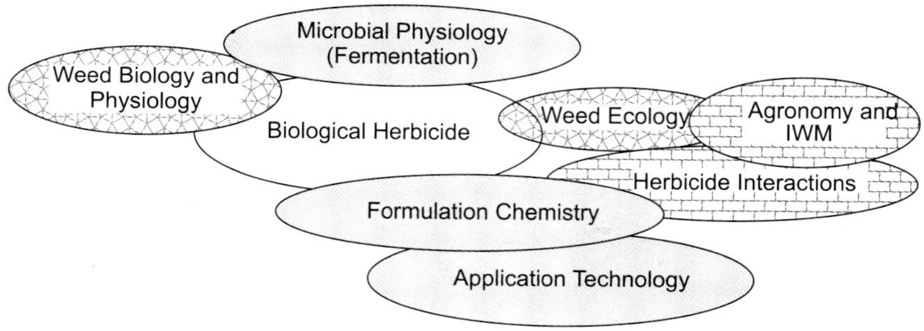

Figure 1.3 The integration of scientific disciplines and technology platforms required for bioherbicide development.

Advancing the Science of Biological Control in the 21st Century

Our Philosophy

A successful bioherbicide was defined by Charudattan (2005) as the one that has been registered and subsequently used with some regularity. In order to achieve this level of success, there must be a champion who leads the research and development through a convoluted, intertwining pathway of science, engineering, and regulatory demands, while balancing key marketplace and commercial considerations. Success follows a lengthy process of making small incremental steps that are unique for each weed-pathogen system. We call this pathway the "innovation chain" (Fig. 1.4). In the early stages, the research skips back and forth between the links connecting "Discovery & Selection–Proof of Concept–Technology Development" using a series of "Go vs. No-Go" criteria to force decisions on the feasibility of the organism and the target. Basic assessments of biology, environment, biochemistry, and lab scale fermentation and formulation are confirmed under laboratory, greenhouse and field conditions, with the emphasis on characterization, safety, and practicality (Boyetchko and Rosskopf, 2006). Regulatory and market considerations form the backbone of the chain from which the scientific studies branch (Bailey and Mupondwa, 2006). This ultimately fosters development of novel technology platforms that can be expanded to other potential applications, whether this is for multiple patterns of use, application methods, broadening weed targets, new production systems, or integration into other biological product chains for use with different biopesticide organisms (Hynes and Boyetchko, 2006).

The later stages of the chain test the robustness of earlier decisions by actively working with committed stakeholders (industry partner, other collaborators) and regulators to develop the data required for registration. Commercial scale-up can still challenge the eventual success, often requiring significant investment back into technology development to move from the bench to pilot scale and commercial manufacturing. Although commercial scale-up, registration and technology adoption are usually directed by the industry partner, the science champion

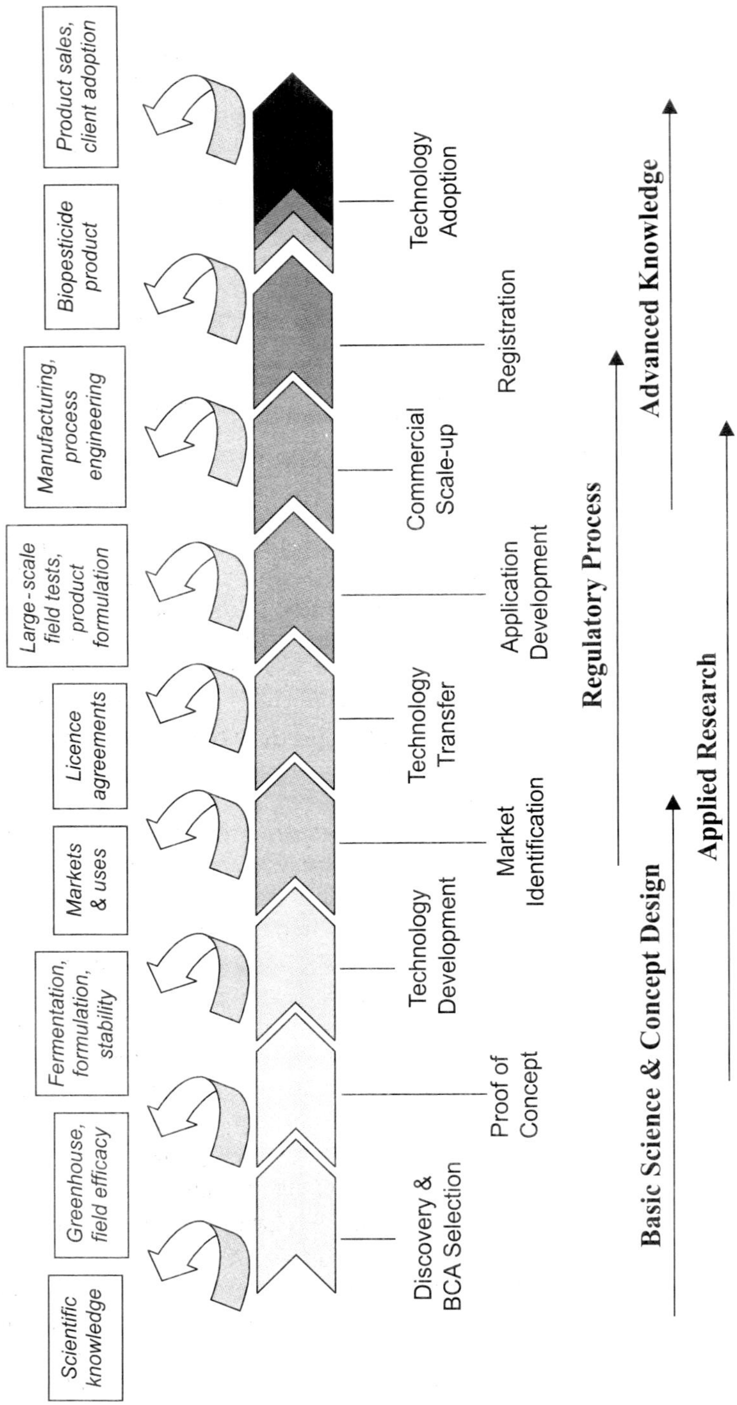

Figure 1.4 Innovation chain for bioherbicide product development.

should not abandon the project until technology adoption has taken place. Besides the direct scientific contributions, the scientist's corporate memory or "know-how", "show-how", and intuition should not be discounted by the commercial partner as this can ease the transition phase and fast track the development process. The successful development of a biopesticide is a combination of science, art, and serendipity taking place over a period of 10–15 years.

Microbial Exploration and Discovery

The discovery of the right agent for the right host is the key for successful bioherbicide development. However, deciding which comes first, the selection of the host (chicken) or the agent (egg) remains problematic. Boyetchko (2005) has postulated that the wrong weeds were targeted in some bioherbicide projects as a result of the accidental discovery of a disease-causing agent. For example, BioMal® was the result of a fortuitous discovery on round-leaved mallow during a greenhouse-screening programme aimed at developing synthetic herbicides. However, despite being a good pathogen, market forces prevented full commercialization and adoption due to limited market size (Cross and Polonenko, 1996; Boyetchko *et al.*, 2006). Hallett (2005) suggested that more amenable targets for bioherbicides would be found with "niche systems" such as parasitic weeds, urban or allergenic weeds, and narcotic plants. Charudattan (2005) argued that the target selection is dependent on either stakeholder demand or need-driven targets. While the latter is more closely aligned with our philosophy, survey work should be focused on predetermined weed targets with collections being done at different weed growth stages in a variety of environments. Sampling from plants should include roots, stems, leaves, and seed to obtain both pathogens and endophytes, while the biology, physiology and genetics of the weed species should be considered when selecting target species (Boyetchko *et al.*, 2002).

After defining the target weed, hundreds to thousands of fungi collected from a broad range of environment (both spatially and temporally) must be evaluated to narrow the selection down to those with desirable traits. Pharmaceutical and other chemical companies often screen 2,000 – 20,000 isolates in order to obtain the one strain that meets their criteria for efficacy, safety, novelty, chemical dereplication, mass production, and cost-effectiveness (Campbell, 1994). Discovery of new synthetic herbicides usually employs a cookbook mentality comprised of exacting recipes (bioassays and procedures) and is conducted with a systematic approach because the market opportunities are well understood. However, the process for biopesticides is relatively immature, and because the companies are usually smaller and do not have the financial resources, much of the discovery work continues to be done by government-supported researchers with programs seeking novel strains and bioactive molecules (Meneley, 2000).

With the prototype models, the criteria for selection initially focused on fulfilling Koch's postulates for pathogenicity, early taxonomic identification, artificially culturing the fungus, and creating maximum disease severity under optimal conditions. Renwick and Powell (1990) suggested that a better approach would be to develop early screening procedures with a broader range of criteria encompassing pathogen-host-environment interactions. As bioherbicide discoveries matured, early screening strategies encompassed *in vitro* and *in vivo* assays with different hosts, environmental conditions and multiple microbial strains. Bailey and Mupondwa (2006) described the standard primary-level screening used for fungi at Agriculture & Agri-Food Canada in Saskatoon. The major steps include confirmation of Koch's postulates, assessing fungal growth characteristics on different cultural media to predict mass production potential, determining preferred sites of infection on hosts (i.e., foliar, stem, root, seed), and appraising efficacy and consistency of response under variable environmental conditions. A modification of this protocol that involves early screening against multiple weed targets and a limited number of important crop species in an attempt to broaden the fields of use was recently employed by Peng *et al.* (2004) to select pathogens for control of green foxtail. These researchers tested 133 fungal isolates for Koch's postulates to ensure some degree of pathogenicity to green foxtail, screened isolates on several common agar media (potato dextrose at full- and half-strength, V-8, minimal, and oatmeal) to assess potential for mass production of spores, inoculated green foxtail plants at different growth stages to compare efficacy, and then screened all of those isolates causing greater than 50% disease on green foxtail against 27 other plant species from 5 families to determine host range specificity and crop tolerance. Only one candidate, *Pyricularia setariae* Nisikado emerged as the prime candidate causing severe damage to the target weed without adverse effects to potential crop species.

Often the final step taken in the discovery stage is to assign a taxonomic designation to the fungal candidate. Early designations often support identification only to the genus level and greater expertise is required to elucidate the actual species, which may or may not have been described previously. Preliminary identification alerts researchers to the general traits that the fungus may possess and often requires considerable time and expense depending upon the number of isolates involved. In reality, the name and hence the nomenclature associated with the fungus is less important than the biology of the organism, although, accurate taxonomic identification to the species level is required for registration, and increasingly involves molecular characterization in addition to traditional cultural and morphological studies (Samson 1995).

Advanced Biological Characterization and Development

As mentioned in our philosophy, regulatory considerations form the backbone for the scientific studies. Hence, the advanced characterization is guided by criteria outlined in the "Guidelines for the Registration of Microbial Pest Control Agents and Products" as documented by the Pest Management Regulatory Agency (PMRA 2001). Since the Canadian data requirements are generally harmonized with those prepared by the Environmental Protection Agency (EPA) in the United

States, considerable time and money can be saved by planning the studies to meet the data requirements for registration.

Developing a product profile and use patterns helps to focus the types of characterization studies needed to supplement what is currently available in the literature. The profile defines the vision of the final product form and enables researchers to work towards achieving it by identifying gaps in basic knowledge on the host and pathogen that must be fulfilled prior to conducting characterization studies. Key regulatory elements that influence decisions to develop a particular fungus as a bioherbicide include: (i) origin, derivation, and identification (complete and accurate taxonomic identification to the species level), (ii) biological properties of the fungus (such as natural distribution, pathogenicity, toxic dose, transmission, mode of action, pest host range and closely related species, life cycle and various forms or states, physiological growth parameters, unusual traits, history of use and relationships to other similar organisms, and adverse effects to humans and environment), (iii) environmental fate (i.e., dispersal, persistence, survival, carry over effects), (iv) environmental toxicity to non-target species, and (v) potential risk to human health either by the degree of the hazard or level of exposure (PMRA, 2001). Prioritizing various studies not only depends on the fungus selected and the current state of knowledge about the pathogen, but also depends on quickly identifying the conditions needed to acquire evidence for proving the concept under field conditions. To warrant further investigation, the pathogen must elicit a predetermined level of disease on the host plant following artificial inoculation in the field.

Biological factors that affect efficacy and influence the consistency of the response need to be studied in detail prior to conducting field trials to confirm the "proof-of-concept". Some of the important studies include: (i) defining environmental parameters under constant and intermittent conditions to identify the circumstances that could severely limit weed control, (ii) exploring within-species biodiversity in the pathogen to select the most virulent isolate with the widest tolerance to the environment in which it will be applied, and (iii) assessing pathogenicity, dose and host range (multiple host targets and crop tolerance) to address patterns of use and environmental safety.

Moisture and temperature are the key parameters influencing plant disease development and can be particularly limiting for foliar applied bioherbicides in Western Canada which has a semi-arid temperate environment. *Pyricularia setariae* and *Drechslera gigantea* (Heald & F.A. Wolf) Ito were examined for their potential to control green foxtail (Green *et al.*, 2004). Disease severity increased with longer periods of leaf wetness and increasing temperatures from 15–32°C. Germination was minimal at 10°C but more than 80% at 15°C. Using a thermogradient plate to study variable dew temperature, Peng and Boyetchko (2006) found that *D. gigantea* caused more disease in a shorter period of time and under cooler dew temperatures (15/10°C and 20/5°C day/night regimes) than *P. setariae*. However, both fungi required immediate and prolonged wet periods in order to cause disease and it was concluded that these requirements would severely limit their effectiveness in the Canadian prairies.

Similarly, Zhang *et al.* (2002a) assessed the potential of *Plectosporium tabacinum* (van Beyma) M.E. Palm, W. Gams et Nirenberg as a bioherbicide for false cleavers. A single isolate was screened for pathogenicity to herbicide-resistant and herbicide-susceptible false cleavers in addition to a closely related crop species, canola, where this weed is particularly problematic. Host range testing was expanded using the modified centrifugal phylogenetic and varietal strategy of Wapshere (1974) such that the susceptibility of 34 plant species in 26 genera and 12 families was investigated. However, following a second study on the effect of inoculum concentration, weed growth stage and dew period temperature and duration, the requirement for a 16 hours dew period and temperature minimum of 15°C severely limited the success of *P. tabacinum* for control of false cleavers (Zhang *et al.*, 2002b).

In another study, Peng *et al.* (2005) compared 14 isolates of *Colletotrichum truncatum* (Schwein) Andrus & W.D. Moore from European and Canadian origins for virulence to scentless chamomile, crop safety to 14 plant species in 5 families, and the influence of leaf wetness duration. In Canada, scentless chamomile is an introduced weed from Europe that has attained the status of a noxious weed with legislation requiring landowners to remove it from their properties. Peng *et al.* (2005) found that isolate 00-003B1 originating from Canada had similar virulence to the European isolates, which came from the weed's centre of origin. Isolate 00-003B1 was five times more virulent than the least virulent Canadian isolate 00-215A. One isolate (00-193C) had greater virulence, but dew period studies showed that the high level of virulence could only be achieved with 24 h of dew. However, isolate 00-003B1 had greater virulence with fewer hours of dew, and thus, was the best indigenous strain for developmental work in Canada. A differential reaction among isolates to the interaction of disease severity and dew period was also observed among European isolates.

Deciding on the appropriate time to prove that the fungal candidate will successfully initiate disease and/or provide weed control under field conditions may depend on the results of early biological testing. For example, *Colletotrichum truncatum* 00-003B1 only had moderate virulence and primarily caused reduction in weed biomass as opposed to weed mortality (Peng *et al.*, 2005; Graham *et al.*, 2006a). Therefore, it was important to explore ways that could enhance efficacy before undergoing field testing. In this case, efficacy was improved by creating a synergy with herbicides that increased damage and broadened the window of application to older weeds (Graham *et al.*, 2006b, 2006c). Yet, in the case of *Plectosporium tabacinum*, preliminary field testing showed disease initiation on false cleavers 2–3 days after inoculation when night temperatures were 9–15°C, despite earlier greenhouse conditions that showed the temperature greater than 15°C, were required for spore germination (Zhang *et al.*, 2002b).

Generally, we believe that it is better to initiate and continue field testing sooner rather than later to determine efficacy and consistency under a wide range of conditions. With BioMal®, 20 field trials showed 13 of them with efficacy greater than 90%, 3 with 60% weed control, and 4 failing to provide control (Boyetchko *et al.*, 2006). However, determining why trials fail to incite disease can often be more enlightening than successful trials as was the case for *Alternaria cirsinoxia* Simmons & Mortensen which appeared to be a good candidate for control of Canada thistle (Bailey, 2004). The pathogen was biologically and ecologically adapted to the weed causing severe leaf spots and stem lesions under natural disease epidemics. The fungus also demonstrated restricted host range, tolerance to a wide range of temperatures and survived multiple cycles of wet and dry conditions. Yet, the disease was not induced from artificial inoculations under field conditions. It required detailed studies addressing disease development and infection process, mechanisms of attack, and host defence to uncover the reasons why this pathogen could not be developed as a bioherbicide (Green and Bailey, 2000a; 2000b; 2001, Green *et al.*, 2001a; 2001b).

Target Weed: Biological and Ecological Considerations

The pathogen and its potential as a bioherbicide agent generally receive the most attention in biological weed control programmes without much consideration being given to fundamental knowledge of the target weed, its biology, physiology, ecology, and population structure. While Koch's postulates may prove the association, infection, and pathogenicity of the fungal pathogen on a particular weed species, the severity of disease and thus, the bioherbicidal potential may, in part, be controlled by biological factors inherent within the weed host.

Architecture and morphology of the target weed can partially determine the success or failure of a bioherbicide (Auld and Morin, 1995). Pre-existing plant barriers such as leaf hairs and thick epicuticular waxes can interfere with the infection process and retention on the leaf surface. Green and Bailey (2000) observed that infection of *Alternaria cirsinoxia* on Canada thistle was highest on older leaves than younger ones. It was postulated that as a plant matures, the structure and thickness of epicuticular waxes and the cuticle may be altered, thus affecting the pathogenicity of *A. cirsinoxia*. In addition, Canada thistle was found to develop a variety of structural defence mechanisms such as lignins, and mobilization of callose and silicon in response to *A. cirsinoxia* infection (Green *et al.*, 2001b). Lignification of host cell walls acts as a mechanical barrier, thereby making the cell walls resistant to fungal enzymes and toxins. Callose deposition is a general plant defence response involving the alteration and thickening of the cell wall that prevents pathogen infection, while accumulation of silicon at the site of infection has been known to increase mechanical strength of plants and thus act as another plant defence response.

On the other extreme, green foxtail is an annual grassy weed with leafy hairs as a physical barrier and an upright architecture that may affect spray retention and deposition of the fungal propagules (Peng *et al.*, 2005b; Byer *et al.*, 2006). The majority of weed species selected as the targets for biological control using foliar fungal pathogens have been broadleaved weeds (Charudattan, 1991; 2001).

Grass meristems are covered by leaf sheaths, thus making it more difficult to control these weeds at the growing point than with broadleaved weeds (Greaves and MacQueen, 1992). In addition, the vertical orientation of the leaves affect the retention of the fungal propagules unless the trajectory of the spray application method is considered. Use of a twin-nozzle by varying the angle of the spray trajectory both forward and backward was found to improve spray retention and biocontrol efficacy of *Pyricularia setariae* on green foxtail (Peng *et al.*, 2001). Preliminary research also revealed differences in plant size and width and length of leaves among different accessions of green foxtail (Sarah Green, unpublished data), but the variation in amount of leaf hairs among accessions is not fully understood. This may affect the deposition of spray droplets onto the leaf surfaces. Peng *et al.* (2005b) reported that application of *P. setariae* to green foxtail through delivery of finer droplet sizes provided higher retention efficiency and thus an incremental increase in biocontrol efficacy over coarser sprays. However, one must consider the size of the fungal propagule in relation to the droplet size and optimum dose transfer of the active ingredient (i.e., the fungal pathogen).

Byer *et al.* (2006a, 2006b) reported greater retention of *C. gloeosporioides* f.sp. *malvae* and *Colletotrichum truncatum* on round-leaved mallow and scentless chamomile, respectively, using finer sprays. An application of *C. gloeosporioides* f.sp. *malvae* with finer droplets onto the weed did not necessarily improve efficacy because the fungus is primarily a stem pathogen where spraying till runoff will result in deposition of droplets into leaf axils and adjoining stems. Inundative application of high inoculum rates have been used to overcome some of these barriers, however, these high doses also represent impractical application rates and/or water volumes. High application rates representative of spraying till runoff results in excessive spray volumes which carry the fungal propagules to the leaf edges and washing them off the leaf surface, particularly if the fungus is a leaf pathogen rather than a stem pathogen (Greaves *et al.*, 1998; Byer *et al.*, 2006a; 2006b).

The physiological status and age of the plant affects the efficacy of the weed pathogen and susceptibility of the plant to the pathogen. Results of *A. cirsinoxia* infection on Canada thistle revealed greater infection on older, senescing leaf tissue (Green and Bailey, 2000). Although infection on the older leaf tissue was high, Canada thistle, a perennial weed, recovered from the disease resulting in regrowth after initial infection, and thus limited its potential as a bioherbicide. This example underscores an important factor with development of bioherbicides for perennial weed control: regeneration of weeds via rhizomes, stolons, and underground stems can make long-term weed control more difficult to achieve (Greaves and MacQueen, 1992). Although these weeds may outgrow the effects of foliar-applied pathogens, there are opportunities for exploiting soil-borne pathogens as pre-emergent bioherbicides for control of perennial and *Poaceae* weed species (Boyetchko *et al.*, 2002).

Weed communities are often inherently genetically diverse in nature, and unlike crops that are bred for homogeneous, agronomic traits, discovery of a single pathogen with consistent field performance to control an entire weed population over a variety of geographical locations may be a little more challenging (Boyetchko *et al.*, 2002; Charudattan, 2005; Boyetchko and Rosskopf, 2006). Numerous biotypes of Canada thistle possessing differences in leaf morphology and lipid deposition have been reported (Hodgson 1973), and these variations may have a significant impact on the infection process (Green *et al.* 2001a). It has further been suggested that introduced, invasive weed species that reproduce asexually will generate homogeneous populations and have a high probability of being infected by highly virulent pathogens that are locally adapted (Charudattan, 2005). Conversely, a more genetically diverse population of pathotypes will co-evolve with a genetically diverse weed population. This could result in the discovery of a pathotype(s) that controls a heterogeneous weed population.

Application of reduced herbicide rates combined with a bioherbicide has been shown to act synergistically and thus enhance efficacy (Graham, 2004; Peng and Byer, 2005; Boyetchko, 2005; Boyetchko and Rosskopf, 2006). Split applications of several herbicides in combination with *C. gloeosporioides* f.sp. *malvae* resulted in improved efficacy on round-leaved mallow than with the pathogen alone (Grant *et al.*, 1990b). Tests for compatibility between the pathogen and the chemicals showed that more than 50% of the chemicals were not detrimental to the growth of the pathogen when applied at 0.1 to 2 times of the recommended rate (Grant *et al.*, 1990a). In an another study, application of the herbicide sethoxydim plus *P. setariae* significantly improved control of green foxtail, especially when compared to either the herbicide or pathogen alone (Peng and Byer, 2005). Higher rates of propanil and quinclorac in combination with *P. setariae* resulted in 100% weed mortality. Similarly, Graham (2004) reported an enhanced bioherbicidal effect of *Colletotrichum truncatum* when applied with metribuzin. In certain cases, co-application reduced the rates of the chemical herbicide and the dose of the pathogen that was normally required if applied alone. Although the mechanism of the synergies between the pathogens and the chemicals were not identified in these studies or are unknown, it has been reported that co-application may be the result of lowered plant defence responses (Hoagland, 1996). When plant defence mechanisms are impaired, there is a corresponding increase in the weed's susceptibility to the pathogen and thus enhanced efficacy. Therefore, the synergistic interaction appears to be an attractive system for enhancing the efficacy of bioherbicide agents, especially where application of the pathogen alone is not as effective, while simultaneously reducing the application rates of the chemical herbicides.

One major issue critical to the selection of an appropriate bioherbicide candidate is an understanding of the weed ecology. A researcher must be aware of the dynamics of the weed in its own environment prior to or during the candidate selection process as it may predetermine the selection approach. The two main ecological factors are the time of weed emergence in relation to the competitive ability of the crop and the critical period for weed control or suppression to prevent significant crop loss (Weaver, 1984; O'Donovan *et al.*, 1985; Swanton and Weise, 1991; Boyetchko, 1997). The critical period consists of two components:

(i) the duration period during which weeds must be controlled in order to prevent crop yield loss; and (ii) the length of time a crop can tolerate the presence of weeds before experiencing detrimental effects on crop growth and yield. For example, round-leaved mallow is a highly competitive weed and its competitive ability can greatly affect yield loss in crops such as wheat and lentil (Makowski, 1995a). Lentil is a poor competitive crop compared to wheat and in the presence of the weed the relative yield loss in lentil is greater than that in wheat. Application of the BioMal® (bioherbicide containing *C. gloeosporioides* f.sp. *malvae*) controlled round-leaved mallow in both crops, however, the speed of weed control was not sufficient in the lentil crop to prevent economic yield reductions as compared to insignificant yield reductions in wheat. Therefore, the selection of this bioherbicide is less suitable for lentil. On the other hand, control of the same weed in a highly competitive cultivar of strawberry resulted in a concomitant increase in crop yield (Makowski, 1995b). These examples suggest that the critical period and relative time of emergence could provide a basis for deciding whether a bioherbicide will be an effective weed control option. If the bioherbicide takes too long to control the weed in a particular crop, then it may be difficult to justify using it as a weed control strategy, particularly in crops that have little or no tolerance for weeds early in the growing season. While foliar-applied bioherbicides would be more appropriate for post-emergence application, it may be more suitable to apply soil-applied bioherbicides for pre-emergence application, especially if the target weed species is a strong competitor and emerges early in the growing season when compared to the crop.

Molecular Diagnostic Tools for Identification, Characterization, and Monitoring

Registration of a biopesticide requires an evaluation of the risks associated with the large-scale deployment of a specific microorganism on non-target species, people and the environment. Regulators are concerned with the identification for classification and nomenclature of the fungus, as well as any novel biological traits. Environmental toxicology and fate address the issues of potential harm to nontarget species, the likelihood of genetic exchange with existing microbial populations, modes of reproduction, and unexpected dispersal, persistence and survival. Although this information may be obtained by traditional mycological techniques, the use of molecular diagnostic tools for strain identification, detection, and monitoring is becoming routine because these tools are more rapid, sensitive and accurate. Molecular tools can also confirm strain authenticity, thus ensuring the quality and genetic integrity of commercial products as well as providing protection of intellectual property (Jenkins and Grzywacz, 2000; Avis *et al.*, 2001).

The nuclear ribosomal DNA (rDNA) region is one target used for fungal diagnostics (Lievens and Thomma, 2005). By exploiting conserved regions of the 18S, 5.8S and 28S rRNA genes, comparative studies of the intervening internal transcribed spacer (ITS) regions are commonly used to establish phylogenetic and taxonomic relationships within and between species (White *et al.*, 1990). *Rhynchosporium alismatis* (Oudem.) Davis, a hyphomyceteous fungus for control of alismataceous weeds in Australian rice fields, was thought to be closely related to the barley scaled and orchard grass pathogens, *Rhynocosporium secalis* (Oudem.) Davis and *R. orthosporum* Caldwell, respectively. However, ITS sequence comparison showed that *R. alismatis* was unrelated to these species, and was congeneric with *Plectosporium* Palm, Gams & Nirenberg, prompting the introduction of the new combination *Plectosporium alismatis* (Oudem.) Pitt, Gams & Braun (Pitt *et al.*, 2004) and redescription of the species (Pitt and Gams, 2005).

In the case of *Colletotrichum truncatum*, ITS sequence analysis and exploitation of sequence differences facilitated the development of taxon-specific primers with the capacity to differentiate *C. truncatum* isolates originating from different hosts, previously impossible based on morphological characters (Liu *et al.*, 2005). Strains isolated from scentless chamomile caused disease on the weed, but were not pathogenic to lentil (Forseille *et al.*, 2006), and PCR-assays with host-specific primers provided accurate and rapid detection of different *C. truncatum* strains both from scentless chamomile and legume crops. These strain-specific primers could also be used for field-based monitoring to show that strains from scentless chamomile would not pose a risk to lentil crops.

PCR-based molecular markers have been used to generate randomly amplified fragments, which are then separated by electrophoresis, compared between related fungi and potential diagnostic bands sequenced to design sequence characterized amplified region markers or SCARs (McCartney *et al.*, 2003). Ramsfield *et al.* (1999) developed scar assays based on mitochondrial DNA (mtDNA) for screening a world collection of *Chondrostereum purpureum* (Pers.:Fr.) Pouzar, which has since been developed as a bioherbicide for vegetation management in forestry. Restriction enzyme digestion of an ATPase VI – cytochrome *b* SCAR produced a single polymorphism, with both mitochondrial haplotypes being present in all surveyed regions. This data suggests that the introduction of novel mitochondrial genes as a result of bioherbicide application is highly unlikely and hence contributed significantly to the decision to pursue registration and commercialization.

Despite the abundance of molecular markers, multilocus PCR fingerprinting techniques such as amplified fragment length polymorphism (AFLP) and random fragment length polymorphism (RFLP) provide complex banding patterns that have excellent value for strain identification and authentification (Vos *et al.*, 1995; Ramsfield *et al.*, 1999). Methods have been developed to convert complex AFLP fingerprints to simple single-locus markers for locus- and allele-specific assays (Brugmans *et al.*, 2003). Comparisons of DNA fingerprints between related fungi can differentiate closely related species (Abeln *et al.*, 2002), assess genetic diversity within and between species (Kistler, 1997), and determine population structures (McDonald, 1997). Such studies are often used to address biosafety concerns prior to registration (Hintz *et al.*, 2001) and examine the residual effect of evolutionary forces on population structure (McDonald, 1997), including the potential for unexpected dispersal (Carbone and Kohn, 2004), risk to nontarget species (Gosselin *et al.*, 1999), and the likelihood of genetic exchange with existing microbial populations, either through recombination, migration (Brown, 2000; Wilkin *et al.*, 2005), or the introduction of novel genes (Ramsfield *et al.*, 1999).

In Canada, where an ecozone concept is imposed, the release of a microorganism into an ecozone other than where it originated is prohibited because there is potential risk to non-target species, people and the environment. Encroachment of the ecozone boundaries can only be accomplished if biosafety requirements can be met or specific attributes are shown to be present in multiple zones (PMRA, 2001). Random amplified polymorphic DNA (RAPD) and pulse-field gel electrophoresis (PFGE) were used to characterize bioherbicidal strains of *P. macrostoma* into two biotypes (Zhou *et al.*, 2005). Because both biotypes were found in each of the surveyed ecozones, these strains were considered indigenous to each region, thus satisfying biosafety concerns for deploying strains across the country.

Similar studies conducted on populations of the *Alismataceae* pathogen *P. alismatis* from three field sites throughout Southern Australia revealed a similar story, with populations largely shaped by clonal reproduction, but likely supplemented with intermittent rounds of recombination or gene flow over considerable distances. Migration via the passage of infected seed and/or plant materials was suspected as a means of gene flow between populations, but the low levels of diversity and limited means of dissemination suggested that fungal applications posed minimal risk to the aforementioned groups (Pitt *et al.*, 2005b).

As research on biocontrol pathogens has become more intensive, and biosafety assessment for registration more stringent, 'identification' of biocontrol strains has become synonymous with 'detection' for the purposes of monitoring the strains post-application. As a consequence, detection involves differentiation of biocontrol strains from other native strains and species present in the environment, whilst monitoring of strains now comprises quality control measures for strain maintenance, contamination and genetic stability, and quantification to assess fate of infective propagules. Based largely on the approaches previously outlined, strain-specific regions of the genome are identified either by assessing large numbers of hypervariable loci through multilocus fingerprinting or identifying specific genomic regions of biological significance. Multiple primer sets designed and incorporated into a single reaction provide unmatched specificity for identification and rapid detection of strains released into the environment, but also serve to provide tools for quality control and environmental management of antagonistic strains, necessary for registration and commercialization of biopesticides (Fravel *et al.*, 1999).

Caron *et al.* (2005) recently developed a multiplex assay for molecular quality control of *Pseudozyma flocculosa* (Traquair, L.A. and Shaw & Jarvis) Boekhout

& Traquair, the active ingredient of Sporodex, a biofungicide for control of powdery mildew. Multiplex-PCR fingerprints differentiated biocontrol strains of the fungus from closely related species and genera, identified contamination otherwise undetectable by traditional culture techniques and provided a means for the detection of potential mutations for strain authentication and maintenance of genetic stability of the product.

Zhou *et al.* (2004) used traditional PCR techniques to measure the presence, mobility and persistence of strain-specific fungal DNA in soils inoculated with differing concentrations of *P. macrostoma*. The DNA was detected when the technical product was applied at rates between 4 and 1000g/m^2, but was observed to decline over time, often becoming undetectable after 4 months. Biocontrol strains had limited mobility and were rarely detected beyond the site of inoculation, suggesting that strains had minimal dispersal potential and were unlikely to undergo self dissemination in the field. Whilst successful in determining the presence or absence of the fungal DNA and establishing detection limits for diagnostic PCR, these studies often fail to accurately quantify DNA as a result of the non-linear nature of PCR amplification (Wang *et al.*, 1989). Future studies with biocontrol fungi may employ the use of real-time PCR to remove many of the earlier limitations of PCR-based quantification. Real-time PCR measures the accumulation of PCR products after each cycle and calibrates the accumulation of product in comparison to the amount of target DNA (Schena *et al.*, 2004).

Production and Utilization of Microbial Metabolites with Phytotoxic Activity

Most plant pathogens use chemical forces to attack their hosts by producing enzymes to soften the cuticle for mechanical penetration, growth regulators to distort cell development and growth, polysaccharides to block the vascular system, or phytotoxins to kill the cells (Agrios, 1978). The virulence of a plant pathogen is often dependent on producing one or more compounds which contribute to the bioactivity. Some microbial metabolites are very host specific with regard to their activity, while others may have broad-spectrum activity. Venkatasubbaiah and Chilton (1992a) showed that an unidentified species of *Phoma* (a rhubarb pathogen) produced four phytotoxins in liquid culture. Culture filtrates of *Phoma sorghina* (Sacc.) Boerema, Dorenbosch, and van Kesteren, caused necrosis when spotted on pokeweed leaves and eight other weed species, indicating a non-specific nature (Venkatasubbaiah and Chilton,1992b). The isolation of phytotoxins from culture filtrates of *Phoma putaminum* Speg., the causal agent of leaf necrosis of annual fleabane, gave putaminoxin as the main phytotoxin as well as the closely related but inactive putaminoxin B, and a polyketide putaminoxin C (Evidente *et al.*, 1995; Evidente *et al.*, 1997). Putaminoxin C showed broad-spectrum toxic effects on leaves of weedy and cultivated host plants that were similar to those produced by putaminoxin itself.

Among fungi being studied for biological control of weeds, many have been reported to produce phytotoxins which function like herbicides to kill plants

(Kenfield *et al.*, 1989; Strobel *et al.*, 1991; Abbas and Duke, 1995; Evidente and Motta, 2001; Saxena and Pandey 2001; Singh *et al.*, 2003). Historically, the pesticide industry has concentrated on developing synthetic herbicides, but rising costs of development and public pressure for reduced risk products have increased research activities on microbial metabolites with phytotoxic activity as templates for new herbicide chemistry. These phytotoxins, often of low molecular weight, are difficult to synthesize because of their complex chemical structure. Phytotoxins may be the product of invasion and defence compounds resulting from the co-evolution of the organism and its biotic environment. These compounds have reduced environmental impact because of a shorter environmental half-life than synthetics (Rimando and Duke, 2006).

Research has shown that microbial metabolites have different bioactivities and often present novel modes of action. Pedras *et al.* (2003) discovered that the bacterium, *Pseudomonas fluorescens* strain BRG100, produced two cyclic lipodepsipeptides which had differing bioactivities: pseudophomin B demonstrated antifungal activity and pseudophomin A exhibited herbicidal activity by inhibiting germination and root growth of green foxtail. Rivero-Cruz *et al.* (2000) found two new phytotoxic nonenolides, herbarumin I and herbarumin II in extracts of *Phoma herbarum* (West.). These compounds, each containing a 10-membered heterocyclic ring, caused significant inhibition of radicle growth of *Amaranthus hypochondriacus*. The fungus *Phoma exigua* Desm., which was investigated for the control of dandelion (Neumann and Boland, 2002; Stewart-Wade and Boland, 2004), was shown to produce cytochalasins, a large group of compounds containing a heterocyclic dehydro oxindole ring (Vurro *et al.*, 1997). Cytochalasins, however, possess various bioactivities and are probably not solely responsible for weed control by *Phoma exigua*. Rothweiler and Tamm (1970) identified the cytotoxic antibiotics phomine (shown to be cytochalasin B) and 5-dehydrophomin (cytochalasin A) from cultures of *Phoma* sp. S-298. Bousquet and Barbier (1972), claimed cytochalasin B was found in culture media of three strains of *Phoma exigua* and that this material was phytotoxic on chicory, potato and periwinkle.

In our laboratory and in collaboration with scientists at Dow AgroSciences in Indianapolis, Indiana, we studied microbial metabolites from *Phoma macrostoma* that showed phytotoxic activity against several broadleaved weeds. Bioassay-directed isolation from liquid cultures led to the discovery of macrocidin A (Fig. 1.5), a new tetramic acid derivative, as the main component (Graupner *et al.*, 2003). Macrocidin B, also an active metabolite, was tentatively identified as 13-hydroxymacrocidin A. Macrocidins are defined not only by the heterocyclic tetramic acid moiety (2, 4-pyrrolidinedione) but also by an epoxy group at position 16. The macrocidins were isolated by reversed-phase liquid chromatography and characterized by high resolution NMR spectroscopy. The structure of macrocidin A was solved by single crystal X-ray diffraction measurements. In another publication (Graupner *et al.*, 2006), structures of other naturally-occurring macrocidin derivatives were presented, including one with an olefinic double bond replacing the epoxide. Methods of synthesis of substituted deoxymacrocidins and relation between structure and herbicidal activity were also reported for both cyclic and acyclic derivatives. The epoxy group was not necessary for the herbicidal activity. The mode of action of herbicidal macrocidins is currently unknown.

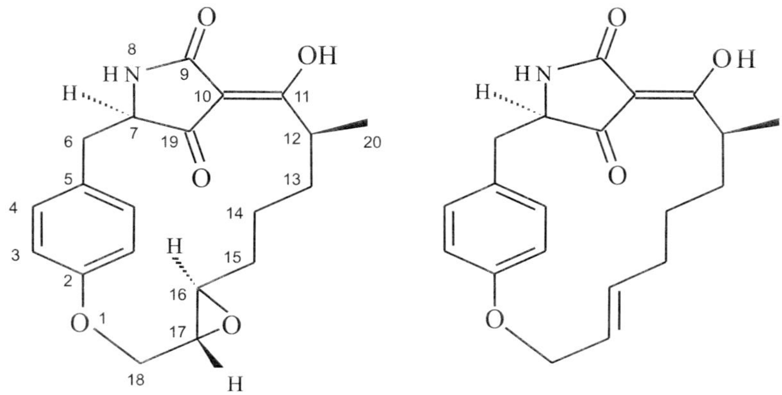

Figure 1.5 Structure of macrocidin A and deoxymacrocidin A.

Industry mostly views these natural products as substitutes for synthetic herbicides and wants to either make synthetic derivatives or genetically modify the organisms into super producers, thereby having the fungus manufacture the phytotoxins *in vitro* and then separating the phytotoxins from the fungus for application as a biochemical (examples being glufosinate and bialaphos derived from *Streptomyces* spp.). This is different from a microbial bioherbicide which retains the living component in the final product so the fungus can naturally produce the phytotoxins *in situ*. The bioherbicide approach reinforces the principles of co-evolution and ecological balance allowing nature to be responsible for fine tuning the system over time. The derivative and biochemical approaches have significant regulatory considerations because they are both classified as synthetic herbicides in Canada, although in the United States a biochemical is recognized as a reduced risk product and reviewed as a bioherbicide (PMRA, 2001; Braverman *et al.*, 2006).

Fermentation

One of the major hindrances to commercial success of bioherbicides is the inability to produce the microorganisms in an efficient and economical manner. Different methods, in particular, submerged and solid substrate fermentations, have been studied for mass production of bioherbicides (Morin *et al.*, 1990; Daigle *et al.*, 1998; Dokken *et al.*, 2004), and there are several reviews on the general strategies and challenges in producing microbial inocula (Stowell, 1991; Auld, 1993; Jackson *et al.*, 1996; Holker *et al.*, 2004). In the following section, we will highlight key technologies used for fungal mass production focusing on aspects which are pertinent to addressing the challenges with scale-up.

Solid state fermentation (SSF) is defined as the growth of microorganisms on the solid materials containing sufficient moisture adsorbed within the matrix, but in the absence of free water (Pandey, 1992). This technology is particularly useful for growing filamentous fungi and for the production of secondary metabolites (Robinson *et al.*, 2001). The process uses low cost ingredients and inexpensive facilities to produce a stable end product with high yield. The end product usually requires minimal downstream processing, especially when the fungal biomass and solid matrix is used as an integrated product. However, difficulties may be encountered due to contamination, limitations in gas exchange, and poor control of temperature and moisture. Additionally, production cycles may be time-consuming and a large surface area is required for adequate aeration (Stanbury *et al.*, 1995). Nevertheless, the technology is less expensive than submerged fermentation (SmF) for many fungi and yeasts with key factors for growth being temperature, moisture content, humidity, aeration, agitation, and pH. Although this process has not been readily adopted by western industrialized countries, companies such as Sylvan Bioproducts Inc. in the USA and Prophyta in Germany have emerged as toll manufacturers of biopesticides using solid state systems. SSF is not suitable for all fungal species, but has been used successfully with the following bioherbicide candidates: *Alternaria destruens* E.G. Simmons (Hopen *et al.*, 1997), *Sclerotinia minor* Jagger (Schnick *et al.*, 2002), *Phoma macrostoma* (Zhou *et al.*, 2004), and *Alternaria alternata* (Fr.) Keissler (Babu *et al.*, 2004).

Submerged fermentation (SmF) is a more familiar technology for microbial mass production and is still preferred by industry for large-scale commercial production of fungal biomass (Charudattan, 2001). While many fungi may not sporulate sufficiently in liquid shake culture, some have shown great potential for spore production using SmF Ogel *et al.* 1994; (Dokken *et al.*, 2004). Some advantages of SmF include maintenance of homogenous cultures with less risk of contamination, ease of monitoring and manipulating critical parameters, and simple recovery of propagules using centrifugation or filtration (Davis and Blevins, 1979), all of which help to reduce production costs.

The third type of production system is a two-stage process involving submerged fermentation of mycelia followed by drying the biomass on a solid matrix to induce sporulation (Stowell, 1991). Two fungal agents, *Pyricularia setariae* and *Alternaria cirsinoxia* developed by our team for control of green foxtail and Canada thistle, do not sporulate in liquid culture. A process adopted from the production of *Alternaria macrospora* Zimm. (Walker, 1980) was used to produce mycelia of *P. setariae* and *A. cirsinoxia* in oatmeal broth and modified Richard's medium, respectively. Mycelia were harvested from liquid cultures after one week, homogenized and spread on trays to sporulate. Production was more successful with *A. cirsinoxia* as sporulation was profuse as the mycelia dried slowly and the dry spores were easily detached by vacuum suction. Sporulation of *P. setariae* proved more difficult as its initiation was triggered by a drop in humidity.

This was achieved by covering the trays with transparent polyethylene to reduce evaporation and then making slits in the plastic covering to accelerate moisture release. Timing of the moisture drop was critical because newly produced spores germinated under the high humidity if the drying process was initiated too late. This production method, however, provided sufficient inoculum only for small-scale trials and scale up appeared difficult.

Regardless of the production system used, fungi are inherently unpredictable during fermentation. Determination of the media, specific conditions and processes for each fungal species presents new challenges despite prior experiences, as even closely related species/strains may have vastly different nutritional and environmental requirements for growth, sporulation, virulence, and metabolite production. Initial inoculum concentration may affect fermentation yield. In our experience with the scentless chamomile strain of *Colletotrichum truncatum*, a lower initial concentration of 10^2 spores/ml yielded twice as many conidia as cultures inoculated with 10^4 or 10^5 spores/ml. Cultures with higher initial spore concentrations showed more dispersed growth, whereas at the low concentration the fungus formed larger pellets that were conducive to sporulation likely due to greater nutritional stress at the centre of the pellet (Stanbury *et al.*, 1995). Nutritionally poor conditions often trigger differentiation from vegetative growth (Vega *et al.*, 2003).

Medium selection is the first crucial step in fermentation. The basic components often include water, inorganic salts, vitamins, carbon (C) and nitrogen (N) sources. In early stages of research defined media, with a known quality and quantity of each ingredient, may be more suitable than undefined, complex media. Adopted from Jackson *et al.* (1996), a semi-defined salt solution supplemented with glucose and casamino acids as C and N sources was used as a medium that could be manipulated with a range of elements for maximum conidial yields and weed-control efficacy by *Colletotrichum truncatum* (Dokken *et al.*, 2004). In shake flask cultures at relatively low aeration levels, this medium yielded more conidia when 40 g glucose/L were added but spores from high-sugar media were less efficacious for control of scentless chamomile when compared to those produced at 5–10 g glucose/L (Dokken, 2003). In the same study, the type of C and N sources also affected fungal sporulation, with glucose, sucrose, hydrolyzed casein and casamino acids being superior to other C or N sources tested, including molasses, fructose, cottonseed hydrolysate and ammonium nitrate. Often selection of medium ingredients involves balancing of several criteria: inoculum yield, potency, cost, and materials with a consistent quality and abundant supply. Complex media may be employed once the key ingredients required for the production of the desired propagules are identified. For example, a simple plant protein, prolamin, found in oatmeal stimulated sporulation of *Pyricularia oryzae* Cavara (Namai and Yamanaka, 1983). Similarly, ground oatmeal was used as an ingredient on a solid substrate to produce up to 10^7 spores per cm^2 surface of *P. setariae* (Peng and Byer, 2005). This ingredient can also be mixed with inert granular carriers to enhance the surface area for greater sporulation (Peng, unreported data).

Environmental conditions, particularly temperature, moisture, aeration, and light, are important factors for liquid culturing of fungi. Light may stimulate sporulation in some fungi, others may require complete darkness, and yet others may need specific light regimes (Urquhart *et al.*, 1994; Amsellem *et al.*, 1999; Montazeri, 2002). Conidial generation of *P. setariae* was promoted by 14 h near-UV light (Peng *et al.*, 2004), whereas a range of light regimes showed no difference for sporulation of *C. truncatum* (Dokken, 2003). Optimum temperatures can be different for vegetative growth and sporulation, and often the latter has a narrower range and may require fluctuating rather than constant temperatures. The preferred temperature range for sporulation may be related to the environment from which the organism originated. An isolate of *P. setariae* obtained from Southern Ontario, one of the warmest regions in Canada, had greater sporulation at a constant temperature of 26°C whereas lower temperatures slowed its growth and conidial initiation. The chamomile strain of *C. truncatum* isolated in Saskatchewan, Canada (a much cooler region than Ontario) produced fewer conidia at 28°C than at 16°C, and lowering the fermentation temperature to 12°C after an initial growth period at 20°C for one week, stimulated the fungus to produce the highest number of conidia (Dokken *et al.*, 2004). The higher temperature favoured mycelial growth, but the low temperature stress triggered fungal sporulation.

Another important factor for fermentation is to provide sufficient oxygen transfer through liquid or solid media. Oxygen diffusion into liquid media in shake-flask cultures is usually achieved by agitation. Flasks with side indentations (baffled) encourage greater turbulence and, consequently aeration. More oxygen may be dissolved in media through increased agitation, which minimizes boundary layer resistance and maximizes the surface area for oxygen transfer (Auld, 1993). Conidial production for the chamomile strain of *C. truncatum* was favoured by higher levels of aeration. At the same shaker speed of 100 rpm, cultures in baffled flasks yielded an average of 4 times more conidia than those cultured in standard Erlenmeyer flasks (Dokken, unpublished data). Oxygen needs can be influenced by nutritional state of the medium. Higher carbon concentrations can encourage biomass increase (Slininger *et al.*, 1993), but the process would require a greater oxygen supply. The effect of interactions between these two factors on fungal sporulation is more complex. Large-scale manufacturing often uses forced air and agitation to overcome aeration deficiency in liquid fermentation. Oxygen transfer is regulated by measuring dissolved oxygen (dO) concentration and adjusting agitation speed and/or forced-air pressure (Hosobuchi, 1999).

Fermentation development is usually carried out in three stages: shake flask cultures, bench top and pilot scale fermenters for optimization of operating conditions, and then to the manufacturing plant for industrial yields and economy of scale. Physical, chemical, and process factors may need to be adjusted at each step as the scale increases. To produce larger quantities of bioherbicide inoculum for more elaborate trials and to fulfil commercialization potential, more precise control of the process is required and a scale up to bench-top fermenters is essential. These models are capable of monitoring and controlling key environmental factors during fermentation. As the scale increases, maintaining the rate of oxygen transfer becomes more important and challenging due to differences in dissolved oxygen at varying depths of the medium. The same medium used in flask cultures may also be used for fermenters, but ingredient strength may need to be adjusted to compensate for differing oxygen-transfer capabilities between vessels of different sizes (Hilton, 1999). Nevertheless, economics often dictate changes in medium as most defined media are too costly to use at a commercial scale. The efficiency of nutrient utilization may change when fermentation is scaled up, where agitation and aeration conditions are more intense, and nutrient requirements may increase (Dahod, 1999). Seed cultures need to be produced in steps depending on the fermenter size, and the inoculum should be in an appropriate physiological condition before being transferred to the next stage. Cultural vigour, indicated by oxygen uptake rates, may be used to determine the timing for transfer but subculturing should be kept minimum since the productivity and stability of an organism can be affected negatively (Monaghan, 1999). The efficacy must be ensured with seed cultures prior to scale-up production. *Alternaria alternata* lost bioherbicidal effectiveness against *Amaranthus retroflexus* L. (redroot pigweed) after being subcultured only a few times (Montazeri, 2002).

A study was conducted to scale up production of the chamomile strain of *C. truncatum* using 20 L fermenters. Three levels of dO were used at varying agitation speeds; 10% (low), 30% (medium), and 60% (high). Based on previous flask-culture studies at low aeration levels (100 rpm shaker speed), glucose was used as the C source at approximately 10 g/L for all dO levels and fermentation was carried out at 22°C for first week followed by a drop to 12°C for the second week. Data showed that glucose in the liquid medium was exhausted more quickly under high and medium dO conditions and, correspondingly, the fungal biomass reached at peak at least 4 days earlier than with the cultures under the low dO level. However, higher dO levels did not enhance sporulation substantially and the conidial yield in fermenter productions was about 100 times lower than those in flask cultures. To understand the cause of the phenomenon, a further study was conducted using flask cultures and results suggested that benefits of higher aeration to sporulation of *C. truncatum* could be further enhanced by increasing the amount of glucose in the medium (Fig. 1.6). At 40 g glucose/L, the conidial yield was 18 times higher than cultures supplied with only 5 g glucose/L despite the same level of high aeration was used. Although the ability to increase dO concentration in liquid cultures is much greater with state-of-the-art equipment, the optimal dO may vary between biomass increase and sporulation and can be influenced by the nutritional level of media. The impact of these factors may sometimes need to be assessed interactively.

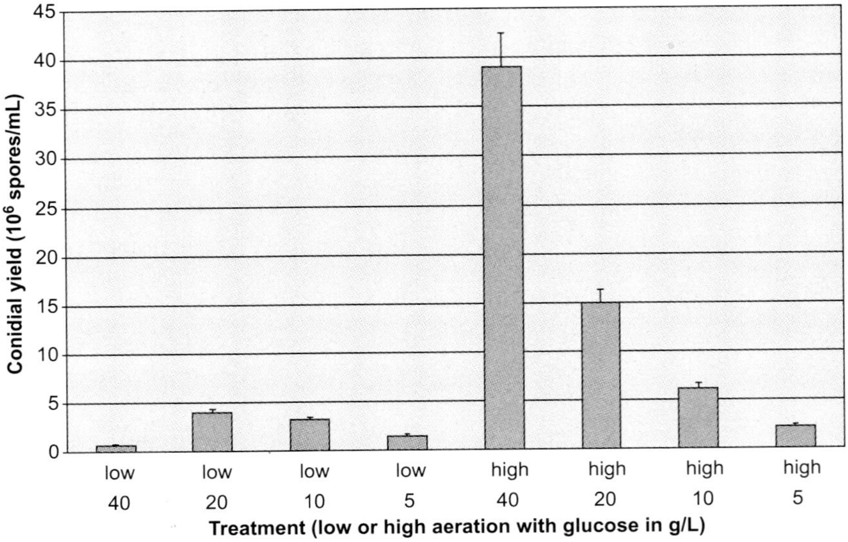

Figure 1.6 Effect of aeration and glucose concentrations on spore production by *Colletotrichum truncatum* in liquid flask culture. The low and high aeration was created by setting shake speeds at 100 rpm and 200 rpm, respectively.

Formulation

Development of formulations and application protocols are crucial for advancing bioherbicides as a viable option in crop production systems (Zhang *et al.*, 2003). Formulation ingredients are chosen for their ability to: (i) shelter and protect microbial biopesticides from environmental stresses during storage and following application onto the target weed, (ii) promote ease of delivery to the target and (iii) support efficacy on the target. Foliar spray and soil-applied granular inoculants are being developed to attract commercial interest in bioherbicides as viable solutions for weed management.

Development of products that include living microorganisms add an element of complexity and consequently, require a multi-disciplinary team approach to optimize fermentation processes, develop formulation strategies and clarify the interactions among the bioherbicide, pest and environment (Deacon, 1991; Whipps, 1997; Butt *et al.*, 2001). Fermentation and formulation are closely linked processes; the success and ability of the formulation to extend the survival and efficacy of the fungal propagules is highly dependent on a fermentation process that produces robust spores and mycelium. For example, fungal spores should be harvested at the peak of maturity when melanin, a UV radiation protective by-product and glycerol or other osmotic protective metabolites are abundant and can defend the spores from environmental stresses. Additional UV radiation protectants and compounds promoting osmotic tolerance can be supplemented in the formulation; however, fungi have evolved mechanisms that shield them from many environment stresses and it would only be prudent to take advantage of these inherent protective mechanisms.

There are several fundamental details that must be accounted for while formulating a bioherbicide that relates to the needs of commercial producers, distributors, retailers, end-users and government regulatory agencies. Assuming that the fungus can be mass produced economically, the formulation process and ingredients must be inexpensive, protect the bioherbicide from environmental stresses such as temperature fluctuations during transportation and storage, sustain an efficacious bioherbicidal population and ensure a long shelf-life. Shelf-life duration, i.e., one or two crop growing seasons, is frequently determined by the distribution and retail industries and is motivated by a need to manage their product inventories. Formulation stringency for promoting long shelf-life of bioherbicides frequently becomes an on-going study, with periodic improvements of the formulation and entry of new products into the marketplace. Technology transfer to industry is more likely to be successful if cost-effective constituents are used and if the formulation processes can be scaled-up in standard industrial equipment. The end-users want the formulated bioherbicide to (i) be easy to handle, (ii) be applied with conventional equipment, (iii) be compatible with the other integrated pest management and fertility programs, and (iv) have no adverse effects on seed germination or plant growth, or on livestock and human health. Government agencies have regulations in place that control biopesticide products advancing into the marketplace. They insist that formulation ingredients have no adverse effects on seed germination, plant growth, livestock, or human health and do not alter the microbial diversity of the soil.

A weed management strategy that includes bioherbicidal fungi requires a comprehensive understanding of the weed-bioherbicide interaction and mode of action, optimum temperature for activity, site of infection and mode of entry. With these things in place, the options for formulation type and means of application can then be considered. The formulation types include granules (0.5–2 mm), dusts, liquid emulsions (oil or aqueous), wettable and dispersible granules and powders. Soil-applied granular inoculants may be efficient for root infecting bioherbicides and a pre-emergent weed management strategy. Soil placement of a granular formulation, i.e., sweeps, side-banding or with the seed, and addition of dispersants to promote slow or fast release of a bioherbicide, can be optimized when mobility and competitiveness of the bioherbicide in the rhizosphere are known. Foliar application of liquid emulsions or wettable and dispersible powders and dusts are required for stem and leaf infection and a post-emergent target weed management strategy (Deacon, 1991). For foliar application of bioherbicides, knowledge of the chemistry and surface structure of the targeted plant surface is required so that formulation ingredients can promote spreading of the fungus over the surface and promote infection. Regardless of the formulation type, the optimized rate, timing and mode of application must be determined to ensure that the bioherbicide is targeted effectively at the most susceptible growth stage of the weed. Experimentation with the bioherbicide-weed interaction under a variety of environmental conditions, examining time before or after rainfall, effect from dew period (duration of leaf wetness), UV radiation from sunlight, air and soil temperatures and wind induced desiccation will reduce field performance variability and advance our knowledge of the plant microbe association.

Formulations are composed of: (i) an active ingredient, i.e., hypha, spore, fungal metabolites; (ii) carriers, often an inert material used to support and deliver the densely populated active ingredient to the target; and (iii) adjuvants, compounds that promote and maintain active ingredient efficacy. Examples of formulations, carriers and adjuvants are described by Burges (1998) and summarized in Table 1.1. Formulation ingredients for fungal bioherbicides: (i) ensure bioherbicide stability during processing and storage, (ii) assist application, (iii) protect the bioherbicide from unfavourable environmental conditions, and (iv) promote bioherbicide activity at the target weed (Jones and Burges, 1998). The wettable powder formulations and oil-liquid emulsions for foliar application may require clay, titanium dioxide or Tinopal to act as a sunscreen and protect the bioherbicide from UV radiation; polyacrylamide, Waterlock®, or psyllium to absorb moisture and protect against desiccation; vegetable gums to ensure rain fastness on the target; and emulsified oils to promote spreading and dispersal on the plant surface, reduce loss of moisture and promote penetration of the hydrophobic cuticular layer by the bioherbicide (Burges, 1998, Boyetchko *et al.*, 2002). The effect of destructive shear forces from conventional spray equipment must be considered when the application strategy requires an emulsified liquid formulation. Several rheology-modifying compounds (e.g., starches, gelatin), are available that can assist with buffering the emulsified carrier of the microorganisms from these shear forces.

Table 1.1 Examples of formulation types, additives and the function of the additives.

Formulation type	*Formulation additives*	*Function*
Water-oil emulsions,	Vegetable, Mineral oil	Carrier, spreader
Wettable powders, water dispersible granules	Clay, Tinopal	Sunscreen ultraviolet radiation protection
	Glycerol	Humectant, osmotic protectant
	Lecithin, SPAN-85	Emulsifiers
	Xanthan gum	Thickener
	Gelatin	Sticker
Granule, dusts	Kaolin, clay	Carrier; dispersant
	Diatomaceous earth	Free-flow agent
	Microcrystalline cellulose	Dispersant
	Vegetable gum, corn syrup,	Binder, nutrient
	Methyl cellulose, polyethylene glycol	Humectants

Pre-emergent weed management by bioherbicides applied to soil in a granular "Pesta" formulation has been reported for *Colletotrichum truncatum* for hemp sesbania (Connick *et al.*, 1996), *Fusarium oxysporum* Schlechtend:Fr. f. sp. *orthoceras* for sunflower broomrape (Shabana *et al.*, 2002), *Fusarium oxysporum* Foxy 2 for *Striga* spp. (Elzein *et al.*, 2004), and soil bacteria for green foxtail (Daigle *et al.*, 2002). Post-emergent management of green foxtail by *Pyricularia setariae* has been reported by Peng *et al.* (2005). Water-in-oil-in-water (W/O/W) emulsions have been studied to serve as carriers for spores of bioherbicidal fungi (Auld *et al.*, 2003). The advantage of W/O/W multiple emulsions are that they typically contain a low concentration of oil, 5% or less, which may not elicit a phytotoxic reaction on non-target plants, and they carry encapsulated water and water soluble nutrients to promote fungal spore germination and reduce dependency on dew (Auld *et al.*, 2003). Fig. 1.7 illustrates the W/O/W emulsion in which *Pyricularia setariae* spores are suspended in aqueous drops which are encapsulated by a layer of oil while suspended in an aqueous phase (Hynes, unpublished). Compatibility of emulsifiers, oils and other ingredients with the active ingredient is a crucial step in the development of the soil and foliar applied bioherbicides (Zhang *et al.*, 2003).

There are similarities between the food processing, pharmaceutical and cosmetic industries with formulation development of bioherbicides (Gaugler, 1997). There has been extensive screening of ingredients to find those which are non-toxic, safe for consumption and inexpensive. Strategic collaborations with these industries may identify formulation ingredients and potential formulation know-how for maintaining survival and efficacy of a bioherbicide.

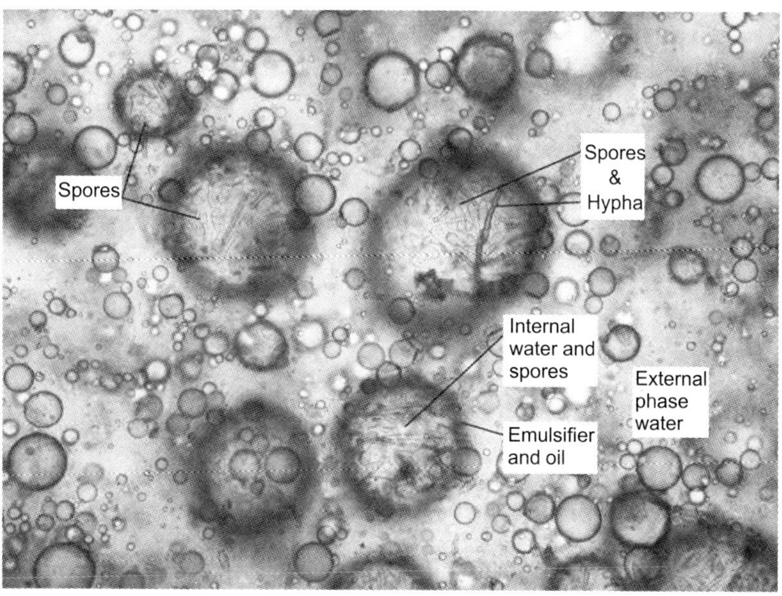

Figure 1.7 Water-in-oil-in-water (W/O/W) emulsion of *Pyricularia setariae* spores for post-emergent management of green foxtail.

The Industry Component: A Shareholders' Dilemma

Perceptions and Expectations

From industry's perspective, the expectations for a bioherbicide are straightforward; it must equal or come close to a chemical benchmark in terms of efficacy, reliability, ease of use, and storage (Meneley, 2000). This is because the *raison d'être* of a business is to increase the customer base and retain existing customers by delivering products and services with perceived or real value. From the public perspective, the customers have similar expectations but their decisions to use a product may also be guided by their social values. In the search for alternatives to conventional herbicides, bioherbicides are becoming more notable in the public domain because they are seen as being compatible with an integrated management philosophy and as having fewer potential health and environmental effects. A survey conducted by the Biocontrol Network and the World Wildlife Fund-Canada interviewed 1000 random members of the public to gauge the attitude towards biological control in Canada (Anonymous, 2006a). The survey showed that 63% of people believed biological control (including the use of biopesticides) was preferable to using pesticides and 60% felt there was less risk of food contamination. More people believed that using insects for pest control was safer than using beneficial microbes when it was being applied directly to food. Only 3% of the respondents said they were very well-informed about biological control, 49% felt they were well-informed and the remainder were poorly informed. The greatest barriers to adoption (according to the respondents) were perceptions that biologicals have poor efficacy, inconsistent results, are more expensive, are not available, and are complicated to use (Anonymous, 2006b). These results demonstrate three important points for scientists and industry partners: (i) there will be adoption if we make products that work and are cost-effective, (ii) there is a need for more public awareness and education programmes, and (iii) we need to get more products into the general marketplace.

The Economic Crunch: Money, Money and Money

10 years ago, optimistic projections speculated that biopesticides would sequester 40–50% of the total pesticide market (Meneley, 2000). In reality, there has been a more modest growth in the biopesticide market increasing from <2% in 1988 to 3% by 1998 based on the percentage of global end-user sales. One reason for the slow growth rate may be attributed to money, either the potential of money as it relates to market share, the real cost of production using current technologies, or the expectation of profit and return on investment.

Market Size and Customer Demand

Market conditions directly influence profitability and can make or break the innovation chain for bioherbicide development. The market size must be large enough to recover the costs of investment and return some profitability. The percentage share of the market must be determined. Cross and Polonenko (1996) suggest that in the year of market launch, 1% market share would be considered as a typical expectation, but 10% market share after 5 years of introduction is considered satisfactory growth and 25% being exceptional.

Customer demand can change market conditions if there are no effective products available, if the available products are too expensive, or if the products available carry some real or perceived risk (Cross and Polonenko, 1996). However, for new technologies where there is poor public awareness or where comparable products do not currently exist, customer demand will have limited power in directing new development. Consequently, without customer demand, companies have difficulty in judging the potential market for new products and are unaware of possible economic returns (Auld, 1991).

Meneley (2000) found that research institutes often spend too much time on the characterization of the organism without having closely looked at what the market will bear. Frequently, companies may decide that the market opportunity is too small, complicated, or competitive for successful commercialization. This came to pass for BioMal® where a detailed market study was completed after registration. The study showed that the market potential was too small to bear the production and packaging costs at the set consumer price (Cross and Polonenko, 1996). To reduce the chances of this reoccurring, our research programme now consults with an economist and potential industry partners prior to target selection.

Manufacturing: Process and Infrastructure Economics

A major challenge in developing a bioherbicide is moving production from the laboratory bench to pilot scale and commercial manufacturing. Keeping the costs of production and formulation low are essential, if a biological product is to succeed. For example, control of weeds such as Canada thistle, leafy spurge, knapweed, and St. John's wort only costs $3.90/acre with the herbicide atrazine, but this rises to $70.00/acre when releasing reared beneficial insects (Uri, 1999). Small changes in production practices can rapidly alter growth and costs (Zorner *et al.*, 1993). A large component of product pricing is determined by recovering the production costs and adding a profit margin to account for an amortized value for R&D, returns to shareholders, and new R&D investment (Cross and Polonenko, 1996). Stowell (1991) estimated that the manufacturer's selling price is double the actual production cost. Beyond the manufacturer, costs must be added for the wholesaler, distributor, and retailer, such that a product that cost $5.00 to produce could end up retailing for $40.00.

Another key challenge in developing bioherbicides for North America has been a poor level of investment towards developing expertise and infrastructure for manufacturing microbial products for non-pharmaceutical uses. Countries such as South Africa and Europe have more microbial-based products than Canada or the United States, partially because they have learned that it can be economical to produce lower volumes on a regional basis for local use with simple systems. Van Lenteren *et al.* (1997) noted that in Europe, commercial use of biological control (both insects and microbials) developed rapidly during the past 30 years with 29 companies producing products compared to only 10 companies in North America. Zorner *et al.* (1993) noted that the production of microbial bioherbicides requires very critical processes and that less than 10% of the people working on them in North America focus their research in this area. Ten years later, this situation has improved modestly in terms of research on liquid and solid state production, but not in terms of national capacity for manufacturing or in attitude towards adopting different concepts and ideas for manufacturing.

Risk vs. Return

Registration constitutes a significant portion of the total investment of a bioherbicide. It costs between $250,000–$650,000 USD and takes 1–2 years to register a bioherbicide as compared to $5–20M USD and 3–4 years for a conventional herbicide (Panetta, 1999; Meneley, 2000; Cuddeford, 2005). The difference in cost is primarily due to the fact that most biopesticide products have been demonstrated to be of low toxicity to mammals and other non-targets species, thus requiring fewer toxicology and environmental safety tests to be conducted. The vast majority of microbial pest control products reviewed by PMRA and EPA have not triggered testing beyond Tier II (Cuddeford, 2005). Regulatory requirements in Canada, United States and Europe are similar but there are some significant differences. For example, efficacy data is required by Canada and Europe but not by the USA (Cuddeford, 2005). The actual registration fees charged by regulatory agencies are quite variable; costs range from $450 USD (only charge for label review) in Canada, $15,000–20,000 USD in the United States, $42,000 USD in the United Kingdom, and may be as high as $212,000 USD in Sweden. In addition, there are other costs associated with regulatory requirements for studies on toxicology ($150,000–180,000 USD), ecotoxicology ($130,000–175,000), and field efficacy testing ($130,000–175,000 USD).

Company size influences the tolerance for risk and the expectation of return. Larger companies may be more equipped to afford the development, registration, and commercialization of a new product, but they also have much greater expectations of return. For example, investing in a bioherbicide that will be registered for use against multiple weeds and on multiple major and minor crops generates higher revenue than one for control of a single weed in one or two major crops (Bailey and Mupondwa, 2006).

Smaller companies have limited cash to invest and raising large amounts for registration can be a burden, especially if they are also involved in the research and development. The use of public funding to generate toxicology and environmental data for products deemed to have environmental benefits or other public good attributes would increase participation of smaller companies to bring forward new products. In the United States, the IR-4 program assists small companies with the registration process by developing research protocols and good laboratory practice standards, funding field trials for efficacy and residue studies, preparing and submitting petitions for tolerance exemptions, and developing efficacy data to expand registrations to additional crops and uses (Anonymous, 2006a). In Canada, there is no comparable programme in place yet, so small and medium companies rely on doing registration themselves with limited manpower or hire private consultants at considerable cost, sometimes upwards of $60,000 USD.

Influence of Government Policy Making and Regulations

The implementation of the Food Quality Protection Act in 1996 by the United States Congress set new health-based standards to reduce risks associated with all pesticide products (Uri, 1999). With regard to biopesticides this served to accelerate development and commercialization of new biological pest approaches by expediting the review of minor use products and approval of new reduced–risk products (i.e., those that have less risk to human health and non-target organisms, reduced potential for contamination of environmental resources, broaden the adoption of integrated pest management, or are more effective when compared to existing alternatives). This programme has resulted in 110 registrations of biopesticide active ingredients from 1996–2005 (EPA, 2005), and of these, four are bioherbicides. Similar initiatives in Canada were established at the same time under a Joint Review Programme of the North American Free Trade Agreement with the initiative further expanded in 2002 to encourage manufacturers to apply for Canadian registration of reduced-risk products currently available in the United States. As of 2005, there were 46 registered biopesticides in Canada, of which four were bioherbicides (PMRA, 2005), including BioMal® which passed a mandatory review process to ensure continuing acceptability of products registered prior to 1994 using current scientific approaches and standards (PMRA, 2006).

Although these regulatory initiatives are laudable, government still has a larger role to play in the research and development of new biopesticides. Many funding programs focus on modifying currently available active ingredients as opposed to developing new active components, despite the fact that new intellectual property generates more wealth for a country than modifying the knowledge created by others. Currently, the infrastructure is not in place to support a strong manufacturing base. The cash flow in many small to medium enterprises cannot sustain the company through the registration and commercialization process, especially when they do not have other products to sell or access to large markets. Government policies that affect price responsiveness, expected returns, technology development, and adoption will impact on the development of biopesticides (Uri, 1999). Some examples include reducing risks associated with basic research by creating new funding sources for research, aggressively supporting product development and technology transfer by encouraging local manufacturing industries for new bioproducts, creating new funding and tax incentives, encouraging the development of minor use biological products, finding ways to offset the costs of development, taking a stronger role in public education, and providing public incentives (environmental or economic) to facilitate adoption. In Canada, some policies have been developed in support of biological control. The Agricultural Policy Framework offers the National Farm Stewardship Programme that provides incentives to producers using "beneficial management practices" which includes a 30% cost-sharing for the purchase of microbial pesticides and companies may also access the Scientific Research and Experimental Development Programme, which provides tax incentives for research performed and technology developed in Canada (Anonymous, 2005). Unfortunately, smaller companies do not have the cash flow to access these tax based incentives that generally require matching funds. There is a need for increased dialogue between the smaller players and government to establish programmes that are more accessible. For example, changing the regulatory requirements to permit a phased-in approach to registration could allow a company to manufacture small lots of product and sell to test markets on limited and regulated scale to generate revenue while gaining valuable insight as to the value and demand for the product.

Conclusion

The diversity of traits found in fungal species presents unique opportunities to researchers who are interested in novel aspects to the fungal functions and seek ways to utilize these selected traits. The development of bioherbicides from fungi is one opportunity that continues to challenge us. The concept of using fungi for weed control was shown to be feasible by the bioherbicide prototypes developed in the 1990s and we continue to discover large numbers of fungal species and strains with bioherbicidal traits. However, past experiences have taught us that there is much more to bioherbicide development than finding a fungus that can cause damage to a weed target.

The first set of challenges that became apparent after prototype developments concentrated on learning how to select the right strain for the right target, accumulating knowledge on pathogen-weed-crop interactions and modes of action, as well as putting more effort into developing mass production and formulation systems for application of the technology. Although many of these challenges remain a high priority, researchers began to realize that to achieve successful commercialization, there is more to the process than what can be delivered by conducting smaller-scale research trials, whether in the laboratory or in the field.

The future of bioherbicides as an emerging industry will depend on our ability to create a smooth transition through the various stages from discovery and basic science to application, development, registration, commercial scale-up, and technology adoption (See Integration of Science and Technology in Fig. 1.3 and Innovation Chain in Fig. 1.4). This will require finding new ways to reduce the risks to industry partners, especially small and medium size enterprises. Researchers need to do a better job at transferring the basic technology from the "laboratory" scale to an intermediate stage or pilot scale process that delivers a final product form suitable for manufacturing by proving cost-effectiveness, comparable product quality, and commercial feasibility. Industry partners must be willing to take the risks of commercializing an under-developed, unfamiliar technology and work closely with the scientist, and possibly other partners to move the technology through the innovation chain. Unfortunately, this becomes a "catch-22" situation for the inventor and the industry entrepreneur. The "laboratory" ideas generated by scientists may be shaken when confronted by marketing and costing realities presented by industry partners. The entrepreneur may not have experience, resources or personnel to devote to development, especially when a potential product still has too many technological risks. Both partners are taken out of their areas of expertise and training, and this uncertainty may lead to inaction, whereby government policy and incentive programs may be needed to encourage the partners to work through these barriers. The timeframe for the development of a single bioherbicide technology may be 10–15 years, but the evolution of a new industry takes even longer.

References

Abeln, E.C.A., Stax, A.M., de Gruyter, J., and van der Aa, H.A. (2002). Genetic differentiation of *Phoma exigua* varieties by means of AFLP fingerprints, *Mycological Research* 106:419–427.

Abbas, H.A., and Duke, S.O. (1995). Phytotoxins from plant pathogens as potential herbicides, *Journal of Toxicology: Toxin Reviews* 14(4):523–543.

Agrios, G.N. (1978). Plant Pathology, 2nd Edition, Academic Press, New York, 48–63.

Amsellem, Z., Zidack, N.K., Quimby, P.C., and Gressel J. (1999). Long-term dry preservation of viable mycelia of two mycoherbicidal organisms, *Crop Protection* 18:643–649.

Anonymous (2002). Legislation to modernize pesticide legislation and protect Canadians receives Royal Assent. Health Canada News Release 2002-84, December 23. [www.hc-sc.gc.ca/ahc-asc/media/nr-cp/2002/2002_84_e.html].

Anonymous (2005). Canadian tax incentives for biocontrol research. Biocontrol Files, Issue 3, July, World Wildlife Fund Canada, Biocontrol Network, Agriculture and Agri-Food Canada, Montreal, Quebec, 8. [www.biocontrol.ca.]

Anonymous (2006a). Surveys gauge attitude of Canadian public. Biocontrol Files, Issue 5, January, World Wildlife Fund Canada, Biocontrol Network, Agriculture and Agri-Food Canada, Montreal, Quebec, 8. [www.biocontrol.ca.]

Anonymous (2006b). Barriers to wider adoption of biocontrol and strategies to address them-final results of a *Biocontrol Files* reader survey. Biocontrol Files, Issue 6, April, World Wildlife Fund Canada, Biocontrol Network, Agriculture and Agri-Food Canada, Montreal, Quebec, 1–3. [www.biocontrol.ca.]

Auld, B.A. (1993). Mass production of fungi for biopesticides, *Plant Protection Quarterly* 8:7–9.

Auld, B.A. (1991). Economic aspects of biological control. *In:*Microbial Control of Weeds, TeBeest, D.O. Ed.. Chapman and Hall, London, 262–273.

Auld, B.A. and Morin, L. (1995). Constraints in the development of bioherbicides, Weed Technology 9:638–652.

Auld, B.A., Hetherington, S.D., and Smith, H.E. (2003). Advances in bioherbicide formulation, Weed Biology and Management 3:61–67.

Avis, T.J., Hamelin, R.C., and Belanger, R.R. (2001). Approaches to molecular characterization of fungal biocontrol agents: some case studies, *Canadian Journal of Plant Pathology* 23:8–12.

Babu, R.M., Sajeena, A., and Seetharaman, K. (2004). Solid substrate for production of *Alternaria alternata* conidia: a potential mycoherbicide for the control of *Eichhornia crassipes* (water hyacinth), *Weed Research* 44:298–304.

Bailey, K.L. (2004). Microbial weed control: an off-beat application of plant pathology, *Canadian Journal of Plant Pathology* 26:239–244.

Bailey, K.L., and Mupondwa, E.K. (2006). Developing microbial weed control products: commercial, biological, and technological considerations. In: Handbook of Sustainable Weed Management, Singh H.P, Batish D.R., and Kohli R.K. Eds., The Haworth Press, Binghamton, New York, 431–473.

Bousquet, J.F., and Barbier, M. (1972). Phytotoxical activity of three strains of *Phoma exigua* and the presence of cytochalasine B (or phomine) in the cultural medium, *Phytopathologische Zeitschrift* 75:365–367.

Bowers, R.C. (1986). Commercialization of Collego®–An industrialist's view, *Weed Science* 34:24–25.

Boyetchko, S.M. (1997). Principles of biological weed control with microorganisms, *Hort Science* 32:201–205.

Boyetchko, S.M. (2005). Biological herbicides in the future. In: Weed Management in Transition. Topics in Canadian Weed Science, Volume 2, Ivany J.A. Ed., Canadian Weed Science Society—Société canadienne de malherbologie, Sainte-Anne-de-Bellevue, Quebec, 29–47.

Boyetchko, S.M., and Peng, G. (2004). Challenges and strategies for development of mycoherbicides. *In:* Fungal Biotechnology in Agricultural, Food, and Environmental Applications, Arora, D.K., Ed., Marcel Dekker Inc., New York, 111–121.

Boyetchko, S.M., and Rosskopf, E.N. (2006). Strategies for developing bioherbicides for sustainable weed management. *In*: Handbook of Sustainable Weed Management, Singh, H.P., Batish, D.R., and Kohli, R.K., Eds., The Haworth Press, Binghamton, NewYork, 393–430.

Boyetchko, S.M., Rosskopf, E.N., Caesar, A.J., and Charudattan, R. (2002). Biological weed control with pathogens: search for candidates to applications. *In*: Applied Mycology and Biotechnology, Volume 2. Agriculture and Food Production, Khachatourians, G.G, and Arora, D.K., Eds. Elsevier B.V., the Netherlands, 239–274.

Boyetchko, S.M., K.L. Bailey, R. H. Hynes, and G. Peng. 2007. Development of an inundative mycoherbicide: BIOMAL®. *In* Biological Control: Global perspective. Vincent, C., Goettel, M.S., and Lazarovits, G., eds. CABI Publishing, Wallingford, UK, 274–283.

Braverman, M.P., Kunkel, DL., Baron, J.J., and Holm, R.H. (2006). Regulatory considerations for researchers of natural products in the United States *In*: Natural Products for Pest Management (ACS Symposium Series 927), Rimando, A.M., and Duke, S.O., Eds., American Chemical Society, Washington, D.C., 278–288.

Brown, J.K.M. (2000). Estimation of rates of recombination and migration in populations of plant pathogens, *Phytopathology* 90:320–323.

Brugmans, B., van der Hulst, R.G.M., Visser, R.G.F., Lindhout, P., and van Eck, H.J. (2003). A new and versatile method for the successful conversion of AFLP markers into simple single locus markers, *Nucleic Acids Research* 31(10):e55.

Burges, H.D. (1998). Formulation of Microbial Biopesticides: Beneficial Microorganisms, Nematodes and Seed Treatments, Kluwer Academic, Dordrecht, the Netherlands 1–6.

Burnett, H.C., Tucker, D.P.H., Patterson, M.E., and Ridings, W.H. (1973). Biological control of milkweed vine (*Morrenia odorata*) with a race of *Phytophthora cirophthora*, Proceedings of Florida State Horticultural Society 86:111–115.

Butt, T.M., Jackson, C., and Magan, N. (2001). Introduction – Fungal biological control agents: progress, problems and potential. *In*: Fungi as Biocontrol Agents: Progress, Problems and Potential, Butt, T.M., Jackson, and C., Magan, N., Eds., CABI Publishing, Wallingford, 1–8.

Byer, K.N., Peng, G., Wolf, T.M., and Caldwell, B.C. (2006a). Spray retention and its effect on weed control by mycoherbicides, *Biological Control* 37:307–313.

Byer, K.N., Peng, G., Wolf, T.M., and Caldwell, B.C. (2006b). Spray retention for liquid and mycoherbicide inoculum in three weed-biocontrol systems, *Biocontrol Science and Technology* 16:815–823.

Caccia, C. (2000). Pesticides: making the right choice for the protection of health and the environment. Report of the Standing Committee on Environment and Sustainable Development, May, 2000. Government of Canada, Ottawa. (www.parl.gc.ca/InfoComDoc/36/2/ENVI/Studies/Reports/envi01/04-toc-e.html)

Campbell, R. (1994). Biological control of soil-borne disease: some present problems and different approaches, *Crop Protection* 13: 4–13.

Carbone, I., and Kohn, L. (2004). Inferring process from pattern in fungal population genetics. *In*: Fungal Genomics, Applied Mycology and Biotechnology Series, Volume 4, Arora, D.K., and Khachatourians, G.G., Eds., Elsevier Science B.V., The Netherlands, 29–58.

Caron, S.J., Avis, T.J., Boekhout, T., Hamelin, R.C., and Belanger, R.R. (2005). Fingerprinting techniques as tools towards molecular quality control of *Pseudozyma flocculosa*, *Mycological Research* 109: 335–341.

Charudattan, R. (2005). Ecological, practical, and political inputs into selection of weed targets: what makes a good biological control target?, *Biological Control* 35:183–196.

Charudattan, R. (2001). Biological control of weeds by means of plant pathogens: significance for integrated weed management in modern agro-ecology, *BioControl* 46:229–260.

Charudattan, R. (2000). Current status of biological control of weeds. *In:* Emerging Technologies for Integrated Pest Management, Kennedy, G.C., and Sutton, T.B., Eds., APS Press, St. Paul, Minnesota, 269–288.

Charudattan, R. (1991). The mycoherbicide approach with plant pathogens. *In*: Microbial Control of Weeds, TeBeest D.O. Ed., Chapman and Hall, Inc., New York, 24–57.

Connick Jr., W.J., Daigle, D.J., Boyette, C.D., Williams, K.S., Vinyard, B.T., Quimby Jr, P. C. (1996). Water activity and other factors that affect the viability of *Colletotrichum truncatum* conidia in wheat flour-kaolin granules ('Pesta'), *Biocontrol Science and Technology* 6:277–284.

CropLife Canada. (2003). Industry Statistics: Industry performance 2003 Archives. (www.croplife.ca/english/aboutcpi/industrystatistics.html)

CropLife Canada. (2002a). TAB1: History of crop protection and pest control in our society. In: Analyzing the risks, balancing the benefits: the facts on pesticides and human safety, a crop protection industry information guide. (www.croplife.ca/english/resourcescentre/analyzing2003.html)

CropLife Canada. (2002b). Industry Statistics: Industry performance 2002 Archives. (www.croplife.ca/english/aboutcpi/industrystatistics_2002.html)

CropLife Canada. (2001). Industry Statistics: Industry performance 2001 Archives. (www.croplife.ca/english/aboutcpi/industrystatistics_2001.html)

Cross, J.V., and Polonenko, D.R. (1996). An industry perspective on registration and commercialization of biocontrol agents in Canada, *Canadian Journal of Plant Pathology* 18: 446–454.

Crump, N.S., Cother, E.J., and Ash, G.J. (1999). Clarifying the nomenclature in microbial weed control, *Biocontrol Science and Technology* 9: 89–97.

Cuddeford, V. (2005). Pesticide risk reduction program BCA workshop: Critical regulatory path for biocontrol agents, February 28-March 1, 2005, Saskatoon. *Agriculture and Agri-Food Canada*, 61. (www.agr.gc.ca/env/pest)

Dahod, S.K. (1999). Raw materials selection and medium development for industrial fermentation processes. *In:* Manual of Industrial Microbiology and Biotechnology, 2nd Edition, Demain, A.L., and Davies, J.E., Eds., ASM Press, Washington, D.C., 213–220.

Daigle, D.J., Connick Jr., W.J., and Boyetchko, S.M. (2002). Formulating a weed-suppressive bacterium in "Pesta", *Weed Technology* 16: 407–413.

Daigle, D.J., Connick Jr., W.J., Boyette, C.D., Jackson, M.A., and Dorner, J.W. (1998). Solid-state fermentation plus extrusion to make biopesticide granules, *Biotechnology Techniques* 12:715–719.

Daniel, J.T., Templeton, G.E., Smith Jr., R.J., and Fox, W.T. (1973). Biological control of northern jointvetch in rice with an endemic fungal disease, *Weed Science* 21:303–307.

Davis, N.D., and Blevins, W.T. (1979). Methods for laboratory fermentations. *In:* Microbial Technology, Vol. II, 2nd Edition, Peppler, H.J., and Perlman, D., Eds., Academic Press, Toronto, Canada, 303–329.

Deacon, J.W. (1991). Significance of ecology in the development of biocontrol agents against soil-borne plant pathogens, *Biocontrol Science and Technology* 1: 5–20.

Delaplane, K.S. (2000). Pesticide usage in the United States: history, benefits, risks, and trend, Bulletin 1121, Cooperative Extension Service, College of Agricultural and Environmental Sciences, University of Georgia, 1–8. (http://pubs.caes.ugs.edu/caespubs/pubcd/B1121.htm)

Dokken, F., Peng, G., Bailey, K.L., and Nelson, L.M. (2004). Submerged fermentation of Colletotrichum truncatum for biological control of scentless chamomile, *Canadian Journal of Plant Pathology* 26:224–225.

Dokken, F. (2003). Submerged fermentation of *Colletotrichum* sp. for the biological control of scentless chamomile. Undergraduate Thesis, Department of Plant Sciences, University of Saskatchewan, Saskatoon, pp 54.

Elzein, A., Kroschel, J., and Muller-Stover, D. (2004). Effects of inoculum type and propagule concentration on shelf life of Pesta formulations containing *Fusarium oxysporum* Foxy 2, a potential mycoherbicide agent for *Striga* spp., *Biological Control* 30:203–211.

EPA (2005). New Biopesticide Active Ingredients–2005. (http://www.epa.gov/pesticides/biopesticides/products_list/new_ai_2005.htm)

Evidente, A. and Motta, A. (2001). Phytotoxins from fungi, pathogenic for agrarian, forestal and weedy plants. *In*: Bioactive compounds from natural sources: isolation, characterization, and biological properties, Tringali, C. Ed., Taylor and Francis, London, 474–525.

Evidente, A., Lanzetta, R., Capasso, R., Andolfi, A, Vurro, M., and Zonno, M.C. (1997). Putaminoxins B and C from *Phoma putaminum*, *Phytochemistry* 44(6):1041–1045.

Evidente, A., Lanzetta, R., Capasso, R., Andolfi, A., Bottalico, A., Vurro, M. and Zonno, M.C. (1995). Putaminoxin, a phytotoxic nonenolide from *Phoma putaminum*, *Phytochemistry* 40(6):1637–1641.

Forseille, L., Wei, Y.D., and Peng, G. (2006). Differentiation and detection of *Colletotrichum truncatum* isolates by strain specific PCR markers, *Canadian Journal of Plant Pathology* 28:331 (Abstr.).

Fravel, D.R., Rhodes, D.J., and Larkin, R.P. (1999). Production and commercialization of biocontrol agents. *In*: Integrated Pest and Disease Management in Greenhouse Crops, Albajes, R., Lodovica-Guillino, M., van Lenteren, J.C., Elad, Y., Eds., Kluwer and Academic Publishers, Dordrecht., the Netherlands, 365–376.

Gaugler, R. (1997). Alternative paradigms for commercializing biopesticides, *Phytoparasitica* 25:179–182.

Gosselin, L., Jobidon, R., and Bernier, L. (1999). Genetic variability and structure of Canadian populations of *Chondrostereum purpureum*, a potential biophyticide, *Molecular Ecology* 8:113–122.

Graham, G.L. (2004). Synergism of *Colletotrichum truncatum* with herbicides for control of scentless chamomile (*Matricaria perforata*). M.Sc. Thesis, Department of Plant Sciences, University of Saskatchewan, Saskatoon, Saskatchewan, pp 121.

Graham, G.L., Peng, G., Bailey, K.L., and Holm, F.A. (2006a). Effect of dew temperature, post-inoculation condition, and pathogen dose on suppression of scentless chamomile by *Colletotrichum truncatum*, *Biocontrol Science and Technology* 16: 271–280.

Graham, G.L., Peng, G., Bailey, K.L., and Holm, F.A. (2006b). Effect of plant stage, *Colletotrichum truncatum* dose and use of herbicide on control of *Matricaria perforate*, BioControl 51: (In Press), available on line: (http://www.springerlink.com/content/e7057w13514r7055/fulltext.pdf)

Graham, G.L., Peng, G., Bailey, K.L., and Holm, F.A. (2006c). Interactions of *Colletotrichum truncatum* with herbicides for control of scentless chamomile (*Matricaria perforata*), *Weed Technology* 20:877-884.

Graham, G.L., G. Peng, K.L. Bailey, and F.A. Holm. 2007. Effect of plant stage, *Colletotrichum truncatum* dose, and use of herbicide on control of *Matricaria perforata*. BioControl 52:573–589.

Grant, N.T., Prusinkiewicz, E., Makowski, R.M.D., Holmstrom-Ruddick, B., and Mortensen, K. (1990a). Effect of selected pesticides on survival of *Colletrotrichum gloeosporioides* f.sp. *malvae*, a bioherbicide for round-leaved mallow (*Malva pusilla*), *Weed Technology* 4:701–717.

Grant, N.T., Prusinkiewicz, E., Mortensen, K., and Makowski, R.M.D. (1990b). Herbicide interactions with *Colletrotrichum gloeosporioides* f.sp. *malvae*, a bioherbicide for round-leaved mallow (*Malva pusilla*) control, *Weed Technology* 4:716–723.

Graupner, P.R., Carr, A., Clancy, E., Gilbert, J., Bailey, K.L., Derby, J. and Gerwick, B.C. (2003). The macrocidins: Novel cyclic tetramic acids with herbicidal activity produced by *Phoma macrostoma*, *Journal of Natural Products* 66(12):1558–1561.

Graupner, P.R., Gerwick, B.C., Siddall, T.L., Carr, A.W., Clancy, E., Gilbert, J.R., Bailey, K.L., and Derby, J. (2006). *In:* Natural Products for Pest Management (ACS Symposium Series 927), Rimando, A.M., and Duke, S.O., Eds., American Chemical Society, Washington, DC, 37–47.

Greaves, M.P., and MacQueen, M.D. (1992). Bioherbicides: their role in tomorrow's agriculture. In: Resistance, Achievements and Developments in Combating Pesticide Resistance, Science Symposium, Denholm, I., Devonshire, A. L., and Hollomon, D. W., Eds., Elsevier Applied Science, Harpenden, England, 295–306.

Greaves, M.P., Holloway, P.J., and Auld, B.A. (1998). Formulation of microbial herbicides. *In:* Formulation of Microbial Biopesticides: Beneficial Microorganisms, Nematodes, and Seed Treatments, Burges, H.D., Ed., Kluwer Academic Publishers, Dordrecht, The Netherlands, 203–233.

Green, S. and Bailey, K.L. (2001). A combination of glyphosate and *Alternaria cirsinoxia* for control of Canada thistle (*Cirsium arvense*), *Canadian Journal of Plant Pathology* 23:198

Green, S. and Bailey, K.L. (2000a). Effects of leaf maturity, infection site, and application rate of *Alternaria cirsinoxia* conidia on infection of Canada thistle (*Cirsium arvense*), *Biological Control* 19:167–174.

Green, S. and Bailey, K.L. (2000b). The influence of moisture and temperature on infection of Canada thistle by *Alternaria cirsinoxia*, Plant Disease 84:1126–1132.

Green, S., Bailey, K.L., and Tewari, J.P. (2001a). The infection process of *Alternaria cirsinoxia* on Canada thistle (*Cirsium arvense*) and host structural defense responses, *Mycological Research* 105: 344–351.

Green, S., Mortensen, K., and Bailey, K.L. (2001b). Host range, temperature response, survival, and overwintering of *Alternaria cirsinoxia, Biological Control* 20:57–64.

Green, S., Peng, G., Connolly, T., and Boyetchko, S.M. (2004). Effect of moisture and temperature on disease of green foxtail caused by *Drechslera gigantea* and *Pyrcularia setariae, Plant Disease* 88: 605–612.

Hallett, S.G. (2005). Where are the bioherbicides?, *Weed Science* 53: 404–415.

Hilton, M.D. (1999). Small-scale liquid fermentations. In: Manual of Industrial Microbiology and Biotechnology, 2nd Edition, Demain, A.L., and Davies, J.E., Eds., ASM Press, Washington, D.C., 49–60.

Hintz, W.E., Becker, E.M., and Shamoun, S.F. (2001). Development of molecular markers for risk assessment of biological control agents, *Canadian Journal of Plant Pathology* 23: 13–18.

Hoagland, R.E. (1996). Chemical interactions with bioherbicides to improve efficacy, *Weed Technol.* 10: 651–674.

Holker, U., Hofer, M., and Lenz, J. (2004). Biotechnological advantages of laboratory-scale solid-state fermentation with fungi, Applied Microbiological Biotechnology 64:175–186.

Hopen, H.J., Caruso, F.L., Bewick, T.A., Yarborough, D.E., and Smagula, J.M. (1997). Control of dodder in cranberry (*Vaccinium macrocarpon*) with a pathogen-based bioherbicide, *Acta Horticulture* 446: 427–428.

Hosobuchi, M. and Yoshikawa, H. (1999). Scale-up of microbial processes. In: Manual of Industrial Microbiology and Biotechnology, Demain, A.L., and Davies, J.E., Eds., ASM Press, Washington, D.C., 236–239.

Hynes, R.K. and Boyetchko, S.M. (2006). Research initiatives in the art and science of biopesticide formulations, Soil Biology & Biochemistry 38:845–849.

Jackson, M.A., Schisler D.A., Slininger, P.J., Boyette, C.D., Silman, W.R., and Bothast, R.J. (1996). Fermentation strategies for improving the fitness of a bioherbicide, *Weed Technology* 10:645–650.

Jenkins, N.E. and Grzywacz, D. (2000). Quality control of fungal and viral biocontrol agents – assurance of product performance, *Biocontrol Science and Technology* 10: 753–777.

Jones, K.A. and Burges, H.D. 1998. Technology of formulation and application. In: Formulation of Microbial Biopesticides: Beneficial Microorganisms, Nematodes and Seed Treatments, Burges, H.D., Ed., Kluwer Academic, Dordrecht, The Netherlands, 7–30.

Kenfield, D., Bunkers, G., Strobel, G., and Sugawara, F. (1989). Fungal phytotoxins – potential new herbicides. *In:* Phytotoxins and plant pathogenesis (NATO ASI series, Volume H27), Graniti, A., Ed., Springer-Verlag, Berlin, 319–335.

Kenney, D.S. (1986). DeVine® –The way it was developed – an industrialist's view, *Weed Science* 34 (Suppl.1):15–16.

Kiely, T., Donaldson, D., and Grube, A. (2004). Pesticides Industry Sales and Usage: 2000 and 2001 market estimates. U.S. Environmental Protection Agency, Washington, D.C., (www.epa.gov/oppbead1/pestsales/market_estimates2001.pdf).

Kistler, H.C. (1997). Genetic diversity in the plant-pathogenic fungus *Fusarium oxysporum*, *Phytopathology* 87: 474–479.

Leath, K.T. (1992). Phoma. *In*: Methods for research on soilborne phytopathogenic fungi, Singleton, L.L., Mihail, J.D., and Rush, C.M., Eds., APS Press, St. Paul, Minnesota, 142–144.

Lievens, B. and Thomma, B.P.H.J. (2005). Recent developments in pathogen detection arrays: Implications for fungal plant pathogens and use in practice, *Phytopathology* 95: 1374–1380.

Liu, L., Peng, G., Wei, Y.D., Gossen, B.D., and Bailey, K.L. (2005). Differentiation of *Colletotrichum* spp. from scentless camomile and lentil based on infection characteristics and ITS sequences, *Canadian Journal of Plant Pathology* 27: 472.

Makowski, R.M.D. (1995a) Round-leaved mallow (*Malva pusilla*) interference in spring wheat (*Triticum aestivum*) and lentil (*Lens culinaris*) in Saskatchewan, *Weed Science* 43:381–388.

Makowski, R.M.D. (1995b). Tolerance of strawberries to *Colletotrichum gloeosporioides* f.sp. *malvae*, a mycoherbicide for control of round-leaved mallow (*Malva pusilla*), *Weed Science* 43:429–433.

Makowski, R.M.D. and Mortensen, K. (1992). The first mycoherbicide in Canada: *Colletotrichum gloeosporioides* f. sp. *malvae* for round-leaved mallow control. *In:* Proceedings of the First international Weed Control Congress, 17–21 February, 1992, Volume 2, Weed Science Society of Victoria Inc., Melbourne, Australia, 298–300.

McDonald, B.A. (1997). The population genetics of fungi: Tools and techniques, *Phytopathology* 87:448–453.

McCartney, H.A., Foster, S.J., Fraaije, B.A., Ward, E. (2003). Molecular diagnostics for fungal plant pathogens, *Pest Management Science* 59: 129–142.

Meneley, J.C. (2000). Challenges to the commercialization of biological control technologies for IPM. *In*: Emerging Technologies for Integrated Pest Management, Kennedy, G.C., and Sutton, T.B., Eds., APS Press, St. Paul, Minnesota, 289–304.

Monaghan, R.L., Gagliardi, M.M., and Streicher, S.L. (1999). Culture preservation and inoculum development. *In:* Manual of Industrial Microbiology and Biotechnology, Demain, A.L., and Davies, J.E., Eds., ASM Press, Washington, D.C., 29–48.

Montazeri, M. and Greaves, M.P. (2002). Effects of culture age, washing and storage conditions on desiccation tolerance of *Colletotrichum truncatum* conidia, *Biocontrol Science and Technology* 12: 95–105.

Morin, L., Watson, A.K., and Reeleder, R.D. (1990). Production of conidia by *Phomopsis convolvulus*, *Canadian Journal of Microbiology* 36:86–91.

Mortensen, K. (1998). Biological control of weeds using microorganisms. *In:* Plant-Microbe Interactions and Biological Control, Boland G.J, and Kuykendall, L.D., Eds., Marcel Dekker Inc., New York, 223–248.

Mortensen, K. (1988). The potential of an endemic fungus, *Colletotrichum gloeosporioides*, for biological control of round-leaved mallow (*Malva pusilla*) and velvetleaf (*Abutilon theophrastis*), *Weed Science 36*: 473–478.

Mortensen, K. and Bailey, K.L. (2002). *Malva pusilla* Smith, Round-leaved mallow (*Malvaceae*). *In*: Biological Control Programmes in Canada, 1981–2000, Mason, P.G., and Huber, J.T., Eds. CABI Publishing, Wallingford, UK, 391–394.

Namai, T. and Yamanaka. S. (1983). A component contained in oatmeal that favors the sporulation of rice blast fungus, *Pyricularia oryzae* Cavara prolamin, Nihon Shokubutsu Byori Gakkai-Ho 49:274–276.

Neumann, S. and Boland, G.J. (2002). Influence of host and pathogen variables on the efficacy of *Phoma herbarum*, a potential biological control agent of *Taraxacum officinale*, *Canadian Journal of Botany* 80:425–429.

O'Donovan, J.T., de St. Remy, E.A., O'Sullivan, P.A., Dew, D.A., and Sharma. A.K. (1985). Influence of the relative time of emergence of wild oat (*Avena fatua*) on yield loss of barley (*Hordeum vulgare*) and wheat (*Triticum aestivum*), *Weed Science* 33:498–503.

Ogel, Z.B., Brayford, D., and McPherson, M.J. (1994). Cellulose-triggered sporulation in the galactose oxidase-producing fungus *Cladobotryum* (*Dactylium*) *dendroides* NRRL 2903 and its re-identification as a species of *Fusarium, Mycological Research* 98:474–480.

Pandey, A. (1992). Recent process developments in solid-state fermentation, *Process Biochemistry* 27:109–117.

Panetta, J.D. (1999). Environmental and regulatory aspects: Industry view and approach. *In*: Methods in Biotechnology, Volume 5, Biopesticides: Use and Delivery, Hall, F.R., and Menn, J.J., Eds., Humana Press Inc., Totowa, New Jersey, 473–484.

Pedras, M.S.C, Ismail, N., Quail, J.W., and Boyetchko, S.M. (2003). Structure, chemistry, and biological activity of pseudophomins A and B, new cyclic lipodepsipeptides isolated from the biocontrol bacterium *Pseudomonas fluorescens*, Phytochemistry 62: 1105–1114.

Peng, G. and S.M. Boyetchko. 2006. Effect of variable dew temperatures on infection of green foxtail by *Pyricularia setariae*, *Drechslera gigantea*, and *Exserohilum rostratum*, Bio. Control 39:539–546.

Peng, G. and Byer, K.N. (2005) Interactions of *Pyricularia setariae* with herbicides for control of green foxtail (*Setaria viridis*), *Weed Technology* 19:589–598.

Peng, G., Wolf, T.M., Byer, K.N., and Caldwell, B. (2001). Spray retention on green foxtail (*Setaria viridis*) using airbrush and broadcast sprayers and its impact on the efficacy of a mycoherbicide agent. In: Proceedings of 18th Asian-Pacific Weed Science Conference, May 28–June 2, 2001, Ni, H.W., and You, Z.G., Eds., Beijing, China, 699–706.

Peng, G., Byer, K.N., and Bailey, K.L. (2004). *Pyricularia setariae*: a potential bioherbicide agent for control of green foxtail (*Setaria viridis*), *Weed Science* 52:105–114.

Peng, G., Bailey, K.L., Hinz, H.L., and Byer, K.N. (2005a). *Colletotrichum* sp: A potential candidate for biocontrol of scentless chamomile (*Matricaria perforata*) in western Canada, *Biocontrol Science and Technology* 15:497–511.

Peng, G., Wolf, T.M., Byer K.N., and Caldwell, B. (2005b). Spray retention on green foxtail (*Setaria viridis*) and its effect on weed control efficacy by *Pyricularia setariae, Weed Technology* 19:86–93.

Pitt, W.M. and Gams, W. (2005). A redescription of *Plectosporium alismatis* (hyphomycetes with glomerellaceous affinities), *Nova Hedwigia*, 81: 311–323.

Pitt, W.M., Goodwin, S.B., Ash, G.J., Cother, N.J., and Cother, E.J. (2004). *Plectosporium alismatis* comb. nov., a new placement for the *Alismataceae* pathogen *Rhynchosporium alismatis. Mycological Research* 108: 775–780.

Pitt, W.M., Ash, G.J., and Cother, E.J. (2005b). A preliminary population structure analysis of *Plectosporium alismatis*, a potential biocontrol agent for *Alismataceae* weeds in Australian rice crops, *Canadian Journal of Plant Pathology* 27(1):167–168.

PMRA (2001). Guidelines for the registration of microbial pest control agents and products., DIR 2001–02, March 30, 2001. Health Canada, Pest Management Regulatory Agency, Ottawa., 1-99. (http://www.pmra-arla.gc.ca)

PMRA (2005). Update on reduced-risk pesticides in Canada, RR2005-02, July 19, 2005. Health Canada, Pest Management Regulatory Agency, Ottawa, 1–50. (http://www.pmra-arla.gc.ca)

PMRA. (2006). Re-evaluation of *Colletotrichum gloeosporioides* f. sp. *malvae* {CGM}, REV2006–10, June 27, 2006. Health Canada, Pest Management Regulatory Agency, Ottawa. 1–7. (http://www.pmra-arla.gc.ca.)

Ramsfield, T.D., Shamoun, S.F., Punja, Z.K., and Hintz, W.E. (1999). Variation in the mitochondrial DNA of the potential biological control agent *Chondrostereum purpurem*, *Canadian Journal of Botany* 77:1490–1498.

Renwick, A., and Powell, K. (1990). Biological control of soil-borne plant pathogens, Brighton Crop Protection Conference: *Pests and Diseases* 1:227–231.

Riches, C.R. (2001). The World's Worst Weeds, BCPC Symposium proceedings No.77. British Crop Protection Council, Surrey, UK, 1–118.

Rimando, A.M. and Duke, S.O. (2006). Natural products for pest management. *In*: Natural Products for Pest Management (ACS Symposium Series 927), Rimando, A.M., and Duke, S.O., Eds., American Chemical Society, Washington, D.C., 2–21.

Rivero-Cruz, J.F., Aguirre, G.G., Rojas, C.M.C.G., and Mata, R. (2000). Conformational behaviour and absolute stereostructure of two phytotoxic nonenolides from the fungus *Phoma herbarum*, Tetrahedron 56:5337–5344.

Robinson, T., Singh, D., and Nigam, P. (2001). Solid-state fermentation: a promising microbial technology for secondary metabolite production, *Applied Microbiological Biotechnology* 55:284–289.

Rothweiler, W. and Tamm, C. (1970). Isolation and structure of the antibiotics phomin and 5-dehydrophomin, *Helvetica Chimica Acta* 53(79):696–724.

Samson, R.A. (1995). Constraints associated with taxonomy of biocontrol fungi, *Canadian Journal of Botany* 73:S83–S88.

Saxena, S. and Pandey, A.K. (2001). Microbial metabolites as eco-friendly agrochemicals for the next millennium, *Applied Microbiology and Biotechnology* 55:395–403.

Schena, L., Nigro, F., Ippolito, A., and Gallitelli, D. (2004). Real-time quantitative PCR: a new technology to detect and study phytopathogenic and antagonistic fungi, *European Journal of Plant Pathology* 110: 893–908.

Schnick, P.J., Stewart-Wade, S.M., and Boland, G.J. (2002). 2, 4–D and *Sclerotinia minor* to control common dandelion, *Weed Science* 50:173–178.

Shabana, Y.M., Muller-Stover, D., and Sauerborn, J. (2002). Granular pesta formulation of *Fusarium oxysporum* f. sp. *orthoceras* for biological control of sunflower broomrape: efficacy and shelf-life, *Biological Control* 26: 189–201.

Slininger, P.J., Silman, R.W., and Jackson, M.A. (1993). Oxygen delivery requirements of *Colletotrichum truncatum* during germination, vegetative growth, and sporulation, *Applied Microbiological Biotechnology* 39:744–74.

Singh, H.P., Batish, D.R., and Kohli, R.K. (2003). Allelopathic interactions and allelochemicals: New possibilities for sustainable weed management, *Critical Reviews in Plant Sciences* 22:239–311.

Stanbury, P. F., Whitaker, A., and Hall, S.J. (1995). Media for industrial fermentations. *In:* Principles of Fermentation Technology, 2nd Edition, by Stanbury, P.F., Whitaker, A., and Hall, S.J. Elsevier Science Ltd., Pergamon Press, New York, 93–122.

Stewart-Wade, S.M. and Boland, G.J. (2004). Selected cultural and environmental parameters influence disease severity of dandelion caused by the potential bioherbicidal fungi, *Phoma herbarum* and *Phoma exigua*, *Biocontrol Science and i* 14(6):561–569.

Stowell, L.J. (1991). Submerged fermentation of biological herbicides. *In:* Microbial Control of Weeds, TeBeest, D.O. (Ed.). Chapman and Hall, New York, 225–261.

Strobel, G., Kenfield, D., Bunkers, G., Sugawara, F., and Clardy, J. (1991). Phytotoxins as potential herbicides, *Experientia* 47:819–826.

Swanton, C.J. and Weise, S.F. (1991). Integrated weed management: the rationale and approach, *Weed Technology* 5:657–663.

Templeton, G.E., Smith, R.J., and TeBeest, D.O. (1989). Perspectives on mycoherbicides two decades after discovery of the Collego® pathogen. In: Proceedings of the VII International Symposium on Biological Control of Weeds, 6-11 March, 1988, Delfosse, E.S.,Ed. Ist. Sper.Patol. veg. (MAF), Rome, Italy, 553–558.

Uri, N.D. (1999).Pesticide policy influences on biopesticide technologies. In: Methods in Biotechnology, Volume 5, Biopesticides: Use and Delivery, Hall F.R., and Menn, J.J., Eds., Humana Press Inc., Totowa, New Jersey, 55–73.

Urquhart, E.J., Menzies, J.G., and Punja, Z.K. (1994). Growth and biological control activity of *Tilletiopsis* species against powdery mildew (*Sphaerotheca fuliginea*) on greenhouse cucumber, *Phytopathology* 84:341–351.

van Lenteren, J.C., Roskam, M.M., and Timmer, R. (1997). Commercial mass production and pricing of organisms for biologiocal control of pests in Europe, *Biological Control* 10:143–149.

Vega, F.E., Jackson, M.A., Mercadier, G., and Poprawski, T.J. (2003). The impact of nutrition on spore yields for various fungal entomopathogens in liquid culture, *World Journal of Microbiology and Biotechnology* 19: 363–368.

Venkatasubbaiah, P. and Chilton, W.S. (1992a). An epoxydon-derived ester from a *Phoma* sp. Pathogenic to rhubarb, *Journal of Natural Products* 55(5):639–643.

Venkatasubbaiah, P. Van Dyke, C.G., and Chilton, W.S. (1992b). Phytotoxic metabolites of *Phoma sorghina*, a new foliar pathogen of pokeweed, *Mycologia* 84(5):715–723.

Vos, P., Hogers, R., Bleeker, M., Rijans, M., Van der Lee, T., Hornes, M., Frijters, A., Pot, J., Peleman, J., Kuiper, M., and Zabeau, M. (1995). AFLP: a new technique for DNA fingerprinting, *Nucleic Acids Research* 23: 4407–4414.

Vurro, M., Bottalico, A., Capasso, R., and Evidente, A. (1997). Cytochalasins from phytopathogenic *Ascochyta* and *Phoma* species. *In*: Toxins in Plant Disease Development and Evolving Technology, Upadhyay, R.K., and Mukerji, K.G., Eds., Oxford and IBH Publishing Co. Pvt. Ltd., New Delhi, 127–147.

Walker, H.L. (1980). *Alternaria macrospora* as a potential biocontrol agent for spurred anoda: production of spores for field studies, *Advances in Agricultural Technology* 12:1–5.

Wang, A.M., Doyle, M.V., and Mark, D.F. (1989). Quantification of mRNA by the polymerase chain reaction, Proceedings of the National Academy of Sciences of the USA 24: 9717–9721.

Wapshere, A.J. (1974). A strategy for evaluating the safety of organisms for biological weed control, *Annals of Applied Biology* 77:201–211.

Weaver, SE. (1984). Critical period of weed competition in three vegetable crops in relation to management practices, *Weed Research* 24:317–325.

Whipps, J.M. (1997). Developments in the biological control of soil-borne plant pathogens, *Advances in Botanical Research* 26:1–134.

White, T.J., Bruns, T., Lee, S., and Taylor, J.W. (1990). Amplification and direct sequencing of fungal ribosomal RNA genes for phylogenetics. *In:* PCR Protocols: A guide to methods and applications, Innis, M.A., Gelfand, D.H., Sninsky, J.J., and White, T.J., Eds., Academic Press, New York, 315–322.

Wilkin, J.E., Shamoun, S.F., Ritland, C., Ritland, K., and El-Kassaby, Y.A. (2005). Genetic diversity and population structure of *Valdensinia heterodoxa*, a potential biocontrol agent for salal in coastal British Columbia, *Canadian Journal of Plant Pathology* 27: 59–571.

Will, K.W., and Rubinoff, D. (2004). Myth of the molecule: DNA barcodes for species cannot replace morphology for identification and classification, *Cladistics* 20: 47–55.

Zhang, W.M., Sulz, M., and Bailey, K.L. (2002a). Evaluation of *Plectosporium tabacinum* for control of herbicide-resistant and herbicide-susceptible false cleavers, *Weed Science* 50:79–85.

Zhang, W.M., Sulz, M., Bailey, K.L., and Cole, D.E. (2002b). Effect of epidemiological factors on the impact of the fungus *Plectosporium tabacinum* on false cleavers (*Gallium spurium*), *Biocontrol Science and Technology* 12:183–194.

Zhang, W., Wolf, T.M., Bailey, K.L., Mortensen, K., and Boyetchko, S.M. (2003). Screening of adjuvants for bioherbicide formulations with *Colletotrichum* spp. and *Phoma* spp., *Biological Control* 26:95–108.

Zhou, L., Bailey, K.L., and Derby, J. (2004). Plant colonization and environmental fate of the biocontrol fungus *Phoma macrostoma*, *Biological Control* 30: 634–644.

Zhou, L., Bailey, K.L., Chen, C.Y., and Keri, M. (2005). Molecular and genetic analyses of geographic variation in isolates of *Phoma macrostoma* used for biological weed control, *Mycologia* 97: 612–620.

Zorner, P.S., Evans, S.L., and Savage, S.D. (1993). Perspectives on providing a realistic technical foundation for the commercialization of bioherbicides. *In*: Pest Control with Enhanced Environmental Safety, Duke, S.O., Menn, J.J., and Plimmer, J.R., Eds., American Chemical Society, Washington, D.C., 79–86.

2

Management of Plant Pathogenic Fungi by Mycoviruses

László Radócz

Department of Plant Protection, University of Debrecen
138. Böszörményi str. H-403, Telephone: + 36 52 508659, Fax: +36 52 413385,
E-mail: radocz@agr.unideb.hu

INTRODUCTION

Viruses are recognized as common in fungi and have been detected in species belonging each of the major taxonomic classes of the fungi (Aist and Williams, 1971; Moffitt and Lister, 1975; Lemke, 1977; Radócz, 1998). The presence of viruses in fungi adds a new dimension to experimental mycology. These viruses are applied in biological control methods against the host fungus. Today it is not surprising that quite a few of the known fungi contain virus-like particles (VLPs) or viruses. Many of them responding unique host phenotype (e.g., hypovirulence) related specifically to the presence of the mycovirus (Fig. 2.1)

Figure 2.1 Abnormal canker caused by the hypovirulent (dsRNA-containing) strain of the blight fungus *Cryphonectria parasitica* on European chestnut tree.

A virus-linked disease exists in the *Agaricus brunnescens* (Lemke, 1977) and "killer" phenotypes related to the viral double-stranded RNA (dsRNA) in the corn smut *Ustilago maydis* (Kotlin and Day, 1976) and in the yeast *Saccharomyces cerevisiae* (Wickner, 1976). Among plant pathogenic fungi belonging to *Ascomycota* one of the best example for the cytoplasmatically induced phenotype for hypovirulence is the chestnut blight fungus *Cryphonectria parasitica* (syn.:*Endothia parasitica*). Another good example of this phenomenon can be found in the phyto-pathogenic teleomorph *Thanatephorus cucumeris* (Castanho *et al.*, 1978). Therefore, a few important species of plant pathogenic fungi are very interesting too as experimental models for study of mycoviruses.

Conidia are efficient transporters of mycoviruses for cell transmission. This is parallel in transmissibility of host cell and virus and rather reflect a considerable degree of coevolutional relationships between these fungi and their viruses. Extensive vegetative reproduction with the formation of different types of propagation are well-known among fungi. These are frequently accompanied by a reduction or loss in sexuality. Therefore, the most of the fungal viruses are non-infectious as isolated virions and can be transmitted from one fungal cell to another only by cell fusion. Cell-limited transmission of such a virus might lead to considerable loss of classical integrity in function and organization of a virus, and this is rather proved from the organizational and physico-chemical analyses of mycoviruses (Lemke, 1981). A coadaptation is very well represented by the extensive loss of sexuality among fungi and of infectivity and structure among mycoviruses.

DISCOVERY OF FUNGAL VIRUSES

In 1950, a disease of the cultivated mushroom *Agaricus bisporus* was described (Sinden and Hauser, 1950). The symptoms of the disease were a significant decrease in production of mushrooms, distorted morphology in the development and premature deterioration of mushroom tissue. This early study on a disease of a commercial mushroom resulted in the first observation of a mycovirus by using electron microscope (Gandy and Hollings, 1962). Between 1962 and 1965 considerable evidence for fungal viruses in *Agaricus bisporus* accumulated and indicated by immunological tests that at least two serological distinct fungal viruses existed (Hollings *et al.*, 1965).

A transmissible disease of the plant pathogenic fungus *Helminthosporium (Cochliobolus) victoriae* was described in 1959 (Lindberg, 1959). Cellular extracts from abnormally stunted colonies of this fungus induced normal colonies to yield abnormal sectors. The infective agent was isolated from abnormal cells and concentrated by differential centrifugation and precipitation with ammonium-sulphate. Although Lindberg (1959) postulated that this transmissible agent might be a virus but the cellular extracts from infected fungal strains were not examined for virus by electron microscopy. Later, it was confirmed that the diseased isolates of the plant pathogenic fungus *Helminthosporium (Cochliobolus) victoriae* harbour two isometric dsRNA viruses. The diseased isolates produced little or no victorin, and were hypovirulent.

Studying the antibiotic activity associated with two conidial fungi, *Penicillium funiculosum* (Lampson *et al.*, 1967) and *Penicillium stoloniferum* (Kleinschmidt *et al.*, 1968) led to the discovery of viral dsRNA in fungi. Since these earlier studies provide the evidence of viruses in fungi (mainly dsRNA viruses) has grown progressively (Table 2.1).

Table 2.1 Distribution of fungal viruses among plant pathogenic fungi.

Fungal species	*Reference(s)*
Plasmodiophoromycota	
Plasmodiophora brassicae	Aist and Williams, 1971
Oomycota	
Phytophthora infestans	Corbett and Styer, 1977
Pythium butleri	Moffitt and Lister, 1975
Zygomycota	
Mucor sp.	Bozarth, 1972
Rhizopus sp.	Bozarth, 1972
Ascomycota	
Cryphonectria parasitica	Hansen *et al.*, 1985
Diplocarpon rosae	Bozarth *et al.*, 1972
Erysiphe (Blumeria) graminis	Yarwood *et al.*, 1973
Gaeumannomyces graminis	Rawlinson *et al.*, 1973
Nectria radicicola	Il-Pyung and Yong-Hwan, 2001
Neurospora crassa	Tuveson and Peterson, 1972
Rosellinia necatrix	Ikeda *et al.*, 2004
Sclerotinia sclerotiorum	Xie *et al.*, 2001
Basidiomycota	
Helicobasidium mompa	Ikeda *et al.*, 2004
Puccinia graminis	Mussell *et al.*, 1973
Puccinia recondita, P. striiformis	Yarwood *et al.*, 1973
Thanatephorus cucumeris	Castanho *et al.*, 1978
Ustilago maydis	Kotlin and Day, 1976
Deuteromycota	
Alternaria tenuis	Hollings *et al.*, 1965
Aspergillus niger	Banks *et al.*, 1970
Botrytis cinerea	Robyn *et al.*, 2001
Chalara elegans	Bozarth and Goenaga, 1977
Colletotrichum lindemuthianum	Rawlinson, 1973
Coniothyrium minitans	Cheng *et al.*, 2002
Fusarium oxysporum	Kilic, 1997
Helminthosporium maydis	Bozarth, 1977
Helminthosporium (Cochliobolus) victoriae	Lindberg, 1959
Penicillium funiculosum	Lampson *et al.*, 1967
Pyricularia oryzae	Yamashita *et al.*, 1971
Sclerotium cepivorum	Spire, 1971
Stemphylium botryosum	Hollings and Stone, 1969
Trichotecium sp.	Bozarth, 1972
Verticillium sp.	Bozarth, 1972

The dsRNA-containing virus-like particles were characterized as small, spherical parts in the fungal cell with a rather simple structure (Wood, 1973). These VLPs are single-shelled or they have simple capsids and the diameters of the particles range from 15 to 50 nanometres. The heterogenicity of their evolutionary origin are represented by the complexity and distribution of their dsRNA genomes, despite their small size and relatively simple structure. Most of the dsRNA fungal viruses examined so far are serologically unrelated, and sometimes serologically unrelated virus particle exist within a same host fungus. A typical example is strains of *Chalara elegans*, which has shown to have at least five different VLPs, each containing dsRNA (Bozarth and Goenaga, 1977).

COMPONENT MYCOVIRUSES

Many dsRNA-containing mycoviruses are component viruses (i.e., the viral genome is segmented and distributed among a population of serologically identical particles). The virus of the chestnut blight fungus *Cryphonectria parasitica* is a good example of this. The *Cryphonectria hypovirus* has been characterized in great detail (Shapira *et al.*, 1991). The entire genome of this virus is encoded on the large dsRNA (L-dsRNA, 12.712 bp) which contains two open reading frames (ORF A and ORF B). ORF A encodes for a proteinase, whereas ORF B contains a RNA polymerase and a RNA helicase motif. Transformation tests have shown that ORF A causes reduction of pigmentation and sporulation in the host fungus but not hypovirulence (Choi and Nuss, 1992).

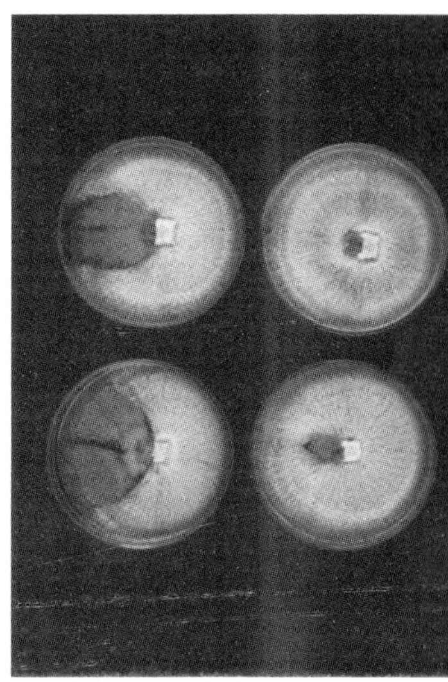

Figure 2.2 *In vitro* conversion of virulent (orange-dsRNA-free) isolates with hypovirulent (white, L-dsRNA-containing) strain of the chestnut blight fungus *Cryphonectria parasitica*.

Besidses L-dsRNA, smaller dsRNA parts (M- and S-dsRNA) which are internally deleted forms of the L-dsRNA, may occur. Typically, these multisegmented dsRNAs consist of a larger 12-13 kb segment along with smaller segments that range in size from 1 to 10 kb. These smaller segments are derived by internal deletions from the largest segment so that dsRNA segments within isolates examined have conserved regions and share terminal sequences (Shapira *et al.*, 1991). The largest segment contains all genetic information required for virus replication and maintenance (Choi and Nuss, 1992; Fig. 2.2).

As for most fungal viruses, there is no known extracellular phase of *Cryphonectria hypoviruses*. Members of the

Hypoviridae mycovirus family are not encapsulated in a protein coat but in lipid vesicles derived from host membranes (Hansen *et al.*, 1985). The only members characterized to date belong to the recently described *Hypoviridae* family (Hillman *et al.*, 1995), which are ancestrally related to the *Potyviridae* family of plant viruses.

Dissemination of the hypovirus may occur vertically through asexually produced conidia but not through sexually produced ascospores (Anagnostakis, 1987). The virus is transmitted horizontally between different strains of the host fungus via hyphal anastomosis (Anagnostakis and Day, 1979). A vegetative incompatibility (vic) system in *Cryphonectria parasitica* involving at least six vic loci, controls the formation of hyphal anastomosis and limits the transmission of the hypovirus (Cortesi and Milgroom, 1998).

Other mycoviruses have their genome as a single large dsRNA molecule. The 6,3 megadalton dsRNA of a *Helminthosporium maydis* virus is an example and represents the largest dsRNA molecule described to date from any virus (Bozarth, 1977).

PRACTICAL ASPECTS

Isolates of *Fusarium oxysporum* were recovered from healthy and necrotic Essex soyabean seedlings grown in naturally infested soil. These were tested for pathogenicity in artificially infested, autoclaved field soil. Highly pathogenic, moderately pathogenic, and hypovirulent isolates of this species were identified. Fifty-seven *F. oxysporum* isolates were tested for the presence of dsRNA. The presence of dsRNA was associated with hypovirulence in *F. oxysporum.* Four segments of dsRNA, with sizes of 4.0, 3.1, 2.7, and 2.2 kb, were detected in extracts of all *F. oxysporum* isolates. No morphological differences were found between dsRNA-containing and dsRNA-free *F. oxysporum* isolates. Prior inoculation of Essex soyabean seeds with conidia of dsRNA-containing hypovirulent *F. oxysporum* isolates, plus metalaxyl seed treatment, significantly ($p<0.05$) reduced disease severity on both cotyledons and hypocotyls and increased the rate of seedling emergence in field soil, compared to the control plants treated with metalaxyl alone or not treated with metalaxyl (Kilic, 1997).

It was previously reported that three dsRNA segments, designated L, M and S, were isolated from *Sclerotinia sclerotiorum* and that the M dsRNA segment was coincident with hypovirulence and debilitation of the fungal host (Xie *et al.*, 2001). The M dsRNA-encoded putative replicase protein contained the conserved methyl transferase, helicase and RNA-dependent RNA polymerase (RdRp) domains characteristic of the replicases of potcx-like plant viruses (flexiviruses) and *Botrytis virus F* (BVF), a flexuous rod mycovirus infecting the phytopathogenic fungus *Botrytis cinerea.* It was demonstrated that the debilitation phenotype was transmitted from the parental debilitated strain to its normal ascospore progeny via hyphal anastomosis (Robyn *et al.*, 2001).

Double-stranded RNAs (dsRNAs) are widespread in plant pathogenic fungi, but their functions in fungal hosts remain mostly unclear, with a few exceptions. dsRNAs from *Nectria radicicola*, the causal fungus of ginseng root rot were analyzed (Il-Pyung and Yong-Hwan, 2001). Four distinct sizes of dsRNAs, 6.0, 5.0, 2.5, and 1.5 kbp, were detected in 24 out of the 81 strains tested. Curing tests of individual dsRNAs suggested that the presence of 6.0 kbp dsRNA was associated with high levels of virulence, sporulation, laccase activity, and pigmentation in this fungus. The 6.0 kbp dsRNA-cured strains completely lost virulence-related phenotypes. This 6.0 kbp dsRNA was reintroduced by hyphal anastomosis to a dsRNA-cured strain marked with hygromycin resistance, which resulted in the restoration of virulence-related phenotypes. These results strongly suggest that 6.0-kbp dsRNA up regulates fungal virulence in *N. radicicola*. Phylogenetic analysis showed that this gene is closely related to those of plant cryptic viruses.

The complete nucleotide sequence, 4975 bp, of the double-stranded RNA (dsRNA) mycovirus infecting the sclerotial parasite *Coniothyrium minitans* (CmRV) was also determined (Cheng *et al.*, 2002). Sequence analysis revealed the occurrence of two overlapping open reading frames (ORFs). Based on genome organization and sequence analysis encoded proteins, the virus infecting *C. minitans* strain Chy-1, designated *C. minitans* RNA virus (CmRV), belongs to the family *Totiviridae*. Pairwise sequence comparisons of the deduced amino acid sequences encoded by CmRV as well as phylogentic analysis indicated that it is more closely related to the totiviruses that infect filamentous fungi than to those infecting protozoa, yeast and smut fungi.

The diversity and vertical transmission of double-stranded (ds) RNA in *Helicobasidium mompa* and *Rosellinia necatrix* was examined by electrophoresis and Northern hybridization (Ikeda *et al.*, 2004). These two fungi share the similar niche as root-rot pathogens of trees in forests and orchards, and had diverse dsRNAs. The detection frequency of dsRNA in both fungi was different; in *H. mompa*, 68.4% (132 out of 193 MCGs; mycelial compatibility groups) had dsRNA, whereas 20.9% (53 out of 254 MCGs) in *R. necatrix*. dsRNA banding patterns and Northern blot analyses revealed the presence of various dsRNA elements in both fungi. Hyphal tip isolation was mostly unsuccessful to remove dsRNA with some exceptions. Sexual reproduction functioned to remove dsRNA in both fungi since dsRNA was not detected from single sexual spore cultures.

CHESTNUT BLIGHT AND HYPOVIRULENCE

The chestnut blight fungus *Cryphonectria* (syn: *Endothia*) *parasitica,* an ascomycete and causal agent of chestnut blight, virtually eliminated the American chestnut trees (*Castanea dentata*) within their natural range following introduction into the USA at the beginning of this century (Anagnostakis, 1987). A similar development was expected when chestnut blight fungus was first detected on European chestnut trees (*Castanea sativa*) near Genova (Northern Italy) in 1938 (Fig. 2.3).

Figure 2.3 A chestnut blight canker caused by the normal (virulent, dsRNA-free) strain of *Cryphonectria parasitica* fungus on young European chest nut tree

Indeed, the disease spread rapidly, and by 1969, most of the chestnut growing areas in Europe were affected by the *C. parasitica* fungus. Transmissible hypovirulence, a viral disease in the pathogen population, has been associated with the decline of the chestnut blight epidemic which was observed in many chestnut growing regions of Europe after 1956 (reviewed by Heiniger and Rigling, 1994).

Hypovirulence of *C. parasitica* is caused by an unencapsidated dsRNA virus of the genus Hypovirus which is located in the cytoplasm of the fungus (Hillman *et al.*, 1995). Infection by the *Cryphonectria hypovirus* (CHV) causes a reduction of fungal virulence (hypovirulence) ranging from avirulence to almost normal virulence (Rigling *et al.*, 1989). In comparison with virus-free isolates, fungal isolates containing the European–type hypovirus (CHV1) exhibit reduced sporulation and pigmentation, reducing in a typical white cultural appearance. One CHV1 subtype (the Italian CHV1-I) was widespread throughout most of the chestnut blight infested areas in Europe (Italy, Slovenia, Croatia, Greece, Hungary, Corsica, Switzerland) and dominates the *Cryphonectria hypovirus* in Europe. In contrast to the Italian CHV1, that occurrence of the four other CHV1 subtypes was geographically restricted to Western Europe. Two subtypes were found in France and one each in Germany and Spain, both near the French border (Allemann *et al.*, 1999).

The original *C. parasitica* strain harbouring hypovirus CHV2 was isolated quite recently in North America. However, this hypovirus is speculated to be of distant European origin (Hillman *et al.*, 1995). Some variants of *Cryphonectria hypovirus* 3 and 4 (CHV3 and CHV4), the species most commonly found in Eastern North America. CHV4 is distinguished from other hypoviruses by having little effect on fungal virulence and colony morphology. The 9.1 kb genome of strain CHV4 is the smallest of any member of the family characterized to date (Linder-Basso *et al.*, 2005).

FUTURE PERSPECTIVES

Elliston (1985) noted that hypovirulence-associated phenotypes exhibited by *Cryphonectria parasitica* isolates are most dependent on dsRNAs rather than host genotype. Studies demonstrate the similarity of European hypovirulent isolates to each other and the differences from North-American strains used in the study. Examination of Hungarian hypovirulent strains confirms these observations (Radócz, 2001). Hungarian hypovirulent strains showed homogenicity in electrophoretic characteristics of their dsRNAs and similarity in their effect on the phenotype of the host fungus (Fig. 2.4).

Figure 2.4 Sampling from abnormal canker caused by hypovirulent strain of the chestnut blight fungus *Cryphonectria parasitica* in Hungary.

Hypovirulence is a potentially important factor in limiting destruction in European chestnut populations and the control of virulent cankers caused by the chestnut blight fungus. Observations of genetic diversity and population dynamics of the host fungus are therefore important. In some countries (like France, Italy, and Slovakia) mixtures of hypovirulent strains from different vegetative compatibility groups (multiconverter inocula) are used. The introduction of a new genotype of the fungus into small, scattered chestnut stands (with regularly one or two VCGs) could be potentially dangerous and requires continuous monitoring. An isolation and *in vitro* conversion of virulent, local fungal strains (belonging to the same VCG) and used them as a hypovirulent inocula is a better solution to control the pathogen in the small chestnut stands.

CONCLUSION

Mycoviruses are probably widespread in fungi in nature, despite the fact that relatively few have been isolated and characterized, and still fewer have been experimentally transmitted to test fungi and demonstrated to be infectious.

Typical mycovirus particles are isometric in shape and have a diameter between 15–50 nm. Some mycoviruses are rod-shaped or have some other form. This nucleic acid usually consists of dsRNA and seldom contains dsDNA. Some mycoviruses appear to be multipartite. Mycoviruses appear to be transmitted by means of cytoplasm and through mainly asexual spores. Mycoviruses appear to be retained in the cytoplasm of a fungal strain indefinitely.

Many mycoviruses do not cause symptoms in their hosts, despite the fact that cells may harbour large numbers of virus particles. Viral infection in the edible mushroom *Agaricus brunnescens* causes a degeneration of the mycelium and development of malformed basidiocarps, resulting in a reduction of yield. Hypovirulent strains of the chestnut blight fungus (*Cryphonectria parasitica*) carry double-stranded RNA viruses, located in the cytoplasm of the fungal cells. Hypovirulent fungal strains produce nonlethal, superficial cankers that are restricted to the outer parts of the bark and do not destroy the vascular cambium of the host tree. Besides reducing fungal virulence, *Cryphonectria hypovirus* affects other fungal features, such as sporulation and pigmentation, laccase and cutinase expression, and other proteins and mRNAs normally produced. Acquisition of dsRNA often affects the fungus cultural phenotype, resulting in slow growth, lobate margins or reduced aerial mycelium. The mycovirus can be transmitted via hyphal anastomosis, resulting in conversion of the recipient fungal isolate to the hypovirulent morphology.

References

Aist, J.R. and Williams, P.H. (1971). The cytology and kinetics of cabbage root hair penetration by *Plasmodiophora brassicae*. *Can. J. Bot.* 49:2023–2034.

Allemann, C., Hoegger, P., Heiniger, U. and Rigling D. (1999). Genetic variation of *Cryphonectria hypoviruses* (CHV1) in Europe, assessed using restriction fragment length polymorphism (RFLP) markers. *Mol. Ecol.* 8: 843–854.

Anagnostakis, S.L. (1987). Chestnut blight: the classical problem of an introduced pathogen. *Mycologia* 79:23–37.

Anagnostakis, S.L. and Day, P.R. (1979). Hypovirulence conversion in *Endothia parasitica*. *Phytopathology* 69:1226–1229.

Banks, G.T., Buck, K.W., Chain, E.B., Darbyshire, J.E., Himmelweit F., Ratti, G., Sharpe, T.J. and Planterose, D.N. (1970). Antiviral activity of double-stranded RNA from a virus isolated from *Aspergillus foetidus*. *Nature* 227:505–507.

Bozarth, R.F. (1972). Mycoviruses: a new dimension in microbiology. In "Environmental health Perspectives". US Dept. of Health Education and Welfare.Washington D.C., 23–29.

Bozarth, R.F. (1977). Biophysical and biochemical characterization of virus-like particles containing a high molecular weight dsRNA from *Helminthosporium maydis. Virology* 80:149–157.

Bozarth, R.F. and Goenaga, A. (1977). Complex of virus-like particles containing double-stranded RNA from *Thielaviopsis basicola. J. Virol.* 24:846–849.

Bozarth, R.F., Wood, H.A. and Goenaga, A. (1972). Virus-like particles from a culture of *Diplocarpon rosae. Phytopathology* 62: 493.

Castanho, B., Butler, E.E. and Shepherd, R.J. (1978). The association of double-stranded RNA with *Rhizoctonia* decline. *Phytopathology* 68:1515–1519.

Cheng, J., Jiang, D., Fu, Y., Li, G., Peng, Y. and Ghabrial, S.A. (2002). Molecular characterization of a dsRNA totivirus infecting the sclerotial parasite *Coniothyrium minitans. Virus Res.* 93:41–50.

Choi, G.H. and Nuss, D.L. (1992). Hypovirulence of chestnut blight fungus conferred by an infectious viral cDNA. *Science*, 257:800–803.

Corbett, M.K. and Styer, E.L. (1977). Intranuclear virus-like particles in *Phytophthora infestans. Proc. Am. Phytopathol. Soc.* 3:332.

Cortesi, P. and Milgroom, M.G. (1998). Genetics of vegetative incompatibility in *Cryphonectria parasitica. Appl. Environ. Microbiol.* 64:2988–2994.

Elliston, J.E. (1985). Characteristics of dsRNA-free and dsRNA-containing strains of *Endothia parasitica* in relation to hypovirulence. *Phytopathology* 75:151–158.

Gandy, D. and Hollings, M. (1962). Viruses associated with a die-back disease of cultivated mushroom. *Nature* 196:962–965.

Hansen, D.R., Van Alfen, N.K., Gilles, K. and Powell W.A. (1985). Naked dsRNA associated with hypovirulence of *Endothia parasitica* is packaged in fungal vesicles. *J. Gen. Virol.* 66:2605–2614.

Heiniger, U. and Rigling, D. (1994). Biological control of chestnut blight in Europe. *Ann. Rev. Phytopathol.* 32:581–599.

Hillman, B.I., Fulbright, D.W., Nuss D.L. and Van Alfen N.K. (1995). *Hypoviridae*. In "Virus Taxonomy: Sixth Report of International Committee for the Taxonomy of Viruses" (F.A. Murphy, C.M. Fauquet, D.H.L. Bishop, S.A. Ghabrial, A.W. Jarwis, G.P. Martel, M.P. Mayo and M.D. Summers eds.). Springer-Verlag, New York, 261–264.

Hollings, M., Gandy, D. and Stone, O.M. (1965). Viruses that infect fungi. *Ann. Rev. Phytopathol.* 9:93–118.

Hollings, M. and Stone, O.M. (1969). Viruses in fungi. *Sci. Progr. London* 57:371–391.

Ikeda, K.I., Nakamura, H., Arakawa, M. and Matsumoto, N. (2004). Diversity and vertical transmission of double-stranded RNA elements in root rot pathogens of trees, *Helicobasidium mompa* and *Rosellinia necatrix. Mycol. Res.* 108:626–634.

Il-Pyung, A. and Yong-Hwan, L. (2001). A viral double-stranded RNA upregulates the fungal virulence of *Nectria radicicola. Mol. Plant-Microbe Interact.* 14:496–507.

Kleinschmidt, W.J., Ellis, L.F., van Frank, R.M. and Murphy, E.B. (1968). Interferon stimulation by a double-stranded RNA of a mycophage in statolon reparations. *Nature* 220:167–168.

Kilic, O. (1997). Effect of dsRNA-containing and dsRNA-free hypovirulent isolates of *Fusarium oxysporum* on severity of Fusarium seedling disease of Essex soybean. MSc. Thesis. Virginia Polytechnic Institute and State University. pp. 57.

Kotlin, Y. and Day, P.R. (1976). Suppression of the killer phenotype in *Ustilago maydis. Genetics* 82:629–637.

Lampson, G.P., Tytell, A.A., Field A.K., Nemes, M.M. and Hilleman, M.R. (1967). Inducers of interferon and host resistance, I. Double-stranded RNA from extracts of *Penicillium funiculosum*. *Proc. Natl. Acad. Sci. USA* 58:782–789.

Lemke, P.A. (1977). Fungal viruses in agriculture. In "Virology in Agriculture" (J.A. Romberger, J.D. Anderson and R.L. Powell eds.) Allaheld, Osmun & Co., Montclair, New Jersey, USA, 159–175.

Lemke, P.A. (1981). Viruses of Conidial Fungi. In "Biology of Conidial Fungi" (P.A. Lemke and R.L. Powell eds.) Academic Press Inc. New York, USA, 395–415.

Lindberg, G.D. (1959). Reduction in pathogenicity and toxin production in diseased *Helminthosporium victoriae*. *Phytopathology* 50:457–461.

Linder-Basso, D., Dynek, J.N. and Hillman, B. (2005). Genome analysis of *Cryphonectria hypovirus 4*, the most common hypovirus species in North America. *Virology* 337: 192–203.

Moffitt, E.M. and Lister, R.M. (1975). Application of a serological test for detecting double-stranded RNA mycoviruses. *Phytopathology* 65:851–859.

Mussel, H.W., Wood, H.A., Adler, J.P. and Bozarth, R.F. (1973). *Puccinia graminis*: virus-like particles in mycelium from axenic cultures. 2nd Int. Cong. Plant Pathol., September 1973, Minneapolis. Abstract 0908.

Radócz, L. (1998). Chestnut blight *Cryphonectria parasitica* and its biological control in Hungary. *Acta Phytopathologica et Entomologica Hungarica* 33:131–145.

Radócz, L. (2001). Study of subpopulations of the chestnut blight (*Cryphonectria parasitica*) fungus in the Carpathian-basin. *Forest, Snow and Landscape Research* 76:368–372.

Rawlinson, C.J. (1973). Virus-like particles (VLP) in fungi. *Rep. Rothamsted Exp. Station* 1972, 123–124.

Rawlinson, C.J., Hornby, D., Pearson, V. and Carpenter, J.M. (1973). Virus-like particles in the take-all fungus *Gaeumannomyces graminis*. *Ann. Appl. Biol.* 74:197–209.

Rigling, D., Heiniger, U. and Hohl, H.R. (1989). Reduction of laccase activity in dsRNA-containing hypovirulent strains of *Cryphonectria (Endothia) parasitica*. *Phytopathology* 79:219–223.

Robyn, L.J., Ross, H., Beever, E., Pearson, M.N. and Forster, R.L. (2001). Genome characterization of *Botrytis* virus F, a flexuous rod-shaped mycovirus resembling plant 'potex-like' viruses. *J. Gen. Virol.* 82:67–78.

Shapira, R., Choi, G.H., Hillman, B.I. and Nuss, D.L. (1991). The contribution of defective RNAs to the complexity of viral-encoded double-stranded RNA populations present in hypovirulent strains of the chestnut blight fungus *Cryphonectria parasitica*. *EMBO J.* 10:741–746.

Sinden, J.W. and Hauser, E.M. (1950). Transmission of a disease of mushrooms by infected spores. *Phytopathology* 53:888.

Spire, D. (1971). Virus des champignons. *Physiol. Vég.* 9:555–567.

Tuveson, R.W. and Peterson, J.F. (1972). Virus-like particles in certain slow-growing strains of *Neurosora crassa*. *Virology* 47:527–531.

Wickner, R.B. (1976). Killer of *Saccharomyces cerevisiae*, a double-stranded ribonucleic acid plasmid. *Bacteriol. Rev.* 40:757–773.

Wood, H.A. (1973). Viruses with double-stranded RNA genomes. *J. Gen. Virol.* 20, 61–85.

Xie, J., Wei, D., Jiang D., Fu, Y., Li, G., Ghabrial, S. and Youliang, P. (2001). Characterization of debilitation-associated mycovirus infecting the plant-pathogenic fungus *Sclerotinia sclerotiorum. J. Gen. Vir.* 87:241–249.

Yamashita, S., Doi, Y. and Yora, K. (1971). A polyhedral virus found in rice blast fungus *Pyricularia oryzae. Ann. Phytopathol. Soc. Jap.* 37:356–359.

Yarwood, C.E. and Hecht-Poinar, E. (1973). Viruses from rusts and mildews. *Phytopathology* 63:1111–1115.

3

Biotechnology: A New Tack on Fungal Resistant Crops

Neetu Dahiya

Department of Biotechnology, Panjab University, Chandigarh, Pin-160014, India

INTRODUCTION

With the rapid population growth, the resulting food and energy storage anticipated and the numerous ecological problems partly due to intensive agricultural practices of our present society necessitate the major limitations of classical breeding to be overcome. Improving the yield and nutritional quality of crops will be necessary to meet the caloric and protein needs of a growing world population. Earlier the alteration of crops to improve their production was performed through selection. In those days, variation occurred through induced mutation or hybridization where two or more plants were crossed. Selection occurred through nature, using a "*selection of the fittest*" concept, where only the seeds best adapted to that environment succeeded. For example, farmers selected only the biggest seeds with non-shattering seed heads, assuming these to be the best. It was possible to produce some improved products, although it was very difficult to do and many things could not be improved or changed. An improved product may involve changing the color or the size of a plant, making it more appealing to the buyer's eye or it may be more practical, allowing increased tolerance to cold, frost, or drought; all making a crop easier to grow in a constantly changing environment. By all accounts classical plant breeding has been most successful in the last couple of decades.

However, there are several limitations of classical plant breeding. First, classical breeding is limited by sexual compatibility. The result is that a very limited gene pool is available for breeding processes involving sexual crosses. Second, plants often exchange large, unregulated chunks of their genomes. This can lead to both useful and unwanted traits in the offspring. Sometimes these unwanted traits can be unsafe. One example would be potato varieties created using conventional plant breeding that inadvertently produced excessive levels of naturally occuring glycoalkaloids. These glycoalkaloids cause gastrointestinal, circulatory, neurological and dermatological problems associated with alkaloid poisoning. An extensive time and labour consuming effort is consequently required to eliminate the undesired characteristics by out crossing while still maintaining the desired one.

The tools of genetic engineering and the creation of genomics (the study of the structural and functional nature of genes) though pioneered in medicine have quickly spread to agriculture and now offer enormous potential to invest plants and livestock with desirable traits that, heretofore, were unimaginable.

Both traditional plant breeding methods and transgenic techniques create new gene combinations with novel traits such as resistance to pests, disease and herbicides. Although both types of techniques generate a similar end product, that is, crop plant with certain novel traits, there may be large qualitative differences in degree of "genetic novelty" brought into transgenic plants compared to traditional cultivars. For example, almost all traditionally improved varieties of corn are the result of the inbreeding of corn varieties. In contrast transgenic corn may contain transgenes from an organism as distantly related as a bacterium. For example, "Bt corn" contains a gene from soil bacterium, *Bacillus thuringiensis*. The bacterial gene is capable of producing a class of proteins called "Cry proteins," and each one of these proteins is toxic to a specific group of insects. Thus, transgenic crop plants hold great promise for the world agriculture, as novel insect, pest and disease resistant crop plants (http://www.ccst.us/ccst/pubs/gmf/ FB_Ch3.pdf).

A transgenic crop plant contains a gene or genes which have been artificially inserted instead of the plant acquiring them through pollination. The inserted gene sequence (known as the transgene) may come from another unrelated plant, or from a completely different species: transgenic Bt corn, for example, which produces its own insecticide, contains a gene from a bacterium. Plants containing transgenes are often called genetically modified or GM crops, although in reality all crops have been genetically modified from their original wild state by domestication, selection and controlled breeding over long periods of time. Through the use of transgenics, one can produce plants with desired traits and even increase yields to allow for more crops that last longer and withstand pests and diseases. The idea being that we could produce the traits we want without throwing the dice as with selection techniques, allowing us to better feed a growing population and produce more desirable products. In this chapter, we address several aspects concerning role of biotechnology in development of transgenic crops and the potential future for this rapidly advancing technology.

Thus, transgenics is now preferred when trying to come up with new products to improve sales. Today, scientists can not only select, but also create crops by inserting genes to make seeds with desired traits. GM techniques allow new traits to be introduced one at a time without complications from extra genes and extensive crossbreeding. The advances made in the field of plant biotechnology have contributed significantly for development of genetically modified crops. The major breakthrough includes:

1. The ability of transferring isolated and well-defined genes into plant cells,
2. The identification and characterization of regulatory similarities present in plant genes,
3. The regeneration of intact and fertile plants starting from a single transformed cell, and

4. Generation of transgenic crop plants which show improved agronomic traits.

Since genes are transferred by ways different from that of sexual processes, there is no limitation with respect to the gene pool available. In addition, since only well-defined single gene or small group of genes encoding the desired traits are transferred, gene technology should enable the starting elite plant to keep all its good characteristics while being specifically improved for the desired trait.

The leap to genetic engineering was made possible only because sustained research over the past five decades has disclosed many secrets of gene structure and function. Biotechnology is only the most recent of a continuum of scientific advancement that is contributing its solutions to agriculture. One can anticipate that biotechnology will incrementally be enhanced and then be supplemented by other presently unknown technologies (http://www.pesticidebook. com/pdfs/chapter25_pages317-319.pdf).

In order to make a transgenic crop, there are five main steps: extracting DNA, cloning a gene of interest, designing the gene for plant infiltration, transformation, and finally plant breeding. In order to obtain the DNA comprising a gene, DNA is first extracted from cells and put into a bacterial plasmid. A plasmid is a molecular biological tool that allows any segment of DNA to be put into a carrier cell (usually a bacterial cell) and is capable of autonomous replication it.

Once the gene of interest has been amplified, it is time to introduce it into the plant species we are interested in. The nucleus of the plant cell is the target for the new transgenic DNA. The transfer of isolated DNA into plant cell has become a routine procedure which is applicable to many different plants, including crop species. Several methods exist to transfer DNA into cell like, use of Gram negative soil bacterium *Agrobacterium tumefaciens* to introduce genetic material into the nucleus of plant cell of most dicotyledonous plants, transformation of wall less protoplasts of plant by naked DNA in the presence of polyethylene glycol or calcium chloride, gene gun method, liposome fusion (Caboche, 1990) and the biolistic approach (Klein *et al.*, 1987). Among the different methods developed, most commonly used are, the "Gene Gun" and *Agrobacterium* method.

The "Gene Gun" method, also known as the micro projectile bombardment method, is most commonly used in species such as corn and rice. As its name implies, this procedure involves high velocity micro-projectiles to deliver DNA into living cells using a gun. It involves sticking DNA to small micro-projectiles and then firing these into a cell. This technique is clean and safe. It enables scientists to transform organized tissue of plant species and has a universal delivery system common to many tissue types from many different species. It can give rise to unwanted side effects, such as the gene of interest being rearranged upon entry or the target cell sustaining damage upon bombardment. Nevertheless, it has been quite useful for getting transgenes into organisms when no other options are available. The *Agrobacterium* method involves the use of a soil-dwelling bacteria known as *Agrobacterium tumefaciens*, which has the ability to infect plant cells with a piece of its DNA. The piece of DNA that infects a plant is integrated into a plant chromosome through a tumor-inducing plasmid (T_i plasmid), which can take control of the plant's cellular machinery and use it to make many copies of its own bacterial DNA. The T_i plasmid is a large circular DNA particle that replicates independently of the bacterial chromosome.

One important aspect of plant gene technology and the production of transgenic plants is the controlled and predetermined expression of foreign genes. The regulatory structures of plant genes which control their transcription are very similar to those from other eukaryotic sources. It has been observed that the sequence located at the 5'-end of the RNA coding sequences of the gene contains many signals necessary for its correct expression. One of the best studied and most commonly used promoter is that of the 35sRNA of Cauliflower Mosaic Virus (CaMV). It contains several different domains. It has a strong transcriptional enhancer which when fused to other promoters can stimulate the level of expression up to tenfold without interfering with the specificity of expression which remains controlled by the basal promoter (Nagy *et al.*, 1987). Plant derived promoters maintain their expression specificity on heterologous plants and can therefore be widely used. In addition, an artificial promoter is reported which is composed of the basic promoter 35sRNA of CaMV, the operator sequence of the bt gene of the transposable element Tn10 and the repressor of Tn10 (Gatz *et al.*, 1991).

One of the hottest areas right now is the modification of crops to increase the resistance of plants to fungi. Fungi are responsible for a range of serious plant diseases such as blight, grey mould, bunts, powdery mildew, and downy mildew. Some of the most devastating and universal crop diseases are caused by fungal pathogens (Table 3.1). For instance, the rust fungi are the most widespread and generally cause the largest crop losses per season. Crop losses can be considerable due to fungal pathogens. For example, the fungal agent of rice blast disease (*Magnaporthe grisea)* destroys 157 million tonnes of cultivated rice each year, enough rice to feed 60 million people worldwide (Pennisi, 2001). Significant crop losses due to fungal attacks occur in most of the agricultural and horticultural crops. Contribution of fungal diseases toward yield loss in some important crops in India has been summarized in Table 3.2 (Grower and Gowthaman, 2003).

Table 3.1. Fungal diseases of some important crops worldwide.

Cereals	Powdery mildew, rusts, leaf-spot diseases, common bunt of wheat, loose smuts
Rice	Blast, sheath blight and other leaf spot diseases
Vegetables	Leaf rot, brown rot, grey mould, powdery mildew and downy mildew, leaf spot and fruit spot diseases (e.g., *Alternaria*)
Potatoes	Late blight, early blight (*Alternaria*), black scab (*Rhizoctonia*), silver scurf
Grapevine	Powdery mildew (*Odium*) and downy mildew (*Peronospora*), grey mould (*Botrytis*)
Peanuts	White mould (*Rhizoctonia*), *Sclerotia* stem rot, leaf spot, rust

(http://www.gmo-compass.org/eng/agri_biotechnology/gmo_planting/countries_ growing_gmos.html)

Table 3.2. Contribution of fungal diseases towards yield loss in some major crops of India.

Crop	*Pathogen*	*Disease*	*Total yield loss (%)*
Rice	*Pyricularia oryzae*	Blast	21
Wheat	*Puccinia recondiata*	Leaf rust (Brown rust)	30
Maize	*Helminthosporium*	Leaf blight	30
Sorghum	*Sphacelotheca reiliaria maydis* and *H. turcium*	Grain mould	18
Pigeonpea	*Fusarium udum*	Wilt	24
Chickpea	*Fusarium oxysporum*	Wilt	23
Brassica	*Alternaria brassiceae*	Blight	30
Soyabean	*Phakospora packyrhizi*	Rust	23
Potato	*Phytophthora infestans*	Late blight	31

Fungal plant diseases are usually managed with applications of chemical fungicides or heavy metals. In some cases, conventional breeding has provided fungus resistant cultivars. Besides combatting yield losses, preventing fungal infection keeps crops free of toxic compounds produced by some pathogenic fungi. These compounds, often referred to as mycotoxins, can affect the immune system and disrupt hormone balances. Some mycotoxins are carcinogenic. Late blight of potato is caused by the fungus *Phytophtora infestans*, the disease responsible for the great Irish potato famine of 1846-47. It led to the deaths of more than one million people (http://www.gmo-compass.org/eng/agri_biotechnology /breeding_aims/148.disease_resistant_crops.html).

The negative effects of some fungal pathogens can be limited by the use of chemical fungicides. Demand for fungicides amongst farmers is high, indicating that for many farmers there are few available alternatives. The world market for agricultural and non-crop fungicides amounted to an estimated US$6 billion at the end-user level in 1999. The United States, Western Europe and Japan together accounted for 75% of the total world market. Small grains constitute the largest market for fungicides worldwide. This sector accounted for an estimated 27% of the total world market in 1999, followed closely by tree and vine crops (24%), rice (16%), and vegetables and potatoes (10%). Other crop markets accounted for 17% of the world fungicide market, and non-crop markets accounted for 6 %.

In many countries, fungicides as crop protection products are subject to strict legislative regulation and undergo a rigorous and expensive process of registration for public sale. While fungicides can provide a level of control, this chemical option is often limited for many farmers, particularly in developing countries, by high costs and lack of knowledge about application. In addition, the negative effects of fungicide applications on human health, with special reference to the laborers and the environment can be considerable. There is a need to find more environmentally benign alternatives to fungicides to control fungal diseases of crops.

Development of Transgenic Crop Varieties Resistant to Fungal Diseases

Genes encoding of many antifungal proteins which can inhibit fungal growth *in vitro* has been exploited to make fungal resistant transgenic plants. Different strategies can be used for making plants resistant against fungi. A few are mentioned below:

1. Gene encoding enzymes involved in formation of small organic compounds which act on fungus-like phytoalexins,
2. Gene encoding enzymes which attack fungal cell wall, like chitinases and glucanases,
3. Gene encoding low molecular weight proteins with a direct antifungal action.

By changing their physiological conditions, higher plants protect themselves from various stresses such as pathogen attacks, wounding, application of chemicals including phytohormone and heavy metals, air pollutants like ozone, ultraviolet rays, and harsh growing conditions. These protective reactions are known as "defense responses" of higher plants and the proteins actively synthesized in accordance with this reaction are called "defense-related proteins". In particular, protective plant proteins specifically induced in pathological or related situations have been intensively studied from an agricultural perspective and are called "pathogenesis-related proteins" (PR proteins). Enzymes indispensable for biosynthesizing low-molecular-weight antibiotics (phytoalexin), isoflavone reductase, for example, can also be considered to be defense-related proteins. On the other hand, many of the reserve proteins accumulated in seeds and fruits are considered to have a constitutive defence function against microbial pathogens and invertebrate pests in addition to their storage function. These inducible or constitutive defense mechanisms of higher plants are relatively conserved during the course of evolution. Accordingly, most plants produce or accumulate structurally and functionally similar protective proteins under certain situations, irrespective of their morphological differences (http://dmd.nihs.go.jp/latex/defense-e.html).

Thirteen classes of antifungal proteins have been reported: PR-1 proteins, (1,3)β-glucanases, chitinases, chitin-binding proteins, thaumatin-like (TL) proteins, defensins, cyclophilin-like protein, glycine/histidine-rich proteins, ribosome-inactivating proteins (RIPs), lipid-transfer proteins (LTPs), killer proteins (killer toxins), protease inhibitors, and other proteins. These proteins have been named primarily on the basis of either their mechanism of action, (e.g., glucanases), their structure (e.g., cysteine rich), or their similarity to a known "type" protein (Selitrennikoff, 2001).

(i) PR-1 proteins. PR-1 proteins are accumulated to high levels after pathogen infection and are antifungal both in plants (transgenic plants overexpressing tobacco PR-1) and *in vitro*. PR-1 proteins have been found in rice, wheat, maize, tobacco, *Arabidopsis thaliana,* barley, and many other plants (Muradov *et al.*, 1993; Bryngelsson *et al.*, 1994; Molina *et al.*, 1999; Agrawal *et al.*, 2000). Although these proteins are from diverse sources, they are remarkably similar (at least 35% identity). PR-1 proteins have antifungal activity at the micromolar level against a number of plant pathogenic fungi, including *Uromyces fabae, Phytophthora infestans*, and *Erysiphe graminis* (Niderman *et al.*, 1995). PR-1 proteins have molecular masses of ~15 to 17 kDa and have homology to the superfamily of cysteine-rich proteins. Although the precise mechanism of antifungal activity is not understood for plant PR-1 proteins, a PR-1-like protein, helothermine, from the Mexican banded lizard interacted with membrane-channel proteins of target cells, inhibiting the release of Ca^{2+} (Morrisette *et al.*, 1995). Whether antifungal plant PR-1 proteins act by this mechanism is not known but is suspected.

(ii) PR-2 proteins (*β*-glucanases). PR-2 proteins have (1,3) *β*-endoglucanase activity *in vitro* and have been grouped into three classes on the basis of amino acid sequence analysis. Class I glucanases are basic proteins of ~ 33 kDa and are found in the plant vacuole. Classes II and III include acidic, extracellular proteins of about 36 kDa. The major structural difference between class I proteins and the other two classes is that class I proteins are synthesized as preproteins that are processed prior to being enzymatically active. PR-2 proteins have been found in a wide variety of plants, including tobacco, *A. thaliana*, peas, grains, and fruits (Cote *et al.*, 1991; Kim and Hwang, 1997; Rezzonico *et al.*, 1998), the proteins are active *in vitro* at micromolar levels (~50 μg/ml) against a wide number of fungi, including human and plant pathogens (e.g., *Rhizoctonia solani, C. albicans* and *Aspergillus fumigatus*). The antifungal activity of PR-2 proteins has been convincingly demonstrated by a number of *in vitro* enzyme and whole-cell assays (Stintzi *et al.*, 1993) as well as in plants using transgenic plants overexpressing a PR-2 protein (Jach *et al.*, 1995).

The antifungal activity of plant (1,3) *β*-glucanases is thought to occur by PR-2 proteins hydrolyzing the structural (1,3) *β*-glucan present in the fungal cell wall, particularly at the hyphal apex of filamentous molds where glucan is most exposed, resulting in a cell wall that is weak. This weakened cell wall results in cell lysis and cell death.

(iii) PR-3 proteins (chitinases). A number of enzymatic assays have shown PR-3 proteins to have *in vitro* chitinase activity. Most PR-3 proteins have molecular masses between 26 and 43 kDa. Chitinases (both plant PR-3 chitinases and chitinases from other sources) have been divided into five groups. Class I chitinases contain an N-terminal cysteine-rich domain of ~40 amino acids (also known as the wheat germ agglutinin domain), a chitin-binding hevein-like domain, a highly conserved central portion, and a hinge region; most class I proteins have molecular masses of ~32 kDa. Class II proteins are similar in amino acid sequence to class I proteins, but they lack the N-terminal cysteine-rich domain and have molecular masses of 27 to 28 kDa. Class III proteins do not share amino acid sequence homology to any other class and have molecular masses of ~28 to 30 kDa. Class IV proteins resemble class I chitinases but are significantly smaller due to four major deletions. Class V chitinases show sequence similarities to bacterial exochitinases and have molecular masses of 41 to 43 kDa. In addition to chitinases, a chitosanase (chitosan is deacetylated chitin) from *Streptomyces* strain N174 with antifungal activity has been isolated (Money and Harold, 1992) and its X-ray structure has been determined.

The first report on developing fungus resistant transgenic came in 1991. Broglie *et al.* (1991), constitutively expressed bean chitinase gene in tobacco and *Brassica napus* and the plant showed enhanced resistance to *Rhizoctonia solani*. Many papers and reviews have discussed the advantages of using chitinases for plant protection because these enzymes are fungicidal, part of the plant defense system, and non-toxic to plants, animals, and higher vertebrates. Transgenic plant expressing chitinase genes has been shown to improve disease resistance in various crops, and patents have been issued on related methods (Lorito *et al.*, 1998). The assumption that chitinases are involved in the plant defense reaction was deduced from the following observation: 1. Chitin is the major component of the cell wall of plant pathogens, 2. Chitinase is one of the pathogensis related proteins and 3. Some plant chitinases have antifungal activity *in vitro*. This idea has been supported by the observation that transgenic plants with over expression of the plant chitinase gene showed increased resistance to fungal diseases (Itoh *et al.*, 2003). Chitinase, capable of degrading the cell walls of invading phytopathogenic fungi, plays an important role in plant defense response, particularly when this enzyme is overexpressed through genetic engineering.

The plant defense system against microbial pathogens may be modified to produce high constitutive levels of antimicrobial compounds (Staskawicz *et al.*, 1995). Plant genes encoding cell wall degrading enzymes, especially chitinases, have been used to alter plant resistance to fungal pathogens, but no single genes have produced an adequate level of resistance (Broglie *et al.*, 1991; Zhu *et al.*, 1994; Jach *et al.*, 1995; Punja and Raharjo, 1996), and almost no paper report resistance to multiple pathogens. Disease resistance in transgenic plants has been improved, for the first time, by the insertion of a gene from a biocontrol fungus. The gene encoding a strongly antifungal endochitinase from the mycoparasitic fungus *Trichoderma harzianum* was transferred to tobacco and potato. High expression levels of the fungal gene were obtained in different plant tissues, which had no visible effect on plant growth and development. Substantial differences in endochitinase activity were detected among transformants. Selected transgenic lines were highly tolerant or completely resistant to the foliar pathogens *Alternaria alternata*, *A. solani*, *Botrytis cinerea*, and the soil-borne pathogen *Rhizoctonia solani*. The high level and the broad spectrum of resistance obtained with a single chitinase gene from *Trichoderma* overcome the limited efficacy of transgenic expression of chitinase genes in plants, from plants as well as from bacteria (Lorito *et al.*, 1998).

Chitinase gene (chi) of bean which has been cloned in recombinant binary plasmid vector, pBI121 with 35S promoter of Cauliflower Mosaic Virus (CaMV) were used for transformation of soybean using strain LBA4404 of *Agrobacterium*. Results showed that some lines of soybean contained Gus and CHN genes. Moreover, chitinase activity in the leaf extract of transformed soybean lines was significantly more than the untransformed soybean, except in one sample. Bioassay of chitinase activity of transgenic lines on *in vitro* condition prevented mycelial growth of *Rhizoctonia solani* in comparison with untransformed control leaf extract (Salehi *et al.*, 2005).

The ch5B gene encodes for a chitinase from *Phaseolus vulgaris*, were transformed in plants of strawberry (cultivar Pajaro). Phytopathological tests showed that two transgenic lines expressing only the ch5B gene displayed high levels of resistance to gray mold disease (*Botrytis cinerea*) (Vellicce *et al.*, 2006). When a rice chitinase (Cht-2; RCC2) gene was introduced into calli of Italian ryegrass (*Lolium multiflorum* Lam) the bioassay of detached leaves indicated increased resistance to crown rust (*Puccinia coronata*) in transgenic plants, which exhibited higher chitinase activity than a nontransgenic plant (Takahashi *et al.*, 2005). *Brassica* plant (*Brassica juncea* L.) was transformed with chitinase gene tagged with an overexpressing promoter 35S CaMV. The putative transgenics were assayed for their inhibitory activity against *Alternaria brassicae*, the inducer of *Alternaria* leaf spot of *Brassica* both *in vitro* and under polyhouse conditions. In *in vitro* fungal growth inhibition assays, chitinase inhibited the fungal colony size by 12-56% over the non-transgenic control. The bioassay under artificial epiphytotic conditions revealed the delay in the onset of disease as well as reduced lesion number and size in 35S-chitinase *Brassica* as compared to the untransformed control plants (Mondal *et al.*, 2003).

Transgenic tobacco plants were generated by Agrobacterium-mediated transformation using *Saccharomyces cerevisiae* chitinase, encoded by the *CTS1-2* gene. Leaf extracts from the transgenic lines inhibited *Botrytis cinerea* spore germination and hyphal growth by up to 70% in a quantitative *in vitro* assay, leading to severe physical damage on the hyphae. The plant lines that showed increased disease tolerance were also shown to have higher chitinase activities (Carstens *et al.*, 2003). Transgenic broccoli (cv. Green Comet) expressing *Trichoderma harzianum* endochitinase genes were used to assess the interaction between endochitinase and the fungicide Bayleton in the control of *Alternaria brassicola*. Thus endochitinase-transgenic plants alone provided a significant reduction of disease severity, comparable to the protection by fungicide on non-transgenic plants (Mora and Earle, 2001). The *chiC* gene of *Streptomyces griseus* HUT6037 was introduced into rice plants. Transgenic plants were morphologically normal. Resistance to leaf blast disease caused by *Magnaporthe grisea* was evaluated in R_1 and R_2 generations using a spray method. 90% percent of transgenic rice plants expressing *chiC* had higher resistance than non-transgenic plants. Disease resistance of sibling plants within the same line was correlated with the *chiC* expression levels (Itoh *et al.*, 2003). The two disease-resistance genes chitinase and glucanase, which were introduced into cotton genome via *Agrobacterium tumefaciens* showed resistant to *Verticillium dahliea* (Wu *et al.*, 2004). Not surprisingly, PR-2 (β-glucanases) and PR-3 (chitinases) proteins act synergistically in inhibiting fungal growth, both *in vitro* and in plants (Jach *et al.*, 1995).

(iv) PR-4 proteins. PR-4 proteins are chitin-binding proteins, have molecular masses of 13–14.5 kDa, and have been classified into two groups. Class I proteins have amino acid sequence similar to hevein (a chitin-binding polypeptide) and belong to the superfamily of chitin-binding lectins. Class II proteins lack the chitin-binding domain. PR-4 proteins have been isolated from potato, tobacco, barley, tomato, and many other plants. Note that the PR-4 proteins from the diverse sources share common sequences. Both classes of proteins have potent antifungal activity against a wide variety of human and plant pathogens (e.g., *Trichoderma harzianum, Fusarium culmorum, F. graminearum*, and *B. cinerea*). The antifungal activity of class I proteins is likely the result of protein binding to nascent fungal cell wall β-chitin. By mechanisms not understood, this resulted in disrupted cell polarity, with concomitant inhibition of growth (Bormann *et al.*, 1999). The mechanism of action of class II proteins (which lack the chitin-binding hevein domain but are antifungal nonetheless) is not understood.

Chitin-binding proteins and peptides that have antifungal activity but are not PR proteins. They have been isolated from a number of sources, including bacteria (Bormann *et al.*, 1999), plants, insects, and crustaceans (Khardori, 1989; De Bolle *et al.*, 1993; Osborn *et al.*, 1995; Nielsen *et al.*, 1997; Kolbe *et al.*, 1998; Chae *et al.*, 1999; Huang *et al.*, 2000;). These non-PR-4 chitin-binding proteins include the tachystatins (Kawabata *et al.*, 1996; Osaki *et al.*, 1999) (horseshoe crab, 6.8 to 7.4 kDa), the penaeidins (Destoumieux *et al.*, 2000) (penaeid shrimp, 5.5 to 6.6 kDa), antifungal protein 1 (AFP1) (Bormann *et al.* 1999) (*Streptomyces tendae*, 9.8 kDa) and others. Fungi inhibited by these proteins include plant and human pathogens, e.g., *Paecilomyces variotii, Aspergillus* spp., *F. oxysporum, N. crassa, B. cinerea*, and *Alternaria brassicola*. It is likely that these proteins act by a mechanism similar to that of the class I PR-4 proteins, namely, binding to cell wall chitin and disrupting cell polarity, thus leading to inhibition of fungal growth.

(v) PR-5 (TL) proteins. PR-5 proteins share significant amino acid homology to thaumatin (a sweet-tasting [to humans] protein from the South African ketemfe berry bush) and are known as TL proteins. TL proteins have been isolated from *A. thaliana* (Hu and Reddy, 1997), corn (Huynh *et al.*, 1992), soybeans, rice, wheat, tobacco (Koiwa *et al.*, 1997), tomato (Singh *et al.*, 1987), beans (Ye *et al.*, 1999), barley (Hejgaard *et al.*, 1991), flax (Borgmeyer *et al.*, 1992), and many other plants. Majority of PR-5 proteins have molecular masses of ~22 kDa and are stabilized by eight disulfide bonds. This highly stabilized structure allows PR-5 proteins to be very resistant to protease degradation (Roberts and Selitrennikoff, 1990).

Although the precise mechanism of action of PR-5 proteins is not completely understood, there are a number of interesting observations that may eventually lead to a unified hypothesis for how these proteins function to kill fungi. First, several TL proteins cause cell permeability changes in fungal cells with a cell wall but have no or little effect on protoplasts (Roberts and Selitrennikoff, 1990). For example, zeamatin (a TL protein from corn) caused very rapid cell lysis of *N. crassa*, even at 4°C; lysis occurred primarily at subapical regions (Roberts and Selitrennikoff, 1990). Second, a number of PR-5 proteins bind (1,3) β-glucan and have detectable *in vitro* (1,3) β-glucanase activity (Grenier *et al.*, 1993, Trudel *et al.*, 1998). Third, zeamatin inhibits insect β-amylase and mammalian trypsin activities *in vitro* (Schimoler –O'Rourke *et al.*, 2001). Fourth, osmotin, a TL protein from tobacco, causes perturbations in the regulation of fungal cell wall assembly (Yun *et al.*, 1997; Yun *et al.*, 1998). Fifth, zeamatin and nikkomycin act in synergy, reducing the amount of zeamatin required for cell killing up to 1,000-fold (Roberts and Selitrennikoff, 1990).

(vi) Defensins. Defensins are a diverse group of low-molecular-mass cysteine-rich proteins found in mammals, fungi, insects, and plants. The insect and mammalian defensins are quite small (3 to 5 kDa) and form voltage-dependent ion channels in plasma membranes (Thevissen *et al.*, 1996; Landon *et al.*, 1997, 2000,). Thionins are also small (3 to 5 kDa) cysteine-rich peptides that are toxic to fungi (Thevissen *et al.*, 1996).

Plant and fungal defensins are cysteine-rich proteins ranging from 45 to 54 amino acids, are positively charged, and in most cases contain four disulfide bonds that stabilize each protein in solution. In addition, most defensins are highly oligomeric (many subunits of 4 to 5 kDa) *in situ* (Terras *et al.*, 1993). Defensins are classified into four groups—Group I defensins cause morphological changes in susceptible fungi and are known as morphogenic defensins, Group II proteins inhibit fungal growth but do not cause morphological changes (non-morphogenic group), Group III are inactive against test fungi but inhibit α-amylases *in vitro*, and Group IV are unique in terms of antifungal specificity and structure (Segura *et al.*, 1998). In addition, the positions of the cysteine residues are conserved in group I, II, and III proteins. No other significant homology exists between groups. In contrast to mammalian and insect defensins, plant defensins do not form channels either in artificial bilayers or in artificial liposomes, and they do not show significant hyphal permeabilization activity. In studies using *N. crassa*, Theviseen and coworkers have shown that treatment of hyphae with the defensins from radish (Rs-AFP2) or dahlia (DM-AMP1) caused K^+ efflux and Ca^{2+} uptake through binding to specific cell membrane receptors (Thevissen *et al.*, 1999). Although not tested with other fungi, it is likely that fungal inhibition occurs through this mechanism, i.e., ion efflux. Defensins are broadly active, inhibiting a large number of human and plant fungal pathogens, including *B. cinerea, Alternaria brassicola, F. culmorum, F. oxysporum, F. solani*, and *C. albicans* at micromolar levels (Ostoa-Saloma *et al.*, 2000).

(vii) Cyclophilin-like protein. Cyclophilins are a highly conserved group of proteins that are the intracellular receptors for cyclosporin; they have been found in a wide variety of organisms, including bacteria, plants, animals, and fungi (Ostoa-Saloma *et al.*, 2000). Recently, an 18-kDa protein was isolated from mung bean (*Phaseolus mungo*) with activity against *R. solani, F. oxysporum, B. cinerea*, and *Coprinus comatus* (Ye and Ng, 2000). This protein, called mungin, showed significant homology to cyclophilins and inhibited α- and β-glucosidases *in vitro*. However, the antifungal mechanism of action of mungin is not known.

(viii) Glycine/histidine-rich proteins. Insects synthesize a number of glycine/histidine-rich antifungal proteins and polypeptides, including those from *Holotrichia diomphalia* larvae (holotrichin, 84 amino acids (Lee *et al.*, 1995), *Sarcophaga peregrina* (flesh fly), AFP, 67 amino acids (Iijima *et al.*, 1993), and *Tenebrio molitor* (tenecin), 49 amino acids (Dae-Hee *et al.*, 1998; Lee *et al.*, 1999). They are extremely rich in glycine and histidine, which comprise as much as 80% of the amino acids. Importantly, fungi inhibited included *C. albicans*, the most common human pathogen (e.g., the 50% inhibitory concentration of tenecin is ~8 μg/ml (Dae-Hee *et al.*, 1998). The mechanism of action of these proteins is not understood.

(ix) Ribosome-inactivating proteins (RIPs). Ribosome-inactivating proteins (RIPs) are a group of cytotoxic N-glycosidases that depurinate the universally conserved α-sarcin loop of large rRNAs. This depurination inactivates the ribosome, thereby blocking its further participation in protein synthesis. RIPs may be induced by mechanical wounding, jasmonic acid, abscisic acid, salt stress, virus and fungi (Leah *et al.*, 1991; Logemann *et al.*, 1992; Song *et al.*, 1994; Desvoyes *et al.*, 1997). RIPs are generally viewed as pathogen-related proteins because of their toxicity towards animals, bacteria, fungi or viruses (Lam *et al.*, 1992; Barbieri *et al.*, 1993).

Plant RIPs inhibit mammalian, bacterial, fungal, and plant protein synthases *in vitro* and *in vivo* (Iglesias *et al.*, 1993). As an aside, how plants protect themselves from the action of their own RIPs is a subject of very interesting research. RIPs are widely distributed among different plant genera and within a variety of different tissues. RIPs have been classified into three groups. Type 1 RIPs are single-chain *N*-glycosidases with molecular masses of 11 to 30 kDa. Type 2 RIPs contain two chains, a cell-binding lectin (B chain) and an *N*-glycosidase (A chain), with molecular masses of ~60 kDa (Zhang *et al.*, 1999); type 2 RIPs include toxic members such as ricin and nontoxic members such as ebulin 1 (Girbes *et al.*, 1993) and nigrin b. Type 3 RIPs consist of four chains organized as two dimers of type 2 RIPs. Although RIPs were first identified more than 100 years ago, their biological function(s) still remains open to speculation. Besides their known activity of depurinating ribosomes at the sarcin/ricin (S/R) loop *in vitro*, there is no unequivocal answer to the question of why plants synthesize and accumulate RIPs. A critical reason for our ignorance of RIPs is the lack of information about this group of enzymes from a well-studied model plant system, such as *Arabidopsis thaliana* or *Nicotiana tobacum*. It was originally thought that RIPs do not occur universally among plants, evidenced by the failure to identify RIPs in such model plant systems. RIPs have intrinsic antifungal activity due to their ability to inactive fungal ribosomes *in vitro* and, presumably, *in situ*. Recent studies with a type 2 RIP showed that the cell-binding B chain (lectin) binds to fungal cells, forming a channel allowing the *N*-glycosidase A-chain entry into cells, resulting in RNA damage (Zhang *et al.*, 1999; Xia and Sui, 2000). Precisely, how the type 1 RIPs, which do not have a cell-binding chain, inhibit fungi (i.e., how they are internalized), is not known. Both type 1 and type 2 RIPs show broad activity against a number of plant and human pathogenic fungi as well as toxicity against mammalian cells (some type 2 RIPs are highly toxic to animals, likely because of the presence of the cell-binding B chain) (Oda *et al.*, 1997; Pohi *et al.*, 1998).

Recently, a 26-kDa RIP-like protein termed TRIP, possessing N-glycosidase activity, was isolated and characterized from *N. tobacum* (Sharma *et al.*, 2004). TRIP was found to be expressed at very low levels in the leaf (0.01% of total starting material), which could be the reason behind the long-standing failure to identify an RIP from tobacco. The expression of several RIPs (PAP, trichosanthin from *Trichosanthes kirilowii*, and antiviral protein from *Phytolacca insularis*) in transgenic plants successfully led to resistance against various viral and/or fungal pathogens (Lodge *et al.*, 1993; Lam *et al.*, 1996; Moon *et al.*, 1997). Ng and Prakash (2002) reported antifungal activity of a novel ribosome inactivating protein, Hispin from hairy melon seeds (Ng and Prakash, 2002). The two novel type 1 ribosome-inactivating proteins (RIPs), named ME1 and ME2, were isolated from the storage roots of *Mirabilis expansa*, an underutilized Andean root crop. ME1 and ME2 were active against several fungi, including *Pythium irregulare*, *Fusarium oxysporum solani*, *Alternaria solani*, *Trichoderma reesei*, and *Trichoderma harzianum* (Vivanco *et al.*, 1999). Tobacco plants expressing barley ribosome inactivating protein (RIP) under control of wound inducible promoter of the potato wun1gene, showed resistance againt *Rhizoctonia solani* (Logemann *et al.*, 1992). Another RIP is, Saporin that depurinate the major rRNA, thus leading to inactivation of ribosomes (Poma *et al.*, 1999). In spite of having been long investigated as a potentially useful toxin, the role and function of RIPs in plants and their distribution in nature remain enigmatic.

(x) LTPs. LTPs have the ability to transfer phospholipids between membranes. LTPs are small proteins (~8.7 kDa) of ~90 amino acids stabilized by four disulfide bonds with a central tunnel-like hydrophobic cavity. They have been isolated from a number of sources, including mammals, plants, fungi, and bacteria, and may play several *in vivo* roles, including lipid exchange between cytoplasmic organelles and, importantly, defense against pathogens (Guerbette *et al.*, 1999). Note that although the proteins are from diverse sources, they have striking homologies (between 37 and 90% identity). LTPs are active *in vitro* against a number of bacteria and fungi; however, the mechanism of action is not known. It may be that these proteins insert themselves into the fungal cell membrane, and the central hydrophobic cavity forms a pore, allowing efflux of intracellular ions and thus leading to fungal cell death. How this is related to their lipid transfer function is not clear.

(xi) Killer proteins (killer toxins). A number of yeasts secrete proteins that are lethal to sensitive fungal cells. These proteins, called killer proteins or killer toxins, are encoded either by double-stranded RNA, linear double-stranded plasmid DNA, or nuclear genes. Fungal cells secreting a killer toxin are resistant to their own toxin but are sensitive to other toxins. *Saccharomyces cerevisiae, Ustilago maydis, Hanseniaspora uvarum, Zygosaccharomyces bailii, Phaffia rhodozyma, Kluveromyces lactis*, and several *Pichia* species secrete a number of killer proteins. Over 20 individual killer toxins have been identified, varying in molecular mass from 10.7 to 156.5 kDa (Hiraga *et al*., 1999; Komiyama *et al*., 1998). The killer toxins have broad, potent antifungal activity against a number of human and plant pathogens (including *Pneumocystis carinii* (Seguy *et al*., 1998). MICs vary from 20μg/ml to far less. Although they have varied mechanisms of action, the first step of killer protein activity involves binding of the protein to specific cell surface receptors. Once bound, killer proteins are internalized and can disrupt cell wall synthesis, DNA synthesis, and K^+ channel activity, inhibit (1,3) β-glucan synthesis, or arrest the cell cycle. Any one of these effects leads to inhibition of fungal growth and to fungal cell death.

(xii) Protease inhibitors. Protein inhibitors of serine (e.g., trypsin and chymotrypsin) and cysteine proteases have emerged as a class of antifungal proteins that have potent activity against plant and animal pathogens. Cysteine protease inhibitors have been isolated from a number of plants and form a fourth group of cystatins, the phytocystatins (Blanco-Labra and Iturbe-Chinas, 1980; Joshi *et al*., 1998; Kouzuma *et al*., 2000; Park *et al*., 2000). The phytocystatins are single polypeptides of 10 to 12 kDa and share common structural motifs. Although phytocystatins are active against plant pathogens such as *F. solani* (MIC of 20 μg/disk in a disk agar diffusion assay) and *Trichoderma reesei* (250 ng/disk) (Joshi *et al*., 1998), the mechanism of antifungal activity is not understood.

Serine protease inhibitors that have antifungal activity also have the interesting property of inhibiting α-amylase activity from insects but not from bacterial or mammalian sources (Schimoler-O'Rourke *et al*., 1994). These proteins are bifunctional, inhibiting enzymes as well as inhibiting insect and fungal growth. Blanco-Labra and Iturbe-Chinas (1980) identified a bifunctional α-amlyase/trypsin inhibitor from corn (Blanco-Labra and Iturbe-Chinas, 1980); later it was found that this protein was identical to zeamatin (Richardson *et al*., 1987; Roberts and Selitrennikoff, 1990). We have recently confirmed that at high trypsin/zeamatin and α-amylase/zeamatin molar ratios, zeamatin inhibits trypsin and insect α-amylase activities *in vitro* (Schimoler-O'Rourke *et al*., 2001). Other bifunctional proteins from ragi (*Eleusine coracana*), wheat and barley have been isolated and characterized (Odani *et al*., 1983; Singh *et al*., 1987; Sanchez de la

Hz *et al.*, 1994; Behnke *et al.*, 1998; Gourinath *et al.*, 2000). Only a few of these proteins have been tested for *in vitro* antifungal activity, with zeamatin being the most extensively characterized. The mechanism of antifungal activity of these proteins is not fully understood.

(xiii) Other proteins. New proteins that have antifungal activity but do not neatly fall into any of the above classes are being discovered at a rapid pace. Only a few can be mentioned here. Viridin, a novel protein isolated from the culture medium of *Trichoderma viride*, has a molecular mass of 65 kDa and is active against sensitive fungi at 6 μM (Hao *et al.*, 1999; Hao *et al.*, 2000). Snakin-1 isolated from potato has a molecular mass of 6.9 kDa and is active at 10 μM (Segura *et al.*, 1999). A 30-kDa protein with very potent antifungal activity (50 ng/disk in an agar diffusion assay) was isolated from Engelmann's daisy (*Engelmannia pinnatifida*); this protein showed 35-50% identity to self-incompatibility glycoproteins, not previously known to be antifungal (Huynh *et al.*, 1996). The mechanism of action of none of these proteins is known.

However, these novel transgenic plants may also bring a set of new problems to the environment, since such transgenic plant genotypes have never previously occurred in the environment before. One of the reasonable steps after the creation of transgenic products, therefore, is to evaluate its potential benefits and risks to the environment (http://www.ccst.us/ccst/pubs/gmf/FB_Ch3.pdf).

Advantages and Disadvantages of Transgenic Crops

The use of transgenic crops has been an issue for many years. Many concerns have been raised and these generally fall into two categories:

1. A concern about what affect genetically modified material could have on human health. For example, transgenic crops have been suggested to cause allergies in some people, although it is uncertain whether transgenic crops are the source of this reaction. Furthermore, the antibiotic resistance genes placed in these crops have been suggested to cause resistance to antibiotics leading to super bugs that cannot be killed with antibiotic treatments (Ferber, 1999). The idea of a population being uncomfortable with ingesting DNA that originated from another source, such as a virus or bacterium, must also be considered when thinking about producing transgenic crops. However, to date, there is no evidence of the DNA from transgenic crops being any different from the DNA ingested from conventional crops.
2. A concern about whether transgenic crops cause damage to the natural environment. One example includes pollen from transgenic corn, which has been suggested to kill the Monarch butterfly larvae. It has been shown that hybrid corn expresses a bacterial toxin in its pollen, which is then dispersed over 60 metres by wind. In this range, the corn pollen is deposited on other plants near cornfields where it can be ingested by non-target organisms including the monarch butterfly. These butterflies have been found to eat less, have a slower growth rate and higher death rate (Losey *et al.*, 1999). A

second example is the hybridization of crops with nearby weeds. This could cause these weeds to attain resistance to herbicides or other things that we have been trying to avoid for many years. Genes that provide resistance to viral disease or other traits allowing them to survive in their environment could end up benefitting weed populations around a crop field. This trait could make that population more difficult to control. To date, there has been little evidence to support this theory (Crawley *et al.*, 2001).

On the other side of the coin are the notions that support the use of transgenic crops. The potential benefits of which are quite obvious, including such things as increased yields (to feed a growing population), decreasing the use of pesticides (to save the environment and the cost of pesticides), and the production of novel crops (such as providing crops with increased nutritional value) (Ferber, 1999). Being able to retrofit any crop to our desires is a powerful concept, especially with the changing climates of today.

Actual and potential access by farmers to traditional fungal control measures such as fungicides, and the absence of durable transgenic genetic resistance strategies, may explain why there are few reported efforts to develop transgenic varieties which are resistant to fungal pathogens. In relation to fungi, the current stage of successful research worldwide on identification/isolation of genes conferring durable resistance to fungal diseases of crops, is probably not encouraging developing countries with scarce research resources to embark on transgenic approaches to fungal resistance.

In the African region, only two initiatives for fungal resistant transgenic varieties are reported, a field trial of transgenic strawberry with phytoalexin synthesis genes (e.g., *Vst1*, *Vst2*) and laboratory work on transgenic maize for resistance to cob rot (*Stenocarpella maydis*), both in South Africa. The three initiatives for the development of fungal resistant transgenic varieties reported in Eastern Europe were all in Bosnia and Herzegovina where laboratory testing of transgenic potato for resistance to *Fusarium, Verticilium* and *Rhizoctonia* has been initiated. In the Asian region, there is a field trial underway in China for transgenic cotton with resistance to *Verticilium* and *Fusarium*. Other R&D initiatives reported include the involvement of India, Malaysia and Pakistan in research on sheath blight of rice, and of Indonesia on rice blast and leaf rust of coffee.

A few countries in Latin America, mainly Argentina, Brazil and Cuba, are carrying out a number of activities on transgenic resistance to fungi, particularly on tropical fruit trees, with some results already being tested in the field. In this region, most of the activities for transgenic fungal resistance are reported in Cuba, in particular involving field trials of transgenic potato for late blight resistance, and fungal-resistant sugar cane. Other field trials in the Latin America region for transgenic fungal resistance are reported for maize, sunflower and wheat in Argentina, and tobacco in Mexico. Other crops subject to transgenic R&D for fungal resistance in Cuba are banana, plantain, pineapple, tomato, papaya, citrus and rice. Other countries involved in transgenic fungal resistance research are Argentina on alfalfa, Brazil on rice, barley and cocoa, Chile on grape and apple,

Colombia on tree tomato, Peru on potato for late blight resistance and Venezuela on sugar cane.

Countries Growing GMOs. The world's leading producers of GM crops are the United States, Argentina, Brazil, Canada, and China. In 2005, GM crop production also reached noteworthy levels in Paraguay, South Africa, Uruguay and Australia, Table 3.3. In the EU, GM crops have remained uncommon. Appreciable GM maize production in the EU only took place in Spain on an area of about 53,000 hectares. In Portugal, Germany, France and the Czech Republic, transgenic crops were primarily grown for small-scale field trials. Iran and the Czech Republic were also added to the list of countries commercially growing transgenic crops. As of 2005, 38% of GM crops are grown in developing countries.

Table 3.3 Global Area of Genetically Engineered Crops, 1996 to 2005: By Country (Million Hectares).

Country	*1996*	*1997*	*1998*	*1999*	*2000*	*2001*	*2002*	*2003*	*2004*	*2005*
USA	1.5	8.1	20.5	28.7	30.3	35.7	39.0	42.8	47.6	49.8
Argentina	0.1	1.4	4.3	6.7	10.0	11.8	13.5	13.9	16.2	17.1
Brazil	-	-	-	-	-	-	-	3.0	5.0	9.4
Canada	0.1	1.3	2.8	4.0	3.0	3.2	3.5	4.4	5.4	5.8
China	-	0.0	<0.1	0.3	0.5	1.5	2.1	2.8	3.7	3.3
Paraguay	-	-	-	-	-	-	-	-	1.2	1.8
India	-	-	-	-	-	-	<0.1	0.1	0.5	1.3
South Africa	-	-	<0.1	0.1	0.2	0.2	0.3	0.4	0.5	0.5
Uruguay	-	-	-	-	<0.1	<0.1	<0.1	0.1	0.3	0.3
Australia	<0.1	0.1	0.1	0.1	0.2	0.2	0.1	0.1	0.2	0.3
Mexico	<0.1	<0.1	0.1	<0.1	<0.1	<0.1	<0.1	<0.1	0.1	0.1
Romania	-	-	-	<0.1	<0.1	<0.1	<0.1	<0.1	0.1	0.1
Philippines	-	-	-	-	-	-	-	<0.1	0.1	0.1
Spain	-	-	<0.1	<0.1	<0.1	<0.1	<0.1	<0.1	0.1	0.1
Colombia	-	-	-	-	-	-	<0.1	<0.1	<0.1	<0.1
Iran	-	-	-	-	-	-	-	-	-	<0.1
Honduras	-	-	-	-	-	-	<0.1	<0.1	<0.1	<0.1
Portugal	-	-	-	<0.1	-	-	-	-	-	<0.1
Germany	-	-	-	-	<0.1	<0.1	<0.1	<0.1	<0.1	<0.1
France	-	-	<0.1	<0.1	<0.1	-	-	-	-	<0.1
Czech Republic	-	-	-	<0.1	-	-	-	-	-	<0.1
Bulgaria	-	-	-	-	<0.1	<0.1	<0.1	<0.1	-	-
Indonesia	-	-	-	-	-	<0.1	<0.1	<0.1	-	-
Ukraine	-	-	-	<0.1	-	-	-	-	-	-
Total	1.7	11.0	27.8	39.9	44.2	52.6	58.7	67.7	81.0	90.0

Source: ISAAA, Clive James, 2005. January 16, 2006.

Conclusion

The control of disease is an essential part of crop production strategy. Antifungal proteins and polypeptides have been isolated from diverse groups of organisms, including plants, fungi, bacteria, insects, and animals (both vertebrates and invertebrates). The mechanisms of action of these proteins are as varied as their sources and include fungal cell wall polymer degradation, membrane channel and pore formation, damage to cellular ribosomes, inhibition of DNA synthesis, and inhibition of the cell cycle. The mode of action of many proteins remains unknown and is the subject of active research. The range of fungi inhibited by antifungal proteins is extremely broad, with plant pathogens and human pathogens being sensitive at micromolar levels; in some cases, even more potent inhibition was found. The genes encoding many antifungal proteins are currently being used by agribusiness to create genetically modified plants that have increased fungal resistance in the field. Whether these transgenic plants and the crops derived from them gain acceptance in the marketplace remains to be seen. Novel methods of disease control will continue to emerge as the limitation of older methods are revealed and new techniques discovered in response to new demands on cropping system to produce food in ever more hostile environments. The skills of plant breeder, coupled with present day technology will ensure an even greater supply of cultivars with more resistance to diseases.

References

Agrawal, G.K., Jwa, N.S. and Rakwal, R. (2000). A novel rice (*Oryza sativa L.*) acidic PR1 gene highly responsive to cut, phytohormones, and protein phosphatase inhibitors. *Biochem. Biophys. Res. Commun.* 274:157–165.

Barbieri, L., Battelli, M.G. and Stirpe, F. (1993). Ribosome inactivating proteins from plants. *Biochim. Biophys. Acta.* 1154(3-4):237–282.

Behnke, C.A., Yee, V.C., Trong, I.L., Pedersen, L.C., R.E. Stenkamp, Kim, S.S., Reeck, G.R. and Teller, D.C. (1998). Structural determinants of the bifunctional corn Hageman factor inhibitor: x-ray structure at 1.95 Å resolution. *Biochem.* 37:15277–15288.

Blanco-Labra, A. and Iturbe–Chinas, F.A. (1980). Purification and characterization of an α-amylase inhibitor from maize (*Zea maize*). *J. Food Biochem.* 5:1–17.

Borgmeyer, J.R., Smith, C.E. and Huynh, Q.K. (1992). Isolation and characterization of a 25 kDa antifungal protein from flax seeds. *Biochem. Biophys. Res. Commun.* 187: 480–487.

Bormann, C., Baier, D., Horr, I., Raps, C., Berger, J., Jung, G. and Schwartz, H. (1999). Characterization of a novel, antifungal, chitin-binding protein *Streptomyces tendae* Tu901 that interferes with growth polarity. *J. Bacteriol.* 181:7421–7429.

Broglie, K., Chet, I., Holliday, M., Cressman, R., Biddle, P., Knowlton, S., Mauvais, C.J. and Broglie, R. (1991). *Science* 254:1194–1197.

Bryngelsson, T., Sommer-Knudsen, J., Gregersen, P.L., Collinge, D.B., Ek, B. and Thordal-Christensen, H. (1994). Purification, characterization, and molecular cloning of basic PR-1-type pathogenesis-related proteins from barley. *Mol. Plant Microbe Interact.* 7:267–275.

Caboche M. (1990). Liposome mediated transformation of nucleic acids into plant cells. Physiol. *Plant* 79:173–176.

Carstens, M., Vivier, M.A. and Pretorius, I.S.(2003). The *Saccharomyces cerevisiae* chitinase, encoded by the CTS1-2 gene, confers antifungal activity against *Botrytis cinerea* to transgenic tobacco. *Transgenic Res.* 12(4):497–508.

Chae, K.S., Lee, I.H., Choi, C.S. and Kim, H.R. (1999). Purification and characterization of chitin-binding proteins from the hemolymph of sweet potato hornworm, *Agrius convolvuli. Comp. Biochem. Physiol. B* 124:475–481.

Cote, F., Cutt, J.R., Asselin, A. and Klessig, D.F. (1991). Pathogenesis-related acidic β-1,3-glucanase genes of tobacco are regulated by both stress and developmental signals. *Mol. Plant Microbe Interact.* 4:173–181.

Crawley, M.J., Brown, S.L., Hails, R.S., Kohn, D.D. and Rees, M. (2001). Transgenic crops in natural habitats. *Nature* 409:682–683.

Dae-Hee, K., Lee, Y.T., Lee, Y.J., Chung, J.H., Lee, B.L., Choi, B.S. and Younghoon, L. (1998). Bacterial expression of tenecin 3, and insect antifungal protein isolated from *Tenebrio molitor*, and its efficient purification. *Mol. Cells* 8:786–789.

De Bolle, M.F., David, K.M., Rees, S.B., Vanderleyden, J., Cammue, B.P. and Broekaert, W.F. (1993). Cloning and characterization of a cDNA encoding an antimicrobial chitin-binding protein from *Amaranthus, Amaranthus caudatus. Plant Mol. Biol.* 22:1187–1190.

Destoumieux, D., Munoz, M., Bulet, P. and Bachere, E. (2000). Penaeidins, a family of antimicrobial peptides from penaeid shrimp (Crustacea, Decapoda). *Cell Mol Life Sci.* 57:1260–1271.

Desvoyes, B., Poyet, J.L., Schlick, J.L., Adami, P., Jouvenot, M. and Dulieu, P. (1997). Identification of a biological inactive complex form of pokeweed antiviral protein. *FEBS Lett.* 410(2-3):303–308.

Ferber, D. (1999). Risks and benefits: GM crops in the cross hairs. *Science* 286:1662–1666.

Gatz, C., Kaiser, A. and Wendenburg, R. (1991). Regulation of a midified CaMV 35SRNA by the Tn 10 encoded tet repressor intransgenic tobacco. *Mol. Gen. Genet.* 227: 229–237.

Girbes, T., Citores, L., Iglesias, R., Ferreras, J.M., Munoz, R., Rojo, M.A., Arias, F.J., Garcia, J.R., Mendez, E. and Calonge, M. (1993). Ebulin 1, a nontoxic novel type 2 ribosome-inactivating protein from *Sambucus ebulus* L. leaves. *J. Biol. Chem.* 268:18195–18199.

Gourinath, S., Alam., N., Srinivasan, A., Betzel, C. and Singh, T.P. (2000). Structure of the bifunctional inhibitor of trypsin and α-amylase from ragi seeds at 2.2 Å resolution. *Acta Crystallogr. D* 56:287–293.

Grenier, J., Potvin, C. and Asselin, A. (1993). Barley pathogenesis-related proteins with fungal cell wall lytic activity inhibit the growth of yeasts. *Plant Physiol.* 103:1277–1283.

Grower, A. and Gowthaman, R. (2003). Strategies for development of fungus-resistant transgenic plants. *Curr. Sci.* 84(3):330–340.

Guerbette, F., Grosbois, M., Jolliot-Croquin, A., Kader, J.C. and Zachowski, A. (1999). Lipid-transfer proteins from plants: structure and binding properties. *Mol. Cell. Biochem.* 192:157–161.

Hao, J.J., Geng, C., Xie, W., Gong, Z., Liu, W.Y. and Wang, E. (1999). Isolation and characterization of viridin, a new 65 kDa antifungal protein from the mould *Trichoderma viride*. *Biol. Chem.* 380:1243–1245.

Hao, J.J., Ye, J.Q., Yang, Q., Gong, Z.Z., Liu, W.Y. and Wang, E. (2000). A silent antifungal protein (AFP)-like gene lacking two introns in the mould *Trichoderma viride*. *Biochim. Biophys. Acta* 1475(2):119–124.

Hejgaard, J., Jacobsen, S. and Svendsen, I. (1991). Two antifungal thaumatin-like proteins from barley grain. *FEBS Lett.* 291:127–131.

Hiraga, K., Hayashi, S., Kitazawa, M. and Oda, K. (1999). Isolation and some properties of a novel killer toxin like protein produced by *Streptomyces* sp. F-287. *Biosci. Biotechnol. Biochem.* 63(6):1037–1044.

http://dmd.nihs.go.jp/latex/defense-e.html

http://www.ccst.us/ccst/pubs/gmf/FB_Ch3.pdf

http://www.gmo-compass.org/eng/agri_biotechnology/breeding_aims/148.disease_resistant_ crops.html

http://www.gmo-compass.org/eng/agri_biotechnology/gmo_planting/142.countries_growing _gmos.html

http://www.gmo-compass.org/eng/agri_biotechnology/gmo_planting countries_growinggmo s . html)

http://www.pesticidebook.com /pdfs/chapter25_pages317–319.pdf

Hu, X., and Reddy, A.S. (1997). Cloning and expression of a PR5-like protein from *Arabidopsis*: inhibition of fungal growth by bacterially expressed protein. *Plant Mol. Biol.* 34:949–959.

Huang, X., Xie, W. and Gong, Z. (2000). Characteristics and antifungal activity of a chitin binding protein from *Ginkgo biloba*. *FEBS Lett.* 478:123–126.

Huynh, Q.K., Borgmeyer, J.R. and Zobel J.F. (1992). Isolation and characterization of a 22 kDa protein with antifungal properties from maize seeds. *Biochem. Biophys. Res. Commun.* 182:1–5.

Huynh, Q.K., Borgmeyer, J.R., Smith, C.E., Bell, L.D. and Shah, D.M. (1996). Isolation and characterization of a 30 kDa protein with antifungal activity from leaves of *Engelmannia pinnatifida*. *Biochem. J.* 15:723–727.

Iglesias, R., Arias, F.J., Rojo, M.A., Escarmis, C., Ferreras, J.M. and Girbes, T. (1993). Molecular action of the type 1 ribosome-inactivating protein saporin 5 on *Vicia sativa* ribosomes. *FEBS Lett.* 325:291–294.

Iijima, R., Kurata, S. and Natori, S. (1993). Purification, characterization, and cDNA cloning of an antifungal protein from the hemolymph of *Sarcophaga peregrina* (flesh fly) larvae. *J. Biol. Chem.* 268:12055–12061.

Itoh, Y., Takahashi, K., Takizawa, H., Nikaiduo, N., Tanaka, H., Nishihashi, H., Watanabe, T. and Nishizawa, Y. (2003). Family 19 chitinase of *Streptomyces grieus* HUT6037 increase plant resistance to fungal disease. *Biosci. Biotech. Biochem.* 67(4):847–855.

Jach, G., Gornhardt, B., Mundy, J., Logemann, J., Pinsdorf, E., Leah, R., Schell, J. and Maas, C. (1995). Enhanced quantitative resistance against fungal disease by combinatorial expression of different barley antifungal proteins in transgenic tobacco. *Plant J.* 8:97–109.

Joshi, B.N., Sainani, M.N., Bastawade, K.B., Gupta, V.S. and Ranjekar, P.K. (1998). Cysteine protease inhibitor from pearl millet: a new class of antifungal protein. *Biochem. Biophys. Res. Commun.* 246:382–387.

Kawabata, S., Nagayama, R., Hirata, M., Shigenaga, T., Agarwala, S.L., Saito, T., Cho, J., Nakajima, H., Takagi, T. and Iwanaga, S. (1996). Tachycitin, a small granular component in horseshoe crab hemocytes, is an antimicrobial protein with chitin-binding activity. *J. Biochem.* 120:1253–1260.

Khardori, N. (1989). Host-parasite interaction in fungal infections. *Eur. J. Clin. Microbiol. Infect. Dis.* 8:331–351.

Kim, Y. J., and Hwang, B.K. (1997). Isolation of a basic 34 kilo Dalton α-1,3-glucanase with inhibitory activity against *Phytophthora capsici* from pepper stems. *Physiol. Mol. Plant Pathol.* 50:103–115.

Klein, T.M., Wolf, E.D., Wu. R.D. and Standar, J.C. (1987). High velocity microinjection for delivering nucleic acids into living cells. *Nature* 327:70–73.

Koiwa, H., Kato, H., Nakatsu, T., Oda, J., Yamada, Y. and Sato, F. (1997). Purification and characterization of tobacco pathogenesis-related protein PR-5d, an antifungal thaumatin-like protein. *Plant Cell Physiol.* 38:783–791.

Kolbe, S., Fischer, S., Becirevic, A., Hinz, P. and Schrempf, H. (1998). The *Streptomyces reticuli* alpha-chitin-binding protein CHB2 and its gene. *Microbiol.* 144:1291–1297.

Komiyama, T., Shirai, T., Ohta, T., Urakami, H., Furuichi, Y., Ohta, Y. and Tsukada, Y. (1998). Action properties of HYI killer toxin from *Williopsis saturnus* var. *saturnus* and antibiotics, aculeacin A and papulacandin B. *Biol. Pharm. Bull.* 21:1013–1019.

Kouzuma, Y., Inanaga, H., Doi-Kawano, K., Yamasaki, N. and Kimura, M. (2000). Molecular cloning and functional expression of cDNA encoding the cysteine proteinase inhibitor with three cystatin domains from sunflower seeds. *J. Biochem.* 128:161–166.

Lam YH, Wong YS, Wang B, Wong RNS, Yeung HW, Shaw PC(1996) Use of trichosanthin to reduce infection by turnip mosaic virus. *Plant Sci.* 114:11–117.

Lam, S.K. and Ng, T.B. (2001). Hypsin, a novel thermostable ribosome-inactivating protein with antifungal and antiproliferative activities from fruiting bodies of the edible mushroom *Hypsizigus marmoreus*. *Biochem. Biophys. Res. Commun.* 285:1071–1075.

Landon, C., Pajon, A., Vovelle, F. and Sodano, P. (2000). The active site of drosomycin, a small insect antifungal protein, delineated by comparison with the modeled structure of Rs-AFP2, a plant antifungal protein. *J. Pept. Res.* 56:231–238.

Landon, C., Sodano, P., Hetru, C., Hoffmann, J. and Ptak, M. (1997). Solution structure of drosomycin, the first inducible antifungal protein from insects. *Protein Sci.* 6:1878–1884.

Leah, R., Tommerup, H., Svendsen, I. and Mundy, J. (1991). Biochemical and molecular characterization of three barley seed proteins with antifungal properties. *J. Biol. Chem.* 266:1564–1573.

Lee, K. M., D. H. Kim, Y. H. Lee, B. S. Choi, J. H. Chung, and B. L. Lee. (1999). Antifungal activities of recombinant antifungal protein by conjugation with polyethylene glycol. *Mol. Cells* 9:410–416.

Lee, S. Y., Moon, J.H., Kurata, S., Natori, S. and Lee, B.L. (1995). Purification and cDNA cloning of an antifungal protein from the hemolymph of *Holotrichia diomphalia* larvae. *Biol. Pharm. Bull.* 18:1049–1052.

Lodge, J.K., Kaniewski, W.K. and Tumer, N.E. (1993). Broad-spectrum virus resistance in transgenic plants expressing pokeweed antiviral protein. *Proc. Natl. Acad. Sci. USA* 90:7089–7093.

Logemann, J, Jach, G., Tommerup, H., Mundy, J. and Schell, J. (1992). Expression of a Barley Ribosome-Inactivating Protein Leads to Increased Fungal Protection in Transgenic Tobacco Plants. *Biotechnol.* 10:305–308.

Lorito, M., Woo, S.L., Fernandez, I.G., Colucci, G., Harman, G.E., Pintor-Toro, J.A., Filippone, E., Muccifora, S., Lawrence, C.B., Zoina, A., Tuzun, S. and Scala, F. (1998). Genes from mycoparasitic fungi as a source for improving plant resistance to fungal pathogens. *Proc. Natl. Acad. Sci. USA.*. 95:7860–7865.

Losey, J.E., Rayor, L.S. and Carter, M.E. (1999). Transgenic pollen harms monarch larvae. *Nature* 399: 214.

Molina, A., Gorlach, J., Volrath, S. and Ryals, J. (1999). Wheat genes encoding two types of PR-1 proteins are pathogen inducible, but do not respond to activators of systemic acquired resistance. *Mol. Plant Microbe Interact.* 12:53–58.

Mondal, K.K., Chatterjee, S.C., Viswakarma, N., Bhattacharya, R.C. and Grover, A. (2003). Chitinase-mediated inhibitory activity of Brassica transgenic on growth of *Alternaria brassicae. Curr Microbiol.* 47:171–173.

Money, N. and Harold, F. (1992). Extension growth of the water mold *Achyla*: interplay of turgor and wall strength. *Proc. Natl. Acad. Sci. USA* 89:4245–4249.

Moon, Y.H., Song, S.K., Choi, K.W. and Lee, J.S. (1997). Expression of a cDNA encoding *Phytolacca insularis* antiviral protein confers virus resistance on transgenic potato plants. *Mol. Cells* 7:807–815.

Mora, A. and Earle, E.D.(2001). Combination of *Trichoderma harzianum* endochitinase and a membrane-affecting fungicide on control of Alternaria leaf spot in transgenic broccoli plants. *Appl. Microbiol. Biotechnol.* 55:306–310.

Morrisette, J., Kratzschmar, J., Haendler, B., El-Hayek, R., Mochca-Morales, J., Martin, B. M., Patel, J.R., Moss, R.L., Schleuning, W.D., Coronado, R. and Possani, L.D. (1995). Primary structure and properties of helothermine, a peptide toxin that blocks ryanodine receptors. *Biophys. J.* 68:2280–2288.

Muradov, A., Petrasovits, L., Davidson, A. and Scott, K.J. (1993). A cDNA clone for a pathogenesis-related protein 1 from barley. *Plant Mol. Biol.* 23:439–442.

Nagy, F., Bartry, M., Hsu-Wong, M. and Chua, N. (1987). The 5' proximal region of the wheat Cab-1 gene contains a 286 bp enhancer like sequence for phytochrome response. *EMBO J.* 6: 2537–2542.

Ng, T.B. and Parkash, A. (2002). Hispin, a novel ribosome inactivating protein with antifungal activity from hairy melon seeds. *Protein Expr. Purif.* 26(2):211–217.

Niderman, T., Genetet, I., Buryere, T., Gees, R., Stintzi, A., Legrand, M., Fritig, B. and Mosinger, E. (1995). Pathogenesis-related PR-1 proteins are antifungal. Isolation and characterization of three 14-kilodalton proteins of tomato and of a basic PR-1 of tobacco with inhibitory activity against *Phytophthora infestans. Plant Physiol.* 108:17–27.

Nielsen, K. K., Nielsen, J.E., Madrid, S.M. and Mikkelsen, J.D. (1997). Characterization of a new antifungal chitin-binding peptide from sugar beet leaves. *Plant Physiol.* 113:83–91.

Oda, T., Komatsu, N. and Muramatsu, T. (1997). Cell lysis induced by ricin D and ricin E in various cell lines. *Biosci. Biotechnol. Biochem.* 61:291–297.

Odani, S., Koide, T. and Ono, T. (1983). The complete amino acid sequence of barley trypsin inhibitor. *J. Biol. Chem.* 258:7998–8003.

Osaki, T., Omotezako, M., Nagayama, R., Hirata, M., Iwanaga, S., Kasahara, J., Hattori, J., Ito, I., Sugiyama, H. and Kawabata, S. (1999). Horseshoe crab hemocyte-derived antimicrobial polypeptides, tachystatins, with sequence similarity to spider neurotoxins. *J. Biol. Chem.* 274:26172–26178.

Osborn, R. W., De Samblanx, G.W., Thevissen, K., Goderis, I., Torrekens, S., Van Leuven, F., Attenborough, S., Rees, S.B. and Broekaert, W.F. (1995). Isolation and characterization of plant defensins from seeds of *Asteraceae, Fabaceae, Hippocastanaceae* and *Saxifragaceae. FEBS Lett.* 368:257–262.

Ostoa-Saloma, P., Carrero, J.C., Petrossian, P., Herion, P., Landa, A. and Laclette, J.P. (2000). Cloning, characterization and functional expression of a cyclophilin of *Entamoeba histolytica. Mol. Biochem. Parasitol.* 107:219–225.

Park, K.S., Cheong, J.J., Lee, S.J., Suh, M.C. and Choi, D. (2000). A novel proteinase inhibitor gene transiently induced by tobacco mosaic virus infection. *Biochim. Biophys. Acta* 1492:509–512.

Pennisi, E. (2001). The push to pit genomics against fungal pathogens. *Science.* 292: 2273–2274.

Pohi, P., Antonenko, Y.N., Evtodienko, V.Y., Pohl, E.E., Saparov, S.M., Agapov, I.I. and Tonevitsky, A.G. (1998). Membrane fusion mediated by ricin and viscumin. *Biochim. Biophys. Acta* 1371:11–16.

Poma, A., Zarivi, O., Bianchini, S. and Spano, L. (1999). The plant ribosome inactivating protein saporin induces micronucleus formation in peripheral human lymphocytes *in vitro. Toxicol Lett.* 105:67–73.

Punja Z.K., Raharjo S.H.T. (1996). Response of transgenic cucumber and carrot plants expressing different chitinase enzymes to inoculation with fungal pathogens. *Plant Dis.* 80:999–1005.

Rezzonico, E., N. Flury, F. Meins, Jr., and R. Beffa. 1998. Transcriptional down-regulation by abscisic acid of pathogenesis-related α-1,3-glucanase genes in tobacco cell cultures. *Plant Physiol.* 117:585–592.

Richardson, M., Valdes-Rodriquez, S. and Blanco-Labra, A. (1987). A possible function for thaumatin and a TMV-induced protein suggested by homology to a maize inhibitor. *Nature* 327:432–434.

Roberts, W. and Selitrennikoff, C.P. (1990). Zeamatin, an antifungal protein from maize with membrane-permeabilizing activity. *J. Gen. Microbiol.* 136:1771–1778.

Salehi, A., Mohammadi, M., Okhovvat, S.M. and Omidi, M. (2005). Chitinase gene transformation through Agrobacterium and its explanation in soybean in order to induce resistance to root rot caused by *Rhizoctonia solani. Commun. Agric. Appl. Biol. Sci.*70:399–406.

Sanchez de la Hoz, P., Castagnaro, A. and Carbonero, P. (1994). Sharp divergence between wheat and barley at loci encoding novel members of the trypsin/α-amylase inhibitor family. *Plant Mol. Biol.* 26:1231–1236.

Schimoler-O'Rourke, R., Richardson, M. and Selitrennikoff, C.P. (2001). Zeamatin inhibits trypsin and α-amylase activities. *Appl. Environ. Microbiol.* 67:2365-2366.

Segura, A., Moreno, M., Madueno, F., Molina, A. and Garcia-Olmedo, F. (1999). Snakin-1, a peptide from potato that is active against plant pathogens. *Mol. Plant Microbe Interact.* 12:16–23.

Segura, A., Moreno, M., Molina, A. and Garcia-Olmedo, F. (1998). Novel defensin subfamily from spinach (*Spinacia oleracea*). *FEBS Lett.* 435:159–162

Seguy, N., Polonelli, L., Dei-Cas, E. and Cailliez, J.C. (1998). Effect of a killer toxin of *Pichia anomala* on *Pneumocystis*. Perspectives in the control of pneumocytosis. *FEMS Immunol. Med. Microbiol.* 22:145–149.

Selitrennikoff, C.P. (2001). Antifungal Proteins. *Appl. Environ. Microbiol.* 67(7): 2883–2894.

Sharma, N., Park, S.W., Vepachedu, R., Barbieri, L., Ciani, M., Stirpe, F., Savary, B.J. and Vivanco, J.M. (2004). Isolation and characterization of an RIP (ribosome-inactivating protein)-like protein from tobacco with dual enzymatic activity. *Plant Physiol.* 134(1):171–181.

Singh, N.K., Bracker, C.A., Hasegawa, P.M., Handa, A.K., Buckel, S., Hermodson, M.A., Pfankoch, E., Regnier, F.E. and Bressan, R.A. (1987). Characterization of osmotin. *Plant Physiol.* 85:529–536.

Song, H.K., Hwang, K.Y., Kim, K.K. and Suh, S.W. (1994). Crystallization and preliminary X-ray crystallographic study of ribosome-inactivating protein from barley seeds. *Acta Cryst.* D50, 910–912.

Source: ISAAA, Clive James, 2005. January 16, 2006

Staskawicz, B.J., Ausubel, F.M., Baker, B.J., Ellis, J.G. and Jones, J.D.G. (1995). *Science*.268:661–667.

Stintzi, A., Heitz, T., Prasad, V., Wiedemann-Merdioglu, S., Kauffmann, S., Geoffroy, P., Legrand, M. and Fritig, B. (1993). Plant 'pathogenesis-related' proteins and their role in defense against pathogens. *Biochimie* 75:687–706.

Takahashi, W., Fujimori, M., Miura,Y., Komatsu, T., Nishizawa, Y., Hibi, T. and Takamizo, T. (2005). Increased resistance to crown rust disease in transgenic Italian ryegrass (*Lolium multiflorum* Lam.) expressing the rice chitinase gene. *Plant Cell Rep.* 23:811–818.

Terras, F. R., Schoofs, H.M., De Bolle, M.F., Van Leuven, F., Rees, F.B., Vanderleyden, J., Cammue, B.P. and Broekaert, W.F. (1992). Analysis of two novel classes of plant antifungal proteins from radish (*Raphanus sativus L.*) seeds. *J. Biol. Chem.* 267:15301–15309.

Terras, F. R., Torrekens, S., Van Leuven, F., Osborn, R.W., Vanderleyden, J., Cammue, B.P. and Broekaert, W.F. (1993). A new family of basic cysteine-rich plant antifungal proteins from *Brassicaceae* species. *FEBS Lett.* 316:233–240.

Thevissen, K., Ghazi, A., De Samblanx, G.W., Brownlee, C., Osborn, R.W. and Broekaert, W.F. (1996). Fungal membrane responses induced by plant defensins and thionins. *J. Biol. Chem.* 271:15018–15025.

Thevissen, K., Terras, F.T. and Broekaert, W.F. (1999). Permeabilization of fungal membranes by plant defensins inhibits fungal growth. *Appl. Environ. Microbiol.* 65:5451–5458.

Trudel, J., Grenier, J., Potvin, C. and Asselin, A.(1998). Several thaumatin-like proteins bind to α-1,3-glucans. *Plant Physiol.* 118:1431–1438

Vellicce, G.R., Ricci, J.C., Hernandez, L and Castagnaro, A.P. (2006). Enhanced resistance to *Botrytis cinerea* mediated by the transgenic expression of the chitinase gene ch5B in strawberry. *Transgenic Res.* 15:57–68.

Vivanco, J.M., Savary, B.J. and Flores, H.E. (1999). Characterization of Two Novel Type I Ribosome-Inactivating Proteins from the Storage Roots of the Andean Crop *Mirabilis expansa*. *Plant Physiol.* 119: 1447–1456.

Wu, J.H., Zhang, X.L., Luo, X.L., Nie, Y.C., Tian, Y.C. and Chen, Z.H. (2004). Transgenic cotton plants of chitinase and glucanase genes and their performance of resistance to *Verticillium dahliea*. *Yi Chuan Xue Bao.* 31:183–188.

Xia, X.F., and Sui, S.F. (2000). The membrane insertion of trichosanthin is membrane-surface-pH dependent. *Biochem. J.* 349:835–841.

Ye, X.Y., Wang, H.X. and Ng, T.B. (1999). First chromatographic isolation of an antifungal thaumatin-like protein from French bean legumes and demonstration of its antifungal activity. *Biochem. Biophys. Res. Commun.* 263:130–134.

Ye, X.Y., and Ng, T.B. (2000). Mungin, a novel cyclophilin-like antifungal protein from the mung bean. *Biochem. Biophys. Res. Commun.* 273:1111–1115.

Yun, D.J., Ibeas, J.I., Lee, H., Coca, M.A., Narasimhan, M.L., Uesono, Y., Hasegawa, P.M., Pardo, J.M. and Bressan, R.A. (1998). Osmotin, a plant antifungal protein, subverts signal transduction to enhance fungal cell susceptibility. *Mol. Cell* 1:807–817.

Yun, D., Zhao, Y., Pardo, J., Narasimhan, M.L., Damsz, B., Lee, H., Abad, L.R., D'Urzo, M. P., Hasegawa, P.M. and Bressan, R.A. (1997). Stress proteins on the yeast cell surface determine resistance to osmotin, a plant antifungal protein. *Proc. Natl. Acad. Sci. USA* 94:7082–7087.

Zhang, G. P., Shi, Y.L., Wang, W.P. and Liu, W.Y. (1999). Cation channel formed at lipid bilayer by Cinnamomin, a new type II ribosome-inactivating protein. *Toxicon.* 37:1313–1322.

Zhu Q, Maher E A, Masoud A, Dixon R A, Lamb C J. (1994). Enhanced Protection Against Fungal Attack by Constitutive Co-expression of Chitinase and Glucanase Genes in Transgenic Tobacco. *Bio/Technol.* 12:807–812.

4

Properties and Commercial Applications of Xylanases from Fungi

Maria de Lourdes T.M. Polizeli

Departmento de Biologia, Faculdade de Filosofia, Ciências e Letras de Ribeirão Preto – Universidade de São Paulo
Av. Bandeirantes, 3900, Bairro Monte Alegre, 14040-901 Ribeirão Preto – SP, Brazil.
Telephone + 55 16 36024680; Fax +55 16-36024886. E-mail: polizeli@ffclrp.usp.br

INTRODUCTION

Ever since prehistoric times, wood has played a vitally important part in our lives. It is created in the trunks, stems, branches and roots of arboreal gymnosperms and angiosperms. *Softwood* refers to the wood of conifers, such as the genus *Pinus*. On account of its long fibers, softwood is used in the manufacture of types of paper in which strength is essential, for example, wrapping paper. On the other hand, wood from the Angiospermae, including the genus *Eucalyptus*, is called *hardwood*. Possessing short fibers, it can be used to fabricate printing and writing paper. The terms 'softwood' and 'hardwood' must not be understood as referring to physical hardness, but rather to morphological differences. Softwoods are structurally homogeneous, 90% of their axial elements being tracheids, while hardwoods are heterogeneous, containing vessel elements, fibers and sometimes tracheids (Fujita and Harada, 1991).

The organization and composition of the xylem cell wall give rise to the properties of a given wood and thus determine whether and how it will be used in papermaking. Basically, the plant cell wall consists of primary (P) and secondary (S) walls (Fig.4.1). After a cell divides, the daughter cells remain united by the middle lamella (ML) that spreads across the plane of division. The ML is an intercellular plate, made mainly of lignin, on which cellulose is deposited to form the primary wall. In some plant cells, that is the only wall formed, but in most it represents only 7–14% of the final thickness. The secondary wall is laid down after the cell ceases to grow and can be divided into three layers, each one with microfibrils pointing in a different direction: S1 ($\approx$ 0.2 μm thick) is next to the primary wall and forms 5–11% of the total thickness; S2 (2-5 μm) represents 74–84% of the wall, and S3 is similar to S1, contributing 3–4%.

The main polymers constituting hardwood and softwoods and their distribution within the cell wall are described in the Tables 4.1 and 4.2, respectively, where average values have been taken (Timell, 1965; Coughlan *et al.*, 1993). The chemical composition of the cell wall in a tree also depends on the species, soil and climatic conditions, individual genotype, height of the trunk, sampling and even the method of analysis.

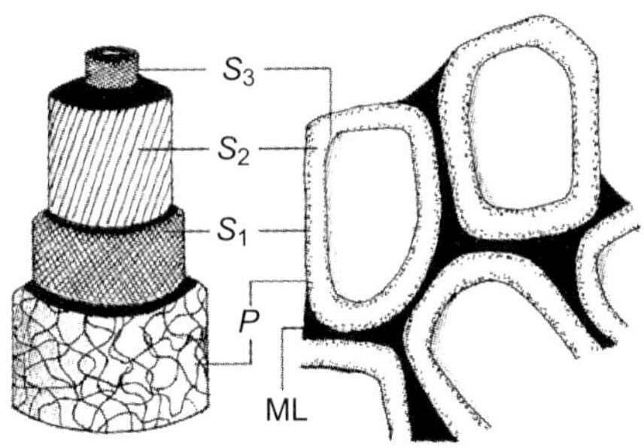

Figure 4.1 Expanded scheme of the basic structure of the cellular wall of softwood tracheids or fiber hardwood. Abbreviations: ML – middle lamella; P primary wall, layers of the secondary wall S1, S2 and S3.

Table 4.1 Chemical composition of softwood and hardwood.

Constituents	*Percentage (%)*
Cellulose	40-50
Lignin	15-35
Hemicellulose	20-35
Extractives	3-10
Mineral compounds	0.5

Table 4.2 Chemical composition of the primary cell wall, secondary cell wall and middle lamella.

Cell wall	*Cellulose %*	*Hemicellulose %*	*Lignin %*
Primary wall	10	20	70
Secondary wall			
S1	35	25	40
S2	55	30	15
S3	55	40	5
Middle lamella	0	10	90

HEMICELLULOSE

The term hemicellulose was introduced by Schulze in 1891 (Filho, 1994) and represents the vegetal cell easily hydrolyzed. Hemicelluloses are found not only in the cell wall but also in the medullar rays, axial parenchyma and middle lamella of plant cells. They are non-cellulose heteropolysaccharides composed in various proportions of the monosaccharides units D-xylose, D-mannose, D-galactose and L-arabinose. In addition, the hemicellulose molecules can be composed of branched chains with D-glucuronic acid, D-galacturonic acid, 4-*O*-methyl-α-D-glucuronopyranosyl acid, *O*-acetyl groups, ferulic acid and *p*-coumaric acid. Classes of hemicellulose are named according to the main sugar unit. Thus, when a polymer is hydrolyzed and yields xylose, it is a xylan; in the same way, hemicelluloses include mannans, glucans, arabinans and galactans (Viikari *et al.,* 1994; Uffen, 1997; Ebringerova and Heinze, 2000; Polizeli *et al.*, 2005).

In nature, of course, wood hemicelluloses are complex structures made of more than one polymer being glucuronoxylans, arabinoglucuronoxylans, glucomannans, arabinogalactans and galactoglucomannans the more frequent. For instance, hydrolysis of an arabinoglucuronoxylan, yields arabinose, glucuronic acid, and mainly, xylose.

Glucuronoxylans and glucuronoarabinoxylans are present mainly in the secondary wall, and act as an adhesive by forming covalent and non-covalent bonds with lignin, cellulose and the other polymers essential to the integrity of the cell wall. Hemicellulose may be extracted by alkaline reagents (Coughlan *et al.*, 1993, Habibi *et al.*, 2002). The amount of each component varies from species to species, tree-to-tree and even among parts of the same plant (Haltrich *et al.,* 1996; Sunna and Antranikian, 1997; Kulkarni *et al.,* 1999a; Subramaniyan and Prema, 2002; Polizeli *et al.*, 2005). For example, bamboo shoots, barley, oat, pangola grass, rice, rye, rye grass, sorghum, wheat, produce satisfactory levels of arabinoxylans (Izydorczyk and Biliaderis, 1995).

Hemicellulose is a class of polymeric components of plant fibers with properties peculiar to each one. They have various effects on paper properties. Considering another area of industrial activity, hemicelluloses are notable for their use in the production of furfural, the raw material in the fabrication of some plastics (furan resins, Bakelite) with a range of applications.

XYLAN

Xylans are the principal class of hemicelluloses in hardwood, and are less abundant in softwood contributing to the total dry weight with 15–30% and 7–12% in angiosperms and gymnosperms, respectively (Haltrich *et al.,* 1996). The two main types of xylan are represented by glucuronoxylans and glucuronoarabinoxylans, but both will be referred to as "xylans" from here on, following an established practice (Filho, 1994; Sunna and Antranikian, 1997; Polizeli *et al.*, 2005).

Xylans are linear polymers of *β*-D-xylopyranosyl units (xylose) linked by *β*-(1-4) glycosidic bonds (Fig. 4.2). Depending on the origin of the xylan the chain has substituent groups in the polysaccharide backbone, these include acetyl groups, 4-*O*-methyl-*α*-D-glucuronopyranosyl units and *α*-L-arabinofuranosyl groups. Angiosperm xylans have a high rate of substitution (70-80%) by acetyl groups, at position 2 and/or 3 of the β-D-xylopyranosyl, conferring on the xylan its partial solubility in water (Coughlan and Hazlewood, 1993). Gymnosperm presents mainly glucuronoarabinoxylans, where for each ten β-D-xylopyranosyl units of the xylan backbone, there are one to three α-L-arabinofuranosyl groups linked at position 2 or 3 by α-(1-3) or α-(1-3) bonds.

Gymnosperms have a higher content of 4-*O*-methyl-α-glucuronopyranosyl units than angiosperms, and these xylans are not acetylated and, because of their furanoside structure, the arabinose side-groups are readily hydrolyzed by acid. Some arabinoxylans of seeds have a uncommon structure, because they present arabinofuranosyl residues linked by α-(1-5) bonds (Kennedy and White, 1988). Other exception is the xylan from the marine algae *Rhodymenia palmate* which has a non-branched structure with approximately 80% of the β-D-xylopyranosyl units linked by β-(1-4) glycosidic bonds and 20% of β-(1-3) glycosidic bonds with irregular form (Filho, 1994).

According to its chemical structure and the substituent groups of the lateral chain, the xylan is covalently bound with the overlapping sheath of lignin (Biely, 1985). The layers of covalently linked xylans and lignin, and their non-covalent interaction with cellulose, are important for the maintenance of the integrity of the cellulose *in situ*, and consequently in the protection of the fibers against the action of cellulase (Uffen, 1997).

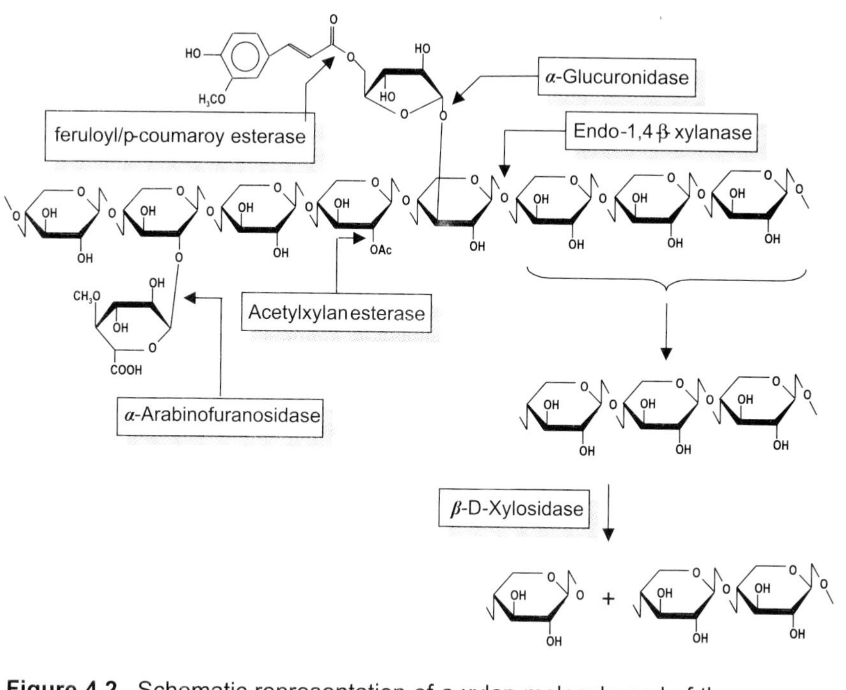

Figure 4.2 Schematic representation of a xylan molecule and of the enzymes of the xylanolytic system. The action of β-xylosidase is represented in more detail below

The lateral chains of xylan determine the solubility, physic conformation and the reactivity of the molecule of xylan with hemicellulosic components and, therefore, greatly influence in the way and extension of its enzymatic hydrolysis. The ramification points play an important role, since they may cause steric hindrance, preventing the formation of the enzyme-substrate complex (Coughlan and Hazlewood, 1993). Furthermore, they can alter the structure of the molecule changing the solubility, the configuration and the linking of phenolic components in the lignin. Therefore, for complete hydrolysis of xylan are necessary enzymes that hydrolyze the branches (lateral chains), and that synergistically contribute to the hydrolysis of the main chain. Therefore, the knowledge of the substrate composition is as important as that of the xylanolytic system.

The Xylanolytic System

Due to the heterogeneous structure of hemicelluloses, the complete degradation of xylan requires an enzymatic system with several enzymes, acting on the main or the lateral chains. Enzymes acting on the main chain are the exohydrolases, which hydrolyze terminal glycosidic bonds, removing monosaccharide units of the non-reducing terminus, and endohydrolases that cleave randomly or at specific glycosidic bonds depending on the type of enzyme. Several enzymes participate in this process (Fig. 4.2): endo-1,4-β-xylanase (1,4-β-D-xylan xylanohydrolase; EC 3.2.1.8), β-D-mannanase (1,4-β-D-mannan mannanohyrolase, EC 3.2.1.78), β-D-xylosidase (1,4-β-D-xylan xylohydrolase; EC 3.2.1.37), β-D-mannosidase (β-D-mannopyranoside hydrolase, EC 3.2.1.25), endo-1,5-α-L-arabinase (EC 3.2.1.99) and β-D- galactanase (β-D- galactoside galactohydrolase, EC3.2.1.23). Other enzymes act on the lateral chains, such as acetylxylan esterase (EC 3.1.1.72), α-D-galactosidase (α-D-galactoside galactohydrolase, EC 3.2.1.22), exo-α-L-arabinofuranosidase (α-L-arabinofuranoside arabinofuranohydrolase, EC 3.2.1.55), α-glucuronidase (xylan-α-D-1,2-(4-O-methyl) glucuronohydrolase, EC 3.2.1.131) and feruloyl/p-coumaroyl esterase (EC 3.1.1.73). The feruloyl/p-coumaroyl esterase cleaves ester bonds on xylan between arabinose and ferulic acid and/or arabinose and *p*-coumaric acid side-groups (Christov and Prior, 1993, Collins *et al.*, 2005). Ferulic acid esterase, *p*-coumaric acid esterase, acetylxylan esterase and α-glucuronidase were discovered only by the end of the 1980s, probably because of the difficulty of obtaining suitable substrates. These enzymes occur in fungi and bacteria, but up to now a few have been purified and analyzed in terms of their physical and chemical properties (Christov and Prior, 1993; Williamson *et al.*, 1998; Crepin *et al.*, 2004).

The conversion of xylan to its monomers requires the synergistic participation of several enzymes, some acting on the main chain, and others on its branched groups. The endo-1,4-β-xylanase (1,4-β-D-xylan xylanohydrolase; EC 3.2.1.8) cleaves glycosidic bonds in the xylan backbone, producing xylooligosaccharides of variable length, bringing about a reduction in the degree of polymerization of the substrate. The xylan is not attacked randomly, since the bonds selected for hydrolysis depend on the nature of the substrate molecule, i.e. on the length of chain, the degree of branching and the presence of substituents (Reilly, 1981; Puls and Poutanen, 1989; Li *et al*, 2000). Initially, the main products of xylanase action are xylose oligomers, but at a later stage, small molecules such as the mono-, di- and trisaccharide of β-D-xylopyranosyl may be liberated. The accumulation of these short oligomers of xylose may inhibit the endoxylanase activity. β-xylosidase then hydrolyses these products, removing the cause of inhibition and increasing the efficiency of xylan hydrolysis (Andrade *et al.*, 2004; Zanoelo *et al.*, 2004). The best substrate for β-xylosidase is xylobiose. Its affinity for xylooligosaccharides is inversely proportional to their degree of polymerization. β-xylosidases also are able to cleave artificial substrates, such as *p*-nitrophenyl- and *o*-nitrophenyl-β-D-xylopyranoside.

Transxylosylation activity has also been detected in fungi, resulting in products of higher molecular weight than the original substrates (Kurakabe *et al.*, 1997). Some properties of β-xylosidases from filamentous fungi were recently reported by Polizeli *et al.* (2005) and a good example of stable β-xylosidase is that of the enzyme from *Aspergillus phoenicis*, which retains 100% of its activity after 4 h at 60°C or 21 days at room temperature (Rizzatti *et al.*, 2001).

Aiming biotechnological applications, the ideal source of xylanase would be one that produces an adequate amount of each of the enzymes of the xylanolytic system. Therefore, the synergistic interactions among the xylan-degrading enzymes enhance the susceptibility of the heteropolymeric xylan to the attack of endoxylanases (van Peij *et al.*, 1997; de Vries *et al.*, 2000). Acetylxylan esterase, for example, removes the *O*-acetyl groups from positions 2 and/or 3 on the xylose residues of acetyl xylan producing acetic acid and a less-acetylated xylan, affording greater susceptibility to hydrolysis by endoxylanases.

On the other hand, small-acetylated polymers, produced by endoxylanase, are the preferred substrates of the esterases. An example is the degradation of the wheat bran, a complex compound that contains large amounts of arabinoxylan, and it cannot be easily degraded by endoxylanases without being previously treated with α-arabinofuranosidase, which removes the L-arabinose residues substituted at positions 2 and 3 of the xylose. There are two types of α-arabinofuranosidase, with distinct modes of action: the most common, exo-α-L-arabinofuranosidase (EC 3.2.1.55) degrades *p*-nitrophenyl-α-L arabinofuranosides and branched arabinans, whereas endo-1,5-α-L-arabinase (EC 3.2.1.99) hydrolyses only linear arabinans (Kaneko *et al.*, 1993).

It has also been noted that acetyl groups close to the glucuronosyl substituents can partially hinder the α-glucuronidase activity, an enzyme that hydrolyses the α-1,2 bonds between the glucuronic acid residues and xylose backbone units found in glucuronoxylan. Some glucuronidases are able to hydrolyze the intact xylan but the substrate specificity varies with the microbial source (Puls and Schuseil, 1993).

The association of xylanases is also possible. Xylanosome, a term analog to cellulosome (a multienzyme complex that may mediate the adhesion of the bacterium to cellulose) (Bayer *et al.*, 1994 and Doi *et al.*, 1994) is applied to structures observed as protein aggregates with many subunits. These complexes have very high molecular weights (500–600 kDa) and consist of more than ten proteins with xylanolytic activity, several of which are endoxylanases. In a few microorganisms, a cellulosome may be associated with a xylanosome, forming large complexes responsible for the hydrolysis of both cellulose and xylan (Sunna and Antranikian, 1997).

Properties of the Xylanases

Microorganisms are undoubtedly the main source of xylanases but these enzymes are also present in marine algae, snails (Sunna and Antranikian, 1997), protozoans (Bera-Maillet *et al,* 2005), crustaceans (Johnston *et al.*, 2005), insects (Brennan *et al.*, 2004) and plant seeds (Caspers *et al.*, 2001). Among microbial sources, the filamentous fungi are especially interesting as they secrete enzymes into the medium, and their xylanase levels are very much higher than those in yeasts and bacteria.

Cloning, sequencing and expression in *Escherichia coli* of xylanase genes has been performed for microorganisms of various genera (Sung *et al.*, 1995; Horikoshi, 1996; Kulkarni *et al.*, 1999b; Liu *et al.*, 2003; Huang *et al.*, 2006). The xylanases in bacteria are restricted to the intracellular or periplasmic fractions. Furthermore, enzymes expressed in bacteria are not subjected to post-translation modifications such as glycosylation. But besides bacteria, xylanases have been also expressed in yeast (Nuyens *et al.*, 2001) and fungi (Dalboge, 1997; Decelle *et al.*, 2004). Recently, Ramchuran *et al.* (2005) expressed in the methylotrophic yeast *Pichia pastoris* a thermostable xylanase produced by *Rhodothermus marinus.*

In general, fungal endoxylanases have optimum activity and stability in the range of 40°C–75°C and pH 4.0–6.5 (Table 4.3) but, although bacteria is not the focus of this chapter, it is important to remember that extremophile bacteria produce endoxylanases with optimal temperature higher than 80°C (Reeves *et al.*, 2000). Endoxylanases are usually small glycoproteins (6 to 50 kDa), with multiplicity of forms, being that in some cases three or more enzymatic isoforms have been separated from a single culture (Rizzatti *et al.*, 2004). A number of factors may be responsible for this phenomenon, such as differential processing of mRNA, post-translation modifications (glycosylation and self-aggregation) or proteolysis (Polizeli *et al.*, 2005). Multiple endoxylanases can also be expressed by distinct alleles of one gene, or even by completely separate genes (Gomez de Segura *et al.*, 1998; Chavez *et al.*, 2002).

The production of thermostable xylanase is desirable for several industrial processes. Apparently, this type of enzyme is a characteristic of thermophilic microorganisms. But despite intensive biochemical and structural investigation, no definitive conclusion for the phenomenon of thermophilia has been reached, since the information obtained for the enzyme of a given microorganism may not necessarily be repeated in another. Endoxylanases from thermophiles can have high

Table 4.3 Properties of some endoxylanases produced by different microorganisms.

Fungus	*Molecular Mass (kDa)*	*Optimal Temperature (°C)*	*Optimal pH*	*pI*	*Km (mg/mL)*	*Reference*
Acrophialophora nainiana	22	55	7.0	n.d.	16.1	Salles *et al.*, 2000
Aspergillus aculeatus	18,26,52	50,50,70	4.0, 4.0, 5.0	5.6,9.0,3.8	n.d.	Fujimoto *et al.*, 1995
Aspergillus awamori	39,23,26	45-55	4.0-5.5	n.d.	n.d	Kormelink *et al.*, 1993
Aspergillus caespitosus	26.3, 17.7	50-55	7-7.5,5.5-6.5	6.7,5.8	2.5,3.9	Sandrim *et al.*, 2005
Aspergillus fischeri	31	60	6.0	n.d.	4.88	Raj and Chandra, 1996
Aspergillus fumigatus	19, 8.5	55	5.5	n.d.	2.19-5.19	Silva *et al.*, 1999
Aspergillus fumigatus	212-253	60	6.0-6.5	n.d.	n.d.	Anthony *et al.*, 2003
Aspergillus giganteus	21,24	50	6.5,6.0	n.d.	n.d.	Fialho and Carmona, 2004
Aspergillus kawachii	35,26,29	60,55,50	5.5,4.5,2.0	3.5-6.7	n.d	Ito *et al.*, 1992
Aspergillus nidulans	34	56	6.0	3.4	0.97	Fernández-Espinar *et al.*, 1994
Aspergillus niger	23	50	5.5	n.d.	7.1	Levasseur *et al.*, 2005
Aspergillus nidulans KK-99	n.d.	55	8.0	n.d.	n.d.	Taneja *et al.*, 2002
Aspergillus niger (IBT-90)	60	50	5.5	n.d.	n.d.	Romanowska *et al.*, 2006
Aspergillus oryzae	35	60	5.0	n.d	n.d.	Kitamoto et *al.*, 1999
Aspergillus phoenicis	6,7,13,17,17.8	40-55	3.0-5.5	4.4,3,4.4,5,3	17,7,6,3.6,7	Rizzatti *et al.*, 2004
Aspergillus sp	26,21	50	5.0	7.2, 4.7	n.d.	Khanna et *al.*, 1995
Aspergillus sojae	33,36	60,50	5.0,5.5	3.5, 3.75	n.d.	Kimura *et al.*, 1995
Aspergillus sydowii	33	50	4.0	n.d.	n.d.	Ghosh and Nanda, 1994
Aspergillus terreus	n.d.	50	7.0	n.d.	16.7	Ghanem *et al.*, 2000
Aspergillus terreus	n.d.	45	4.5	n.d.	0.83 mM	Ghareib and El Dein, 1992
Aspergillus versicolor	19,32	55,55	6.0,6.0-7.0	5.4	2.3	Carmona *et al.*, 1998
Aspergillus versicolor (MKU3)	n.d.	50	6.0	n.d.	n.d.	Jeya *et al.*, 2005

(*Contd.*)

Table 4.3 (*Contd.*)

Fungus	*Molecular Mass (kDa)*	*Optimal Temperature (°C)*	*Optimal pH*	*pI*	*Km (mg/mL)*	*Reference*
Aureobasidium pullulans	25	54	4.4	9.7	7.6	Li *et al.*, 1993
Aureobasidium pullulans	39	70	6.0	8.9	n.d.	Tanaka *et al.*, 2006
Chaetomium cellulolyticum	25,47,57	50	5.0-7.0	8.9,8.4,5.0	n.d.	Baraznenok *et al.*, 1999
Fusarium oxysporum F3	20.8, 23.5	60, 55	6.0	9.5,8.5-8.7	0.41,0.37	Christakopoulos *et al.*, 1996
Humicola grisea var thermoidea	23	70	5.5	n.d.	3.3	Monti *et al.*, 1991
Humicola insolens	6,21	55,60	6.5,7	n.d	n.d.	Düsterhöft *et al.*, 1997
Myceliophthora sp. IMI 387099	53,53	75	6.0	5.2, 4.8	n.d.	Chadha *et al.*, 2004
Penicillium brasilianum	31	n.d.	n.d.	> 9.0	n.d.	Jørgensen *et al.*, 2003
Penicillum capsulatum	22	48	3.8	2.8	7.0-32.0	Ryan *et al.*, 2003
Penicillium sp. 40	25	50	2.0	4.7	0.0083	Kimura *et al.*, 2000
Phlebia radiata	18.6, 15.8	n.d.	n.d.	4.1	2.76	Rogalski *et al.*, 2001
Thermoascus aurantiacus	n.d.	70-75	4.0-5.0	6.8-5.5	n.d.	Kalogeris *et al.*, 1998
Thermomonospora fusca	22	60	6.0	n.d.	n.d.	Weng and Sun 2005
Thermomyces lanuginosus	24.7	70	6.0-6.5	3.9	n.d.	Singh *et al.*, 2000
Thermomyces lanuginosus	n.d.	70-75	6.5	n.d.	3.26	Lin *et al.*, 1999
Thermomyces lanuginosus	25.6	75	6.2	n.d.	n.d.	Jiang *et al.*, 2005
Thermomyces lanuginosus	n.d.	50	6.5	n.d.	n.d.	Sonia *et al.*, 2005
Trichoderma harzianum	20,29	50,60	5.0, 5.0	9.4,9.5	0.58	Tan *et al.*, 1985
Trichoderma lignorum	21,20	45	3.5,6.5	5.1,8.7	n.d.	John *et al.*, 1979
Trichoderma longibrachiatum	37.7	45	5.0-6.0	n.d.	n.d.	Chen *et al.*, 1997

* n.d. = not determined.

degree of structural homology with those from mesophiles. Factors are commonly cited in the literature as being responsible for the enzymatic thermostability: extra disulfide bridges, an N-terminal proline residue causing a reduction in conformational freedom, salt bridges, the presence of hydrophobic side-chains and others (Maheshwari *et al.*, 2000; Turunen *et al.,* 2001).

The xylanases were originally classified into two families of glycosyl hydrolases, called F and G, or 10 and 11 (Henrissat and Bairoch, 1993; Sapag *et al.*, 2002), respectively, both known to occur in bacteria and eukaryotes and assumed to have different physiological roles (Biely *et al.*, 1997). This classification is based in the structural homology of the hydrophobic regions, deduced from 2D representations of the primary sequences. More recently, Collins *et al.* (2005) reported the existence of enzymes with xylanase activity also belonging to families 5, 7, 8 and 43.

The family F/10 consists of a cellulose-binding domain and a catalytic domain (which is a TIM barrel domain), interconnected by a flexible spacer and a substrate-binding domain. The family 10 xylanases found in species of different groups of bacteria and eukaryotes are not so dissimilar to each other, even though a number of subfamilies have been characterized (Jun *et al.*, 2003). The family G/11 is composed by highly specific endoxylanases of low MW, with 40-90% sequence identity, with different *pIs*, thermostability, pH profiles, structure, and catalytic properties. According to Törrönen and Rouvinen (1997), besides these homology-based classes, xylanases could be further divided into those with alkaline and those with acid *pI*, but some microorganisms produce both types, e.g. *Aspergillus* and *Trichoderma* sp. This resulted in a further classification of the xylanases in classes such as A,B, C, D, 1, 2, 3, I, II, III (Kulkarni *et al.*, 1999a,b).

Purification and Assay Methods for Endoxylanases

A wide variety of technical procedures have been described for the isolation and purification of the xylanolytic complex (Belancic *et al.,* 1995; Baitaillon *et al.,* 2000; Monti *et al.,* 2003; Sa-Pereira *et al.*, 2003). These depend on physical-chemical properties peculiar to the protein composition.

These methods are based on (a) electrical charge (ion-exchange chromatography, electrophoresis) and (b) molecular size (gel sieving chromatography). These are referred to as 'classical methods', in contrast to those based on more specific interactions.

One distinctive feature of the enzymes in question is the ability of the endoxylanases to interact specifically with resins having dextrans in their composition (Sandrim *et al.,* 2005). Such enzyme-resin carbohydrate interaction can lead to a slowing down of the elution and, often, to erroneous estimates of molecular weights.

Several procedures have been recommended to measure endoxylanase activity. Many are based on the same principle but differ in assay conditions such as temperature, pH, reaction time and substrate used. The simplest assay detects the rate of liberation of reducing sugars from a substrate, using 3'5' dinitrosalicylic acid (DNS), as for example in the Miller method (Miller, 1959). This is considered a fast technique that gives highly reproducible results. The Somogyi-Nelson method (Nelson, 1944 and Somogyi, 1952) is also used, but less frequently.

Alternative approaches, involving viscosimetry, where the diminishing viscosity of a xylan solution is recorded, and nephelometry, in which the fall in turbidity is monitored, may be used, but they are unsuitable for assays of insoluble enzyme sources and analyses requiring many samples. The viscosimetric method is considered more specific than those based on reducing-sugar assay (Haltrich *et al.,* 1996).

Methods based on the formation of chromogens have also been proposed, chiefly for the assay of xylanase activity on electrophoretic gels. These involve enzymatic hydrolysis of a soluble chromogenic substrate such as Remazol Brilliant Blue R-Xylan (RBB-Xylan), a dye covalently bonded to xylan (Biely, 1985; Haltrich *et al,* 1996; Sandrim *et al.,* 2005).

The xylans used as substrates in enzyme assays are usually obtained from commercial sources such as the birchwood (about 35% of the dry weight being xylan), larchwood, beechwood, and oat spelt, being the last one to contain 10% arabinose and 15% glucose residues. Xylan can also be extracted from *eucalyptus.*

Xylanase Production: Induction and Catabolite Repression

Industrially, xylanases may be produced either in submerged liquid culture or on a solid substrate, but approximately 80–90% are produced in submerged culture. Thereby xylanase is produced in response to xylans from various sources (Gomes *et al.,* 1994; Liu *et al.,* 1999; Rani and Nand, 2000). Xylose can also act as an inducer of the xylanolytic complex (Ghosh and Nanda, 1994; Rizzatti *et al.,* 2001) but in some microorganisms this can give rise to discriminatory control, leading to catabolite repression of endoxylanases (Flores *et al.*, 1996; Mach *et al.,* 1996). Other compound used as a potent inducer is *β*-methyl xyloside, a non-metabolizable structural analog of xylobiose that can be produced at low cost (Morosoli *et al.,* 1987; Simão *et al.,* 1997a, b).

In cultures on solid substrate, wheat bran and rice are regarded as inducers. Alternative substrates have been described for enzyme production, such as sugarcane bagasse, rice husks and wood pulp (Kadowaki *et al.,* 1995; Damaso *et al.,* 2000; Medeiros *et al.,* 2000; Pandey *et al.,* 2000; Singh *et al.,* 2000; Anthony *et al.,* 2003; Sandrim *et al.*, 2005; Krishna, 2005).

The induction of the xylanolytic system is supposed to occur through physical contact between some recognition site on the cell surface, and the inducer. Relatively low levels of constitutive xylanases may be responsible for the initial hydrolysis of xylan, producing xylobiose and xylotriose, among other products (Biely 1985; Kulkarni *et al.,* 1999 a,b) that may be the true inducers of endoxylanase synthesis (Haltrich *et al.,* 1996; Sunna and Antranikian, 1997). These oligosaccharides are transported into the cell with the help of *β*-xyloside permeases, and then trigger the transcription of the xylanolytic system genes (Polizeli *et al.,* 2005). Permease activity is reduced by the presence of glucose, but it is very efficient in the presence of the xylanolytic inducers. In filamentous fungi, carbon catabolite repression is mediated by the protein CREA (transcription repressor) (Dowzer and Kelly, 1991; Fernández-Espinar *et al.,* 1994; Piñaga *et al.,* 1994; de Graaff *et al.,* 1994).

Research on microorganisms that produces xylanase systems is becoming more and more relevant, both in ecological and economic terms. These enzymes are responsible for biodegradation, generating metabolic energy that maintains the flow of carbon, and their importance in the natural recycling of biomass is enormous. Xylanases are synthesized by mesophiles and thermophiles (Smith *et al.,* 1991). Among the mesophilic fungi, the genera *Aspergillus* and *Trichoderma* are pre-eminent in xylanase production and among the thermophiles the most important fungi are *Chaetomium thermophile, Humicola insolens, H. lanuginosa, H. grisea, Melanocarpus albomyces, Paecilomyces variotii, Talaromyces byssochlamydoides, T. emersonii, Thermomyces lanuginosus* and *T. aurantiacus*, as reviewed by Polizeli *et al.* (2005).

Industrial Applications

Xylanases take part in the breakdown of plant cell walls, along with other enzymes that hydrolyze polysaccharides and digest xylan during the germination of some seeds. Industrially, xylanases began to be used in the 1980s, initially for the preparation of animal feed and later in the food, textile and paper industries. Currently, xylanase and cellulase, together with pectinase, account for 20% of the world market in enzymes. This topic has been recently revised by Bhat (2000), Aristidou and Pentilä (2000), Subramaniyan and Prema (2000, 2002), Beg *et al.* (2001), Techapun *et al.* (2003), Polizeli *et al.* (2005), Chavez *et al.* (2006) and others.

World industries have demonstrated great interest in the development of efficient enzymatic processes, in substitution of acid hydrolysis, for the treatment of material containing hemicellulose. Commercial xylanases are made, for example, in Canada, China, Denmark, Finland, India, Japan, and USA. A large number of the fungal species used are from the genera *Aspergillus*, *Trichoderma*, *Humicola* and *Penicillium,* which are active producers of xylanases with important industrial applications. Xylanase can be used in the processing of soybean protein, starch and beer, etc. to accelerate filtration and improve yield. This enzyme can be applied in the industry of food and foodstuff processing, animal feed, manufacture of bread, solvents and other chemicals, xylitol production, textiles, pulp and paper industries.

The use of xylanase in the production of animal feed is destined for poultry, swine and ruminants. This enzyme may be used together with other exogenous hydrolases (glucanases, pectinases, cellulases, proteases, amylases, phytases, galactosidases and lipases) for improving the performance of the food mixture, making the animal feed more digestible, since young animals produce smaller quantities of endogenous digestive enzymes than adults.

The supplementation of enzymes in the animal rations reduces the phosphorus, nitrogen, zinc and copper levels in the excreta, thus decreasing the environmental contamination. Moreover, β-glucanase and xylanase rapidly breaks down the viscosity caused by β-glucan released during digestion of barley or oats.

Xylanase may also reduce the viscosity of the raw material, containing arabinoxylans, also known as pentosans (Twomey *et al.,* 2003), leading to improved digestion of nutrients in the initial part of the digestive tract, providing a more efficient use of energy. Wheat contains soluble or insoluble arabinoxylans. Branny skin is largely insoluble and forms a kind of sludge during separation, but soluble arabinoxylans are reactive and form molecular complexes with soluble starch and protein. Then, those xylanases that has a high level of activity towards soluble arabinoxylan but a low level of activity towards insoluble arabinoxylan break down soluble arabinoxylans releasing starch and proteins and have a good performance in the separation of wheat fractions.

Xylanases may be used in bakery, together with α-amylase, malting amylase, glucose oxidase, and proteases, increasing the bread volume and the absorption of water (Harbak and Thygesen, 2002; Jiang *et al*., 2005; Romanowska *et al*., 2006). Also, a larger amount of arabinoxylooligosaccharides in bread would be beneficial to health. In biscuit industry xylanase is recommended for making cream-crackers lighter, and for improving the texture, palatability and uniformity in wafers.

Xylanases, in conjunction with cellulases, amylases and pectinases have been applied in the juice and wine industries. This enzyme requests high stability and optimum activity at an acid pH. These enzymes are utilized for extraction, clearing and stabilization of the fruit pulp, making easy the liquefaction of fruit and vegetables, reducing the viscosity and cloudiness in the concentrate. The action of these enzymes increases the recovery of aromas, essential oils, vitamins, mineral salts, edible dyes and pigments, and are important on the coffee-bean mucilage (Wong *et al.,*1988; Wong and Saddler, 1993).

Hemicellulases and xylanases can hydrolyze lignocellulosic materials rich in xylan to produce β-D-xylopyranosyl units (xylose) that can be converted in xylitol by yeast or bacteria (Parajó *et al.,* 1998; Nyyssölä *et al*., 2005). Xylitol is a polyalcohol with properties of an artificial low-calorie sweetener (Parajó *et al.,* 1998). It is non-carcinogenic, suitable for diabetic and obese individuals. A variety of commercial products containing xylitol, such as chewing gums, can be found in the market.

Yeasts are generally considered to be more efficient xylitol producers than bacteria or fungi, principally the genus *Candida* (Parajó *et al*., 1998). However, the use of *Candida* in food industry is dangerous because of its pathogenic nature. The construction of recombinant yeasts producing xylitol has been considered for *Saccharomyces cerevisiae* (Govinden *et al*., 2001).

Xylose, the hydrolytic product of xylan, can be also converted by fermentation into combustible liquids (ethanol), furfural and solvents, using mainly the yeasts *Pichia stipitis* and *Candida shehatae* (Slapack *et al.,* 1987; Screenath and

Jeffries, 2000). These end products of xylan degradation have considerable commercial importance (Parajó *et al.,* 1998).

In the textiles industries, the xylanolytic complex can be used to process plant fibers, such as hessian, linen and dried ramie ('china grass') stems. For this purpose and in the paper industry, the xylanase should be free of cellulolytic enzymes, since it is necessary to extract hemicellulose from the natural fibers without damaging the cellulose. Xylanases are specifically synthesized when microorganisms are cultured on xylan, whereas on cellulose the same organisms will produce cellulases in association with xylanases, perhaps because the cellulose substrate contains traces of hemicelluloses (Prade, 1995; Brühlmann *et al.,* 2000; Csiszár *et al.,* 2001). The application of xylanases on the preparation of textile fibers appears to be a market demanding the development of new techniques.

A promissory industrial application of xylanases is for the bleaching of cellulose pulp (biobleaching). Enzymes began to be used in this sector during the last two decades, since peroxidases were applied to the degradation of lignin (Viikari *et al.,* 1991; Wong and Saddler 1993; Araújo *et al.,* 1999; Tenkanen *et al.,* 1997; Christov *et al.,* 1999, 2000; Beg *et al.*, 2001; Whitmire and Miti, 2002; Saha, 2003; Bajpai, 2004; Li *et al.*, 2005; Zhao *et al.,* 2006).

In many countries, including Brazil, the chemical route is employed in paper manufacture, rather than enzymatic hydrolysis. The usual method is known as the Kraft process (strength or force in German). However, industries from Canada have begun to use xylanase as an auxiliary brightening agent for wood pulp in papermaking. The enzyme opens up the lignin structure of the wood pulp so that it takes 10–15% less chlorine dioxide to achieve the desired level of brightness. This process is just beginning, but some industries have reduced the production cost and improved the performance of xylanase by genetically engineering the fungus from which these enzymes are extracted. The use of xylanase has helped to reduce the amount of organically bound chlorine in wastewater by 60% and the cost of bleaching chemicals by 10–15%. Japanese industry produces its own xylanase by fermentation, reducing the production cost. Xylanase has been used to achieve similar reductions in the requirement for bleaching chemicals and in levels of organically bound chlorine in its wastewater.

A schematic representation of the bleaching cellulose pulp is shown in Fig. 4.3. Fibers and lignin from wood shavings of the three species of *Eucalyptus* (*E. grandis, E. saligna* or *E. urophylla*) are pretreated with white liquor in a digester (cooking liquor) to accelerate the delignification of the wood and the recovery of cellulose fibers. At this stage, the cellulose pulp is known as brown mass, appearing dark because of the soluble lignin (black liquor) deposited on the pulp. Approximately 90–95% of the hemicellulose and lignin are dissolved and partially degraded during this process.

Then, the brown mass is washed and the pre-bleaching is carried out, consisting in removal minor impurities and a part of the remaining lignin from the cooking, with the aim of reducing the cost of other reagents employed in the bleaching. The pre-bleached pulp has a pale yellow color due to residual lignin on the

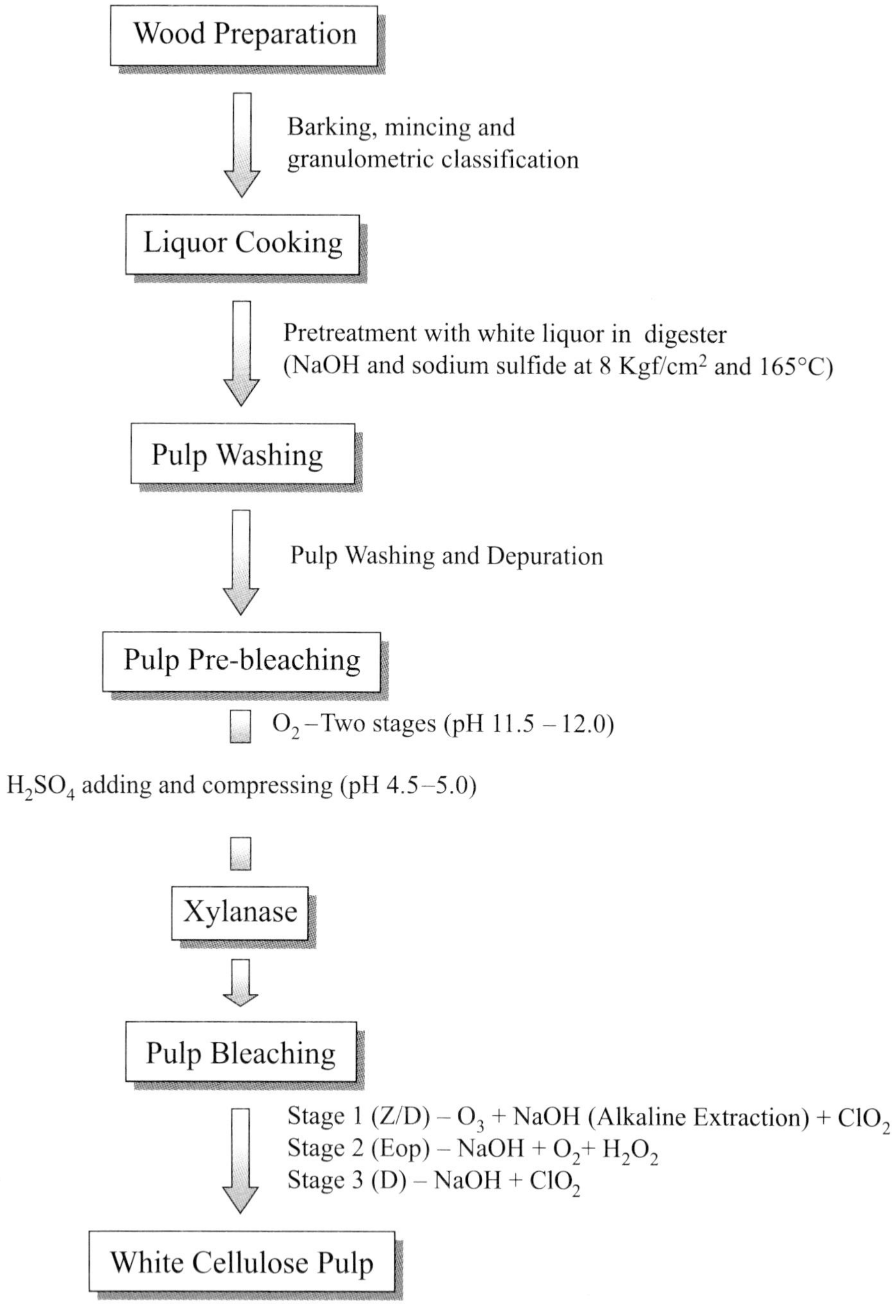

Figure 4.3 Schematic representation of the cellulose pulp bleaching. Xylanases and the other hemicellulases may be added for hydrolyses of reprecipitated xylan or for removal of xylan from lignin carbohydrate complexes and removal of glucomanan.

walls of the fibers. In the subsequent bleaching, constituted of three different treatments (Fig. 4.3), the paste attains a pure white color, through the removal of light-absorbing substances (chromophores) formed by products of the breakdown of lignin. Fig. 4.4 shows stages of the bleaching of cellulose pulp for paper production.

Figure 4.4 Stages of the production and bleaching of cellulose pulp for paper manufacture. Eucalyptus plantlet (A); eucalyptus forest (B), log wood (C), kraft process (D): wood shavings (a), cooked pulp (b), bleaching through two oxygen stages (c and d), addition of sulfuric acid to adjust of pH and compressing (e), bleaching with ozone and alkaline extraction (f), bleached with oxygen, hydrogen peroxide and alkaline extraction (g), and alkaline extraction and chlorine dioxide (h).

There are advantages and disadvantages in the use of the Kraft process (Polizeli *et al.,* 2005). The main advantage is the possibility of industrial recovery of compounds from the black liquor (dissolved wood, lignin and chemical reagents), and as disadvantages can be cited the high initial costs and strong smell of gases exhaled. Treatment with xylanase renders the pulp more permeable to the subsequent chemical extraction of the residual brown lignin and molecules of lignin-carbohydrate from the fibers (Buchert *et al.* 1992; Paice *et al.* 1992; Viikari *et al.* 1994).

CONCLUSION AND FUTURE PERSPECTIVES

Xylanases represent an enzymatic system that acts on the hemicellulose xylan, a heteropolyccharide of the plant cell wall. In nature, endoxylanases are mainly synthesized by several microorganisms to obtain carbon sources and energy. This enzymatic system can be produced in submerged cultures or solid-state fermentation using agroindustrial residues such as sugar cane bagasse, wheat straw and others. Innumerous applications are attributed to these enzymes, such as in the food industry, paper and pulp, textiles, bakery and beer industries, in environmental science (bio-fueling), effluent treatment, agro-waste treatment, etc.

Much attention is being attributed to the use of xylanases in the cellulose-pulp bleaching in the paper industry, but the use in large scale is as yet reduced in the world. Merely some countries are enforcing new environmental regulations that restrict the use of chlorine compounds in the bleaching processes of paper and cellulose industries. In a near future there must be a change in this mentality, making of the production of paper a less harmful process to nature, since in that of the conventional Kraft process the use of chlorine results in the production of organochlorines from the degradation products of lignin, which are highly toxic and mutagenic, requiring treatment of the effluents from the paper-making plant for better preservation of the fauna and flora. Special attention has been given to using xylanase in pre-bleaching, which would lower the amount of chlorine compounds used by up to 30%, so that a 15–20% reduction of organochlorines in the effluents could be achieved. The utilization of xylanases could lead to the replacement of 5 to 7 kg of chlorine dioxide per ton of Kraft pulp and an average fall of 2–4 units in the number kappa, a measure of the lignin content in the cellulose pulp.

Xylanases employed in paper technology must be active at high temperatures and alkaline pH and must not contain cellulolytic enzymes, in order to preserve the cellulose fibers. Although important applications are attributed for xylanases, the progress of the understanding of the regulatory system of these enzymes is still limited. Great hopes are concentrated on technological advance, but there is also a search for new microorganisms, within the immense biodiversity of this planet, which may possess better adaptative characteristics in relation to temperature, pH of the medium and adaptability to low-cost substrates, until now hardly exploited. The enormous potential of extremophiles should be emphasized, considering their impact in several areas of application. On the other hand, we cannot forget the undeniable contribution of the advent of recombinant DNA technology.

References

Andrade, S.V., Polizeli, M.L.T.M., Terenzi, H.F. and Jorge, J.A. (2004). Effect of carbon source on the biochemical properties of the β-xylosidase produced by *Aspergillus versicolor*. *Process Biochemistry* 39: 1931–1938.

Anthony, T., Raj, K.C., Rajendran, A. and Gunasekaran, P. (2003). High molecular weight cellulase-free xylanases from alkali-tolerant *Aspergillus fumigatus* AR1. *Enzyme and Microbial Technology* 32: 647–654.

Araújo, J.H.B., Moraes, F.F. and Zanin, G.M. (1999). Bleaching of kraft pulp with commercial xylanases. *Applied Biochemistry and Biotechnology* 77-79: 713–722.

Aristidou, A. and Pentillä, M. (2000). Metabolic engineering applications to renewable resource utilization. *Current Opinion in Biotechnology* 11: 187–198.

Baitaillon, M., Cardinali, A.P.N., Castillon, N. and Duchiron, F. (2000). Purification and characterization of a moderately thermostable xylanase from *Bacillus* sp. strain SPS-0. *Enzyme and Microbial Technology* 26: 187–192.

Bajpai, P. (2004). Biological bleaching of chemical pulps. *Critical Review in Biotechnology* 24(1): 1–58.

Baraznenok, V.A., Becker, E.G., Ankudimova, N.V. and Okunev, N.N. (1999). Characterization of neutral xylanases from *Chaetomium cellulolyticum* and their biobleaching effect on eucalyptus pulp. *Enzyme and Microbial Technology* 25: 651–659.

Bayer, E.A., Morag, E. and Lamed, R. (1994). The cellulosome–a treasure–trove for biotechnology. *Trends in Biotechnology* 12: 379–386.

Beg, Q.K., Kapoor, M., Mahajan, L. and Hoondal, G.S. (2001). Microbial xylanases and their industrial applications: a review. *Applied Microbiology and Biotechnology* 56: 326–338.

Belancic, A., Scarpa, J., Peirano, A., Diaz, R., Steiner, J. and Eyzaguirre, J. (1995). *Penicillium purpurogenum* produces several xylanases, purification and properties of two of the enzymes. *Journal of Biotechnology* 41: 71–79.

Bera-Maillet, C., Devillard, E., Cezette, M., Jouany, J.P. and Forano, E. (2005). Xylanases and carboxymethylcellulases of the rumen protozoa *Polyplastron multivesiculatum, Eudiplodinium maggii* and *Entodinium* sp. *FEMS Microbiology Letters* 244(1):149–156.

Bhat, M.K. (2000). Cellulases and related enzymes in biotechnology. *Biotechnology Advances* 18: 355–383.

Biely, P. (1985). Microbial xylanolytic systems. *Trends in Biotechnology* 3: 286–290.

Biely, P., Vrsanská, M., Tenkanen, M. and Kluepfel, D. (1997). Endo-α-1,4-xylanase families: differences in catalytic properties. *Journal of Biotechnology* 57:151–166.

Brennan, Y., Callen, W.N., Christoffersen, L., Dupree, P., Goubert, F., Healey, F., Hernandez, M., Keller, M., Li K., Palackal, N., Sittenfeld, A., Tamayo, G., Wells, S.. Hazlewood, G.P., Mathur, E.J., Short, J.M., Robertson, D.E. and Steer, B.A. (2004). Unusual microbial xylanases from insect guts. *Applied and Environmental Microbiology* 70:3609–3617.

Brühlmann, F., Leupin, M., Erismann, K.H. and Fiechter, A. (2000). Enzymatic degumming of ramie bast fibers. *Journal of Biotechnology* 76: 43–50.

Buchert, J., Ranua, M., Kantelinem, A. and Viikari, L. (1992). The role of two *Trichoderma reesei* xylanases in the bleaching of kraft pulp. *Applied Microbiology and Biotechnology* 37: 825–829.

Carmona, E.C., Brochetto-Braga, M.R., Pizzirani-Kleiner, A.A. and Jorge, J.A. (1998.) Purification and biochemical characterization of an endoxylanase from *Aspergillus versicolor. FEMS Microbiology Letters* 166:311–315.

Caspers, M.P., Lok, F., Sinjorgo, K.M., van Zeijl, M.J., Nielsen, K.A. and Cameron-Mills, V. (2001). Synthesis, processing and export of cytoplasmic endo-beta-1,4-xylanase from barley aleurone during germination. *The Plant Journal: for Cell and Molecular Biology* 26:191–204.

Chadha, B.S., Ajay, B.K., Mellon, F. and Bhat, M.K. (2004). Two endoxylanases active and stable at alkaline pH from the newly isolated thermophilic fungus *Myceliophthora* sp. IMI 387099. *Journal of Biotechnology* 109:227–237.

Chavez, R., Schachter, K., Navarro, C., Peirano, A., Aguirre, C., Bull, P. and Eyzaguirre, J. (2002). Differences in expression of two endoxylanase genes (xynA and xynB) from *Penicillium purpurogenum*. *Gene* 26:293(1-2):161–168.

Chavez, R., Bull, P. and Eyzaguirre, J. (2006). The xylanolytic enzyme system from the genus *Penicillium*. *Journal of Biotechnology* 123(4): 413–433.

Chen, C., Chen, J.L. and Lin, T.Y. (1997). Purification and characterization of a xylanase from *Trichoderma longibrachiatum* for xylooligossacharide production. *Enzyme and Microbial Technology* 21: 91–96.

Christakopoulos, P., Nerinckx, W., Kekos, D., Macris, B. and Claeyssens, M. (1996). Purification and characterization of two low molecular mass alkaline xylanases from *Fusarium oxysporum* F3. *Journal of Biotechnology* 51:181–189.

Christov, L., Biely, P., Kalogeris, E., Christakopoulos, P., Prior, B.A. and Bhat, M.K. (2000). Effects of purified endo-β-1,4-xylanases of family 10 and 11 and acetyl xylan esterases on eucalypt sulfite dissolving pulp. *Journal of Biotechnology* 83: 231–244.

Christov, L.P. and Prior, B.A. (1993). Esterases of xylan-degrading microorganisms, production, properties and significance. *Enzyme and Microbial Technology* 15: 460–475.

Christov, L.P., Szakacs, G. and Balakrishnan, H. (1999). Production, partial characterization and use of fungal cellulase-free xylanases in pulp bleaching. *Process Biochemistry* 34: 511–517.

Collins, T., Gerday, C. and Feller, G. (2005). Xylanases, xylanase families and extremophilic xylanases. *FEMS Microbiology Reviews* 29(1):3–23.

Coughlan, M.P. and Hazlewood, G.P. (1993). β-1,4-D-xylan-degrading enzyme systems, biochemistry, molecular biology and applications. *Biotechnology and Applied Biochemistry* 17: 259–289.

Coughlan, M.P., Tuohy, M.G., Filho, E.X.F., Puls, J., Claeyssens, M., Vrsan Ska, M. and Hughes, M.M. (1993). Enzymological aspects of microbial hemicellulases with emphasis on fungal systems. In: *Hemicelluloses and Hemicellulases*. Cougland M.P., Hazlewood G.P., Eds., Portland Press, London 53–84.

Crepin, V.F., Fauld, C.B. and Connerton, I.F. (2004). Functional classification of the microbial feruloyl esterases. *Applied Microbiology and Biotechnology* 63(6):647–652.

Csiszár, E., Urbánszki, K. and Szakás, G. (2001). Biotreatment of desized cotton fabric by commercial cellulase and xylanase enzymes. *Journal of Molecular Catalysis B: Enzymatic*. 11: 1065–1072.

Dalboge, H. (1997). Expression cloning of fungal enzyme genes; a novel approach for efficient isolation of enzyme genes of industrial relevance. *FEMS Microbiology Reviews* 21: 29–42.

Damaso, M.C.T., Andrade, C.M.M.C., and Pereira N. Jr. (2000). Use of corncob for endoxylanase production by thermophilic fungus *Thermomyces lanuginosus* IOC-4145. *Applied Biochemistry and Biotechnology* 84-86:821–834.

de Graaff, L.H., van den Broeck, H.C. and Ooijen, A.J.J. (1994). Regulation of the xylanase-encoding *xlnA* gene of *Aspergillus tubigensis*. *Molecular Microbiology* 12: 479–490.

de Vries, R.P., Kester, H.C., Pousen, C.H., Benen, J.A. and Visser, J. (2000). Synergy between enzymes from *Aspergillus* involved in the degradation of plant cell wall polysaccharides. *Carbohydrate Research* 327: 401–410.

Decelle, B., Tsang, A. and Storms, R.K. (2004). Cloning, functional expression and characterization of three *Phanerochaete chrysosporium* endo-1,4-beta-xylanases. *Current Genetics* 46:166–75.

Doi, R.H., Goldstein, M., Hashida, S., Park, J.S. and Takagi, M. (1994). The *Clostridium cellulovorans* cellulosome. *Critical Reviews in Microbiology* 20: 87–93.

Dowzer, C.E. and Kelly J.M. (1991). Analysis of the *creA* gene, a regulator of carbon catabolite repression in *Aspergillus niger. Molecular and Cell Biology* 7:1256–1266.

Düsterhöft, E.M., Linssen, V.A.J.M., Voragen, A.G.J. and Beldman, G. (1997). Purification, characterization and properties of two xylanases from *Humicola insolens. Enzyme and Microbial Technology* 20:437–445.

Ebringerova, A. and Heinze, T. (2000). Xylan and xylan derivatives – biopolymers with valuable properties. 1. Naturally occurring xylan structures, isolation procedures and properties. *Molecular and Cell Biology* 21:542–556.

Fernández-Espinar, M., Piñaga, F., de Graaff, L., Visser, J., Ramón, D. and Vallés, S. (1994). Purification, characterization and regulation of the synthesis of an *Aspergillus nidulans* acidic xylanase. *Applied Microbiology and Biotechnology* 42:555–562.

Fialho, M.B. and Carmona, E.C. (2004). Purification and characterization of xylanases from *Aspergillus giganteus. Folia Microbiologica* 49: 13–18.

Filho, E.X.F. (1994). The xylan-degrading enzyme system. *Brazilian Journal of Medical and Biological Research* 27: 1093–1109.

Flores, M.E., Perea, M., Rodríguez, O., Malváez, A. and Huitron, C. (1996). Physiological studies on induction and catabolic repression of β-xylosidase and endoxylanase in *Streptomyces* sp. CH-M-1035. *Journal of Biotechnology* 49:179–187.

Fujimoto, H., Ooi, T., Wang, S-L., Takizawa, T., Hidaka, H., Murao, S. and Arai, M. (1995). Purification and properties of three xylanases from *Aspergillus aculeatus. Bioscience, Biotechnology, and Biochemistry* 59:538–540.

Fujita, M. and Harada, H. (1991). Ultrastruture and formation of wood cell wall. In: *Wood and cellulosic chemistry*, David N-S Hon and Shiraishi Nobuo Eds., Marcel Dekker, New York, pp 3–57.

Ghanem, N.B., Yusef, H.H. and Mahrouse, H.K. (2000). Production of *Aspergillus terreus* xylanase in solid-state cultures: application of the Plackett-Burman experimental design to evaluate nutritional requirements. *Bioresource Technology* 73:113–121.

Ghareib, M. and El Dein, M.M.N. (1992). Purification and general properties of xylanase from *Aspergillus terreus. Zentralblatt für Mikrobiologie* 147:569–576.

Ghosh, M. and Nanda, G. (1994). Physiological studies on xylose induction and glucose repression of xylanolytic enzymes in *Aspergillus sydowii* MG49. *FEBS Letters* 117:151–156.

Gomes, D.J., Gomes, J. and Steiner, W. (1994.) Factors influencing the induction of endo-xylanase by *Thermoascus aurantiacus. Journal of Biotechnology* 33:87–94.

Gomez de Segura, B., Durand, R. and Fèvre, M. (1998). Multiplicity and expression of xylanases in the rumen fungus *Neocallimastix frontalis. FEMS Microbiology Letters* 164(1):47–53.

Govinden, R., Pillay, B., van Zyl, WR. and Pillay, D. (2001). Xylitol production by recombinant *Saccharomyces cerevisiae* expressing the *Pichia spitipis* and *Candida shehatae* XYL1 genes. *Applied Microbiology and Biotechnology* 55(1):76–80.

Habibi, Y., Mahrouz, M. and Vignon, M.R. (2002). Isolation and structure of D-xylans from pericarp seeds of *Opuntia ficus-indica* prickly pear fruits. *Carbohydrate Research* 337(17):1593–1598.

Haltrich, D., Nidetzky, B., Kulbe, K.D., Steiner, W. and Zupancic, S. (1996). Production of fungal xylanases. *Bioresource Technology* 58: 137–161.

Harbak, L. and Thygesen, H.V. (2002). Safety evaluation of a xylanase expressed in *Bacillus subtilis*. *Food and Chemistry Toxicology* 40: 1–8.

Henrissat, B.A. and Bairoch, A. (1993). New families in the classification of glycosyl hydrolases based on amino acid sequence similarities. *Biochemistry Journal* 293:781–788.

Horikoshi, K. (1996). Alkaliphiles - from an industrial point of view. *FEMS Microbiology Reviews* 18:259–270.

Huang, J., Wang, G. and Xiao, L. (2006). Cloning, sequencing and expression of the xylanase gene from a *Bacillus subtilis* strain B10 in *Escherichia coli*. *Bioresource Technology* 97(6):802–808.

Ito, K., Iwashita, K. and Iwano, K. (1992). Cloning and sequencing of the xyn C gene encoding acid xylanase of *Aspergillus kawachii*. *Bioscience, Biotechnology, and Biochemistry* 56:1338–1340.

Izydorczyk, M.S. and Biliaderis, C.G. (1995). Cereal arabinoxylans: advances in structure and physiochemical properties. *Carbohydrate Polymers* 28: 33–48.

Jeya, M., Thiagarajan, S. and Gunasekaran, P. (2005). Improvement of xylanase production in solid-state fermentation by alkali tolerant *Aspergillus versicolor* MKU3. *Letters in Applied Microbiology* 41(2):175–178.

Jiang, Z.Q., Yang, S.Q., Tan, S.S., Li, L.T. and Li, X.T. (2005). Characterization of xylanase from the newly isolated thermophilic *Thermomyces lanuginosus* CAU44 and its application in bread making. *Letters in Applied Microbiology* 41: 69–76.

John, M., Schmidt, B. and Schmidt, J.(1979). Purification and some properties of five endo-I, 4-beta-D-xylanases and a β-D-xylosidase produced by a strain of *Aspergillus niger*. *Canadian Journal of Biochemistry* 57: 125–134.

Johnston, M., Johnston, D. and Richardson, A. (2005). Digestive capabilities reflect the major food sources in three species of *talitrid amphipods*. Comparative Biochemistry and Physiology. *Biochemistry and Molecular Biology* 140(2):251–257.

Jørgensen, H., Eriksson, T., Börjesson, J., Tjerneld, F. and Olsson, L. (2003). Purification and characterization of five cellulases and one xylanase from *Penicillium brasilianum* IBT 20888. *Enzyme and Microbial Technology* 32:851–861.

Jun, H.S., Ha, J.K., Malburg, L.M., Verrinder, G.A. and Forsberg, C.W. (2003). Characteristics of a cluster of xylanase genes in *Fibrobacter succinogenes* S85. *Canadian Journal of Microbiology* 49(3):171–180.

Kadowaki, M.K., Pacheco, M.A.C. and Peralta, R.M. (1995). Xylanase production by *Aspergillus* isolates grown on corn cob. *Revista de Microbiologia* 263:219–223.

Kalogeris, E., Christakopoulos, P., Kekos, D. and Macris, B.J. (1998). Studies on solid-state production of thermostable endoxylanases from *Thermoascus aurantiacus*. Characterization of two isozymes. *Journal of Biotechonology* 60:155–163.

Kaneko, S., Shimasaki, T. and Kusakable, I. (1993). Purification and some properties of intracellular α-L-arabinofuranosidase from *Aspergillus niger* 5-16. *Bioscience, Biotechnology, and Biochemistry* 57: 1161–1165.

Kennedy, J.F. and White, C.A. (1988). The plant, algal, and microbial polysaccharides. In: *Carbohydrate Chemistry*. Oxford, Clarendon Press, pp 220–262.

Khanna, P., Sundari, S.S. and Kumar, N.J. (1995). Production, isolation and partial purification of xylanase from an *Aspergillus* sp. *World Journal of Microbiology & Biotechnology* 11: 242–243.

Kimura, I., Sasahara, H. and Tajima, S. (1995). Purification and characterization of two xylanases and an arabinofuranosidase from *Aspergillus sojae*. *Journal of Fermentation and Bioengineering* 804:334–339.

Kimura, T., Ito, J., Kawano, A., Makino, T., Kondo, H., Karita, S., Sakka, K. and Ohmiya, K. (2000). Purification, characterization, and molecular cloning of acidophilic xylanase from *Penicillium* sp. 40. *Bioscience, Biotechnology, and Biochemistry* 64:1230–1237.

Kitamoto, N., Yoshino, S., Ohmiya, K. and Tsukagoshi, N. (1999). Purification and characterization of the overexpressed *Aspergillus oryzae* xylanase, XynF1. *Bioscience, Biotechnology, and Biochemistry* 6310:1791–1794.

Kormelink, F.J.M., Leeuwen, M.G.F.S.L., Wood, T.M. and Voragen, A.G.J. (1993). Purification and characterization of three endo (1,4)-β-D-xylanases and one β-xylosidase from *Aspergillus awamori*. *Journal of Biotechnology* 27:249–253.

Krishna, C. (2005). Solid-state fermentation systems—an overview. *Critical Reviews in Biotechnology* 25(1-2):1–30.

Kulkarni, N., Shendye, A. and Rao, M. (1999a). Molecular and biotechnological aspects of xylanases. *FEMS Microbiology Reviews* 23: 411–456.

Kulkarni, N., Lakshmikumaran, M. and Rao, M. (1999b). Xylanase II from an alkaliphilic thermophilic *Bacillus* with distinctly different structure from other xylanases, evolutionary relationship to alkaliphilic xylanases. *Biochemistry and Biophysical Research Communications* 263: 640–645.

Kurakabe, M., Shinjii, O. and Komaki, T. (1997). Transxylosilation of β-xylosidase from *Aspergillus awamori* K4. *Bioscience, Biotechnology and Biochemistry* 6112: 2010–2014.

Levasseur, A., Asther, M. and Record, E. (2005). Overproduction and characterization of xylanase B from *Aspergillus niger. Canadian Journal of Microbiology* 51(2):177-83.

Li, X.T., Jiang, Z.Q., Li, L.T., Yang, S.Q., Feng, W.Y., Fan, J.Y. and Kusakabe, I. (2005). Characterization of a cellulase-free, neutral xylanase from *Thermomyces lanuginosus* CBS 288.54 and its biobleaching effect on wheat straw pulp. *Bioresource Technology* 96(12):1370–1379.

Li, X.L., Zhang, Z., Dean, I.F.D., Eriksson, K.L. and Ljungdahl, L.G. (1993). Purification and characterization of a new xylanase APX-II from the fungus *Aureobasidium pullulans* Y-2311-1. *Applied and Environmental Microbiology* 59:3212–3218.

Li, K., Azadi, P., Collins, R., Tolan., J., Kim, J.S. and Eriksson Karl-Erik, L. (2000). Relationships between activities of xylanases and xylan structures. *Enzyme and Microbial Technology* 27: 89–94.

Lin, J.J., Ndlovu, L.M., Singh, S. and Balakrishna, P. (1999). Purification and biochemical characteristics of β-D-Xylanase from a thermophilic fungus *Thermomyces lanuginosus*-SSBP. *Biotechnology and Applied Biochemistry* 30:73–79.

Liu, W., Lu, Y. and Ma, G. (1999). Induction and glucose repression of endo-β-xylanase in the yeast *Trichosporon cutaneum* SL409. *Process Biochemistry* 34:67–72.

Liu, C.J., Suzuki, T., Hirata, S. and Sakai, K. (2003). Processing of XynE (110-kDa) of *Aeromonas caviae* ME-1 to 72-kDa xylanase in *Escherichia coli* transformant. *Journal of Bioscience and Bioengineering* 96(4): 406–408.

Mach, R.L., Strauss, J., Zeilinger, S., Schindler, M. and Kubicek, C.P. (1996). Carbon catabolite repression of xylanase I *xyn1* gene expression in *Trichoderma reesei. Molecular Microbiology* 216:1273–1281.

Maheshwari, R., Bharadwaj, G. and Bhat, M.K. (2000). Thermophilic fungi: their physiology and enzymes. *Microbiology and Molecular Biological Reviews* 64:461–488.

Medeiros, R.G., Soffner, M.A.P., Tomé, J.A., Cacais, A.O.G., Estelles, R.S., Salles, B.C., Ferreira H.M., Neto S.A.L., Silva, F.G.Jr and Filho E.X.F. (2000). The production of hemicellulases by aerobic fungi on medium containing residues of banana plant as substrate. *Biotechnology Progress* 16:522–524.

Miller, G.L. (1959). Use of dinitrosalicylic acid reagent for determination of reducing sugar. *Analytical Chemistry* 31: 426–428.

Monti, R., Terenzi, H.F. and Jorge, J.A. (1991). Purification and properties of an extracellular xylanase from thermophilic fungus *Humicola grisea* var. thermoidea. *Canadian Journal of Microbiology* 37:675–681.

Monti, R., Cardello, L., Custódio, M. F., Goulart, A. J., Sayama, A. H. and Contiero, J. (2003). Production and purification of an endo-1,4-β-xylanase from *Humicola grisea* var. thermoidea by electroelution. *Brazilian Journal of Microbiology* 34: 124–128.

Morosoli, R., Durand, S. and Letendre, E. (1987). Induction of xylanase by β-methylxyloside in *Cryptococcus albidus. FEMS Microbiology Letters* 48: 261–266.

Nelson, N. (1944). A photometric adaptation of the Somogyi method for the determination of glucose. *The Journal of Biological Chemistry* 153: 375–381.

Nuyens, F., van Zyl, H.W., Iserentant, D., Verachtert, H. and Michiels, C. (2001). Heterologous expression of the *Bacillus pumilus* endo-β-xylanase (xynA) gene in the yeast *Saccharomyces cerevisiae. Applied Microbiology and Biotechnology* 56: 431–434.

Nyyssola, A., Pihlajaniemi, A., Palva, A., von Weymarn, N. and Leisola, M. (2005). Production of xylitol from D-xylose by recombinant *Lactococcus lactis. Journal of Biotechnology* 118(1): 55–66.

Paice, M.G., Gurnagul, N., Page, D.H. and Jurasek, L. (1992). Mechanism of hemicellulose-directed prebleaching of kraft pulps. *Enzyme and Microbial Technology* 14: 272–276.

Pandey, A., Soccol, C.R., Nigam, P. and Soccol, V.T. (2000). Biotechnological potential of agro-industrial residues. I, sugarcane bagasse. *Bioresource Technology* 74: 69–80.

Parajó, J.C., Domíngues, H. and Domíngues, J.M. (1998). Biotechnological production of xylitol. Part 1, Interest of xylitol and fundamentals of its biosynthesis. *Bioresource Technology* 65: 191–120.

Piñaga, F., Fernández-Espinar, M.T., Vallés, S. and Ramon, D. (1994). Xylanase production in *Aspergillus nidulans*, induction and carbon catabolite repression. *FEMS Microbiology Letters* 115: 319–324.

Polizeli, M.L.T.M., Rizzatti, A.C.S., Monti, R., Terenzi, H.F., Jorge, J.A. and Amorim, D.S. (2005). Xylanases from fungi: properties and industrial applications. *Applied Microbiology and Biotechnology* 65: 577–591.

Prade, R.A. (1995). Xylanases, from Biology to Biotechnology. *Biotechnology & Genetic Engineering Reviews* 13: 101–131.

Puls, J. and Poutanen, K. (1989). Mechanisms of enzymatic hydrolysis of hemicelluloses xylans and procedures for determination of the enzyme activities involved. In: Proc. *3rd Int. Conf. Biotechnology in the Pulp and Paper Industry*. Ericksson K.E.E., Ander, P., Eds., STFI, Stockholm, 93–95.

Puls, J. and Schuseil, J. (1993). Chemistry of hemicelluloses, relationship between hemicellulose structure and enzyme required for hydrolysis. In: *Hemicelluloses and Hemicellulases*. Cougland M.P., Hazlewood G.P., Eds., Portland Press, London,1–27.

Raj, K.C. and Chandra, T.S. (1996). Purification and characterization of xylanase from alkali-tolerant *Aspergillus fischeri* Fxn1. *FEMS Microbiology Letters* 1453: 457–461.

Ramchuran, S.O., Mateus, B., Holst, O. and Karlsson, E.N. (2005). The methylotrophic yeast *Pichia pastoris* as a host for the expression and production of thermostable xylanase from the bacterium *Rhodothermus marinus*. *FEMS Yeast Research* 5: 839–850.

Rani, D.S. and Nand, K. (2000). Production of thermostable cellulase-free xylanase by *Clostridium absonum* CFR-702. *Process Biochemistry* 36: 355–362.

Reeves, R.A., Gibbs, M.D., Morris, D.D., Griffiths, K.R., Saul, D.J. and Bergquist, P.L. (2000). Sequencing and expression of additional xylanase genes from the hyperthermophile *Thermotoga maritima* FjSS3B.1. *Applied and Environmental Microbiology* 66(4): 1532–1537.

Reilly, P.J. (1981). Xylanases, structure and function. In: *Trends in the Biology of Fermentation for Fuels and Chemicals*. Hollaender, A., Ed., Plenum Press, New York, 111–129.

Rizzatti, A.C.S., Jorge, J.A., Terenzi, H.F., Rechia, C.G.V. and Polizeli, M.L.T.M. (2001). Purification and properties of a termostable extracellular β-xylosidase produced by a thermotolerant *Aspergillus phoenicis*. *Journal Industrial Microbiology & Biotechnology* 26: 156–160.

Rizzatti, A.C.S., Sandrim, V.C., Jorge, J.A., Terenzi, H.F. and Polizeli, M.L.T.M. (2004). Influence of temperature on the properties of xylanolytic enzymes of the thermotolerant fungus *Aspergillus phoenicis*. *Journal Industrial Microbiology & Biotechnology* 31: 88–93.

Rogalski, J., Oleszek, M. and Tokarzewska-Zadora, J. (2001). Purification and characterization of two endo-1,4-beta-xylanases and a 3-xylosidase from *Phlebia radiate*. *Acta Microbiologica Polonica* 50: 117–128.

Romanowska, I., Polak, J. and Bielcki, S. (2006). Isolation and properties of *Aspergillus niger* IBT-90 xylanase for bakery. *Applied Microbiology and Biotechnology* 69(6): 665–671.

Ryan, S.E., Nolan, K., Thompson, R., Gubitz, G.M., Savage, A.V. and Tuohy, M.G. (2003). Purification and characterization of a new low molecular weight endoxylanase from *Penicillium capsularium*. *Enzyme and Microbial Technology* 33: 775–785.

Saha, B.C. (2003). Hemicellulose bioconversion. *Journal Industrial Microbiology & Biotechnology* 30(5): 279–291.

Salles, B.C., Cunha, R.B., Fontes, W., Sousa, M.V. and Filho, E.X.F. (2000). Purification and characterization of a new xylanase from *Acrophialophora nainiana*. *Journal of Biotechnology* 81: 199–204.

Sandrim, V.C., Rizzatti, A.C.S., Terenzi, H.F., Jorge, J.A., Milagres, A.M.F. and Polizeli, M.L.T.M. (2005). Purification and biochemical characterization of two xylanases produced by *Aspergillus caespitosus* and their potential for kraft pulp bleaching. *Process Biochemistry* 40: 1823–1828.

Sapag, A., Wouters, J., Lambert, C., de Ioannes, P., Eyzaguirre, J. and Depiereux, E. (2002). The endoxylanases from family 11: computer analysis of protein sequences reveals important structural and phylogenetic relationships. *Journal of Biotechnology* 95:109–131.

Sa-Pereira, P., Paveia, H., Costa-Ferreira, M. and Aires-Barros, M. (2003). A new look at xylanases: an overview of purification strategies. *Molecular Biotechnology* 24(3): 257–281.

Screenath, H.K. and Jeffries, T.W. (2000). Production of ethanol from wood hydrolysate by yeasts. *Bioresource Technology* 72: 253–260.

Slapack, G.E., Russel, I. and Stewart, G.G. (1987). *Thermophilic microbes in Ethanol production.* Library of Congress Card Number 86-9597, CRC Press, Inc., pp 1–50.

Silva, C.H.C., Puls, J., Sousa, M.V. and Ferreira-Filho, E.X. (1999). Purification and characterization of a low molecular weight xylanase from solid-state cultures of *Aspergillus fumigatus* Fresenius. *Revista de Microbiologia* 30: 114–119.

Simão, R.C.G., Souza, C.G.M. and Peralta, R.M. (1997a). The use of methyl β-D-xyloside as a substrate for xylanase production by *Aspergillus tamarii. Canadian Journal of Microbiology* 43: 56–60.

Simão, R.C.G., Souza, C.G.M. and Peralta, R.M (1997b). Induction of xylanase in *Aspergillus tamarii* by methyl β-D-xyloside. *Applied Microbiology and Biotechnology* 47: 267–271.

Singh, S., Reddy, P., Haarhoff, J., Biely, P., Janse, B., Pillay, B., Pillay, D. and Prior, B.A. (2000). Relatedness of *Thermomyces lanuginosus* strains producing a thermostable xylanase. *Journal of Biotechnology* 81: 119–128.

Smith, D.C., Bhat, K.M. and Wood, T.M. (1991) .Xylan-hydrolysing enzymes from thermophilic and mesophilic fungi. *World Journal of Microbiology & Biotechnology* 7:475–484.

Somogyi, M.(1952). Notes on sugar determination. *The Journal of Biological Chemistry* 195: 19–23.

Sonia, K.G., Chadha, B.S. and Saini, H.S. (2005). Sorghum straw for xylanase hyper-production by *Thermomyces lanuginosus* (D_2W_3) under solid-state fermentation. *Bioresource Technology* 96: 1561–1569.

Subramaniyan, S. and Prema, P. (2000). Cellulase-free xylanases from *Bacillus* and other microorganisms. *FEMS Microbiology Letters* 183: 1–7.

Subramaniyan, S. and Prema, P.(2002). Biotechnology of microbial xylanases, enzymology, molecular biology, and application. *Critical Review in Biotechnology* 22: 33–64.

Sung, W.L., Luk, C.K., Chan, B., Wakarchuk, W., Yaguchi, M., Campbell, R., Willick, G., Ishikawa, K. and Zahab, D.M. (1995). Expression of *Trichoderma reesei* and *Trichoderma viride* xylanases in *Escherichia coli. Biochemistry in Cell Biology* 73 (5-6):253–259.

Sunna, A. and Antranikian, G. (1997). Xylanolytic enzymes from fungi and bacteria. *Critical Review in Biotechnology*. 17(1): 39–67.

Tan, L.U.L., Wong, K.K.Y., Yu, E.K.C. and Saddler, J.N. (1985). Purification and characterization of two D-xylanases from *Trichoderma harzianum. Enzyme and Microbial Technology* 7: 425–430.

Tanaka, H., Muguruma, M. and Ohta, K. (2006). Purification and properties of family-10 xylanase from *Aureobasidium pullulans* ATCC 20524 and characterization of the encoding gene. *Applied Microbiology and Biotechnology*: 70(2): 202–211.

Taneja, K., Gupta, S. and Kuhad, R.C. (2002). Properties and application of a partially purified alkaline xylanase from an alkalophilic fungus *Aspergillus nidulans* KK-99. *Bioresource Technology* 85: 39–12.

Techapun, C., Poosaran, N., Watanabe, M. and Sasaki, K. (2003). Thermostable and alkaline-tolerant microbial cellulose-free xylanases produced from agricultural wastes and the properties required for use in pulp bleaching bioprocesses: a review. *Process Biochemistry* 38: 1327–1340.

Tenkanen, M., Viikari, L. and Buchert., J. (1997). Use of acid-tolerant xylanase for bleaching of kraft pulps. *Biotechnology Techniques* 11 (12): 935–938.

Timell, T.E.(1965). "Wood hemicelluloses". In: *Advances Carbohydrate Chemistry* Part II, 20: 409.

Törrönen, A. and Rouvinen, J. (1997). Structural and functional properties of low molecular weight endo-1,4-b-xylanases. *Journal of Biotechnology* 57: 137–149.

Turunen with O., Etuaho, K., Fenel, F., Vehmaanperä, J., Wu, X., Rouvinen, J. and Leisola, M. (2001). A combination of weakly stabilizing mutations with a disulfide bridge in the α-helix region of *Trichoderma reesei* endo-1,4-β-xylanase II increases the thermal stability through synergism. *Journal of Biotechnology*. 88: 37–46.

Twomey, L., N., Pluske, J.R., Rowe, J. B., Choct, M., Brown, W., McConnell, M. F. and Pethick, D.W. (2003). The effects of increasing levels of soluble non-starch polysaccharides and inclusion of feed enzymes in dog diets on faecal quality and digestibility. *Animal Feed Science and Technology* 108 1-4: 71–82.

Uffen, R.L. (1997). Xylan degradation: a glimpse at microbial diversity. *Journal Industrial Microbiology & Biotechnology* 19: 1–6.

van Peij, N.N.M.E., Brinkmann, J., Vrsanská, M., Visser, J. and de Graaff, L.H. (1997). β-xylosidase activity, encoded by *xlnD*, is essential for complete hydrolysis of xylan by *Aspergillus niger* but not for induction of the xylanolytic enzyme spectrum. *European Journal of Biochemistry* 245: 164–173.

Viikari, L., Kantelinen, A., Sundquist, J. and Linko, M. (1994). Xylanases in bleaching, from an idea to the industry. *FEMS Microbiology Reviews* 13: 335–350.

Viikari, L., Sundquist, J. and Kettunen, J. (1991). Xylanase enzymes promote pulp bleaching. *Paper Timber* 73: 384–389.

Weng, X.Y. and Sun, J.Y. (2005). Construction, expression, and characterization of a thermostable xylanase. *Current Microbiology* 51(3): 188–192.

Whitmire, D. and Miti, B. (2002). Xylanase effects on pulp delignification. *Chemical Engineering Communications* 189 (5): 608–622.

Williamson, G., Faulds, C.B. and Kroon, P.A. (1998). Specificity of ferulic acid (feruloyl) esterases. *Biochemical Society transactions* 26(2): 205–209.

Wong, K.K.Y. and Saddler, J.N. (1993). Applications of hemicellulases in the food, feed, and pulp and paper industries. In: *Hemicellulose and Hemicellulases*. Coughlan, M.P. and Hazlewood, G.P., Eds., Portland Press, London, 127–143.

Wong, K.K.Y., Tan, L.U.L. and Saddler, J.N. (1988). Multiplicity of β-1,4-xylanase in microorganisms, functions and applications. *Microbiology Reviews* 52(3): 305–317.

Zanoelo, F.F., Polizeli, M.L.T.M., Terenzi, H.F. and Jorge, J.A. (2004). Purification and biochemical properties of a thermostable xylose-tolerant β-D- xylosidase from *Scytalidium thermophilum. Journal Industrial Microbiology & Biotechnology* 31: 170–176.

Zhao, J., Li, X. and Qu, Y. (2006). Application of enzymes in producing bleached pulp from wheat straw. *Bioresource Technology* 97(13): 1470–1476.

5

Laccases of Fungi in Nature and Biotechnology

Petr Baldrian

Laboratory of Biochemistry of Wood-Rotting Fungi, Institute of Microbiology of the ASCR V.V.i, Vídeňská 1083, 14220 Praha 4, Czech Republic, Tel.: +420241062315, Fax: +420241062384, E-mail: baldrian@biomed.cas.cz

INTRODUCTION

Laccase (benzenediol:oxygen oxidoreductase, EC 1.10.3.2) belongs to multi-copper oxidases—a group of polyphenol oxidases containing copper atoms in the catalytic centre. Other members of this group are ascorbate oxidases of plants and the mammalian plasma protein ceruloplasmin. Laccases typically contain three types of copper, one of which gives it its characteristic blue colour. Similar enzymes lacking the Cu atom responsible for the blue colour are called "yellow" or "white" laccases. Laccases catalyze the reduction of oxygen to water accompanied by the oxidation of a substrate, typically a *p*-dihydroxy phenol or another phenolic compound. Laccases, however, have a very broad substrate range which varies from one enzyme to another. Their substrate specificity also overlaps with the substrate range of the monophenol mono-oxygenase tyrosinase (EC 1.14.18.1). Syringaldazine [N,N'-bis (3,5-dimethoxy-4-hydroxybenzylidene hydrazine)] is often considered to be a unique laccase substrate (Harkin *et al.*, 1974), as long as hydrogen peroxide is avoided in the reaction. Laccase is thus an oxidase that oxidizes polyphenols, methoxy-substituted phenols, aromatic diamines and a range of other compounds but does not oxidize tyrosine as tyrosinases do.

Laccases can be typically found in plants and fungi. Plant laccases participate in the radical-based mechanisms of lignin polymer formation (Sterjiades *et al.*, 1992; Liu *et al.*, 1994; Boudet, 2000; Ranocha *et al.*, 2002; Hoopes and Dean, 2004), whereas in fungi it probably has more different roles including lignin modifications, fungal plant-pathogen/host interaction, stress defence and morphogenesis (Thurston, 1994). There are some reports about laccase activity in bacteria (Alexandre and Zhulin, 2000; Martins *et al.*, 2002; Givaudan *et al.*, 2004), but it is so far not known how common laccases are in prokaryotic microorganisms. Bacterial laccase-like proteins are intracellular or periplasmic proteins (Claus, 2003). Probably the best characterized bacterial laccases are the periplasmic protein from *Sinorhizobium meliloti* (Rosconi *et al.*, 2005) and a spore coat laccase of *Bacillus subtilis* (Enguita *et al.*, 2003).

The structure, catalytic properties and occurrence of laccases has recently been reviewed (Baldrian, 2006). The aspects of laccase structure and function were previously reviewed by Thurston (1994) and Mayer and Staples (2002), whereas Leonowicz *et al.* (2001) focused on the involvement of laccase in lignin transformation. The physico-chemical properties of multicopper oxidases have been comprehensively reviewed by Solomon *et al.* (1996, 2001). An overview of technological applications of oxidases including laccase was published by Durán and Esposito (2000), Durán *et al.* (2002) and most recently by Riva (2006). A comprehensive review about the phylogeny of laccase gene sequences has been published by Hoegger and co-workers (2006). The main aim of the present chapter is to focus on the occurrence of laccases in the natural environment and on their possible use in biotechnology.

LACCASE PRODUCERS AMONG FUNGI

Laccase activity has been demonstrated in many fungal species and the enzyme has already been purified from more than a hundred species (Baldrian, 2006). However, the presence and role of laccase in certain taxonomic species of fungi has not yet been studied in detail. So far, laccase production was not demonstrated in Zygomycetes and Chytridiomycetes.

There are many records of laccase production by ascomycetes. Laccase was purified from several phytopathogenic ascomycetes such as *Gaeumannomyces graminis*, *Magnaporthe grisea* and *Ophiostoma novo-ulmi* (Binz and Canevascini, 1997; Edens *et al.*, 1999; Iyer and Chattoo, 2003), as well as from the ascomycete model organisms *Neurospora crassa* or *Podospora anserina* (Molitoris and Esser, 1970; Froehner and Eriksson, 1974) and many others (Baldrian, 2006). In addition to plant pathogenic species, laccase production was also reported for some soil ascomycete species from the genera *Aspergillus*, *Curvularia* and *Penicillium* (Banerjee and Vohra, 1991; Rodriguez *et al.*, 1996; Scherer and Fischer, 1998) and it seems that it is widely produced by freshwater ascomycetes participating in the transformation of dead plant material (Abdel-Raheem and Shearer, 2002; Lyons *et al.*, 2003; Abdel-Raheem and Ali, 2004; Junghanns *et al.*, 2005). Laccase was recently also reported from lichens from the order *Peltigerinae* where it is probably produced by the ascomycetous mycobiont (Laufer *et al.*, 2006).

Wood-degrading ascomycetes are ecologically closely related to the wood-rotting basidiomycetes producing laccase. Laccase activity was described in the genera *Bothryosphaeria*, *Trichoderma* and *Xylaria*, but whereas *Bothryosphaeria* sp. and *Xylaria polymorpha* produce constitutively oxidases that are probably true laccases that participate in lignin transformation (Vasconcelos *et al.*, 2000; Pointing *et al.*, 2005; Liers *et al.*, 2006), only some strains of *Trichoderma* exhibit a low level of production of a syringaldazine-oxidizing enzyme, associated mainly with spores (Assavanig *et al.*, 1992; Holker *et al.*, 2002).

Yeasts are a physiologically specific group of both ascomycetes and basidiomycetes. Laccase has been so far studied mainly in *Cryptococcus (Filobasidiella) neoformans* where it probably contributes to the resistance to fungicides (Zhu *et al.*, 2001; Ikeda *et al.*, 2003). Homologous genes or proteins have also been found in *Cryptococcus podzolicus* and *Saccharomyces cerevisiae* but not in other heterobasidiomycetous yeasts tested (Petter *et al.*, 2001; Machonkin *et al.*, 2001; Stoj and Kosman, 2003) and a low laccase-like activity has been reported from some yeast species isolated from decaying wood (Jimenez *et al.*, 1991).

Among physiological groups of fungi, laccases are typical for the saprotrophic basidiomycetes causing white-rot of wood or decomposition of plant litter. Almost all species of white-rot fungi were reported to produce laccase at various degree (Baldrian, 2006). Although it is undoubtedly the most widely present ligninolytic enzyme in these fungi, it is not produced by the model white-rot fungus *Phanerochaete chrysosporium* (Martinez *et al.*, 2004). The attempts to find laccase in the basidiomycetes causing brown-rot of wood yielded so far only a DNA sequence with a high similarity to laccases of white-rot fungi in *Gloeophyllum trabeum.* Oxidation of 2,2′-azinobis (3-ethylbenzathiazoline-6-sulfonic acid) (ABTS) as an indirect indication of oxidative activity was also found in this fungus and in a few other species (D'Souza *et al.*, 1996; Schlosser and Höfer, 2002). The oxidation of syringaldazine – a reliable indication of laccase presence – has been detected only in the brown-rotter *Coniophora puteana* (Lee *et al.*, 2004). Significant involvement of laccase in brown-rot decay of wood seems thus improbable.

Recent studies searching for laccase genes and their transcripts in forest soils revealed, that relatively many laccase genes can be found in ectomycorrhizal fungal species (Luis *et al.*, 2004; 2005a) and that these genes are transcribed (Luis *et al.*, 2005b). However, in *Paxillus involutus*, a species containing many putative laccase sequences, oxidation of syringaldazine was never detected (Günther *et al.*, 1998; Timonen and Sen, 1998). Laccases have been purified from a few fungi forming ectomycorrhiza: *Cantharellus cibarius*, *Lactarius piperatus*, *Russula delica* and *Thelephora terrestris* or orchideoid mycorrhiza: *Armillaria mellea* (Baldrian, 2006). However, the actual significance of laccase in mycorrhizal fungi as well as the possible semi-saprotrophic lifestyle of these fungi are still questionable.

Structure and Localization of Laccase Proteins

General conclusions about laccase structures can be made based on the data about purified laccases. This approach, however, discriminates towards laccases of white-rot fungi which are the most often studied. One must always keep in mind that laccases from other groups of basidiomycetes, ascomycetes and imperfect fungi were studied to a much lesser extent although these enzymes often exhibit unique structural properties (Baldrian, 2006). The most typical fungal laccase is a protein with a molecular weight about 60–70 kDa and an acidic isoelectric point below pH 4.5 (Table 5.1).

Table 5.1 Typical properties of fungal laccases (according to Baldrian, 2006).

Molecular weight (Da)		60000–75000
pI		3.5–4.5
Temperature optimum (°C)		50–70
pH optimum	ABTS	2.5–4.0
	2,6-dimethoxyphenol	3.0–5.5
	Syringaldazine	4.5–6.0
K_M	ABTS	10–100
(μM)	2,6-dimethoxyphenol	100–1000
	Syringaldazine	10–150
k_{cat}	ABTS	5000–40000
(s^{-1})	2,6-dimethoxyphenol	800–6000
	Syringaldazine	15000–25000

Several laccase isoenzymes have been detected in many fungal species. More than one isoenzyme is produced in most white-rot fungi. Blaich and Esser (1975) found in their study covering more than 10 wood-rotting basidiomycetes always more than one isoenzyme.

Several species produce a wide variety of isoenzymes. The white-rot fungus *Pleurotus ostreatus* produces at least nine different laccase isoenzymes, seven of which have been isolated and characterized. The enzymes differ in their rate of production, catalytic properties and structure, including both monomers and oligomers. Some of them are only produced under certain situations, e.g. in the presence of higher copper concentrations (Sannia *et al.*, 1986; Palmieri *et al.*, 1993; 1997; 2003; Giardina *et al.*, 1999). Isoenzymes of laccase with different molecular weight and pI were also detected in the litter-decomposing fungus *Marasmius quercophilus* (Dedeyan *et al.*, 2000; Farnet *et al.*, 2000; 2002; 2004). The isoenzyme pattern as well as their inducibility by phenolic compounds were consistent in 17 different isolates of this litter-decomposer (Farnet *et al.*, 1999). Some fungal species, e.g. *Coriolopsis rigida*, *Dichomitus squalens*, *Physisporinus rivulosus* or *Trametes gallica* produce isoenzymes that are closely related both structurally and in their catalytic properties (Baldrian, 2006) and apparently have similar function. The molecular basis for the production of different isoenzymes is the presence of multiple laccase genes in fungi. However, it seems that many putative laccase gene sequences are not transcribed (D'Souza *et al.*, 1996; Luis *et al.*, 2004).

Most fungal laccases are monomeric proteins, but both homodimers and heterooligomers of up to three subunits have been described in both basidiomycetes and ascomycetes (Baldrian, 2006). Like almost all fungal extracellular enzymes, laccases are glycoproteins. The extent of glycosylation usually ranges between 10–25%, but laccases with saccharide content as high as

45–80% have also been described (Gigi *et al.*, 1981; Marbach *et al.*, 1984; Zouari *et al.*, 1987; Slomczynski *et al.*, 1995; de Souza and Peralta, 2003). The most detailed structure of laccase glycan is available for *Rigidoporus lignosus* laccase that is glycosylated with N-bound mannose (Garavaglia *et al.*, 2004). It was proposed that glycosylation plays some role in the protection of laccase from proteolytic degradation (Yoshitake *et al.*, 1993). The glycosylation of fungal laccases poses a serious problem for its heterologous expression.

Laccases belong to the group of blue multicopper oxidases (BMCO) that catalyze a one-electron oxidation concomitantly with the four-electron reduction of molecular oxygen to water (Salomon *et al.*, 1996; 2001; Messerschmidt, 1997). The catalysis carried out by all members of this family is guaranteed by the presence of different copper centres in the enzyme molecule. In particular, all BMCO are characterized by the presence of at least one type-1 (T1) copper, together with at least three additional copper ions: one type-2 (T2) and two type-3 (T3) copper ions, arranged in a trinuclear cluster (Baldrian, 2006). Some enzymes lack the T1 copper and thus the characteristic absorption band around 600 nm and can be designated as "white" or "yellow" laccases (Leontievsky *et al.*, 1997a; b). Although some authors do not regard these enzymes as laccases, their catalytic properties are very similar to "true" laccases.

Three-dimensional structures are available for five fungal laccases: *Coprinus cinereus*, *Trametes versicolor*, *Pycnoporus cinnabarinus*, *Melanocarpus albomyces* and *Rigidoporus lignosus* (Ducros *et al.*, 1998; Bertrand *et al.*, 2002; Piontek *et al.*, 2002; Antorini *et al.*, 2002; Hakulinen *et al.*, 2002; Garavaglia *et al.*, 2004) and for the CoA laccase from *Bacillus subtilis* endospore (Enguita *et al.*, 2003). Despite the amount of information on laccases as well as other blue multicopper oxidases, neither the precise electron transfer pathway nor the details of dioxygen reduction are fully understood (Piontek *et al.*, 2002; Garavaglia *et al.*, 2004). Unlike laccases with 3-D structures available, the enzyme from *Phellinus ribis* with catalytic features typical for laccases does not belong to the blue copper proteins because it lacks Cu1 and contains one Mn atom per molecule (Min *et al.*, 2001). The "white" laccase POXA1 from *Pleurotus ostreatus* contains only one copper atom, together with two zinc and one iron atoms per molecule (Palmieri *et al.*, 1997). Future structural studies thus probably show that laccases are a structurally heterogeneous group of proteins.

Due to the properties of their substrate, the enzymes participating in the breakdown of lignin should be exclusively extracellular. This is without exception true for ligninolytic peroxidases of white-rot fungi. Although most laccases purified to that extent are extracellular enzymes, laccases of wood-rotting fungi are usually also found intracellularly (Blaich and Esser, 1975; Schlosser *et al.*, 1997; Baldrian, 2006). Substantial part of *Trametes versicolor* and *Pleurotus ostreatus* laccase was associated with the cell wall (Valášková and Baldrian, 2006) and almost all activity was associated with cell walls in *Irpex lacteus*, *Cryptococcus neoformans* and *Trichoderma* spp. (Zhu *et al.*, 2001; Hoker *et al.*, 2002; Svobodová, 2005). The intracellular laccases of fungi probably participate in the transformation of phenolic compounds in the cell while the cell wall and spores-associated laccases were linked to the possible formation of protective cell wall compounds (Eggert *et al.*, 1995; Galhaup and Haltrich, 2001).

CATALYTIC PROPERTIES

Laccase catalyzes the reduction of O_2 to H_2O using a range of phenolic compounds as hydrogen donors (Thurston, 1994; Solomon *et al.*, 1996). Unfortunately, laccase shares a number of hydrogen donors with tyrosinase, making it difficult to assign unique descriptions to either enzyme (Baldrian, 2006). A further complication is the overlap in activity between monophenol monooxygenase (EC 1.10.3.1) and catechol oxidase. The oxidation of syringaldazine in combination with the inability to oxidize tyrosine has most often been taken as an indicator of laccase activity (Harkin *et al.*, 1974; Thurston, 1994). In common with catechol oxidase and tyrosinase, laccase catalyzes the four-electron reduction of O_2 to H_2O. This is in the case of laccase coupled to the single-electron oxidation of the hydrogen-donating substrate (Reinhammar and Malmstrom, 1981). Laccases are known to be highly oxidizing with redox potential (E_0) ranging from 450 to over 750 mV and exceeding by far the redox potential of tyrosinases (Baldrian, 2006).

A very wide range of substrates has been shown to be oxidized by fungal laccases (Baldrian, 2006) but the catalytic constants have been reported mostly for a small group of substrates – e.g. ABTS, 2,6-dimethoxyphenol (DMP), guaiacol and syringaldazine. K_M ranges from tens of mM for syringaldazine and ABTS to hundreds of µM for DMP and guaiacol. The catalytic performance expressed as k_{cat} spans several orders of magnitude for different substrates and is usually a characteristic for a specific protein (Table 5.1). Low K_M values are typical for sinapic acid, hydroquinone and syringic acid, whereas relatively high values were found for *para*-substituted phenols, vanillic acid or its aldehyde. For the species capable of oxidizing polycyclic aromatic hydrocarbons or pentachlorophenol, only very low rate constants were detected for these xenobiotic compounds. In general, rate preference for different substrates varies considerably among laccases but the oxidation of ABTS and 2,6 dimethoxyphenol is always relatively fast (Baldrian, 2006).

In addition to the oxidation of phenols, laccases have also been demonstrated to catalyze the oxidation of Mn^{2+} in the presence of chelators. Laccase from the white-rot fungus *Trametes versicolor* oxidized Mn^{2+} to Mn^{3+} in the presence of pyrophosphate (Höfer and Schlosser, 1999). The same was also later demonstrated for the enzyme of the litter-decomposer *Stropharia rugosoannulata* with oxalic and malonic acids as chelators (Schlosser and Höfer, 2002). The chelators probably decrease the high redox potential of the Mn^{2+}/Mn^{3+} couple. It was proposed that laccase and Mn-peroxidase can co-operate: in the presence of Mn^{2+} and oxalate, laccase produces Mn^{3+}-oxalate. The latter initializes a set of follow-up reactions leading to H_2O_2 formation, which may support peroxidase reactions (Schlosser and Höfer, 2002).

Fungal laccases typically exhibit pH optima in the acidic pH range. While the pH optima for the oxidation of ABTS are generally lower than 4.0, phenolic compounds exhibit higher values between 4.0 and 7.0 (Table 5.1). pH optima of different fungal enzymes for hydroquinone and catechol are 3.5–4.0 and 3.5–6.5 respectively (Baldrian, 2006). The products of laccase-catalyzed reactions may differ at different pH values (Xu, 1997). The stability of fungal laccases is usually higher at acidic pH (Baldrian, 2006).

Temperature profiles of laccase activity usually exhibit optima between 50–70 °C (Table 5.1). The temperature stability which is a crucial factor for the biotechnological applications varies but some enzymes show high tolerance laccases from *Lentinus edodes* and *Agaricus bisporus* have a half-life of 2-3 hours at 50°C (Wood, 1980; D'Annibale *et al.*, 1996) and it can be up to 50–70 hours in *Trametes* sp. (Smirnov *et al.*, 2001). The thermostable laccase from *Melanocarpus albomyces* exhibited a half-life over 5 h at 60°C (Kiiskinen *et al.*, 2002).

Biotechnological use of Laccases

Laccase has several properties that make this enzyme potentially useful for biotechnological application. These include the fact that laccase, unlike peroxidases, does not need the addition or synthesis of a low molecular weight cofactors like hydrogen peroxide, since its cosubstrate – oxygen – is usually present in its environment. Purification of laccases from fungal cultures is usually simple due to the fact that most laccases are extracellular enzymes. Laccases thus also generally exhibit a considerable level of stability in the extracellular environment. They can be extracted from lignocellulosic substrates colonized by fungi as well as from soil or forest litter, or used in the form of spent substrate from the cultivation of edible mushrooms (Eggen, 1999; Lau *et al.*, 2003; Law *et al.*, 2003). The possibility of increasing the production of laccase by the addition of inducers to fungal cultures and a relatively simple purification process represent other advantages for their use. Last but not the least, the considerable amount of data concerning the properties of fungal laccases accumulated in the past years allows us to select a protein suitable for a specific application (e.g. temperature-resistant or pH-stable). The low substrate specificity is mainly advantageous for applications like detoxication of mixtures of xenobiotic compounds while it can bring problems when selective transformation is required.

The biotechnological processes involving laccase are variable and reflect the wide substrate spectrum of this enzyme (Table 5.2). The most studied processes include the transformation of xenobiotic compounds, synthetic dyes, pesticides or polycyclic aromatic hydrocarbons (PAH), bleaching of kraft pulp or detoxication of agricultural by-products. For review on different aspects of laccases in biotechnology (see Reid and Paice, 1994; Paice *et al.*, 1995; Durán and Esposito 2000; Durán *et al.*, 2002; Mayer and Staples, 2002; Chiacchierini *et al.*, 2004; Sigoillot *et al.*, 2004; Riva, 2006) Only a few different laccase enzymes – mostly from the model white-rot fungi *Trametes versicolor* and *Pleurotus ostreatus*–have been so far used in biotechnology applications (Table 5.2) while the majority of laccase enzymes are still unexplored in this respect.

Table 5.2 Biotechnological applications involving purified fungal laccases.

Enzyme origin	*Process*	*Reference*
Clavariopsis aquatica	Nonylphenol degradation	Junghanns *et al.*, 2005
Coriolopsis rigida	Anthracene oxidation	Rodriguez-Solar *et al.*, 2005
Lentinula edodes	Dyes decolorization	Nagai *et al.*, 2002
L. edodes	Phenolics removal from olive mill wastewater	D'Annibale *et al.*, 2000
Myceliophthora thermophila	Dyes decolorization	Claus *et al.*, 2002
Panus rudis	Dyes decolorization	Zhang *et al.*, 2006
Pleurotus ostreatus	PAH oxidation	Pozdnyakova *et al.*, 2004; 2006
P. ostreatus	Phenolics removal from coffee pulp	Velazquez-Cedeno *et al.*, 2002
P. ostreatus	Dyes decolorization	Palmieri *et al.*, 2005
P. pulmonarius	Phenolics removal from coffee pulp	Velazquez-Cedeno *et al.*, 2002
Pleurotus spp.	Phenolics removal from olive mill wastewater	Tsioulpas *et al.*, 2002
Polyporus pinisitus	Dyes decolorization	Claus *et al.*, 2002
P. pinsitus	4-hydroxybiphenyl oxidation	Bratkovskaya *et al.*, 2006
P. cinnabarinus	Chlorinated hydroxy-biphenyls dehalogenation	Schultz *et al.*, 2001
P. cinnabarinus	Kraft pulp bleaching – removal of sterols	Gutierrez *et al.*, 2006; Ibarra *et al.*, 2006
Pyricularia oryzae	Azo-dyes oxidation	Chivukula and Renganathan, 1995
Trametes hirsuta	PAH oxidation	Cho *et al.*, 2002
T. hirsuta	Milled wood lignin oxidation	Gronqvist *et al.*, 2005
T. hirsuta	Dyes decolorization	Abadulla *et al.*, 2000; Couto and Sanromán, 2004; Couto *et al.*, 2006
T. hispida	Dyes decolorization	Rodriguez *et al.*, 1999
T. modesta	2,4,6-trinitrotoluene metabolites oxidative coupling to humic compounds	Nyanhongo *et al.*, 2006
T. pubescens	Natural glycosides regioselective oxidation	Baratto *et al.*, 2006
Trametes sp.	Dymron herbicide degradation	Maruyama *et al.*, 2006
T. versicolor	Pentachlorophenol degradation	Ullah *et al.*, 2000

(Contd.)

Enzyme origin	Process	Reference
T. versicolor	Chlorophenols degradation	Royarcand and Archibald, 1991
T. versicolor	Trichlorophenol degradation	Roper *et al.*, 1995
T. versicolor	Chlorophenols oxidative coupling	Bollag *et al.*, 2003
T. versicolor	Dyes decolorization	Wong and Yu, 1999; Claus *et al.*, 2002; Peralta-Zamora *et al.*, 2003
T. versicolor	Phenylurea pesticide degradation	Jolivalt *et al.*, 2000; 2006
T. versicolor	PAH oxidation	Collins *et al.*, 1996; Johannes *et al.*, 1996; Majcherczyk *et al.*, 1998; Majcherczyk and Johannes, 2000
T. versicolor	Kraft pulp delignification (mediators)	Bourbonnais and Paice, 1996; Call and Mucke, 1997
T. versicolor	Kraft pulp bleaching	Monteiro and de Carvalho, 1998; de Carvalho *et al.*, 1999; Balakshin *et al.*, 2001; Lund *et al.*, 2003; Gamelas *et al.*, 2005
T. versicolor	Benzyl alcohol and 1-phenylethyl alcohol oxidation (mediators)	Arends *et al.*, 2006
T. versicolor	Biofuel cell based on laccase and glucose oxidase	Barriere *et al.*, 2006
T. versicolor	Triclosan degradation	Kim and Nicell, 2006a
T. versicolor	Bisphenol-A oxidation	Kim and Nicell, 2006b; c
T. villosa	Aminophenols oxidative coupling	Manda *et al.*, 2005
T. villosa	Azo-dyes degradation	Zille *et al.*, 2005
Commercial enzyme	Dyes decolorization	Soares *et al.*, 2002
Commercial enzyme	Bisphenol-A oxidation	Modaressi *et al.*, 2005
Different sources	Residual Kraft lignin oxidation	Elegir *et al.*, 2005

The low redox potential of laccases (450–800 mV) is the main disadvantage compared to ligninolytic peroxidases (E_0 >1 V). It only allows the direct degradation of low-redox-potential compounds and not the oxidation of more recalcitrant aromatic compounds including some synthetic dyes or PAH (Xu *et al.*, 1996). There is, however, some evidence that PAH can be oxidized by some laccases to a considerable degree—yellow laccase from *Pleurotus ostreatus*

(YLPO) degraded PAH anthracene (95% within 2d) and fluoranthene (14% within 2d) with an optimum pH of 6.0 without redox mediators (Pozdnyakova *et al.*, 2004) while "blue" laccases from other fungi were not capable of efficient oxidation (Johannes *et al.*, 1996; Majcherczyk *et al.*, 1998; Kottermann *et al.*, 1998). The compounds with higher redox potential can only be transformed if the reaction product is subject to an immediately following reaction or when its redox potential is lowered, for instance by chelation.

Another possibility for the oxidation of compounds with high redox potentials is the involvement of redox mediators in the catalyzed reaction (Wells *et al.*, 2006). Since the description of the first laccase mediators like ABTS to the more recent use of the -NOH- type, synthetic mediators such as 1 hydroxybenzotriazole, violuric acid, and N-hydroxyacetanilide or TEMPO, a large number of studies have been performed on the mechanisms of oxidation of nonphenolic substrates (Bourbonnais and Paice, 1990; Bourbonnais *et al.*, 1998; Xu *et al.*, 2000; Fabbrini *et al.*, 2002; Baiocco *et al.*, 2003), the search for new mediators (Bourbonnais *et al.*, 1997; Fabbrini *et al.*, 2002), and their applications in the degradation of aromatic xenobiotics (Bourbonnais *et al.*, 1997; Johannes *et al.*, 1998; Kang *et al.*, 2002; Keum and Li, 2004). Nevertheless, the laccase-mediator system cannot be applied on the process scale so far due to the cost of mediators and the lack of studies that guarantee the absence of toxic effects of these compounds or their derivatives. The use of naturally occurring laccase mediators would present environmental and economic advantages. Their capability to act as laccase mediators has recently been demonstrated. The possibility of obtaining mediators from natural sources and the new mediators applicable at a low mediator/substrate ratios enhanced the feasibility of laccase-mediator system for the practical use (Eggert *et al.*, 1996; Camarero *et al.*, 2005; Baldrian, 2006).

In addition to substrate oxidation, laccase can also perform immobilization of soil pollutants by their coupling to soil humic substances – a process analogous to humic acid synthesis in soils (Bollag, 1991; Bollag and Myers, 1992). The xenobiotics like phenolic compounds and anilines like 3,4-dichloroaniline, 2,4,6-trinitrotoluene or chlorinated phenols can be immobilized in this way (Tatsumi *et al.*, 1994; Dawel *et al.*, 1997; Dec and Bollag, 2000; Ahn *et al.*, 2002) The immobilization decreases biological availability and thus the toxicity of these xenobiotics.

Laccases in the Natural Environment

The mechanisms involved in lignocellulose degradation by laccases have been studied in detail and a comprehensive review is available on this topic (Leonowicz *et al.*, 2001). Far less is known about the occurrence, properties and roles of laccases occurring in other environments. Soil or dead plant litter are a more complex and heterogeneous environments compared to wood. This may hamper detection and estimation of enzyme activities. It is usually also difficult to link activity in a complex substrate to single species of a laccase producer. Several works studied the production of ligninolytic enzymes by fungi introduced into soils and a number of protocols for laccase activity measurements were proposed (Lang *et al.*, 1997; 1998; Baldrian *et al.*, 2000). These include direct incubation with guaiacol as laccase substrate (Nannipieri *et al.*, 1991), extraction with surfactants or calcium chloride (Criquet *et al.*, 1999) or the most widely used extraction with phosphate buffer (Lang *et al.*, 1997) depending on the nature of the substrate (forest litter, agricultural soil, compost).

Relatively high activities of laccase – compared to agricultural or meadow soils – can be detected in forest litter and soils where laccase is always the dominant ligninolytic enzyme (Rosenbrock *et al.*, 1995; Criquet *et al.*, 2000; Carreiro *et al.*, 2000; Ghosh *et al.*, 2003). Laccase activity was found to increase during leaf litter degradation in mediterranean broadleaved litter while the pattern of produced isoenzymes varied during decomposition (Fioretto *et al.*, 2000; Di Nardo *et al.*, 2004; Hao *et al.*, 2006). In evergreen oak litter, laccase activity is driven by the changes of litter humidity (Criquet *et al.*, 2000). The association of laccase to fungal fruit bodies can be documented by the increased activity in the soil under fairy rings of *Marasmius oreades* (Gramms *et al.*, 2005). Along with the vertical gradient of fungal biomass distribution in soil profiles, also the laccase activity decreases with increasing depth. The decrease of laccase activity is also reflected in the decrease of laccase gene diversity with soil depth (Luis *et al.*, 2004). From recent studies it seems that laccase activity in soil has a patchy distribution with a high small-scale variability. Moreover, several laccase enzymes are usually responsible for the measured activity and the distribution of mRNA transcripts not always reflects the laccase gene pool in the soil (Luis *et al.*, 2005a; b).

The fact that laccases are dominant ligninolytic enzymes in soils implies their involvement in carbon cycle and their regulation has been studied in this respect. Several studies documented a significant decrease of laccases in forest soils subjected to elevated nitrogen doses (Carreiro *et al.*, 2000; Gallo *et al.*, 2004), with the simultaneous increase in the litter layer (Saiya-Cork *et al.*, 2002). It seems that nitrate deposition directs the flow of carbon through the heterotrophic soil wood web (DeForest *et al.*, 2004a; b). Not surprisingly, the increase of phenolic compounds availability in forest soil after burning increased laccase activity (Boerner and Brinkman, 2003).

It is also generally accepted that laccases are able to react with soil humic substances that can be directly formed from lignin although the character of catalyzed reactions can vary (Yavmetdinov *et al.*, 2003). It was found that humic acids induce laccase activity and mRNA expression (Scheel *et al.*, 2000). The interaction of laccases with humic substances probably leads both to depolymerization of humic substances and their synthesis from monomeric precursors and the balance of these two processes can be influenced by the nature of the humic compounds (Zavarzina *et al.*, 2004). Decolorization and decrease of molecular weight of humic acids along with the production of fulvic acids was observed during the growth of *Trametes versicolor* cultures producing mainly laccase, while humic acid synthesis was also documented *in vitro* using the same enzyme (Katase and Bollag 1991, Fakoussa and Frost 1999). Adsorption of laccases to soil humic substances or inorganic soil constituents changes their temperature and activity profiles and inhibit its activity (Claus and Filp, 1990; Criquet *et al.*, 2000). Inhibition constants for humic acids towards *Panus tigrinus* laccase ranged from 0.003 $\mu g\ ml^{-1}$ to 0.025 $\mu g\ ml^{-1}$ depending on soil type (peat to chernozem) (Zavarzina *et al.*, 2004). Humic substances do not strongly bind laccase and the inactivation is thus due to the dissociation of copper that is chelated by humic substances (Keum and Li, 2004). This brings another difficulty for the determination of laccase activities in soil or forest litter, because extraction methods that do not co-extract proteins (enzymes) with humic acids are not available (Criquet *et al.*, 1999).

Some low-molecular weight compounds including phenolic compounds that can be oxidized by laccase to stable radicals can act as redox mediators and oxidize other compounds that are individually not substrates of laccase due to its low redox potential. Several compounds involved in the natural degradation of lignin by white-rot fungi may be derived from oxidized lignin units or directly from fungal metabolism and act as mediators (Eggert *et al.*, 1996; Johannes and Majcherczyk, 2000; Camarero *et al.*, 2005). The potential mediating compounds syringaldehyde and acetosyringone belong to the main products of biological degradation of lignin and are thus present in the decayed wood (Kirk and Farrell, 1987). Several phenols, including syringaldehyde and acetosyringone, have been described as laccase mediators for indigo decolorization, transformation of the fungicide cyprodinil and degradation of polycyclic aromatic hydrocarbons (Johannes and Majcherczyk, 2000; Campos *et al.*, 2001; Kang *et al.*, 2002).

Laccases are also actively produced during the composting process. Out of thirty-four isolates of fungi from woody compost, eleven were able to oxidize syringaldazine (Chamuris *et al.*, 2000). Laccase was isolated both from compost-specific fungi and the compost itself (Chefetz *et al.*, 1998a; b) and it seems that it participates in both degradation of lignin and humic acids and humic acid formation during composting (Chefetz *et al.*, 1998a; Kluczek-Turpeinen *et al.*, 2003; 2005). In water-saturated environments, laccase activity is driven by the concentration of oxygen. Laccase activity in peatlands is thus usually low due to low oxygen availability but increases dramatically upon increase in oxygen concentration (Pind *et al.*, 1994; Williams *et al.*, 2000; Freeman *et al.*, 2004). The oxygen-regulated laccase activity plays apparently an important role in carbon cycling in this environment. In water environment, laccase was demonstrated to participate in the degradation of wood as well as humic substances (Claus and Filip, 1998; Hendel and Marxsen, 2000). Its activity is dependent on the succession step of substrate decomposition and it can exhibit a seasonal pattern of activity dependent on the input of its substrate (Artigas *et al.*, 2004).

Another roles of laccase in natural conditions include the interactions of fungi with different microorganisms including soil fungi and bacteria, a process usually accompanied by a strong induction of laccase (Freitag and Morrell, 1992; Savoie *et al.*, 1998; Savoie, 2001; Velazquez-Cedeno *et al.*, 2004) that is probably general for laccase-producing basidiomycetes (Iakovlev and Stenlid, 2000; Baldrian, 2004; Gregorio *et al.*, 2006) but was also demonstrated in *Rhizoctonia solani* challenged with *Pseudomonas* strains producing antifungal compounds (Crowe and Olsson, 2001). Laccase and their products do not have a direct effect on soil bacteria or fungi (Baldrian, 2004) but it is probably involved in the formation of melanins or similar defensive compounds (Eggert *et al.*, 1995; Baldrian, 2003). The role of laccases in the defence against heavy metals was also proposed in spite of the fact that different heavy metals induce the synthesis of melanins (Galhaup and Haltrich, 2001; Baldrian, 2003).

Conclusions and Future Perspectives

The broad substrate specificity of laccases, no requirements of cofactors and other properties contribute to a great potential of these fungal enzymes in biotechnology. Current knowledge about the physico-chemical properties of laccases makes it possible to choose the proper enzyme for the intended application. The current development in laccase catalysis research and the design of mediators along with the research on its heterologous expression opens a wide spectrum of possible applications in the near future. Moreover, laccase can also offer a simple and convenient alternative to supplying peroxidases with H_2O_2, because laccases are available at an economically feasible scale.

Despite many studies, the questions about the roles of laccase in the natural environment are answered only partially. It seems clear that laccases have many different roles depending on the situation and conditions of production. Future research is also needed to completely understand the contribution of laccase to the transformation of lignocellulose including lignin degradation and their contribution formation and/or transformation of humic components in the soil and aquatic environments. The dominance of laccase among the oxidative enzymes in different environments will hopefully also more attention that can help to formulate its role in carbon transformation during the turnover of soil organic matter.

ACKNOWLEDGEMENTS

This work was supported by the Grant Agency of the Czech Academy of Sciences (B600200516), by the Czech Science Foundation (526/05/0168), by the Ministry of Education, Youth and Sports of the Czech Republic (LC06066), and by the Institutional Research Concept No. AV0Z50200510 of the Institute of Microbiology, of the ASCR, V.V.i.

References

Abadulla, E., Tzanov, T., Costa, S., Robra, K.H., Cavaco-Paulo, A. and Gubitz, G.M. (2000). Decolorization and detoxification of textile dyes with a laccase from *Trametes hirsuta*, *Applied and Environmental Microbiology* 66: 3357–3362.

Abdel-Raheem, A.M. and Ali, E.H. (2004). Lignocellulolytic enzyme production by aquatic hyphomycetes species isolated from the Nile's delta region, *Mycopathologia* 157: 277–286.

Abdel-Raheem, A. and Shearer, C.A. (2002). Extracellular enzyme production by freshwater ascomycetes, *Fungal Diversity* 11: 1–19.

Ahn, M.Y., Dec, J., Kim, J.E. and Bollag, J.M. (2002). Treatment of 2,4-Dichlorophenol polluted soil with free and immobilized laccase, *Journal of Environmental Quality* 31: 1509–1515.

Alexandre, G. and Zhulin, I.B. (2000) Laccases are widespread in bacteria. *Trends in Biotechnology* 18: 41–42.

Antorini, M., Herpoel-Gimbert, I., Choinowski, T., Sigoillot, J.C., Asther, M., Winterhalter, K. and Piontek, K. (2002). Purification, crystallisation and X-ray diffraction study of fully functional laccases from two ligninolytic fungi, *Biochimica et Biophysica Acta-Protein Structure and Molecular Enzymology* 1594: 109-114.

Arends, I.W.C.E., Li, Y.X., Ausan, R. and Sheldon, R.A. (2006). Comparison of TEMPO and its derivatives as mediators in laccase catalysed oxidation of alcohols, *Tetrahedron* 62: 6659–6665.

Artigas, J., Romani, A.M. and Sabater, S. (2004). Organic matter decomposition by fungi in a Mediterranean forested stream: contribution of streambed substrata, *Annales de Limnologie-International Journal of Limnology* 40: 269–277.

Assavanig, A., Amornkitticharoen, B., Ekpaisal, N., Meevootisom, V. and Flegel, T.W. (1992). Isolation, Characterization and Function of laccase from *Trichoderma*, *Applied Microbiology and Biotechnology* 38: 198–202.

Baiocco, P., Barreca, A.M., Fabbrini, M., Galli, C. and Gentili, P. (2003). Promoting laccase activity towards non-phenolic substrates: a mechanistic investigation with some laccase-mediator systems, *Organic & Biomolecular Chemistry* 1: 191–197.

Balakshin, M., Capanema, E., Chen, C.L., Gratzl, J., Kirkman, A. and Gracz, H. (2001). Biobleaching of pulp with dioxygen in the laccase-mediator system—reaction mechanisms for degradation of residual lignin, *Journal of Molecular Catalysis B-Enzymatic* 13: 1–16.

Baldrian, P. (2003). Interactions of heavy metals with white-rot fungi, *Enzyme and Microbial Technology* 32: 78–91.

Baldrian, P. (2004). Increase of laccase activity during interspecific interactions of white-rot fungi, *FEMS Microbiology Ecology* 50: 245–253.

Baldrian, P. (2006) Fungal laccases – occurrence and properties, *FEMS Microbiology Reviews* 30: 215–242.

Baldrian, P., in der Wiesche, C., Gabriel, J., Nerud, F. and Zadrazil, F. (2000). Influence of cadmium and mercury on activities of ligninolytic enzymes and degradation of polycyclic aromatic hydrocarbons by *Pleurotus ostreatus* in soil, *Applied and Environmental Microbiology* 66: 2471–2478.

Banerjee, U.C. and Vohra, R.M. (1991). Production of laccase by *Curvularia* sp, *Folia Microbiologica* 36: 343–346.

Baratto, L., Candido, A., Marzorati, M., Sagui, F., Riva, S. and Danieli, B. (2006). Laccase-mediated oxidation of natural glycosides, *Journal of Molecular Catalysis B-Enzymatic* 39: 3–8.

Barriere, F., Kavanagh, P. and Leech, D. (2006). A laccase-glucose oxidase biofuel cell prototype operating in a physiological buffer, *Electrochimica Acta* 51: 5187–5192.

Bertrand, T., Jolivalt, C., Briozzo, P., Caminade, E., Joly, N., Madzak, C. and Mougin, C. (2002). Crystal structure of a four-copper laccase complexed with an arylamine: insights into substrate recognition and correlation with kinetics, *Biochemistry* 41: 7325–7333.

Binz, T. and Canevascini, G. (1997). Purification and partial characterization of the extracellular laccase from *Ophiostoma novo-ulmi*, *Current Microbiology* 35: 278–281.

Blaich, R. and Esser, K. (1975) Function of enzymes in wood destroying fungi. 2. Multiple forms of laccase in white rot fungi, *Archiv für Mikrobiologie* 103: 271–277.

Boerner, R.E.J. and Brinkman, J.A. (2003). Fire frequency and soil enzyme activity in southern Ohio oak-hickory forests, *Applied Soil Ecology* 23: 137–146 .

Bollag, J.M. (1991). Enzymatic Binding of Pesticide Degradation Products to Soil Organic-Matter and their Possible Release, *ACS Symposium Series* 459: 122–132.

Bollag, J.M. and Myers, C. (1992). Detoxification of aquatic and terrestrial sites through binding of pollutants to humic substances, *Science of the Total Environment* 118: 357–366.

Bollag, J.M., Chu, H.L., Rao, M.A. and Gianfreda, L. (2003). Enzymatic oxidative transformation of chlorophenol mixtures, *Journal of Environmental Quality* 32: 63–69.

Boudet, A.M. (2000) Lignins and lignification: selected issues, *Plant Physiology and Biochemistry* 38: 81–96.

Bourbonnais, R. and Paice, M.G. (1990) Oxidation of non-phenolic substrates. An expanded role for laccase in lignin biodegradation, *FEBS Letters* 267: 99–102.

Bourbonnais, R. and Paice, M.G. (1996). Enzymatic delignification of kraft pulp using laccase and a mediator, *Tappi Journal* 79: 199–204.

Bourbonnais, R., Paice, M.G., Freiermuth, B., Bodie, E. and Borneman, S. (1997). Reactivities of various mediators and laccases with kraft pulp and lignin model compounds, *Applied and Environmental Microbiology* 63: 4627–4632.

Bourbonnais, R., Leech, D. and Paice, M.G. (1998). Electrochemical analysis of the interactions of laccase mediators with lignin model compounds, *Biochimica et Biophysica Acta-General Subjects* 1379: 381–390.

Bratkovskaya, I., Ivanec, R. and Kulys, J. (2006). Mediator-assisted laccase-catalyzed oxidation of 4-hydroxybiphenyl, *Biochemistry-Moscow* 71: 550–554.

Call, H.P. and Mucke, I. (1997). History, overview and applications of mediated lignolytic systems, especially laccase-mediator-systems (Lignozym®-process), *Journal of Biotechnology* 53: 163–202.

Camarero, S., Ibarra, D., Martínez, M.J. and Martínez, A.T. (2005) Lignin-derived compounds as efficient laccase mediators for decolorization of different types of recalcitrant dyes, *Applied and Environmental Microbiology* 71: 1775–1784.

Campos, R., Kandelbauer, A., Robra, K.H., Cavaco-Paulo, A. and Gübitz, G.M. (2001). Indigo degradation with purified laccases from *Trametes hirsuta* and *Sclerotium rolfsii*, *Journal of Biotechnology* 89: 131–139.

Carreiro, M.M., Sinsabaugh, R.L., Repert, D.A. and Parkhurst, D.F. (2000). Microbial enzyme shifts explain litter decay responses to simulated nitrogen deposition, *Ecology* 81: 2359–2365.

Chamuris, G.P., Koziol-Kotch, S. and Brouse, T.M. (2000). Screening fungi isolated from woody compost for lignin-degrading potential, *Compost Science & Utilization* 8: 6–11.

Chefetz, B., Chen, Y. and Hadar, Y. (1998a). Purification and characterization of laccase from *Chaetomium thermophilium* and its role in humification, *Applied and Environmental Microbiology* 64: 3175–3179.

Chefetz, B., Kerem, Z., Chen, Y. and Hadar, Y. (1998b) Isolation and partial characterization of laccase from a thermophilic composted municipal solid waste, *Soil Biology and Biochemistry* 30: 1091–1098.

Chiacchierini, E., Restuccia, D. and Vinci, G. (2004). Bioremediation of food industry effluents: Recent applications of free and immobilised polyphenoloxidases, *Food Science and Technology International* 10: 373–382.

Chivukula, M. and Renganathan, V. (1995). Phenolic azo-dye oxidation by laccase from *Pyricularia oryzae*, *Applied and Environmental Microbiology* 61: 4374–4377.

Cho, S.J., Park, S.J., Lim, J.S., Rhee, Y.H. and Shin, K.S. (2002). Oxidation of polycyclic aromatic hydrocarbons by laccase of *Coriolus hirsutus*, *Biotechnology Letters* 24: 1337–1340.

Claus, H. (2003). Laccases and their occurrence in prokaryotes, *Archives of Microbiology* 179: 145–150.

Claus, H. and Filip, Z. (1990). Effects of clays and other solids on the activity of phenoloxidases produced by some fungi and actinomycetes. *Soil Biology and Biochemistry* 22: 483–488.

Claus, H. and Filip, Z. (1998). Degradation and transformation of aquatic humic substances by laccase-producing fungi *Cladosporium cladosporioides* and *Polyporus versicolor*, *Acta Hydrochimica et Hydrobiologica* 26: 180–185.

Claus, H., Faber, G. and Konig, H. (2002). Redox-mediated decolorization of synthetic dyes by fungal laccases, *Applied Microbiology and Biotechnology* 59: 672–678.

Collins, P.J., Kotterman, M.J.J., Field, J.A. and Dobson, A.D.W. (1996). Oxidation of anthracene and benzo[a]pyrene by laccases from *Trametes versicolor*, *Applied and Environmental Microbiology* 62: 4563–4567.

Couto, S.R., Rosales, E. and Sanroman, M.A. (2006). Decolorization of synthetic dyes by *Trametes hirsuta* in expanded-bed reactors, *Chemosphere* 62: 1558-1563.

Couto, S.R. and Sanroman, M.A. (2004). Continuous decolorisation of a leather azo dye by *Trametes hirsuta*, *Afinidad* 61: 460–463.

Criquet, S., Tagger, S., Vogt, G., Iacazio, G. and Le Petit, J. (1999). Laccase activity of forest litter, *Soil Biology and Biochemistry* 31: 1239–1244.

Criquet, S., Farnet, A.M., Tagger, S. and Le Petit, J. (2000). Annual variations of phenoloxidase activities in an evergreen oak litter: influence of certain biotic and abiotic factors, *Soil Biology and Biochemistry* 32: 1505–1513.

Crowe, J.D. and Olsson, S. (2001). Induction of laccase activity in *Rhizoctonia solani* by antagonistic *Pseudomonas fluorescens* strains and a range of chemical treatments, *Applied and Environmental Microbiology* 67: 2088–2094.

D'Annibale, A., Celletti, D., Felici, M., Dimattia, E. and Giovannozzi-Sermanni, G. (1996) Substrate specificity of laccase from *Lentinus edodes*, *Acta Biotechnologica* 16: 257–270.

D'Annibale, A., Stazi, S.R., Vinciguerra, V. and Sermanni, G.G. (2000). Oxirane-immobilized Lentinula edodes laccase: stability and phenolics removal efficiency in olive mill wastewater, *Journal of Biotechnology* 77: 265–273.

Dawel, G., Kästner, M., Michels, J., Poppitz, W., Günther, W. and Fritsche W. (1997). Structure of a laccase-mediated product of coupling of 2,4-diamino-6-nitrotoluene to guaiacol, a model for coupling of 2,4,6-trinitrotoluene metabolites to a humic organic soil matrix, *Applied and Environmental Microbiology* 63: 2560–2565.

de Carvalho, M.E.A., Monteiro, M.C. and Sant'Anna, G.L. (1999). Laccase from *Trametes versicolor*–Stability at temperature and alkaline conditions and ifs effect on biobleaching of hardwood kraft pulp, *Applied Biochemistry and Biotechnology* 77-9: 723–733.

de Souza, C.G.M. and Peralta, R.M. (2003). Purification and characterization of the main laccase produced by the white-rot fungus *Pleurotus pulmonarius* on wheat bran solid state medium, *Journal of Basic Microbiology* 43: 278–286.

Dec, J. and Bollag, J.M. (2000). Phenoloxidase-mediated interactions of phenols and anilines with humic materials, *Journal of Environmental Quality* 29: 665–676.

Dedeyan, B., Klonowska, A., Tagger, S., Tron, T., Iacazio, G., Gil, G. and Le Petit, J. (2000). Biochemical and molecular characterization of a laccase from *Marasmius quercophilus*, *Applied and Environmental Microbiology* 66: 925–929.

DeForest, J.L., Zak, D.R., Pregitzer, K.S. and Burton, A.J. (2004a). Atmospheric nitrate deposition and the microbial degradation of cellobiose and vanillin in a northern hardwood forest, *Soil Biology and Biochemistry* 36: 965–971.

DeForest, J.L., Zak, D.R., Pregitzer, K.S. and Burton, A.J. (2004b). Atmospheric nitrate deposition, microbial community composition, and enzyme activity in northern hardwood forests, *Soil Science Society of America Journal* 68: 132–138.

Di Nardo, C., Cinquegrana, A., Papa, S., Fuggi, A. and Fioretto, A. (2004). Laccase and peroxidase isoenzymes during leaf litter decomposition of *Quercus ilex* in a Mediterranean ecosystem, *Soil Biology and Biochemistry* 36: 1539–1544.

D'Souza, T.M., Boominathan, K. and Reddy, C.A. (1996) Isolation of laccase gene-specific sequences from white rot and brown rot fungi by PCR, *Applied and Environmental Microbiology* 62: 3739–3744.

Ducros, V., Brzozowski, A.M., Wilson, K.S., Brown, S.H., Ostergaard, P., Schneider, P., Yaver, D.S., Pedersen, A.H. and Davies, G.J. (1998). Crystal structure of the type-2 Cu depleted laccase from *Coprinus cinereus* at 2.2 angstrom resolution, *Nature Structural Biology* 5: 310–316.

Duran, N. and Esposito, E. (2000). Potential applications of oxidative enzymes and phenoloxidase-like compounds in wastewater and soil treatment: a review, *Applied Catalysis B-Environmental* 28: 83–99.

Duran, N., Rosa, M.A., D'Annibale, A. and Gianfreda, L. (2002). Applications of laccases and tyrosinases (phenoloxidases) immobilized on different supports: a review, *Enzyme and Microbial Technology* 31: 907–931.

Edens, W.A., Goins, T.Q., Dooley, D. and Henson, J.M. (1999). Purification and characterization of a secreted laccase of *Gaeumannomyces graminis* var. *tritici*, *Applied and Environmental Microbiology* 65: 3071–3074.

Eggen, T. (1999). Application of fungal substrate from commercial mushroom production —Pleuorotus ostreatus—for bioremediation of creosote contaminated soil, *International Biodeterioration & Biodegradation* 44: 117–126.

Eggert, C., Temp, U., Dean, J.F.D. and Eriksson, K.E.L. (1995). Laccase-mediated formation of the phenoxazinone derivative, cinnabarinic acid. *FEBS Letters* 376: 202–206.

Eggert, C., Temp, U. and Eriksson, K.E.L. (1996). The ligninolytic system of the white rot fungus *Pycnoporus cinnabarinus*: Purification and characterization of the laccase, *Applied and Environmental Microbiology* 62: 1151–1158.

Elegir, G., Daina, S., Zoia, L., Bestetti, G. and Orlandi, M. (2005). Laccase mediator system: Oxidation of recalcitrant lignin model structures present in residual kraft lignin, *Enzyme and Microbial Technology* 37: 340–346.

Enguita, F.J., Martins, L.O., Henriques, A.O. and Carrondo, M.A. (2003). Crystal structure of a bacterial endospore coat component - a laccase with enhanced thermostability properties. *Journal of Biological Chemistry* 278: 19416–19425.

Fabbrini, M., Galli, C. and Gentili, P. (2002). Comparing the catalytic efficiency of some mediators of laccase, *Journal of Molecular Catalysis B-Enzymatic* 16: 231–240.

Fakoussa, R.M. and Frost, P.J. (1999). *In vivo*-decolorization of coal-derived humic acids by laccase-excreting fungus *Trametes versicolor*, *Applied Microbiology and Biotechnology* 52: 60–65.

Farnet, A.M., Criquet, S., Tagger, S., Gil, G. and Le Petit, J. (2000). Purification, partial characterization, and reactivity with aromatic compounds of two laccases from *Marasmius quercophilus* strain 17, *Canadian Journal of Microbiology* 46: 189–194.

Farnet, A.M., Criquet, S., Pocachard, E., Gil, G. and Ferre, E. (2002). Purification of a new isoform of laccase from a *Marasmius quercophilus* strain isolated from a cork oak litter (*Quercus suber* L.), *Mycologia* 94: 735–740.

Farnet, A.M., Criquet, S., Cigna, M., Gil, G. and Ferre, E. (2004). Purification of a laccase from *Marasmius quercophilus* induced with ferulic acid: reactivity towards natural and xenobiotic aromatic compounds, *Enzyme and Microbial Technology* 34: 549–554.

Farnet, A.M., Tagger, S. and Le Petit, J. (1999). Effects of copper and aromatic inducers on the laccases of the white-rot fungus *Marasmius quercophilus*. *Comptes Rendus de la Academie des Sciences, Serie III* 322: 499–503.

Fioretto, A., Papa, S., Curcio, E., Sorrentino, G. and Fuggi, A. (2000). Enzyme dynamics on decomposing leaf litter of *Cistus incanus* and *Myrtus communis* in a Mediterranean ecosystem, *Soil Biology and Biochemistry* 32: 1847–1855.

Freeman, C., Ostle, N.J., Fenner, N. and Kang, H. (2004). A regulatory role for phenol oxidase during decomposition in peatlands, *Soil Biology and Biochemistry* 36: 1663–1667.

Freitag, M. and Morrell, J.J. (1992) Changes in selected enzyme activities during growth of pure and mixed cultures of the white-rot decay fungus *Trametes versicolor* and the potential biocontrol fungus *Trichoderma harzianum*, *Canadian Journal of Microbiology* 38: 317–323.

Froehner, S.C., and Eriksson, K.E.L. (1974). Purification and properties of *Neurospora crassa* laccase. *Journal of Bacteriology* 120: 458–465.

Galhaup, C. and Haltrich, D. (2001). Enhanced formation of laccase activity by the white-rot fungus *Trametes pubescens* in the presence of copper, *Applied Microbiology and Biotechnology* 56: 225–232.

Gallo, M., Amonette, R., Lauber, C., Sinsabaugh, R.L. and Zak, D.R. (2004). Microbial community structure and oxidative enzyme activity in nitrogen-amended north temperate forest soils, *Microbial Ecology* 48: 218–229.

Gamelas, J.A.F., Tavares, A.P.M., Evtuguin, D.V. and Xavier, A.M.B. (2005). Oxygen bleaching of kraft pulp with polyoxometalates and laccase applying a novel multi-stage process, *Journal of Molecular Catalysis B-Enzymatic* 33: 57–64.

Garavaglia, S., Cambria, M.T., Miglio, M., Ragusa, S., Lacobazzi, V., Palmieri, F., D'Ambrosio, C., Scaloni, A. and Rizzi, M. (2004). The structure of *Rigidoporus lignosus* laccase containing a full complement of copper ions, reveals an asymmetrical arrangement for the T3 copper pair, *Journal of Molecular Biology* 342: 1519–1531.

Ghosh, A., Frankland, J.C., Thurston, C.F. and Robinson, C.H. (2003). Enzyme production by *Mycena galopus* mycelium in artificial media and in *Picea sitchensis* F-1 horizon needle litter, *Mycological Research* 107: 996–1008.

Giardina, P., Palmieri, G., Scaloni, A., Fontanella, B., Faraco, V., Cennamo, G. and Sannia, G. (1999). Protein and gene structure of a blue laccase from *Pleurotus ostreatus*, *Biochemical Journal* 341: 655–663.

Gigi, O., Marbach, I. and Mayer, A.M. (1981). Properties of gallic acid-induced extracellular laccase of *Botrytis cinerea*, *Phytochemistry* 20: 1211–1213.

Givaudan, A., Effosse, A., Faure, D., Potier, P., Bouillant, M.L. and Bally, R. (2004). Polyphenol oxidase in *Azospirillum lipoferum* isolated from rice rhizosphere—evidence for laccase activity in nonmotile strains of *Azospirillum lipoferum*, *FEMS Microbiology Letters* 108: 205–210.

Gramss, G., Voigt, K.D. and Bergmann, H. (2005). Factors influencing water solubility and plant availability of mineral compounds in the tripartite fairy rings of *Marasmius oreades* (BOLT.: FR.) FR. *Journal of Basic Microbiology* 45: 41–54.

Gregorio, A.P.F., Da Silva, I.R., Sedarati, M.R. and Hedger, J.N. (2006) Changes in production of lignin degrading enzymes during interactions between mycelia of the tropical decomposer basidiomycetes *Marasmiellus troyanus* and *Marasmius pallescens*, *Mycological Research* 110: 161–168.

Gronqvist, S., Viikari, L., Niku-Paavola, M.L., Orlandi, M., Canevali, C. and Buchert, J. (2005). Oxidation of milled wood lignin with laccase, tyrosinase and horseradish peroxidase, *Applied Microbiology and Biotechnology* 67: 489–494.

Günther, H., Perner, B. and Gramss, G. (1998) Activities of phenol oxidizing enzymes of ectomycorrhizal fungi in axenic culture and in symbiosis with Scots pine (*Pinus sylvestris* L.), *Journal of Basic Microbiology* 38: 197–206.

Gutierrez, A., del Rio, J.C., Ibarra, D., Rencoret, J., Romero, J., Speranza, M., Camarero, S., Martínez, M.J. and Martínez, A.T. (2006). Enzymatic removal of free and conjugated sterols forming pitch deposits in environmentally sound bleaching of eucalypt paper pulp, *Environmental Science & Technology* 40: 3416–3422.

Hakulinen, N., Kiiskinen, L.L., Kruus, K., Saloheimo, M., Paananen, A., Koivula, A. and Rouvinen, J. (2002). Crystal structure of a laccase from *Melanocarpus albomyces* with an intact trinuclear copper site, *Nature Structural Biology* 9: 601–605.

Hao, J.J., Tian, X.J., Song, F.Q., He, X.B., Zhang, Z.J. and Zhang, P. (2006) Involvement of lignocellulolytic enzymes in the decomposition of leaf litter in a subtropical forest, *Journal of Eukaryotic Microbiology* 53:193–198.

Harkin, J.M., Larsen, M.J. and Obst, J.R. (1974). Use of syringaldazine for detection of laccase in sporophores of wood rotting fungi, *Mycologia* 66: 469–476.

Hendel, B. and Marxsen, J. (2000). Extracellular enzyme activity associated with degradation of beech wood in a Central European stream. *International Review of Hydrobiology* 85: 95–105.

Hoegger, P.J., Kilaru, S., James, T.Y., Thacker, J.R. and Kües, U. (2006) Phylogenetic comparison and classification of laccase and related multicopper oxidase protein sequences, *FEBS Journal* 273: 2308–2326.

Höfer, C. and Schlosser, D. (1999) Novel enzymatic oxidation of Mn^{2+} to Mn^{3+} catalyzed by a fungal laccase, *FEBS Letters* 451: 186–190.

Holker, U., Dohs,e J. and Hofer, M. (2002). Extracellular laccases in ascomycetes *Trichoderma atroviride* and *Trichoderma harzianum*, *Folia Microbiologica* 47: 423–427.

Hoopes, J.T. and Dean, J.F.D. (2004). Ferroxidase activity in a laccase-like multicopper oxidase from *Liriodendron tulipifera*, *Plant Physiology and Biochemistry* 42: 27–33.

Iakovlev, A. and Stenlid, J. (2000). Spatiotemporal patterns of laccase activity in interacting mycelia of wood-decaying basidiomycete fungi, *Microbial Ecology* 39: 236–245.

Ibarra, D., Camarero, S., Romero, J., Martínez, M.J. and Martínez, A.T. (2006). Integrating laccase-mediator treatment into an industrial-type sequence for totally chlorine-free bleaching of eucalypt kraft pulp, *Journal of Chemical Technology and Biotechnology* 81: 1159–1165.

Ikeda, R., Sugita, T., Jacobson, E.S. and Shinoda, T. (2003). Effects of melanin upon susceptibility of *Cryptococcus* to antifungals, *Microbiology and Immunology* 47: 271–277.

Iyer, G. and Chattoo, B.B. (2003). Purification and characterization of laccase from the rice blast fungus, *Magnaporthe grisea*, *FEMS Microbiology Letters* 227: 121–126.

Jimenez, M., Gonzalez, A.E., Martínez, M.J., Martínez, A.T. and Dále, B.E. (1991) Screening of yeasts isolated from decayed wood for lignocellulose degrading enzyme activities, *Mycological Research* 95: 1299–1302.

Johannes, C., Majcherczyk, A. and Hüttermann, A. (1996). Degradation of anthracene by laccase of *Trametes versicolor* in the presence of different mediator compounds, *Applied Microbiology and Biotechnology* 46: 313–317.

Johannes, C. and Majcherczyk, A. (2000). Natural mediators in the oxidation of polycyclic aromatic hydrocarbons by laccase mediator systems, *Applied and Environmental Microbiology* 66: 524–528.

Johannes, C., Majcherczyk, A. and Hüttermann, A. (1998). Oxidation of acenaphthene and acenaphthylene by laccase of *Trametes versicolor* in a laccase-mediator system. *Journal of Biotechnology* 61: 151–156.

Jolivalt, C., Brenon, S., Caminade, E., Mougin, C. and Pontie, M. (2000). Immobilization of laccase from *Trametes versicolor* on a modified PVDF microfiltration membrane: characterization of the grafted support and application in removing a phenylurea pesticide in wastewater, *Journal of Membrane Science* 180: 103–113.

Jolivalt, C., Neuville, L., Boyer, F.D., Kerhoas, L. and Mougin, C. (2006). Identification and formation pathway of laccase-mediated oxidation products formed from hydroxyphenylureas, *Journal of Agricultural and Food Chemistry* 54: 5046–5054.

Junghanns, C., Möder, M., Krauss, G., Martin, C. and Schlosser, D. (2005). Degradation of the xenoestrogen nonylphenol by aquatic fungi and their laccases, *Microbiology* 151: 45–57.

Kang, K.H., Dec, J., Park, H. and Bollag, J.M. (2002). Transformation of the fungicide cyprodinil by a laccase of *Trametes villosa* in the presence of phenolic mediators and humic acid, *Water Research* 36: 4907–4915.

Katase, T. and Bollag, J.M. (1991). Transformation of trans-4-hydroxycinnamic acid by a laccase of the fungus *Trametes versicolor*—Its significance in humification, *Soil Science* 151: 291–296.

Keum, Y.S. and Li, Q.X. (2004). Copper dissociation as a mechanism of fungal laccase denaturation by humic acid, *Applied Microbiology and Biotechnology* 64: 588–592.

Kiiskinen, L.L., Viikari, L. and Kruus, K. (2002). Purification and characterisation of a novel laccase from the ascomycete *Melanocarpus albomyces*, *Applied Microbiology and Biotechnology* 59: 198–204.

Kim, Y.J. and Nicell, J.A. (2006a). Laccase-catalysed oxidation of aqueous triclosan, *Journal of Chemical Technology and Biotechnology* 81: 1344–1352.

Kim, Y.J. and Nicell, J.A. (2006b). Impact of reaction conditions on the laccase-catalyzed conversion of bisphenol A, *Bioresource Technology* 97: 1431–1442.

Kim, Y.J. and Nicell, J.A. (2006c). Laccase-catalyzed oxidation of bisphenol A with the aid of additives, *Process Biochemistry* 41: 1029–1037.

Kirk, T.K. and Farrell, R.L. (1987). Enzymatic "combustion": the microbial degradation of lignin, *Annual Reviews of Microbiology* 41: 465–505.

Kluczek-Turpeinen, B., Tuomela, M., Hatakka, A. and Hofrichter, M. (2003). Lignin degradation in a compost environment by the deuteromycete *Paecilomyces inflatus*, *Applied Microbiology and Biotechnology* 61: 374–379.

Kluczek-Turpeinen, B., Steffen, K.T. , Tuomela, M., Hatakka, A. and Hofrichter, M. (2005). Modification of humic acids by the compost-dwelling deuteromycete *Paecilomyces inflatus*, *Applied Microbiology and Biotechnology* 66: 443–449.

Kottermann, M.J.J., Rietberg, H.J., Hage, A. and Field, J.A. (1998). Polycyclic aromatic hydrocarbon oxidation by the white rot fungus *Bjerkandera* sp. strain BOS55 in the presence of nonionic surfactants, *Biotechnology and Bioengineering* 57: 220–227.

Lang, E., Eller, G. and Zadrazil, F. (1997). Lignocellulose decomposition and production of ligninolytic enzymes during interaction of white-rot fungi with soil microorganisms, *Microbial Ecology* 34: 1–10.

Lang, E., Nerud, F. and Zadrazil, F. (1998). Production of ligninolytic enzymes by *Pleurotus* sp. and *Dichomitus squalens* in soil and lignocellulose substrate as influenced by soil microorganisms, *FEMS Microbiology Letters* 167: 239–244.

Lau, K.L., Tsang, Y.Y. and Chiu, S.W. (2003). Use of spent mushroom compost to bioremediate PAH-contaminated samples, *Chemosphere* 52: 1539–1546.

Laufer, Z., Beckett, R.P., Minibayeva, F.V., Lüthje, S., Böttger, M. (2006) Occurrence of laccases in lichenized ascomycetes of the *Peltigerineae*, *Mycological Research* 110: 846–853.

Law, W.M., Lau, W.N., Lo, K.L., Wai, L.M. and Chiu, S.W. (2003). Removal of biocide pentachlorophenol in water system by the spent mushroom compost of *Pleurotus pulmonarius*, *Chemosphere* 52: 1531–1537.

Lee, K.H., Wi, S.G., Singh, A.P. and Kim, Y.S. (2004). Micromorphological characteristics of decayed wood and laccase produced by the brown-rot fungus *Coniophora puteana*, *Journal of Wood Science* 50: 281–284.

Leonowicz, A., Cho, N.S., Luterek, J., Wilkolazka, A., Wojtas-Wasilewska, M., Matuszewska, A., Hofrichter, M., Wesenberg, D. and Rogalski, J. (2001). Fungal laccase: properties and activity on lignin, *Journal of Basic Microbiology* 41: 185–227.

Leontievsky, A., Myasoedova, N., Pozdnyakova, N. and Golovleva, L. (1997a). 'Yellow' laccase of *Panus tigrinus* oxidizes non-phenolic substrates without electron-transfer mediators, *FEBS Letters* 413: 446–448.

Leontievsky, A.A., Vares, T., Lankinen, P., Shergill, J.K., Pozdnyakova, N.N., Myasoedova, N.M., Kalkkinen, N., Golovleva, L.A., Cammack, R., Thurston, C.F. and

Hatakka, A. (1997b). Blue and yellow laccases of ligninolytic fungi, *FEMS Microbiology Letters* 156: 9–14.

Liers, C., Ullrich, R., Steffen, K.T., Hatakka, A. and Hofrichter, M. (2006) Mineralization of ^{14}C-labelled lignin and extracellular enzyme activities of the wood-colonizing ascomycetes *Xylaria hypoxylon* and *Xylaria polymorpha*, *Applied Microbiology and Biotechnology* 69: 573–579.

Liu, L., Dean, J.F.D., Friedman, W.E. and Eriksson, K.E.L. (1994). A laccase-like phenoloxidase is correlated with lignin biosynthesis in *Zinnia elegans* stem tissues, *Plant Journal* 6: 213–224.

Luis, P., Kellner, H., Zimdars, B., Langer, U., Martin, F. and Buscot, F. (2005a). Patchiness and spatial distribution of laccase genes of ectomycorrhizal, saprotrophic, and unknown basidiomycetes in the upper horizons of a mixed forest cambisol, *Microbial Ecology* 50: 570–579.

Luis, P., Kellner, H., Martin, F. and Buscot, F. (2005b). A molecular method to evaluate basidiomycete laccase gene expression in forest soils, *Geoderma* 128: 18-27.

Luis, P., Walther, G., Kellner, H., Martin, F. and Buscot, F. (2004). Diversity of laccase genes from basidiomycetes in a forest soil, *Soil Biology and Biochemistry* 36: 1025–1036.

Lund, M., Eriksson, M. and Felby, C. (2003). Reactivity of a fungal laccase towards lignin in softwood kraft pulp, *Holzforschung* 57: 21–26.

Lyons, J.I., Newell, S.Y., Buchan, A. and Moran, M.A. (2003). Diversity of ascomycete laccase gene sequences in a southeastern US salt marsh, *Microbial Ecology* 45: 270–281.

Machonkin, T.E., Quintanar, L., Palmer, A.E., Hassett, R., Severance, S., Kosman, D.J. and Solomon, E.I. (2001). Spectroscopy and reactivity of the type 1 copper site in Fet3p from *Saccharomyces cerevisiae*: Correlation of structure with reactivity in the multicopper oxidases, *Journal of the American Chemical Society* 123: 5507–5517.

Majcherczyk, A. and Johannes, C. (2000). Radical mediated indirect oxidation of a PEG-coupled polycyclic aromatic hydrocarbon (PAH) model compound by fungal laccase, *Biochimica et Biophysica Acta-General Subjects* 1474: 157–162.

Majcherczyk, A., Johannes, C. and Hüttermann, A. (1998). Oxidation of polycyclic aromatic hydrocarbons (PAH) by laccase of *Trametes versicolor*, *Enzyme and Microbial Technology* 22: 335–341.

Manda, K., Hammer, E., Mikolasch, A., Niedermeyer, T., Dec, J., Jones, A.D., Benesi, A.J., Schauer, F. and Bollag, J.M. (2005). Laccase-induced cross-coupling of 4-aminobenzoic acid with para-dihydroxylated compounds 2,5-dihydroxy-N-(2-hydroxyethyl)-benzamide and 2,5-dihydroxybenzoic acid methyl ester, *Journal of Molecular Catalysis B-Enzymatic* 35: 86–92.

Marbach, I., Harel, E. and Mayer, A.M. (1984). Molecular properties of extracellular *Botrytis cinerea* laccase, *Phytochemistry* 23: 2713–2717.

Martins, L.O., Soares, C.M., Pereira, M.M., Teixeira, M., Costa, T., Jones, G.H. and Henriques, A.O. (2002). Molecular and biochemical characterization of a highly stable bacterial laccase that occurs as a structural component of the *Bacillus subtilis* endospore coat, *Journal of Biological Chemistry* 277: 18849–18859.

Martinez, D., Larrondo, L.F., Putnam, N., Gelpke, M.D.S., Huang, K., Chapman, J., Helfenbein, K.G., Ramaiya, P., Detter, J.C., Larimer, F., Coutinho, P.M., Henrissat, N.,

Berka, R., Cullen, D. and Rokhsar, D. (2004). Genome sequence of the lignocellulose degrading fungus *Phanerochaete chrysosporium* strain RP78, *Nature Biotechnology* 22: 695–700.

Maruyama, T., Komatsu, C., Michizoe, J., Ichinose, H. and Goto, M. (2006). Laccase-mediated oxidative degradation of the herbicide dymron, *Biotechnology Progress* 22: 426–430.

Mayer, A.M. and Staples, R.C. (2002). Laccase: new functions for an old enzyme, *Phytochemistry* 60: 551–565.

Messerschmidt, A. (1997). Multi-copper Oxidases, World Scientific, Singapore.

Min, K.L., Kim, Y.H., Kim, Y.W., Jung, H.S. and Hah, Y.C. (2001). Characterization of a novel laccase produced by the wood-rotting fungus *Phellinus ribis*, *Archives of Biochemistry and Biophysics* 392: 279–286.

Modaressi, K., Taylor, K.E., Bewtra, J.K. and Biswas, N. (2005). Laccase-catalyzed removal of bisphenol-A from water: Protective effect of PEG on enzyme activity, *Water Research* 39: 4309–4316.

Molitoris, H.P. and Esser, K. (1970). The phenoloxidases of the ascomycete *Podospora anserina*. V. Properties of laccase I after further purification, *Archiv für Mikrobiologie* 72: 267–296.

Monteiro, M.C. and de Carvalho, M.E.A. (1998). Pulp bleaching using laccase from *Trametes versicolor* under high temperature and alkaline conditions, *Applied Biochemistry and Biotechnology* 70-2: 983–993.

Nagai, M., Sato, T., Watanabe, H., Saito, K., Kawata, M. and Enei, H. (2002). Purification and characterization of an extracellular laccase from the edible mushroom *Lentinula edodes*, and decolorization of chemically different dyes, *Applied Microbiology and Biotechnology* 60: 327–335.

Nannipieri, P., Gelsomino, A. and Felici, M. (1991). Method to determine guaiacol oxidase activity in soil, *Soil Science Society of America Journal* 55: 1347-1352.

Nyanhongo, G.S., Couto, S.R. and Gübitz, G.M. (2006). Coupling of 2,4,6-trinitrotoluene (TNT) metabolites onto humic monomers by anew laccase from *Trametes modesta*, *Chemosphere* 64: 359–370.

Paice, M.G., Bourbonnais, R., Reid, I.D., Archibald, F.S. and Jurasek, L. (1995). Oxidative bleaching enzymes—a review, *Journal of Pulp and Paper Science* 21: J280–J284.

Palmieri, G., Cennamo, G., Faraco, V., Amoresano, A., Sannia, G. and Giardina, P. (2003). Atypical laccase isoenzymes from copper supplemented *Pleurotus ostreatus* cultures, *Enzyme and Microbial Technology* 33: 220–230.

Palmieri, G., Giardina, P., Bianco, C., Scaloni, A., Capasso, A. and Sannia, G. (1997). A novel white laccase from *Pleurotus ostreatus*. *Journal of Biological Chemistry* 272, 31301–31307.

Palmieri, G., Giardina, P., Marzullo, L., Desiderio, B., Nitti, G., Cannio, R. and Sannia, G. (1993). Stability and activity of a phenol oxidase from the ligninolytic fungus *Pleurotus ostreatus*, *Applied Microbiology and Biotechnology* 39: 632–636.

Palmieri, G., Giardina, P. and Sannia, G. (2005). Laccase-mediated Remazol Brilliant Blue R decolorization in a fixed-bed bioreactor, *Biotechnology Progress* 21: 1436–1441.

Peralta-Zamora, P., Pereira, C.M., Tiburtius, E.R.L., Moraes, S.G., Rosa, M.A., Minussi, R.C. and Durán, N. (2003). Decolorization of reactive dyes by immobilized laccase, *Applied Catalysis B-Environmental* 42: 131–144.

Petter, R., Kang, B.S., Boekhout, T., Davis, B.J. and Kwon-Chung, K.J. (2001). A survey of heterobasidiomycetous yeasts for the presence of the genes homologous to virulence factors of *Filobasidiella neoformans*, CNLAC1 and CAP59, *Microbiology* 147: 2029–2036.

Pind, A., Freeman, C. and Lock, M.A. (1994). Enzymatic degradation of phenolic materials in peatlands—measurement of phenol oxidase activity, *Plant and Soil* 159: 227–231.

Piontek, K., Antorini, M. and Choinowski, T. (2002). Crystal structure of a laccase from the fungus *Trametes versicolor* at 1.90-angstrom resolution containing a full complement of coppers, *Journal of Biological Chemistry* 277: 37663–37669.

Pointing, S.B., Pelling, A.L., Smith, G.J.D., Hyde, K.D. and Reddy, C.A. (2005) Screening of basidiomycetes and xylariaceous fungi for lignin peroxidase and laccase gene-specific sequences, *Mycological Research* 109: 115–124.

Pozdnyakova, N.N., Rodakiewicz-Nowak, J. and Turkovskaya, O.V. (2004). Catalytic properties of yellow laccase from *Pleurotus ostreatus* D1, *Journal of Molecular Catalysis B-Enzymatic* 30: 19–24.

Pozdnyakova, N.N., Rodakiewicz-Nowak, J., Turkovskaya, O.V. and Haber, J. (2006). Oxidative degradation of polyaromatic hydrocarbons and their derivatives catalyzed directly by the yellow laccase from *Pleurotus ostreatus* D1, *Journal of Molecular Catalysis B-Enzymatic* 41: 8–15.

Ranocha, P., Chabannes, M., Chamayou, S., Danoun, S., Jauneau, A., Boudet, A.M. and Goffner, D. (2002). Laccase down-regulation causes alterations in phenolic metabolism and cell wall structure in poplar, *Plant Physiology* 129: 145–155.

Reid, I.D. and Paice, M.G. (1994). Biological bleaching of kraft pulps by white-rot fungi and their enzymes, *FEMS Microbiology Reviews* 13: 369–376.

Reinhammar, B., Malmstrom, B.G. (1981) Blue copper-containing oxidases. In: Copper proteins. Metal ions in biology, vol. 3 (Editor: Spiro, T.G.) pp. 109–149, Wiley, New York, USA.

Riva, S. (2006) Laccases: blue enzymes for green chemistry, *Trends in Biotechnology* 24: 219–226.

Rodriguez, A., Falcon, M.A., Carnicero, A., Perestelo, F., delaFuente, G. and Trojanowski, J. (1996). Laccase activities of *Penicillium chrysogenum* in relation to lignin degradation, *Applied Microbiology and Biotechnology* 45: 399–403.

Rodriguez, E., Pickard, M.A. and Vazquez-Duhalt, R. (1999). Industrial dye decolorization by laccases from ligninolytic fungi, *Current Microbiology* 38: 27–32.

Rodriguez-Solar, D., Sieiro, J., Alcantara, T., Longo, M.A. and Sanromán, M.A. (2005). Degradation of anthracene by laccase of *Coriolopsis rigida*, *Journal of Biotechnology* 118: S176–S176.

Roper, J.C., Sarkar, J.M., Dec, J. and Bollag, J.M. (1995). Enhanced enzymatic removal of chlorophenols in the presence of co-substrates, *Water Research* 29 : 2720–2724.

Rosconi, F., Fraguas, L.F., Martínez-Drets, G. and Castro-Sowinski, S. (2005). Purification and characterization of a periplasmic laccase produced by *Sinorhizobium meliloti*, *Enzyme and Microbial Technology* 36: 800–807.

Rosenbrock, P., Buscot, F. and Munch, J.C. (1995). Fungal succession and changes in the fungal degradation potential during the initial stage of litter decomposition in a black alder Forest [*Alnus glutinosa* (L) Gaertn], *European Journal of Soil Biology* 31: 1–11.

Royarcand, L. and Archibald, F.S. (1991). Direct dechlorination of chlorophenolic compounds by laccases from *Trametes (Coriolus) versicolor*, *Enzyme and Microbial Technology* 13: 194–203.

Saiya-Cork, K.R., Sinsabaugh, R.L. and Zak, D.R. (2002). The effects of long term nitrogen deposition on extracellular enzyme activity in an *Acer saccharum* forest soil, *Soil Biology and Biochemistry* 34: 1309–1315.

Sannia, G., Giardina, P., Luna, M., Rossi, M. and Buonocore, V. (1986). Laccase from *Pleurotus ostreatus*, *Biotechnology Letters* 8: 797–800.

Savoie, J.M. (2001). Variability in brown line formation and extracellular laccase production during interaction between white-rot basidiomycetes and *Trichoderma harzianum* biotype Th2, *Mycologia* 93: 243–248.

Savoie, J.M., Mata, G. and Billette, C. (1998). Extracellular laccase production during hyphal interactions between *Trichoderma* sp and Shiitake, *Lentinula edodes*, *Applied Microbiology and Biotechnology* 49: 589–593.

Scheel, T., Höfer, M., Ludwig, S. and Holker, U. (2000). Differential expression of manganese peroxidase and laccase in white-rot fungi in the presence of manganese or aromatic compounds, *Applied Microbiology and Biotechnology* 54: 686–691.

Scherer, M. and Fischer, R. (1998). Purification and characterization of laccase II of *Aspergillus nidulans*, *Archives of Microbiology* 170: 78–84.

Schlosser, D., Grey, R. and Fritsche, W. (1997). Patterns of ligninolytic enzymes in *Trametes versicolor*. Distribution of extra- and intracellular enzyme activities during cultivation on glucose, wheat straw and beech wood, *Applied Microbiology and Biotechnology* 47: 412–418.

Schlosser, D. and Höfer, C. (2002). Laccase-catalyzed oxidation of Mn^{2+} in the presence of natural Mn^{3+} chelators as a novel source of extracellular H_2O_2 production and its impact on manganese peroxidase, *Applied and Environmental Microbiology* 68: 3514–3521.

Schultz, A., Jonas, U., Hammer, E. and Schauer, F. (2001). Dehalogenation of chlorinated hydroxybiphenyls by fungal laccase, *Applied and Environmental Microbiology* 67: 4377–4381.

Sigoillot, C., Record, E., Belle, V., Robert, J.L., Levasseur, A., Punt, P.J., van den Hondel, C.A.M.J., Fournel, A., Sigoillot, J.C. and Asther, M. (2004). Natural and recombinant fungal laccases for paper pulp bleaching, *Applied Microbiology and Biotechnology* 64: 346–352.

Slomczynski, D., Nakas, J.P. and Tanenbaum, S.W. (1995). Production and characterization of laccase from *Botrytis cinerea* 61-34, *Applied and Environmental Microbiology* 61: 907–912.

Smirnov, S.A., Koroleva, O.V., Gavrilova, V.P., Belova, A.B. and Klyachko, N.L. (2001). Laccases from basidiomycetes: Physicochemical characteristics and substrate specificity towards methoxyphenolic compounds, *Biochemistry-Moscow* 66: 774–779.

Soares, G.M.B., Amorim, M.T.P., Hrdina, R. and Costa-Ferreira, M. (2002). Studies on the biotransformation of novel disazo dyes by laccase, *Process Biochemistry* 37: 581–587.

Solomon, E.I., Sundaram, U.M. and Machonkin, T.E. (1996). Multicopper oxidases and oxygenases, *Chemical Reviews* 96: 2563–2605.

Solomon, E.I., Chen, P., Metz, M., Lee, S.K. and Palmer, A.E. (2001). Oxygen binding, activation, and reduction to water by copper proteins, *Angewandte Chemie-International Edition* 40: 4570–4590.

Sterjiades, R., Dean, J.F.D. and Eriksson, K.E.L. (1992). Laccase from sycamore maple (*Acer pseudoplatanus*) polymerizes monolignols, *Plant Physiology* 99: 1162–1168.

Stoj, C. and Kosman, D.J. (2003). Cuprous oxidase activity of yeast Fet3p and human ceruloplasmin: implication for function, *FEBS Letters* 554: 422–426.

Svobodová, K. (2005). The implications of ligninolytic enzymes in the decolorization of synthetic dyes by the white-rot fungus *Irpex lacteus*, PhD Thesis, Charles University, Prague, Czech Republic.

Tatsumi, K., Freyer, A., Minard, R.D. and Bollag, J.M. (1994). Enzyme-mediated coupling of 3,4-dichloroaniline and ferulic acid—a model for pollutant binding to humic materials, *Environmental Science & Technology* 28: 210–215.

Thurston, C.F. (1994). The structure and function of fungal laccases, *Microbiology* 140: 19–26.

Timonen, S. and Sen, R. (1998). Heterogeneity of fungal and plant enzyme expression in intact Scots pine *Suillus bovinus* and *Paxillus involutus* mycorrhizospheres developed in natural forest humus, New Phytologist 138: 355–366.

Tsioulpas, A., Dimou, D., Iconomou, D. and Aggelis, G. (2002). Phenolic removal in olive oil mill wastewater by strains of *Pleurotus* spp. in respect to their phenol oxidase (laccase) activity, *Bioresource Technology* 84: 251–257.

Ullah, M.A., Bedford, C.T. and Evans, C.S. (2000). Reactions of pentachlorophenol with laccase from *Coriolus versicolor*, *Applied Microbiology and Biotechnology* 53: 230–234.

Valášková, V. and Baldrian, P. (2006). Estimation of bound and free fractions of lignocellulose-degrading enzymes of wood-rotting fungi *Pleurotus ostreatus*, *Trametes versicolor* and *Piptoporus betulinus*, *Research in Microbiology* 157: 119–124.

Vasconcelos, A.F.D., Barbosa, A.M., Dekker, R.F.H., Scarminio, I.S. and Rezende, M.I. (2000). Optimization of laccase production by *Botryosphaeria* sp in the presence of veratryl alcohol by the response-surface method, *Process Biochemistry* 35: 1131–1138.

Velazquez-Cedeno, M.A., Mata, G. and Savoie, J.M. (2002). Waste-reducing cultivation of *Pleurotus ostreatus* and *Pleurotus pulmonarius* on coffee pulp: changes in the production of some lignocellulolytic enzymes, *World Journal of Microbiology & Biotechnology* 18: 201–207.

Velazquez-Cedeno, M.A., Farnet, A.M. , Ferre, E. and Savoie, J.M. (2004). Variations of lignocellulosic activities in dual cultures of *Pleurotus ostreatus* and *Trichoderma longibrachiatum* on unsterilized wheat straw, *Mycologia* 96: 712–719.

Wells, A., Teria, M. and Eve, T. (2006). Green oxidations with laccase-mediator systems, *Biochemical Society Transactions* 34: 304–308.

Wesenberg, D., Kyriakides, I. and Agathos, S.N. (2003). White-rot fungi and their enzymes for the treatment of industrial dye effluents, *Biotechnology Advances* 22: 161–187.

Williams, C.J., Shingara, E.A. and Yavitt, J.B. (2000). Phenol oxidase activity in peatlands in New York State: Response to summer drought and peat type, *Wetlands* 20: 416–421.

Wong, Y.X. and Yu, J. (1999). Laccase-catalyzed decolorization of synthetic dyes, *Water Research* 33: 3512–3520.

Wood, D.A. (1980). Production, purification and properties of extracellular laccase of *Agaricus bisporus*, *Journal of General Microbiology* 117: 327–338.

Xu, F. (1997). Effects of redox potential and hydroxide inhibition on the pH activity profile of fungal laccases, *Journal of Biological Chemistry* 272: 924–928.

Xu F., Kulys J.J., Duke K., Li K.C., Krikstopaitis K., Deussen H.J.W., Abbate E., Galinyte V. and Schneider P. (2000). Redox chemistry in laccase-catalyzed oxidation of N-hydroxy compounds, *Applied and Environmental Microbiology* 66: 2052–2056.

Xu, F., Shin, W.S., Brown, S.H., Wahleithner, J.A., Sundaram, U.M. and Solomon, E.I. (1996). A study of a series of recombinant fungal laccases and bilirubin oxidase that exhibit significant differences in redox potential, substrate specificity, and stability, *Biochimica et Biophysica Acta-Protein Structure and Molecular Enzymology* 1292: 303–311.

Yavmetdinov, I.S., Stepanova, E.V., Gavrilova, V.P., Lokshin, B.V., Perminova, I.V. and Koroleva, O.V. (2003). Isolation and characterization of humin-like substances produced by wood-degrading white-rot fungi, *Applied Biochemistry and Microbiology* 39: 257–264.

Yoshitake, A., Katayama, Y., Nakamura, M., Iimura, Y., Kawai, S. and Morohoshi, N. (1993). N-Linked carbohydrate chains protect laccase-III from proteolysis in *Coriolus versicolor*. *Journal of General Microbiology* 139: 179–185.

Zavarzina, A.G., Leontievsky, A.A., Golovleva, L.A. and Trofimov, S.Y. (2004). Biotransformation of soil humic acids by blue laccase of *Panus tigrinus* 8/18: an in vitro study. *Soil Biology and Biochemistry* 36, 359–369.

Zhang, M., Wu, F., Wei, Z.Y., Xiao, Y.Z. and Gong, W.M. (2006). Characterization and decolorization ability of a laccase from *Panus rudis*, *Enzyme and Microbial Technology* 39: 92–97.

Zhu, X.D., Gibbons, J., Garcia-Rivera, J., Casadevall, A. and Williamson, P.R. (2001). Laccase of *Cryptococcus neoformans* is a cell wall-associated virulence factor, *Infection and Immunity* 69: 5589–5596.

Zille, A., Gornacka, B., Rehorek, A. and Cavaco-Paulo, A. (2005). Degradation of azo dyes by *Trametes villosa* laccase over long periods of oxidative conditions, *Applied and Environmental Microbiology* 71: 6711–6718.

Zouari, N., Romette, J.L. and Thomas, D. (1987). Purification and properties of 2 laccase isoenzymes produced by *Botrytis cinerea*, *Applied Biochemistry and Biotechnology* 15: 213–225.

6

Enzymes from Mushrooms and their Industrial Applications

Deepak K. Rahi[1], Sonu Rahi[2], A.K. Pandey[3] and R.C. Rajak[4]

[1]Department of Microbiology, Panjab University, Chandigarh
deepakrahi10@rediffmail.com / dkrio@pu.ac.in
[2]Department of Botany & Microbiology, M. H. College of Home Science & Science for Women, Jabalpur
[3]Department of Bioscience, R.D. University, Jabalpur
[4]Seedling Acadamy of Science, Design Technology & Management, Jaipur

INTRODUCTION

Fungi thrive in harsh environments shunned by higher organisms. Fungi can exploit marginal living conditions in large part because they produce unusual enzymes capable of performing chemically difficult reactions. These fungal enzymes can convert wood, plastic, paints and jet fuel, among other materials, into nutrients. Some of these enzymes have already been harnessed in pulp and paper processing (biopulping and biobleaching), and in the synthesis of fine chemicals (biocatalysis). Using enzymes permits these industrial processes to be performed under milder conditions, using less energy and producing fewer toxic by-products. In addition, many fungal enzymes are capable of breaking down a broad range of complex compounds, making them potentially useful for destruction of persistent pollutants (bioremediation and biodegradation). Thus, one of the important industrial uses of fungi is in the production of enzymes. Mushrooms are important for the commercial production of certain enzymes that are involved in the various useful degradatory activities.The capacities and capabilities of mushroom mycelium to biodegrade the substrate are enormous and multitudinal. On the other hand, the chemical, biological, and biochemical properties of the fruiting bodies are numerous. In this article, a concerted picture of the degradative enzymatic capacities of the mushrooms in relevance with industrial utilization have been made.

WHAT IS MUSHROOM?

Mushrooms are the fruiting bodies produced by basidiomycetous fungi, a classs of fungi whose spores are produced by basidia. Not all fruit bodies are true mushrooms. The word 'mushroom' is usually thought to be derived from the French 'mousseron' (muceron), 'mousse' or 'moss' which comprise a large heterogenous group having various shapes, sizes, colours, all quite different in characters, appearance and edibility and can be defined as the macrofungi with fleshy, subfleshy or sometimes leathery, umbrella-like sporophores, that bear their fertile surfaces either on lamellae or lining the tubes, opening out by means of pores. The lamellate members are commonly called 'mushrooms' (Fig. 6.1) or 'toadstools' depending upon whether they are edible or poisonous and the tube bearing or poroid members as 'boletes'.The function of this visible part of some fungi is to produce and disperse the largest possible number of spores in the shortest possible time. Spores create new individuals after being carried away on the wind and landing in a good place for growth.True mushrooms typically look like umbrellas. They consist of a stalk topped by a flat or cup-shaped cap. Their spores are produced on special cells called basidia, located on the underside of the cap.

Fig. 6.1 A typical mushroom.

Saprophytic Mushrooms

Mushrooms differ from the higher plant by the absence of the green pigment chlorophyll, which enables plants to photosynthesize. By this process green plants are able to obtain their carbohydrates, as mushrooms are unable to do this, they have to obtain their carbohydrates from decomposed animal or plant tissues. Hence, they are found in habitats rich in rotting vegetation such as woodland, grassland, compost heaps, sawdust piles on dung or manure heaps and on burnt ground colonized by moss. Mushrooms obtaining their food from these sources are known as saprophytic mushrooms. The hyphal growth habit of saprophytic fungi is uniquely suited to the penetration and digestion of substrates, in addition many fungi produce enzymes capable of attacking the most stubborn waste materials such as cellulose and lignin. Important saprophytic species belong to genera *Stropharia, Panaeolus, Coprinus, Conocybe, Lentinus, Pholiota*, etc. Most of the gourmet mushrooms are saprophytic, wood-decomposing fungi.The filamentous mycelial network is designed to weave between and through the cell walls of plants. The enzymes and acids they secrete degrade large molecular complexes into simpler compounds. All ecosystems depend upon fungi's ability to decompose organic plant matter soon after it is rendered available. The end result of their activity is the return of carbon, hydrogen, nitrogen, and minerals back into the ecosystem in forms usable to plants, insects, and other organisms. As decomposers, they can be separated into three key groups as described below. Some mushroom species cross over from one category to another depending upon prevailing conditions.

Primary Decomposers: These are the fungi first to capture a twig, a blade of grass, a chip of wood, a log or stump. Primary decomposers are typically fast-growing, sending out ropy strands of mycelium that quickly attach to and decompose plant tissue. Most of the decomposers degrade wood. Hence, the majority of these saprophytes are woodland species, such as Oyster mushrooms (*Pleurotus* species), Shiitake (*Lentinula edodes*), and King Stropharia (*Stropharia rugoso-annulata*). However, each species has developed specific sets of enzymes to break down lignin-cellulose, the structural components of most plant cells. Once the enzymes of one mushroom species have broken down the lignin-cellulose to its fullest potential, other saprophytes utilizing their own repertoire of enzymes to reduce this material even further.

Secondary Decomposers: These mushrooms rely on the previous activity of other fungi to partially break down a substrate to a state wherein they can thrive. Secondary decomposers typically grow from composted material. The actions of other fungi, actinomycetes, bacteria, and yeasts all operate within compost. As plant residue is degraded by these microorganisms, the mass, structure, and composition of the compost is reduced, and proportionately available nitrogen is increased. Heat, carbon dioxide, ammonia, and other gases are emitted as by-products of the composting process. Once these microorganisms (especially actinomycetes) have completed their life cycles, the compost is susceptible to invasion by a selected secondary decomposer. A classic example of a secondary decomposer is the Button Mushroom, *Agaricus brunnescens*, the most commonly cultivated mushroom. Another example is *Stropharia ambigua*, which invades outdoor mushroom beds after wood chips have been first decomposed by a primary saprophyte.

Tertiary Decomposers: The fungi represented by this group are typically soil dwellers. Fungi existing in these reduced substrates are remarkable in that the habitat appears in hospitable for most other mushrooms. A classic example of a tertiary decomposer is *Aleuria aurantia*, the Orange Peel Mushroom. *Panaeolus subbalteatus* is yet another example. Although one can grow it on composted substrates, this mushroom has the reputation of growing prolificially in the discarded compost from button mushroom farms. Other tertiary decomposers include species of *Conocybe, Agrocybe, Pluteus*, and some *Agaricus* species. The floor of a forest is constantly being replenished by new organic matter. Primary, secondary, and tertiary decomposers can all occupy the same location. In the complex environment of the forest floor, and over time, as each habitat is being transformed, successions of mushrooms occur. This model becomes infinitely complex when taking into account the interrelationships of not only the fungi to one another, but also the fungi to other microorganisms (yeasts, bacteria, protozoa), plants, insects, and mammals. Primary and secondary decomposers afford the most opportunities for cultivation.

Mycorrhizal Mushrooms

Some mushrooms enter into a special symbiotic relationship with the roots of certain green plants whereby both partners benefit. This condition is described as a mycorrhizal association (meaning fungus root). 'Myco' means mushroom while 'rhizal' means roots. In the case of many forest trees, the fungus partner, often a gill bearing mushroom or Bolete, forms a sheath of threads around the trees ultimate rootlets, some of the threads running out into the soil and others penetrating between the cells of the root. In this way substances absorbed from the soil by the fungus, after the breakdown of rotting tree or plant material, are diffused through the fungal mycelium and ultimately into the tree root. In effect the sheath of fungal hyphae surrounding the tree and roots greatly increases the area of absorption of food materials in the soil available to the tree and confers on it a great benefit, especially when growing in poor soils where competition for available food supplies is intense. The fungus partner in turn receives its carbohydrates and a constant supply of water from the tree via the roots. This relationship, in which the fungal threads are mostly outside the plant root is referred to as an ectotrophic mycorrhiza. If the threads of the fungal partner are inside the host cells, the condition is described as an endotrophic mycorrhiza. Mycorrhizal mushrooms form a mutually dependent, beneficial relationship with the roots of host plants, ranging from trees to grasses.The collection of filament of cells that grow into the mushroom body is called the mycelium. The mycelia of these mycorrhizal mushrooms can form an exterior sheath covering the roots of plants and are called ectomycorrhizal. When they invade the interior root cells of host plants they are called endomycorrhizal. In either case, both organisms benefit from this association. Plant growth is accelerated. The resident mushroom mycelium increases the plant's absorption of nutrients, nitrogenous compounds, and essential elements (phosphorus, copper, and zinc). By growing beyond the immediate root zone, the mycelium channels and concentrates nutrients from afar. Plants with mycorrhizal fungal partners can also resist diseases far better than those without. Most ecologists now recognize that the environment's health is directly related to the presence, abundance, and variety of mycorrhizal associations.

Parasitic Mushrooms

Mushrooms obtaining their food material directly from living plants or animals are known as parasites for example, bracket fungi *Polyporus squamosus* commonly infects the host through wounds. The mycelium grows into the heart wood where it attacks the timber and eventually kills the tree. The fungus then continues to grow as a saprophyte on the dead tree and produces the fruting bodies on the dead wood. Parasitic mushrooms are the bane of foresters. They do immeasurable damage to the health of resident tree species, but in the process create new habitats for many other organisms of environmental significance.

ENZYMES AND CO-ENZYMES PRODUCED BY MUSHROOMS

Mushrooms display a wide capacity to secrete a number of enzymes of valid properties and distinct functions. They are of different types: (1) secreted by the mycelium to biodegrade the growth substrate; (2) those involved in the transformation of myceloid phase into fructification; (3) those responsible for autolysis; besides, there are also other (4) constitutive/inductive enzymes characteristic of mycelium and fruiting body. The ability of mushroom species to grow and yield over a wide spectrum of plant wastes is realized only by the secretion and function of these enzymes. It is not surprising to state that the great diversity in shape, size, colour, texture, structure, and flavour of the fruiting bodies is an ultimate expression of the physiology and biochemistry of these enzymes (Rahi *et al.*, 2004).The genetic system of each species/strain, to regulate and condition the production and formation of its own type of mushroom fruiting bodies, despite the availability of a large divergent spectrum of chemical substances in the growth substrate, is beset with the complex coordination of sequential production of these enzymes, that also renders completion of the life cycle in nature.The capacity of mushroom mycelium to grow on a wide spectrum of lignocellulosic materials have been attributed to its ability to secrete a range of saccharifying and oxidizing enzymes. The enzymes produced from the metabolism of mushrooms have many industrial applications.

Most of the mushrooms species have been recognized for their property of degradation of the natural lignocellulosic wastes, in their original form or preformed (composted) form. In general, the plant residues like cereal straws contain about 35% cellulose, 28% hemicellulose, and 25% lignin. These values vary with the plant material, and maturity. The lignin content is invariably high in the wood waste materials. In simple terms, the fungi having preference to utilize cellulose and hemicellulose leave behind a brown residue of lignin and such species are called "brown-rots" while the other group degrade strikingly the lignin component besides holo-cellulose, leaving behind a white residuc (eventually cellulose) and such fungi are termed the "white-rots". White-rots are the best lignin degraders which completely metabolize the complex polymer, and have been the most studied. Most of them, however, have the enzymatic capacity to use cellulose, hemicellulose, and other components of lignocellulosic matter as a source of carbon and energy; hence, total biomass breakdown usually occurs and lignin removal is accompanied by removal of polysaccharides (Kirk and Moore, 1972; Rahi and Soni, 2007; Soni and Rahi, 2007). Since most of the mushrooms species have been recognized for their degradative efficiencies which are utilize in various industrial purposes whether food, textile, agriculture, pharmaceutical or environmental processes, therefore, in this review a detailed discussion has been made on degradative enzymes of mushrooms (Table 6.1).

Table 6.1 Degradatory Enzymes of Mushrooms.

Mushroom species	*Enzyme(s)*	*Properties*	*Ref.*
Coprinus cinereus	Peroxidase	Acidic protein, molecular weight 41,600 exhibited spectra of a hemoprotein	Morita *et al.* (1988)
Pleurotus ostreatus	Peroxidase	Glycoprotein containing 35.7% carbohydrate; enzyme is a dimer of identical subunits molecular weight = 2,400); contains iron protoporphyrin; optimum activity at pH 3.5–4.0 and at 40°C.	Bae *et al.* (1989)
Trametes versicolor	Peroxidase	Heme-containing inducible intracellular; isolated affinity chromatography on vanillin-activated controlled porosity glass column; polymerizes as well as depolymerizes soluble lignin derivatives; can cooperate with glucose-oxidase in degradation of lignocellulosics	Loparzeski (1987)
Coriolus versicolor	Laccase	Has 2 Cu atoms per protein molecule believed to be necessary for catalysis; molecular weight of 66,000; acts on lignins to loosen their three-dimensional structure, permitting depolymerization and solubilization	Morohoshi (1991)
Coriolus hirsutus	Laccase	Molecular mass = 55 kDa; optimum pH for activity 3.5–4.5; catalyzes the electro-reduction of O_2 to H_2O with a direct electron transfer from the enzyme active center to the electrode	Gindilis *et al.* (1988)
Coriolus versicolor	Laccase	Optimum pH 4.6 and 40°C for activity; stable at 40°C for 30 min; Cu, Fe and Ca activated; Mn and Hg were inhibitory; totally inhibited by 1 mM NaN_3 and KCN	Hong *et al.* (1987)
Coprinus 7 N	Proteinase	Stable at pH 6–9; optimum activity at 37°C; completely inactivated by the specific inhibitors of serine proteinases; hydrolyzes azocasein, azoalbumin, hemoglobin fibrinin, and synthetic chromogenic peptide substrates	Shaginyan *et al.*(1990)
Coprinus cinereus	Proteinases	Chromatographed into 2 fractions — EL and A; both had an apparent molecular weight of 31 kDa; pH 9 and temperature 30°C favored the activity of EL; AD was active at pH 5 and 9 and 56–62°C,	Kalisz *et al.* (1989)
Agaricus bisporus	Protease	Optimum activity at pH 6 and stable at pH 4–7; active at 50°C and instantly inactivated at 60°C; inhibited by Ag, Cu, Ba, Fe, Co, Ca, Pb, Mg, and Mn	Eun *et al.* (1989)

(*Contd.*)

Mushroom species	*Enzyme(s)*	*Properties*	*Ref.*
Bjerkandera adjusta	Aryl alcohol oxidase	Molecular weight = 78,000; oxidized several aromatic alcohols to the corresponding aldehydes, simultaneously reducing O_2 to H_2O_2; pH optimum 3.7, enzyme stable for weeks at 4°C	Muheim *et al.* (1990)
Pleurotus eryngii	Aryl alcohol oxidase	Catalyzes conversion of primary aromatic alcohols to the corresponding aldehydes, and H_2O_2; showing no activity with aliphatic and secondary aromatic alcohols; enzyme stable at pH 4–9, temperature 45–50°C; and pH 6–6.5 optima; inhibited by Ag, Pb and NaN_3	Guillen *et al.* (1990)
Trametes versicolor	Oxidizing enzyme	Isolated from the culture filtrates; induction of synthesis late in the growth cycle and does not require an inducer; pH optimum 3.5; ferulic acid and o-dianisidine are oxidized	Khan and Overend, 1991)
Pleurotus ostreatus	Veratryl alcohol oxidase	A glycoprotein with FAD as a prosthetic group; oxidizes cinnamyl alcohol faster; oxidation of cinnamyl alcohol is slower	Sannia *et al.* (1991)

LIGNINOLYTIC ENZYMES OF MUSHROOMS

Lignin Molecule

Lignin is the material that confers the qualities of rigidity and durability that make woody plants 'woody'. It makes up about 20–30% of wood, and is found in cell walls, in a complex with cellulosic and hemicellulosic polysaccharides. In this natural composite material, the cellulose fibrils provide tensile strength, and the hemicellulose and lignin provide cross-linking, binding the structure together. Lignin is a very large, highly cross-linked and stereochemically complex polymer that is biosynthesized by the polymerization of phenylpropanoid precursors. There are three of these precursors, differing in the number of methoxyl groups on the aromatic ring (Fig. 6.2).

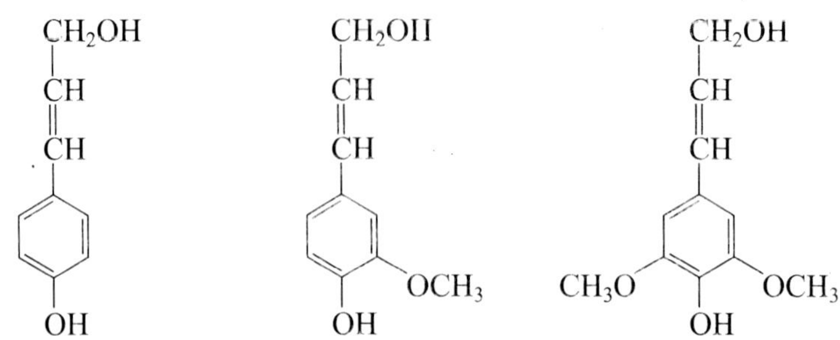

(1) (2) (3)

Fig. 6.2 Structure of the three lignin precursors: (1) P-coumaryl alcohol, (2) Coniferyl alcohol and (3) Sinapyl alcohol.

Lignin polymerization takes place in cell walls after the polysaccharides have been deposited, and is initiated by enzymatic oxidation of the precursors to phenoxy radicals. These radicals can couple with each other and with the growing lignin polymer in numerous ways to form a complex cross-linked network. This random polymerization of lignin gives it a very complex and irregular structure (Fig.6.3).

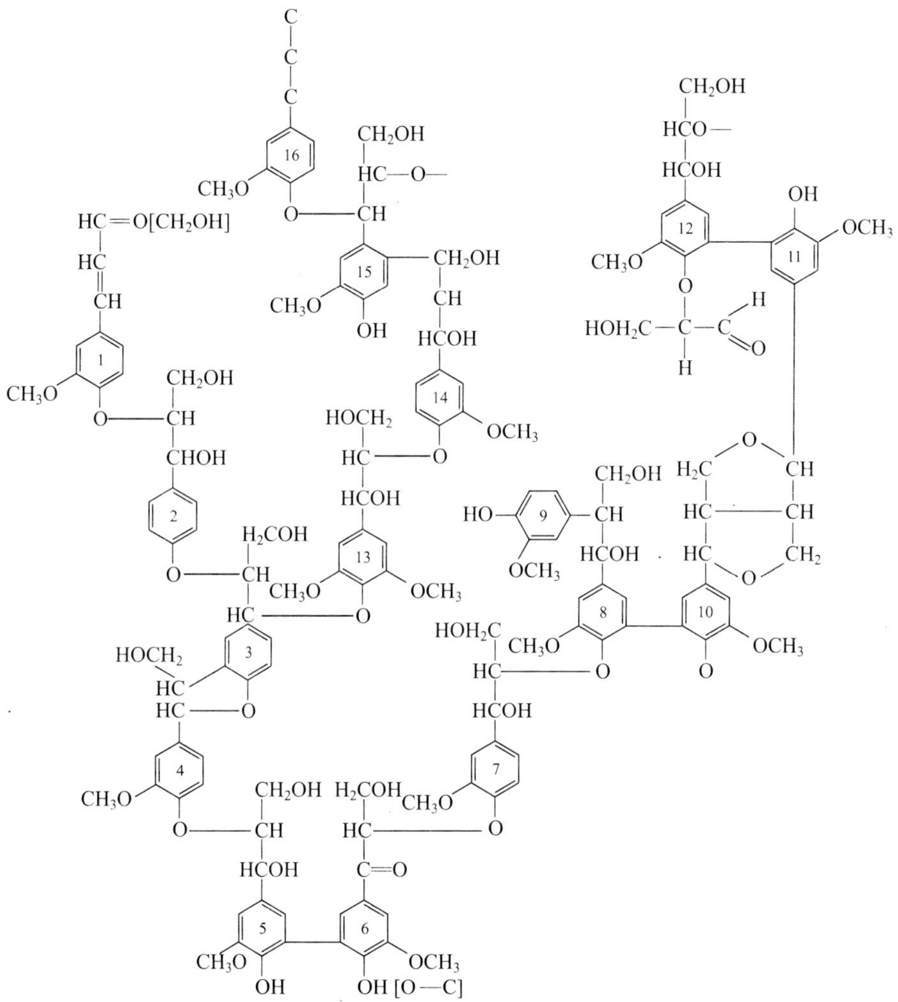

Fig. 6.3 Schematic structure of a portion of a lignin molecule.

Lignin Degrading Mushrooms

Lignin degrading fungi are ubiquitous. The most familiar are those that form mushrooms, brackets (Conks) and other sporophores on decaying trees, wood, forest litter, and other lignocellulosics. These are fungi that cause the white-rot type of wood decay, and the closely related litter-decomposing fungi. The most vigorous lignin-degraders are white-rot wood decay fungi, which are mainly basidiomycetes, and which belong to the orders Agaricales, Aphyllophorales, and Tremellales (Table 6.2).

Table 6.2 Important mushrooms and their ligninolytic enzymes with decay types.

Fungus	*Decay type enzymes*	*Ligninolytic*	*Reference*
Agaricus bisporus, A., brunnescens A. bitorquis, Cuitivated mushroom: Button mushroom	w- r	Laccase, MnP, LiP	Bonnen *et al.,* 1994,
Agrocybe praecox	l	MnP, LiP, laccase	Steffen *et. al.*, 2002c
Auricularia auricula Wood ear, Jew's ear	w-r	MnP, LiP, laccase	Hofrichter and Fritch 1997
Bjerkandera adusta	w-r	MnP, LiP	Wang *et al.*, 2001
Cerrena maxima	w-r	Laccase	Koroleva *et al.*, 2002a
Clitocybula dusenii	w-r	MnP	Nuske *et al.*, 2002
Collybia dryophila	l	MnP, LiP, laccase	Steffen 2002a
Coriolopsis gallica, C. rigida	w-r	Laccase	Pickard *et al.*, 1999b
Flammulina velutipes Enoki, golden needle, velvet shank	w-r	n. r.	
Fomes sclerodermeus	w-r	MnP, laccase	Papinutti *et al.*, 2003
Ganoderma lucidum	w-r	MnP, LiP, laccase	D'Souza *et al.*, 1999
Lentinula edodes (shiitake)	w-r	MnP, LiP, laccase	Forrester *et.al.*, 1990 Nagai *et al.,* 2002
Lepista irina	l	VP	Zorn *et al.*, 2003
Marasmius quercophilus	l	Laccase	Klonowska *et al.*, 2001
Mycena galopus	l	MnP, LiP, laccase	
Nematoloma frowardii	w-r	MnP, LiP, laccase	Nuske *et al.*, 2002
Panaeolus sphinctrinus	l	MnP, LiP, laccase	Heinzkill *et. al,* 1998
Panus tigrinus	w-r	MnP	Fenice *et al.*, 2003
Phaenerochaete chrysosporium	w-r	MnP, LiP	Herpoel *et al.,* 1999; Anadurai *et al.*, 2000; Cabaleiro *et al.*, 2002; Gill & Arora, 2003
Phlebia floridensis, P. brevispora, P. radiate, P. fascicularia	w-r	Laccase	Arora & Rampal 2002
Pleurotus ostreatus, P. eryngii, P. florida, P.	w-r	Laccase,VP, MnP	Reddy *et al.*, 2003; Munoz *et al.*,

(*Contd.*)

Fungus	Decay type enzymes	Ligninolytic	Reference
pulmonarius, P. Sajor caju			1997a,b; Das *et al.*, 1999
Pycnoporus cinnabarinus, P. sanguineus	w-r	Laccase	Lomascolo *et al.*, 2002, 2003
Rigidoporus lignosus	w-r	Laccase	Cambria *et al.*, 2002
Trametes versicolor, T. hirusta, T. modesta, T. multicolour, T. pubescens, T. trogii	w-r	MnP, LiP, laccase	Rosales *et al.*, 2002; Nyanhongo *et al.*, 2002a; Hess *et al.*, 2002; Levin *et al.*, 2002
Stropharia coronilla	w-r	MnP, LiP, laccase	Steffen *et. al.*, 2002c
Stropharia rugosoannulata	w-r	MnP, LiP, laccase	Steffen, 2003
Volvariella volvacea	1	Laccase	

l : litter degrading, **MnP** : manganese peroxidase, **LiP** : Lignin Peroxidase, **n. r.** : not reported, **w-r** : white rot, **VP** : versatile peroxidase

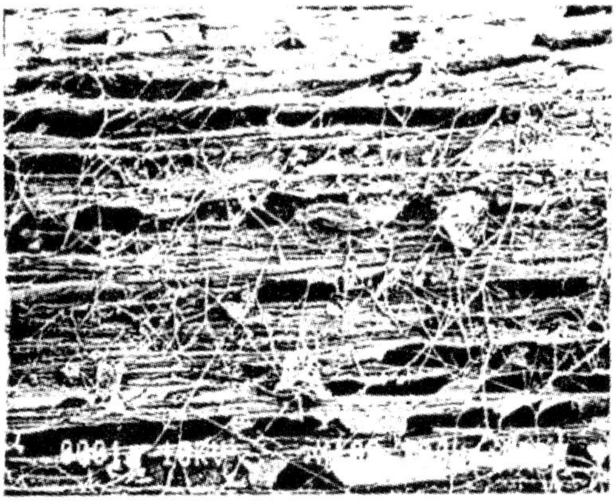

Fig. 6.4 Scaning electron microscopic view of hyphae of *Phaenerochaete chrysosporium* growing in cells of aspen (*Poplus* sp.) Wood.

A few ascomycetes belonging to the order Speriales also cause white-rot wood decay. It is white-rot fungi that have been most intensively studied for bioremediation, and it is their lignin-degrading system that seems to be important in such applications. Fig. 6.4 shows the hyphae of the most studied white-rot fungus, *Phaenerochaete chrysosporium,* growing in cells of aspen wood. Lignin degrading fungi are classified into three major categories based on the type of wood decay caused by these organisms: white-rot fungi, brown-rot fungi and soft-rot fungi. Of these three groups, white-rot fungi are the most effective lignin degraders and have been the most extensively studied group. In this group the most commonly used model organism in lignin biodegradation is strain of *Phaenerochaete chrysosporium* (Boominathan and Reddy, 1992). White-rot fungi comprise a heterogeneous collection of several hundreds of species of basidiomycetes.They are able to completely mineralize both the lignin and carbohydrate components of wood. Despite the fact that lignin is rich in carbon, it is not a growth substrate. White-rot fungi metabolize lignin only in the presence of an alternate energy source, such as cellulose, hemicellulose or simple carbohydrates. This is called co metabolism. Most white-rot fungi degrade lignin only during secondary metabolism, which is triggered by a limitation of nutrients. However, some fungal species have been reported to degrade lignin during primary metabolism also under conditions of nutrient sufficiency. There are many factors involved in the lignin degradation process, but very few of them are completely understood.

Mechanism of Lignin Biodegradation

Due to random polymerization process of lignin formation it has a complex and irregular structure. The diversity of the interunit linkages and the irregularity of their arrangement make it difficult for a ligninolytic fungus to produce enzymes that could recognize and cleave all of them. The solution that has evolved in the white rot fungi is to produce enzymes of low specificity that initiate, but do not direct, oxidative reactions in lignin. Kirk and Farrell (1987) have termed this process as enzymatic combustion in which the enzyme activates the lignin to overcome an energy barrier and begin a thermodynamically favoured oxidative fragmentation without further control of the reaction pathway by the enzyme. The chemical changes produced by white-rot fungi in lignin include oxidative cleavage of the propanoid side chains and also demethylation and oxidation cleavage of aromatic rings. Lignin biodegradation does not proceed by an orderly removal of the peripheral subunits. Instead it also involves oxidation of the aromatic rings and side chains in the interior of the polymer, increasing the solubility of the polymer core at the same time as fragments of varying size are set free. The disorderly nature of this degradation agrees with the concept of enzymatic combustion.

Lignin Degrading System of White-rot Fungi

The ligninolytic system of *P. chrysosporium* is illustrated schematically in Fig. 6.5. The key extracellular enzymes are thought to be lignin peroxidase and glyoxal oxidase. The latter oxidizes and metabolizes glyoxal and methyl glyoxal with reduction of O_2 to H_2O_2, which activates lignin peroxidase. Lignin peroxidase oxidizes nonphenolic aromatic nuclei in lignin by one electron to generate aryl cation radicals, which degrade non enzymatically via many reactions. Most of those reactions result in polymer cleavages, generating both aromatic and aliphatic products which are taken up by the hyphae and mineralized.

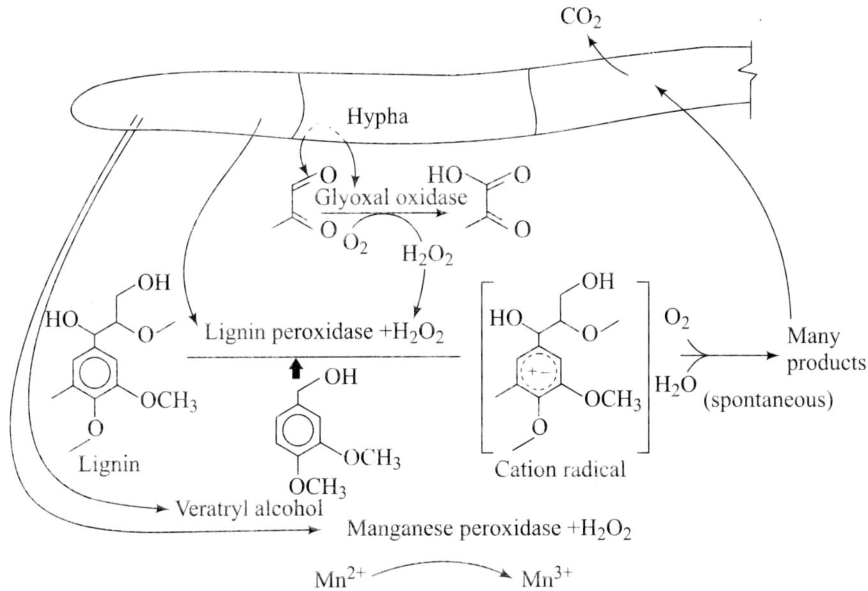

Fig. 6.5 Schematic illustration of the ligninolytic system of *Phaenerochaete chrysosporium*.

Little is known about the intracellular system. A second kind of peroxidase, manganese peroxidase, in the presence of H_2O_2, oxidizes Mn^{2+} to Mn^{3+}, which in turn can oxidize phenolic units in lignin.The role of such oxidation, if any, and that of manganese peroxidase, are not yet clear. The aromatic metabolize veratryl alcohol (Fig. 6.5) seems to play multiple roles, including stimulation of the production of the enzymes and electron transfer reactions during substrate degradation. Both the extracellular and intracellular components of this system apparently plays roles in xenobiotic degradation.

Table 6.3 Ligninolytic enzymes and their main reactions (Hatakka, 1994).

Enzyme and abbreviation	*Cofactor*	*Substrate, mediator*	*Reaction*
Lignin peroxidase. Lip	H_2O_2	Veratry alcohol	Aromatic ring oxidized to cation radicial
Manganese Peroxidase MnP	H_2O_2	Mn, organic acid as chelators, thiols, unsaturated fatty acids	Mn(II) oxidized to Mn (III); chelated Mn(III) oxidizes phenolic compounds to phenoxyl radicals; other reactions in the presence of additional compounds

(Contd.)

Enzyme and abbreviation	*Cofactor*	*Substrate, mediator*	*Reaction*
Versatile peroxidase VP	H_2O_2	Mn, veratryl alcohol, compounds similar to LiP and MnP	Mn(II) oxidized to Mn (III), oxidation of phenolic and non-phenolic compounds, and dyes
Laccase	O_3	Phenols, mediators e.g. hydroxybenzotriazole or ABTS	Phenols are oxidized to phenoxyl radicals; other reactions in the presence of mediators
Glyoxal oxidase, GLOX		Glyoxal, methyl glyoxal	Glyoxal oxidized to glyoxal acid; H_2O_2 production
Aryl alcohol Oxidase, AAO		Aromatic alcohols (anisyl, veratryl alcohol)	Aromatic alcohols oxidized to aldehydes: H_2O_2 production
Other H_2O_2-producing enzymes		Many organic compounds	O_2 reduced to H_2O_2

Lignin Degrading Enzymes

A set of enzymes (Table 6.3) that will extensively degrade lignin and reproduce the effects of a white-rot fungus have been studied by various workers. However, three important oxidative enzymes are commonly found extracellularly in ligninolytic cultures of white-rot fungi (Table 6.4). It is clear that different combinations of the known enzymes are produced by various lignin degrading fungi, suggesting that there is more than one successful strategy for lignin biodegradation (Hatakka, 1994).

Table 6.4 Distribution of extracellular enzyme, implicated in lignocellulosics degradation, in the white-rot species (Hatakka, 1994).

Species	*Laccase*	*Lignin peroxidase*	*Manganese peroxidase*	*References*
Agaricus bisporus	+	–	+	Bonnen *et al.* (1994)
Phanerochaete Chrysosporium	–	+	+	Ruttimann-Johnson (1994)
Phanerochaete sordida	–	–	+	Ruttimann-Johnson (1994)
Dichomitus squalens	+	–	+	Perie and Gold (1991)
Cereporiopsis subvermispora	+	–	+	Ruttimann *et al.* (1992)
Phlebia brevispora	+	–	+	Ruttimann *et al.* (1992)

(Contd.)

Species	Laccase	Lignin peroxidase	Manganese peroxidase	References
Rigidoporus lignosus	+	–	+	Galliano *et al.* (1991)
Panus tigrinus	+	–	+	Maltseva *et al.* (1991)
Lentinus edodes	+	–	+	Forrester *et al.* (1990)
Ganoderma colossum	+	–	+	Horvath *et al.* (1993)
Phlebia radiata	+	+	+	Kantelinen *et al.* (1989) Niku-Paavola *et al.* (1988)
Coriolus versicolor	+	+	+	Jonsson *et al.* (1987)
Phlebia ochraceofulva	+	+	–	Vares *et al.* (1993)
Junghunia separabilima	+	+	–	Vares *et al.* (1993)

Lignin peroxidase : Lignin peroxidase (LiP) was first discovered in *Phanerochaete chrysosporium* (Tien and Kirk, 1984) and it is produced by many white-rot fungi. This enzyme is an extracellular hemeprotein, dependent of H_2O_2, with an unusually high redox potential and low optimum pH. It shows little substrate specificity, reacting with a wide variety of lignin model compounds and even unrelated molecules (Barr and Aust, 1994). It has the distinction of being able to oxidize methoxylated aromatic rings without a free phenolic group (Fig.6.6), generating cation radicals that can react further by a variety of pathways, including Cα-Cβ cleavage, ring opening, demethylation and phenol dimerisation.

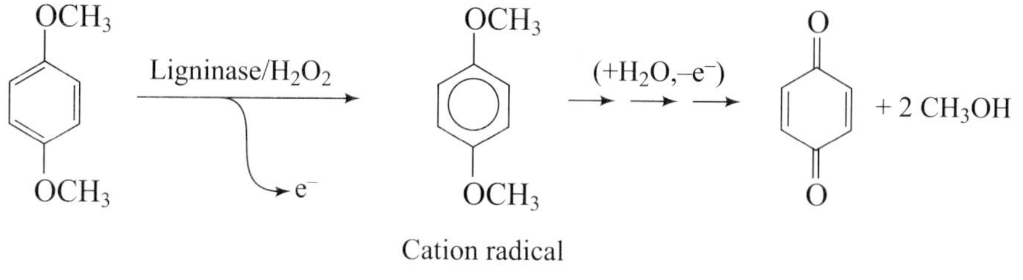

Fig. 6.6 The action mechanism of lignin peroxidase was discovered from studies of metoxylated benzenes. They were oxidized by the enzyme to unstable molecules called cation radicals. In the figure the cation radical 1,4-dimetoxibenzene, decomposes in the reaction with H_2O_2 producing methanol and benzenequinone.

Manganese peroxidase : Another type of heme peroxidase, manganese peroxidase (MnP), was later found in the culture media of the same fungi. This enzyme shows a strong preference for Mn(II) as its reducing substrate. The product

Mn(III) forms a complex with organic acids and diffuses away from the enzyme to oxidize other materials, such as lignin (Fig. 6.7).

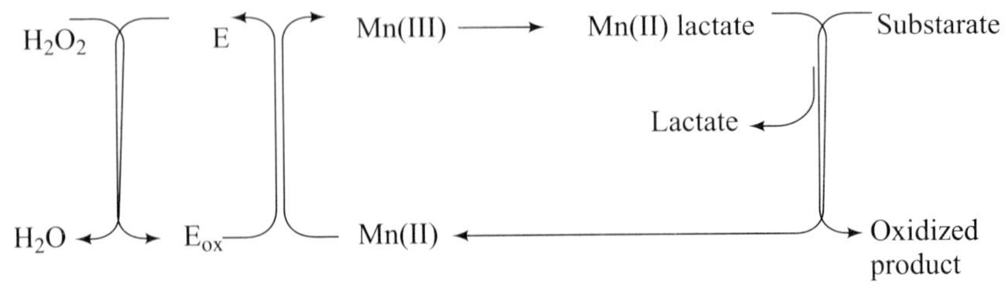

Fig. 6.7 The action mechanism of manganese peroxidase.

The redox potential of the Mn peroxidase-Mn system is lower than that of lignin peroxidase and it has only shown capacity for oxidize *in vitro* phenolic substrates (Vares, 1996). Phenolic substrates are oxidized to phenoxy radicals, which can react further by demethylation, alkyl-phenyl cleavage, Cα-oxidation or Cα-Cβ cleavage. Manganese peroxidase seems to be more widespread among white-rot fungi than lignin peroxidase (Hatakka, 1994) and is readily detected in cultures on real lignocellulosic substrates.

Laccases : This enzyme is a copper-containing oxidase and it does not require peroxide, H_2O_2. Like manganese peroxidase, it normally oxidizes only those lignin model compounds with a free phenolic group, forming phenoxy radicals. However, in the presence of the artificial substrate ABTS (2,2´-azinobis(3-ethylbenzthiazoline- 5-sulphonate) or some other synthetic mediators, laccase can also oxidize certain non-phenolic compounds, veratryl alcohol and Mn (II). Laccase is produced by most white-rot fungi (Hatakka, 1994), but normally not in *Phanerochaete chrysosporium* (Kirk and Farrell, 1987). Laccase has the capability to both depolymerize and polymerize lignin model compounds (Eriksson *et al.*, 1990).

Aryl-Alcohol Oxidases : Two veratryl alcohol oxidase (VAO) enzymes has been found in the cultured medium of *Pleurotus sajor-caju* that mineralizes ring-14C labeled lignin (dehydrogenative polymer) when grown in mycological broth. Both the enzymes are glycoproteins and contain flavin prosthetic groups, with pH optima around 5.0. 4-Methoxybenzyl alcohol was oxidized most rapidly, followed by veratryl alcohol. It does not oxidize either of the alcohol, i.e. aromatic and non-aromatic. Cinnamyl alcohol is oxidized at the gamma-position. The VAO enzymes aptly represent a significantly different route for veratryl alcohol oxidation from that catalyzed by LiPs from *Phanerochaete chrysosporium*. The role of the oxidases in biodegradation might be to produce H_2O_2 during oxidation of lignin fragments. If VAO plays a role in lignin biodegradation, it could be either through two electron redox cycling or veratryl alcohol or lignin fragments, with H_2O_2 production from O_2 or in combination with laccase, as a dehydrogenase of aromatic compounds (Bourbonnais and Paice, 1988). Sannia *et al.* (1991) have purified and characterized a VAO from the cultures of *Pleurotus ostreatus*. This enzyme oxidized veratryl alcohol to veratraldehyde with a concomitant reduction of O_2 to H_2O_2. Its production is not induced by the presence of veratryl alcohol, olive oil, sawdust, induline or Tween 80 in the culture broth. The possible role of a laccase and FAD-dependent aryl alcohol oxidase excreted by *Pleurotus ostreatus* was investigated by Marzullo *et al.* (1995). They found that the VAO was able to reduce synthetic quinones, laccase-generated quinonoids, and phenoxy radicals with concomitant oxidation of veratryl alcohol to veratraldehyde. This cooperative action of laccase and VAO also prevented the polymerization of phenolic compounds and reduced the molecular weight of soluble lignosulphonates to a significant extent The production of an extracellular aryl-alcohol oxidase from the white-rot fungus *Bjerkandera adusta* has also been studied under limitations of carbon and nitrogen by various workers. The purified enzyme consisted of two oxidases.

The white-rot fungi are known to employ a variety of extracellular oxidative enzymes to cleave lignin. The best characterized of these are the lignin peroxidases (Li-ps) (Edwards *et al.*, 1993; Gold *et al.*, 1989; Kirk and Farrell, 1987; Piontek *et al.*, 1993), which cleave polymeric lignin between C and C of its propyl side chain to give oxidized lignin oligomers (Hammel *et al.*, 1993; Hammel and Moen, 1991). Other ligninolytic mechanisms exist, however, perhaps allowing certain Li-p-negative fungi to degrade lignin (Ruttimann-Johnson *et al.*, 1993), Li-ps are common to many white-rot species, and the evidence is strong that they play a role in ligninolysis (Hammel *et al.*, 1993). In LiP-producing fungi, lignin degradation coincides with LiP secretion, and in most cases both processes are functions of secondary (idiophasic) metabolism (Gold *et al.*, 1989; Kirk and Farerell, 1987). White-rot species which produce LiPs generally secrete a secondary metabolite, veratryl alcohol (VA), which is a substrate for LiP. VA protects LiP against H_2O_2-mediated inactivation and has been proposed to act *in vivo* as a stabilizer for the enzyme. This stabilizing effect is probably responsible for the observation that exogeneous addition of VA enhances LiP secretion in certain white-rot species. VA also promotes the LiP catalyzed oxidation of other molecules that are either inaccessible to the enzyme or recalcitrant, by themselves, to enzymatic attack (Akamatsu *et al.*, 1990; Hammel and Moen 1991). Some researchers have proposed that VA participates in these reactions by acting as a redox shuttle between LiP and substrate (Akamatsu *et al.*, 1990; Harvey *et al.*, 1986). Others have presented evidence that only one of the redox states in the LiP catalytic cycle is capable of oxidizing certain substrates and that VA is required in these cases as a co-substrate that allows the enzyme to complete its turnover. VA thus plays diverse roles in white-rot fungal metabolism.

Industrial Potential of Ligninolytic Fungi

The preparations such as LiP from *Phanerochaete chrysosporium* and *Phlebia radiata* and MnP from *P. chrysosporium* and *Nematoloma frowardii* are on the market for research purposes. An engineered form of *Coprinus cinereus* peroxidase was successfully produced in a heterologous system and sold as a bleaching agent for laundry detergents. Pure enzymes, crude enzyme preparations and cultures of ligninolytic fungi have been widely studied for biopulping, degradation of toxic compounds (Kremer and Anke, 1997) and dye removal. These potential applications are based on the concomitant and nonspecific action of the fungal ligninolytic system (Pointing 2001). Pure enzymes have proved to be less effective for applications based on the use of ligninolytic activity than ligninolytic culture preparations. Table 6.5 lists some ligninolytic fungi and their enzymes undertaken to determine their use in industrial applications. Bleaching of industrial effluent and dye colours by white-rot fungi and their ligninolytic enzymes is a promising industrial application. Laccase was reported to be the enzyme responsible for decolorization of industrial dyes (Rodriguez *et al.,* 1999, Mayer and Staples 2002). LiP was reported to be effective in decolorizing kraft pulp mill effluents (Ferrer *et al.,* 1991). Moreira *et al.,* (1998) considered MnP as the most important enzyme in biobleaching experiments with *Bjerkandera* sp. strain BOS55. Biopulping by lignin-degrading fungi has been studied at the pilot scale and development of commercial applications is in progress. Mechanical pulping includes all process that fiberize the wood chips with little or no application of chemicals. Organopollutants such as 2,4,6-trinitrotoluene (TNT), polychlorinated biphenyls (PCBs), organochlorines, PAHs and wood preservatives were shown to be degraded by white-rot fungi (Pointing 2001). The use of elemental chlorine and hypochlorite in pulp bleaching has caused environmental problems in the pulp and paper industry. High-molecular weight chlorolignins are formed when active chlorine reacts with lignin during pulp bleaching (Kringstad and Lindström 1984). These chlorolignins are degraded extremely slowly under natural conditions. White-rot fungi and their ligninolytic enzymes decolorize and degrade organic chlorine-containing material from pulp mill wastewaters (Eriksson and Kirk, 1985). In the 1980s methods called MYCOR and MYCOPOR (Messner *et.,* 1989) were developed to remove colour but they also decreased the total organic chlorine from pulp mill wastewaters. These processes were based on the white-rot fungus *P. chrysosporium*. Dehalogenation of organohalogen compounds was also reported in anaerobic pulp mill wastewater treatment (Jokela *et al.,* 1993).

The potential for white-rot fungi and their ligninolytic enzymes has not yet been exploited. Factors limiting the biotechnological use of ligninolytic enzymes, especially the peroxidases, have included their low production level and rather low stability (Ikehata *et al.,*2004). Peroxidases also need the addition of H_2O_2 for their catalytic cycle. Heterologous expression of ligninolytic enzymes has been attempted, using several recombinant systems with various degrees of success. The most successful approaches have been those using laccase; moreover, laccase is produced constitutively by many fungi and the homologous production of laccase may also be economically feasible. Laccase works with oxygen and a mediator, a low-molecular weight redox compound that generate stable radicals. These laccase-mediator systems (LMS) can achieve 40–60% delignification of kraft pulps, has enlarged the potential of laccases in different applications including the paper pulp industry. Laccases have also been used for reducing the energy of mechanical pulping. Peroxidases and laccases have also been used to polymerize and copolymerize materials with lignocellulosic fibres. Laccases have also been used for reducing the energy of mechanical pulping. The treated wood chips are more readily broken apart during subsequent refining, saving as much as 30% of the electrical energy while producing more flexible intact fibres compared to refining untreated wood chips.The large-scale production of ligninolytic peroxidases is still under development (Lomascolo *et al.*,2003). Extracellular lignin peroxidases and/or manganese peroxidases that are able to initiate the depolymerization and degradation of lignin and its oligomers and monomers of white-rot fungi have broad substrate specificity and have been implicated in the transformation and mineralization of organopollutants with structural similarities to lignin. These enzymes involves in degradation of munitions waste, pesticides, polychlorinated biphenyls, polycyclic aromatic hydrocarbons, bleach plant effluent, synthetic dyes, synthetic polymers, and wood preservatives. Much bioremediation work has concentrated on organic pollutants, although the range of substances that can be transformed or detoxified by mircroorganisms includes both natural and synthetic organic materials and inorganic pollutants, such as toxic metals. Managanese peroxidase (MnP) has been implicated as an important enzyme in delignification of kraft pulps by white rot fungi. Treating kraft pulps with MnP can lower their kappa numbers and increase their bleachability. In the presence of the surfactant Tween 80, the enzyme can increase the brightness of hardwood kraft pulp.

Table 6.5 Potential industrial applications of ligninolytic mushrooms and their enzymes.

Application	*Fungus*	*Reference*
Biopulping	*Ceriporiopsis subvermispora* *Physisporinus rivulosus* T241i	Messner and Srebotnik 1994; Akhtar *et al.*, 1998 Maijala *et al.*, 2002, Hakala *et al.*, 2004
Decolorization of synthetic dyes and dye industry wastewaters	*Phanerochaete chrysosporium*, *Trametes versicolor, T. cingulata, T. hispida, T. pocas, Datronia concentrica, Bjerkandera sp.* BOS55, *Phanerochaete beiidinus, Bjerkandsfii adusta, Pycnoporus cinnabarinus, P. sanguineus, Coriolus hirsutus, Funalia trogii, Irpex lacteus, Pleurotus ostreatus, P. Sajor-caju, P. pulmonarius*	Review by Shah and Nerud 2002 and review by Wesenberg *et al.*, 2003., Trupkin *et al.*, 2003
Biobleaching	*Phanerochaete sordida* *Trametes versicolor* *Bjerkandera sp.* BOS55	Kondo *et al.*, 1994 Archibald *et al.*, 1997 Moreira *et al.*, 1998

(*Contd.*)

Application	*Fungus*	*Reference*
Treatment of pulp mill wastewaters (decrease of colour and organic halogen)	*Trametes elegans, Phlebia radiata, Panus crinitus, Trametes villosa, Phanerochaete chrysosporium*	Duran and Esposito 2000. Soares and Duran 2001, Lara *et al.*, 2003
Biodegradation of toxic compounds	*Pleurotus pulmonarius, Pleurotus ostreatus*	Review by Pointing 2001, Lau *et al.*, 2003, Hestbjerg 2003,
Improvement of digestibility of lignocellulosic animal feed	*Pleurotus ostreatus, Ganoderma applanatum*	Zadrazil 1977

CELLULOLYTIC ENZYMES OF MUSHROOMS

Structure of Cellulose

Cellulose is a linear polymer made of glucose subunits linked by β-1,4 bonds (Fig. 6.8). Each glucose residue is rotated by 180° relative to its neighbours, so that the basic repeating unit is in fact cellobiose. The degree of polymerization (DP) of native cellulose is in the range of 7,000–15,000.

$$DP = \frac{\text{Molecular weight of cellulose}}{\text{Molecular weight of one glucose unit}}$$

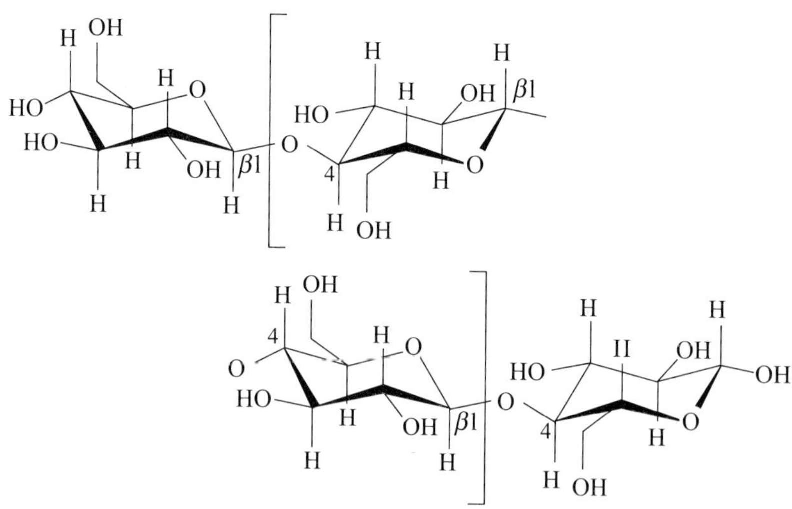

Fig. 6.8 Three-dimensional structure of cellulose.

The cellulose structure favours organization into bundles with the crystalline order held together by hydrogen bonds (Fig. 6.9).

Fig. 6.9 Cellulose consists of crystalline regions (blue) and disordered or amorphous regions (red).

Native crystalline cellulose is generally found as a mixture of cellulose Iα and cellulose Iβ polymorphic forms (Fig. 6.10). The native, crystalline form of cellulose has a structure designated as type 1, which can be converted into type II by alkali treatment. The two types differ in their intra chain hydrogen bonding pattern. Furthermore, most native celluloses are composed of two slightly different forms of type I cellulose, termed I_α, and I_β which differ in their intermolecular hydrogen bonding pattern. The two forms occur in different proportions, depending on the source of the cellulose (Atalla *et al.*, 1984; Kuga and Brown 1991; Richmond, 1991). The proportions of the two allomorphs of crystalline cellulose vary depending on the source of the cellulose. Higher plants such as trees and corn believed to have a higher proportion of cellulose I_β. Studies have determined cellulose I_β is more stable than cellulose I_α and the conversion from type I_α to I_β is irreversible. Electron diffraction patterns show type I_α to have a triclinic unit cell and type I_β to have a monoclinic unit cell. Chain length varies between 100 and 14000 residues. Cellulose chains form numerous intra- and intermolecular hydrogen bonds, which account for the formation of rigid, insoluble microfibrils. Microfibrils range in lateral dimension from 3–4 nm in higher plants up to 20 nm for the microfibrils of the alga *Valonia rnacrophysa,* which contain up to several hundred cellulose chains. The chains are oriented in parallel and form highly ordered, crystalline domains interspersed by more disordered, amorphous regions. Raman spectroscopy and solid-state ^{13}C-NMR have shown that crystalline cellulose occurs in various forms. The diversity and heterogeneity of cellulosic substrates contribute to a large extent to the difficulty of enzymological studies. Practically, a variety of substrates is available, whose recalcitrance to hydrolysis is directly correlated with their degree of similarity with natural cellulose. Pure cellulose is commercially available in several forms (cotton, filter paper, avicel). These forms are generally used to assess the efficiency of complete cellulase systems.

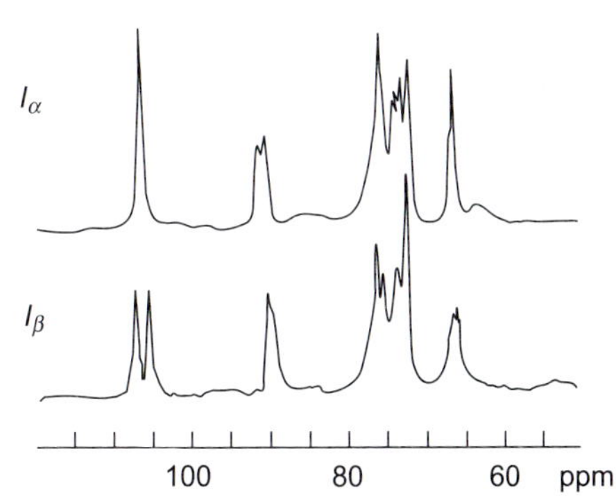

Fig. 6.10 Solid State NMR spectrum of cellulose I_α and I_β.

Cellulose Degrading Fungi

The important genera of cellulose degrading mushrooms include : *Phanerochaete chrysoporium, Pycnoporus cinnabarinus, Trametes versicolor, T. pubesecens, T. villosa, Ceriporiopsis subvermispora, Agaricus bisporus, Volvariella volvacea, Pleurotus florida, Phellijus ignlanus, species of Poria, Schizophyllum and Serpula.* Other members of higher fungi that have received considerable study with respect to their cellulolytic enzymes and/or wood degrading capability include *Bulgaria, Helotium* (Ascomycetes).

Cellulose Biodegradation

The cellulose polymer is degraded to low molecular weight soluble products, particularly glucose by the action of a multienzyme cellulose complex. A cellulosic enzyme system consists of three major components: endo-ß-glucanase (EC 3.2.1.4), exo-ß-glucanase (EC 3.2.1.91) and ß-glucosidase (EC 3.2.1.21) on the basis of substrate specificity.

(i) Endo-1,4-β-glucanase (EG:EC 3.2.1.4) act randomly at 1,4- β-linkages. It prefers to degrade amorphous (non-crystalline) cellulose and not the crystalline cellulose.

(ii) Exo-1,4- β-glucanase (exo-1,4- β-D-glucan cellobiohydrolase; CBH; EC. 3.2.1.91) releases both glucose and cellobiose units from the non-reducing ednds of cellulose chains.

(iii) 1,4- β-glucosidases (EC 3.2.1.21) hydrolyze cellobiose to glucose, and cellobionic acids to glucose and gluconolactone.

Wood and Wood (1992) have demonstrated the essential role of three enzymes-CBH I, CBH II and trace EG for rapid hydrolysis of microcrystalline cellulose. Reese *et al.* (1950) proposed that exo-ß-glucanase causes a disruption in cellulose hydrogen bonding, followed by hydrolysis of the accessible cellulose with endo-ß-gucanase. According to this hypothesis, in a synergistic sequence of events, endo-ß-glucanase acts randomly on the cellulose chain, while exo-ß-glucanase acts on exposed chain ends by splitting off cellobiose or glucose. Cellobiose is subsequently hydrolysed by p-glucosidase to glucose. This hypothesis is however the opposite of that proposed by Reese *et al.,* (1950) and indicates that three, rather than two enzymes are essential for the decomposition of cellulosic biomass.

In addition to hydrolytic enzymes, oxidative enzymes from *Phanerochaete chrysoporium* have also shown to play in cellulose degradation. The oxidative enzymes cellobiose: quinone oxidoreductase (EC 1.1.5.1), a flavoprotein, reduces quinines and phenoxy radicals (produced during lignin degradation) in the presence of cellobiose which is oxidized to cellobio-α-lactone. Cellobiose oxidase (EC 1.1.3.25), a flavoheme protein, is another enzyme, oxidizing cellobiose and higher cellodextrins to their corresponding-onics acids using molecular oxygen. Two possible functions have been suggested for this enzyme, i.e. to oxidize the reducing end group formed when β-1,4 glucosidic linkage is cleaved by an endoglucanase in order to prevent the linkage reforming and the release of hydrogen peroxide (H_2O_2) for lignin degradation. Both enzymes have been shown to bind microcrystalline cellulose while retaining enzyme activity. Wood and Wood (1992) have suggested that the cellobiose quinone oxidoreductase is not a separate enzyme but a breakdown product of cellulose oxidase. This enzyme might act as a link between the process of lignin and cellulose degradation. Cellulose is the most abundant renewable organic resource. It comprises approximately 45% of dry wood weight. Crystalline cellulose is composed of long ß-1,4-glucan chains that are hydrogen bonded into water-insoluble microfibrils. The crystalline cellulose coexists with less ordered amorphous regions (Fig. 6.11). The fibres are embedded in a matrix of hemicellulose and lignin, but the precise organization is poorly understood.Cellulose is often characterized by the degree of polymerization (DP), which is the average number of glucan residues per molecule. The average DP of native cellulose is 3500.

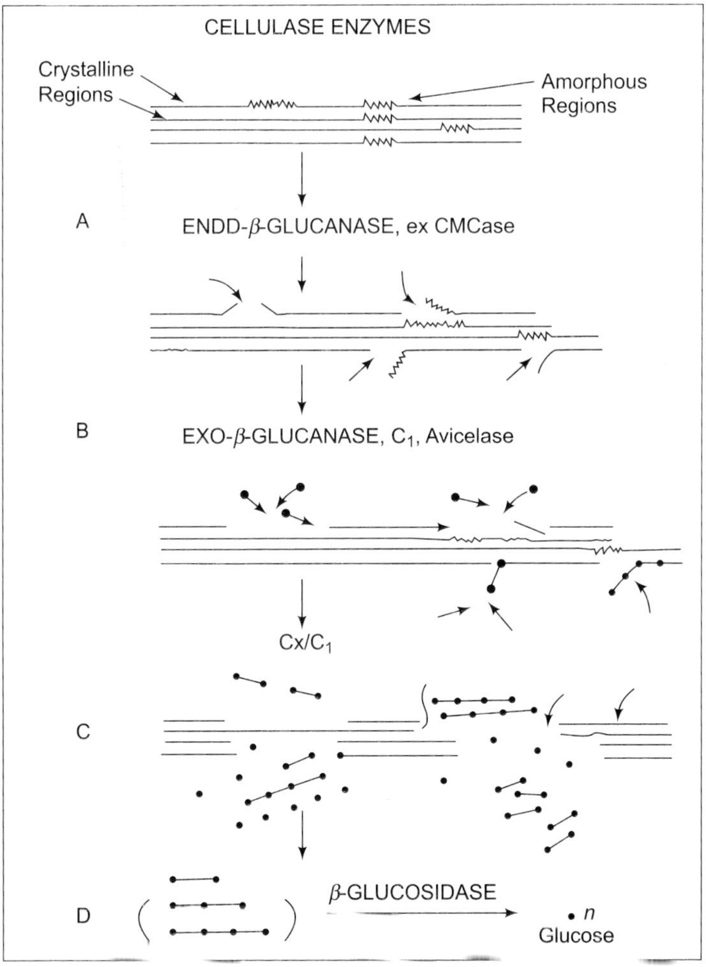

Fig. 6.11 Schematic representation of sequential stages in cellulolysis.

Applications of Cellulolytic Enzymes

Cellulose finds a wide application in various industrial processes. Cellulose and hemicellulose preparations are used to clear fruit juices from remaining pulp particles and also help the extraction of juice and oil from fruit and seed pulp. These enzymes are also expected to have an impact on the processing of animal feed. Treatment of silage with cellulases, for partial hydrolysis of the plant cell wall to speed up the release of sugars to be fermented by lactic acid bacteria. In addition, cellulases improve the digestibility of silage. Fungal cellulases helps in hydrolysis of glucans and xylans present in low grade feedstuffs, which increase the viscosity of bowel content, reduce nutrient uptake and lead to 'sticky faeces' problems in chicken. Besides, fungal cellulases are also employed in various bioremediation processes to clean the environmental pollutants. For instance, in the treatment of cellulosic wastes, which are produced in increasing amounts either as municipal waste or as agricultural waste.

HEMICELLULOLYTIC ENZYMES OF MUSHROOMS

Hemicelluloses are most abundant natural polymers next to cellulose and are present in association with lignin and cellulose in plant cell wall. The term hemicellulose was first introduced by Schuleze (1891) for the fractions extracted from plant material with dilute alkali. Hemicelluloses include xylan, mannan, galactan and arabinan as the main heteropolymer depending on the types of sugar monomer moieties present such as D-xylose, D-mannose, D-galactose and L-arabinose respectively. Unlike cellulose, hemicelluloses consist of different monosachharide units.

In addition, the polymer chains of hemicelluloses have short branches and are amorphous. Because of the amorphous morphology, hemicelluloses are partially soluble or swellable in water. The backbone of the chains of hemicelluloses can be a homopolymer (generally consisting of single sugar repeat unit) or a heteropolymer (mixture of different sugars). Formulas of the sugar component of hemicelluloses are listed in Fig. 6.12. Among the most important sugar of the hemicelluloses component is xylose.

MANNAN AND GLUCOMANNAN SUBSTANCES

Mannan exists in nature in two forms, galactomannan and glucomannan. Mannans are present in some plants such as endosperm of copra and ivory palm nuts, seeds of guar, beans of locust and coffee and roots of Konjac (Tamaru *et al.*, 1995; Braithwaite *et al.*, 1995; Tipson and Horton 1976). Mannan consists of a backbone of β-1,4 linked mannose residues whereas glucomannan comprises a heterogeneous polymer of β-1,4 linked glucose and mannose sugar (Fig. 6.13a). The backbone of both mannan and glucomannan can also have α-1,6 linked galactosyl residues and thus these polysaccharides are often referred to as galactomannan and galactoglucomannan. Galactoglucomannan is generally found in the cell walls of seed such as carob, where its main function is as a storage

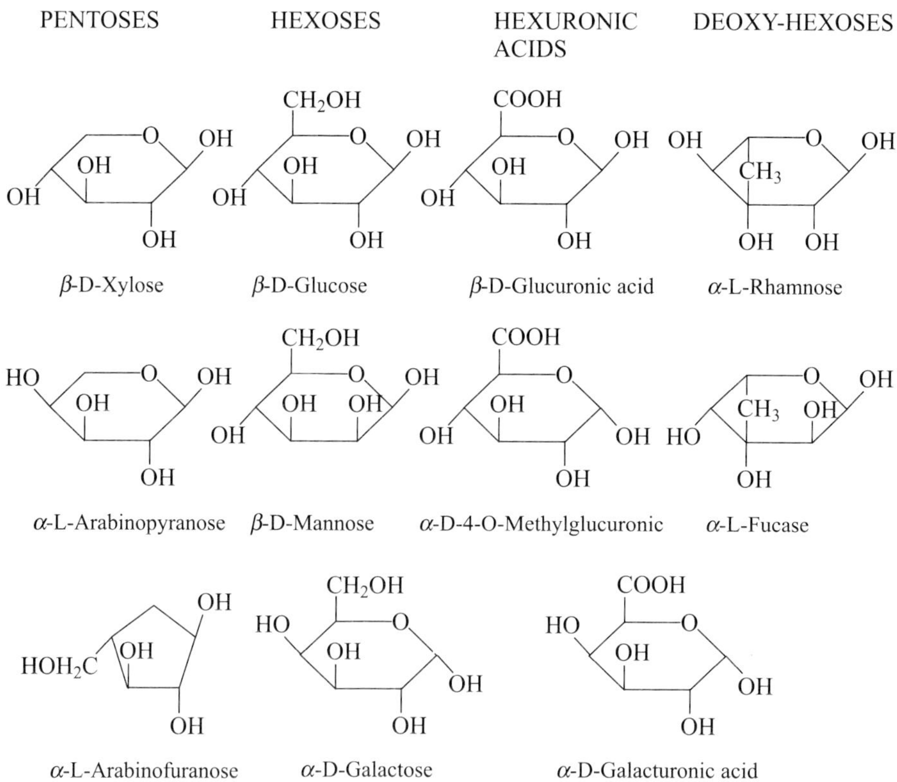

Fig. 6.12 Schematic illustration of sugar units of hemicelluloses.

polysaccharide. Mannan is an important structural component of marine algae and terrestrial plants such as ivory nuts and coffee plants (Jones 1950). It is also present as principal hemicellulose in softwood, accounting for up to 25% of dry weight, are O-acetylgalactoglucomanan in which the backbone comprises mannose and glucose in the ratio 3:1, the glucose residues may be distributed randomly (Stoll *et al.,* 1999; Puls and Schuseil,.1996; Tenkanen *et al.,*1997).

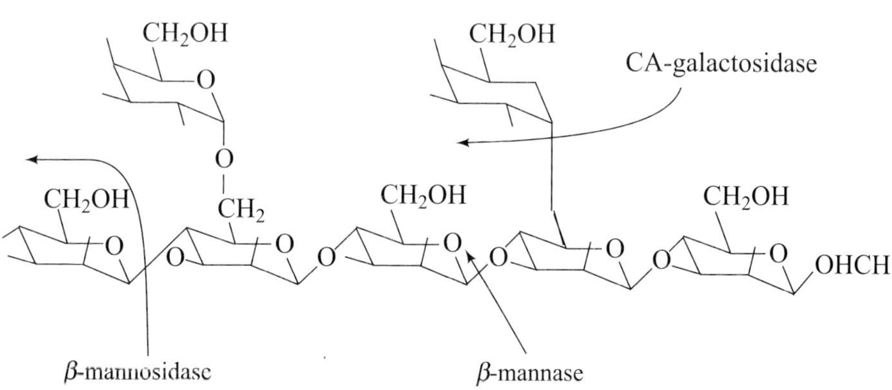

Fig. 6.13a Structure of galactomannan with possible site for enzyme hydrolysis.

Glucomannan is a dietary fiber obtained from tubers of *Amorphophallus konjac* cultivated in Asia. Glucomannan is used as a hunger suppressant because it produces a feeling of fullness by creating very viscous solutions that retard absorption of the nutrients in food. One gram of this soluble polysaccharide can absorb up to 200 ml of water, so it is also used for absorbent articles such as disposable diapers and sanitary napkins. The polysaccharide consists of glucose (G) and mannose (M) in a proportion of 5:8 joined by 1β → 4 linkages (Fig. 6.13b). The basic polymeric repeating unit has the pattern: GGMMGMMMMMGGM. Short side chains of 11–16 monosaccharides occur at intervals of 50–60 units of the main chain attached by 1β → 3 linkages. Also, acetate groups on carbon 6 occur at every 9–19 units of the main chain. Hydrolysis of the acetate groups favors the formation of intermolecular hydrogen bonds that are responsible for the gelling action.

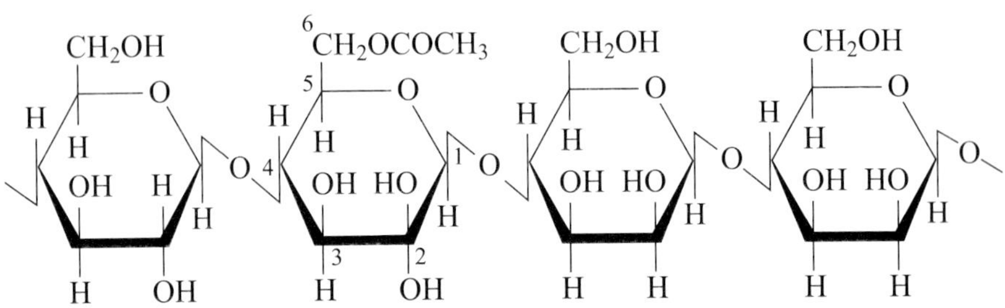

Fig. 6.13b A portion (GGMM) of the glucomannan repeating unit.The second glucose has an acetate group.

Mannanase Producing Fungi

Mannanase has most widely been reported to occur in microorganisms, although it is also found in plants and animals. Mannanase (mannan endo-1,4 β-mannosidase; EC 3.2.1.7.8) is produced by a number of fungi, yeast, bacteria and marine algae. Fungal β-mannanase have been described by Hashimoto and Funkumoto (1969a) and yeast mannanase have been reported by Kremnicky *et al.* (1996) and Feng *et al.* (2003) Mannanase possess different characteristic in terms of molecular weight, pI, pH and temperature (Hossain *et al.,* 1996). Importanat mannanase producing mushroom species include *Pleurotus spp, Lentinus spp., Trametes,* etc. (Ishihara and Shimizu, 1980).

Mannan Degrading Enzymes

Three different types of mannanases are involved in degradation of mannan which are as follows:

β-1,4-Endomannanase (1,4-β-D-mannan mannohydrolase. EC 3.2.1.78) : β-D mannanase (1,4-β D mannan mannohydrolase. EC 3.2.1.78) is an enzyme capable of hydrolyzing 1,4-β D mannopyranosyl linkages of mannans, glucomannas, galactomannan and galactoglucomannanans yielding mannooligosaccharides (Zhang *et al.,* 2002). It is one of the key enzyme groups that attack the plant cell wall. There are abundant natural sources of 1,4-β-D mannopyranosyl linkages of polysaccharides for examples konjac powder, locust bean gum, gaur gum etc. Gaur gum is a natural product derived from the leguminous plant, *Cyamopsis tetragonolobus*. It consists of a backbone of O-mannose residues bound to each other by a β-1, 4 linkages of galactose in a ratio of 1:2 are also randomly attached to the backbone by α-1,6 linkage (McCleary *et al.*, 1985). The repeating unit and structure of guar gum is illustrated in Fig. 6.14. Locust bean gum is a polysaccharide made of a long chain of the sugar galactose and mannose. Locust bean gum is extracted from the endosperm of the seeds of the carob tree (*Ceratonia siliqua*). It consist of only mannose, with galactose monomer linked to it randomly by α-1,6 bonds. The hydrolysis of locust bean gum requires endo β-mannanase, β-mannosidase and α-galactosidase for the complete hydrolysis. Locust bean gum is off white in colour and soluble in hot water. The chemical structure of locust bean gum is illustrated in Fig. 6.15.

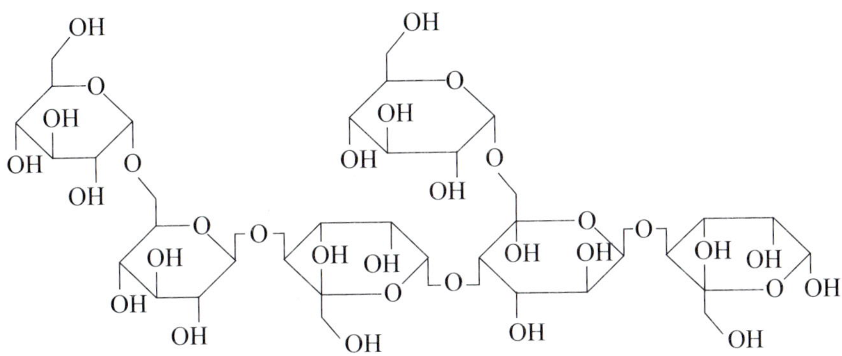

Gal	Gal		Gal	Gal	Gal
1	1		1	1	1
↓	↓		↓	↓	↓
6	6		6	6	6

4Man 1 → 4Man 1 → 4Man 1 → 4Man 1 → 4Man 1 → 4Man 1 → 4Man 1
→ 4Man 1 → 4Man 1 → 4Man 1 →

(Man = β-D-mannopyranose, Gal = α-D-galactopyranose)

Fig. 6.14 Chemical structure of gaur gum.

These endomannanases are the most widely studied and most characterized mannan-degrading enzymes :

β-Mannosidase (1,4-β mannoside mannohydrolase; EC 3.2.1.25) : β-mannosidases are exoglycosidases that hydrolyze short oligosaccharides like mannotriose and the higher mannooligosaccharide from the non-reducing end to liberate mannose. Among the oligomers, mannobiose is usually the best substrate. The affinity of the enzyme towards manooligosaccharides decreases with increasing chain length. True β-mannosidases is able to cleave artificial substrates like p-nitrophenyl-β-D-mannopyranoside.

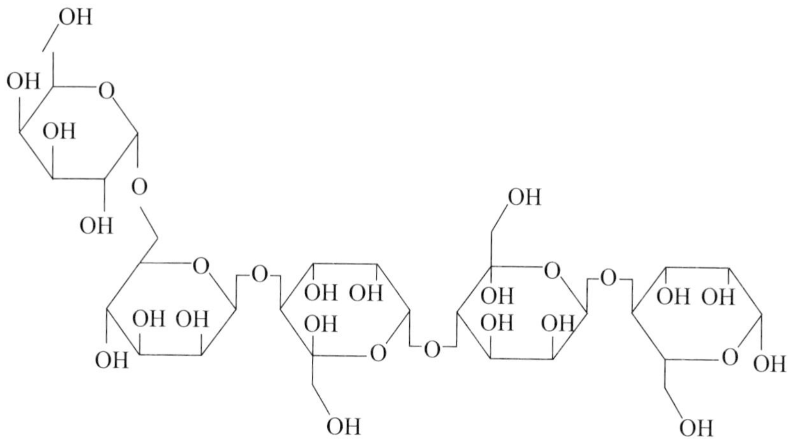

Gal 1 ↓ 6   Gal 1 ↓ 6

4Man 1 → 4Man 1 → 4Man 1 → 4Man 1 → 4Man 1 → 4Man 1 → 4Man 1 → 4Man 1 →

(Man = β-D-mannopyranose, Gal = α-D-galactopyranose)

Fig. 6.15 Chemical structure of locust bean gum.

α-Galactosidases (melibase or α-D-galactoside galactohydrolase; EC 3.21.22) : α -Galactosidase efficiently clears galactose from O-nitrophenyl-α-D-galactoside, and a galactomannan from *Aspergillus tamari and Thermotoga neopolitana* 5068 (Duffand *et al.,*1997).

XYLAN

Xylan are the major renewable hemicellulose polysaccharides consisting principally of D-xylose as its monomeric unit and traces of L-arabinose (representing 30–35% of the total dry weight of higher plants (Joseleau *et al.,*1992; Puls and Poutanen, 1989) with a high potential for degradation to useful end products. It occurs as heteropolysaccharides, containing different substituent groups in the backbone and in the side chain. The common substituents found on the backbone of xylan are acetyl, arabinosyl and glucuronosyl residues (Biely, 1985). Xylans having β-1,3-linked backbone are found only in marine algae (Dekkar and Richards, 1976) while β-1,4-linked xylans are characteristics of hardwoods, softwood and grasses (Timmel, 1967). The mixed link β-1,3- and β-1,4-xylans are found in seaweed such as *Palmeria palmate*.

Substituted Xylan

Xylans, including arabinoxylan, glucuronoxylans, and glucuronoarabinoxyalans, are quantitatively minor components of the primary walls of dicots and non-graminaceous monocots but are abundant in the walls of the Poaceae and secondary walls of woody plants. These polysaccharides all have a backbone composed of 1,4-linked β-D-Xyl residues. Many of the Xyl residues are substituted at O-2 and/or O-3 with α-Ara, β-GlcA, 4-O-Me β-GlcA, and β-Xyl residues. The backbone Xyl residue may also be *O*-acetylated. The side chain Ara residues may be esterified with ferulic or coumaric acid (Fig. 6.16).

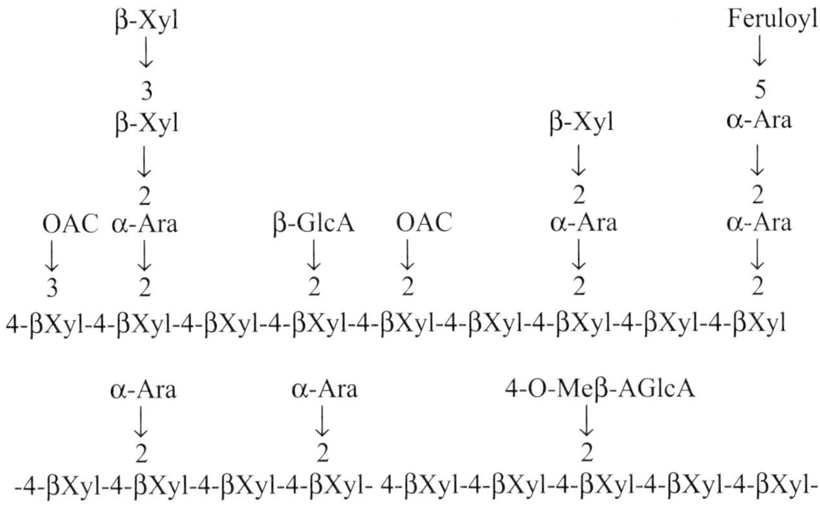

Fig. 6.16 Structure of the substituted xylan.

Hardwood Xylan

The xylan of hardwood, which accounts for 10–35 % of dry weight (Dekkar and Richards, 1976), is typically acetyl-4-O-methylglucronoxylan (DP 200). Approximately 10% of the backbone xylose units are α-1,2-linked to a 4-O-methyl-α-D-glucuronic acid residues, while 70% of xylose residues are acetylated at C-2 or C-3 or both. Most hardwoods contain small amounts of rhamanose and galacturonic acid as integral component of the main chain. In hardwood xylan, the backbone chain consists of xylose units which are linked by β-1,4-glycosidic bonds and branched by α-1,2-glycosidic bonds with 4-O methylglucuronic acid groups. In addition, O-acetyl groups sometime replace the OHgroups in position C2 and C3 (Fig. 6.17).

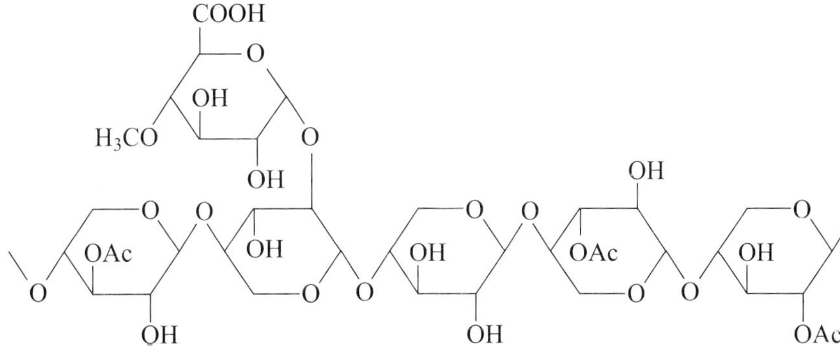

Fig. 6.17 Partial xylan structure from hardwood chain.

Softwood Xylan

The xylan of softwoods (10–15% of dry weight) is an arabino-4-O-methylglucronoxylan (DP>120) and differs from that of hardwood in not being acetylated. Substituents are 4-O-methylglucuronosyl residues attached to C-2 and L-arabinofuranosyl residues attached to C-3 of the relevant xylose backbone unit (Puls and Pountanen, 1989). For softwood xylan, the acetyl groups are fewer in the backbone chain. However, softwood xylan has additional branches consisting of arabinofuranose units linked by α-(1,3)-glycosidic bonds to the backbone (Fig.6.18).

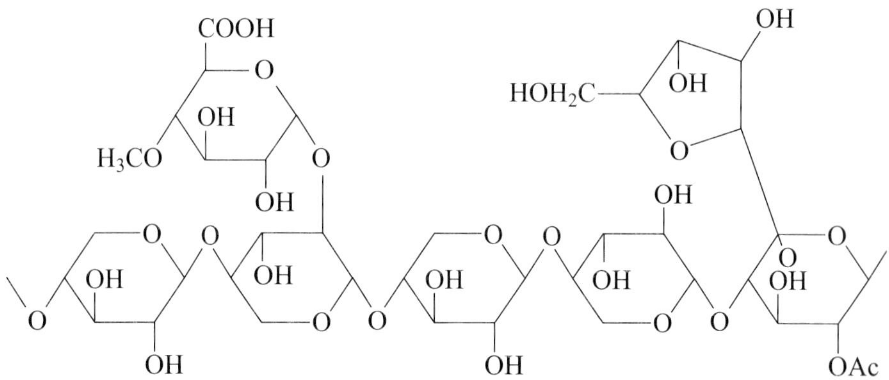

Fig. 6.18 Partial xylan structure from softwood chain.

Xylan of Grasses and Cereals

The xylan of grasses and cereals is also an arabino-4-O-methylglucronoxylan (DP 70), but it has a lower 4-O-methylglucuronic acid content than hardwood xylan. A larger proportion of L-arabinofuranosyl side chains are linked to C-2 or C-3 (Fig. 6.19), or both of the xylose main residues. These xylans also contain 2-5% by weight of O-Acetyl groups linked to C-2 or C-3 of the xylose units (Bacon *et al.*,1975). Approximately 6% of the arabinosyl side chain residues are substituted at C-5 with feruloyl group, while approximately 3% are substituted at C-5 with p-coumaroyl residues (Mueller-Harwey *et al.*,1986). The relative proportion of thc substituents vary from species to species. Feruloyl and courmaroyl groups have been implicated in the formation of covalent linkages between xylans, xylan and lignin and between xylan and other cell wall polysaccharides particularly galacturonans (Joseleau *et al.*,1992). In wheat flour, the distribution of the type of substitution is not random (contiguous similarly substituted residues being preferred due to the enzymatic mechanism) but the distribution of substituted (irrespective of the substitution type) residues along the chain appears random. The arabinose residues may also be linked to other groups attached such as glucuronic acid residues and ferulic acid crosslinks.

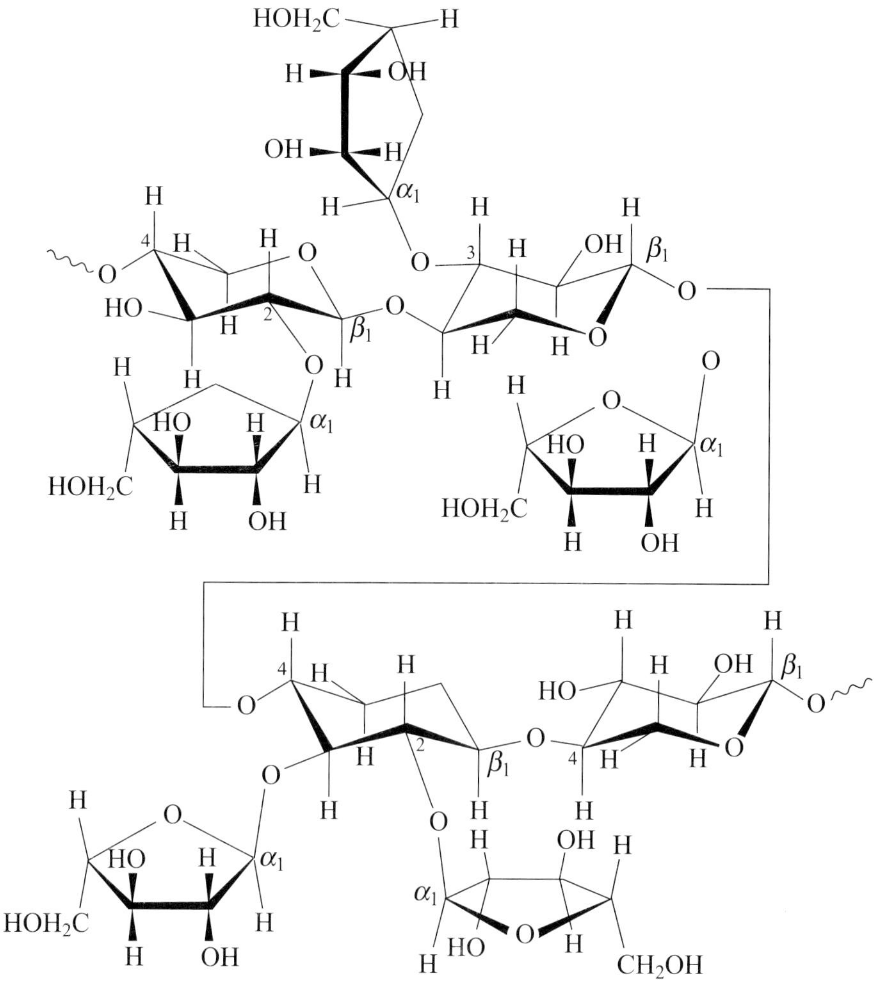

Fig. 6.19 Structure of arabinoxylan in the bran of grasses (*Graminiae*).

Xylan Degrading Fungi and their Enyzymes

Xylan is hydrolyzed by three groups of hydrolytic enzymes that act cooperatively to convert xylan to its constituent sugars (Gilbert and Hazlewood, 1993 & Sunna and Antranikian, 1997). Xylanases are widespread in nature and have been reported in marine and terrestrial bacteria, rumen and ruminant bacteria, fungi, actinomycetes, marine algae, protozoa, snails, crustaceans, insects and seeds of terrestrial plant (Dekkar and Richards, 1976). According to the mode of action, these enzymes have been grouped in three classes.

Group I:

Endo-β-1,4-D-xylanases (β-1,4-D-xylan xylanohydrolase;EC 3.2.1.8) : β-1,4-endoxylanase cleaves the internal glycosidic linkages of the heteroxylan backbone resulting in a decreased DP of the substrate. The attack of the substrate is not random and the bonds to be hydrolyzed depend on the nature of substrate (e.g. length and degree of branching of substrate or the presence of substiuents (Reilly,1981). During the early course of hydrolysis of xylan, the main products formed are xylo-oligosaccharides which are further hydrolysed to xylotriose, xylobiose and xylose. The endo-acting xylanases have been differentiated according to the end products released from the hydrolysis of xylan (e.g. xylose, xylobiose and xylotriose, and/or arabinose) as non-debranching (arabinaose non liberating) or debranching (arabinose liberating) enzymes (Dekker and Richards, 1976). Endoxylanase could be grouped into those that are low molecular weight (below 30,000 Da) basic proteins with high pI values and high molecular weight (above 30,000Da) acidic proteins with low pI values.

β-1,4-D-xylosidase (β-1,4-D-xyloside xyhydrolase;EC 3.2.1.37) : β-1,4-D-xylosidase hydrolyze short oligosaccharides and xylobiose from the non-reducing ends to liberate xylose. These are larger enzymes with molecular weight between 60 and 360 kDa and may be monomeric or dimeric proteins. The affinity of the enzyme towards xylo-oligosacchharides decreases with increasing DP.

Group II:

Acetylan esterase (EC 3.1.1.6) : These estaerase break the ester linkages between xylose unit of the xylan polymer and acetic acid i.e. acetyl xylan, and it removes O-actyl groups from C-2 and/or C-3 of xylose residues in xylan. The production of acetylxylan esterase was first reported in fungi (Biely *et al.*, 1985). The main reason for the late discovery of this enzyme was due to the lack of suitable substrate (Sunna and Antranikian, 1997).

α-L-arabinofuranosidase(α-L-arabinofuranosidase arbinofuranohydrolase; EC 3.2.1.55) :

α -L-arabinofuranosidases are characterized by their ability to cleave artificial (p-nitropheyl- α -L-arabinofuranosidases) and natural (branched arabinan) substrates, their specificity is towards the glycosidic linkages viz. α -1,2, α -1,3 or α -1,5 and to effect the release of arabinose from singly or doubly substituted xylose residues of arabinoxylan (Coughlan and Hazlewood,1993 & Sunna and Antranikian, 1997). An endo-1,5- α -L-arabinase (EC 3.2.1.99) has been found to be active only on linear arabinans. The size of native arabinofuranosidases may reach up to 495 kDa and are found in mono-, di-, tetra-, hexa- and octameric forms (Coughlan and Hazlewood, 1993).

α -Glucuronidase (α-4-O-methyl-D-glucuronidase; EC 3.2.1.1): α-D-glucuronidase hydrolyzes the α -1, 2 linkage between glucuronic acid and xylose residues in glucuronoxylan. So far, very few enzymes having molecular weight above 100 kDa have been purified but the substrate specificity differs according to the enzyme source (Sunna and Antranikian, 1997).

Group III

Ferulic acid esterase (EC 3.1.1.1.1) and p-Coumaric acid esterase (EC 3.1.1.1) : Ferulic acid esterase cleaves the ester linkages between arabinose side chains and ferulic acid in xylan and p-coumaric acid esterase cleaves the ester linkage between arabinose side chain and p-coumaric acid. The purified p-coumaroyl esterase is a dimmer with a molecular weight of 11 kDa and the feruloyl esterase (two in number) are monomeric with molecular weights of 68 and 24 kDa, respectively (Mackenzie *et al.,* 1987 and Borenman *et al.,*1993).

Applications of Hemicellulolytic Enzymes

Xylanases have shown an immense potential to create a technological breakthrough in most of the industrial processes. Enzymes application in the prebleaching of kraft pulp improves pulp fibrillation and water retention, reduction of beating times in virgin pulps, restoration of bonding and increases freeness in recycled fibres and selective removal of xylans from dissolving pulps (Bajpai, 1999). It is useful in yielding cellulose from dissolving pulps for rayon production and biobleaching of wood pulps (Vikari *et al.,* 1994a; Srinivasan and Rele,1995). Enzymatic degradation of the hemicellulose-lignin complexes present in pulps leaves the cellulose fibres intact and strongly reduces the amount of bleaching chemicals (e.g. chlorine) required. This not only results in reduction of costs of chemicals but also reduces the environmental problems caused by the use of chlorine. Xylanolytic enzyme preparations are widely used in the bakery industry as solubilizing agent and solubilize arabinoxylan fraction of the dough resulting in increased bread volume and an improved quality of the dough (Petit-Benvegnen *et al.,*1998; Poutanen, 1997). Xylanases are used concurrently with cellulose and pectinase for clarifying must and juices and for liquefying fruits and vegetables, and in the pretreatment of forage crops to improve the digestibility of ruminant feeds and to facilitate composting (Gilbert and Hazzlewood, 1993; Biely, 1985). Alkyl glycosides are one of the most promising candidates for new surfactants. Commercially, they are produced from monomeric sugars such as D-glucose and a fatty alcohol. But the direct glycosylation using polysaccharide can be feasible for their industrial production because hydrolysis of polysaccharides and subsequent steps can be omitted. Xylanase from *Aurobasidium pullulans* have been used for direct transglycosylation of xylan, 1-octanal and 2-othyl hexanol into octyl-β-xylobioside, xyloside and 2-ethylhexyl-β-xylobioside, respectively. Xylanases can also be used to improve cell wall maceration for the production of plant protoplasts (Wong *et al.,* 1998). An important application of xylanolytic enzyme system in conjuction with pectinolytic enzyme is in the degumming of bast fibres such as flax, hemp, jute and ramie (Puchart *et al.,* 1999). Xylanase-pectinase combination is also used in debarking process, which is the first step in wood processing (Wong and Saddler, 1992; Bajpai, 1999). The fibre liberation from plants is affected by retting i.e. the removal of binding material present in plant tissues using enzymes produced *in situ* by microorganisms. According to Sharma (1987) pectinases are believed to play main role in this process, but xylanases may also be involved.

Mannanases recently attracted the attentions of industrial sector due to their positive role in the pulp and paper industry to remove the hemicelluloses from the dissolving pulps (Gubitz *el al.,*1997). Their effective role in pulp bleaching process minimized the use of environmentally harmful bleaching chemicals in pulp and paper industry (Lantinen *et al.,* 1995; Cuevas *et al.,* 1996). Mannanases have been also used in the food industry for the extraction of vegetable oils from leguminous seeds and the clarification of fruit juices (Christgau *et al.,* 1994). They also used in the reduction of the viscosity of extracts during manufacturing of instant coffee chocolate and cacao liquor (Belitz and Grosch 1987; Francoise *et al.,* 1996) to lower the cost for subsequent evaporation and drying (Wong and Saddler, 1993). In addition, mannanases are potentially applied in the pharmaceutically interesting oligosaccharides (Christgau *et al.*, 1994).

PROTEOLYTIC ENZYMES OF MUSHROOMS

Industrially available proteolytic enzymes produced by microorganisms are usually mixtures of endopeptidases (proteinases) and exopeptidases.Most fungal proteases tolerate and act effectively over a wide range of pH (about 4 to 8). Fungal protease has been used for centuries in the Orient for the production of soy sauce, tamari sauce, and miso, a breakfast food. In these usages, soya beans or other grains are steamed and inoculated with spores of *Aspergillus flavus-oryzae or Aspergillus tamari.* After maximum enzyme production has taken place, the koji is covered with brine and enzymatic digestion allowed to take place. In this process enzymes were not separated from the producing organisms. One of the largest uses for fungal protease is in baking bread and crackers. The proper amount of protease action reduces mixing time and increases extensibility of doughs, and improves grain, texture, and loaf volume. Cereal foods are also treated with proteolytic enzymes to modify their proteins, resulting in better processing, including improved product handling, increased drying capacity, and lower power requirements. To prevent development of undesirable haze in beer and ale when these beverages are cooled, proteolytic enzymes are added during the finishing operation to "chillproof" these beverages. Chillproofing agents contain pepsin, papain, bromelin, fungal and bacterial proteases in various combinations, and digest enough of the protein to prevent formation of haze. Fungal proteolytic enzymes are used for tenderizing meats, and animal casings for processed meats. Recent work at the American Meat Institute has shown that various proteolytic enzymes preferentially attack different meat tissues. Recent practical tests have indicated that combinations of plant, fungal, and bacterial proteases have an advantage over any single enzyme for meat tenderizing. Pharmaceutical and clinical applications for fungal proteases include their use in digestive aids, and in debridement of wounds and by injection to relieve inflammation, bruises, and blood clots. *Agaricus bisporus*, *Coprinus cinereus,* and *Volvariella volvacea* are prominent producers of proteinases (Kalisz *et al.*, 1987). *Absidia blakesleeana* and *Morchella esculenta* are the most promising strains for production of protease in liquid cultures on synthetic medium. *Clitocybe*, *Boletus*, *Coprinus* possess high fibrinolytic and low thrombolytic activity. Denisova *et al.* (1989) used two stable strains of *Coprinus* for their potential as industrial producers of fibrinolytic proteases. Increase in the intracellular activity of mycelial proteinases has been observed during morphogenesis in *Flammulina velutipes* (Chao and Gruen, 1987). Protease activity increased steadily during cultivation of *Pleurotus ostreatus* on rice and sugar cane straw, without a marked jump at any stage; the activity stayed low in the spent straw. Terashita *et al.* (1995) while working on *Flammulina velutipes*, found metal proteinase having the isoelectric point pH 6.0 as important for the growth of fruiting body and basidiospore formation. Similarly, Murao *et al.* (1986) found carboxyl proteinase(s) in the caps of *Lentinus edodes* as essential for the maturation of basidiospores.

OXIDO-REDUCTASES AND PYRANOSE-OXIDASES OF MUSHROOMS

Tsuyl *et al.* (1989) observed the stereospecificity of NAD(P) H oxidation catalyzed by 4-aminobenzoate hydroxylase from *Agaricus bisporus*. The action of glucose oxidase in reducing quinone improved the efficiency of lignin depolymerization by *Trametes versicolor* (Szklarz and Leonowicz, 1986). Khan and Overend (1990) found an oxygenase enzyme system from *Trametes versicolor* was capable of attacking lignin and a large number of di- and tri-substituted benzene rings containing at least one hydroxy group. This enzyme was produced late in the growth cycle without the requirement for any inducer. The complex of oxido-reductases of *Coriolus hirsutus* was immobilized on fibrous supports by Vol'f *et al.*, in 1986. similarly the conditions for production and properties of pyranose oxidase from *Polyporus* was evaluated by Uchida *et al.* in 1995. Nakanishi and Machida in 1987 found a glucose-specific pyranose oxidase produced by *Coriolus*, *Daedaleopsis*, *Gloeophyllum* or *Pleurotus* as useful for enzymic determination of glucose and of substances or enzymes yielding glucose as a reaction product.

OTHER ENZYMES

Studies conducted from 58 accessions of *Lentinus edodes* revealed quite different pattern of isoenzyme in its mono- and dikaryotic mycelia. Besides intercompatibilities and morphological features, employing an index of similarity based on relative mobilities, the enzyme bands are also useful in analysis of species relation (Magae *et al.*, 1990). The mannitol dehydrogenase purified from *Agaricus bisporus* was used for the determination of D-mannitol and also the D-mannose content of glycoproteins and polysacchrides following the liberation of hexose and its reduction to D-mannitol (Berenzenko and Sturgeon, 1991). The white-rot fungus *Trametes versicolor* degrades lignocellulosic materials at least in part by oxidizing the lignin via a number of secreted oxidative and peroxidative enzymes. An extracellular reductive enzyme, cellobiose dehydrogenase (CDH), oxidizes cellobiose and reduces insoluble Mn (IV) O_2, commonly found as dark deposits in decaying wood, to form Mn (III), a powerful lignin-oxidizing agent. CDH also reduces ortho-quinones and produces sugar acids that can promote Mn-P and therefore exhibit ligninolytic activity. Two allelic variants are known to encode cellobiose dehydrogenase from *Phanerochaete chrysosporium* (Li *et al.,* 1997). Activities of trehalase and glycogen phosphorylase were determined for stage 1 (pin) sporophores of *Agaricus bisporus* harvested daily throughout several fruiting cycles (flushes). Both enzymes exhibited maximum activity, co-incident with peak flush as defined by maximum yield of stage 2 to 4 sporophores, and minimum activity during the interflush periods. Glycogen phosphorylase activity sometimes peaked on more than one occasion during a flush, whereas trehalase showed only a single peak of activity per flush. These results are discussed in relation to metabolic events associated with carbohydrate changes during the periodic fruiting cycle of this mushroom by Wells *et al.* (1987). A novel type of trehalose phosphorylase is also reported from *Flammulina velutipes*. Comparison of the specific activity of trehalose phosphorylase with that of trehalase suggested that the function of the former enzyme is more important in the fruiting bodies (Kitamoto *et al.*, 1988). Similarly, a novel type of ascorbate oxidase has also been purified from the cytosolic fraction of the mycelia of *Pleurotus ostreatus* with an overall yield of 13%. The enzyme is a hemoprotein, quite similar to β type cytochrome, and contains 2 mol of heme per molecule. The reaction is catalyzed by the enzyme L-ascorbic acid + O_2 α-dehydro- L-ascorbic acid + H_2O_2 (Kim *et al.,* 1996b). The RNA-ase and DNAase enzymes are also known in fresh fruiting bodies of *Flammulina velutipes* (Kurosawa *et al.,* 1990). Application of recombinant DNA technology for production of wood-degrading enzymes of *Coriolus versicolor* is reviewed with 19 references (Williams *et al.,* 1990). Evolution of 3 species of *Pleurotus,* i.e. *Pleurotus florida*, *Pleurotus ostreatus,* and *Pleurotus sapidus*, has been traced by comparing their isoenzymes of malate dehydrogenase, malic enzyme and tetrazole oxidase (Haung *et al.*, 1987). Proline iminopeptidase from *Pleurotus ostreatus* can be used for removing the bitterness from food, such as soy proteins (Yoshimoto *et al.*, 1991). *Agaricus bisporus* has the enzymic potential to assimilate ammonia by the activities of glutamine synthetase, NAD-, and NADP-dependent glutamate dehydrogenases. It also contains glutamate synthetase and a number of transaminating activities such as glutamate-oxaloacetate transaminase, glutamate-pyruvate transaminase, and alanine-glyoxylate transaminase. The specific activities of the ammonia assimilating enzymes showed no variation during maturation of the fruiting bodies (Baars *et al.*, 1994). Oxalate decarboxylase was detected both intra- and extracellularly in liquid cultures of *Coriolus versicolor*. Immunogoldcytochemical labeling of ultrathin sections of the degraded beach wood showed that the enzyme was located close to the plasma membrane and in intracellular vesicles (Dutton *et al.,* 1994). Highest levels of ammonia assimilating enzymes *viz*, NAD-dependent glutamate dehydrogenase and glutamine synthetase has been found in fruiting bodies of *Pleurotus ostreatus* (Mikes *et al.*, 1994). A guanine nucleotide-specific RNA-ase has been isolated from the fruiting bodies of *Pleurotus ostreatus*. An oxalate decarboxylase has been purified from the fungus *Collybia velutipes,* and the gene encoding it is also cloned for expression in plants. Several other enzymes of mushroom species are depicted in Table 6.6.

Table 6.6 Other enzymes of mushrooms.

Mushroom Species	*Enzyme(s)*	*Properties*	*References*
Agaricus bisporus	Glutamate dehydrogenase	NADP-dependent; key enzyme in NH_3 assimilation; molecular weight 330 kDa;Isoelectric point at pH 4.8; pH optima for amination and deamination were 7.6 and 9.0; optimum activity at 33°C	Yamamoto *et al.* (1986)
Agaricus bisporus	DNA polymerase	Purification yielded 3 bands of 72, 65, and 36 kDa; optimum activity at pH 7.8 in the presence of 5 mM Mg and 50 mM KCL at 28–30°C; found only in the meiotic cell rich fraction	Llerena *et al.* (1987)
Agaricus bisporus	Propyl endopeptidase	Enzyme was most active at pH 7.5 and stable in the range of pH 5–9, when checked with Z–Gly–Pro–napthylamide (Z = N-benzoxycarbonyl)	Sattar *et al.* (1990)
Collybia velutipes	Oxalate decarboxylase	Fraction A of the 2 purified fractions was 1670-fold; had a Km of 4.5 mM for potassium oxalate; inhibited by PO_4 (K i for PO_4 = 9 mM); optimum pH at 3.0	Dodson *et al.* (1987)
Coprinus cinereus	Endonuclease	Purified from basidiocarp tissues; its peak activity appears during the meiotic prophase; optimal pH 8.3 and temperature 50°C; contains a single 43 kDa polypeptide; converts the supercoiled DNA to relaxed DNA	Sakaguchi (1991)
Flammulina velutipes	Trehalose phosphorylase	Catalyzes both the reversible phosphorlysis of trehalose to form d-glucose-l-phosphate + glucose and also synthesis of trehalose (the latter is more important in the fruiting bodies).	Kitamoto *et al.* (1988)
Flammulina velutipes	Nicotinamidase	Activity optimum at pH 7.3 and at 45°C; activation energy 10.96 Kcal molecule^{-1} nicotinaldehyde, 3-cyanopyridine, nicotinic acid-hydrazide; pyrazinamide, benzamide are competitive inhibitors	Taguchi *et al.* (1989)

(*Contd.*)

Mushroom Species	*Enzyme(s)*	*Properties*	*References*
Lentinus edodes	DNA topoisomerase	Molecular weight 71,500; no energy factor was required; activity enhanced 10 folds by Mg++ and 8 folds by KCL	Kono *et al.* (1986)
Lentinus edodes	Mitochondrial ATP-ase	Stimulated by 680 nm illumination; optimum pH 7.5; , temperature 59°C; enzyme was inhibited by 5 m mol Na+	Min *et al.*, (1987a, b)
Lentinus edodes	Acid phosphatase	Stable at pH 4; 60°C; activity markedly inhibited by NaF, $K_2Cr_2O_7$, $Na_2M_0O_4$, and Na_2NO_4	Sawada & Endo, (1987)
Lentinus edodes	RNA-ase	Molecular weight 21,000; one of the base nonspecific and adenylic-acid preferential RNA-ases such as RNA-ase T2, RNA-ase Rh and RNA-ase M	Shimada *et al.* (1991)
Lentinus edodes	Aryl-alkyl oxidase	Purified from wheat bran culture; molecular weight 65,000 oxidizes guaiacol, O- and p-phenylene-diamines, syringaldazine, catalyzes alkyl-aryl cleavage	Kofujita *et al.* (1991)
Lentinus edodes	Branched chain amino acid tranferase	Isoenzyme II; mol wt.=72,000; N-terminal amino acid glycine; optimum pH 8.2 and temperature 38°C; stable at 20°C for 30 min; activity on L-leucine as substrate; inhibited by hydroxylamine, N-ethylamaleimide, and P-chloromercuribenzoate, and by Cu, Hg, and Fe	Min *et al.* (1992)
Panaeolus papillonaceus	α-fructo-furanosidase (inulinase)	Molecular weight – 1,16,000; most active on sucrose; Ag and Hg are highly inhibitory, sucrose; inulinase = 5.7; optimum temperature = 60–65°C, and pH = 6.0; Fructose was detected as liberated sugar from raffinose, stachyose, inulin	Mukherjee & Sengupta (1987)
Pleurotus ostreatus	Proline Iminopeptidase	Extracted from mushrooms; optimum pH 7.2 and temperature 42°C	Yoshimoto *et al.* (1991)

Conclusions and Future Outlook

The practical application and industrial use of enzymes to accomplish certain reactions apart from the cell dates back many centuries and was practiced long before the nature or function of enzymes was understood. Use of barley malt for starch conversion in brewing, and of dung for bating of hides in leather making, are examples of ancient use of enzymes. It was not until nearly the turn of this century that the causative agents or enzymes responsible for bringing about such biochemical reactions became known. Then crude preparations from certain animal tissues such as pancreas and stomach mucosa, or from plant tissues such as malt and papaya fruit, were prepared which found technical applications in the textile, leather, brewing, and other industries. Once the favorable results of employing such enzyme preparations were established, a search began for better, less expensive, and more readily available sources of such enzymes. It was found that certain microorganisms produce enzymes similar in action to the amylases of malt and pancreas, or to the proteases of the pancreas and papaya fruit. This led to the development of processes for producing such microbial enzymes on a commercial scale. Microbial cultivated enzymes have replaced the animal or plant enzymes. At present, only a relatively small number of microbial enzymes have found commercial application and there is a need to search some more potent microbial group/s efficient in producing enzymes of industrial significance. Fungi, especially mushrooms (a basidiomycetous fungi) in this regard are the most important organisms in the whole world not only because of their vital roles in ecosystem functioning but also because of their influence on human and human related activities. Due to their various inbuild enzymatic machinery they play an important role in environmental bioremediation, food, feed, pharmaceutical, textile, paper-pulp, chemical and in agricultural industries. However, the number of applications are increasing and the field will undoubtedly be much expanded in the future. Among different mushroom species, the functional approach of enzymes of white-rot fungi towards these applications are unparallel. Industrial uses of enzymes of white-rot fungi have increased greatly during the past few years. Prospects are excellent for continuing increased usage of presently available enzymes in above mentioned applications, and in new uses for many purposes. Many more mushroom species and their enzymes with unique functions have been discovered by means of extensive screening and are now commonly used in industrial applications. Screening for such enzymes, in combination with current biotechnologies, such as genetic engineering, protein engineering (including directed and random mutagenesis), metabolic engineering and so on, will offer further exciting possibilities for much better and efficient applications of industrial enzymes from mushrooms.

References

Akamatsu, Y., Ma, D. B., Higuchi, T., and Shimada, M. (1990). A novel enzymatic decarboxylation of oxalic acid by the lignin peroxidase system of white-rot fungus, *Phanerochaete chrysosporium*. *FEBS Lett.* 269: 261–263.

Akhtar, M., Blanchette, R.A., Myers, G. & Kirk, T.K. (1998). An overview of biomechanical pulping research In:Young, R.A. & Akhtar, M. (eds.): Environmentally Friendly Technologies for the Pulp and Paper Industry, John Wiley & Sons, New York, pp. 309–340

Annadurai, G., Babu, S.R., Nagarajan, G., and Ragu, K. (2000). Use of Box-Behnken design of experiments in the production of manganese peroxidase by *Phaenerochate chrysosporium* (MTCC 767) and decolorization of crystal violet. *Bioprocess Eng.* 23 : 715–719.

Archibald, F.S., Bourbonnais, R., Jurasek, L., Paice, M.G. and Reid, I.D. (1997). Kraft pulp bleaching and delignification by *Trametes versicolor. Journal of Biotechnology*, 53: 215–336.

Arora, D.S. and Rampal, P. (2002). Laccase production by some *Phlebia* species. *J. Basic Microbiol.* 42(5): 295–301.

Atalla, R.H. and VanderHart, D.L. (1984). Native cellulose: A composite of two distinct crystalline forms. *Science* 223: 283–284.

Bacon, J.S.D., Gordon, A.H., Morris, E.J. and Farmer, V.C. (1975). Acetyl group in cell wall preparations from higher plants. *J. Biotechnol.* 23: 257–270.

Bae, S. H., Shin, K.S., Kang, S.O., Hah, Y.C., Chae, S.J., Kim, K.J., and Chol, H.T. (1989). Characterization of extracellular peroxidase from *Pleurotus ostreatus. Misaengumi Hakhoechi* 27(4): 348–356.

Bajpai, P. (1999). Application of enzymes in the pulp and paper industry. *Biotechnol. Progress.* 15: 147–157.

Baars, J.J.P., OP den Camp, H.J.M. Hermans, J.M.H., Mikes, V., Vander Drift, C., Van Griensven, L.J.L.D., and Vogels, G.D. (1994). Nitrogen assimilating enzymes in the white button mushroom, *Agaricus bisporus. Microbiology* 140(5): 1161–1168.

Barr, D.P. and Aust, S.D. (1994). Mechanism of whiterot fungi used to degrade pollutants. *Environ. Sci.Technol.* 28: 78–87.

Belitz H.D. and W. Groch, (1987). Enzymes In: Food Chemistry, Eds, Hadziev, D. Translator, Springer verlag Berline. pp: 73–127.

Berenzenko, S., and Sturgeon, R.J. (1991). The enzymic determination of D-mannitol with mannitol dehydrogenase from *Agaricus bisporus. Carbohydr. Res.* 216: 505–509.

Biely, P. (1985). Microbial xylanolytic systems. *Trends Biotechnol.* 3: 286– 290.

Boominathan, K. and Reddy, C.A (1992). Fungal degradation of lignin: biotechnological applications. In : Handbook of applied mycology. (Eds., D.K. Arora, R.P. Elander and K.G. Mukerji), Vol. IV, Marcel Dekker, New York. pp: 822.

Bonnen, A.M., Anton, L.H., and Orth, A.B. (1994). Lignin degrading enzymes of the commercial button mushroom, *Agaricus bisporus. Appl. Environ. Microbiol.* 60(3): 960–965.

Borenaman, W.S., Ljungdhal, L.G., Hartley, R.D. and Akin, D.E. (1993). Feruloyl and p-coumaroyl esterases from the anaerobic fungus *Neocallimastix* strain MC-2: Properties and functions in plant wall degradation. *In*: Hemicelluloses and Hemicellulases. Coughlan. M.P. and Hazlewood, GP. (Edr). Portland Press. London : 85–102.

Bourbonnais, R. and Paice, M. G. (1988). Veratryl alcoholoxidases from the lignin degrading basidiomycete, *Pleurotus sajor-caju. Biochem. J.* 255:445–450.

Braithwaite, K.L., Black, G.W., Hazlewood, G.P., Ali, B.R.S., and Gilbert, H.J. (1995). A non-modular endo β1,4-mannanase from *Pseudomonas fluorescens subspecies cellulosa. Biochem. J.* 305: 1005–1010.

Cabalerio, D.R., Rodriguez-Couto, S., Sanroman, A., and Longo, M.A. (2002). Comparison between the protease production ability of ligninolytic fungi cultivated in solid-state media. *Process Biochem.* 37 : 1017–1023.

Cambria, M.T., Cambria, A., Rasusa, S., and Rizzarelli, E. (2002). Production, purification, and properties of an extrcellular laccase from *Rigidiporus lignosus. Protein Expr. Purif.* 18 : 141–147.

Chao, E.E. and Gruen, H.E. 1987. Intracellular activity of mycelial proteinases during fruit body development in *Flammulina velutipes. Can. J. Bot.* 65(3): 518–525.

Christgau, S., L.N. Andersen, S. Kauppinen, H.P. Heldt-Hansen and H. Dalboege, (1994).Purified Enzyme Exhibiting Mannanase Activity; Application in Oil, Paper, Pulp, Fruit and Vegetables Juice Industry and in Carrageenan Extraction. Patent Novo-Nordisk, 9425576, 10 November (1994).

Coughlan, M.P. and Hazlewood, G.P. (1993). β-1,4-Dxylan-degrading enzyme systems: biochemistry, molecular biology and applications. *Biotechnol. Appl. Biochem.* 17: 259–289.

Cuevas, W.A. Kantelinen, P. Tanner, B. Bodie and S. Leskinen, (1996). Purification. In; Strebotnik, E. and K. Mesner (eds.), Biotechnology in the Pulp and Paper Industry, pp: 123–126. Facultas-Universitatsverlag, Vienna, Austria.

Das, N., Chakraborty, T.K. and Mukherjee, M. (1999). Role of potato extract in extracellular laccase production of *Pleurotus florida. J. Basic Microbiol.* 39(5-6) : 2999–303.

Dekker, R. F., and Richards, G.N. (1976). Hemicellulases : their occurrence, purification, properties and mode of action. *Adv. Carbohydr. Chem. Bio-Chem.* 32: 277–352.

Denisova, N.P., Semenova, I.R., and Sukharevich, V.I. (1989). Fibrinolytic proteinase biosynthesis by higher basidiomycetes in submerged culture. *Mikol. Fitopatol.* 23(4): 378–381.

Dodson P., Evans C., Harvey P. and Palmer M. (1987). Production and properties of extracellular peroxidases from *Coriolus versicolor* which catalises Cα-Cβ cleavage in a lignin model compound. *FEMS Microbiology Letters*; 42: 17–22.

D'Souza, T.M., Merritt, C.S. and Reddy, C.A. (1999). Lignin-modifying enzymes of the White-rot basidiomycete *Ganoderma lucidum. Appl. Environ. Microbiol.* 65(12) : 5307–5313.

Duffand, G.D., Carol, M., McCutchen, P., Leyduc, K.N., and Kelly, R.M. (1997). Purification and characterization of extremely thermostable enzyme α-mannanase, α-mannosidase and α-galactosidase from the hyperthermophilic eubacterium *Thermotoga neopolitana* 5068. *Appl. Environ. Microbiol.* 63: 169–177.

Durán, N. and Esposito, E. (2000). Potential applications of oxidative enzymes and phenoloxidase-like compounds in wastewater and soil treatment: a review. *Appl. Catalysis.* 28:83–99.

Dutton, M.V., Kathira, M., Gallagner, I.M., and Evans, C.E. (1994). Purification and characterization of oxalate decarboxylase from *Coriolus versicolor. FEMS Microbiol. Lett.* 116(3): 321–325.

Edwards, S.L., Raag, R., Wariishi, H., Gold, M.H., and Poulos, T.L. (1993). Crystal structure of ligninperoxidase. *Proc. Natl. Acad. Sci.* U.S.A. 90: 750–754.

Eriksson K-EL, Blanchette RA, Ander P (1990). Microbial and enzymatic degradation of wood components. Springer-Verlag, Berlin.

Eun, J.S., Yang, J.H., Cho, D.Y., Lee, T.K., andPark, I.H. (1989). Studies on higher fungi in Korea.(II) Protcolytic enzyme of *Agaricus bisporus*(Lange) Sing. Yakche Hakhoechi 19(1): 565–800.

Feng, Y., Zhimin H., Say, L.O., Jiangyong, H., Zhang, Z., and Wun, J.N. (2003). Optimization of agitation, aeration and temperature conditions for maximum b-mannanase production. *Enz.. Microbiol Technol.* 32: 285–289.

Ferrer, I., Dezotti, M. and Duran, N. (1991). Decolorization of kraft effluent by free and immobilized lignin peroxidases and horseradish peroxidase. *Biotechnol. Lett.* 13:577–582.

Forrester, I.T., Grabski, A.C., Mishra, C., Kelley, B.D., Strickland, W.N., Leatham, G.F. and Burgess, R.R. (1990). Characteristics and N-terminal amino acid sequence of a manganese peroxidase purified from *Lentinula edodes* cultures grown on a commercial wood substrate. *Appl. Microbiol. Biotechnol.* 33:359–365.

Francoise, M., C. Ghakis, C. Dupont, R. Morosoli and D. Kluepfel, (1996). Improved production of mannanase by *Streptomyces lividans. Appl. Environ. Microbiol.*, 62: 4656–58.

Gilbert, H.J. and Hazlewood, G.P. (1993). Bacterial cellulases and xylanases. *J. of General Microbiol.* 61:170–174.

Gill, P.K. and Arora, D.S. (2003). Effect of culture conditions on manganese peroxidase production and activity by some white-rot fungi. *Journal of Industrial Microbiology & Biotechnology* 30 : 28–33.

Gindllis, A.L., Zhazhina, E.O., Baranov, Y.A., Karyakin, A.A., Gavrilova, V.P., and Yaropolov, A.I. (1988). Isolation and properties of laccase from the basidial fungus, *Coriolus hirsutus* (Fr.) Quel. *Biokhimiva (Moscow)* 53(5): 735–739.

Gold, M., Wariishi, H. and Valli, K. (1989). Extracellular peroxidases involved in lignin degradation by the white rot Basidiomycete *Phaneroehaete chrysosporium.* In: ACS Symp. Series No. 389. Biocatalysis in Agricultural Biotechnology , Eds, Whitaker, J. and Sonnet, P.E., Am. Chem. Soc., Washington, pp. 127–1411.

Guillen, F., Martinez, A.T., and Martinez, M.J. (1990).Production of hydrogen peroxide by aryl-alcoholoxidase from the ligninolytic fungus, *Pleurotuseryngii. Appl. Microbiol. Biotechnol.* 32(4): 465–469.

Hakala, T.K., Maijala P., Konn J. and Hatakka A. (2004). Evaluation of novel wood rotting polypores and corticioid fungi for the decay and biopulping of Norway spruce (*Picea abies*) wood. *Enzyme Microb. Technol.* 34:255–263.

Hammel, K.E. and Moen, M.A. (1991). Depolymerization of a synthetic lignin *in vitro* by lignin peroxidase.*Enzyme Microb. Technol.* 13: 15–18.

Hammel, K.E., Jensen, K.A., Mozuch, M.D., Landucci, L.L., Tien, M., and Pease, E.A. (1993). Ligninolysis by a purified lignin peroxidase. *J. Biol. Chem.* 268: 12274–12281.

Harvey, P. J. Schoemaker, H. E., and Palmer, J. M. (1986). Veratryl alcohol as a mediator and the role of radical cations in lignin degradation by *Phanerochaete chrysosporium. FEBS Lett.* 195: 242–246.

Hashimoto, Y. and Funkumoto, J. (1969a). Studies on the enzyme treatment of coffee beans. Purification of mannanase of *Rhizopus niveus* and its action on coffee mannan. *Nippon Nogei kagaku Kaishi.* 43 : 317–322.

Hattaka, A. (1994). Lignin modifying enzymes from selected white-rot fungi: production and role in lignin degradation. *Microbiol. Rev.* 13 : 125–135.

Haung, Z., Zhou, J., Shen, Y., and Liu, H. (1987). Isoenzymes of *Pleurotus* species. Shiyong Jun 4: 8–9.

Heinzkill, M., Bech. L., Halkier. T., Schneider, P. & Anket. (1998). Characterization of laccase and peroxidase from wood-rotting fungi (family *Coprinaceae*). *Appl. Environ. Microbiol.* 64(5): 1601–1606.

Herpoel, I., Asther, M., and Sigoillot, J.C. (1999). Design and scale up of a process for manganese peroxidase production using the hypersecretory strain *Phaenerochaete chrysosporium* I–1512. *Biotechnol. Bioeng.* 65(4) : 468–473.

Hess, J., Leitner, C., Galhaup, C., Kulbe, K.D., Hinterstoisser, B., Steinwender, M., and Haltrich, D. (2002) Enhanced formation of extracellular laccase activity by the white rot fungus *Trametes multicolor. Appl. Biochem. Biotechnol.* 24 : 23–63.

Hestbjerg, H., Willumsen, P.A., Christensen, M., Andersen, O. and Jacobsen, C. S. (2003). Bioaugmentation of tar-contaminated soils under field conditions using *Pleurotus ostreatus* refuse from commercial mushroom production. *Environ. Toxicol. Chem.* 22: 692–698.

Hofrichter, M. and Fritsche, W. (1997). Depolymerization of low-rank coal by extracellular fungal enzyme systems. II. The ligninolytic enzymes of the coal-humicacid-depolymerizing fungus *Nematoloma frowardii* b 19. *Appl. Microbiol. Biotechnol.* 47:419–424.

Hong, J.S., Kim, M.K., Kim, Y.H., and Lee, Y.B. (1987a). Production and properties of laccase from *Coriolus versicolor. Han'guk Kyunhakhoechi* 15(2): 99–107.

Hossain, M.Z., Abe, J. and Hizukuri, S. (1996). Multiple forms of α- mannanase from *Bacillus* sp. KKO1. *Enzy. Microbial. Technol.* 18: 95–98.

Ikehata, K., Buchanan, I. and Smith, D. (2004). Recent developments in the production of extracellular fungal peroxidases and laccases for waste treatment. *J. Environ. Engin. Science* 3:1–19.

Ishihara, M., and Shimizu, K. (1980). *Mokuzai Gakkaishi* 26: 811–818.

Jokela, J., Laine, M., Ek, M. and Salkinoja-Salonen, M. (1993) Effect of biological treatment on halogenated organics in bleached kraft pulp mill effluents studied by molecular weight distribution analysis. *Environ. Sci. Technol.* 27:547–557.

Jones, J.K.N. (1950). The structure of the mannan present in *Porphyra umbilicalis. J. Chem. Soc.* 50: 3292–3295.

Joseleau, J.P., Conltat, J. and Ruel, K. (1992). Chemical structure of xylans and their interaction in the plant cell walls. In: Progress in Biotechnology, Vol. 7, Xylans and Xylanases (Visser, J., Beldman, G., Kusters-van Someren. M.A. and Voragen, A.G.J., Eds.), pp. 1–15. *Elsevier*. Amsterdam.

Kalisz, H.M., Wood, D.A., and Moore, D. (1987). Production, regulation and release of extracellular proteinase activity in basidiomycete fungi. *Trans. Br. Mycol. Soc.* 88(2): 221–227.

Kalisz, H.M. Wood, D.A., and Moore, D. (1989). Some characteristics of extracellular proteinase from *Coprinus cinereus. Mycol. Res.* 92(3): 278–285

Khan, A.W., Lamb, K.A. and Overend, R.P. (1990). Comparison of natural hemicellulose and chemically acetylated xylan as substrates for the determination of acetyl-xylan esterases activity in *Aspergillus. Enzyme and Microbiol. Technol*. 12: 127–131.

Khan, A.W. and Overend, R.P. (1991). An oxygenase enzyme system from *Trametes versicolor. FEMS Microbiol. Lett.* 66(1–3): 215–220.

Klonowska, A., Le Petit, J., and Torn, T. (2001). Enhancment of minor laccases production in the basidiomycete *Marasmius quercophilus* C30. *FEMS Microbiol. Lett.* 200: 25–30.

Koroleva, O.V., Gavrilova, V.P., Stepanova, E.V., Lebedeva, V.I., Sverdlova, N.I., Landesman, E.O., Yavmetdinov, I.S., and Yaropolov, A.I. (2002a). Production of lignin modifying enzymes by co-cultivated white rot fungi *Cerrena maxima* and *Coriolus hirsutus* and characterization of laccase from *Cerrena maxima*. *Enzyme Microb. Technol.* 30 : 573–580.

Kim, Y. R., Yu, S.W., Lee, S.R., Hwang, Y.Y., and Kang, S.O. (1996). A heme-containing ascorbate oxidase from *Pleurotus ostretus*. *J. Biol. Chem.* 271(6): 3105–3111.

Kim, H.M. Han, S.B., Oh, G.T., Kim, Y.H., Hong, D.H., Hong, N.D., and Yoo, I.D. (1996). Stimulation of humoral and cell mediated immunity by polysaccharide from mushroom *Phellinus linteus*. *Int. J. Immunopharmacol.* 18(5): 295–303.

Kirk, T.K. and Farrell, R.L. (1987). Enzymatic 'combustion': the microbial degradation of lignin. *Annual review of Microbiology*, 41: 465–505.

Kirk, T.K., and Moore, W.E., (1972). Removing lignin from wood with white-rot fungi and digestibility of resulting wood. *Wood Fibre.*, 4:72–79.

Kitamoto, Y., Akashi, H., Tanaka, H., and Mori, N. (1988). Alpha-Glucose–1-phosphate formation by a novel trehalose phosphorylase from *Flammulina velutipes*. *FEMS Microbiol. Lett.* 55(2): 147–149.

Kofujita, H., Ohta, T., Asada, Y., and Kuwahara, M. (1991). Purification and characterization of laccase from *Lentinus edodes*. *Mokuzai Gakkaishi* 37(6): 562–590.

Kondo, R., Kurashiki, K., and Sakai, K. (1994). *In vitro* bleaching of hardwood kraft pulp by extracellular enzymes extracted from white-rot fungi in a cultivation system using a membrane filter. *Appl. Environ. Microbiol.* 60(3): 921–926.

Kono, H., Habuka, N., and Shishida, K. (1986). Purification and characterization of type I DNA topoisomerase from *Lentinus edodes*. *FEMS Microbiol. Lett.* 37(2): 169–172.

Kremnicky, L., Slavikova, E., Mislovicova, D. and Biely, P. (1996). Production of extracelluar β-mannanase by yeast and yeast like micro-organisum. *Folia. Microbiol.* (Praque) 41: 43–7.

Kremer, S. and Anke, H. (1997). Fungi in bioremediation. In: Fungal Biotechnology. Anke, T., Eds, Chapman & Hall, Weinheim, Germany. p. 275–295.

Kringstad, K.P. and Lindström, K. (1984). Spent liquors from pulp bleaching. *Environ. Sci. Technol.* 18:236A–248A.

Kuga, S. and Brown, R.M. Jr. (1991). Physical structure of cellulose microfibrils: implications for biogenesis. In: Biosynthesis and Degradation of Cellulose.Haigler, C.H. and Weimer, P.J., Eds., Marcel Dekker, New York, NY. pp. 125–142.

Kurosawa, S., Kawal, T., Akahane, N., and Sen, K. (1990). Purification and general properties of a nuclease from the fruit body of *Flammulina velutipes*. *Agric. Biol. Chem.* 54(10): 2733– 2735.

Lahtinen, T.P. Kristoo and M. Paloheimo, (1995). Mannanases in Softwood kraft Pulp Bleaching in the pulp and paper industry, 11–15 June 1995, Vienna, Austria Abstract.

Lara, M.A., Rodríquez-Malaver, A.J., Rojas, O.J., Holmquist, O., González, A.M., Bullón, J. Peñaloza, N. and Araujo, E. (2003). Black liquor lignin biodegradation by *Trametes elegans*. *Int. Biodeterior. Biodegrad.* 52:167–173.

Lau, K.L., Tsang, Y.Y. and Chiu, S.W. (2003). Use of spent mushroom compost to bioremediate PAH-contaminated samples. *Chemosphere* 52:1539–1546.

Levin, L., Forchiassin, F., and Ramos, A.M. (2002). Copper induction of lignin-modifying enzymes in the white rot fungus *Trametes trogii*. *Mycologia.* 94(3) : 377–383.

Li, B., Nagalla, S.R., and Renganathan, V. (1997). Cellobiose dehydrogenase from *Phanerochaete chrysosporium* is encoded by two allelic variants. *Appl. Environ. Microbiol.* 63(2): 796–799.

Llerena, E., Joseph, I.N., Diaz, P.J., Pacheco, E., Rosado, A., and Jimenez, L.E. (1987). Biosynthesis of fragrances by microorganisms I. production of monoterpenes by a basidiomycete. *Rev. Cubana Quim.* 3(4): 31–35.

Lomascolo, A., Cayol, J.L., Roche, M., Guo, L., Record, E., Lesage-Meessen, L., Olliver, B., Sigoillot, J.C. and Asther, M. (2002). Molecular clustering of *Pycnoporus* strains from various geographic origins and isolation of monokaryotic strain of *Pycnoporus cinnabarinus* strains for laccase hyperproduction. *Mycol Res.* 106(10): 1193–1203.

Lomascolo, A., Record, E., Herpoel-Gimbert, I., Delattre, M., Robert, J., Georis, J., Dauvrin, T., Sigoillot, J-C. and Asther, M. (2003). Overproduction of laccase by a monokaryotic strain of *Pycnoporus cinnabarinus* using ethanol as inducer. *J. Appl. Microbiol.* 94:618–624.

Loparzewski, J. (1987). A lignin-biotransforming peroxidase from *Trametes versicolor* and its use in immobilized form. Collq. INRA, 40 (Lignin Enzymic Microb. Degrad.).

Rotterdam. (1992). Xylanases and their applications in bakery, p. 349–360. *In* G. B. J. Visser, M. A. Kustersvan Someren, and A. G. J. Voragen (ed.), Xylans and xylanases, vol. 7. Elsevier Science, Amsterdam, the Netherlands.

Mackenzie, C.R. Bilons, D., Schneider, H. and Johnson, K.G. (1987). Induction of cellulolytic and xylanolytic enzyme systems in *Streptomyces* sp. *Appl. and Env. Microbiol.* 53: 2835–2839.

Magae, Y., Haga, K., Taniguchi, H., and Sasaki, T. (1990). Enzymes of strains of *Pleurotus species (Basidiomycetes)* compared by electrophoresis. *J. Gen. Appl. Microbiol.* 36(2): 69–80.

Maijala, P., Hakala, T.K., Salo, V., Lundell, T. and Hatakka, A. (2002). Characteristics of a new white-rot fungus with potential in biopulping. In: Conference Proceedings of the Workshop of COST Action E23 Biotechnology in the Pulp and Paper Industry, "Biotechnology for Improving Pulp & Paper Processing", 28–29 November, 2002, Centre Technique du Papier, Grenoble, France (no. 13).

Marzullo, L., Cannio, R., Giardina, P., Santini, M. T., and Sannia, G. (1995). Veratryl alcohol oxidase from *Pleurotus ostreatus* participates in lignin biodegradation and prevents polymerization of laccase-oxidized substrates. *J. Biol. Chem.* 270: 3823–3827.

Mayer, A.M. and Staples, R.C. (2002).Laccase: new functions for an old enzyme. *Phytochemistry* 60:551–565.

McCleary, B., Clark, A., Dea. I. and Rees, D.(1985). The fine structure of carob and guar galactomannans. *Carbo. Res.* 139: 237.

Messner, K., Srebotnik, E., Ertler, G., Foisner, R. Petterson, B., and Stavhelberger, H. (1989). Cell wall systems, extracellular membraneous structure and ligninase of wood rotting fungi. *Lignin Enzymic Microb. Degrad., Colloq. INRA* 40: 243– 248.

Messner, K. and Srebotnik, E. (1994) Biopulping: An overview of developments in an environmentally safe paper-making technology. *FEMS Microbiol. Rev.* 13:351–364.

Mikes, V., Zofall, M., Chytil, M., Fulnecek, J., and Schand, L. (1994). Ammonia-assimilating enzymes in the basidiomycetes fungus, *Pleurotus ostreatus*. *Microbiology* 140(4): 977–982.

Min, T.J., Cho, S.W., and Park, S.S. (1987a). Studies on the development of photoreceptor in the nonchromatophore organisms. I. Light-induced mitochondrial ATPase in *Lentinus edodes* (Berk.) Sing. *Han'guk Kyunhakhoechi* 15(4): 217–223.

Min, T.J., Cho, S.W., Kim, Y.S., Kim, J.W., and Mheen, T.I. (1987b). Studies on the development of photoreceptor in the nonchromatophore organisms. II. Effects of organic compound and metal ion influx on light-induced mitochondrial ATPase in the *Lentinus edodes* (Berk.) Sing. *Han'guk Kyunhakhoechi* 15(4): 224–230.

Moreira, M.T., Feijoo, G., Mester, T., Mayorga, P., Sierra-Alvarez, R. and Field, J.A. (1998). Role of organic acids in the manganese-independent biobleaching system of *Bjerkandera* sp. strain BOS55. *Appl. Environ. Microbiol.* 64: 2409–2417.

Morita, Y., Yamashita, H., Mikami, B., Iwamoto, H., Aibara, S., Terada, M. and Minami J. (1988) Purification, crystallization, and characterization of peroxidase from *Coprinus cinereus*. *J. Biochem.* 103:693–699.

Morohoshi, N. 1991. Laccases of the ligninolytic fungus *Coriolus versicolor*. In: Enzyme in Biomass Conversion. pp. 207–224.

Mueller-Harvey, I., Hartly, R.D., Harries, P.J. and Curzon E.N. (1986). Linkage of p-coumoaroyl and feruloyl groups to cell walls polysaccharides of barley straw. *Carbohydrate Res.* 148: 71–85.

Muheim, A., Waldner, R., Leisola, M. S. A., and Fiechter, A. (1990). An extracellular aryl-alcohol oxidase from the white-rot fungus, *Bjerkandera adusta. Enzyme Microb. Technol.* 12: 204–209.

Munoz, C., Guillen, F., Martinez, M.J. (1997a). Laccase isoenzymes of *Pleurotus eryngii*: characterization, catalytic properties, and participation in activation of molecular oxygen and Mn^{2+} oxidation. *Appl. Environ. Microbiol.* 63(6) : 2166–2174.

Munoz, C., Guillen, F., Martinez, M.J. (1997b). Induction and characterization of laccase in the ligninolytic fungus *Pleurotus eryngii*. *Curr. Microbiol.* 34: 1–5.

Murao, S., Oda, K., Tsuji, M., and Terashita, T. (1986). Carboxyl proteinase essential for maturation of basidiospores of *Lentinus edodes* (shiitake). *Agric. Biol. Che.* 50(12): 3225–3226.

Nagai, M., Sato, T., Watanabe, H., Saito, K., Kawata, M. and Enei, H. (2002). Purification and characterization of an extracellular laccase from the edible mushroom *Lentinula edodes*, and decolorization of chemically different dyes. *Appl. Microbiol. Biotechnol.* 60: 327–335.

Nakanishi, T. and Machida, Y. (1987). Microbial glucose-specific pyranose oxidase, its production by fermentation, and its use in enzymic determination of glucose. U.S. 4636464 (Cl. 435–14) (IPC C12Q–001/54), 13 Jan. 1987, Appl. U.S. 474310, 11 Mar. 1983, 9 pp.

Nuske, J., Scheibner, K., Dornberger, U., Ullrich, R., and Hofrichter, M. (2002). Large-scale production of manganese-peroxidase using agaric white rot fungi. *Enzyme Microb. Technol.* 30: 556–561.

Nyanhongo, G.S., Gomes, J., Gubitz, G., Zvauya, R., Read, J,S., and Steiner, W. (2002a). Production of laccase by a newly isolated strai of *Trametes modesta. Bioresour. Technol*. 84(3) : 259–263.

Papinutti, V.L., Diorio, L.A., and Forchiassin, F. (2003). Production of laccase and manganese peroxidase by *Fomes sclerodermens* grown on wheat bran. *J. Ind. Microbiol. Biotechnol*. 30: 157–160.

Petit-Benvegnen, M.-D., L. Saulnier, and X. Rouau. (1998). Solubilization of arabinoxylans from isolated water-unextractable pentosans and wheat flour doughs by cell wall degrading enzymes. *Cereal Chem.* 75:551–556.

Pickard, M.A., Vandertol, H., Roman, R., and Vazquez-Duhalt, R. (1999b). High production of ligninolytic enzymes from white rot fungi in cereal bran liquid medium. *Can. J. Micrbiol.* 45: 627–631.

Piontek, K., Gulumoff, T., and Winterhalter, K. (1993). Low pH crystal structure of glycosylated lignin peroxidase from *Phanerochaete chrysosporium* at 2.5 AO reolution. *FEBS Lett.* 315: 119–124.

Pointing, S.B. (2001). Feasibility of bioremediation by white-rot fungi. *Appl. Microbiol. Biotechnol.* 57:20–33.

Poutanen, K.J. (1997). Enzymes. An important tool in the improvement of the quality of cereal foods. *Trends Food Sci. Technol.* 8:300–306.

Puchart, V., Katapodis, P., Biely, P., Kreminicky, L., Christakopoulos, P., Vrsanka, M., Kekos, D., Marcis, B.J. and Bhat, M.K. (1999). Production of xylanases, mannanases, and pectinase by the thermophilic fungus *Thermomyces lanuginosus. Enz. Microbial Technol.* 24: 355–361.

Puls, J. and Poutanen, K. (1989). In: Enzyme Systems for lignocellulose degradation (Coughlan, M.P. Editor) Elsevier, London.

Puls, J. and Schuseil, J. (1996). Chemistry of hemicellulose: relationship between hemicellulose structure and enzymes required for hydrolysis, pp. 1–28. In M.P. Coughlan and G.P. Hazlewooa Eds., Hemicelluloses and hemicellulases. Portland Press. London, England.

Rahi, D.K., Shukla, K.K., Rajak, R.C. & Pandey, A.K. (2004). Mushrooms and their sustainable utilization. *Everymans' Science* 38(6) : 357–365.

Rahi, D.K. & Soni, S.K. (2007). Applications of microorganisms and their commercial uses. In: Microbes : A Source of Energy for 21st Century. New India Publishing Agency, New Delhi, India, 69–128.

Reddy, G.V., Babu, P.R., Komaraih, P., Roy, K.R.R.M., and Kothari, I.L. (2003). Utilization of banana waste for the production of lignolytic and cellulolytic enzymes by solid substrate fermentation using two *Pleurotus species* (*P. ostreatus* and *P. sajor caju*). *Process. Biochem.* 38: 1457–1462.

Reilly, P. (1981). Xylanases: Structure and function. In: Trends in the biology of fermentation for fuel and chemicals. A. Hollaender, Ed. P. 111, New York: Plenum Press.

Reese, E.D., Siu, R.G.H. and Levinson, H.S. (1950). The biological degradation of soluble cellulose derivatives and its relationship to the mechanisms of cellulose hydrolysis. *J. Bact.* 59 : 485–497.

Richmond, P.A. (1991). Occurrence and functions of native cellulose. In: Biosynthesis and Biodegradation of Cellulose. Liaigler, C.H. and Weimcr, P.J., Eds., pp.5–23. Marcel Dekker, New York, NY.

Rodriguez E., Pickard M. and Duhalt R. (1999). Industrial dye decoloration by laccase from ligninolytic fungi. *Current Microbiology;* 38: 27–32.

Rosales, E., Couto, S.R., and Sanroman, A. (2002). New uses of food waste: application to laccase production by *Trametes hirsute. Biotechnol. Lett.* 24: 701–704.

Ruttimann-Johnson, C., Salas, L., Vicuna, R., and Kirk, T. K. (1993). Extracellular enzyme production and synthetic lignin mineralization by *Ceriporiopsis subvermispora. Appl. Environ. Microbiol.* 59: 1792–1797.

Sakaguchi, K., Nakajima, T., Okada, T., and Iseda, H. (1991). Recovery of actinoid series elements by sorption with fungi. Jpn. Kokai Tokkyo Koho. JP and characterization of an extracellular prolyl endopeptidase from *Agaricus bisporus. J. Biochem.* (Tokyo) 107(2): 256–261.

Sannia, G., Limongi, P., Cocca, E., Buonocore, F., Nitti, G., and Giardina, P. (1991). Purification and characterization of a veratryl alcohol oxidase enzyme from the lignin degrading basidiomycete, *Pleurotus ostreatus. Biochimicaet Biophysica Acta* 1073: 114–119.

Sattar, A.K., Abdus, K.M., Yamamoto, N., Yoshimoto, T., Tadashi, T., and Tsuru, D. (1990). Purification and characterization of an extracellular prolyl endopeptidase from *Agaricus bisporus. J. Biochem.* (Tokyo) 107(2): 256–261.

Sawada, T. and Endo, K. (1987). Purification and some properties of acid phosphatases from 'Shiitake' mushroom. *Nippon Shokuhin Kogyo Gakkaishi* 34(12): 801–808.

Schulze, E. (1891). Information regarding chemical composition of plant cell membrane. Berichte der Deutschen Chemischem Gesellshaft. 24: 2277–2287.

Shaginyan, K.A., Alekhina, I.A., and Denisova, N.P. (1990). Serine proteinase from the higher basidiomycete genus *Coprinus. Biokhimiya (Moscow)* 55(8): 1387–1395.

Shah, M.M., Grover, T.A., and Sharma, H.S.S. (1987). Enzymatic degradation of residual non-cellulosic polysaccharides present on dew-retted flax fibres. *Appl. Micrbiol. and Biotechnol.* 26: 358–362.

Shah, V. and Nerud, F. (2002). Lignin degrading system of white-rot fungi and its exploitation for dye decolorization. *Can. J. Microbiol.* 48:857–870.

Shimada, H., Inokuchi, N., Okuwaki, H., Koyama, T., and Irie, M. (1991). Purification and characterization of a base non-specific and adenylic acid preferring ribonuclease from the fruit bodies of *Lentinus edodes. Agric. Biol. Chem.* 55(4): 1167–1169.

Soares, C.H. and Duran, N. (2001). Biodegradation of chlorolignin and lignin-like compounds contained in E1-pulp bleaching effluent by fungal treatment. *Appl. Biochem. Biotechnol.* 95:135–149.

Soni, S.K., Rahi, D.K. & Soni, R. (2007). Microbial ecology and pollution management. In : Microbes : A Source of Energy for 21st Century. New India Publishing Agency, New Delhi, India, 127–172.

Srinivasan, M.C. and Rele, M.V. (1995). Microbial xylanases for paper industry. *Current Science.* 77: 137–142.

Steffen, K. (2003).Degradation of recalcitrant biopolymers and polycyclic aromatic hydrocarbons by litter-decomposing basidiomycetous fungi. Dissertationes Biocentri Viikki Universitatis Helsingiensis, 23/2003. Ph.D. Thesis. Department of Applied Chemistry and Microbiology, University of Helsinki, Helsinki. 68 p.

Steffen, K.T., Hatakka, A. and Hofrichter, M. (2002a) Degradation of humic acids by the litter-decomposing basidiomycete *Collybia dryophila. Appl. Environ. Microbiol.* 68: 3442–3448.

Steffen, K.T., Hofrichter, M. and Hatakka, A. (2002c) Purification and characterization of manganese peroxidases from the litter-decomposing basidiomycetes *Agrocybe praecox* and *Stropharia coronilla. Enzyme Microb. Technol.* 30:550–555.

Stoll, D., Henrik, S. and Warren, R.A.J. (1999). Mannan degrading enzymes from *Cellulomonas fimi. Appl. Environ. Microbiol.* 65: 2598–2605.

Sunna, A. and Antranikian, G. (1997). Xylanolytic enzymes from fungi and bacteria. *Critical Rev. in Biotechnol.* 17: 39–67.

Szklarz, G. and Leonowicz, A. (986). Co-operation between fungal laccase and glucose oxidase in the degradation of lignin derivatives. *Phytochemistry* 25(11): 2537–2539.

Taguchi, H., Omura, D., Nishitani, H., Okumura, K., and Shimabayashi, Y.(1989). Properties of nicotinamidase in "Enokitake" mushroom, (*Flammulina velutipes*). *Bitamin.* 63(10): 491–496.

Tamaru, Y., Araki, T., Morishita, T., Kimura, T., Sakka, K. and Ohmiya, K. (1995). Cloning, DNA sequencing and expression of the β-1,4-mannanase gene from a marine, *Vibrio* sp. strain MA-138. *J. Fermen. Bioeng.* 83: 201–205.

Tein M; and Kirk, T.K., (1984). Lignin degrading enzyme from *Phanerochaete chrysosporium* : purification, characterization and catlalytic properties of unique H_2O_2 requiring oxygenase. *Proc. Natt. Acad. Sci. USA* 81: 2280–2284.

Tenkanen, M., Makonen, M., Perttula, M., Viikari, L. and Teleman, A. (1997). Action of *Trichoderma reesei* on galactomannan in pine kraft pulp *J, Biotechnol.* 57: 191–204.

Terashita, T., Nawama, M., Yoshikawa, K., and Shishiyama, J. (1995). Proteinases from the fruit body of *Flammulina velutipes. Nippon Shokuhin Kagaku Kogaku Kaishi.* 42(11): 907–912.

Timmel, T.E. (1967). Recent progress in chemistry of wood hemicellulose. *Wood Science Technol.* 1: 45–70.

Trupkin, S., Levin, L., Forchiassin, F. and Viale, A. (2003). Optimization of a culture medium for ligninolytic enzyme production and synthetic dye decolorization using response surface methodology. *J. Ind. Microbiol. Biotechnol.* 30:682–690.

Tsuyl, H., Ogawa, T., Banodo, N., and Susaoka, K. (1989). Stereo-specificity of NAD(P)H oxidation catalyzed by 4-amino-benzoate hydroxylase from *Agaricus bisporus. Biochim. Biophys. Acta* 991(2): 380–382.

Uchida, T., Matsui, S., and Uhayasni, A. (1985). Manufacture of pyranose oxidase. Jpn. Kokai Tokyo.

Vares, T. (1996). Ligninolytic enzymes and lignin-degrading activity of taxonomically different white-rot fungi. Ph.D. Thesis. Department of Applied Chemistry and Microbiology, Division of Microbiology. University of Helsinky, Finland.

Vikari, L., Kantelinen, A., Sandquist, J. and Linko, M. (1994a). Xylanases in bleaching: From an idea to the industry. *FEMS Microbiol. Reviews.* 13 : 338–350.

Vol'f, L.A., Shamolina, I.I., Goncharova, N.A., Lobova, A.B., and Gavrilova, V.P. (1986). Immobilization of enzymes on fibrous supports. *Priki. Biokhim. Mikrobiol.* 22(5): 664–668.

Wang, Y., Vazquez-Duhalt, R., and Pickard, M.A. (2001). Effect of growth conditions on the production of manganese peroxidase by three strains of *Bjerkandera adusta. Can. J. Microbiol.* 47: 277–282.

Wells, T.K., Hammond, J. B. W., and Dickerson, A.G. (1987). Variations in activities of glycogen phosphorylase and trehalase during the perodic fruiting of the edible mushroom *Agaricus bisporus.* (Lange) *Imbach. New Phytol.* 105(2): 273–280.

Wesenberg, D., Kyriakides, I. and Agathos, S.N. (2003). White-rot fungi and their enzymes for the treatment of industrial dye effluents. *Biotechnol. Adv.* 22:161–187.

Williams, A.C., Moore, N.L., Dashek, W.V., and Williams A.L. (1990). *Coriolus versicolor*, a model system to investigate the biotechnology of wood-deteriorating enzymes. F , Eds., 3 (*Proc. Meet. Pun Am.* 405–17). Plenum Press, New York.

Wood, J.D. and Wood, P.M. (1992). Evidence that cellobiose: quinine oxidoreductase from Phaenerochaete chrysosporium is a breakdown product of cellobiose oxidase. *Biochem. Biophys. Acta.* 1119(1): 90–96.

Wong, K.K.Y. and Saddler, J.N. (1992). *Trichoderma* xylanases, their properties and purification. *Crit. Rev. in Biotechnol.* 12: 413–435.

Wong, K.K.Y., Tan, L.U.L. and Saddler, J.N. (1998). Multiplicity of β-1,4-xylanase in microorganisms: function and applications. *Microbiol. Rev.* 52: 305–317.

Wong, K.K.Y. and J.N. Saddler, (1993). Application of hemicellulases in the food, feed, pulp and paper industries. In: Coughlan, M.P. and G.P. Hazlewood (Eds.), Hemicellulose and Hemicellulases, pp; 127–143. Portland Press, London, England.

Yamamoto, K., Matsuoka, S., and Hiroi, T. (1986). Effects of nutrient nitrogen impregnated in wood on lignin degradation by white-rot fungi. I. Comparison Among Fungus Species. *Ringyo Shikenjo, Kenkyu Hokoku* 340: 197–204.

Yoshimoto, T., Ee, K. E., Abudasu, S., and Tsuru, O. (1991). Proline iminopeptidase and its manufacture with mushrooms. Jpn. Kokai Tokyo Koho. Jp. 03108483 (91108483) (IPC C12N–009/48), 8 May 1991, Appl. 89/245016, 22 Sept. 1989, 6 pp.

Zadrazil, F. (1977). The conversion of straw into feed by basidiomycetes. *Eur. J. Appl. Microbiol.* 4:37–44.

Zhang, J., Mingria, H. and Zhimin, H. (2002). Operational and storage ability of neutral β-mannanase from *Bacillus licheniformis*. *Biotechnol. Lett.*

Zorn, H., Langhoff, S., Scheibner, M., Nimtz, M. and Berger, R. G. (2003). A peroxidase from *Lepista irina* cleaves beta,beta-carotene to flavor compounds. *Biol. Chem.* 384: 1049–1056.

7

Pigments from Fungi: Industrial Perspective

Nelson Durán[1], Roseli De Conti[1,2] and Maria Francisca S. Teixeira[3]

[1]Instituto de Química, Universidade Estadual de Campinas-UNICAMP, C.P. 6154, Campinas, CEP 13083-970, S.P., Brazil, [2]Centro Universitário de Espirito Santo do Pinhal-UNIPINHAL, Espirito Santo do Pinhal, S.P., Brazil, and [3]Mycology Department, Faculty of Science, Universidade Federal de Amazonias-UFAM, Manaus, Brazil.

INTRODUCTION

Pigments in biological materials have been discussed by us in 2002 (Duran *et al.*, 2002) and in 2004 (Duran *et al.*, 2004), by Gill (2003) and most recently by Medentsev *et al.* (2005), Dufossé *et al.* (2005), Mapari *et al.* (2005).

The black pigment frequently encountered in microbial systems is usually considered to be melanin or melanoprotein. Melanin and melanin-like pigmentation have been described in various organisms, which may be dark brown, black only in certain organelles, or may be totally pigmented. Dark brown or black pigments occur widely in fungi. Except for reactions to wounding, melanogenesis in fungi is restricted to certain developmental stages in special structures such as chlamydospores or microsclerotia (Thielaviopsis, Verticillium), conidia (*Aspergillus niger*), and hyaline mycelium.

In the Dematiaceae, both hyphae and conidia are heavily pigmented (Alternaria, Curvularia, Drechslera). Many Ascomycetes and imperfect fungi appear to produce melanin by the pentaketide pathway (Durán *et al.*, 2002). However, Basidiomycetes seems to employ a different route in which special substrates are used for the phenoloxidase system that leads to the formation of melanin. The biological significance of melanin reported involves from resistance to a variety of adverse environmental factors to exposure at extreme temperature. A protective role in desiccation and irradiation has also been suggested, where melanin deposited in the cell wall may act as a shield against immunologically active cells (Durán *et al.*, 2002).

More than 500 different molecules of the carotenoid family have been described. Some 200 carotenogenic fungal species have been described so far, and β-carotene (Fig. 7.1) is usually the major pigment, but xanthophyll (Fig. 7.2) may also be present, especially in the Basiodiomycetes and Deuteromycetes. During recent years, many observations have been published with regards to the antioxidant activity of carotenoids, inhibition of mutagenesis, enhancement of the immune response, and inhibition of tumor development. Photosensitizing pigments as hypericin (Fig. 7.3) and as hypericin-like molecules have also been described in microbial organisms (*Penicilliopsis clavariaeformis*, *Dermocibe austroveneta*). The most important flavin compounds produced by microorganisms and organisms are riboflavin (Fig. 7.4). Fungi synthesize and excrete riboflavin into their growth media (Durán *et al.*, 2002 and references therein).

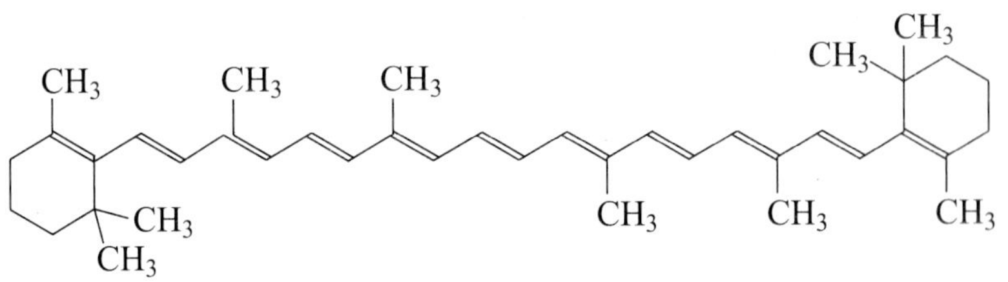

Fig. 7.1 β-Carotene

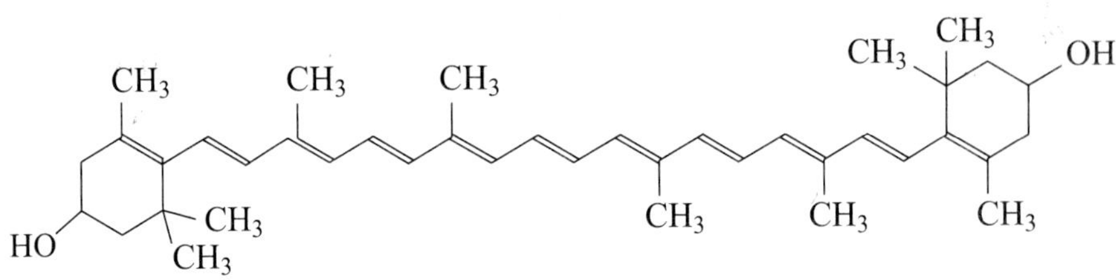

Fig. 7.2 Xanthophyll

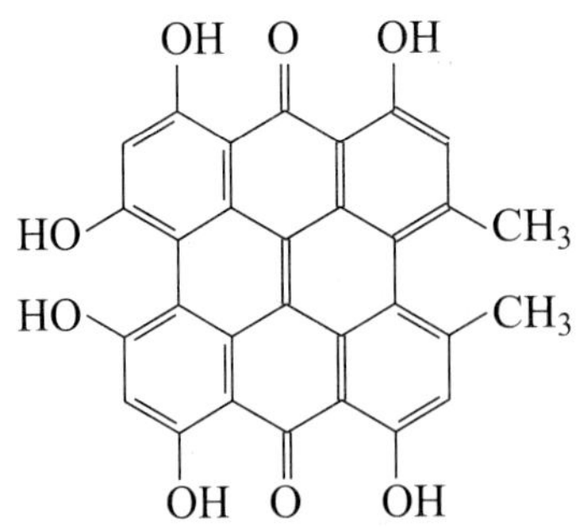

Fig. 7.3 Hypericin

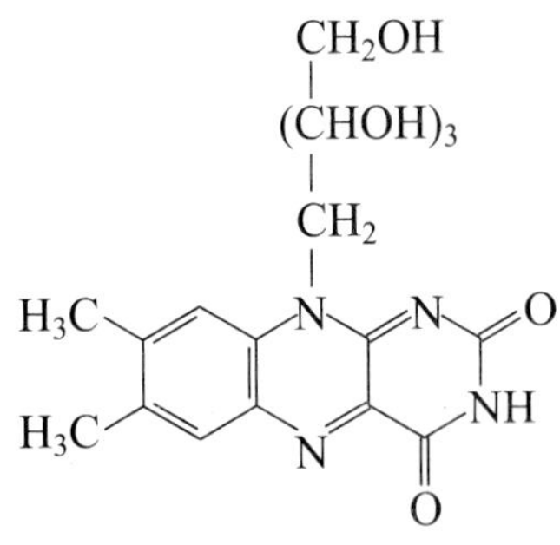

Fig. 7.4 Riboflavin

Quinones display an array of colors: yellow, orange, or red according to the position of the keto groups. In fungi, the naphthoquinone pigments are very common and have been the subject, not only in the research of their chemical structure, but also into their biosynthesis and biological significance. Naphthoquinone pigments related to fusarubin (Fig. 7.5) obtained from the fungus *Fusarium solani*

(Mart) (Parisot *et al.*, 1990) have been described in details (Durán *et al.*, 2002 and references there in).

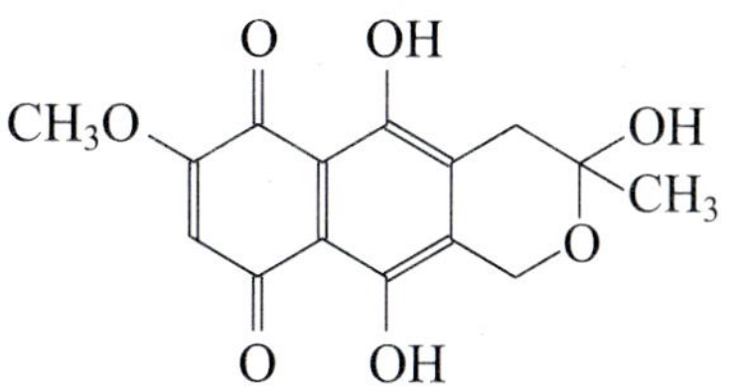

Fig. 7.5 Fusarubin

Monascus pigments, which are water-soluble pigments, are of great biotechnological interest. Monascus is a typical Ascomycete that produces a cleistothecium, a closed fruiting body containing eight ascospores, but reproduces asexually by the formation of conidiospores and a vegetative mycelium. The pigments may appear both in the mycelium and in the fermentation broth. It was suggested earlier that microbial pigments having no obvious function should be classified as secondary metabolites, but many data published show in especial cases that the pigments are certainly not a secondary metabolite (Durán *et al.*, 2002).

Because of its potential as food color additives, there is a high interest in pigments produced by fungi from the genus Monascus. These pigments are well known in eastern countries in the traditional rice ang-kak and in red rice wines. Monascus metabolites comprise six known azaphilone pigments (e.g. monascine (Fig. 7.6), rubropuctatine (Fig. 7.7), monascorubrine (Fig. 7.8), etc.), with color ranging from yellow to red. These substances may be extracted from fermentation medium and used as a mixture or individually after purification. The review by De Carvalho *et al.* (2003) presents the information presently available of Monascus biopigments production.

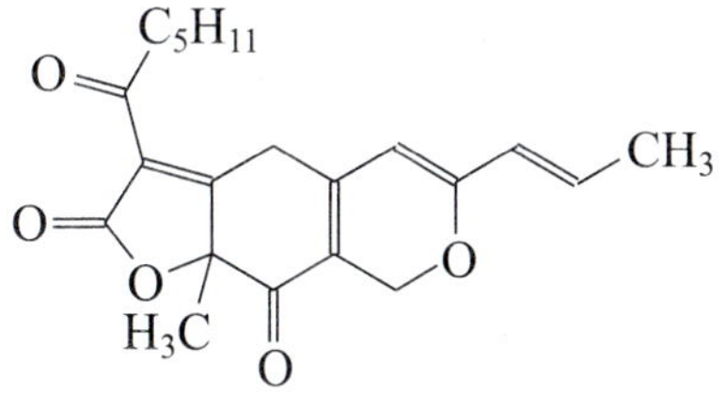

Fig. 7.6 Monascine

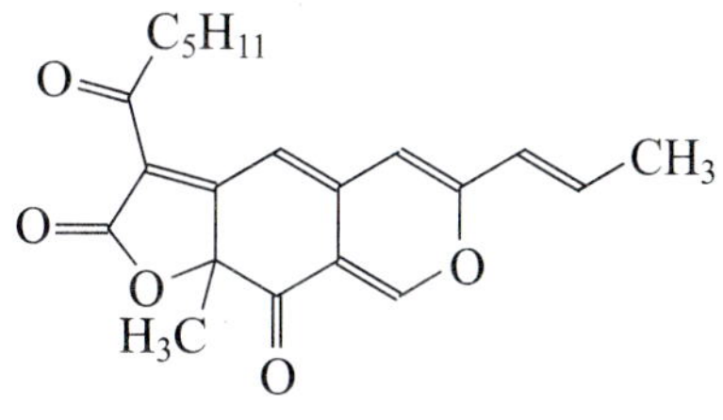

Fig. 7.7 Rubropuctatine

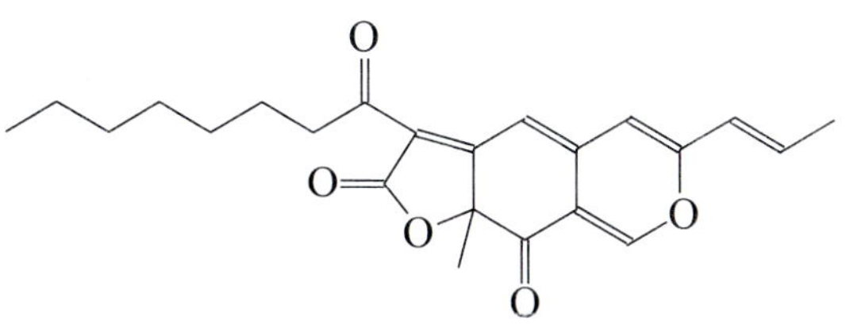

Fig. 7.8 Monascorubrine

The production of many currently authorized natural food colorants has a number of disadvantages, including a dependence on the supply of raw materials and variations in pigment extraction. Fungi provide a readily available alternative source of naturally derived food colorants that could easily be produced in high yields. The recent authorization of a fungal food colorant has induced research to explore the extraordinary chemical diversity and biodiversity of fungi for the biotechnological production of pigments as natural food colorants (Mapari *et al.*, 2005; Dufossé *et al.*, 2005).

Pigments and other secondary metabolite produced in the fruiting bodies of mushrooms and toadstools exhibit a great variety of structures and their chemical constitution often provides a reasonable indication of their biosynthetic origins (Gill, 2001).

Then, the aim of this review is to discuss the most important advances in the fungal pigment area and its interest in biotechnological applications.

Fungal Pigments

Characteristic pigments are produced by a wide variety of fungi. Some species give hydroxyanthraquinones (e.g. helminthosporin (Fig. 7.9) (maroon, brown), catenarin (Fig. 7.10) (red), cynodontin (Fig. 7.11) (bronze), tritisporin (Fig. 7.12) (red brown) and erytroglaucin (Fig. 7.13) (red)), or phenolics (e.g. auroglaucin (Fig. 7.14 (orange), aurantin (Fig. 7.15) (yellow) and flavoglaucin (Fig. 7.16) (yellow). Other chemical structure in many pigments are the conjugated double bonds (e.g. corticrocin (Fig. 7.17) (yellow), mycorradicin (Fig. 7.18) (yellow), β-carotene (Fig. 7.1), xanthophyll (Fig. 7.2) (yellow). Among other pigments, investigations have been made on naphthenics (e.g. rubrofusarin (Fig. 7.19) (red) javanicin (Fig. 7.20) (red orange), marticin (Fig. 7.21) (red) and solaniol (Fig. 7.22) (dark red)), and more complex structures (e.g. viopurpurin (Fig. 7.23) (blue), aurofusarin (Fig. 7.24) (orange yellow), xerocomic (Fig. 7.25) (blue), variegatic acid (Fig. 7.26) (blue) and variegatorubin (Fig. 7.27) (red) (Kahner *et al.*, 1998; Durán *et al.,* 2002). The melanin pigments (black) were isolated from *Phellinus robustus* and *Inonotus obliquus.* Special reference must be done with Monascus pigments (e.g. monascorubrine (Fig. 7.8) (orange), rubropunctatine (Fig. 7.7) (orange) and monascurubramine (Fig. 7.28) (red)) (Durán *et al.*, 2002 and references therein).

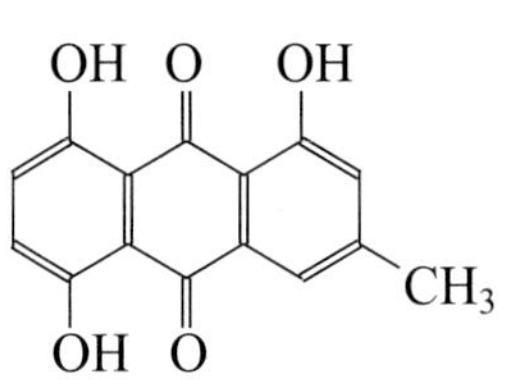

Fig. 7.9 Helminthosporin

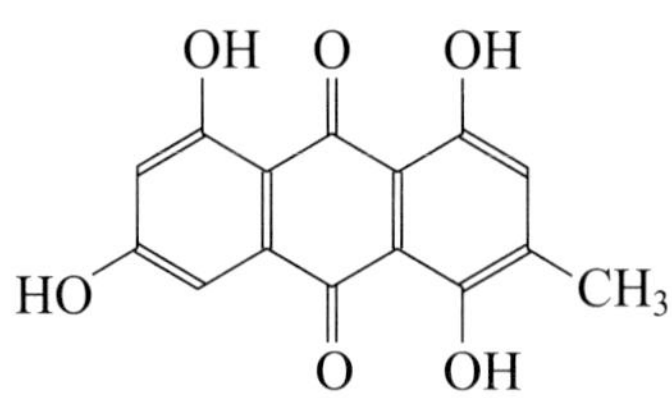

Fig. 7.10 Catenarin

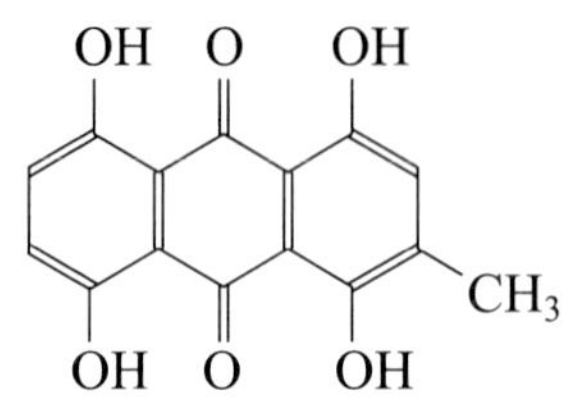

Fig. 7.11 Cynodontin

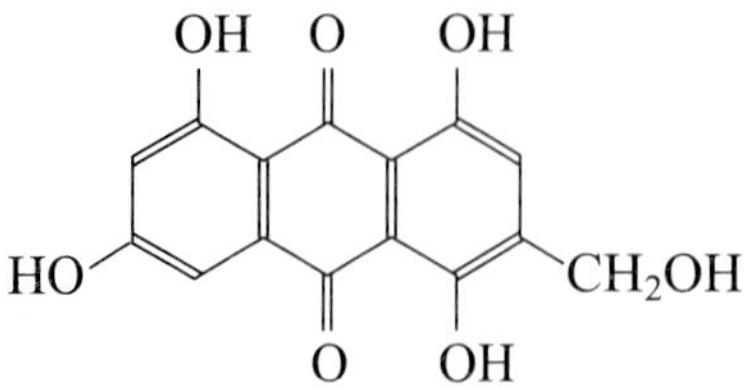

Fig. 7.12 Tritisporin

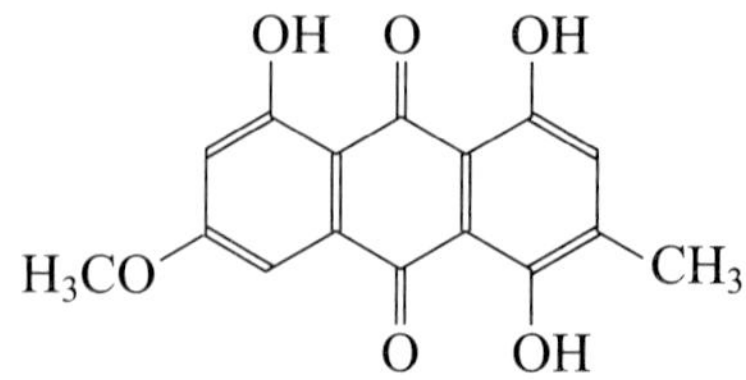

Fig. 7.13 Erytroglaucin

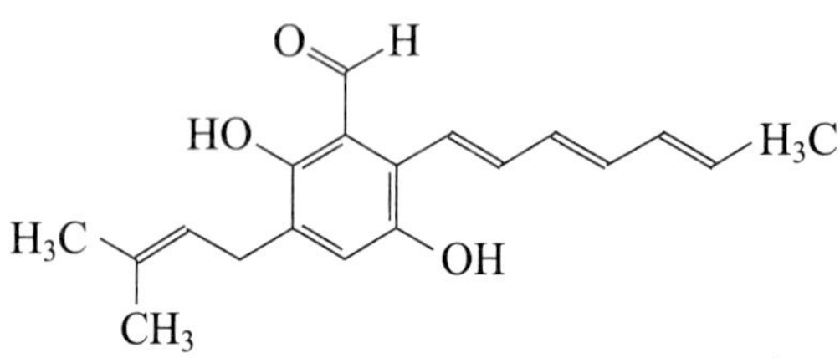

Fig. 7.14 Auroglaucin

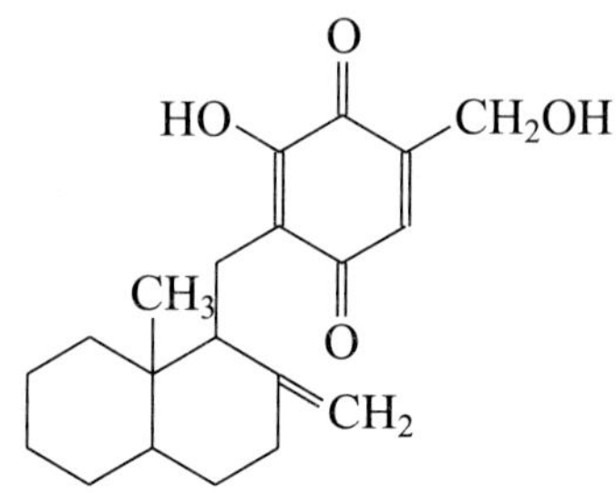

Fig. 7.15 Aurantin

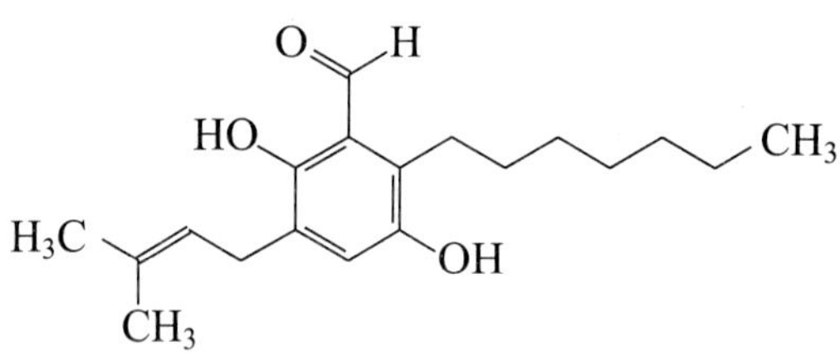

Fig. 7.16 Flavoglaucin

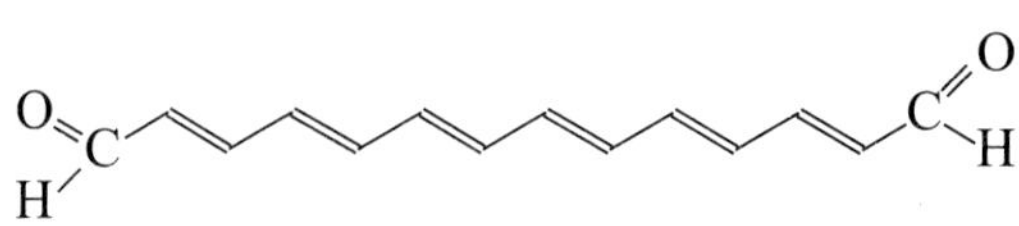

Fig. 7.17 Corticrocin

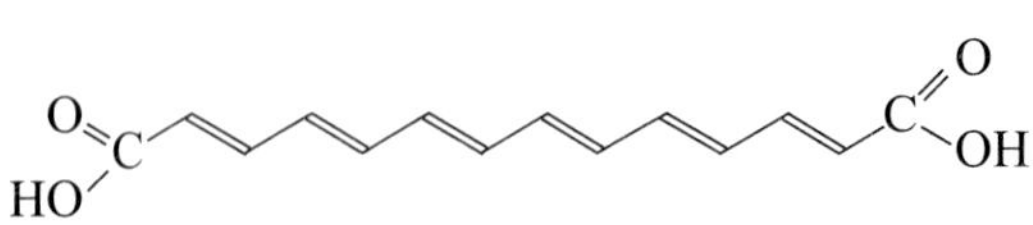

Fig. 7.18 Mycorradicin

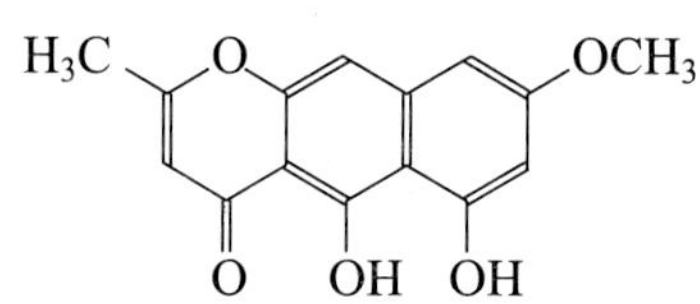

Fig. 7.19 Rubrofusarin

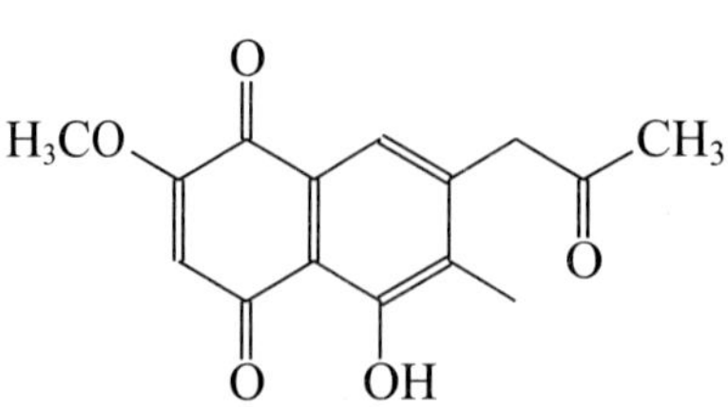

Fig. 7.20 Javanicin

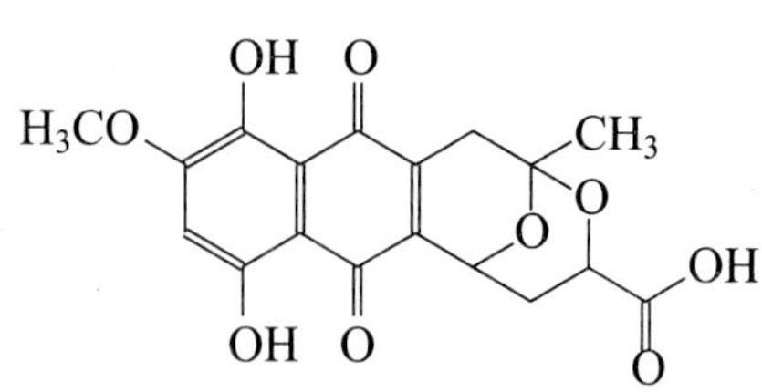

Fig. 7.21 Marticin

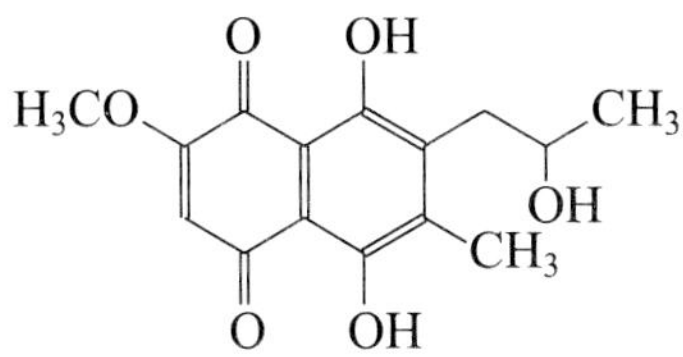

Fig. 7.22 Solaniol

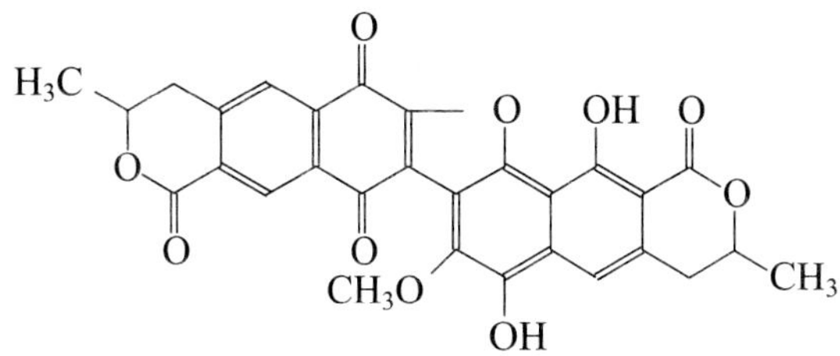

Fig. 7.23 Viopurpurin

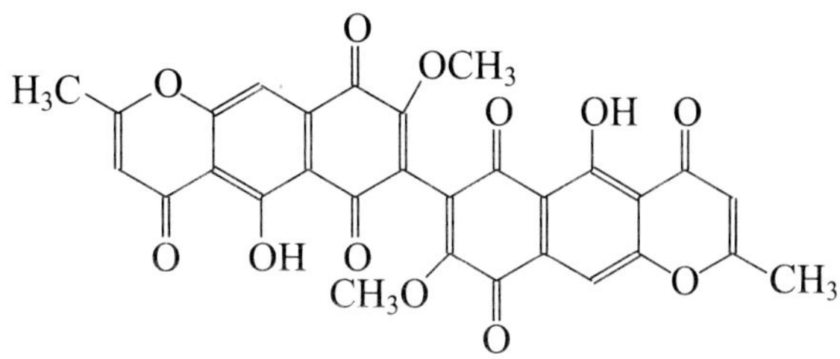

Fig. 7.24 Aurofusarin

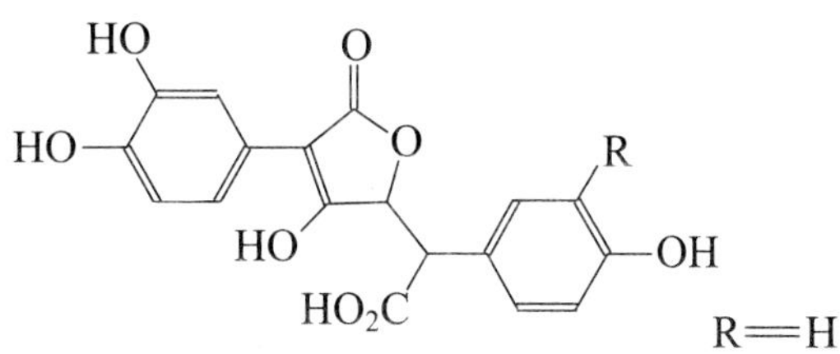

Fig. 7.25 Xerocomic

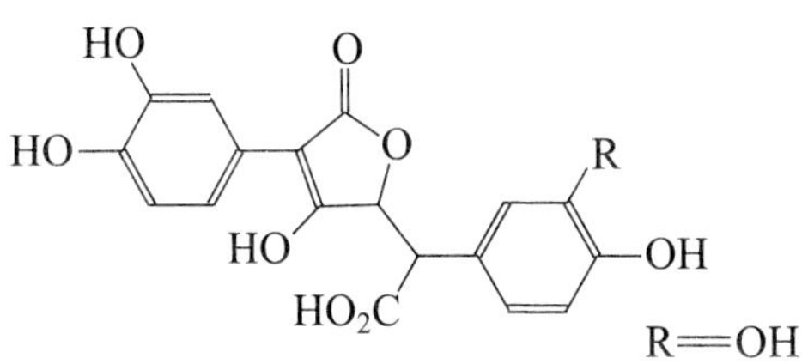

Fig. 7.26 Variegatic Acid

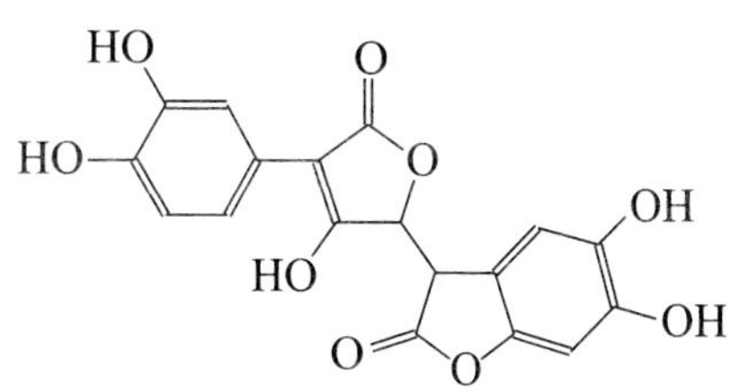

Fig. 7.27 Variegatorubin

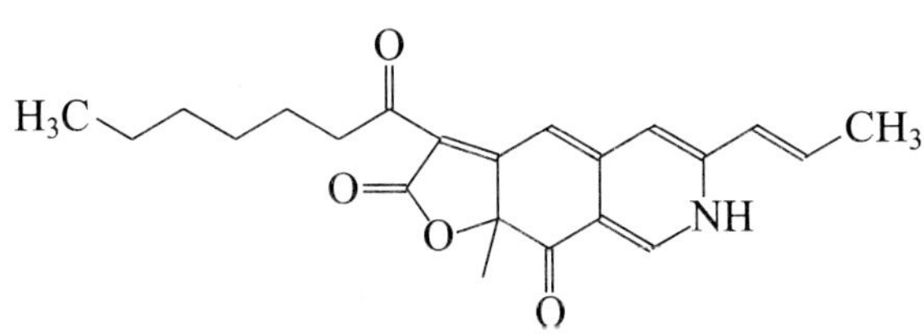

Fig. 7.28 Monascurubramine

Sarcodon imbricatus is appreciated for its bluish-greenish pigment, which is related to atromentin (Fig. 7.29), thelephoric acid (Fig. 7.30) and gyrocyanin (Fig. 7.31) (Johannesson *et al.*, 1999).

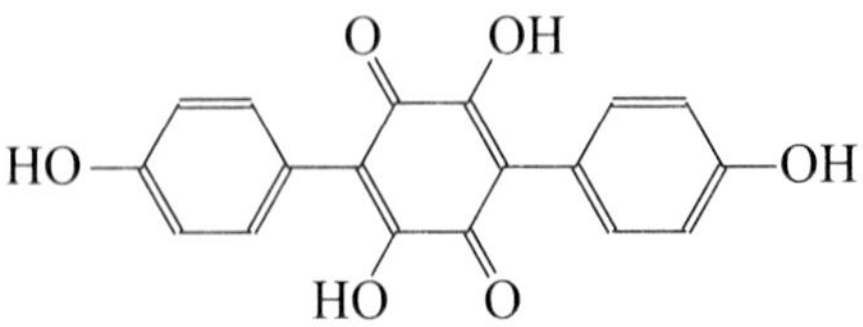

Fig. 7.29 Atromentin

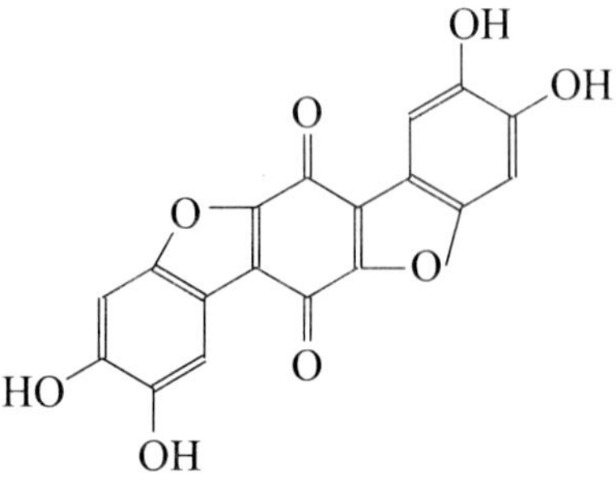

Fig. 7.30 Thelephoric Acid

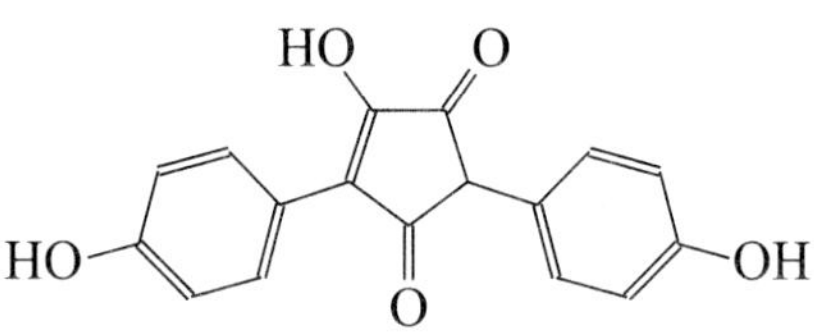

Fig. 7.31 Gyrocyanin

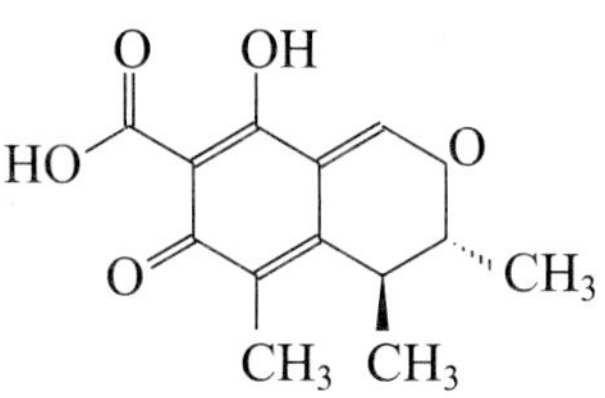

Fig. 7.32 Citrinin

The biosynthesis of naphthoquinone pigments was shown to be the main response of the fungi to stress, observed under conditions of growth inhibition or arrest (Medentsev *et al.* 2005).

Many mycotoxins are pigmented, as the naphthoquinones from *Penicillium* and *Aspergillus* (Hawksworth *et al.*, 1995). *Monascus* red pigment and/or yellow pigments are efficiently produced by several strains and are commercially important (Blanc, 1998; Shin, 1998). *Monascus purpureus* was selected to study the bioconversion of whole wheat flour for the production of food colorants (Espinoza and Webb, 1998). The growth and metabolic behavior of the filamentous fungus *Monascus ruber* were studied in submerged cultures under various aeration and agitation conditions. It was observed that the improvement of the oxygen supply increase the production of secondary metabolites (red pigment and citrinin (Fig. 7.32). However, a low oxygen transfer coefficient was required to improve the red pigment/citrinin (Hajjaj *et al.*, 1999). Fungi form the genus Monascus are a promising source for natural color additives. However, before effectively applying Monascus to foods, it is important to select strains which produce large amounts of biopigments but little or no citrinin, a mycotoxin usually produced by these fungi. Also, color stability of these pigments should be properly investigated in order to compare Monascus strains for biopigment production in solid substrate fermentation (SSF). Thus, four strains were cultivated over potatoes-dextrose-agar (PDA) in Petri dishes, and compared for radial

growth velocity. These strains were cultivated over cooked rice, and were compared in relation to their capacity to produce biopigments and citrinin too. These results showed that one of the strains (LPB 31) was the best strain for biopigment production in SSF, giving both higher pigment and lower mycotoxin (citrinin (Fig. 7.32) concentration on the extracts, showing that it is a promising strain for production of this bioproduct. Biopigment assays for heat and pH stability, showed that these biopigments were unstable at low pH range and high temperatures, but may be successfully used at near-neutral pH's and in non-thermal processed foods (De Carvalho *et al.*, 2005).

In general, high broth viscosity is a key factor to be considered in a submerged fermentation of filamentous fungi. High broth viscosity was also observed in a batch fermentation of *Monascus* sp. J101 at 30°C. In a batch culture at 30°C, most cell growth was accomplished within 48 h. The resultant high viscosity induced heterogeneity inside the fermentor, poor oxygen transfer, and low pigment yield. However, these problems could be overcome by reducing fungal growth rate through culture at low temperature (25°C). Cell growth was moderate and continued for 120 h, and the low viscosity was maintained. The dissolved oxygen (DO) levels remained at 50% or higher with good stirring. As a result, the pigment yield at 25°C was 10 times higher than at 30°C (Ahn *et al.*, 2006).

It was studied the red pigments production by spectrophotometry in semi-synthetic medium by rotary shaker cultures, utilizing seven strains of *Monascus* sp. The strains *M. ruber* isolated from the local soil had an absorbance average of extracellular red pigments higher than the others ATCC strains (Hamano *et al.*, 2005).

In nature, light is one of most crucial environmental signals for developmental and physiological processes in various organisms, including filamentous fungi. Very recently it was found that both red light and blue light affect development of Monascus, influencing the processes of mycelium and spore formation, and the production of secondary metabolites such as gamma-aminobutyric acid, red pigments as monacolin K (Fig. 7.33) and citrinin (Fig. 7.32). Additionally, it was observed that the wavelength of light affects these developmental and physiological processes in different ways. These findings suggest that Monascus possesses a system for differential light response and regulation (Miyake *et al.*, 2005).

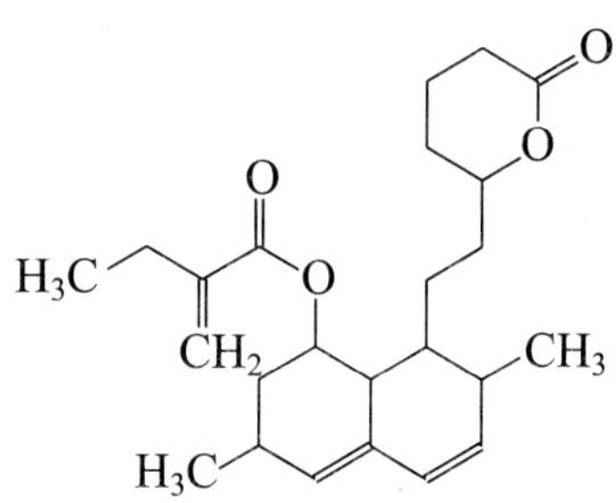

Fig. 7.33 Monacoline

Citrinin (Fig. 7.32) was extracted by a liquid-liquid extraction method from *Penicillium citrinun* (Araujo *et al.*, 1998). *Penicillium* species that produces patulin (Fig. 7.34), citrinin (Fig. 7.32), palitantin (Fig. 7.35), and arthographol (Fig. 7.36), were identified from 149 different microorganisms screened (Yamaji *et al.*, 1999). A red-violet pigment was isolated from *Rissula vinosa* (Li *et al.*, 1999). Also, the yellow rice refers to the color of rice infected by pigment and toxins

producing Penicillia. The important fungal sex hormones such as trisporic acid (Fig. 7.37) from *Blakeslea trispora* are carotenoid pigments. A review summarizes the development in carotenoid biosynthesis from a molecular genetic standpoint (Armastrom and Hearst, 1996). Although astaxanthin (salmon pigmentation) is mainly produced by the yeast *Phaffia rhodozyma,* other fungal sources were discussed, that is, *Neurospora crassa* (Johnson and An, 1991). Astaxanthin is used primarily as a pigment source in aquaculture and the market for it is probably greater than USD 100 million (Johnson and Schroeder, 1995; Dufossé *et al.*, 2005). The production of phenolic compounds by the basidiomycetous fungi *Pleurotus otreatus* and *Crinipellis schevczenkovi* was most efficient stimulated by aromatic amino acids, naphthylacetic acid, indolacetic acid, kinetin and flavonic pigments (Babitzkaya *et al.*, 1998; Babitskaya and Shcherba, 1999). The fruiting bodies of some macromycetes (e.g. agaric) have been used less and not historically, but they can give a range of pinks, blues, yellows, reds, and browns (e.g. *Boletus, Cortinarius, Hydnellum, Hygrocybe* spp (Gripenberg, 1981; Hawksworth *et al.*, 1995; Kahner *et al.,* 1998; Gill and Morgan, 2004; Griffith *et al.*, 2004). The most important pigments, the anthraquinones, were, for a long time, of crucial importance in the dyestuff industry (Hobson and Wales, 1998). These metabolites presumably are secondary ones and posses no obvious function in cell growth. They are synthesized for a finite period by cells that are no longer undergoing balanced growth. On the contrary, primary metabolism includes all the anabolic and catabolic processes that are finely balanced to keep the organism alive. It is essentially an identical process for all living organisms.

There are a lot of secondary metabolites that are used by the modern biotechnology and were mainly obtained from plants (e.g. indigo (Fig. 7.38), carminic acid (Fig. 7.39), alizarin, linalol, cinnamaldehyde, salicin, coninne, morphine, lawsone (Fig. 7.40), etc.), which were applied in dyes, perfume, spices, poisons, pharmacological agents, and cosmetics (Durán *et al.*, 2002).

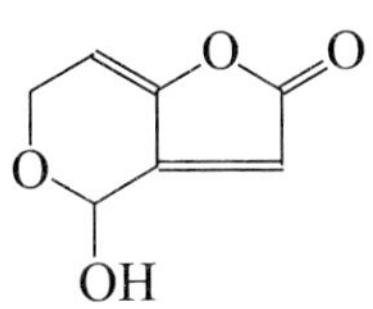

Fig. 7.34 Patulin

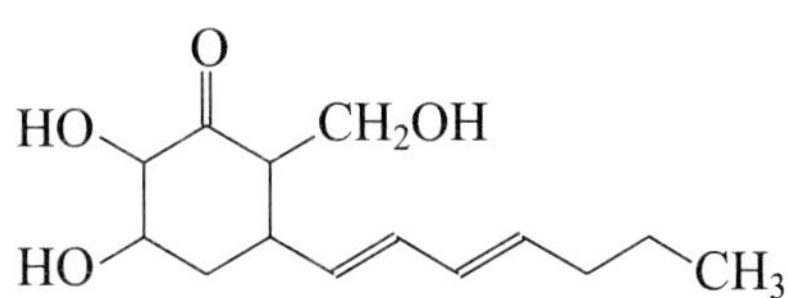

Fig. 7.35 Palitantin

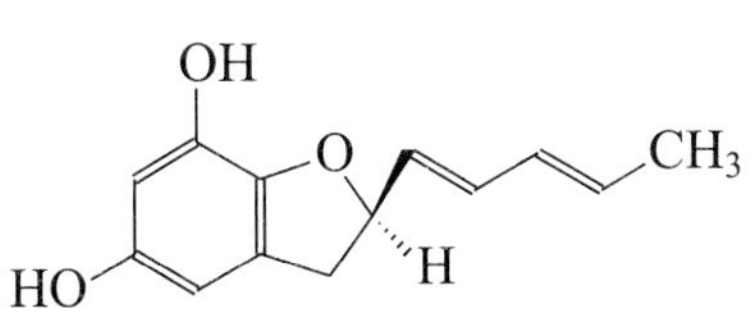

Fig. 7.36 Artographol

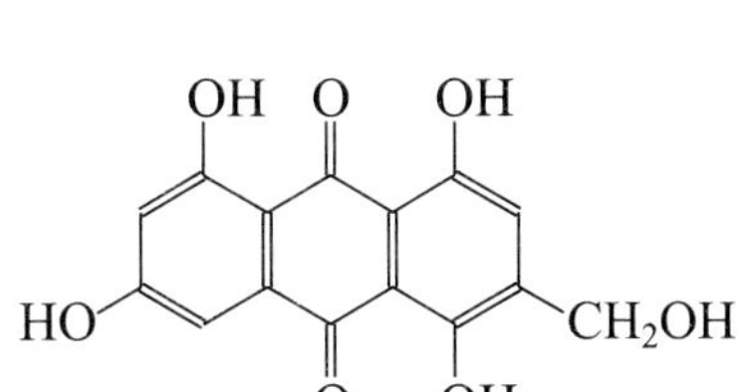

Fig. 7.37 Trisporic Acid

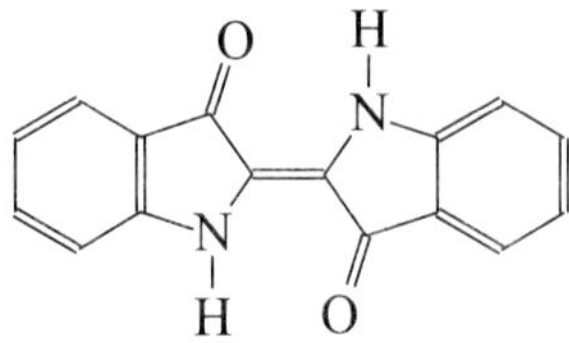

Fig. 7.38 Indigo

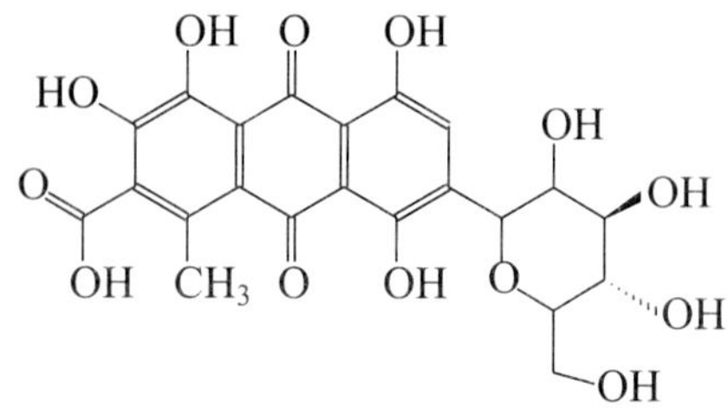

Fig. 7.39 Carminic Acid

The dyestuff industry is suffering from the increases in costs of feedstock and energy for dye synthesis and they are under increasing pressure to minimize the damage to the environment caused by the process and effluents. Thus, the industries are continuously looking for cheaper, more environmen-friendly routes to existing dyes.

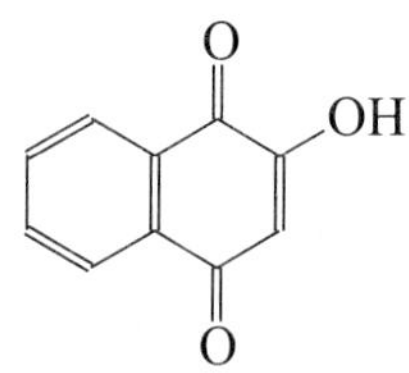

Fig. 7.40 Lawsone

Through the biotechnology techniques anthraquinones have been isolated from a number of fungi, that is, *Trichoderma, Drechslera, Aspergillus,* and *Curvularia* strains (Hobson and Wales, 1998; Gill, 1996, 1999). A strain of *Curvularia* lunata (Hobson *et al.*, 1997) has been identified that produced three anthraquinones: cynodontin (Fig. 7.11), helminthosporin (Fig. 7.9), and chysophanol (Fig. 7.41). Cynodontin (Fig. 7.11) extracted from the biomass of *C. lunata* has been converted successfully to two anthraquinone biodyes similar to Disperse blue 7 (Fig. 7.42) and Acid Green 28 (Fig. 7.43) (blue biodye (Fig. 7.44) and green biodye (Fig. 7.45). The properties of these biodyes applied to knitten polyamides were compared with those of conventional dyes and found to be identical to all-important aspects (Hobson *et al.*, 1997).

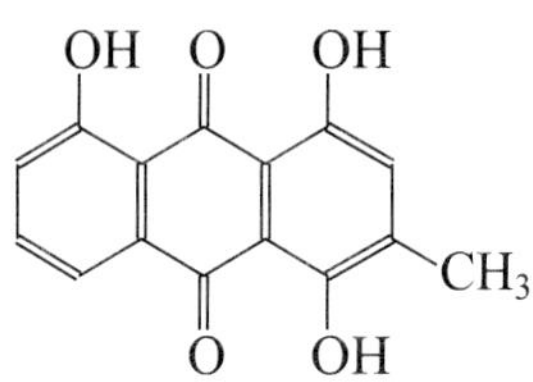

Fig. 7.41 Chysophanol

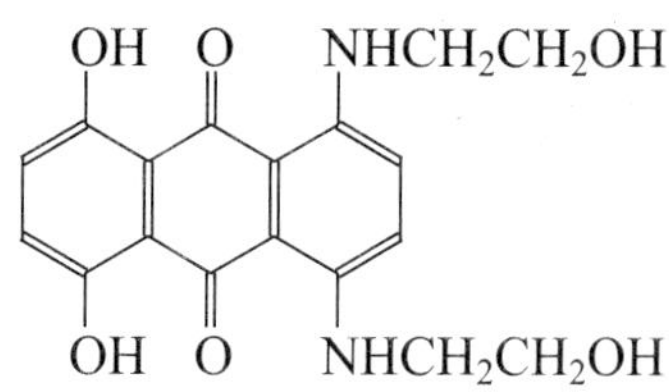

Fig. 7.42 Disperse Blue 7

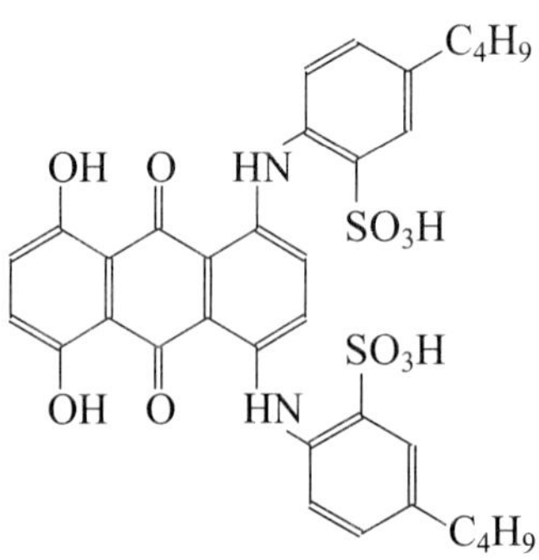

Fig. 7.43 Acid Green 28

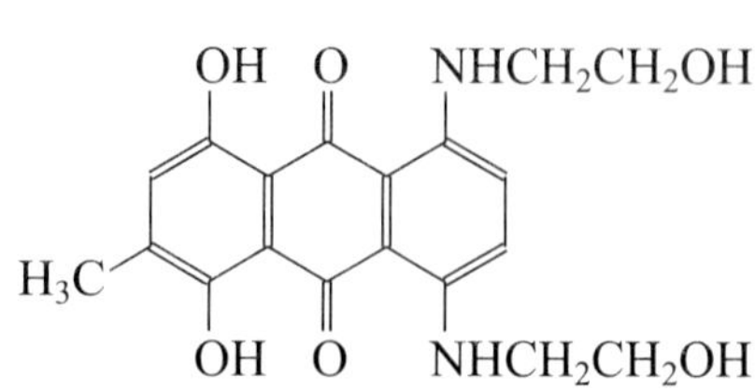

Fig. 7.44 Blue Biodye

Fusarium oxysporum isolated from roots of diseased citrus trees produced anthraquinones with no hydroxyl substituents at the 1,4-position, and they have an acetyl or 1-hydroxyethyl group at the 2 or 3 positions (Baker and Tatum, 1998; Durán *et al.*, 2005). Many anthraquinones and naphthoquinones were isolated and characterized from *Dermocybe cardinalis* (Buchanan *et al.*, 1999; Gill *et al.,* 1998a; Gill, 2001, 2003) and from *Cortinarius sinapicolor* (Buchanan *et al.*, 1998; Gill *et al.*,-1998b; Elsworth *et al.*, 1999).

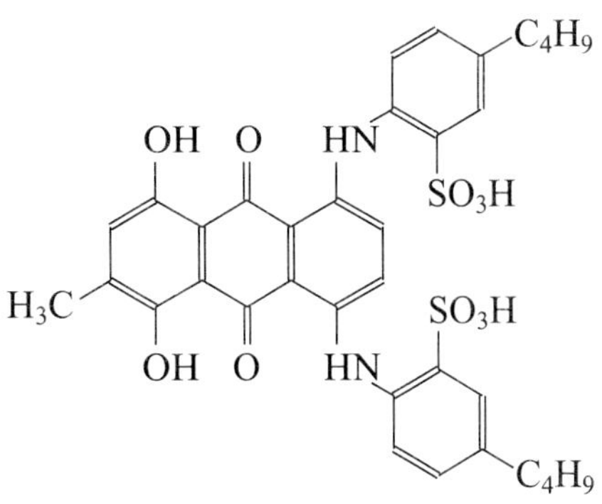

Fig. 7.45 Green Biodye

Exploiting the fungal synthesis of anthraquinones has several advantages over chemical methods. Expensive fuel-consuming high-temperature and environmental unfriendly strong acids and alkalis of the chemical synthesis are not required. Naphthoquinones are widely spread in nature and have been found also in fungi and in actinomycetes (Durán *et al.*, 2002 and references therein). Excellent reviews of the naphthoquinone metabolites of fungi were published recently (Medentsev and Akimenko, 1996, 1998) and the knowledge of the pigments formation was mainly obtained from *Fusarium.* Falconensones A (Fig. 7.46) and B (Fig. 7.47), which are new types of yellow pigments. They were isolated recently from Ascomycete fungi, *Emericella falconensis,* or *Emericella fruticulosa*, whose structures are similar to retinoic acid (Fig. 7.48) (Takahashi *et al.*, 1998).

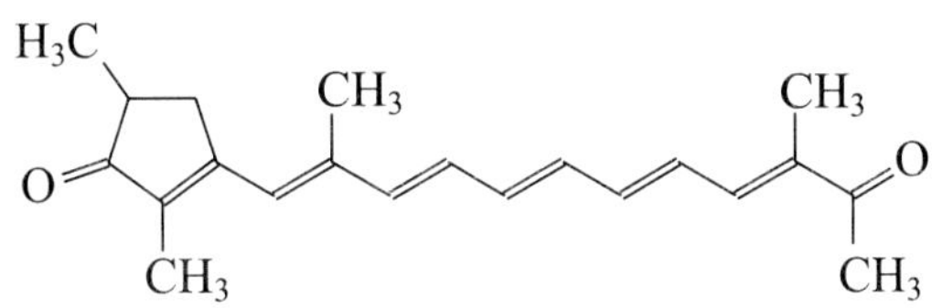

Fig. 7.46 Falconensones A

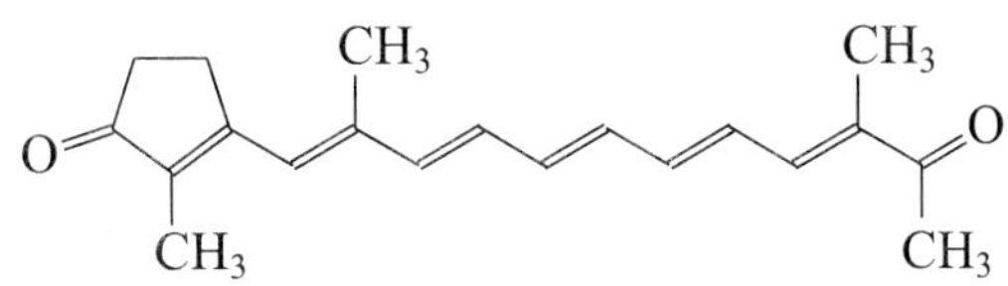

Fig. 7.47 Falconensones B

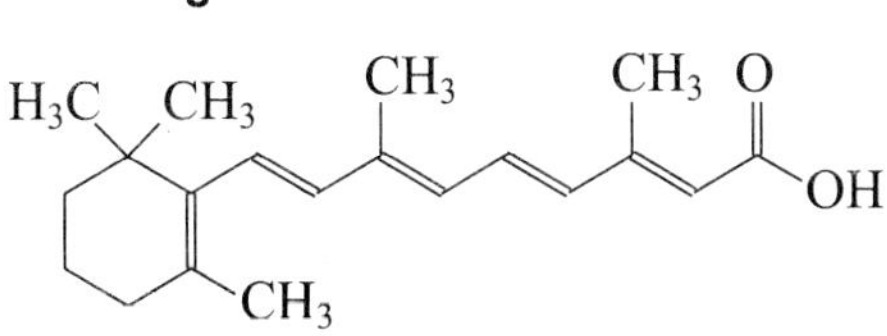

Fig. 7.48 Retinoic Acid

A new pigment, laetiporic acid (49-*Cis* and *Trans*), has been isolated from fruit-bodies of the wood-rotting fungus *Laetiporus sulphureus* (sulfur shelf). Structure elucidation by application of extensive 2D NMR techniques permitted its identification as a polyene of non-soprenoid origin. Laetiporic acid (Fig. 7.49), which represents the main pigment in *L. sulphureus* basidiocarps, bears an unprecedented decaene skeleton as part of its chromophore and, interestingly, contains double bonds with a stable cis configuration (Weber *et al.*, 2004).

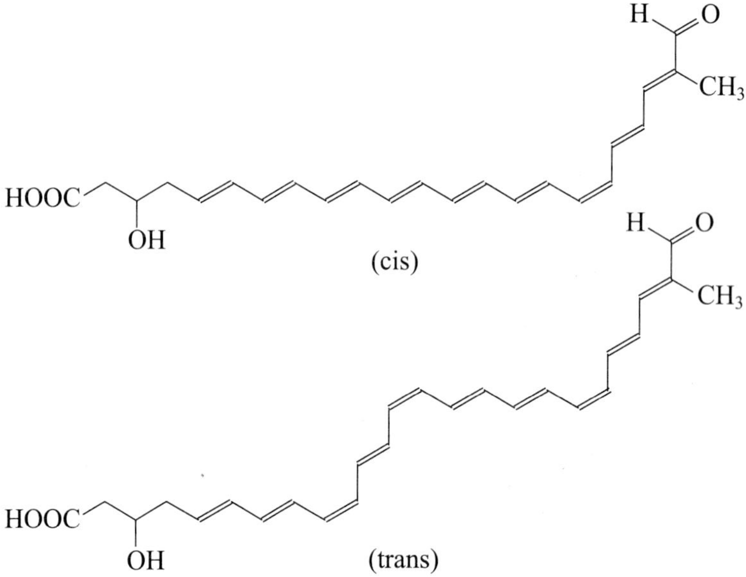

Fig. 7.49 Laetiporic Acid (cis and trans)

A red pigment that accumulates in cultures of a *Drechslera avenae* pathotype with specificity for *Avena sterilis* was isolated and identified as the anthraquinone cynodontin (3-methyl-1,4,5,8-tetrahydroxyanthraquinone) (Fig. 7.11). Satisfactory yield of the compound was obtained with 20-60 day incubations at temperatures between 20 and 27 °C. (Chrysayi-Tokousbalides and Kastanias, 2003).

Pycnoporus sanguineus (Polyporaceae) appeared with a great industrial potentiality due to the metabolites production (Esposito *et al.*, 1993; Castro e Silva *et al.*, 1993). Cinnabarin (Fig. 7.50) is an antibiotic substance produced by *P. sanguineus*. This compound is an orange pigment which has a basic phenoxazin-3-one structure (Gripenberg, 1951; Achenbach and Blumm, 1991). Production of this substance *in vitro* occurs between the 18th and the 23rd day of incubation (Smânia, *et al.*, 1995ab). Moreover, the synthesis of cinnabarin (Fig. 7.50) is increased significantly when the pH of the broth media is adjusted to 9.0, with incubation at 25°C, under light (Smânia, *et al.*, 1997). As this substance has a cromophore orange color and can transfer this color to food products, the substance was tested against bacteria obtained from foods.

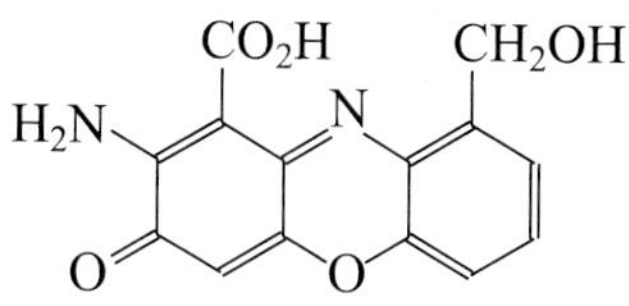

Fig. 7.50 Cinnabarin

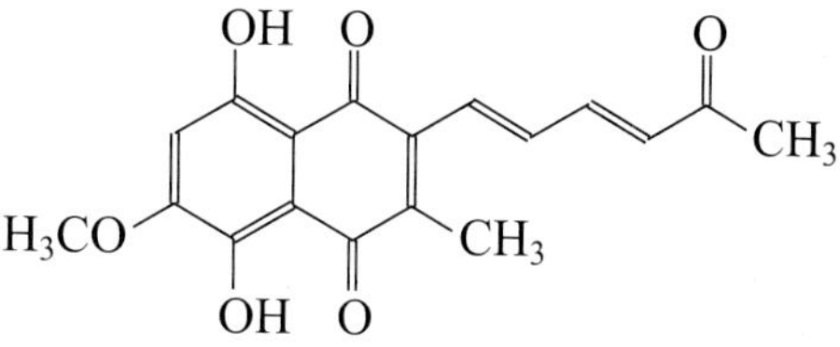

Fig. 7.51 3,5,8-TMON

Carotenoids represent one of the most widely distributed and structurally diverse classes of natural pigments, with important functions in photosynthesis, nutrition, and protection against photooxidative damage. In the eubacterial community, yellow, orange, and red carotenoids are produced by anoxygenic photosynthetic bacteria, cyanobacteria, and certain species of nonphotosynthetic bacteria. Many eukaryotes, including all algae and plants, as well as some fungi (*Phycomyces blakesleeanus*, and *Neuroposra crassa*)(Cerdá-Olmedo *et al.*, 1970; Carattoli *et al.*, 1991), also synthesize these pigments (Armstrong, 1997).

A number of fungi, including *Aspergillus nidulans*, produce pigments related or identical to melanin, which are located on cell walls or exist as extracellular polymers. It was studied the antioxidant activity of synthetic melanin and of the pigment extracted from the mycelium and culture medium after growth of the highly melanized strain (MEL1) from *A. nidulans*. The ability of the melanin pigment to scavenge the oxidants HOCl and H_2O_2 was evaluated by inhibition of the oxidation of 5-thio-2-nitrobenzoic acid (TNB) using several melanin concentrations. The results showed that the pigment of the MEL1 strain competes with TNB for H_2O_2 and HOCl, inhibiting TNB oxidation in a concentration-dependent manner. For the HOCl oxidant, this inhibition was comparable to that of synthetic melanin, whose IC_{50} values were quite close for both pigments. Thus, these results suggest that the melanin from *A. nidulans* is a potential HOCl scavenger and may be considered a promising material for the cosmetic industry for the formulation of creams that protect the skin against possible oxidative damage (Gonçalves *et al.*, 2005).

Production of red pigments (naphthoquinones) by the insect pathogenic fungus *Cordyceps unilateralis* BCC 1869 was investigated and cultivation conditions, including temperature, intitial pH of medium, and aeration, were optimised to improve the yield of total naphthoquinones in shake-flask culture of *C. unilateralis*. The highest yield of total naphthoquinones (3 g L^{-1}) was obtained from a 28-day culture grown in potato dextrose broth with an initial pH of 7.0, at 28°C with shaking-induced aeration at 200 rpm. An extraction process for isolation of the targeted naphthoquinone, 3,5,8-trihydroxy-6-methoxy-2-(5-oxohexa-1,3-dienyl)-1,4-naphthoquinone (3,5,8-TMON)(51), from a culture of *C. unilateralis*, was also developed. The stability of 3,5,8-TMON was very high, even upon exposure to strong sunlight, high temperature, and acid and alkali solutions (Unagul *et al.*, 2005).

The non-isoprenoid polyene laetiporic acid A (Fig. 7.49), recently described from fruit bodies of the wood-rotting fungus *Laetiporus sulphureus*, was found to be the major orange pigment also in mycelium grown in liquid culture. Its formation was variable, ranging from 0.1 to 6.7 mg/g dry weight in three strains, all of which were identified as *L. sulphureus* by ITS rDNA sequence analysis. A second pigment, 2-dehydro-3-deoxylaetiporic acid A, is also described and fully characterized by NMR spectroscopy. Two further minor pigments, laetiporic acids B and C, were produced in liquid culture. These resemble laetiporic acid A (Fig. 7.49) but are enlarged by two and four carbon atoms, respectively, resulting in chromophores with 11 or 12 instead of 10 conjugated double bonds as described for laetiporic acid A. Since fruit-bodies of *L. sulphureus* are edible, laetiporic acids might hold potential as food colorants (Davoli *et al.*, 2005).

Betalains (e.g. red-violet betacyanins (Fig. 7.52) and the yellow betaxanthins (Fig. 7.53), replace the anthocyanins in flowers and fruits of plants of most families of the Caryophyllales. Unexpectedly, they were also found in some higher fungi (*Amanita muscaria*). Whereas the anthocyanin-analogous functions of betalains in flower and fruit coloration are obvious, their role in fungi remains obscure (Strack *et al.*, 2003). Secodopas (Fig. 7.54) were shown to be present in extracts from the colored peels of *Amanita muscaria* and *Hygrocibe conica*. They were cyclized to muscaflabin (Fig. 7.55) and betalamic acid (Fig. 7.56) (Terrads and Wyler, 1991; Strack *et al.*, 2003).

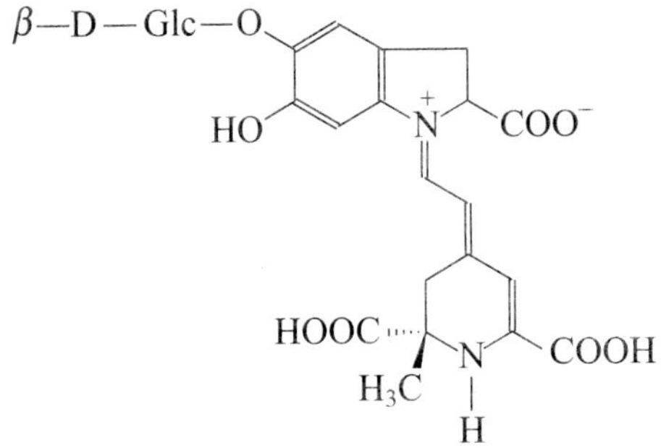

Fig. 7.52 Betacyanins

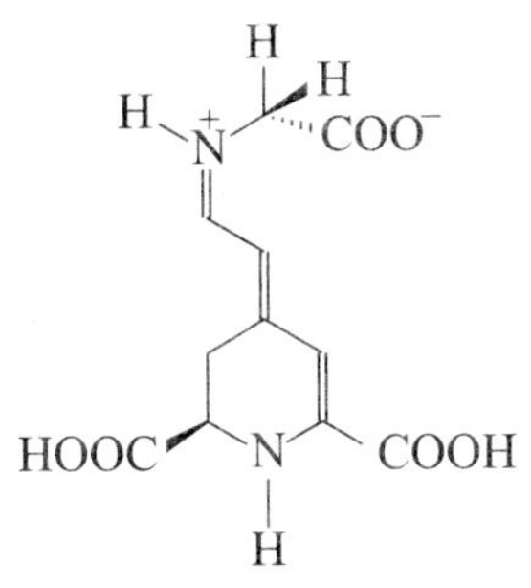

Fig. 7.53 Betaxanthins

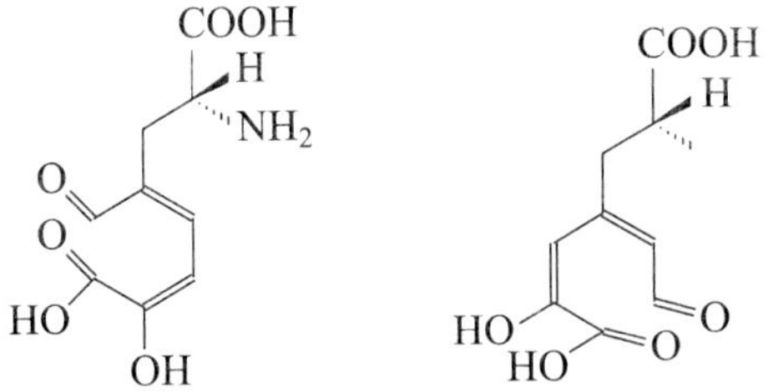

Fig. 7.54 Secodopas

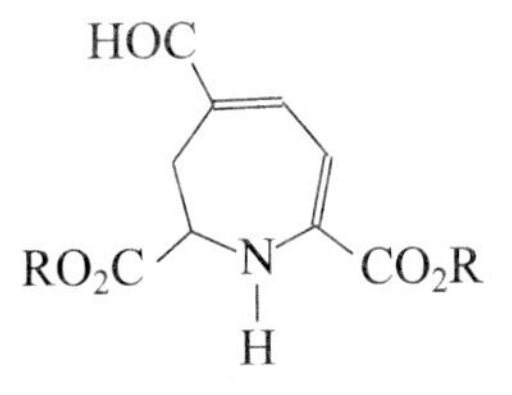

Fig. 7.55 Muscaflavin

Filamentous fungi included in the class of Mucorales produce the pigment β-carotene. This production can be enhanced by a number of stimuli, such as sexual interaction and light. Structural genes involved in the biosynthesis of β-carotene in *Phycomyces blakesleeanus* and *Mucor circinelloides* were cloned and characterized (Ruiz-Vazquez *et al.*, 2003).

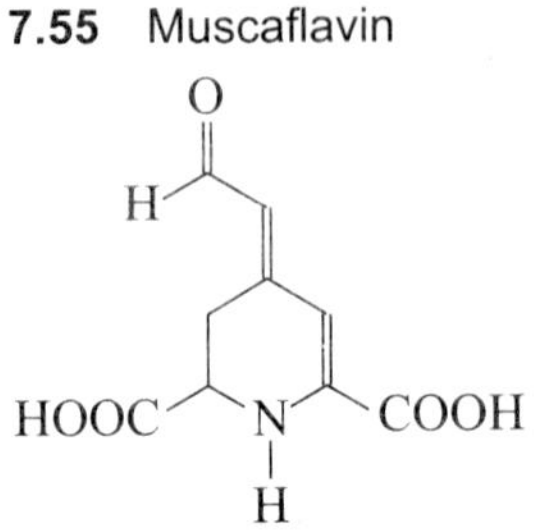

Fig. 7.56 Betalamic acid

Three previously unknown pentaketides, (+)-formylanserinone B (Fig. 7.57) (I), (–)-epoxyserinone A (II) (Fig. 7.58), and (+)-epoxyserinone B (III) (Fig. 7.59), along with 2 known fungal pigments, anserinones A (Fig. 7.60) and B (Fig. 7.61), were isolated and identified from a deep water, marine-derived saltwater fungal culture. Two other minor constituents, hydroxymethylanserinone B (Fig. 7.62) and deoxyanserinone (Fig. 7.63), were also isolated, but not completely purified. The structures of the compounds, each expanding the dense functionalization of the anserinones, were determined by dereplication and spectroscopic analises. Bioactivity was explored in 2 separated cell-based assays (Gautschi *et al.*, 2004).

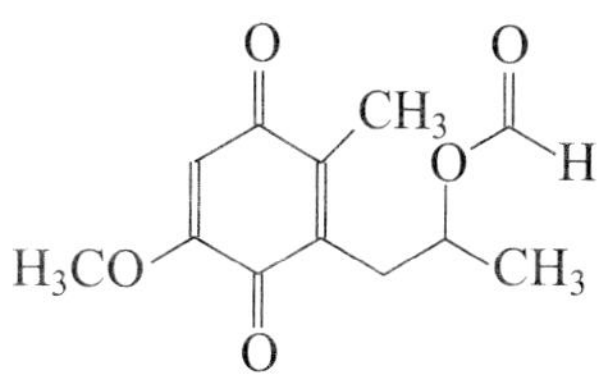

Fig. 7.57 (+)-Formylanserinone B

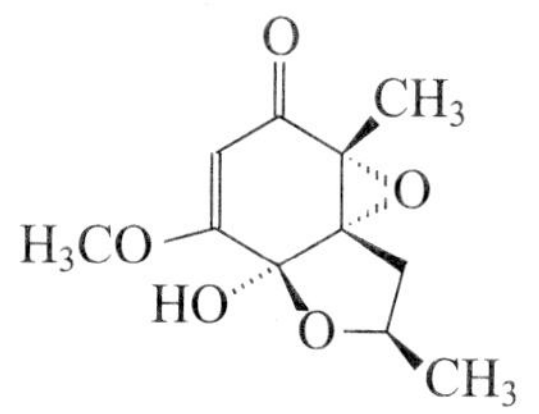

Fig. 7.58 (–)-Epoxyserinone A

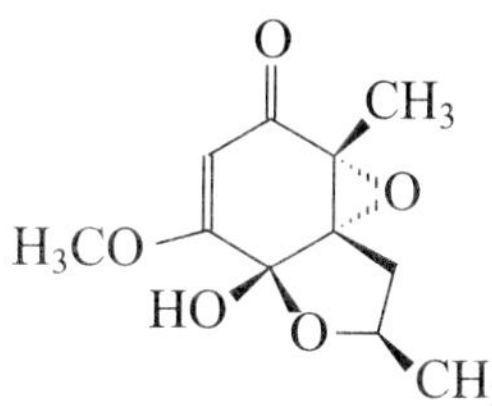

Fig. 7.59 (+)-Epoxyserinone B (III)

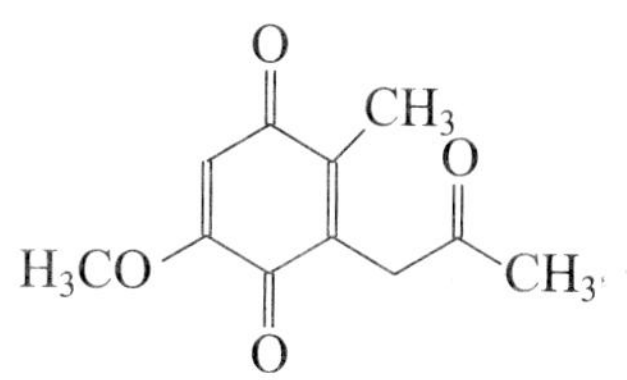

Fig. 7.60 Anserinone A

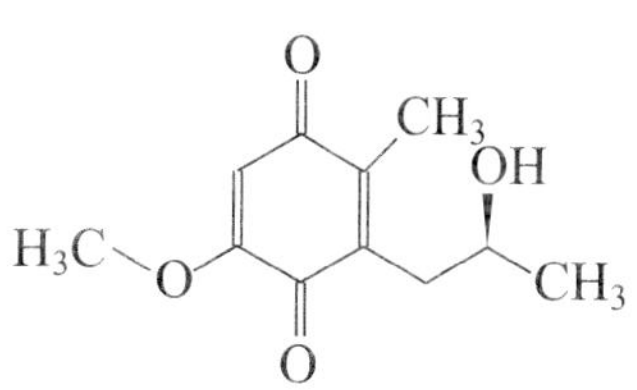

Fig. 7.61 Anserinone B

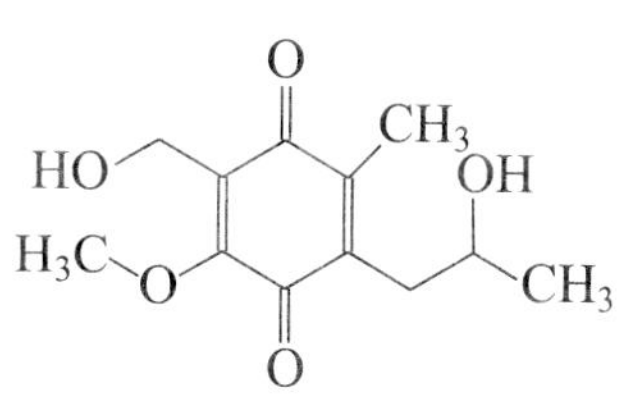

Fig. 7.62 Hydroxymethyl Lanserinone B

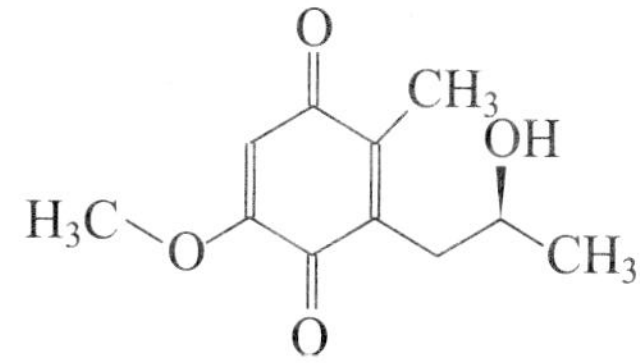

Fig. 7.63 Deoxyanserinonc

The bright color of *Amanita muscaria* nowadays, most of the structural features of the pigments have been elucidated. Muscarufin (Fig. 7.55) is the principal component; it is probably a putative derivative of a terphenylquinone; but efforts by different groups to reproduce these evidently premature results failed. It was proposed a quite novel structure, muscaflavin (Fig. 7.55) (Michelot and Melendez-Howell, 2003), for a yellow pigment, a result corroborated by a biosynthetic pathway that includes prior divergent oxidative cleavages of L-DOPA, further cyclization steps generating muscaflavin together with betalamic acid (Fig. 7.56), and probably also muscazone (Fig. 7.65), stizolobic acid (Fig. 7.66), and stizolobinic acid (Fig. 7.67). The structure has been confirmed by synthesis. Muscaflavin (Fig. 7.64) has also been detected in red-fruited Hygrocybe species. Different compounds, the muscaaurins (Fig. 7.68), have been proved to be the pigments responsible for the characteristic red-orange colour of the caps of several of other Amanita species such as *A. caesarea*. An elegant chromatographic procedure enables the separation of these pigments. Interestingly, ibotenic (Fig. 7.69) and stizolobic (Fig. 7.66) acids are essentially elements added to the whole structure of muscaaurins (Fig. 7.68). The muscaaurins (Fig. 7.68) are betalaines and also originate from the transformation of DOPA. Muscapurpurin (Fig. 7.70) (purple) and muscarufin (Fig. 7.64) (red-brown), whose structures are closely related to the muscaaurins (Fig. 7.68), have been isolated from *Amanita muscaria* (Michelot and Melendez-Howell, 2003).

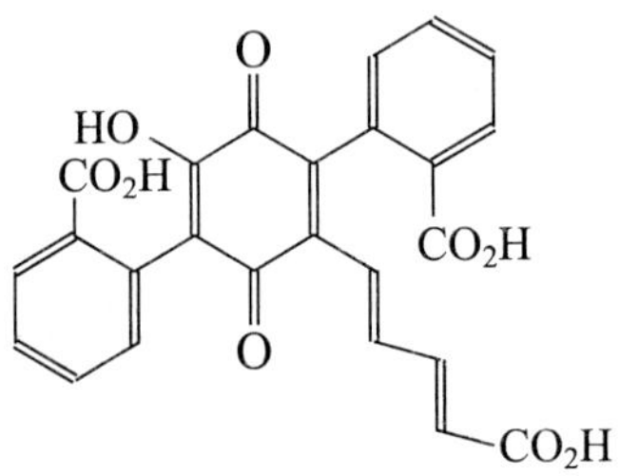

Fig. 7.64 Muscarufin

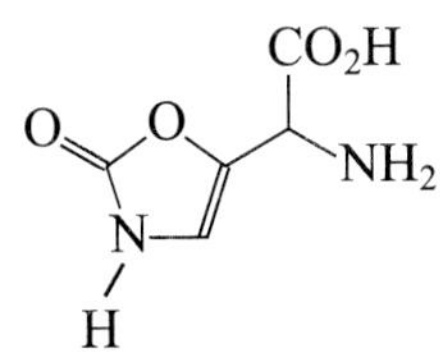

Fig. 7.65 Muscazone

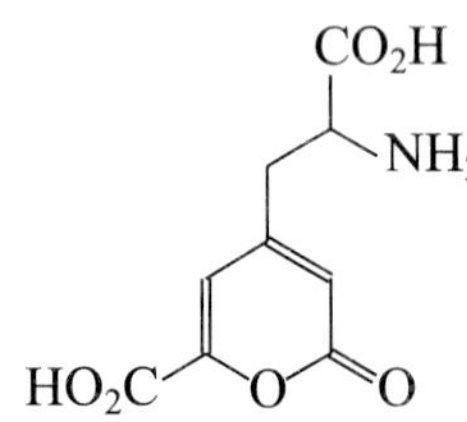

Fig. 7.66 Stizolobic Acid

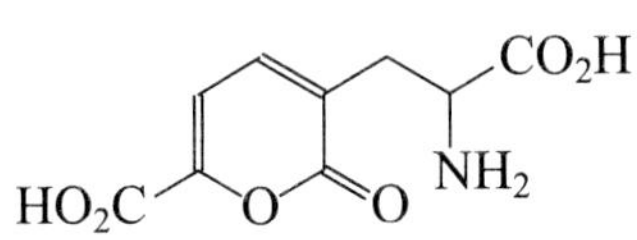

Fig. 7.67 Stizolobinic Acid

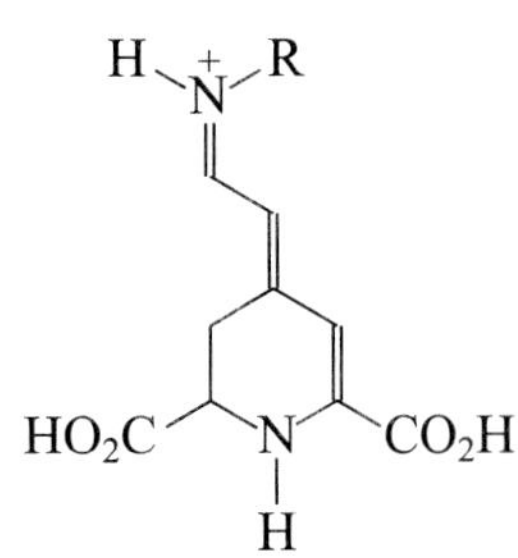

Fig. 7.68 Muscaaurins

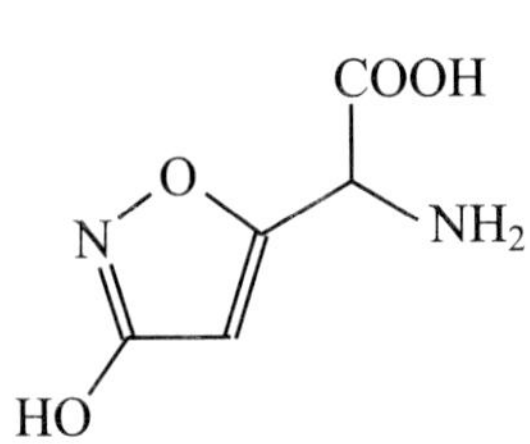

Fig. 7.69 Ibolenic

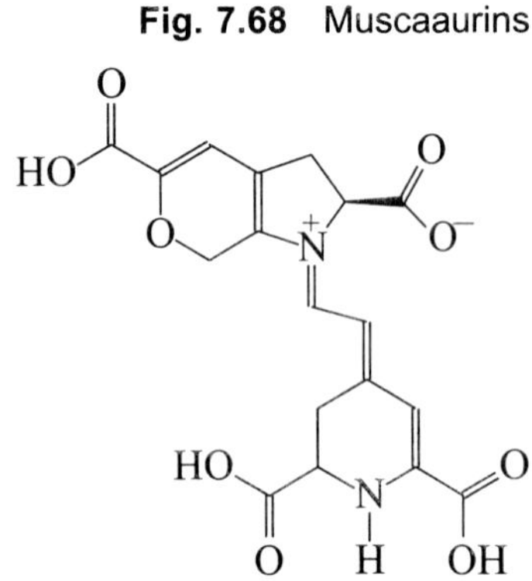

Fig. 7.70 Muscapurpurin

Characterization of Fungal Pigments

The chemical composition of the major pigments was discussed, and new methods for identifying natural dyes used frequently in foods were described (Tricard *et al.*, 1998). Raman and NIR Raman spectroscopy was applied very recently to the characterization of natural dyes (Li-Cham, 1996; Schrader *et al.*, 1999). Structures and physico-chemical properties of 100 naphthoquinone matabolites produced by filamentous fungi were reviewed previously (Medentsev and Akimenko, 1996). Corticrocin (Fig. 7.71) is synthesized by mycelia of *Piloderma croceum* in contrast to mycorradicin (Fig. 7.72), which is deposited into de vacuoles of maize root cells after colonization with arbuscular mycorrhizal fungi. Corticrocin (Fig. 7.71) is formed only when *P. croceum* is grown in darkness. This indicates that the pigment is unlikely to be a corotenoid origin that also follows from the chemical structure (Schreiner *et al.*, 1998). From 476 strains of oligotrophic microorganisms isolated from soil and river water samples, 5 strains capable of reducing the concentration of ytterbium were selected and characterized (Kamijo *et al.*, 1999). A pigment isolated from soil fungi *Cladosporium cladosporioides* was analyzed by pyrolytic gas chromatography and identified by mass spectrometry as samples of dry melanin isolated from the mycelium. It was characterized by increasing content of benzene, pyridine, and indole derivatives as well as by the lack of pyrazine, pyrocatechol, and benzenenitrile derivatives by comparison with a model of synthetic melanin prepared by the autooxidation of dihydroxyphenylalanine or dopamine (Chodurek *et al.*, 1998). Many synthetic pigment structures were described in the SIGMA-ALDRICH handbook (Green, 1991).

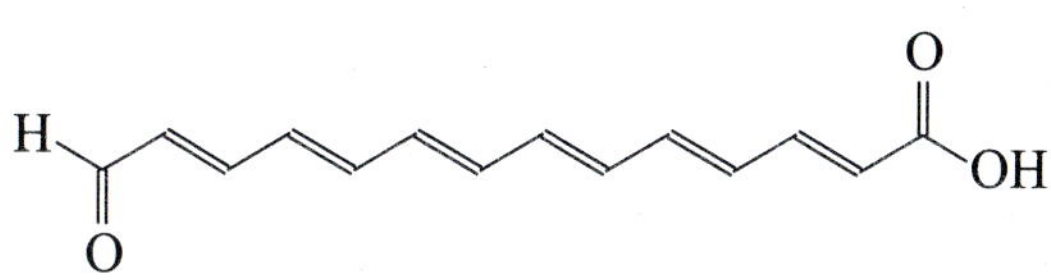

Fig. 7.71 Corticrocin

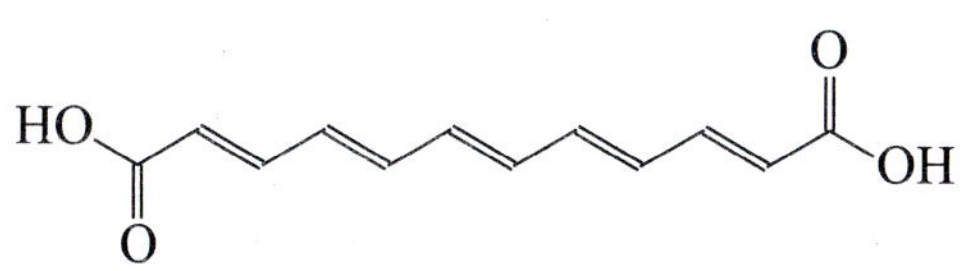

Fig. 7.72 Mycorradicin

An indole derivative, schizocommunin (Fig. 7.73), was isolated along with indigotin (Fig. 7.74) (indigo), indirubin (Fig. 7.75), isatin (Fig. 7.76), and tryptanthrin (Fig. 7.77), from the liquid culture medium in which a culture of *Schizophyllum commune*, isolated from the bronchus of a human patient with allergic bronchopulmonary mycosis, had been grown. The structure of schizocommunin (Fig. 7.73) was established by spectroscopic investigation. The assignments of the I H- and 13C-NMR signals of indigotin were also listed (Hosoe *et al.*, 1999).

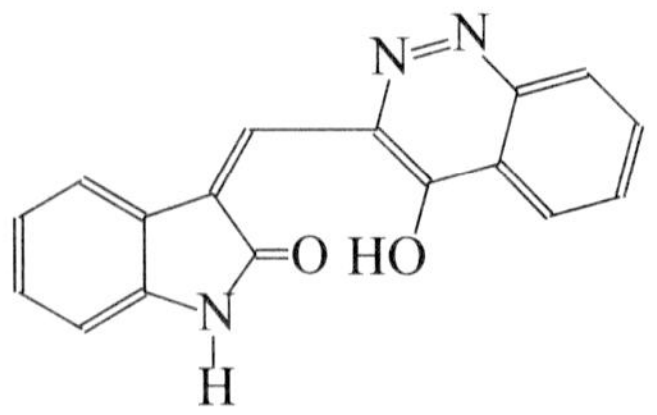

Fig. 7.73 Schizocommunin

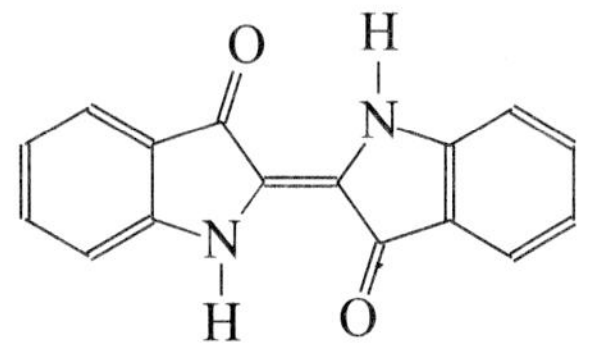

Fig. 7.74 Indigotin

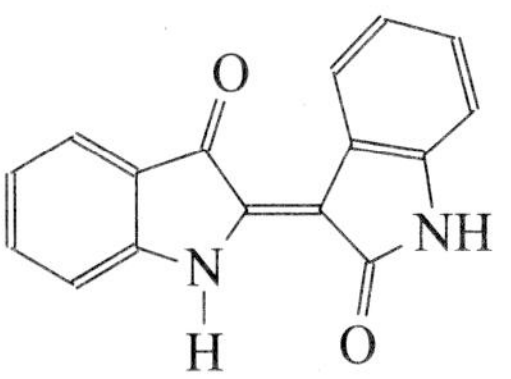

Fig. 7.75 Indirubin

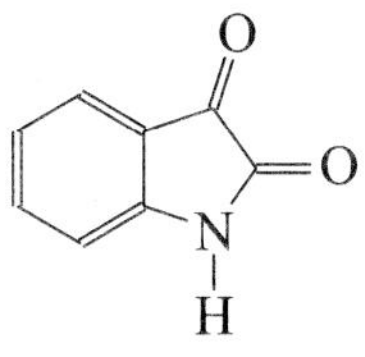

Fig. 7.76 Isatin

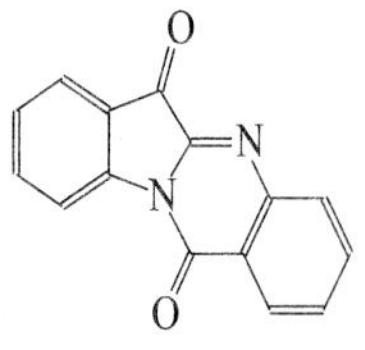

Fig. 7.77 Tryptanthrin

The photostability of 18 amino acid derivatives from monascus pigment was tested under various physical and chemical conditions. Under sunlight, the half-life of derivatives was increased 6-25- fold improvement over a control red pigment. The degradation of pigment derivatives followed a first-order reaction, and the pigment stability increased with an increasing concentration while it decreased with both an increase and decrease in pH from 7. Pigment stability under UV light (365 nm) showed a pattern similar to stability after exposure to sunlight. After 30 days of incubation at 30°C, more than 80% of the initial derivative contents remained while only 29% of the control red remained (Jung *et al.*, 2005).

Azaphilones, named cohaerins A (Fig. 7.78) and B (Fig. 7.79) were isolated from stromata of the xylariaceous ascomycete *Hypoxylon cohaerens*. Their absolute structures were determined by spectroscopic methods (2D NMR, NIS, IR, UV CD), and subsequently confirmed by acetylation. Stromatal metabolite profiles of several taxa of *Hypoxylon* sect. *Annulata* were also generated using analytical HPLC with diode array and MS detection. The cohaerins were neither found in other *Hypoxylon spp*., nor in other Xylariaceae. However, they were present even in holotype material of *H. cohaerens*, collected over 200 years ago. The binaphthalene (BNT) (Fig. 7.80) was also omnipresent in sect Annulata, and its derivatives, the benzofluoranthenes daldinone A (Fig. 7.81) and truncatone (Fig. 7.82), as well as presumably related compounds. These fungi were found devoid of other types of azaphilone pigments of the Xylariaceae, such as mitorubrins (Fig. 7.83) and daldinins (Fig. 7.84), the latter of which are widespread in certain groups of *Hypoxylon* sect. *Hypoxylon*. Hence, chemotaxonomic data largely support the current generic concept. The original source of truncatone (Fig. 7.82) was identified as *Hypoxylon annulatum* (Quang *et al.*, 2005).

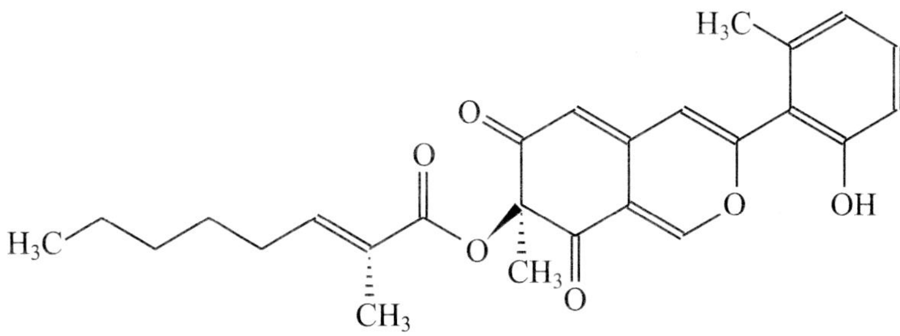

Fig. 7.78 Cohaerins A

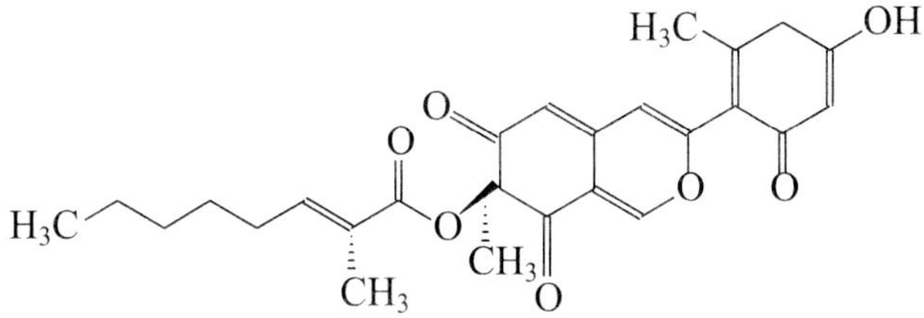

Fig. 7.79 Cohaerins B

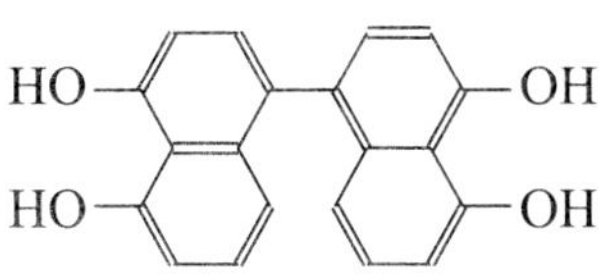

Fig. 7.80 Binaphthalene (BNT)

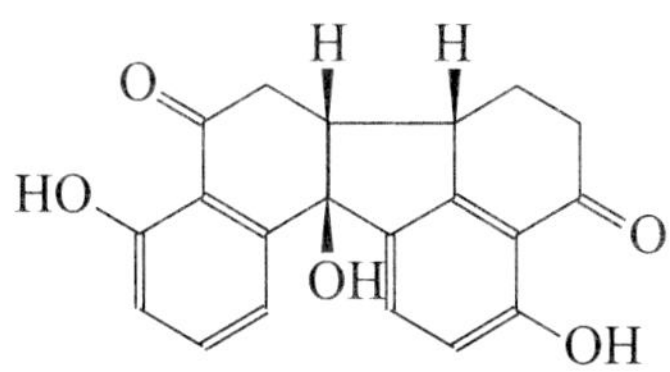

Fig. 7.81 Daldinone A

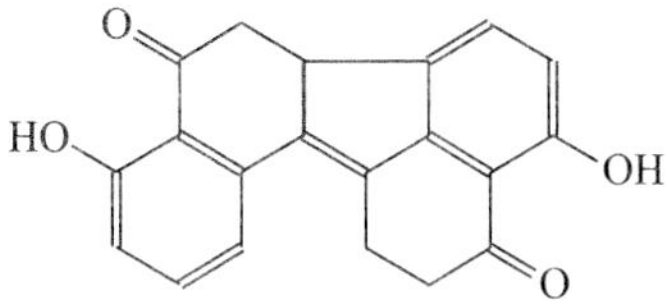

Fig. 7.82 Truncatone

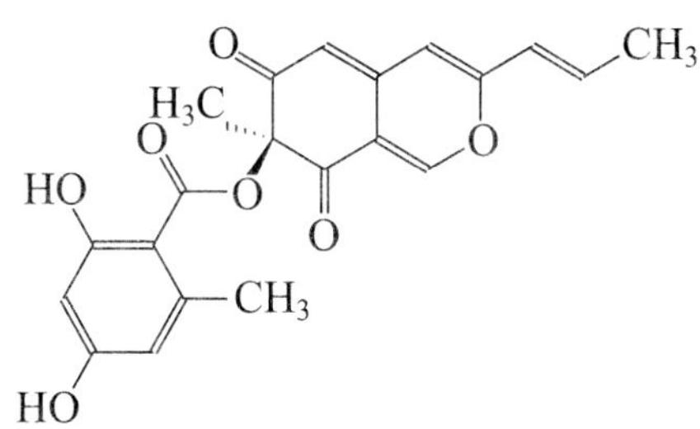

Fig. 7.83 Mitorubrins

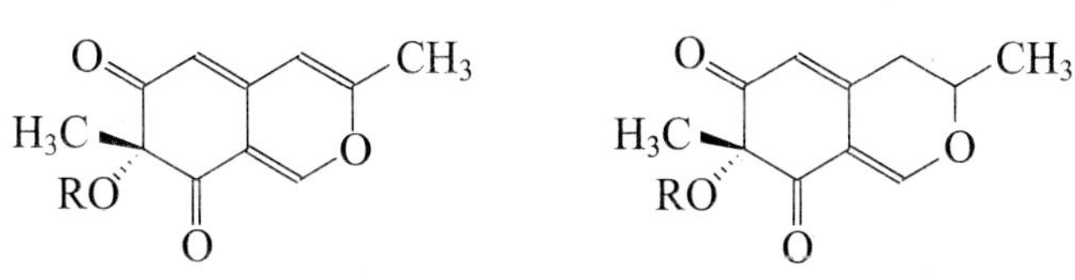

Fig. 7.84 Daldinins

The absolute stereochemistry at the C 3 (and C 3', where appropriate) chiral centre(s) in the coupled dihydroanthracenones, the icterinoidins A(1) (Fig. 7.85) and B-1 (Fig. 7.86) and atrovirin B-2 (Fig. 7.87) (from *Dermocybe icterinoides*), was deduced by application of the 'syn-anti rule', which relies on an empirical relationship between the sign of the Cotton effect couplet centred close to 275 nm in the CD spectrum and the chemical shift of the enantiotopic methylene protons at C 4 in the H-1 NMR spectrum of these pre-anthraquinones. The conclusions also allowed the assignment of central stereochemistry to atrovirin B-1 (Fig. 7.86) (from *Cortinarius atrovirens*). In addition, it was applied Steglich's kinetic resolution method to confirm the (P)-axial configuration of icterinoidin B-1 (Fig. 7.86), previously deduced by using Nakanishi's' exiton chirality' method (Gill and Morgan, 2004).

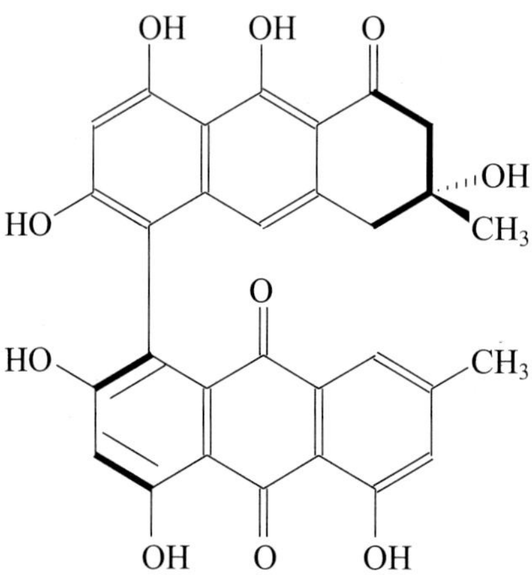

Fig. 7.85 Icterinoidin A1

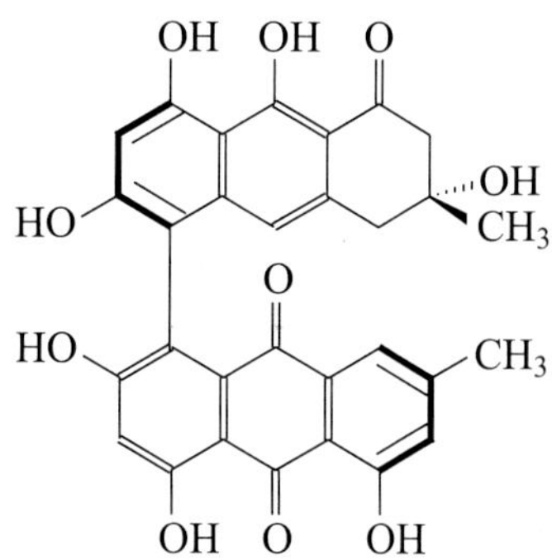

Fig. 7.86 Icterinoidin B1

Monosporascus cannonballus causes severe production losses to muskmelon and watermelon in the United States and other countries. Five colored metabolites from pigmented cultures of *M. cannonballus* isolate TX923038 have now been identified. Two of these, monosporascone (Fig. 7.88) and dehydroxyarthrinone (Fig. 7.89), have been isolated from other fungi, and three, demethylcerdarin (Fig. 7.90), monosporascol A (Fig. 7.91) and azamonosporascone (Fig. 7.92), have not previously been reported. The H-1 NMR and C-13 NMR of all five compounds were reported (Stipanovic *et al.*, 2004).

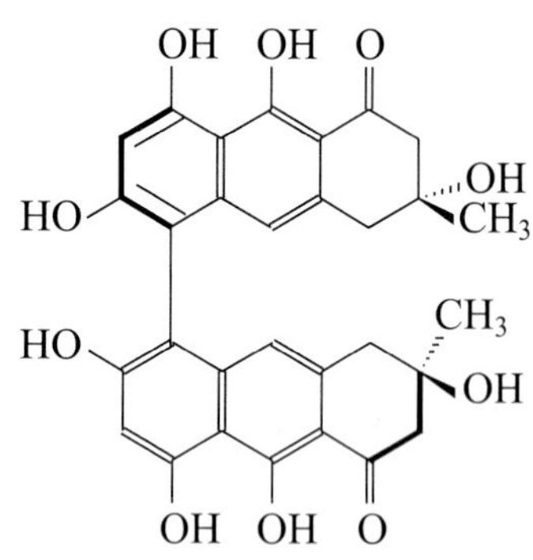

Fig. 7.87 Atrovirin B2

Highly functionalized novel sesquiterpenes, named boletunones A (Fig. 7.93) and B (Fig. 7.94), were isolated from the methanolic extract of the fruiting body of the mushroom *Boletus calopus*. Their structures and relative stereochemistry were established by various spectral analyses (Kim *et al.*, 2004).

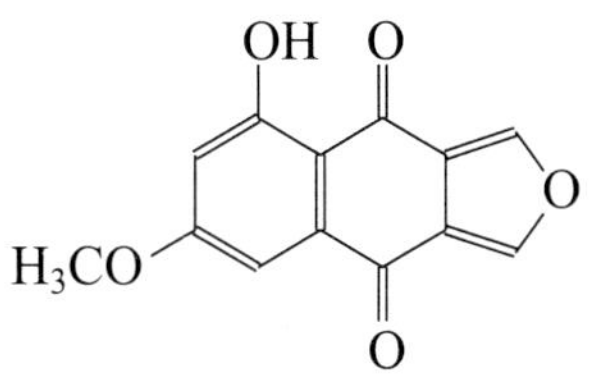

Fig. 7.88 Monosporascone

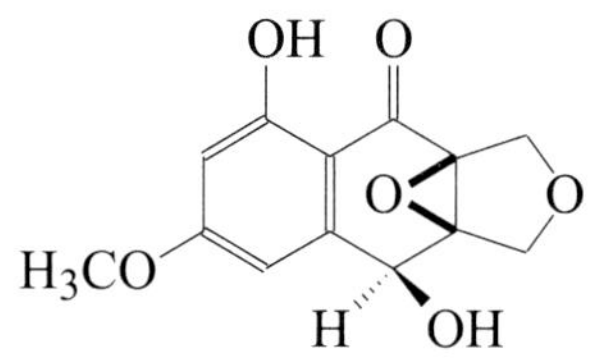

Fig. 7.89 Dehydoxyarthrinone

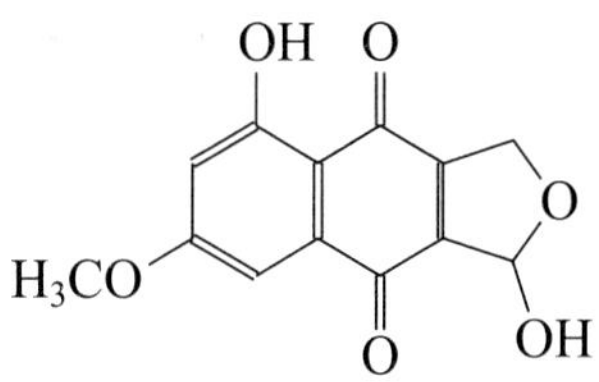

Fig. 7.90 Demethylcerdarin

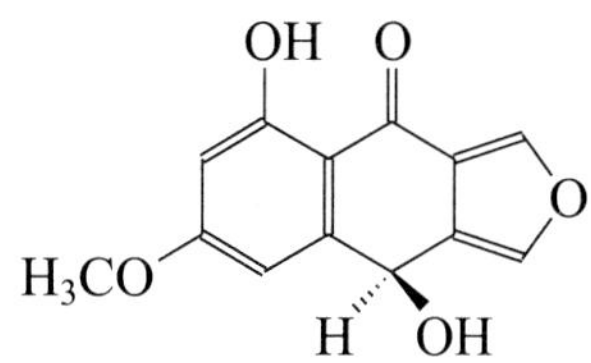

Fig. 7.91 Monosporascol A

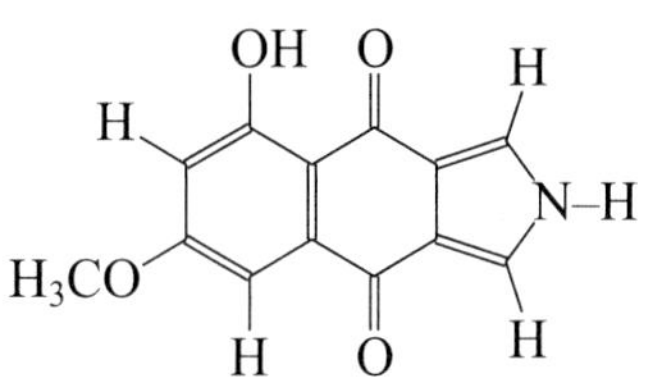

Fig. 7.92 Azamonosporascone

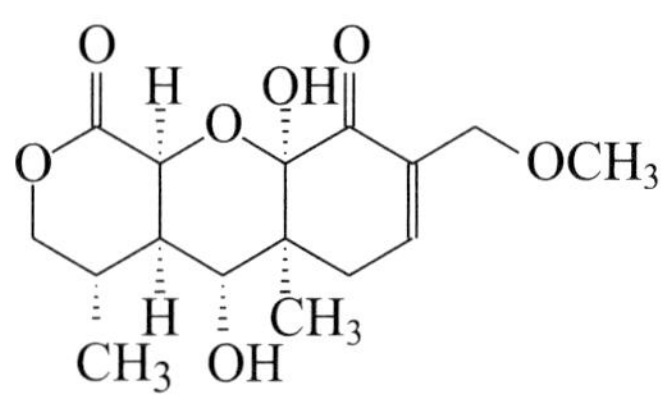

Fig. 7.93 Boletrenone A

Laetiporic acid (Fig. 7.49), from *Laetiporus sulphureus* (sulfur shelf) was analysed by application of extensive 2D NMR techniques and characterized as a polyene of non-isoprenoid origin. Laetiporic acid (Fig. 7.49), which represents the main pigment in *L. sulphureus* basidiocarps, bears an unprecedented decaene skeleton as part of its chromophore and, interestingly, contains double bonds with a stable cis configuration (Weber *et al.,* 2004).

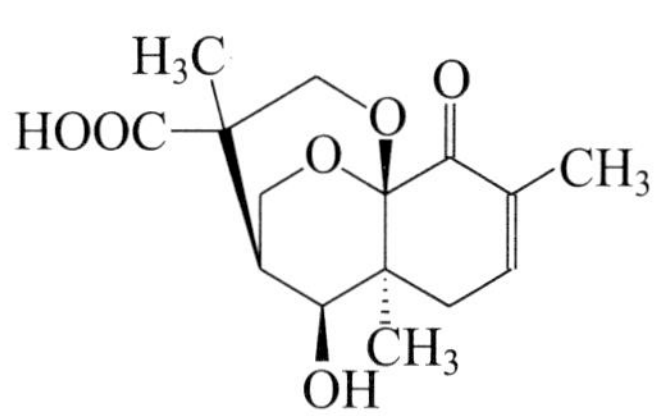

Fig. 7.94 Boletrenone B

A reversed-phase HPLC method had been developed for the analysis of hydroxylated pulvinic acid derivatives which are responsible for the pigmentation of fruit-bodies belonging to the order Boletales (Fungi). Variegatic (Fig. 7.26), xerocomic (Fig. 7.25) and atromentic acid (Fig. 7.29) as well as variegatorubin (Fig. 7.27) were detected and separated in methanolic extracts of *Boletus permagnificus* and *Xerocomus parasiticus*, and the pigment profile of these species was compared. The identity of the pigments was confirmed by means of LC-atmospheric pressure chemical ionization (APCI) mass spectrometry (Davoli and Weber, 2002).

Red yeast rice (angkak, red koji) obtained as cultures of *Monascus purpureus* on rice was extracted and analyzed by HPLC. In addition to the known red, orange, and yellow pigments and the mycotoxin citrinin (Fig. 7.32), a new Monascus metabolite was detected. It was present in the original red yeast rice and formed in higher amounts when red yeast rice was heated. The chemical structure was elucidated by high-resolution mass spectrometry analysis and by NMR data. The new compound, named monascodilone (Fig. 7.95), was characterized by a propenyl group on a pyrone ring, an aromatic ring, and a gamma-lactone group (Wild *et al.*, 2002).

A new and simple enzymatic method was developed for preparative isolation of anthraquinone pigments from *Dermocybe sanguinea*. The endogenous beta-glucosidase of the fungus was used to catalyze the hydrolysis of the O-glycosyl linkage in emodin- (Fig. 7.96) and dermocybin-1-beta-D-glucopyranosides (Fig. 7.97). The developed enzymatic method was found to be effective for the pigment isolation, as the hydrolysis occurred virtually completely, thus leading to a high pigment yield. Two fractions were obtained by the method: Fraction 1 (94% of the total pigment amount: emodin and dermocybin), and Fraction 2 (6%), containing the anthraquinone carboxylic acids. Combined gas chromatography-mass spectrometry was applied to determine the anthraquinone derivatives of Fraction 1 after methylation and acetylation (Hynninen *et al.*, 2002).

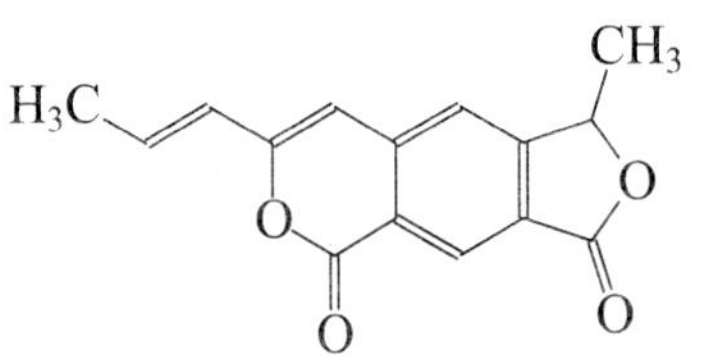

Fig. 7.95 Monascodilone

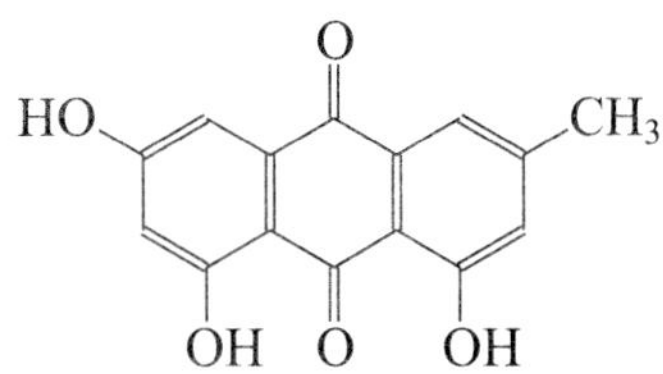

Fig. 7.96 Emodin

The bright orange-red color of the toadstool *Tricholoma aurantium* was due to the benzotropolone pigment aurantricholone (Fig. 7.98). The compound was at least partially present as the calcium salt. Minor pigments were yellow 2H-furo[3,2-b]benzopyran-2-one derivatives aurantricholides A (7a) (Fig. 7.99) and B (7b) (Fig. 7.100), which exhibit strong green fluorescences. Their structures have been established by total syntheses (Klostermeyer *et al.*, 2000).

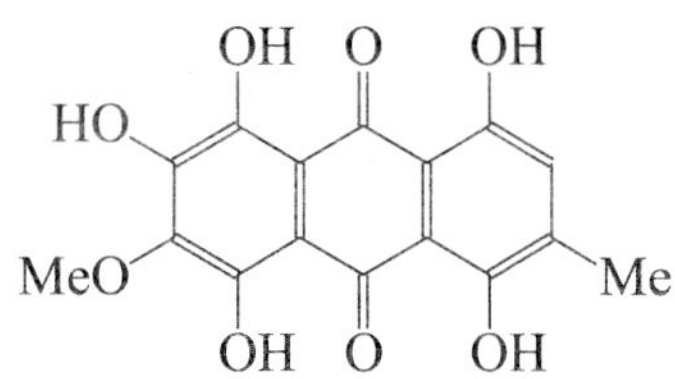

Fig. 7.97 Dermocybin

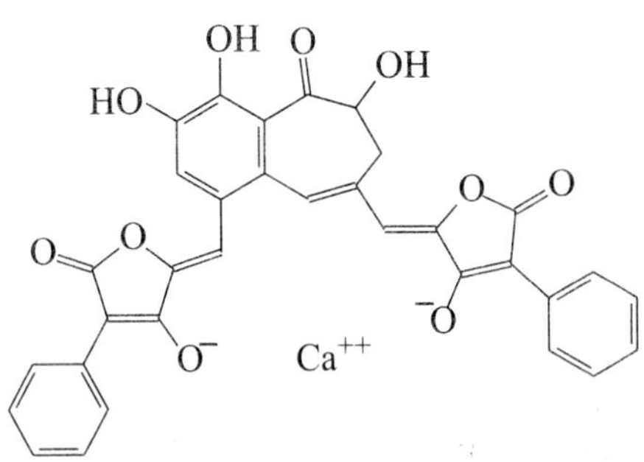

Fig. 7.98 Aurantricolone

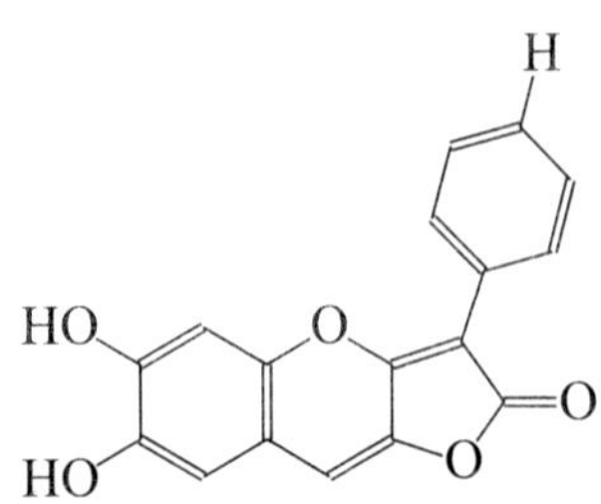

Fig. 7.99 Aurantricholide A

A red pigment, lilacinone (Fig. 7.101) was isolated from fruit bodies of the toadstool *L. lilacinus*. Its structure was established by 2D NMR and APCIMS methods. Lilacinone (Fig. 7.101) is a novel type of fungal aminobenzoquinone pigment and may be biosynthetically derived from 3 molecules of anthranilic acid. Lilacinone (Fig. 7.101) is structurally related to blennione (Fig. 7.102) the green pigment of *Lactarius blennius* and similar to necatorone (Fig. 7.103) (Spiteller *et al.*, 2003).

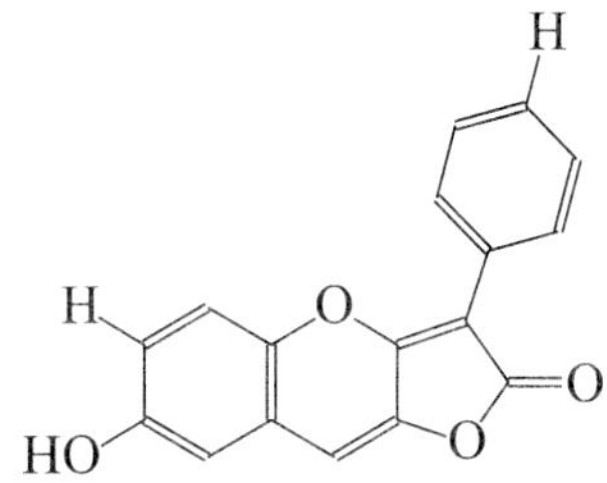

Fig. 7.100 Aurantricholide B

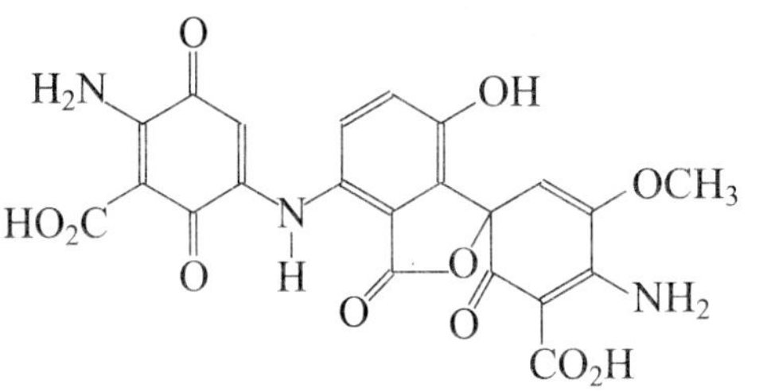

Fig. 7.101 Lilacinone

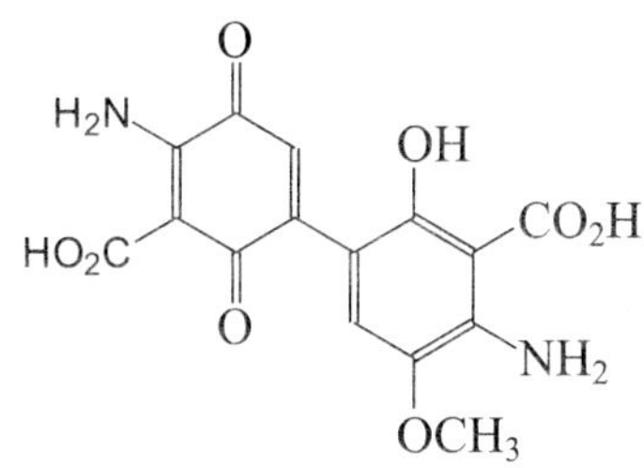

Fig. 7.102 Blennione

Eighty-one compounds of pigmented microorganisms were isolated from natural sources by using spread plate technique with different culture media. From the total, 73 isolates were identified as pigmented yeasts; 3 isolates as black yeasts; 4 isolates as pigmented molds and one isolate was red bacterium. All of pigments were classified in different groups according to their chemical structures by chemical methods. Two isolates of pigmented molds, RF1 and GF produced naphthoquinone and one isolate, RF2 produced anthraquinone. A flavonoid from pigmented mold was also found (Wichai *et al.*, 2001).

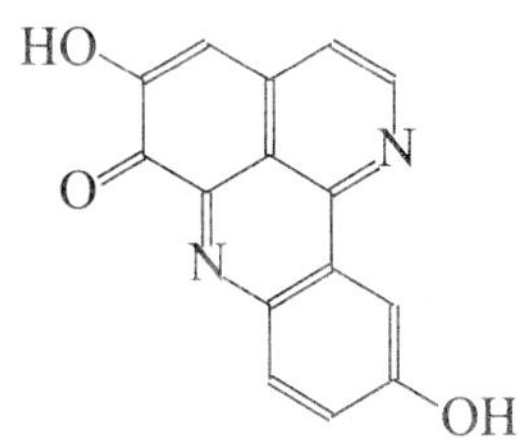

Fig. 7.103 Necatorone

A new polyene pigment, melanocrocin (Fig. 7.104), had been isolated either from fruit bodies or mycelial cultures of the subterranean fungus *Melanogaster broomeianus*. The structure of the pigment was determined by spectroscopic methods and chemical transformations. Melanocrocin (Fig. 7.104) is the N-acyl derivative of L-phenylalanine Me ester with a polyolefinic carboxylic acid (Aulinger *et al.*, 2001).

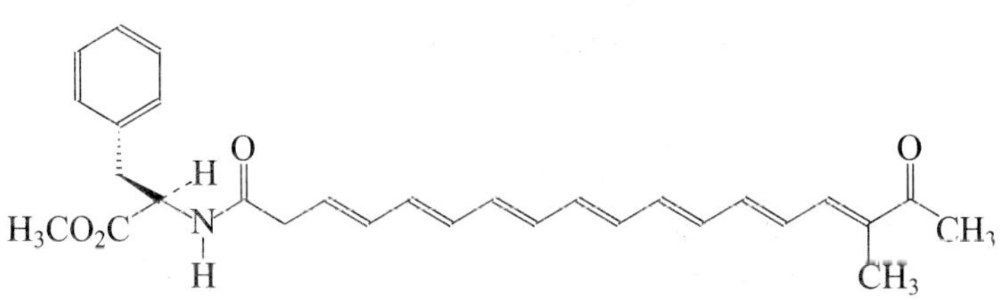

Fig. 7.104 Melanocrocin

Monascus purpureus IB1 produced about 50-fold higher levels of azaphilone pigments than *M. purpureus* NRRL1596. Differently pigmented mutants were obtained from *M. purpureus* IB1 by nitrosoguanidine treatment. A highly pigmented strain, *M. purpureus* HP14, was found to lack the formation of the classical yellow and orange azaphilones and was found to produce only about 10% of the red azaphilone pigments. The intense color was associated with novel pigments as shown by HPLC. The addition of hexanoic acid to *M. purpureus* IB1 resulted in higher volumetric and specific red pigment productivity, but in a complete absence of the classical orange azaphilones, while the classical yellow and red azaphilone pigments were severely reduced; new peaks corresponding to less hydrophobic pigments were found in hexanoic-supplemented cultures by HPLC. Purification of pigments from hexanoic-supplemented cultures showed the presence of five new pigments as indicated by the absorption spectra and HPLC analysis. Two of them, R3 and Y3, were characterized by NMR as 9-hexanoyl-3-(2-hydroxypropyl)-6a-methyl-9,9a-dihydro-6H-furo[2,3-h]isochromene-6,8(6aH)-dione (Fig. 7.105) and 4-[2,4-dihydroxy-6-(3 hydroxybutanethioyloxy)-3-methylphenyl]-3,4-dihydroxy-3,6-dimethylheptanoic acid (Fig. 7.106). These pigments were also found to be present in cultures of the high-producing mutant *M. purpureus* HP14. These new pigments were less hydrophobic than the classical azaphilones and may have better properties as natural colorants in the food industry (Campoy *et al.*, 2006).

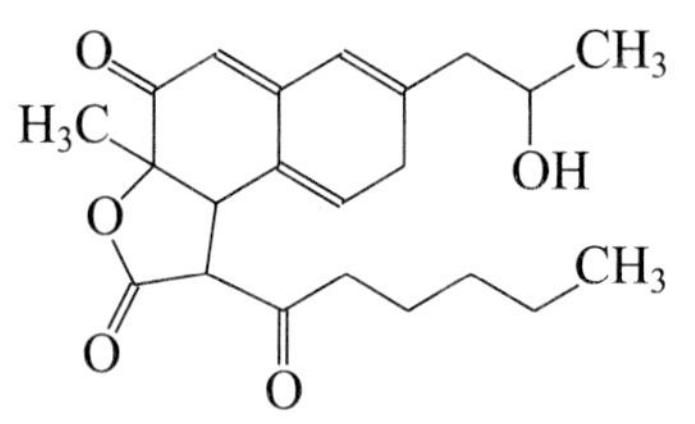

Fig. 7.105 Isochromenedione

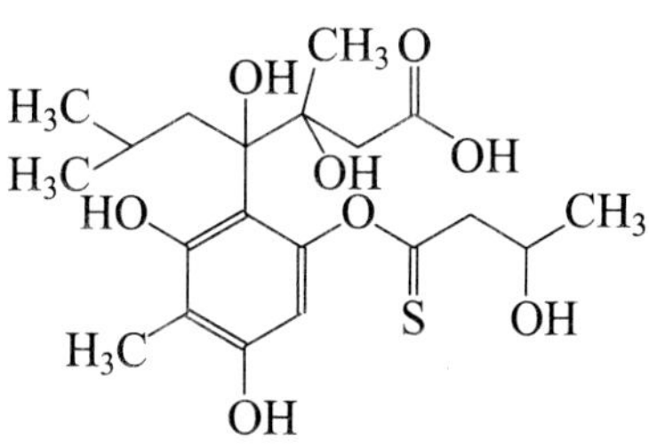

Fig. 7.106 Hydroxybutanethioxyloxy

A new pigment, phlegmacin A 8,8'-di-O-Me ether (Fig. 7.107), was isolated from fruiting bodies of *Xylaria euglossa* (Ascomycota), along with the known fungal pigments (S)-torosachrysone-8-O-Me ether and emodin-6,8-di-O-Me ether. The structure of phlemagcin A was established on the basis of its spectroscopic data and the absolute configuration determined by comparing the CD and 1H NMR spectra of phlegmacin A with those of other phlegmacins. This compound represents the first phlegmacin type pigment isolated from the ascomycete (Wang *et al.*, 2005).

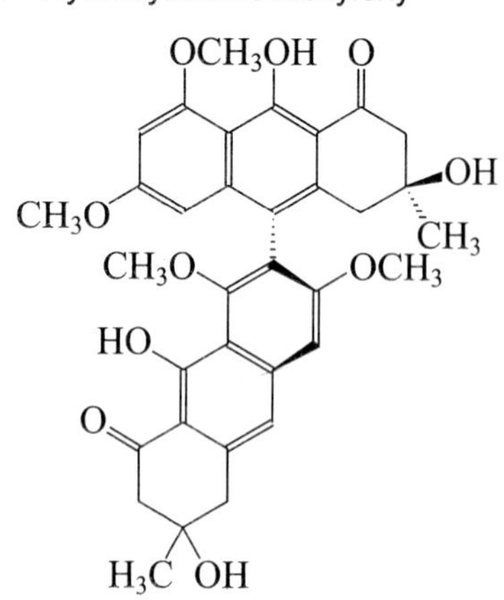

Fig. 7.107 Phlegmacin A

Two new, red-colored azulene pigments, 7-(1,2-dihydroxy-1-methylethyl)-4-methylazulene-1-carbaldehyde (Fig. 7.108) and 7-acetyl-4-methylazulene-1-carbaldehyde (Fig. 7.109), were isolated from the fruiting bodies of the basidiomycete *Lactarius deliciosus*, together with a related, known compound (lactaroviolin (Fig. 7.110). Their structures were established on the basis of spectroscopic evidence including 2D-NMR experiments (Yang *et al.*, 2006).

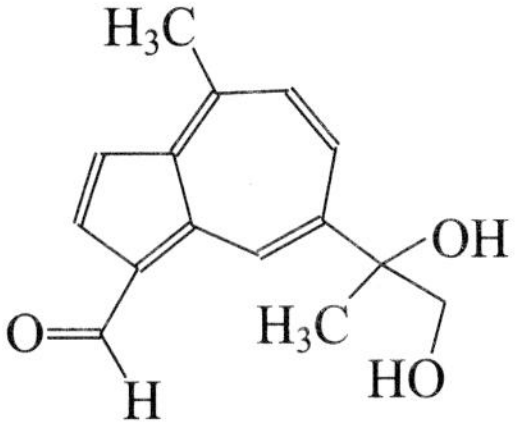

Fig. 7.108 Dihydroxyazulene

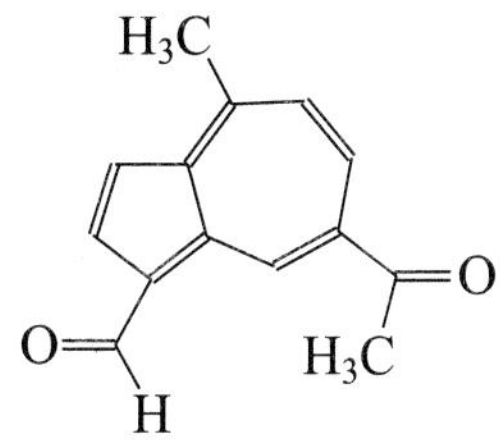

Fig. 7.109 Acetylazulene

A new polyene pigment, melanocrocin, has been isolated either from fruit bodies or mycelial cultures of the subterranean fungus *Melanogaster broomeianus*. The structure of the pigment was determined by spectroscopic methods and chemical transformations. Melanocrocin (Fig. 7.104) is the N-acyl derivative of of L-phenylalanine Me ester with a polyolefinic carboxylic acid (Aulinger *et al.*, 2001).

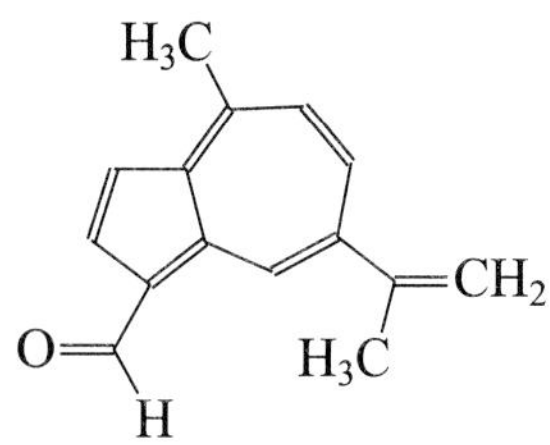

Fig. 7.110 Lactaroviolin

Biological Activities

The fungal pigments are found to be active against bacteria, yeast, fungi, protozoa *Leishmania brasiliensis,* and the insects *Calliphora erythrrocephala.* The cytotoxic activity of naphthoquinones against mouse leukemia and HeLa cells has been reported. Along with the antibiotic and toxic activities, naphthoquinones revealed mutagenic and arcinogenic properties. The mycotoxin activities are expressed in several of these metabolites. Thus, the best known xanthomegnin (Fig. 7.111) and viomellein (Fig. 7.112), synthesized by *Penicillium, Asperigillus, Microsporium, Trichophyton* fungi, affected liver and kidney in laboratory animal (Medentsev and Akimenko, 1998). Pigments produced as secondary metabolites by various isolates of *Monascus* have been used traditionally as colorants in Oriental food. Modern food industry has rediscovered these molds as promising source for natural colorants. However, recent studies indicated that one of the secondary metabolites produced by *Monascus* has an identical structure to the mycotoxin citrinin. Citrinin (Fig. 7.32) and two *Monascus* extracts induced a positive dose depending on the mutagenic response in the Salmonella-hepatocyte assay applying strain TA-98. However, no mutagenecity could be detected in the Salmonella microsome assay, with or without S9-mix, for citrinin (Fig. 7.32) and *Monascus* extracts, applying TA-98, TA-100, TA-1535, TA-1538, and TA-97.

These findings provide further evidence that citrinin (Fig. 7.32) requires complex cellular biotransformation to exert mutagenicity (Fink-Gremmels and Sabater-Vila, 1998; Sabater-Vilar *et al.*, 1999). The biological activity of the pigment extracts from *Monascus purpureus* included their antibiotic action not only against bacteria but also against some species of yeast and filamentous fungi, as well as embryotoxicity and teratogenecity. These activities depended on the presence of the orange components (monascorubrin (Fig. 7.8) and rubropunctatin (Fig. 7.7) (Durán *et al.*, 2002 and references therein).

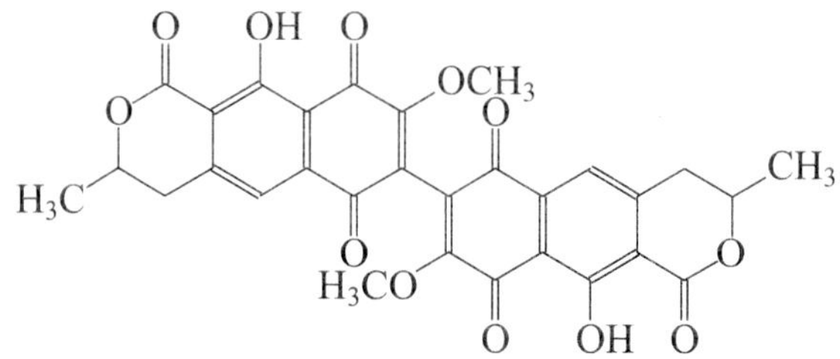

Fig. 7.111 Xanthomegnin

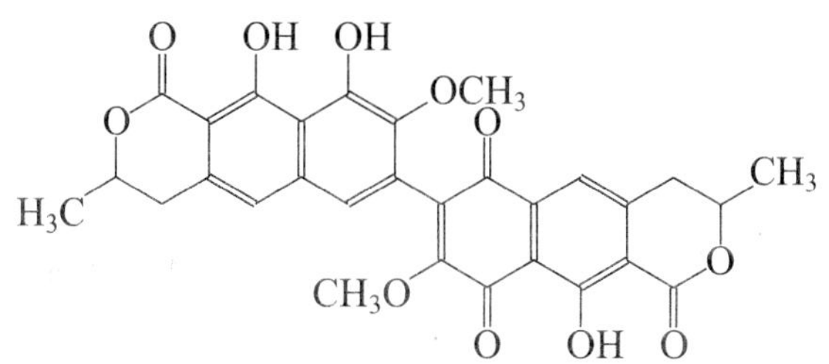

Fig. 7.112 Viomellein

In certain culture conditions purple pigments (monascorubramine (Fig. 7.28) and rubropunctamine (Fig. 7.113) retained some biological activity (Martinkova *et al.*, 1995, 1999). The antibacterial activity of *Monascus* pigments against bacteria, actinomycetes yeast, and fungi was studied. The pigment has a strong activity against *Bacillus subtilis* and *Staphylococcus aureus*, a weak activity against *Escherichia coli* and *Streptomyces griseus,* and no activity against yeast, fungi, or *Sarcina flava* (Tong *et al.*, 1999).

Cynodontin (Fig. 7.11) was tested *in vitro* for fungitoxicity and was found to be a potent inhibitor of the growth of *Sclerotinia minor*, *Sclerotinia sclerotiorum*,

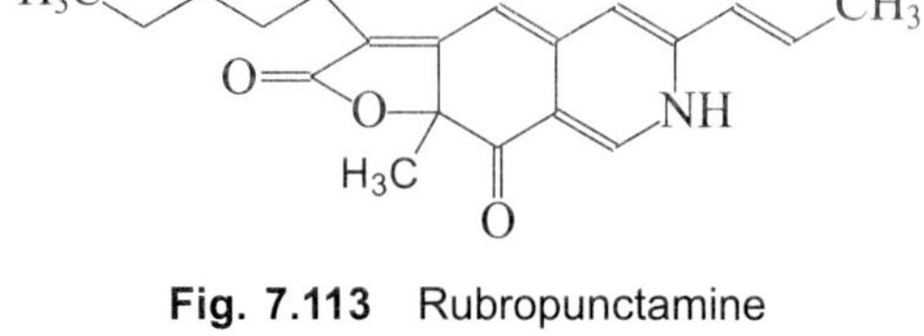

Fig. 7.113 Rubropunctamine

and *Botrytis cinerea* and, to a lesser extent, of *Verticillium dahliae*. The ED50 values obtained with these fungi were of the same order of magnitude as those of the commercial fungicides dicloran and carbendazim, which were used as reference chemicals. In contrast, the growth of a number of other fungi was not significantly inhibited by cynodontin (Fig. 7.11). Anthraquinone and two other anthraquinone derivatives, emodin (Fig. 7.96) and chrysophanol (Fig. 7.41), which were also included in the tests, did not affect the growth of the cynodontin-sensitive fungi. It thus appears that the type and position of the substitutions at the C-ring play a role in the expression of antifungal activity (Chrysayi-Tokousbalides and Kastanias, 2003).

New Monascus metabolites as monoascopyridines A and B (Fig. 7.114) produced by *M. purpureus* displayed cytotoxic effect on human kidney epithelial cells. These compounds did not induce apoptosis, however, they causes a rise of the mitotic index by an accumulation of cells in the metaphase with a simultaneous decrease of cells in the ana- and telophase. These effects are indicating an aneuploidic potential and therefore the monascopyridines (Fig. 7.114) might contribute to tumor formations (Knecht *et al.*, 2006).

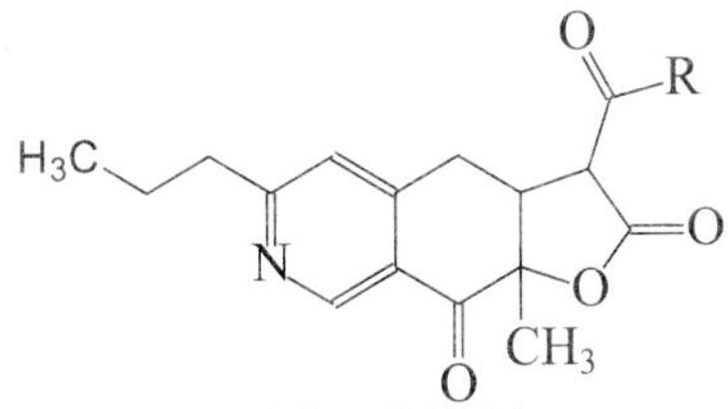

Fig. 7.114 Monascopyridine A & B

Among three strains of *Pycnoporus sanguineus,* MIP 89007 produced more cinnabarin (Fig. 7.50) than MIP 95001 and MIP 95002. The antimicrobial activity of cinnabarin (Fig. 7.50) was tested against 11 species of bacteria isolated from food. *Bacillus cereus* and *Leuconostoc plantarum* were the most sensitive to cinnabarin, being inhibited by 0.0625 mg/ml. *Klebsiella pneumoniae* was the least sensitive (>4.0 mg/ml)(Smania *et al.*, 1998).

The production of cinnabarin (Fig. 7.50) by three *P. sanguineus* strains was investigated, and the antimicrobial activity of this substance against bacterial species from food sources was tested. Its antiviral activity was also studied in this fungus (Smânia *et al.*, 2003).

Luteoskyrin (Fig. 7.115) is a hepatotoxic anthraquinone produced by *Penicillium islandicum Sopp* (Uraguchi *et al.*, 1961; Ueno and Ishikawa, 1969). Long-term feedings of mice with rice infected with *P. islandicum Sopp* and with the isolated luteoskyrin (Fig. 7.115) cause pathological changes of the liver, including adenoma and hepatoma. Takeda *et al.*, (1973) studied the chemistry of this toxin.

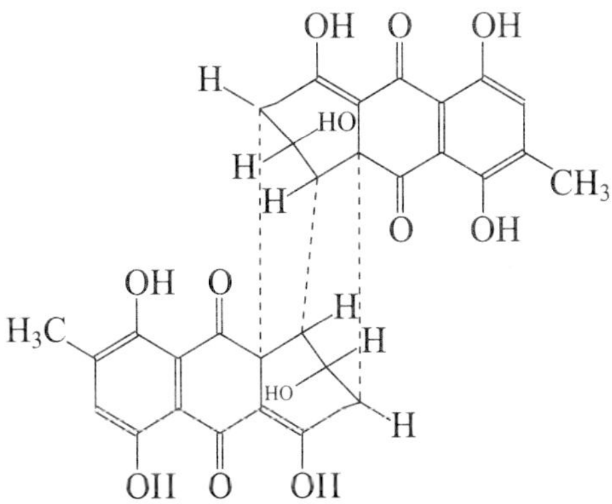

Fig. 7.115 Luteoskyrin

Schizocommunin (Fig. 7.73) had the strong cytotoxicity against murine lymphoma cells: Compound I possessed the inhibitory activity of more than 90% at 39 ng/ml, which corresponded to $1.35 \times 10–7$ M. This cytotoxicity might be concerned as one of the pathogenetic factors in ABPM induced by *S. commune* (Hosoe *et al.*,1999).

Aspergillus niger, Alternaria and Fusarium were isolated from leaves of tomato (*Lycopersicum esculentum*) and fruits of dolichos (*Dolichos lab lab*) and amla (*Emblica officinalis*). These forms were grown on GN medium. The color of the culture filtrate was pink for Fusarium, and yellow for Aspergillus and Alternaria. The development of pigments in the culture filtrates at different pHs was measured. Maximum pigmentation was observed at pH 8.5 for *Aspergillus niger*, pH 3.5 for Fusarium and pH 5.5 and pH 6.5 for Alternaria. The culture filtrates inhibited the germination of seeds of tomato, dolichos and amla (Mawthols *et al.*, 2005).

The pentaketides, (+)-formylanserinone B (Fig. 7.57) and (+)-epoxyserinone B (Fig. 7.59), along with 2 known fungal pigments, anserinones A (Fig. 7.60) and B (Fig. 7.61), were studied. Bioactivity was explored in 2 separated cell-based assays. Leukemia selectivity was greatest with anserinone B (Fig. 7.61) and (Fig. 7.57), while the anserinones and (Fig. 7.57) exhibited modest activity against the MDA-MB-435 cell line (Gautschi *et al.*, 2004).

The pigment isolated from the woody tissue of *Chlorociboria aeruginosa* and similar fungi is useful for control of plant growth, especially algal growth. Isolation of blue xylindein (Fig. 7.116) from *C. aeruginosa* and control of Chlorella growth with the green or blue-green xylindein (Fig. 7.116) were shown (Sakaki *et al.*, 2002).

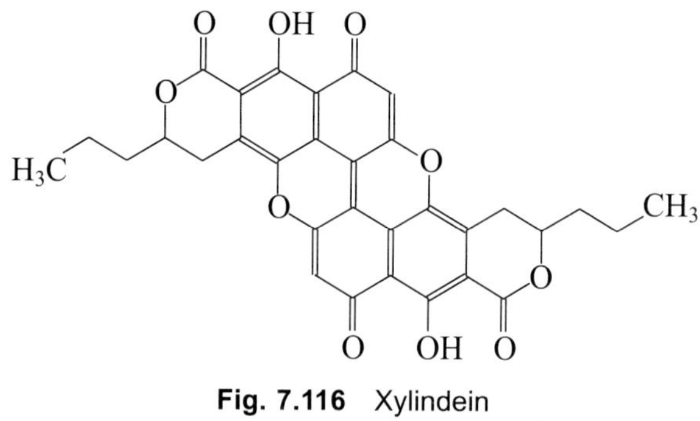

Fig. 7.116 Xylindein

Many toxicological data are available on the Arpink red (Fig. 7.117): acute oral toxicity in mice 90-day subchronical toxicological study, acute dermal irritation/corrosion, acute eye irritation/corrosion, anti-tumor effectiveness, micronucleus test in mice, AMES test (*Salmonella typhimurium* reverse mutation assay), estimation of antibiotic activity, results of estimation of 5 mycotoxins. A new patent

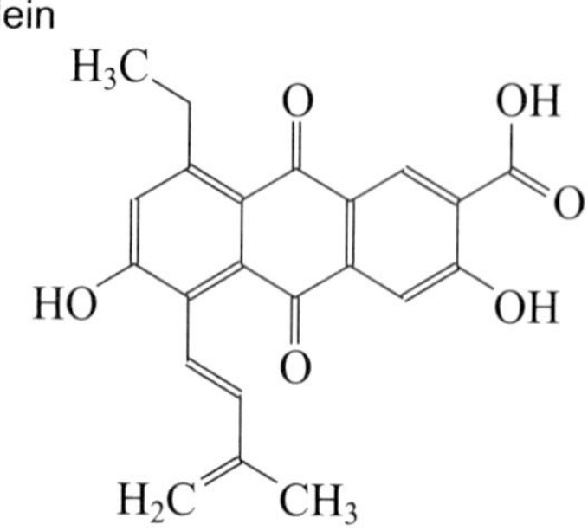

Fig. 7.117 Arpink Red

on Arpink Red (Fig. 7.117) was filed in 2001 (Sardaryan, 2002) with claims of anti-cancer effects of the anthraquinone derivatives and applications within the food and pharmaceutical fields.

Studies on rats had shown no toxicity of astaxanthin (Fig. 7.118) preparations. Dietary administration of astaxanthin (Fig. 7.118) has proved to significantly inhibit carcinogenesis in the mouse urinary bladder, rat oral cavity and rat colon. In addition it was reported to induce xenobiotic-metabolizing enzymes in rat liver.

Astaxanthin has been shown to enhance *in vitro* antibody production by mouse spleen cells stimulated with sheep red blood cells and in human blood cells *in vitro*. Furthermore, it has not exhibited any mutagencity in *in vitro* study at doses up to 14.4 mg/day for two weeks. Astaxanthin (Fig. 7.118) has been reported to have anti-inflammatory effect when administered together with aspirin. There was strong evidence to suggest that astaxanthin was shown to modulate the humoral and non-humoral immune systems. It enhances the release of interleukin-1 and the tumor necrosis factor a in mice, to a greater extent than canthaxanthin or ß-carotene and has the highest cytokinine inducing activity. Astaxanthin has a significant enhancing action on the production of immunoglobulin A, M and G and on T-helper cell antibody production, even when suboptimal amounts of antigen are present (Dufoseé *et al.*, 2005).

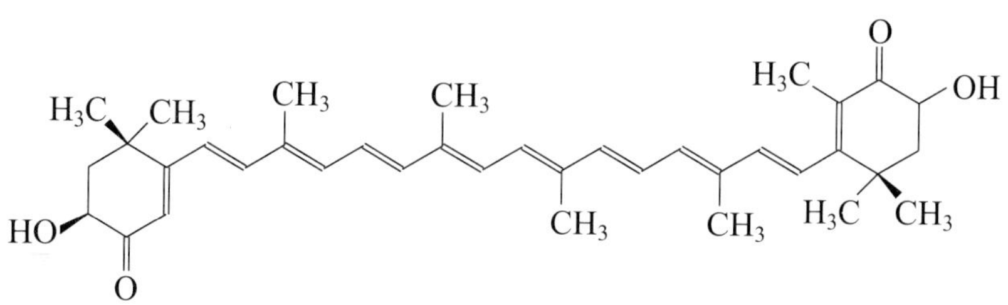

Fig. 7.118 Astaxanthin

Research reports indicate that natural β-carotene (Fig. 7.1) possesses numerous benefits (e.g. antioxidant, cancer, inhibits alveolar lesions, inhibit human lymphocytes proliferation, etc.) for the human body and consistently support the use of β-carotene as part of the human diet (Dufoseé *et al.*, 2005).

Genetics and Molecular Biology of Pigments

Carotenoids are one of the most diverse and widely distributed groups of natural pigments, responsible for the color of many plants and animals. Research into carotenoid biosynthesis has been carried out in many organisms: bacteria, fungi and plants. Among fungi, *Phycomyces blakesleeanus*, *Neurospora crassa*, and more recently *Mucor circinelloides*, have been used to investigate the biosynthesis of carotenoids and the regulation of the pathway. Wild type strains of *P. blakesleeanus* and *M. circinelloides* accumulate β-carotene as the main carotenoid and, depending of the culture conditions, several other compounds can also be found. During the last decade, molecular approaches have led to a deeper knowledge of the pathway and the isolation of a number of genes involved in carotenogenesis. Only a few out of the more than 600 carotenoids identified in nature are industrially important (β-carotene (1), lycopene (Fig. 1.119), lutein (Fig. 7.120) or astaxanthin (Fig. 7.118). These compounds are in an increasing demand by the market, since they are used in pharmaceutical and cosmetics industries and also as human food and animal feed additives. Iturriaga *et al.* (2001) focuse on recent findings on the structure and function of the genes involved in the biosynthesis of carotenoids in Mucor. The possibilities of using this organism as a source for the industrial production. of carotenoids, based on metabolic engineering, were also discussed.

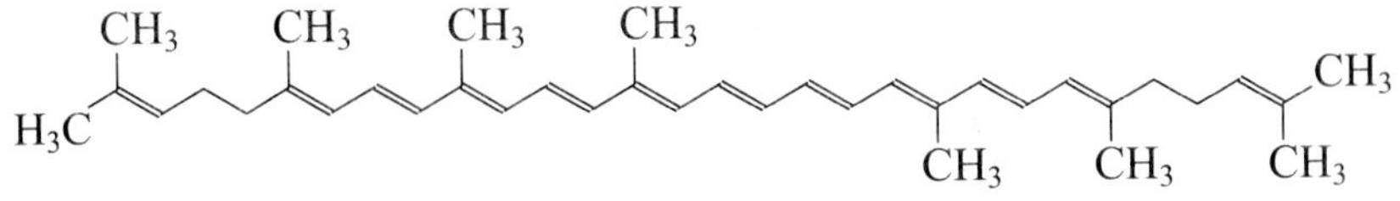

Fig. 7.119 Lycopene

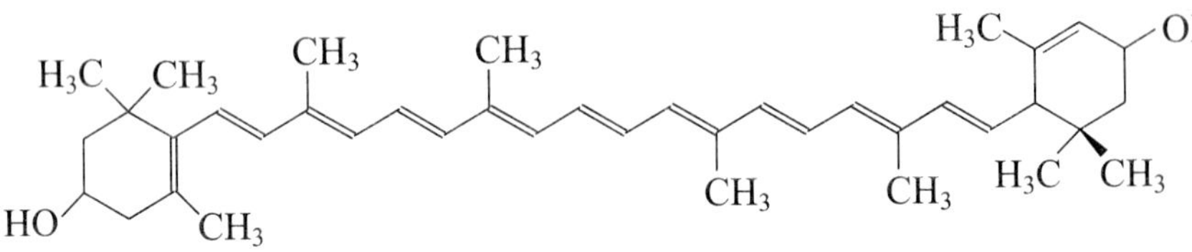

Fig. 7.120 Lutein

A patent in this area was registered. The invention relates to a mutant of *Monascus purpureus*, designated as *Monascus purpureus* M011, which is useful in the production of yellow pigments with a low amount of citrinin. The invention also provides a process for production of yellow pigments, and the products prepared thereby (Chen *et al.*, 2002).

Research projects mixing molecular biology and pigments were investigated all over the world and it seems that current productions are not effective in terms of final yield. Nowadays, combinatorial genetic engineering is being addressed, based on an increasing number of known carotenogenic gene sequences (Mijts *et al.*, 2005). By combining genes some authors were able to obtain more efficient biosynthesis, or new carotenoids, never described in nature, such as multi-hydroxylated ones, which could be very efficient as antioxidants (Dufossé *et al.*, 2005).

The last few years have been a period of great progress using fungi as cell factories. There are four major fronts in which work is currently underway. The first is the development of alternative hosts, especially those that have already been given GRAS status by the FDA and can be used in the food industry. Research is being focused on species such as *Aspergillus sojae*, *Aspergillus japonicus*, *Mortierella alpina* and *Fusarium veneratum*, among others. The second front is the development of better molecular techniques to improve expression and secretion of non-fungal proteins in filamentous fungi. The third major front involves the use of these molecular techniques to carry out metabolic engineering in order to modify and improve particular biosynthetic pathways. The final front will utilize the techniques dealing with the overall analysis of gene expression, i.e. genomics, proteomics and metabolomics. Two fungal genomes have already been sequenced, *A. niger* [http://www.dsm.com] and *Neurospora crassa* [http://wwwgenome. wi.mit.edu], and sequencing of three others are in progress (*A. fumigatus*, *A. nidulans* and *Ustilago maydis*) [http://www.tigr.org]. Initial steps on filamentous fungal genomics and proteomics have recently been published, and undoubtedly much more will become available in the years ahead (Adrio and Demain, 2003).

Industrial Applications

Monascus purpureus C322 was cultivated on well-established production media to yield prevailingly red or orange pigment-rich ethanolic extracts. Once these extracts had been diluted by an overall factor of 50, they were used as such to dye raw wool standard specimens differently premordanted using alum or stannic chloride. Independently of the mordant used, the specimens dyed with the red pigment-rich extracts showed a pale red color tending to pink, whereas the specimens dyed with the orange pigment-rich extracts exhibited a more definite orange color. By carrying out a few color fastness standard tests (manual washing at 40°C, acid and basic perspiration, and hot pressing), SnC14-premordanted wool specimens dyed with the red pigment-rich extracts were found to be less resistant to acid and basic perspiration than their orange counterparts. Since the production of the orange pigment-rich ethanolic extracts appeared to be more cost-effective than that of their red counterparts, the former might support the present demand for colorants of natural origin in the textile sector (De Santis *et al.*, 2005).

There are many fungi that produce carotenoids, but the higher productor are alga and yeast-fungi like microorganims.

Industries producing β-carotenes (*Dunaliella*) are:
Betatene Ltd, Cheltenham, Australia,
Cyanotech Corp., Kailua–Kona, USA,
Inner Mongolia Biological Eng. Co., Inner Mongolia, China,
Nature Beta Technologies (NBT) Ltd, Eilat Israel,
Tianjin Lantai Biotechnology, Ine. Nankai, China,
Western Biotechnology Ltd., Bayswater, Australia,
Aqua Carotene Ltd, Subiaco, Australia,
Small plants are also located in Chile, Mexico, Cuba, Iran, Taiwan, and Japan.

Industries producing astaxanthin (*Haemoptococcus*) are:
Mera Pharmaceuticals, Kailua–Kona, USA,
Cyanotech Corporation, Kailua–Kona, USA,
BioReal Inc., Hawaii, USA,
Parry's Pharmaceuticals, Chennai, India.

A new natural food colorant of fungal origin called Arpink Red^{TM} (Fig. 7.117) is manufactured by ASCOLOR BIOTECH in the Czech Republic (http://www.ascolor-biotec.cz). The red colorant is an extracellular metabolite of the anthraquinone class and is claimed to be produced by a variety of *Penicillium oxalicum*. It is also said to confer anticancer effects when used in food supplements (Mapari *et al.*, 2005). The CODEX Alimentarius Commission (Rotherdam Meeting, 2002) did not may any oblection to use Arpink red (Fig. 7.117) in meats, non-alcoholic drinks, alcoholic drinks, milk product, ice cream and confectionary. JECFA evluation process is in process and Arpink red (Fig. 7.117) situation was discussed during the 63rd meeting of Joint FAO/WHO Expert Commitee on Food Additives in Geneva (Dufossé *et al.*, 2005).

Patents

A novel plant germination inhibitor that without any hazardous effects on cultivated crop and without the danger of environmental pollution, can exert high efficacy with an extremely small dosage. There is provided a plant germination inhibitor comprising as an active ingredient a pigment produced in woody parts by fungi of the chlorosplenium aeruginosum genus or a pigment contained in cells of fungi of the chlorosplenium aeruginosum genus. Through the application of this plant germination inhibitor to a plant growth area during the germination period of the plant, the germination of the plant can be inhibited and the formation of rootstock tissue can be inhibited to thereby enable control of plant tissue formation (Shinnosuke *et al.*, 2005).

Rice is coated with natural functional coloring materials (e.g., from Cordyceps, Monascus, or safflower) with health benefits, including prevention of geriatric diseases. The functional rice is produced by cleaning polished rice and pumping air to distribute the rice grains separation in the air while spraying a mixture of a natural functional coloring matter and a coating solution made of starch, or spraying the coloring matter and the coating solution in turns, and then drying the coated rice. The functional pigment extractions may be derived from animal, plant, or fungal sources (Cha, 2001).

Red and orange pigments, useful as food colorants, are produced from fungi of the genus Monascus on culture media based on milk or whey permeate, with different C and N sources and trace elements added. Through this method a complex, ecological clean product, containing lactose, protein, amino acids, vitamins, and water-sol. pigment is obtained. The method produces no waste, is ecological clean, and uses a nontraditional substrate, waste from the milk industry (Kujumdzieva *et al.*, 1995).

Many patents from Ascolor s.r.o. (Czech Republic) relate to a new fungus strain having the properties to produce a red colorant (Arpink red) (Fig. 7.117) which can be applied in the food and cosmetic industries (Sardaryan, 2002, 2004). The strain *Penicillium oxalicum var. Armeniaca* CCM 8242, obtained from soil, produces a chromophore of the anthraquinone type.

Currently, more than 50 patents have been issued in Japan, the United States, France and Germany, concerning the use of Monascus pigments for food. Annual consumption of Monascus pigments in Japan moved from 100 tons in 1981–600 tons at the end of the nineties and was valued at $1.5 million. New food applications, like the coloration of processed meats (sausage, hams), marine products like fish paste, surimi and tomato ketchup were described (Dufossé *et al.*, 2005).

Asthaxantin-containing preparation for prevention of light-induced aging of skin have been developed by Susuki *et al.* (1996). Astaxanthin (Fig. 7.118) has been shown to enhance *in vitro* antibody production by mouse spleen cells stimulated with sheep red blood cells and in human blood cells *in vitro*. Furthermore it has not exhibited any mutagencity in *in vitro* study at doses up to 14.4 mg/day for two weeks (Miki *et al.*, 1998). Yamashita (1995) has reported anti-inflammatory effect of astaxanthin (Fig. 7.118) when administered together with aspirin. An oral preparation has been developed by Alejung and Wadstroem (1998) for the treatment of Helicobacter infections of the mammalian gastrointestinal tract.

Future Trends

Synthetic pigments traditionally used by food processors continue to be used with success, however with the increasing consumer preference for natural food additives, natural colorants from plants is now a big business and most of the research efforts within the scientific field of colorants are conducted on natural ones (Adrio and Demain, 2003; Dufossé *et al.*, 2005). Regarding bacteria, yeast or fungi, despite common belief about the high production cost of fermentation pigments, two initiatives started in Europe these last few years: β-carotene (Fig. 7.1) from the filamentous fungi, *Blakeslea trispora* (produced by Gist-Brocades now DSM; approved in 2000 by the EU Scientific Committee on Food Safety) and arpink red (Fig. 7.117) from *Penicillium oxalicum* (manufactured by Ascolor). These companies invested significantly to drive an experimental work, process optimization, toxicological studies, and regulatory issues. Another development under process is the production of lycopene (Fig. 7.119) using the filamentous fungus *Blakeslea trispora* by Vitatene, a subsidiary of Spanish penicillin firm Antibioticos. Exploration of fungal biodiversity is still going on, with special interest in water-soluble pigments (Mapari *et al.*, 2005). From these and the previous data on pigment producing fungi, it appears that the versatility and high yield production compare with yeast and bacteria might be feasible the use of fungi to produce new an old pigments (Durán *et al.*, 2002).

Research projects mixing molecular biology and pigments were investigated all over the world and it seems that current productions are not effective in terms of final yield (Dufossé *et al.*, 2005).

A EU Research Framework Program (2004) through the project Fungal Carotenoids (N° QLK1-2001-00780) expired in 2004 was quite successful in the carotinoids production and viability studies.

Blakeslea trispora has been used for the industrial production of β-carotene (Fig. 7.1) in Russia for years. In this fermentation, a fungal mated culture is used with a preferred ratio of minus and plus mating strains. The accumulation of β-carotene (Fig. 1.1) is strongly linked to sexual interaction between the two mating types. A hormone-like substance produced during mating, the major component of which is trisporic acid (Fig. 7.37), stimulates pigment production (Adrio and Demain, 2003).

The Biotechnology Solution

The industry is continuously looking for cheaper, more environment-friendly routes to existing dyes; biotechnology may provide an answer. Anthraquinones have been isolated from a number of fungi including *Drechslera, Trichoderma, Aspergillus* and *Curvularia* strains (Hobson and Wales, 1998; Durán *et al.*, 2002). Most of these fungi produce a mixture of anthraquinones. This is typical of the enzymes of secondary metabolism which are believed to be less specific than those of primary metabolism. *Aspergillus cristatus*, for example, has been reported to produce as many as 15 different anthraquinones. A strain of *Curvularia lunata* has been identified that produces only three anthraquinones: chrysophanol (Fig. 7.41), helminthosporin (Fig. 7.9) and cynodontin (Fig. 7.11), with cynodontin (Fig. 7.11), the 1,4,5,8-tetrahydroxy-3-methylanthraquinone making up over 70% of the mix.

An anthraquinone of 70% purity can be chemically converted to a modern dyestuff, and therefore fermentation of *C. lunata* is the proposed method for producing cheap, environment-friendly anthraquinone dyestuff intermediates. Exploiting the fungal synthesis of anthraquinones has several advantages over chemical methods. The medium in which the fungal culture grows contains no expensive chemicals. The fermentation is carried out at low temperature (30 °C) and neutral pH so that the expensive, fuel-consuming high temperatures and environmentally unfriendly strong acids and alkalis of the chemical synthesis are not required. The purity of the cynodontin (Fig. 7.11) is the key factor making the exploitation of this organism for dye production possible. Cynodontin (Fig. 7.11) extracted from the biomass of *C. lunata* has been successfully converted to two anthraquinone biodyes: Blue biodye and green biodye (similar to CI Disperse Blue 7 and CI Acid Green 28), each with a methyl group in the 3-position. The properties of these biologically derived dyes applied to knitted polyamide were compared with those of conventional dyes, and found to be identical to all important aspects. To date impressive yields of cynodontin have been achieved, although these cannot be considered high enough to warrant full-scale commercial production for which at least a ten-fold increase is required. Yields could be increased by genetically promoting production of the anthraquinone synthase enzyme system.

Cynodontin is a useful anthraquinone since it is tinctorially the strongest and most bathochromic of the hydroxyanthraquinones. Anthraquinones containing three or more substituents are used extensively as commercial dyes. However, other anthraquinones may act as intermediates to a wider range of dyes. Anthraquinone compounds with chloro or amino substitutions would be useful for dye production but are not common among fungal anthraquinones. It is possible that these compounds may be engineered by genetic modification, or by adding to the pathway that is naturally present (Hobson and Wales, 1998).

Acknowledgements

The authors acknolwledge the financial support from FAPESP, CNPq and MCT/CNPq agencies.

References

Achenbach, H. and Blümm, E. (1991). Investigation of the pigments of *Pycnoporus sanguineus* - picnosanguin and new phenoxazin-3-ones. *Archive der Pharmazi.* (Weinheim), 324:3–6.

Adrio, J.L. and Demain, A.L. (2003). Fungal biotechnology. *International Microbiology* 6: 191–199.

Alejung, P. and Wadstroem, T. (1998). Oral preparation for treatment of Helicobacter sp. infections-comprises xanthophylls, especially astaxanthin esterified with a fatty acid and derived from the alga Haematococcus sp. WO Patent 9837874.

Ahn, J., Jung, J., Hyung, W., Haam, S. and Shin, C. (2006). Enhancement of Monascus pigment production by the culture of *Monascus* sp.J101 at low temperature. *Biotechnology Progress* 22: 338-340.

Araujo, A.I.C., Carrazzoni-Pereira, A. R., Porto, A.L.F., Pimentel, M.C.B., Tambourgi, E. B., Lima. J. L. and Melo, E. H. M. (1998). Liquid-liquid extraction of citrinin from fermentation broth using aqueous two phases systems. *XXVII Annual Meet. Braz. Biochem Soc., Caxambu, M.G., Brazil*, G.83: p. 77.

Armstrong, G.A. and Hearst, J.E. (1996). Carotenoids. II. Genetics and molecular biology of carotenoid pigment biosynthesis, *FASEB Journal* 10: 228–237.

Armstrong, G.A. (1997). Genetics of eubacterial caratenoid biosynthesis: A colorful tale. *Annual Review of Microbiology* 51:629–659.

Aulinger, K., Besl, H., Spiteller, P., Spiteller, M. and Steglich, W. (2001). Melanocrocin, a polyene pigment from Melanogaster broomeianus (Basidiomycetes. Zeitschrift fuer Naturforschung, C: *Journal of Biosciences* 56: 495–498.

Babitskaya, V.G., Shcherba, V.V. and Osadchaya, O.V. (1998). Melanin pigments from micromycetes, *Vestsi Nats. Akad. Navuk Belarusi, Ser. Biyal. Navuk.*, 3: 83–88; *Chem. Abstr.* 1999; 31: 29659.

Babitskaya, V.G. and Shcherba, V.V. (1999). Phenolic compounds produced by some basidiomycetous fungi, *Microbiology* 68: 58–62.

Baker, R.A. and Tatum, J.H. (1998). Novel anthraquinones from stationary cultures of *Fusarium oxysporum. Journal of Fermentation and Bioengineering* 85: 359–361.

Blanc. P. (1998). Les pigments rouges de *Monascus. Biofutur* 183: 34–36.

Buchanan, M.S., Gill, M., Millar, P.M., Phonh-Axa, S., Raudies, E. and Yu, J. (1999). Pigments of fungi. Part 51.Structure and stereochemistry of coupled pre-anthraquinones of the seudophlegmacin type from Australian toadstools belonging to the genus *Dermocybe. Journal of Chemical Society Perkin Transaction* 1: 795–801.

Buchanan, M.S., Gill, M., Milanovic, N.M., Phonh-Axa, S. and Steel, P.J. (1998). Pigments of fungi. Part 48. Cavorubin from two Australian toadstools of the genus *Cortinarius* and the X-ray crystal structure of the methyl ester of 6–O-methylclavorubin, *Australian Journal of Chemistry* 51: 347–352.

Campoy, S., Rumbero, A., Martin, J.F. and Liras, P. (2006). Characterization of an hyperpigmenting mutant of Monascus purpureus IB1: Identification of two novel pigment chemical structures. *Applied Microbiology and Biotechnology* 70: 488–496.

Carattoli, A., Romano, N., Ballario, P., Morelli, G., Macino, G. (1991). The *Neurospora crassa* carotenoid biosynthesis gene (albino 3) reveals highly conserved regions among prenyltransferases. *Jounal of Biological Chemistry* 266:5854–59.

Castro e Silva, A., Esposito, E., Ferraz, A., and Durán, N. (1993). Amazonian lignocellulosic materials II. *Pycnoporus sanguineus* acting on *Parkia oppositifolia* wood. Its potentiality, *Holzforschung*. 47: 361–367.

Cerdá-Olmedo, E. and Reau, P. (1970). Genetic classification of the lethal effects of various agents on heterokaryotic spores of Phycomyces. *Mutation Research* 9:369–384

Cha, K.W. (2001). Rice coating with natural functional coloring materials. Repub. Korean Kongkae Taeho Kongbo, Patent KR 2001000266 A 20010105 .

Chen, Y.L., Hwang, I.Er., Lin, M.-Ch., Huang, T.-K., Chen, Ch.-Ch. and Yuan, G.-F. (2002). Monascus pirpureus mutants ans its use in prepraing yellow pigment. U.S. Pat. Appl. 2002061547 A1

Chodurek, E., Latocha, M., Kurkiewicz, S., Buszman, E., Swiatkowska, L. and Wilczok, T. (1998). Chemical characteristics of melanin from *Cladosporium cladosporioides*. *Bulletin Polish Academy of Science and Biological Science* 46: 51–58.

Chrysayi-Tokousbalides M. and Kastanias, M.A. (2003). Cynodontin: A Fungal Metabolite with Antifungal Properties. *Journal of Agricultural and Food Chemistry* 51: 492–493.

Davoli, P., Mucci, A., Schnetti, L. and Weber, R.W.S. (2005). Laetiporic acids, a family of non-carotenoid polyene pigments from fruit-bodies and liquid cultures of *Laetiporus sulphureus* (Polyporales, Fungi). *Phytochemistry* 66: 817–823.

Davoli, P. and Weber, R.W.S. (2002). Simple method for reversed-phase high-performance liquid chromatographic analysis of fungal pigments in fruit-bodies of Boletales (Fungi) *Journal of Chromatography* A 964: 129–135.

De Carvalho J.C., Oishi, B.O., Pandey, A. and Soccol, C.R. (2005). Biopigments from *Monascus*: Strains Selection, Citrinin Production and Color Stability. *Brazilian Archives of Biology and Technology* 48: 885–894.

De Carvalho J.C., Pandey, A., Babitha, S. and Soccol, C.R. (2003). Production of Monascus biopigments: An overview. *Agro Food Industry Hi-Tech* 14: 37–42.

De Santis, D., Moresi, M., Gallo, A.M. and Petruccioli, M. (2005). Assessment of the dyeing properties of pigments from Monascus purpureus. *Journal of Chemical Technology and Biotechnology* 80: 1072–1079.

Dufosse´, L., Galaupa, P., Yaronb, A., Aradb, S.M., Blancc, P., Murthyd, K.N.C. and Ravishankard, G.A. (2005). Microorganisms and microalgae as sources of pigments for food use: a scientific oddity or an industrial reality? *Trends in Food Science & Technology* 16: 389–406.

Durán, N., Teixeira, M.F.S., De Conti, R. and Esposito, E. (2002). Ecological-friendly pigment from fungi. *Critical Review in Food Sciece and Nutrition* 42:53–66.

Durán, N., Teixeira, M.F.S. and Esposito, E. (2004). Pigmentos fúngicos e seu potencial biotecnologico. In Fungos: Uma introdução a biologia, bioquimica e biotecnologia, Esposito, E. and J.L. De Azevedo, Eds., EDUCS Ed. Caxias do Sul, RS, Brasil, 298–307.

Durán, N., Marcato, P.D., Alves, O.L, De Souza, G.I.H. and Esposito, E. (2005). Mechanistic aspects of biosynthesis of silver nanoparticles by several *Fusarium oxysporum* strains, *Journal of Nanobiotechnologia* 3: 1–7.

Elsworth, C., Gill, M., Gimenez, A., Milanovic, N.M. and Raudies, E. (1999). Pigments of fungi. Part 50. Structure, biosynthesis, and stereochemistry of new dimeric dihydroanthracenones of the phlegmacin type from *Cortinarius sinapicolor* Cleland. *Journal of Chemical Society Perkin Transactions* 1: 119–125.

Espinoza, R.M.D. and Webb, C. (1998). Bioconversion of whole wheat flour for production of food colorants, *IchemE Res. Event. Two Day Symp.* 74–81.*Chem. Abstr.* 1999; 129: 80684.

Esposito, E., Innocentini-Mei, L., Ferraz, A., Canhos, V.P. and Durán, N. (1993). Phenoloxidases and hydrolases from *Pycnoporus sanguineus* (UEC-2050 strain): Its applications. *Journal of Biotechnology* 29: 219–224.

Fink-Gremmels, J. and Sabater-Vila, M. (1998). Pharmacological and toxicological aspects of *Monascus* and the role of citrinin. *Proceedings of Symposium of Monascus Culture Applications,,* Toulouse, France, June, France.

Gautschi, J.T., Amagata, T., Amagata, A., Valeriote, F.A., Mooberry, S.L. and Crews, P. (2004). Expanding the strategies in natural product studies of marine-derived fungi: a chemical investigation of Penicillium obtained from deep water sediment. *Journal of Natural Products* 67: 362–367.

Gill, M. (1996). Pigments of fungi (Macromycetes), *Natural Products Report* 13, 513–528.

Gill, M. (1999). Pigments of fungi (Macromycetes), *Natural Products Report* 16: 301–317.

Gill, M. (2001). The biosynthesis of pigments in Basidiomycetes. *Australian Journal of Chemistry* 54: 721–734.

Gill, M. (2003). Pigments of fungi (Macromycetes). *Natural Products Report.* 20: 615–639.

Gill, M., Morgan, P.M., White, J.M. and Yu, J. (1998a). Pigments of fungi. Part. 47. Cardinalic acid, a new anthraquinone carboxylic acid from the New Zealand toadstool *Dermocybe cardinalis* and the synthesis and X-ray crystal structure of methyl 1,7,8–tri-*O*-methylcardinalate, *Australian Journal of Chemistry* 51, 213–218.

Gill, M., Gimenez, A., Jhingran, A.G., Milanovic, N.M. and Palfreyman, A.R. (1998b), Pigments of fungi. Part 49. Structure and biosynthesis of dermocanarin 4, a naphathoquinone-dihydroanthracenone dimer from fungus *Cortinarius sinapicolor* Cleland. *Journal of Chemical Sociey Perkin Transactions* 1: 3431–3436.

Gill, M. and Morgan, P.M. (2004). Pigments of fungi part 73 - Absolute stereochemistry of fungal metabolites: Icterinoidins A1 and B1, and atrovirins B1 and B2 .*Arkivoc* 152–165.

Gonçalves, R.D.R., Pombeiro-Sponchiado, S.R. (2005). Antioxidant activity of the melanin pigment extracted from *Aspergillus nidulans. Biological & Pharmaceutical Bulletin* 28: 1129–1131.

Green, F.J. (1991). The Sigma-Aldrich Handbook of stains, dyes and indicators, *Aldrich Chemical Corporation.* 776p.

Griffith, G. W., Bratton, J. H. and Easten, G. (2004). Charismatic megafungi the conservation of waxcap grassland, *British Wildlife.* (October) 31–43.

Gripenberg, J. (1951). Fungus pigment. I. Cinnabarin, a colouring matter from *Trametes cinnabarina* Jacq. *Acta Chemical Scandinavian* 12: 590–595.

Gripenberg, J. (1981). Fungus pigments.29: The pigments of *Hydnellum ferrogineum* (Fr) Karsten and *Hydnellum zonatum* (Batsch) Krasten. *Acta Chemica Scandinavica Series B-Organic Chemistry and Biochemistry* 35: 513–519 1981.

Hajjaj, H., Blanc, P.J., Groussac, E., Goma, G., Uribelarrea, J.L. and Loubiere, P., (1999). Improvement of red pigment citrinin production ratio as a function of environmental conditions by *Monascus ruber. Biotechnology and Bioengineering* 64: 497–501.

Hamano, P.S., Orozco, S.F.B. and Kilikian, B.V. (2005). Concentration determination of extracellular and intracellular red pigments produced by Monascus sp. *Brazilian Archives of Biology and Technology* 48: 43–49

Hawksworth, D.L., Kirk, P.M., Sutton, B.C. and. Pegler, D.N (Eds).(1995) Ainsworth & Bisby's, *Dictionary of the Fungi,*) 8th ed., Wallingford, CAB International, 616 pp.

Hosoe, T., Nozawal, K., Kawahara, N., Fukushima, F., Nishimura, K., Miyaji, M. and Kawail, K-I. (1999). Isolation of a new potent cytotoxic pigment along with indigotin from the pathogenic basidiomycetous fungus *Schizophyllum commune. Mycopathologia* 146: 9–12.

Hobson, D.K. and Wales, D.S. (1998). Green dyes, *Journal of Society of Dyers Colourists* 114: 42–44.

Hobson, D.K., Edwards, R.L. and Wales, D.S. (1997) Cyanodontin: A secondary metabolite and dyestuff intermediate. *Journal of Chemical Technology and Biotechnology* 70: 343–348.

Hynninen, P.H., Raisanen, R., Elovaara, P. and Nokelainen, E. (2000). Preparative isolation of anthraquinones from the fungus Dermocybe sanguinea using enzymatic hydrolysis by the endogenous beta-glucosidase. *Zeitschrift fur Naturforschung C-A Journal of Biosciences* 55: 600–610.

Iturriaga, E.A., Velayos, A., Eslava, A.P. and Alvarez, M.I. (2001). The genetics and molecular biology of carotenoid biosynthesis in Mucor. *Recent Research Developments in Genetics* 1: 79–92.

Johannesson, H., Rymaqn, S., Lundmark, H. and Danell, E. (1999). *Sarcodon imbricatus* and *S. squamosus*-two confused species. *Mycological Research.* 193: 1447–1452.

Johnson, E.A. and An G.H. (1991). Astaxanthin from microbial sources, *Critical Review in Biotechnology* 11: 297–326.

Johnson, E.A. and Schroeder, W. (1995) Astaxanthin from the yeast *Phaffia rhodozyma, Studies in Mycology* 38: 81–90.

Jung, H., Kim, C. and Shin, C.S. (2005). Enhanced photostability of monascus pigments derived with various amino acids via fermentation. *Journal of Agricultural and Food Chemistry* 53: 7108–7114.

Kahner, L., Dasenbrock, J., Spiteller, P., Steglich, W., Marumoto, R. and Spiteller, M. (1998). Polyene pigments from fruit-bodies of *Boletus laetissimus* and *B. rufo-aureus* (Basidiomycetes). *Phytochemistry* 49: 1693–1697.

Kamijo, M., Suzuki, T., Kawai, K., Fujii, T. and Murase, H. (1999). Ytterbium-decreasing *Streptomyces sp.* and its naphthoquinone-pigment production in the presence of rare-earth elements, *Journal of Bioscience and Bioengineering* 87: 340–343.

Klostermeyer, D., Knops, L., Sindlinger, T., Polbom, K. and Steglich, W. (2000). Novel benzotropolone and 2H-furo[3,2-b]benzopyran-2-one pigments from Tricholoma aurantium (Agaricales). *European Journal of Organic Chemistry 4: 603–609.*

Kim, W.G., Kim, J.W., Ryoo, I.J., Kim, J.P., Kim, Y.H. and Yoo, I.D. (2004). Boletunones A and B, highly functionalized novel sesquiterpenes from *Boletus calopus. Organic Letters* 6: 823–826.

Knecht, A., Cramer, B. and Humpf, H.U. (2006). New Monascus metabolites: Structure elucidation and toxicological properties studied with immortalized human kidney epithelial cells. *Molecular Nutrition & Food Research* 50: 314–321.

Kujumdzieva, A., Savov, V., Wartenberg, A., Davidova, E., Todorov, T. and Rasheva, T. (1995). Food coloring product. Patent WO 9514786 A1 19950601.

Li, H., Huang, D., Xu, X., Zhang, D., Shi, Q. and Wu. S. (1999). Studies on the chemical composition in *Russula vinosa*, *Junwu Xitong*., 17: 68–74; *Chem. Abstr.* **1999;** 129, 227849.

Li-Chan, E.C.Y. (1996). The application of Raman spectroscopy in food science. *Trend in Food Science and Technology* 7: 361–370.

Mapari, S.A.S, Nielsen, K.F., Larsen, T.O. Frisvad, J.C., Meyer A.S. and Thrane, U. (2005). Exploring fungal biodiversity for the production of water-soluble pigments as potential natural food colorants. *Current Opinion in Biotechnology* 16:231–238.

Martinkova, L., Juzlova, P. and Vesely, D. (1995). Biological-activity of polyketide pigments produced by the fungus *Monascus*. *Journal of Applied Bacteriology* 79: 609–616.

Martinkova, L., Juslova, P., Kren, V., Kucerova, Z., Havlicek, V., Olsovsky, P., Hovoka, O., Rihova, B., Vasely, D., Vesela, D., Ulrichova, J. and Prikrylova, V. (1999). Biological activities of oligoketode produced by *Moascus purpureus*. *Food Additives and Contamintans* 16: 15–24.

Mawthols, K.R., Deshpande, R., Ware, D. and Mahajan, M. (2005) Effect of pH on pigment production of fungi and their toxicity on seed germination. *Ecology, Environment and Conservation* 11: 325–326.

Medentsev, A.G. and Akimenko, V.K. (1996). Fungal naphthoquinone metabolites (Review). *Applied Biochemistry and Microbiology* 32: 7–29.

Medentsev, A.G. and Akimenko, V.K. (1998). Naphthoquinone metabolites of the fungi, *Phytochemistry* 47: 935–959.

Medentsev, A.G., Arinbasarova, A.Yu. and Akimenko, V.K. (2005). Biosynthesis of naphthoquinone pigments by fungi of the genus Fusarium. *Applied Biochemistry and Microbiology* 41: 503–507.

Michelot. D. and Melendez-Howell. L. M., (2003) Amanita muscaria: Chemistry, biology, toxicology and ethnomycology. *Mycology Research* 107: 131–146.

Mijts, B.N., Lee, P. C. and Schmidt-Dannert, C. (2005). Identification of a carotenoid oxygenase synthesizing acyclic xanthophylls. Combinatorial biosynthesis and directed evolution. *Chemistry and Biology* 12: 453–460.

Miki,W., Hosada, K., Kondo, K. and Itakura, H. (1998). Astaxanthin -containing drink. Jpn Patent 10155459.

Mitake, T., Mori, A., Kii, T., Okuno, T., Usui, F., Sammoto, H., Watanabe, A. and Kariyama, M. (2005). Light effects on cell development and secondary metabolism in Monascus. *Journal of Industrial Microbiology and Biotechnology* 32: 103–108.

Parisot, D., Devys, M. and Barbier, M. (1990). Naphthoquinone pigments related to fusarubin from the fungus *Fusarium solani (*Mart) Sacc. *Microbios* 64: 31–47.

Quang, D.N., Hashimoto, T., Nomura, Y., Wollweber, H., Hellwig, V., Fournier, J., Stadler, M. and Asakawa, Y. (2005). Cohaerins A and B, azaphilones from the fungus *Hypoxylon cohaerens*, and comparison of HPLC-based metabolite profiles in Hypoxylon sect. Annulata . *Phytochemistry* 66: 797–809.

Ruiz-Vazquez, R.M. and Torres-Martinez, S. (2003). Production of carotenes in Mucorales: structural and regulatory genes In Microorganisms for Health Care, Food and Enzyme Production, Research Signpost, Trivandrum, India, 69–84.

Sabater-Vilar, M., Maas, R.F.M. and Fink-Gremmels, J. (1999). Mutagenicity of commercial *Monascus* fermentation products and the role of citrinin contamination, *Mutation Research, General Toxicology and Environmental Mutagenesis* 444: 7–16.

Sakaki, T., Shibata, M., Mukai, K., Sakai, M., Wakamatsu, K. and Miyauchi, S. (2002). *Chlorociboria aeruginosa* pigment as algicide. Jpn. Kokai Tokkyo Koho JP 2002291493.

Sardaryan, E (2002). Strain of the microorganisms *Penicillium oxalicum* var. Armeniaca and its application. US Patent 6,340,586 B1.

Sardaryan, E. (2004). Food supplement. US Patent 0105864 A1.

Shin, C.S.(1998). Productivity increase of *Monascus* pigments, *Jpn. Kokai Tokkyo Koho JP*, 10004988; *Chem. Abstr.*, 1999; 128: 101172.

Schrader, B., Dippel, B., Erb, I. Keller, S., Lochte, T., Schulz, H., Tatsch, E. and Wessel, S. (1999). NIR Raman spectroscopy in medicine and biology: results and aspects. *Journal of Molecular Structure* 481: 21–32.

Schreiner, T., Hildebrandt, U., Bothe, H., and Marner, F.J. (1998). Chemical and biological characterization of corticrocin, a yellow pigment formed by the ectomycorrhizal fungus *Piloderma croceum. Zeitschrift fur Naturforschung C-A Journal of Biosciences* 53:4–8.

Shinnosuke, M., Masao, S., Tsuyoshi, S., Kunimitsu, W. and Mukae K.M. (2005) Fungal pigment and its use as plant germination inhibitor. Patent WO2005JP05901 20050329.

Smânia, E.F.A., Smânia Jr. A. and Loguercio-Leite, C. (1998). Cinnabarin synthesis by *Pycnoporus sanguineus* strains and antimicrobial activity against bacteria from food products. *Revista de Microbiologia*, 29, 317–320.

Smânia A., Marques, C.J.S., Zanetti, C.R., Carobrez, S.G., Tramonte, R. and Loguercio-Leite, C. (2003). Toxicity and antiviral activity of cinnabarin obtained from *Pycnoporus sanguineus (Fr.) Murr. Phytotherapy Research* 17: 1069–1072.

Smânia, A. Jr., Delle Monache, F., Smânia, E.F.A.; Gil, M.L., Benchetrit, L.C. and Cruz, F.S. (1995a). Antibacterial activity of a substance produced by the fungus *Pycnoporus sanguineus*(Fr.) Murr. *Journal of Ethnopharmacology* 45: 177–181.

Smânia, A. Jr., Smânia, E.F.A., Cruz, F.S. and Benchetrit, L.C. (1995b). Growth and production phases of *Pycnoporus sanguineus. Revista de Microbiologia* (São Paulo), 26: 302–306.

Smânia, E.F.A., Smânia, A. Jr., Loguercio-Leite, C. and Gil, M.L. (1997). Optimal parameters for cinnabarin synthesis by *Pycnoporus sanguineus. Journal of Chemical Technology and Biotechnology* 70:57–59.

Spiteller, P., Arnold, N., Spiteller, M. and Steglich, W. (2003). Lilacinone, a red aminobenzoquinone pigment from *Lactarius lilacinus. Journal of Natural Products* 66: 1402–1403.

Stipanovic, R.D., Zhang, J.X., Bruton, B.D. and Wheller, M.H. (2004). Isolation and identification of hexaketides from a pigmented *Monosporascus cannonballus* isolate. *Journal of Agricultural and Food Chemistry* 52: 4109–4112.

Strack, D., Vogt, T. and Schliemann, W. (2003). Recent advances in betalain research *Phytochemistry* 62: 247–269.

Suzuki, K., Masaki, H. and Takei, M. (1996). External preparation for skin, Jpn. Patent 08733312.

Takahashi, N., Iwahori, A., Kawai, K.-I., and Fukui, T. (1998). Induction of differentiation in human promyelocytic leukemia cell line HL60 by a new type of polyenes, falconensone A and its derivatives, *Archives of Biochemistry and Biophysics.* 360: 113–120.

Takeda, N., Seo, S., Ogihara, Y., Sankawa, U., Iitaka, I., Kitagawa, I. and Shibata, S. (1973). Studies on fungal metabolites .31. anthraquinonoid coloring matters ostuf penicillium-islandicum sopp and some other fungi (–) luteoskyrin, (–) rubroskyrin, (+)rugulosin and their related compounds. *Tetrahedron* 29: 3703–3719.

Terradas, F. and Wyler, H. (1991). The secodopas, natural pigments in Hygrocybe conica and Amanita muscaria. *Phytochemistry* 30, 3251–3253.

Thomson, R.H. (1971). In: *Naturally Occurring Quinones*, Academic Press Publ., London, 198p.

Tong, Q., Gao, K., and Zhou, Z. (1997). Study on antibacterial activity of *Monascus* pigment, *Shipin Gongye Keji* 5: 5–6. *Chem. Abstr.* 1999; 128: 125721.

Tricard, Z.C., Cazabeil, J.M. and Medina, B. (1998). Identification of natural dyes to food products. *Science des Aliments* 18: 25–40.

Ueno, Y. and Ishikawa, I. (1969). Production of Luteoskyrin, a hepatotoxic pigment, by *Penicillium islandicum Sopp. Applied Microbiology* 18: 406–409.

Unagul, P., Wongsa, P., Kittakoop, P., Intamas, S., Srikitikulchai, P. and Tanticharoen, M. (2005). Production of red pigments by the insect pathogenic fungus *Cordyceps unilateralis* BCC 1869. *Journal of Industrial Microbiology & Biotechnology* 32: 135–140.

Uraguchi, K., Tatsuno, T., Sakai, F., Tsukioka, M., Sakai, Y., Onemitsu, O,. Ito, H., Miyake, M., Saito, M., Enomoto, M., Shikata, T. and Ishiko, I. (1961). Isolation of two toxic agents, luteoskyrin and chlorine-containing peptide, from the metabolites of *Penicillim islandicun Sopp*, with some properties thereof. *Japanese Journal Experimental Medicine* 31:19–46.

Wang, X.-N., Tan, R.X., Wang, F., Steglich, W. and Liu, J.K.(2005).The first isolation of a phlegmacin type pigment from the ascomycete Xylaria euglossa. *Zeitschrift fuer Naturforschung, B: Chemical Sciences* 60: 333–336.

Weber, R.W.S., Muccib, A. and Davolib, P. (2004). Laetiporic acid, a new polyene pigment from the wood-rotting basidiomycete *Laetiporus sulphureus* (Polyporales, Fungi). *Tetrahedron Letters* 45: 1075–1078.

Wichai, S., Sriyam, S. and Phutrakul, S. (2001). Isolation of pigmented microorganisms and identification of their pigments. *Chiang Mai Journal of Science* 28: 1–8.

Wild, D., Toth, G. and Humpf, H.U. (2002). New Monascus metabolite isolated from red yeast rice (angkak, red koji). *Journal of Agricultural and Food Chemistry* 50: 3999–4002.

Yamaji, K., Fukushi, Y., Hashidoko, Y., Yoshida, T. and Tahara, S. (1999). Characterization of antifungal metabolites produced by *Penicillium* species isolated from seeds of *Picea glehnii. Journal of Chemical Ecology* 25: 1643–1653.

Yamashita, E. (1995). Anti-inflammatory agent. Jpn Patent 07300421.

Yang, X.L., Luo, D.Q., Dong, Z.J. and Liu, J.K. (2006). Two New Pigments from the Fruiting Bodies of the Basidiomycete Lactarius deliciosus. *Helvetica Chimica Acta* 89:988–990.

8

Microbial Transformation by *Beauveria bassiana*

Jixun Zhan and A. A. Leslie Gunatilaka

Southwest Center for Natural Products Research and Commercialization,
Office of Arid Lands Studies, College of Agriculture and Life Sciences,
The University of Arizona,
250 East Valencia Road, Tucson, Arizona 85706-6800, U.S.A.
E-mail: leslieg@ag.arizona.edu.

INTRODUCTION

Microbial transformation represents a series of biological reactions of xenobiotic substrates catalyzed by whole-cell microorganisms or enzymes obtained from microbial sources. Due to its high regio- and stereo-selectivity, microbial transformation complements organic synthesis and has often been employed in food and pharmaceutical industries (Venisetty and Ciddi, 2003; Panke and Wubbolts, 2005; Suresh *et al.,* 2006). Microbial transformation reactions, especially those utilizing fungi have also been used as *in vitro* models of mammalian metabolism of a variety of bioactive compounds and drugs (Rosazza and Smith, 1979; Davis *et al.*, 1992). *Beauveria bassiana* (Balsamo) Vuillemin. (Clavicipitaceae) is a globally occurring soil-borne filamentous fungus. As an aggressive entomopathogenic and parasitic fungus, it causes a disease called white muscadine in insects. The fungus proliferates throughout the insect's body, producing toxins such as beauvericin that weakens the host's immune system, consuming the internal organs and blood-like fluid, draining the insect of nutrients, and eventually killing it. Unlike bacterial and viral pathogens of insects, *B. bassiana* infects insects on contact and does not need to be consumed by its hosts to cause infection. Once the fungus has killed the insect host, it grows out through the softer portions of the cuticle, covering the insect with a layer of white mold; for this reason *B. bassiana* is also known as "sugar icing fungus". Thus, *B. bassiana* finds applications as a microbial insecticide and has shown a great potential for the management of various insect pests (Liu *et al.*, 2003).

B. bassiana has frequently been employed in biotransformation studies due to its abundant enzyme systems, broad substrate acceptability, and its capability to effect diverse types of reactions. It has been reported that this fungus is capable of biotransformation of over 300 different substrates (Holland *et al.*, 1999). A previous review on reactions catalyzed by *Beauveria* spp. and their applications has dealt with most biotransformation research prior to 1999. As observed in that review, *B. bassiana* ATCC 7159 is an important microbial catalyst and is second only to *Aspergillus niger* as the most frequently used fungal biocatalyst, and has surpassed in applications only by *A. niger*, *Pseudomonas putida*, and baker's yeast (Grogan and Holland, 2000).

The intent of this present chapter is to review biotransformation research employing *B. bassiana* reported since 1999; however, for completeness, some old work not covered in the previous review is also included.

BIOLOGICAL REDUCTIONS

A variety of reductions catalyzed by microorganisms are known, among which most common is the reduction of the carbonyl group. Several studies on carbonyl reductases have revealed that these enzymes belong to a family of short chain NADPH-dependent dehydrogenases/reductases that are mostly monomeric and cytosolic with broad substrate specificity for many endogenous and xenobiotic carbonyl compounds (Forrest and Gonzalez, 2000). *B. bassiana* has displayed its bioreduction ability in a biotransformation study of cadinane sesquiterpenes; the substrate, cadina-4,10(15)-dien-3-one (**1**), on incubation with *B. bassiana* ATCC 7159 at 27°C in liquid culture with shaking for 14 days, has yielded nine new metabolites **2**–**10** (Fig. 8.1), all of which are reduction products of the carbonyl

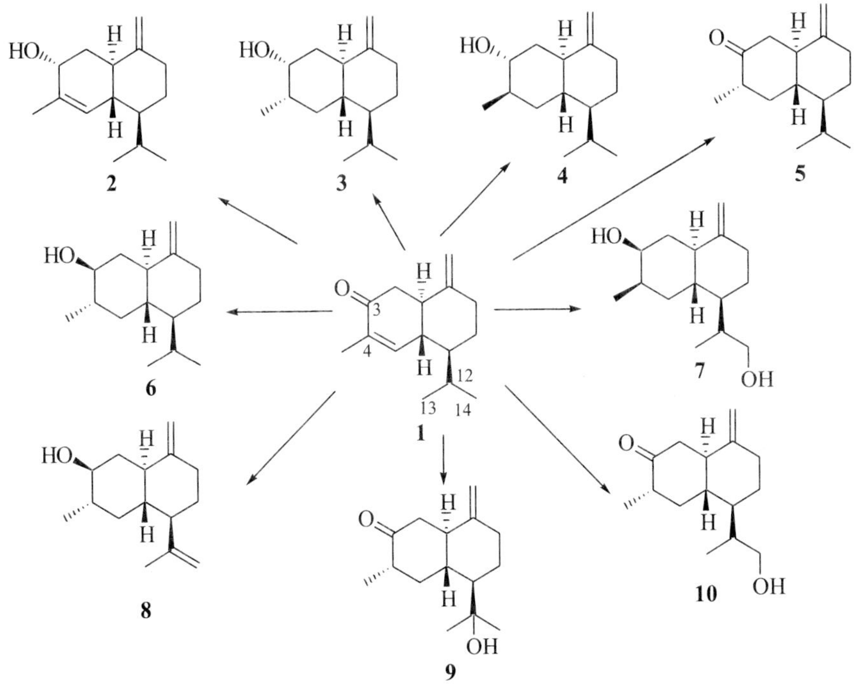

Figure 8.1 Biotransformation of cadina-4,10(15)-dien-3-one (**1**) by *B. bassiana*.

group and/or 4(5)-unsaturation. Interestingly, the 10(15)-unsaturation was not affected, indicating that the reductase enzymes in this microorganism may have some selectivity. Preparative fermentation of 3α-hydroxycadina-4,10(15)-diene (**2**) with the same strain has led to the formation of **1**, **5** and **9** (Buchanan *et al.*, 2000).

Recent interest in the synthesis of biologically active compounds in enantiomerically pure form has led to the investigation of microbial reductions of a variety of aromatic ketones. Barbieri *et al.* (2001) have studied the application of bioreduction of aromatic ketones for the preparation of chiral benzyl alcohols in high enantiomeric purity. The substituted phenacyl chlorides **11** – **13** when reduced with *B. bassiana* ATCC 7159 have yielded (*R*)- or (*S*)-chlorohydrins **14** – **16** in high yields and in high enantiomeric excess (*ee*) (Fig. 8.2) (Barbieri *et al.*, 2001). Asymmetric biological reduction of racemic 5(6),7(8)-tetrahydro-8-methyl-1,3-dimethoxynaphthalen-6-one (**17**) with *B. bassiana* ATCC 7159 has led to the formation of an enantiomeric mixture of *cis*-(6*S*,8*R*)-5(6),7(8)-tetrahydro-8-methyl-1,3-dimethoxynaphthalen-6-ol (**18**) and *trans*-(6*R*,8*R*)-5(6),7(8)-tetrahydro-8-methyl-1,3-dimethoxynaphthalen-6-ol (**19**) in yields of 41% and 59% and with high *ee* of 95% and 42%, respectively, with an overall conversion of 34% (Fig. 8.3) (Aina *et al.*, 2001). In a biotransformation of the synthetic antibacterial agent, cinoxacin (1-ethyl-1,4-dihydro-4-oxo[1,3]dioxolo[4,5-*g*]-cinnoline-3-carboxylic acid) (**20**), *B. bassiana* ATCC 7159 was found to reduce the carboxylic acid group to an alcohol group yielding 1-ethyl-1,4-dihydro-3-(hydroxymethyl)[1,3]dioxolo[4,5-*g*]cinnolin-4-one (**21**) (47.0%), and 1-ethyl-1,4-dihydro-6,7-dihydroxy-3-(hydroxymethyl)-cinnolin-4-one (**22**) (5.6%) (Fig. 8.4a). In order to evaluate the general applicability of this biotransformation to other heterocyclic acids, reduction of quinoline-3-carboxylic acid (**23**) by *B. bassiana* ATCC 7159 has been investigated resulting in the formation of 3-(hydroxymethyl) quinoline (**24**) (Fig. 8.4b) (Parshikov *et al.*, 2002).

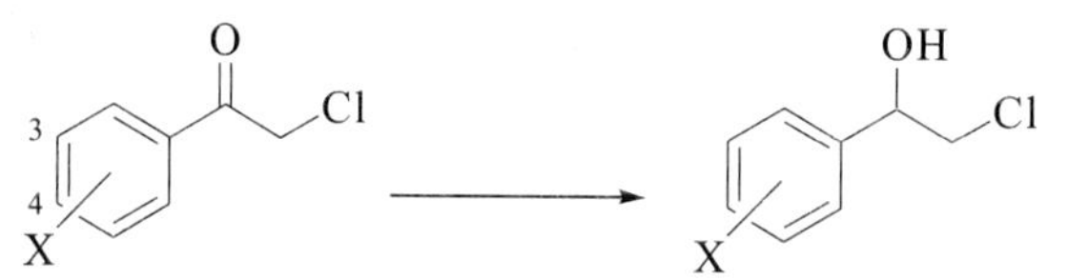

11 X = H　　　　**14** X = H
12 X = 3-Cl, 4-Cl　　**15** X = 3-Cl, 4-Cl
13 X = 4-NO_2　　**16** X = 4-NO_2

Figure 8.2 Enantio-controlled reduction of aromatic ketones by *B. bassiana*.

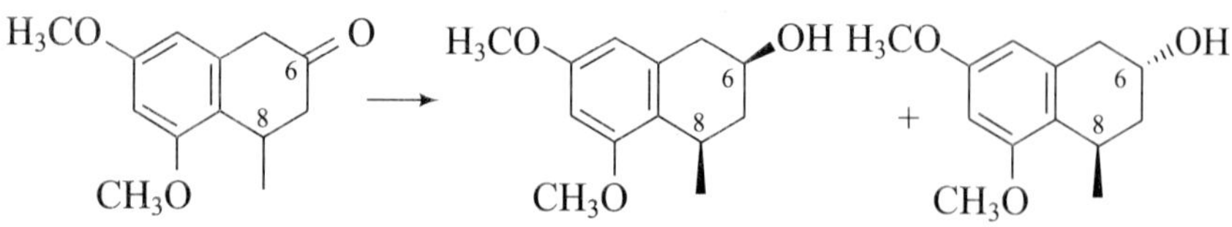

17　　**18**　　**19**

Figure 8.3 Asymmetric reduction of naphthalenan-6-one by *B. bassiana*.

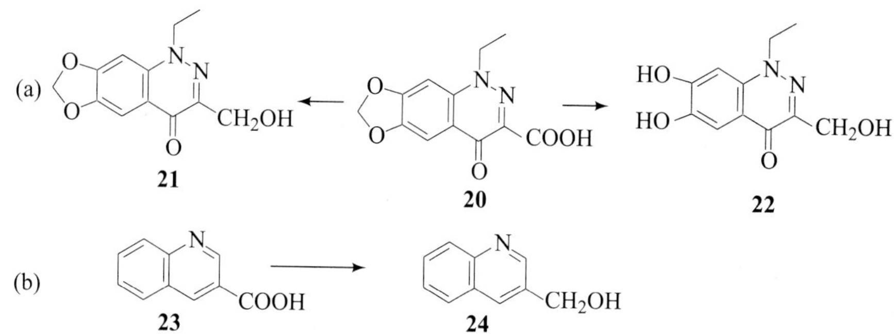

Figure 8.4 Reduction of heterocyclic aromatic carboxylic acids by *B. bassiana*.

Biological reduction of oximes catalyzed by *B. bassiana* has afforded the corresponding alcohols. Thus, indole-3-acetaldoxime (**25**) and 4-hydroxyphenylacetaldoxime (**29**) have yielded tryptophol (**26**) and 2-(4-hydroxyphenyl)-1-ethanol (**30**), respectively (Fig. 8.5) (Pedras and Montaut, 2003).

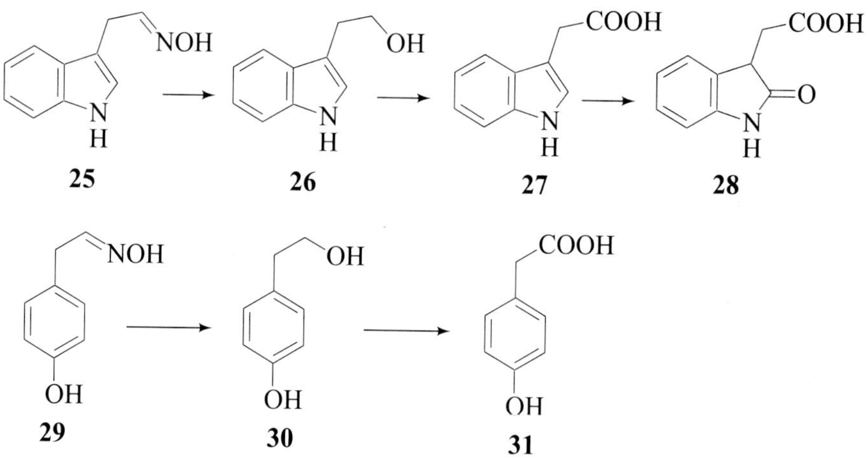

Figure 8.5 Biotransformation of aldoximes by *B. bassiana*.

The steroidal hormone, testosterone (**32**), on biotransformation with *B. bassiana* AM 446, in addition to other bioreactions, has resulted in the reduction of 4(5)-double bond yielding **40** and **42**. However, biotransformation of 1-dehydrotestosterone (**35**) has caused the reduction of the 1(2)-double bond resulting in the formation of **39** and **41** together with other biotransformation products. As seen in the structures of the products **48** and **50**, *B. bassiana* was also found to be capable of reducing the 17-carbonyl group of dehydroepiandrosterone (**38**) (Fig. 8.6) (Huszcza *et al.*, 2005).

Biotransformation of *cis*-2-methyl-3-keto-5-benzyloxymethyl-tetrahydrofuran (**51**) by *B. bassiana* ATCC 7159 has yielded two of its reduction products, the *cis*-alcohol (**52**) and the *trans*-alcohol (**53**) (Fig. 8.7) (Grogan *et al.*, 2001). Terentiev and coworkers (2003) have recently reported that 1(2),5(6)-tetrahydropyridines with different substituents on their *N* atoms, e.g. 1-benzyl-1,2,5,6-tetrahydropyridine (**54**), 1-(1-phenylethyl)-1,2,5,6-tetrahydropyridine (**55**), and 3-methyl-1-propyl-1,2,5,6-tetrahydropyridine (**56**), have resulted in their saturated analogues, 1-benzylpiperidine (**57**), 1-(1-phenylethyl) piperidine (**58**), and 3-methyl-1-propylpiperidine (**59**), respectively. In this biotransformation two additional products, 1-benzyl-3,4-epoxypiperidine (**60**) and 3-methyl-1-propyl-3,4 dihydroxypiperidine (**61**) were also obtained from **54** and **56**, respectively (Fig. 8.8) (Terentiev *et al.*, 2003).

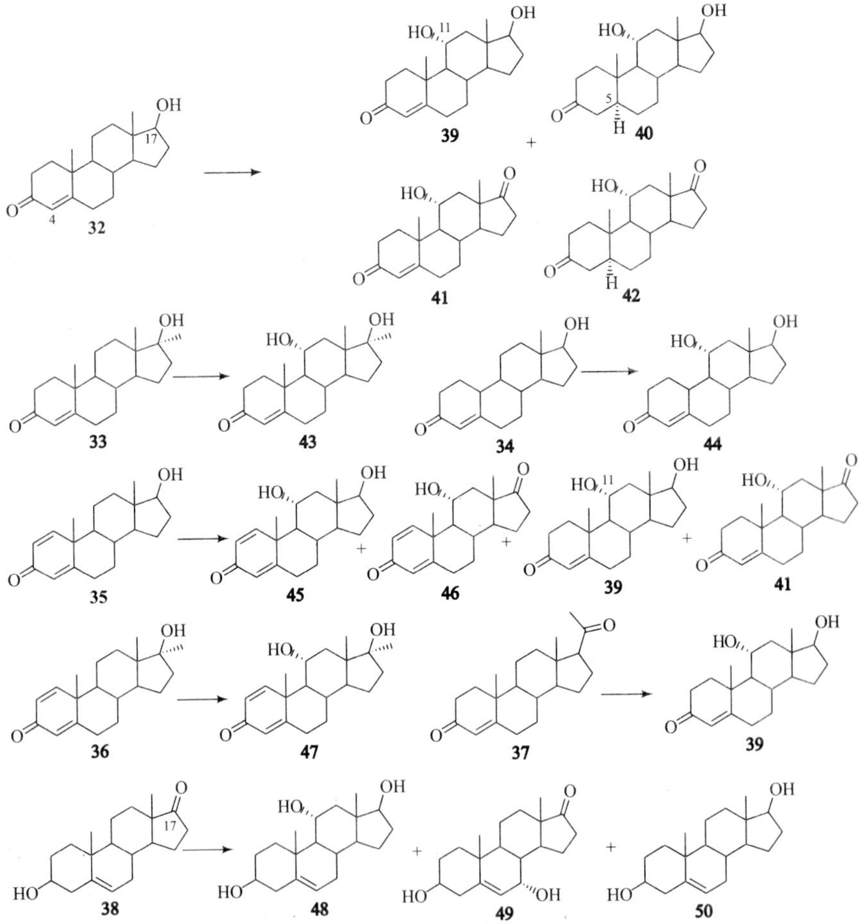

Figure 8.6 Biotransformation of testosterone and its analogues by *B. bassiana*.

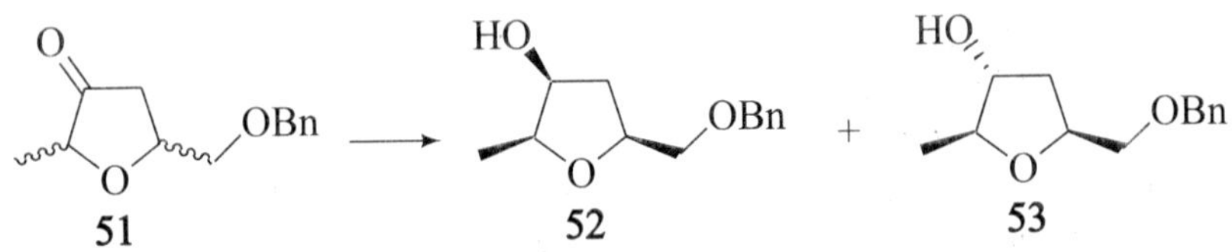

Figure 8.7 Biological reduction of 3-ketofurans by *B. bassiana*.

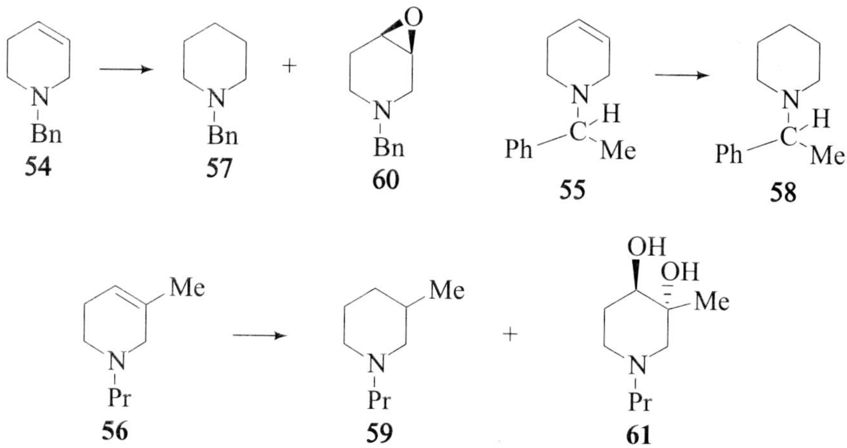

Figure 8.8 Biotransformation of 1(2),5(6)-tetrahydropyridines by *B. bassiana*.

BIOLOGICAL OXIDATIONS

Studies on biocatalytic oxidation by *B. bassiana* in recent years have mainly focused on sulfoxidation reactions. The biotransformation of the *N*-phthaloyl derivatives **62** – **65** of D- and L-methionine and D- and L-ethionine by *B. bassiana* ATCC 7159 have provided their corresponding (S_s) sulfoxides **66** – **69** in good yield and in diastereomeric excess (Fig. 8.9). Pure (S_sS_c) diastereomers have been obtained from the L-series of substrates by crystallization of the biotransformation product mixture, and the corresponding (S_sR_c) products from the D-series of substrates by chromatographic separation of the resulting extract. *N*-Cbz (*N*-carbobenzoxy) and *N*-*t*-Boc (*N*-*t*-butyloxycarbonyl) methionines were also oxidized by *B. bassiana* resulting in sulfoxides with predominantly (S_s) configuration (Holland *et al.*, 1999).

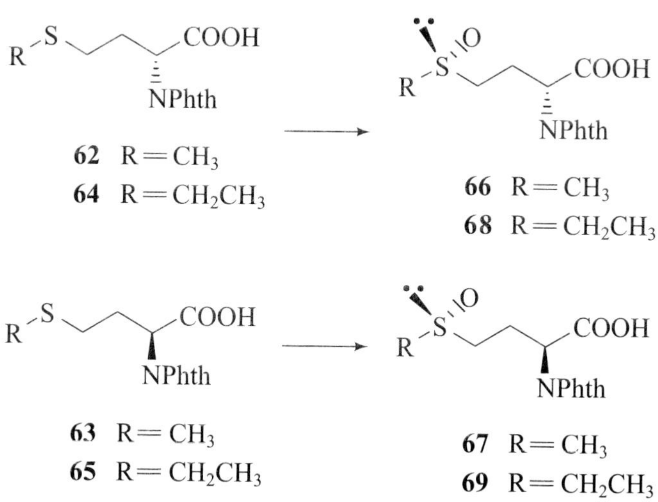

Figure 8.9 Sulfoxidation of sulfides by *B. bassiana*.

Microbial oxidation of *N-t*-Boc-L-4-*S*-morpholine-2-carboxylic acid (**70**) by *B. bassiana* has led to the production of *N-t*-Boc-chondrine (*N-t*-Boc-L-4-*S*-morpholine-2-carboxylic acid sulfoxide) (**71**) (Fig. 8.10a) (Holland *et al.*, 2002). A highly enantioselective oxidation of benzhydrylsulfanyl acetic acid (**72**) to the corresponding (*S*)-sulfinyl carboxylic acid (**73**) was achieved in very good yield (89%) and with high *ee* (99%) by employing *B. bassiana* ATCC 7159. The resulting product **73** has been amidated using the bacterium, *Bacillus subtilis*, to afford (*S*)-modafinil (**74**) (Fig. 8.10b), a psychostimulant agent approved recently by the FDA for the treatment of excessive daytime sleepiness with lack of abuse liability. Interestingly, biotransformation of benzhydrylsulfanyl acetamide (**75**) with *B. bassiana* has afforded the (*R*)-isomer of modafinil (**76**) in good yield (67%) but with low *ee* (Fig. 8.10c), making it possible to have access to both enantiomers of modafinil by microbial transformation (Olivo *et al.*, 2005).

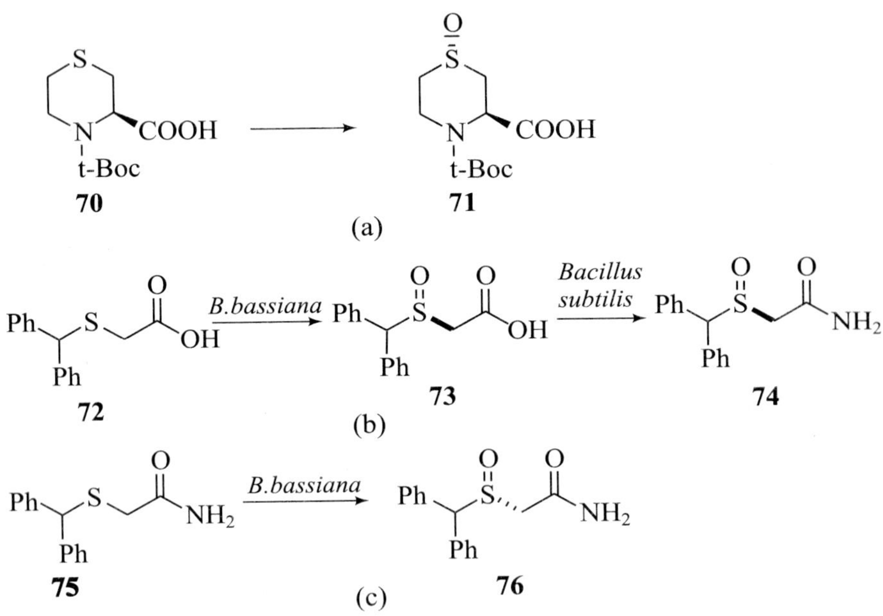

Figure 8.10 Conversion of sulfides to sulfoxides by *B. bassiana*.

B. bassiana ATCC 7159 has also been reported to catalyze *N*-oxidation of tertiary amines. Thus, biotransformation of tamoxifen [*trans*-1-(4-b-dimethylaminoethoxy-phenyl)-1,2-diphenyl but-1-ene] (**77**), a non-steroidal anti-estrogenic drug, has provided tamoxifen-*N*-oxide (**78**) (Fig. 8.11) (El-Sharkawy, 1991).

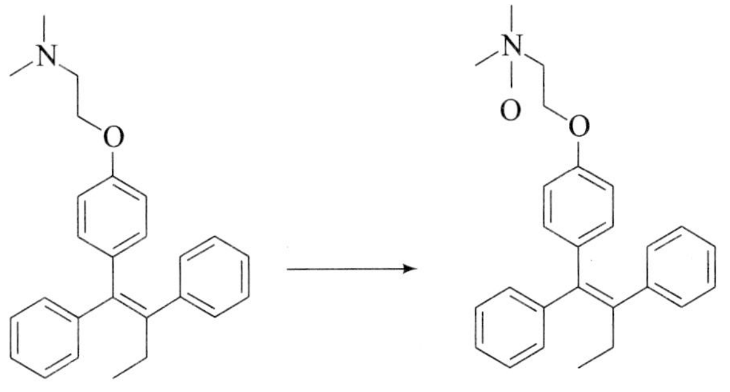

77 78

Figure 8.11 *B. bassiana* catalyzed biotransformation of tamoxifen to its *N*-oxide.

Biological oxidation of indolethyl alcohol, tryptophol (**26**), into 2-*oxo*-indole-3-acetic acid (**28**) *via* indole-3-acetic acid (**27**) by *B. bassiana* has recently been reported; transformation of 4-hydroxyphenylethyl alcohol (**30**) has yielded 4-hydroxyphenylacetic acid (**31**) (Fig. 8.5) (Pedras and Montaut, 2003). Oxidation of alcohols to carboxylic acids by *B. bassiana* has also been reported in the biotransformation of the natural diterpene, stemarin (**89**), into 13-hydroxystemarane-19-carboxylic acid (**91**) (Fig. 8.13) (Buchanan and Reese, 2001). Biological oxidation of 17β-hydroxyl group into 17-oxo function in steroids has been observed in the biotransformation of testosterone (**32**) and 1-dehydrotestosterone (**35**), leading to the metabolites **41**, **42** and **46** (Fig. 8.6) (Huszcza *et al.*, 2005). *B. bassiana* has been found to be capable of oxidizing anthracenes to anthraquinones. Thus, the biotransformation of 1-aminoanthracene (**79**) by *B. bassiana* ATCC 7159 has resulted in the production of 1-acetamidoanthraquinone (**82**) *via* *N*-acetyl-1-aminoanthracene (**80**) (Fig. 8.12) (Zhan and Gunatilaka, 2008).

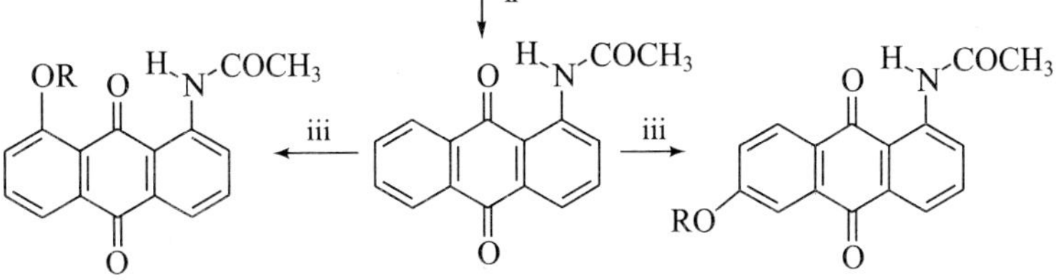

79 80 81 83 82 84

R=4′-O-Methyl-glc

i. Acetylation; ii. Oxidation; iii. 4′-O-Methyl-glucosidation

Figure 8.12 Biotransformation of 1-aminoanthracene by *B. bassiana*.

BIOLOGICAL HYDROXYLATIONS

Microbial hydroxylation is one of the most common and important biotransformation reactions known useful especially in regio- and stereo-selective introduction of a hydroxyl group to a non-activated hydrocarbon position which is hard to achieve by traditional chemical approaches. *B. bassiana* ATCC 7159 has been frequently employed for the hydroxylation of terpenoids and steroids. Thus, incubation of the diterpene stemodin (**85**) with *B. bassiana* ATCC 7159 has resulted in its C-18 hydroxylated product **86**. Similarly, stemodinone (**87**) was found to undergo hydroxylation at the same position to afford 13,18-dihydroxystemodan-2-one (**88**). Several products were detected in the biotransformation of stemarin (**89**), among which the major products were identified as 1β,13,19-trihydroxystemarane (**90**) and 13-hydroxystemarane-19-carboxylic acid (**91**). One of the synthetic derivatives, stemodin *N,N*-dimethylcarbamate (**92**), was transformed to its C-7 mono-hydroxylated derivative, 2α-dimethylcarbamoxy-7β,13-dihydroxystemodane (**93**) (Fig. 8.13) (Buchanan and Reese, 2001).

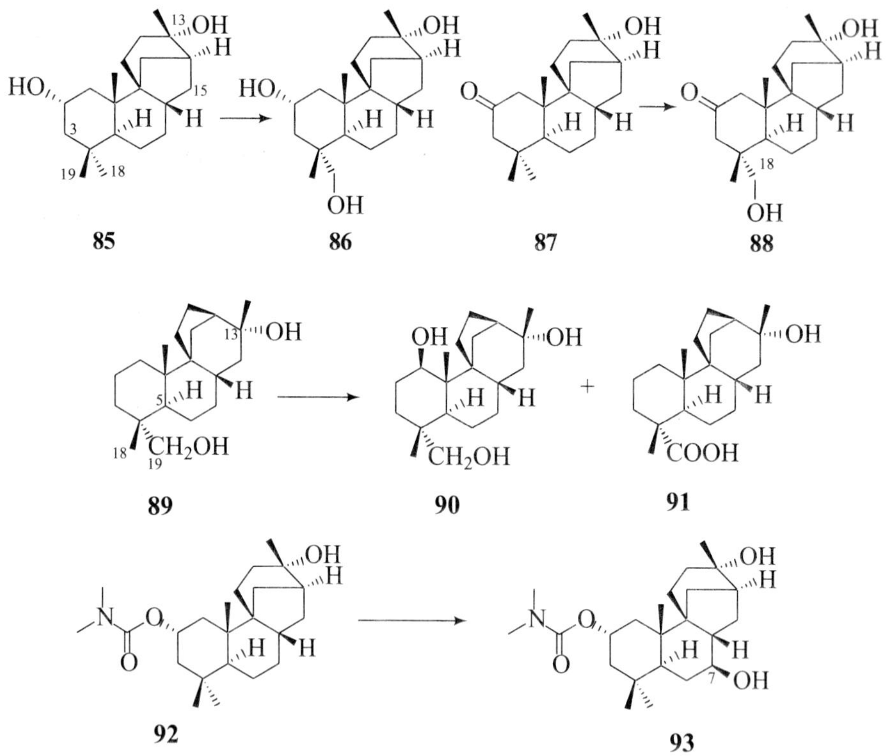

Figure 8.13 Biological hydroxylation of stemodin, stemarin and their derivatives by *B. bassiana*.

Monohydroxylation of 4,5-dianilinophthalimide (**94**) catalyzed by *B. bassiana* DSM 1344 (= ATCC 7159) to yield 4-(4'-hydroxyanilino)-5-anilinophthalimide (**95**) has also been reported (Fig. 8.14a) (Weidner *et al.*,

1999). In an *in vitro* metabolism study of (-)-rhazinilam (**96**), *B. bassiana* ATCC 7159 was found to effect C-3 hydroxylation yielding (3*S*)-hydroxy-rhazinilam (**97**) along with minor amounts of diazaspiroleuconolam (**98**) (Fig. 8.14b) (Décor *et al.*, 2006).

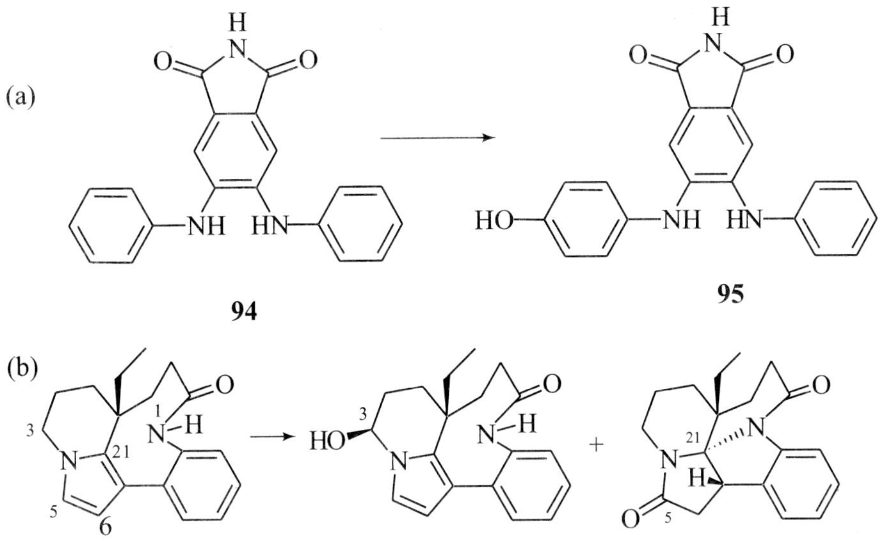

94 **95**

96 **97** **98**

Figure 8.14 Biological hydroxylation catalyzed by *B. bassiana*.

B. bassiana catalyzed biological hydroxylation of fluorinated cyclo-alkyl *N*-phenylcarbamates such as *trans*-2-fluorocyclohexyl *N*-phenyl-carbamate (**99**) and *trans*-2-fluorocycloheptyl *N*-phenylcarbamate (**100**) is also known. As depicted in Fig. 8.15a, hydroxylation of the six-membered ring at C-4 and *p*-hydroxylation of the aromatic ring of **99** have resulted in six products, (1*R*,2*R*,4*R*)-(–)-2-fluoro-4-hydroxycyclohexyl *N*-phenyl-carbamate (**101**), *cis*-2-fluoro-4 hydroxycyclohexyl-*N*-(*p*-hydroxyphenyl) carbamate (**102**), (1*R*,2*R*)-(–)-2-fluorocyclohexyl *N*-(*p*-hydroxyphenyl)-carbamate (**103**), (–)-*trans*-2-fluorocyclohexyl *N*-[*p*-(4'-methyl-*β*-D-glucopyranosyl)]-phenylcarbamate (**104**), (1*S*,2*S*,4*R*)-(+)-2-fluoro-4-hydroxy-cyclohexyl-*N*-phenylcarbamate (**105**) and *trans*-2-fluoro-4-hydroxy-cyclohexyl *N*-(*p*-hydroxyphenyl)carbamate (**106**). Similar biotransformation of the 7-membered ring analogue **100** has led to the production of the six related metabolites **107** – **112** (Fig. 8.15b) (Haufe *et al.*, 2002).

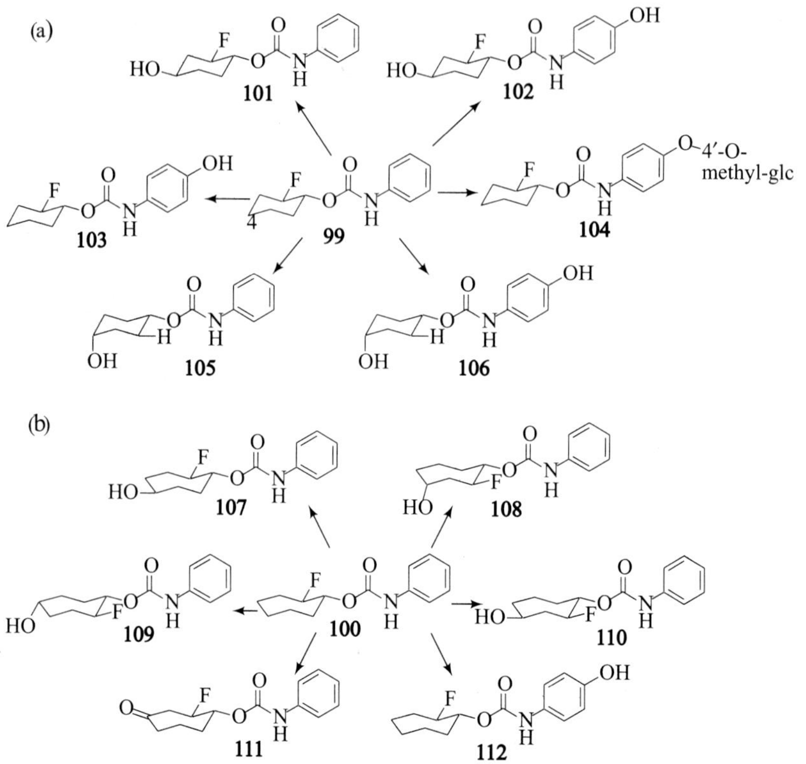

Figure 8.15 Biotransformation of fluorinated cycloalkyl *N*-phenylcarbamates by *B. bassiana*.

B. bassiana has also been reported to be capable of hydroxylating aromatic substrates. Thus, biotransformation of 3-hydroxyflavone (**113**) by *B. bassiana* ATCC 13144 has afforded 3,4'-dihdroxyflavone (**114**) (Fig. 8.16) (Herath *et al.*, 2006). Three anisoyl-7-azanorbornyl amides **116–118** when subjected to biotransformation with *B. bassiana* ATCC 7159 yielded **119–121** resulting in selective hydroxylation of the same position of the norbornyl moiety suggesting that the orientation of the methoxy substituent of the anisoyl group had no effect on this biotransformation reaction (Fig 8.17) (Olivo *et al.*, 2003).

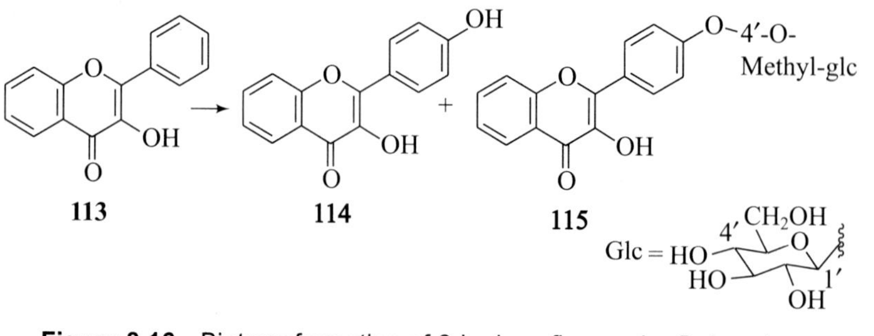

Figure 8.16 Biotransformation of 3-hydroxyflavone by *B. bassiana*.

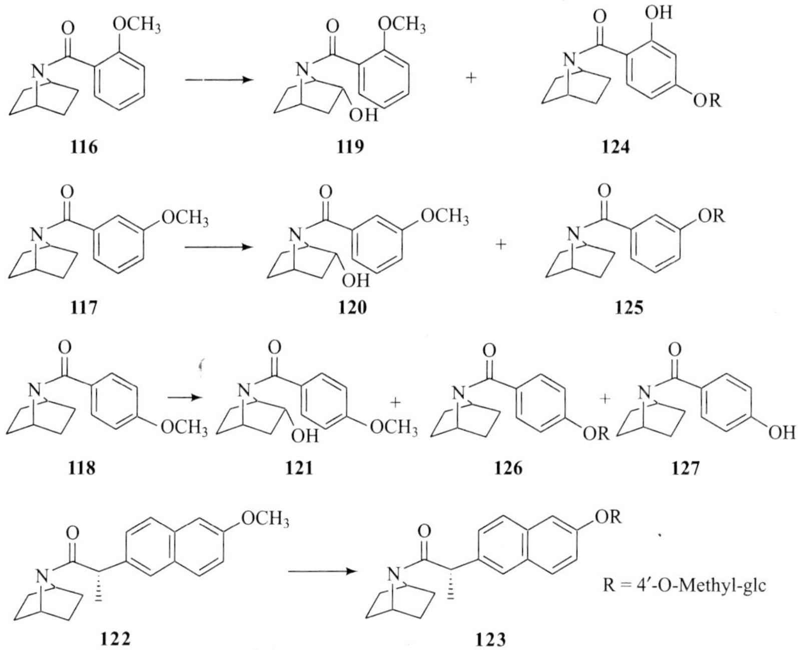

Figure 8.17 Biotransformation of 7-norbornyl amides by *B. bassiana*.

During the metabolism of the eudesmane sesquiterpene, plectranthone (**128**), by *B. bassiana* ATCC 7159, hydroxylation of both methylene and methyl groups were found to occur, yielding 11-hydroxyplectranthone (**129**), 1 hydroxydesacetylplectranthone (**131**), 12-hydroxyisodesacetylplectranthone (**132**), and 15-hydroxydesacetylplectranthone (**133**) (Fig. 8.18) (Orabi, 2000).

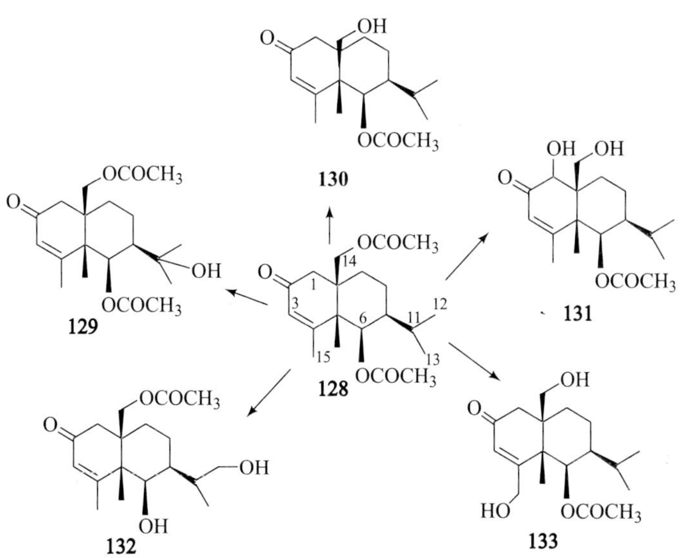

Figure 8.18 Biotransformation of plectranthone (**128**) by *B. bassiana*

The strain AM446 of *B. bassiana* has been shown to cause efficient 11α-hydroxylation of steroids with high regioselectivity and low substrate specificity. As illustrated in Fig. 8.6, seven steroidal substrates namely, testosterone (**32**), 17α-methyltestosterone (**33**), 19-nortestosterone (**34**), 1-dehydrotestosterone (**35**), 1-dehydro-17α-methyltestosterone (**36**), progesterone (**37**), and dehydro-*epi*-androsterone (**38**) have been hydroxylated by this strain to yield their corresponding 11α-hydroxy analogues, **39** - **48**, among other biotransformation products (Huszcza *et al.*, 2005). The cardiovascular steroidal drug , mexrenone (**134**), containing a C-17 spirolactone side chain was hydroxylated at C-11 by *B. bassiana* ATCC 13144 to yield its new derivative, 11α-hydroxymexrenone (**135**) (Fig. 8.19a) (Preisig *et al.*, 2003).

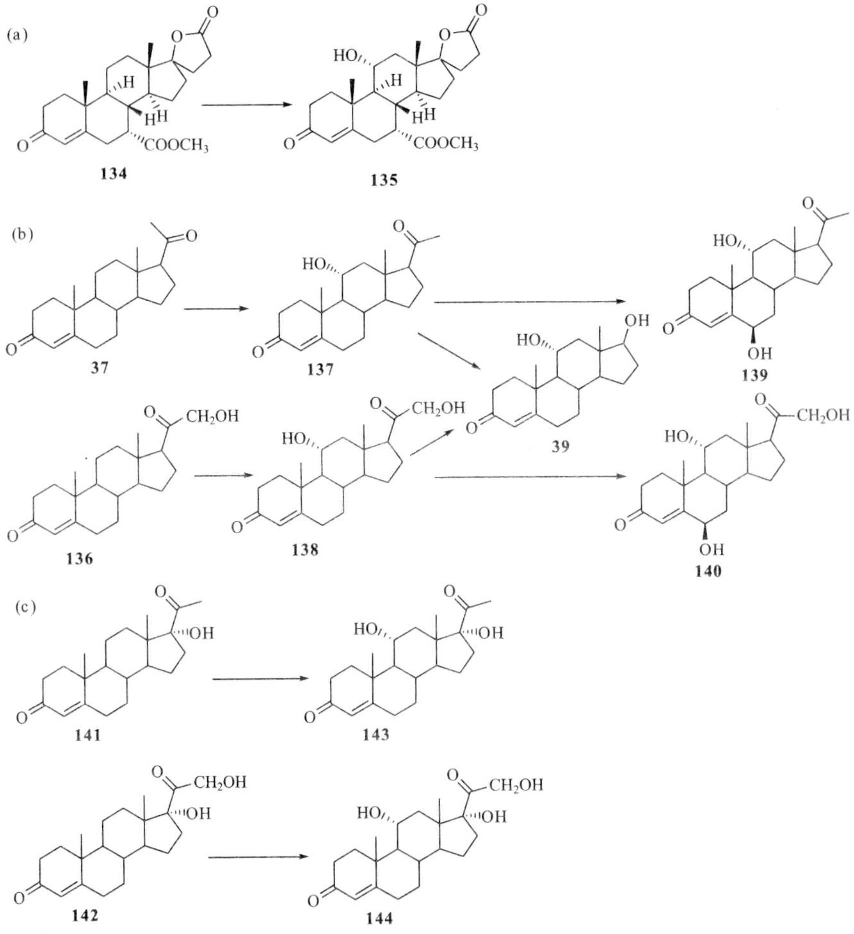

Figure 8.19 Biotransformation of some steroidal substrates by *B. bassiana*.

Čapek and coworkers (1966) have studied the effect of substituents on biotransformation of steroids by *B. bassiana* strain 663. Progesterone (**37**) and cortexone (**136**) were hydroxylated at C-11 to produce 11α-hydroxyprogesterone

(**137**), and epicorticosterone (**138**), respectively. Both these hydroxylated products were found to undergo further biotransformation resulting in the loss of their side-chain to afford 11α-hydroxytestosterone (**39**). Additional products obtained with intact side-chain included those resulting from C-6 hydroxylation [6β,11α-dihydroxyprogesterone (**139**)] and C-6 and side-chain hydroxylation [6β-hydroxyepicorticosterone (**140**)] (Fig. 8.19b). Interestingly, the presence of the 17α-hydroxy group prevented both the 6β-hydroxylation and removal of the side-chain. Thus, the biotransformation of 17α-hydroxyprogesterone (**141**) and cortexolone (**142**) led to the production of 11α-hydroxylated metabolites, 11α,17α-dihydroxyprogesterone (**143**) and epicortisol (**144**), respectively (Fig. 8.19c). Hydroxylation at C-7 and C-11 have been observed during the biotransformation of androsta-1,4-diene-3,17-dione (**145**) by *B. bassiana* i1654 (Fig. 8.20a) which resulted in the formation of 11α-hydroxy-androsta-1,4-diene-3,17-dione (**146**) (40% yield) and 7α,11α-dihydroxy-androsta-1,4-diene-3,17-dione (**147**) (20% yield) (Breiter *et al.*, 1995). Hydroxylation of the sesquiterpene lactone, arteannuin B (**148**), by *B. bassiana* ATCC 7159 has afforded 3β-hydroxyarteannuin B (**149**) (Fig. 8.20b) (Elmarakby *et al.*, 1987).

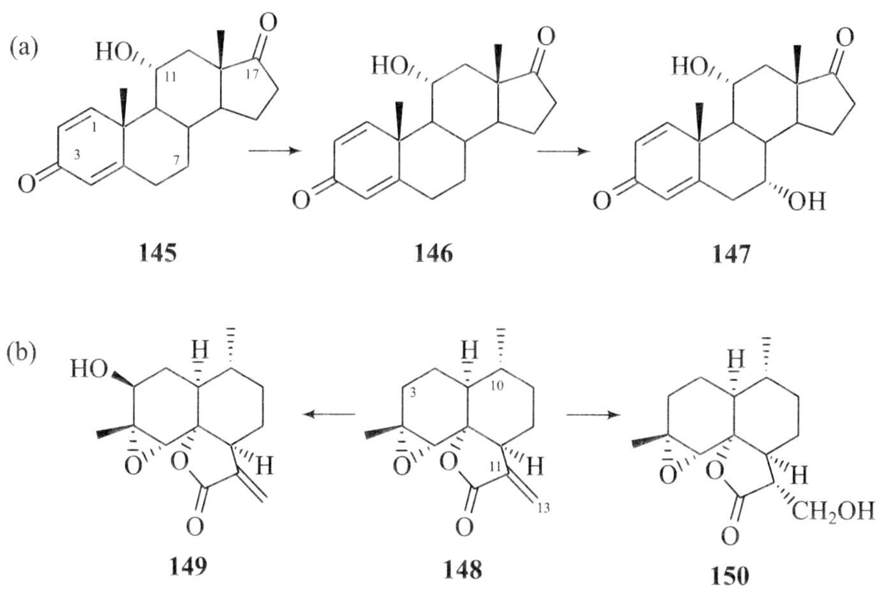

Figure 8.20 Hydroxylation of steroidal and terpenoid substrates by *B. bassiana*.

In a recent study, *B. bassiana* ATCC 7159 was found to catalyze the regio- and diastereo-selective hydroxylation of *trans*-2-methyl-5-benzyloxymethyl-tetrahydrofuran (**151**) to its *cis*-3-hydroxy derivative (**152**) (Fig. 8.21a) isomeric with the products **52** and **53** obtained from the biological reduction of *cis*-2-methyl-3-keto-benzyloxymethyl-tetrahydrofuran (**51**) (Fig. 8.7) by the same fungal strain (Grogan *et al.*, 2001). Shimizu and Fujishima (2001) have developed a method to prepare monohydroxyadamantane esters by bioconversion of the corresponding hydrocarbon esters using *B. bassiana* strains, IFO4848, IFO31676 and IFO31953.

As depicted in Fig. 8.21b, adamantane hydrocarbon ester derivatives represented by **153** were hydroxylated by these three strains in shake culture to yield monohydroxyadamantane esters represented by **154** with high efficiency (Shimizu and Fujishima, 2001). Grotowska and Wawrzeńczyk (2002) have reported hydroxylation of the lactones, δ-iodo-γ-lactone (**155**) and γ-iodo-δ-lactone (**157**), to yield their monohydroxylated derivatives **156** and **158**, respectively (Fig. 8.21c and d).

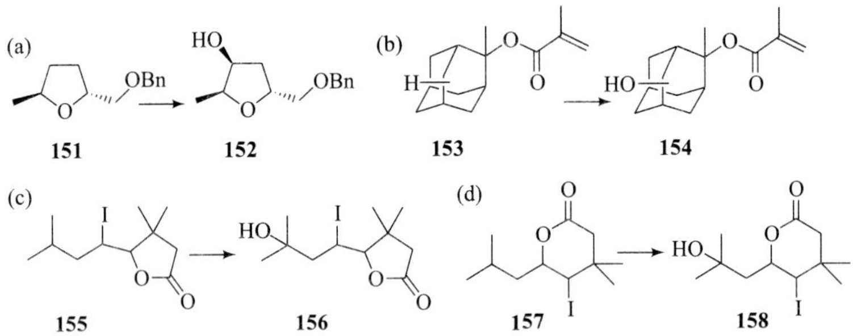

Figure 8.21 Hydroxylation of tetrahydrofuran, adamantane, and iodo-lactone substrates by *B. bassiana*.

The influence of fluorine substitution on biotransformation of nortricyclanyl-*N*-phenylcarbamates have been investigated by Haufe and Wölker (2003) who reported that the biotransformation of tricyclo[2.2.1.$0^{2,6}$]hept-3-yl *N*-phenylcarbamate (**159**) by *B. bassiana* ATCC 7159 led to the formation of a 7:1 mixture of optically active *exo,exo*-5-hydroxytricyclo[2.2.1.$0^{2,6}$]hept-3-yl *N*-phenylcarbamate (**160**) and its *exo*, *endo*-analogue **161** (Fig. 8.22). Interestingly, the fluorine substituted derivatives, 3-*exo*, 5-*exo*-fluorotricyclo[2.2.1.$0^{2,6}$] hept-3-yl-*N*-phenylcarbamate (**162**) and 3-*exo*,5-*endo*-fluorotricyclo-[2.2.1.$0^{2,6}$]hept-3-yl-*N*-phenylcarbamate (**163**) were hydroxylated at *para*-position of the aromatic ring to give the corresponding metabolites **164** and **165**, with no products resulting from the hydroxylation of alicyclic carbon atoms, indicating substitution with a fluorine atom at the 5-*endo*- or 5-*exo*-position prevented aliphatic hydroxylation of the nortricyclane skeleton.

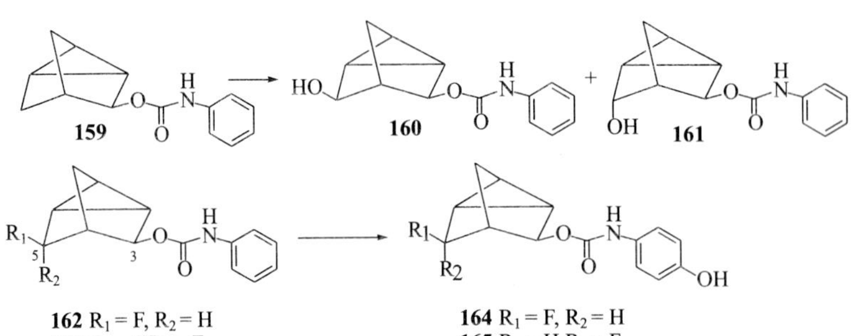

162 $R_1 = F, R_2 = H$
163 $R_1 = HR_2 = F$

164 $R_1 = F, R_2 = H$
165 $R_1 = H R_2 = F$

Figure 8.22 Biotransformation of nortricyclanyl-*N*-phenylcarbamate (**159**) and its fluorinated analogs (**162** and **163**) by *B. bassiana*.

Biotransformation of fluorinated cycloalkyl *N*-phenylcarbamates bearing a fluorine atom in the 4-position of the cycloalkane moiety with *B. bassiana* ATCC 7159 has been reported by Haufe and coworkers (2003) who found that *cis*-4-fluorocyclohexyl *N*-phenylcarbamate (**166**) was transformed into four of its metabolites, 4-oxocyclohexyl *N*-phenylcarbamate (**167**), *cis*-4-hydroxycyclohexyl *N*-phenylcarbamate (**168**), its *trans* isomer **169**, and *cis*-4-fluorocyclohexyl *N*-(4-hydroxyphenyl)carbamate (**170**) (Fig. 8.23a). The related substrate, *cis*-4-fluorocyclooctyl *N*-phenylcarbamate (**171**) was converted into five monohydroxylated products, *cis*-4-fluorooctyl *N*-(4-hydroxyphenyl)carbamate (**173**), (+)-*cis*-4-fluoro-*cis*-5-hydroxycyclooct-1-yl *N*-phenylcarbamate (**174**), (-)-*cis*-4-fluoro-*trans*-6-hydroxyclooct-1-yl *N*-phenylcarbamate (**175**) and (–)-*cis*-4-fluoro-*cis*-6-hydroxycyclooct-1-yl *N*-phenylcarbamate (**176**), and a product in which the fluorine atom is replaced with a carbonyl group [(–)-4-oxocyclooctyl *N*-phenylcarbamate (**172**)] (Fig. 8.23b). The formation of the ketone **172** was explained by hydroxylation of the substrate to form an unstable fluorohydrin, which underwent dehydrofluorination to yield the ketone. In constrast to the results obtained with 5-*exo*- and 5-*endo*-fluoronortricyclan-3-yl *N*-phenylcarbamates, **162** and **163** (Fig. 8.22), this study showed that the presence of a fluorine substituent at a position that is hydroxylated by *B. bassiana* in non-fluorinated parent compounds does not necessarily prevent hydroxylation at positions geminal or vicinal to the fluorine substituent (Haufe *et al.*, 2003).

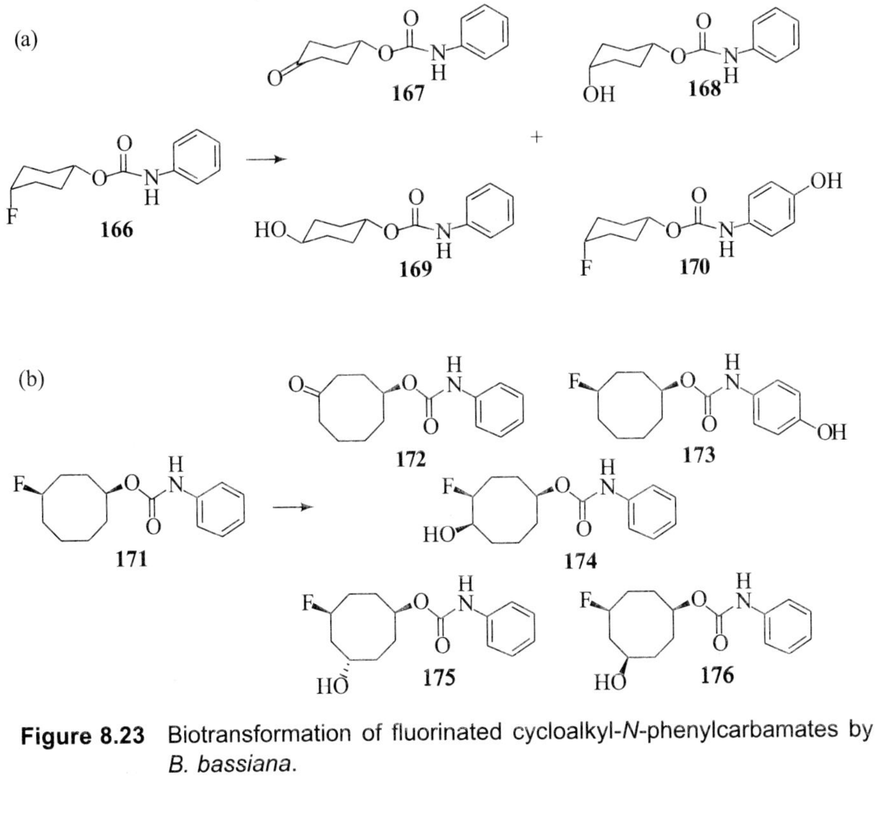

Figure 8.23 Biotransformation of fluorinated cycloalkyl-*N*-phenylcarbamates by *B. bassiana*.

Hydroxylation of a series of *N*-substituted piperidines and alkylpiperidines catalyzed by *B. bassiana* ATCC 7159 has been investigated by Flitsch and coworkers (1999). They found that the *N*-substituted piperidines, **177** and **178**, were hydroxylated at C-4 to yield monohydroxylated products **179** and **180**, respectively (Fig. 8.24a). Biotransformation of *N*-substituted 4-alkylpiperidines, **181** and **182** gave C-4 monohydroxylated products **183** and **184** (Figs. 8.24b and c). Biological transformation of **182** also resulted in **185** formed from hydroxylation of the methyl group at C-4 (Fig. 8.24c). The *N*-benzoyloxycarbonyl 3-methylpiperidine (**186**) was converted into three C-3 or C-4 hydroxylated products **187**, **190** and **191** (Fig. 8.24d), whilst the related substrate *N*-benzoyl 3-methylpiperidne (**188**) was transformed into **189** and **192** (Fig. 8.24e). Biological hydroxylation of *N*-benzoyloxycarbonyl 2-methylpiperidine (**193**) afforded two products, **195** and **196** (Fig. 8.24f), whereas the related substrate **194** yielded the products **197**, **198** and **199** (Fig. 8.24g). Biotransformation of *N*-benzoyloxycarbonyl substrates, **200** and **204**, led exclusively to the 4α-hydroxylated products, **202** and **206**, respectively (Fig. 8.25a and c), whereas the related *N*-benzoyl substrates, **201** and **205**, gave 3- and 4-hydroxylated products **203**, **207** and **208** (Figs. 8.25b and d) (Flitsch *et al.*, 1999). It is significant that the nature of the substituent on *N* atom in alkyl piperidines was found to have an effect on the orientation and sometimes stereochemistry of the newly introduced hydroxyl group during *B. bassiana* catalyzed hydroxylation of these substrates.

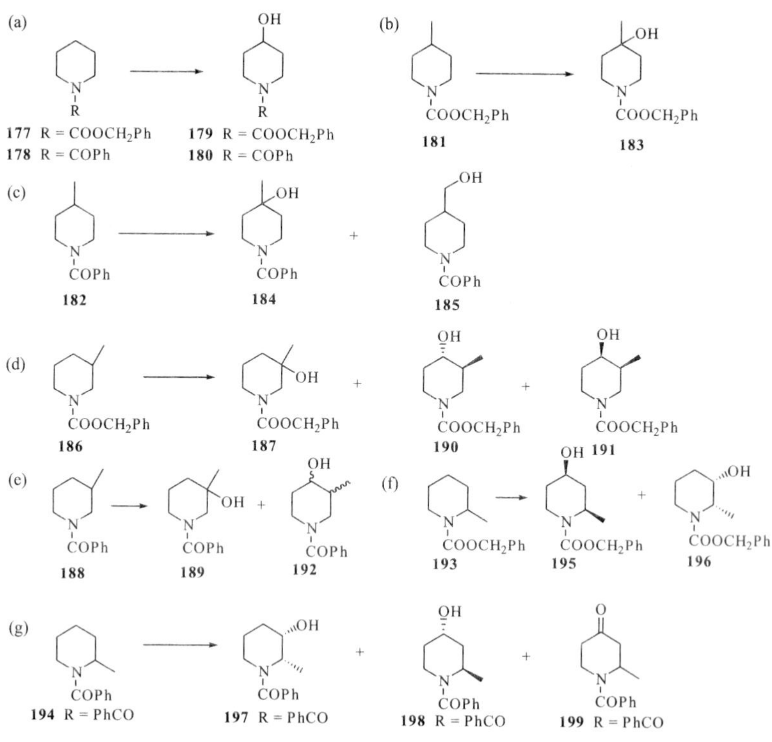

Figure 8.24 Biotransformation of *N*-substituted piperidines by *B. bassiana*.

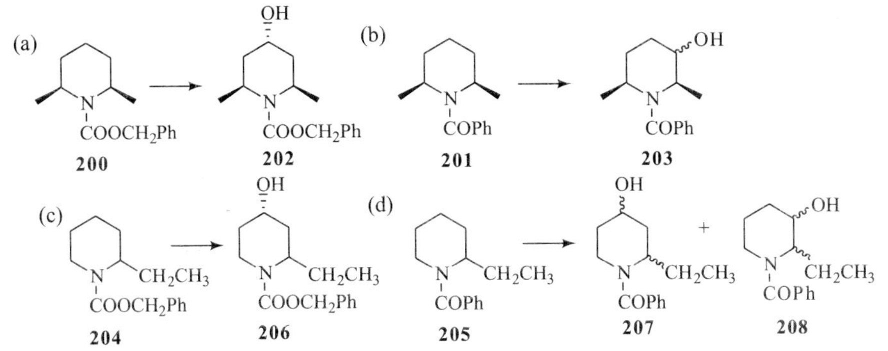

Figure 8.25 *B. bassiana* catalyzed hydroxylation of *N*-substituted piperidines.

In an attempt to broaden the application of biohydroxylation reactions in organic synthesis a concept of docking/protecting groups, which is similar to the common practice of using protecting groups in organic chemistry, has been developed. It is primarily aimed at improving hydroxylation position predictability, preventing undesired side reactions, and aiding substrate detection and product recovery. The process involves: (a) introduction of a docking/protecting group: (b) biotransformation reaction; and (c) the removal of the docking/protecting group (Fig. 8.26). This approach has been successfully applied to a variety of substrates including carboxylic acids, ketones, aldehydes, and alcohols. Protecting groups chosen for evaluation of this strategy included *N*-benzoylspirooxazolidines for aldehydes and ketones, benzoxazoles for carboxylic acids, and isosaccharine derivatives for alcohols. *B. bassiana* ATCC 7159 was employed for biohydroxylation of these due to its broad substrate acceptability (Braunegg *et al.*, 1999; de Raadt *et al.*, 2001a; de Raadt *et al.*, 2001b).

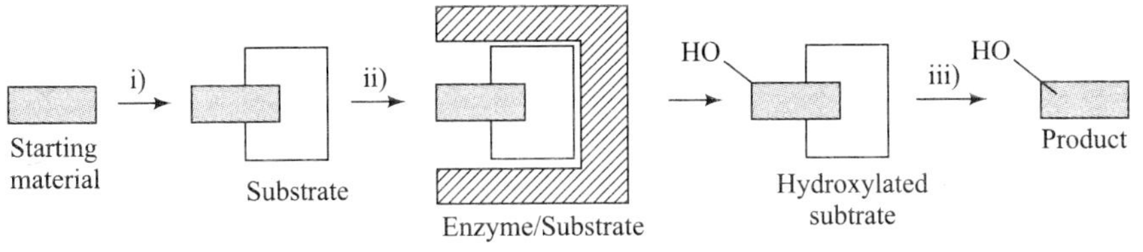

Figure 8.26 Steps involved in the use of docking/protecting (d/p) groups in biohydroxylations. (i) introduction of the d/p group; (ii) microbial biotransformation; (iii) deprotection.

Re-examination of product distribution in biohydroxylation of a series of amide, amine, ketone, and hydrocarbon substrates by *B. bassiana* ATCC 7159 in the presence of cytochrome P-450 inhibitors such as isosafrole, 1-aminobenzotriazole, and phenylacetylene had led to the conclusion that this organism contains a range of hydroxylases with different substrate specificity (Holland *et al.*, 1999). This study has also led to the presentation of a new paradigm for the interpretation of results of microbial hydroxylation and for the application of existing active site models for hydroxylation reactions catalyzed by *B. bassiana*.

4'-*O*-Methyl-Glucosidation and Glucosidation

4'-*O*-Methyl-glucosidation is a typical reaction catalyzed by *B. bassiana*. As depicted in Fig. 8.15a, biotransformation of the *N*-phenylcarbamate **99** had resulted in a 4'-*O*-methyl-*β*-D-glucosidation product identified as (–)-*trans*-2-fluorocyclohexyl *N*-[*p*-(4'-methyl-*β*-D-glucopyranosyl)]-phenylcarbamate (**104**) (Haufe *et al.*, 2002). Fermentation of 3-hydroxy-flavone (**113**) with *B. bassiana* ATCC 13144 has yielded the 4'-*O*-methyl-glucopyranosyloxy derivative **115** (Fig. 8.16) (Herath *et al.*, 2006). Biotransformation of 7-azanorbornylamides **116** – **118** and **122** bearing methylphenyl ether moieties have afforded their corresponding 4'-*O*-methyl-glucosides, **124** – **126** and **123**, respectively (Fig. 8.17) (Olivo *et al.*, 2003), further demonstrating the general applicability of this reaction. Similar transformation has been observed with the EGF-receptor tyrosine kinase inhibitor, CGP-62706 (**209**), but by *B. bassiana* DSM 875 and DSM 1344, giving its 4'-*O*-methyl-glucoside **210** (Fig. 8.27) (Kittelmann *et al.*, 1999). We have recently found that the biotransformation of the pentahydroxy flavonoid, quercetin (**211**), by *B. bassiana* ATCC 7159 afforded a new derivative, quercetin-4'-*O*-methyl-7-*O*-*β*-D-glucopyranoside (**212**) (Fig. 8.28a), in high yield suggesting that 4'-*O*-methyl-*β*-D-glucosidation of the substrate occurred with high selectivity at C-7-OH out of the five hydroxyl groups present in this natural product. Most of the product formed was present in the mycelium and the filtrate of the culture medium did not show any catalytic activity. The mycelium was capable of 4'-*O*-methyl-glucosidation of the substrate when suspended in buffers of pH 2.1 and 7.2, suggesting that intracellular enzymes are involved in this biotransformation and are active at a wide range of pH (Zhan and Gunatilaka, 2006a). The macrolide natural product, curvularin (**213**), on biotransformation with *B. bassiana* ATCC 7159 yielded three glucosidation and 4'-*O*-methyl-glucosidation products, curvularin-7-*O*-*β*-D-glucopyranoside (**214**), curvularin-4'-*O*-methyl-7-*O*-*β*-glucopyranoside (**215**), and 6-hydroxycurvularin-4'-*O*-methyl-6-*O*-*β*-D-glucopyranoside (**216**) resulting from hydroxylation, glucosidation, and methylglucosidation of the substrate (Fig. 8.28b). The isolation of 6-methyl-2,4-dihydroxyphenyl-4'-*O*-methyl-1-*O*-*β*-D-glucopyranoside (**217**) from the fermentation broth of *B. bassiana* ATCC 7159 without any added substrate suggested that hydroxylase, glucosyl transferase, and methylase are constitutive enzymes of this organism (Zhan and Gunatilaka, 2005).

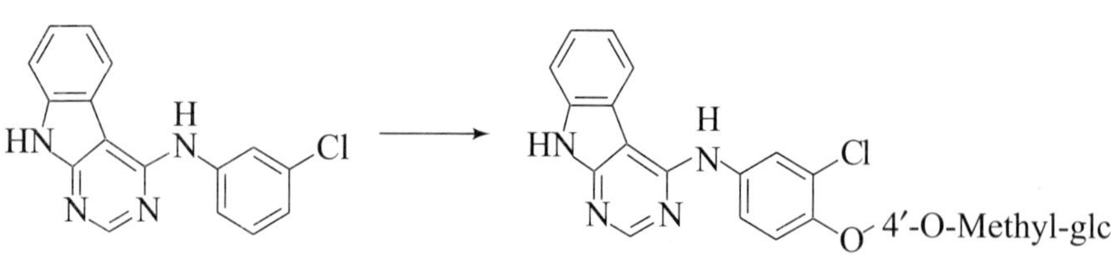

209 **210**

Figure 8.27 Biotransformation of CGP-62706 (**209**) by *B. bassiana*.

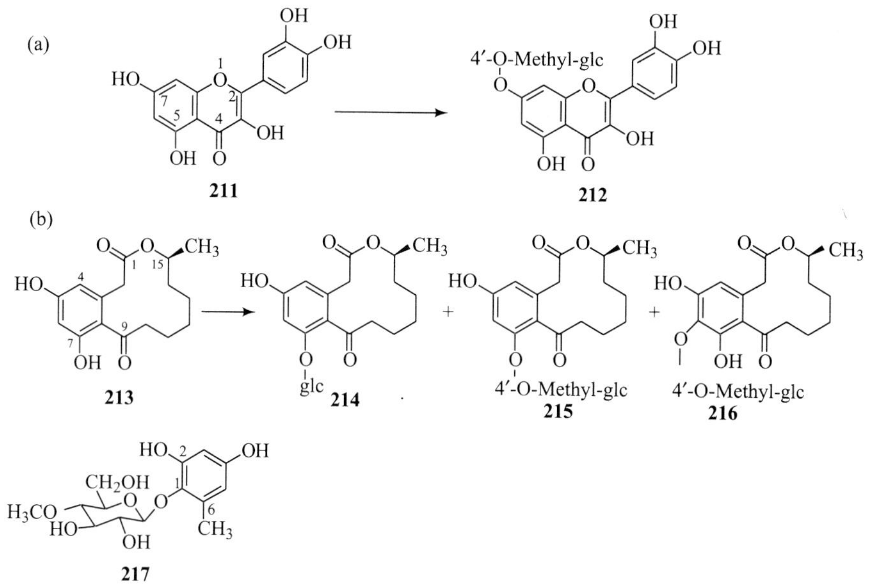

Figure 8.28 4'-O-Methyl-glucosidation of phenolic natural products by *B. bassiana*.

Biotransformation of 1-aminoanthracene (**79**) with *B. bassiana* ATCC 7159 has afforded 1-acetamido-anthracene-4'-*O*-methyl-5-*O*-*β*-D-glucopyranoside (**81**), 1-acetamido-anthraquinone-4'-*O*-methyl-8-*O*-*β*-D-glucopyranoside (**83**) and 1-acetamido-anthraquinone-4'-*O*-methyl-6-*O*-*β*-D-glucopyranoside (**84**) suggesting that the two intermediates 1-acetamidoanthracene (**80**) and 1-acetamido-anthraquinone (**82**) were formed and these underwent ready hydroxylation and 4'-*O*-methyl-glucosidation of the newly introduced hydroxyl groups (Fig. 8.12) (Zhan and Gunatilaka, 2008). When a series of hydroxy-and amino-anthraquinones were subjected to biotransformation by *B. bassiana* ATCC 7159, a novel *N*-4'-*O*-methyl-glucosidation reaction was found to occur with 1,2-diaminoanthraquinone (**218**) yielding 1-amino-2-(4'-*O*-methyl-2*β*-*N*-D-gluco-pyranosylamino)-anthraquinone (**222**). However, *O*-4'-*O*-methyl-glucosidation products, 1-amino-2-(4'-*O*-methyl-2*β*-*O*-D-glucopyranosyloxy)-anthraquinone (**223**), 8-hydroxy-1-(4'-*O*-methyl-1*β*-*O*-D-glucopyranosyloxy)-anthraquinone (**224**), and 1-hydroxy-2-(4'-*O*-methyl-2*β*-*O*-D-glucopyranosyloxy)-anthraquinone (**225**) were obtained from the other substrates [1-amino-anthraquinone (**219**), 1,8-dihydroxyanthraquinone (**220**) and 1,2-dihydro-xyanthraquinone (**221**)] tested (Fig. 8.29). The isolation of *N*-4'-*O*-methyl-gluco-sylation product **222** of 1,2-diaminoanthraquinone (**218**) revealed that this strain is capable of catalyzing selective 4'-*O*-methyl-glucosylation of only some amino groups providing a convenient approach to obtain *N*-glucosides of aromatic amines and therefore broaden the application of this microbial catalyst. During the biotransformation of 1-aminoanthraquinone (**219**) *B. bassiana* was found to catalyze hydroxylation of the substrate followed by selective glucosidation of the newly introduced hydroxyl group to yield **223**. This selectivity was further confirmed by 4'-*O*-methyl-glucosidation of only the hydroxyl group at C-2 of 1,2-dihydroxyanthraquinone (**221**) (Fig. 8.29). The known cytochrome P450 inhibitors, bifonazole and clotrimazole, were shown to have no effect on the above biotransformation reactions suggesting that none of these involved cytochrome P450 enzymes (Zhan and Gunatilaka, 2006b).

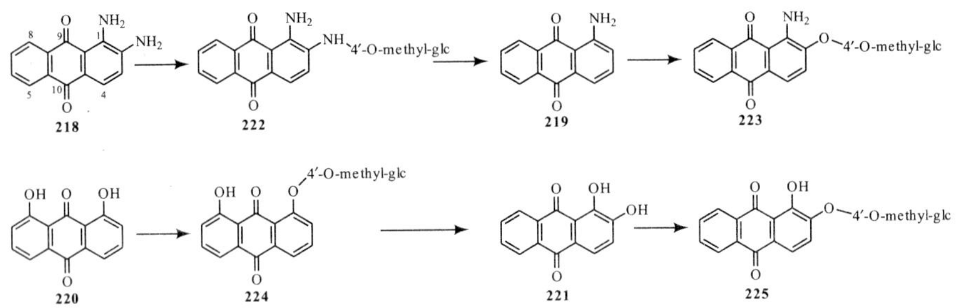

Figure 8.29 Biotransformation of hydroxy- and amino-anthraquinones by *B. bassiana*.

Biotransformation of the antifungal alkaloid, sampangine (**226**), by *B. bassiana* ATCC 7159 has yielded **227** resulting presumably from 1,4-reduction of the iminocarbonyl moiety followed by 4'-*O*-methyl-β-glucosidation (Fig. 8.30) (Orabi *et al.*, 1999).

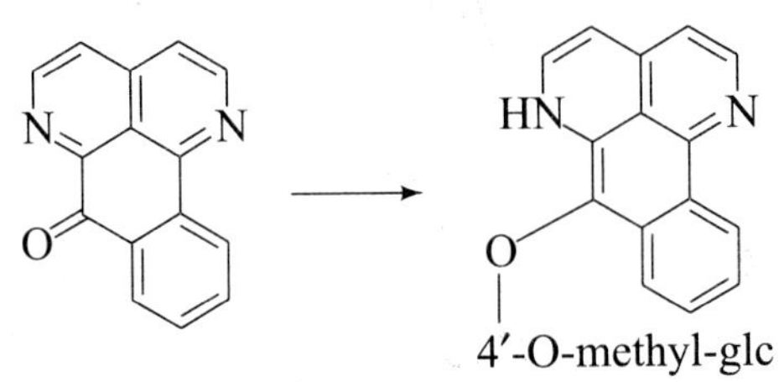

226 **227**

Figure 8.30 Biotransformation of the alkaloid, sampangine, by *B. bassiana*.

BIOLOGICAL HYDROLYSIS

Hydrolysis of esters and oxiranes are reported to be catalyzed by *B. bassiana*. Thus, biotransformation of plectranthone (**128**) has yielded desacetylplectranthone (**130**), 1-hydroxydesacetylplectranthone (**131**), 12-hydroxyisodesacetylplectranthone (**132**), and 15-hydroxydesacetylplectranthone (**133**) (Fig. 8.18). Interestingly, only monodesacetyl products were formed in this biotransformation (Orabi, 2000). *B. bassiana* has also been reported to be useful as an *in vitro* model of mammalian metabolism of an investigational anti-protozoal agent CGP-291 (**228**), biological hydrolysis of which has led to the formation of four deacetylation products including *N*-desacetyl CGP-291 (**229**), dehydro *N*-desacetyl CGP-291 (**230**), and two monohydroxylated deacetylation metabolites, 5'-hydroxy-*N*-desacetyl CGP-291 (**231**) and 4-hydroxy-*N*-desacetyl CGP-291 (**232**) (Fig. 8.31) (Jürgens and Clark, 1990).

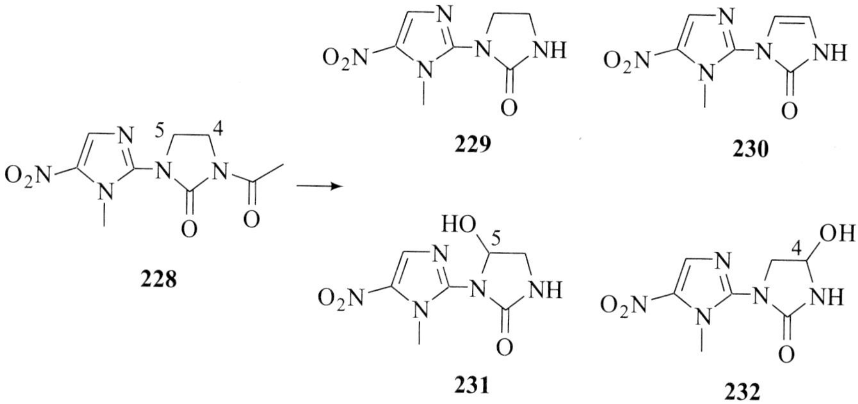

Figure 8.31 Biotransformation of the anti-protozoal drug, GGP-291, by *B. bassiana*.

Moussou and coworkers (2000) have studied the hydrolysis of five synthetic epoxides **233** – **237** (Fig. 8.32) and observed the existence of two different epoxide hydrolase activities in *B. bassiana* ATCC 7159. It was demonstrated that the epoxide hydrolase activity was produced during the exponential and stationary growth phases of the fungus and was totally cell-associated. After cell disruption, epoxide hydrolase activity was recovered in both the cell debris and the soluble fraction but not in the membrane fraction. Activity assays of these fractions with two different substrates have indicated their different substrate specificity and regio-selectivity. Interestingly, opposite absolute configuration for the residual epoxide was also observed indicating that the two epoxide hydrolase activities are substantially different. These workers suggested these enzymes to be complementary biocatalysts that can be used for the preparation of the corresponding epoxides and/or vicinal diols in nearly enantiopure forms.

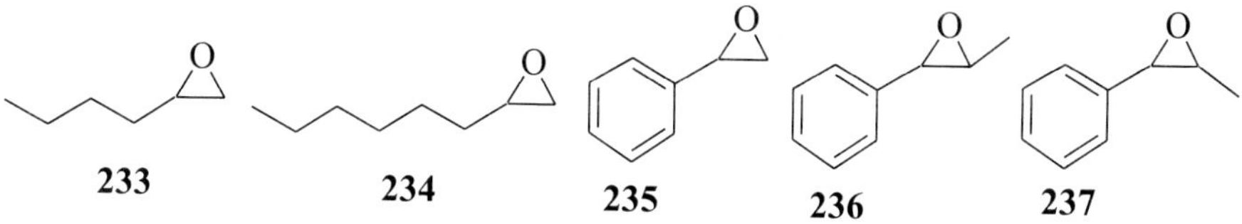

Figure 8.32 Substrates tested for epoxide hydrolase activities in *B. bassiana*.

BIOLOGICAL *N*-ACETYLATION

We have recently found that biotransformation of 1-aminoanthracene (**79**) by *B. bassiana* ATCC 7159 led to the formation of the *N*-acetylation product, 1-acetamidoanthracene (**80**) which underwent further biotransformation (Fig. 8.12) (Zhan and Gunatilaka, 2008). Tixier and coworkers (2002) have reported *B. bassiana* catalyzed *N*-acetylation of 3,4 dichloroaniline (**238**), a biodegradation product of a phenylurea herbicide, diuron (**239**), by the same fungal strain to produce 3,4-dichloroacetanilide (**240**) (Fig. 8.33a) However, on prolonged incubation it was found that the concentration of the product **240** decreased while that of the substrate **238** increased, indicating that the metabolite **240** may have been hydrolysed into the parent compound **238**. This phenomenon was shown to be reproducible and only observed with *B. bassiana* among four selected fungal strains tested, *Aspergillus niger*, *Cunninghamella echinulata* var. *elegans* and *Mortierella isabellina* (Tixier *et al*., 2002).

(a)

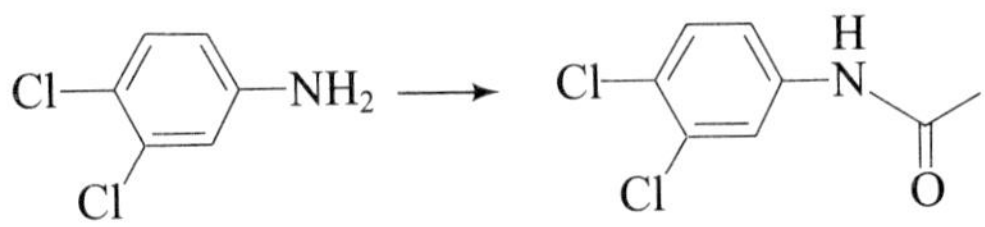

238 **240**

(b)

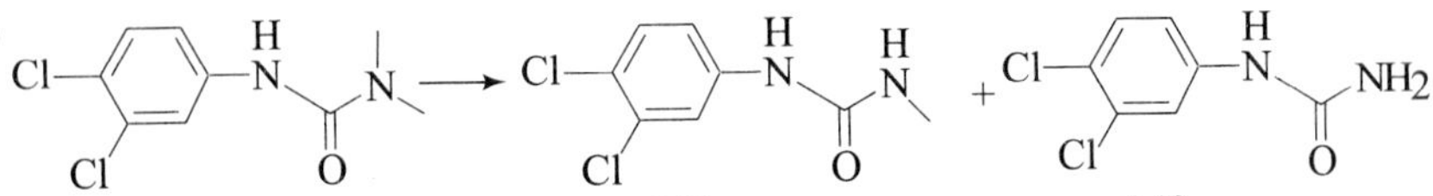

239 **241** **242**

Figure 8.33 Biotransformation of 2,3-dichloroaniline (**238**) and diuron (**239**) by *B. bassiana*.

BIOLOGICAL DEMETHYLATION

O-Demethylation has been reported to occur in the biotransformation of methylbenzamide 7-azanorbornane ethers by *B. bassiana* ATCC 7159. The formation of the phenolic products **124**, **127** and the 4'-*O*-methyl-β-D-glucosidated phenols **125** and **126** from the corresponding *N*-(anisoyl)-7-azanorbornane precursors **116** – **118**, and the 4'-*O*-methyl-β-D-glucosidated phenol **123** from the ethyl ether **122** (Fig. 8.17) suggested that *O*-demethylation of substituted anisoles is independent of the orientation of the substituent (Olivo *et al*., 2003). In a study of biodegradation of the herbicide diuron (**239**) by several soil fungi, *B. bassiana* ATCC 7159 was found to display its enzymatic demethylation ability to yield mono- and di-*N*-demethylation products, *N*-(3,4-dichlorophenyl)-*N*'-methylurea (**241**) and *N*-3,4-dichlorophenylurea (**242**) (Fig. 8.33b). It is noteworthy that these two demethylated products were found to be more toxic than the original herbicide (Tixier *et al*., 2000).

CONCLUSION

As apparent from the above discussion various strains of the fungus, *B. bassiana*, have found applications both as a biocatalyst and as an *in vitro* metabolism model. It was found to have significantly broad substrate acceptability and to display different enzymatic activities including biological reduction, oxidation, hydroxylation, hydrolysis, 4′-*O*-methyl-glucosidation, demethylation, and hydration. Besides these biotransformation reactions, and studies aiming at modification of structures and metabolic fate of substrates, recent investigations have aimed at the applicability of the concept of docking/protecting groups for biological hydroxylation of a variety of substrates. 4′-*O*-Methyl-glucosidation is a typical reaction catalyzed by this fungus, and the enzymes were found to be active in a broad range of pH, which is a desirable characteristic for industrial applications. The recent discovery of the ability of *B. bassiana* ATCC 7159 to effect *N*-4′-*O*-methyl-glucosidation of aromatic amines should find useful synthetic applications. Despite these interesting findings, in order to use this fungus and its abundant enzymes efficiently, more intensive work to determine mechanisms of the biotransformation reactions is required. It may be expected that future research would involve molecular biological aspects of these enzymes and genetic manipulation of the useful strains of *B. bassiana*.

References

Aina, G., Nasini G. and de Pava, O.V. (2001). Asymmetric bioreduction of racemic 5,6,7,8-tetrahydro-8-methyl-1,3-dimethoxynaphthalen-6-one to the corresponding chiral b-tetralols. *Journal of Molecular Catalysis B: Enzymatic* 11(4-6): 367–371.

Barbieri, C., Bossi, L., D'Arrigo, P., Fantoni G.P. and Servi, S. (2001). Bioreduction of aromatic ketones: preparation of chiral benzyl alcohols in both enantiomeric forms. *Journal of Molecular Catalysis B: Enzymatic* 11(4-6): 415–421.

Braunegg, G., de Raadt, A., Feichtenhofer, S., Griengl, H., Kopper, I., Lehmann, A. and Weber, H.-J. (1999). The concept of docking/protecting groups in biohydroxylation. *Angewandte Chemie International Edition* 38(18): 2763–2766.

Breiter, S., Schlosser, D., Weiss, D. and Schmauder, H.-P. (1995). Microbial hydroxylation of androsta-1,4-diene-3,17-dione. *Natural Product Letters* 6(1): 7–14.

Buchanan, G.O. and Reese, P.B. (2001). Biotransformation of diterpenes and diterpene derivatives by *Beauveria bassiana* ATCC 7159. *Phytochemistry* 56(2): 141–151.

Buchanan, G.O., Williams, L.A.D. and Reese, P. B. (2000). Biotransformation of cadinane sesquiterpenes by *Beauveria bassiana* ATCC 7159. *Phytochemistry* 54(1): 39–45.

Čapek, A., Hanè, M., Tadra, M. and Tùma, J. (1966). Microbial transformations of steroids. XXV. The effect of substituents on the direction of the transformation of steroids by *Beauveria bassiana. Folia Microbiologica* 11: 159–162.

Davis, P.J., Yang, W., Hezari, M., Campos, J. and Reddy, C.G. (1992). Applications of whole cell fungal systems as models of mammalian xenobiotic/drug metabolism. *Environmental Science Research* 44: 197–207.

Décor, A., Bellocq, D., Thoison, O., Lekieffre, N., Chiaroni, A., Ouazzani, J., Cresteil, T., Guéritte, F. and Baudoin, O. (2006). *In vitro* oxidative metabolism study of (–)-rhazinilam. *Bioorganic & Medicinal Chemistry* 14(5): 1558–1564.

de Raadt, A., Fetz, B., Griengl, H., Klingler, M.F., Krenn, B., Mereiter, K., Münzer, D.F., Plachota, P., Weber, H. and Saf, R. (2001a). Chiral auxiliaries as docking/protecting groups: biohydroxylation of selected ketones with *Beauveria bassiana* ATCC 7159. *Tetrahedron* 57(38): 8151–8157.

de Raadt, A., Griengl, H. and Weber, H. (2001b). The concept of docking and protecting groups in biohydroxylation. *Chemistry—A European Journal* 7(1): 27–31.

Elmarakby, S.A., El-Feraly, F.S., Elsohly, H.N., Croom, E.M. and Hufford, C.D. (1987). Microbial transformation studies on arteannuin B. *Journal of Natural Products* 50(5): 903–909.

El-Sharkawy, S.H. (1991). Microbial conversion of tamoxifen. *Applied Microbiology and Biotechnology* 35(4): 436–439.

Flitsch S.L., Aitken, S.J., Chow, C.S.-Y., Grogan, G. and Staines, A. (1999). Biohydroxylation reactions catalyzed by enzymes and whole-cell systems. *Bioorganic Chemistry* 27(2): 81–90.

Forrest, G.L. and Gonzalez, B. (2000). Carbonyl reductase. *Chemico-Biological Interactions* 129(1-2): 21–40.

Grogan G.J. and Holland, H.L. (2000). The biocatalytic reactions of *Beauveria spp. Journal of Molecular Catalysis B: Enzymatic* 9: 1–32.

Grogan, G., Sime, J.T. and Turner, N.J. (2001). Diastereoselective hydroxylation and reduction of derivatised tetrahydrofurans by *Beauveria bassiana. Biotechnology Letters* 23(2): 119–124.

Grotowska, A.K. and Wawrzeñczyk, C. (2002). Lactones 13. Biotransformation of iodolactones. *Journal of Molecular Catalysis B: Enzymatic* 19–20: 203–208.

Haufe, G., Wölker, D. and Fröhlich, R. (2002). Selectivity of biohydroxylation with *Beauveria bassiana* of *trans*-2-fluorocycloalkyl *N*-phenylcarbamates. *Journal of Organic Chemistry* 67(9): 3022–3028.

Haufe, G., Pietz, S., Wölker, D. and Fröhlich, R. (2003). Synthesis of fluorinated cycloalkyl *N*-phenylcarbamates and their microbial defluorination/oxygenation by *Beauveria bassiana. European Journal of Organic Chemistry* 11: 2166–2175.

Haufe, G. and Wölker, D. (2003). Blocking fluorine substitution in biotransformation of nortricyclanyl *N*-phenylcarbamates with *Beauveria bassiana. European Journal of Organic Chemistry* 11: 2159–2165.

Herath, W., Mikell, J. R., Hale, A. L., Ferreira, D. and Khan, I. A. (2006). Microbial metabolism. Part 6. Metabolites of 3- and 7-hydroxyflavones. *Chemical & Pharmaceutical Bulletin* 54(3): 320–324.

Holland, H.L., Andreana, P. R. and Brown, F.M. (1999). Biocatalytic and chemical routes to all the stereoisomers of methionine and ethionine sulfoxides. *Tetrahedron: Asymmetry* 10(15): 2833–2843.

Holland, H.L., Brown, F.M., Johnson, D.V., Kerridge, A., Mayne, B., Turner, C.D. and van Vliet, A.J. (2002). Biocatalytic oxidation of *S*-alkylcysteine derivatives by chloroperoxidase and *Beauveria* species. *Journal of Molecular Catalysis B: Enzymatic* 17(6): 249–256.

Holland, H.L., Morris, T.A., Nava, P.J. and Zabic, M. (1999). A new paradigm for biohydroxylation by *Beauveria bassiana* ATCC 7159. *Tetrahedron* 55: 7441–7460.

Huszcza, E., Dmochowska-Gladysz J. and Bartmañska A. (2005). Transformations of steroids by *Beauveria bassiana. Zeitschrift für Naturforschung, C: Journal of Biosciences* 60(1/2): 103–108.

Jürgens, T.M. and Clark A.M. (1990). The metabolism of CGP-291: the use of microorganisms as models of mammalian metabolism. *Pharmaceutical Research* 7(7): 742–745.

Kittelmann, M., Oberer, L., Blum, W. and Ghisalba, O. (1999). Microbial hydroxylation and simultaneous formation of the 4″-*O*-methylglucoside of the tyrosine-kinase inhibitor CGP 62706. *Chimia* 53(12): 594–596.

Liu, H., Skinner, M., Brownbridge, M. and Parker, B.L. (2003). Characterization of *Beauveria bassiana* and *Metarhizium anisopliae* isolates for management of tarnished plant bug, *Lygus lineolaris* (Hemiptera: Miridae). *Journal of Invertebrate Pathology* 82(3): 139–147.

Moussou, P., Archelas, A., Furstoss, R. and Baratti, J.C. (2000). Clues for the existence of two different epoxide hydrolase activities in the fungus *Beauveria bassiana*. *Enzyme and Microbial Technology* 26(5–6): 414–420.

Olivo, H.F., Osorio-Lozada, A. and Peeples, T.L. (2005). Microbial oxidation/amidation of benzhydrylsulfanyl acetic acid. Synthesis of (+)-modafinil. *Tetrahedron: Asymmetry* 16(21): 3507–3511.

Olivo, H.F., Peeples, T.L., Ríos, M.-Y., Velázquez, F., Kim, J.-W. and Narang, S. (2003). Microbial C-hydroxylation and *β*-4-*O*-methylglucosidation of methyl-benzamide 7-azanorbornane ethers with *Beauveria bassiana*. *Journal of Molecular Catalysis B: Enzymatic* 21(3): 97–105.

Orabi, K.Y. (2000). Microbial transformation of the eudesmane sesquiterpene plectranthone. *Journal of Natural Products* 63(12): 1709–1711.

Orabi K.Y., Li, E., Clark, A.M. and Hufford, C.D. (1999). Microbial transformation of sampangine. *Journal of Natural Products* 62(7): 988–992.

Panke, S. and Wubbolts, M. (2005). Advances in biocatalytic synthesis of pharmaceutical intermediates. *Current Opinion in Chemical Biology* 9(2): 188–194.

Parshikov, I.A., Moody, J.D., Heinze, T.M., Freeman, J.P., Williams, A.J. and Sutherland J.B. (2002). Transformation of cinoxacin by *Beauveria bassiana*. *FEMS Microbiology Letters* 214(1): 133–136.

Pedras, M.S.C. and Montaut, S. (2003). Probing crucial metabolic pathways in fungal pathogens of crucifers: biotransformation of indole-3-acetaldoxime, 4-hydroxyphenylacetaldoxime, and their metabolites. *Bioorganic & Medicinal Chemistry* 11(14): 3115–3120.

Preisig, C.L., Laakso, J.A., Mocek, U.M., Wang, P.T., Baez, J. and Byng, G. (2003). Biotransformations of the cardiovascular drugs mexrenone and canrenone. *Journal of Natural Products* 66(3): 350–356.

Rosazza, J. P. and Smith, R.V. (1979). Microbial models for drug metabolism. *Advances in Applied Microbiology* 25: 169–208.

Shimizu, M. and Fujishima, H. (2001). Preparation of monohydroxyadamantane esters by bioconversion. *Jpn. Kokai Tokkyo Koho (Japanese Patent Application JP 99-194354)*, 5 pp.

Suresh, B., Ritu, T. and Ravishankar, G.A. (2006). Biotransformation as applicable to food industries. *Food Science and Technology* 148: 1655–1690.

Terentiev, P.B., Zilberstein, T.M., Borisenko, A.A., Shmorgunov V.A., Piskunkova N.F. and Grishina, G.V. (2003). Transformation of 1,2,5,6-tetrahydropyridines with micellar fungi. *Chemistry of Heterocyclic Compounds* 39 (7): 885–890.

Tixier, C., Bogaerts, P., Sancelme, M., Bonnemoy, F., Twagilimana, L., Cuer, A., Bohatier, J., and Veschambre H. (2000). Fungal biodegradation of a phenylurea herbicide, diuron: structure and toxicity of metabolites. *Pest Management Science* 56(5), 455–462.

Tixier, C., Sancelme, M., Aït-Aïssa, S., Widehem, P., Bonnemoy, F., Cuer, A., Truffaut, N. and Veschambre, H. (2002). Biotransformation of phenylurea herbicides by a soil bacterial strain, *Arthrobacter sp.* N2: structure, ecotoxicity and fate of diuron metabolite with soil fungi. *Chemosphere* 46(4): 519–526.

Venisetty, R.K. and Ciddi, V. (2003). Application of microbial biotransformation for the new drug discovery using natural drugs as substrates. *Current Pharmaceutical Biotechnology* 4(3): 153–167.

Weidner, S., Goeke, K., Trinks, U., Traxler, P., Ucci-Stoll, K. and Ghisalba, O. (1999). Preparation of 4-(4'-hydroxyanilino)-5-anilinophthalimide and 4,5-bis-(4'-hydroxyanilino)-phthalimide by microbial hydroxylation. *Bioscience, Biotechnology, and Biochemistry* 63(8): 1497–1500.

Zhan, J. and Gunatilaka, A.A.L. (2005). Microbial transformation of curvularin. *Journal of Natural Products* 68(8): 1271–1273.

Zhan, J. and Gunatilaka, A.A.L. (2006a). Selective 4'-O-methylglycosylation of the pentahydroxy-flavonoid quercetin by *Beauveria bassiana* ATCC 7159. *Biocatalysis and Biotransformation* 24(5): 396–399.

Zhan, J. and Gunatilaka, A.A.L. (2006b). Microbial transformation of amino-and hydroxyanthraquinones by *Beauveria bassiana* ATCC 7159. *Journal of Natural Products* 69(10): 1525–1527.

Zhan, J. and Gunatilaka, A.A.L (2008). Microbial metabolism of 1-aminoanthracance by *Beauveria bassiana Bioorganic and Medicinal* Chemistry.D01:10.1016 j.bmc.-2008.03.030.

9

Biotransformation of Terpenes by Fungi

Carla C.C.R. de Carvalho

Centro de Engenharia Biológica e Química, Instituto Superior Técnico,
Av. Rovisco Pais, 1049-001 Lisboa, Portugal
* corresponding author. E-mail: ccarvalho@ist.utl.pt
Telephone: +351-21-8419594; Fax: +351-21-8419062

INTRODUCTION

Terpenes are natural unsaturated hydrocarbons derived from isoprene units (C5) and are widely distributed in nature. Isoprene is the main VOC (Volatile Organic Compound) emitted by trees such as oak, poplar, willow, sycamore and eucalyptus. Depending on molecular size, the compounds can be mono-, sesqui-, di- and triterpenes, corresponding to 10, 15, 20 and 30 carbon atoms, respectively. The most important group correspond to monoterpenes because they are responsible for the fragrance and flavour of about 90% of all spices. Monoterpenes can be acyclic (*e.g.*, myrcene), cyclic with one ring (limonene, α-terpinene) or with two rings (α- and β-pinene, camphene, 3-carene). Monoterpenes have mainly ecological roles in plants, acting as deterrent against feeding by herbivores, as antifungal defences and attractants for pollinators (Langenheim, 1994). In mammals they are involved in stabilizing cell membranes (*e.g.*, cholesterol), metabolic pathways (vitamins A, D, K, carotenoids) and in regulating enzymatic reactions (prenilation of proteins). Herbs and higher plants containing their oxidised derivatives, known as terpenoids, have been used for centuries as flavours, fragrances and pharmaceuticals. More than 22,000 terpenoids are known at the present, which makes this the largest group of natural compounds. The availability of terpenes, such as limonene, at large amounts and low cost, favours their application in reactions in which they may be used as substrates. Furthermore, terpenoids are in general GRAS (Generally Recognized As Safe) substances.

Industrial use of terpenes as substitutes of ozone-depleting chlorofluorocarbons is also promising (Kirchner, 1994). As substitutes of chlorinated solvents, they can be used to clean electronic components and cables, to degrease metal and clean aircraft parts (Brown *et al.*, 1992).

Plants have been an important source of essential oils from which terpenes and terpenoids are extracted. However, these compounds are the result of secondary metabolism and are often present in minor quantities. Sometimes they are only present in exotic plants. The isolation and purification of the desired flavours and fragrances is therefore difficult and expensive.

The possibility of producing compounds with high stereo-specificity and selectivity, under mild conditions, is rapidly drawing interest to biotransformations. One of the major advantages is that the compounds produced by microbial systems can be classified as "natural" products by the European and American food legislations (Krings and Berger, 1998). The difference in price between a natural and a chemically synthesised compound can be considerable in a market with increasingly health and nutrition conscious consumers. In addition, the range of natural and engineered biocatalysts that can be used in clean industrial products and processes is widening due to the convergence of biology and chemistry, the discoveries in biodiversity and the results in molecular biology and computational science (Bull *et al.*, 1999). By using microorganisms, it is even possible to perform regiospecific biotransformations at molecule centres that are chemically unreactive. However, the low water solubility of monoterpenes, the chemical instability of some terpenes and terpenoids, their high volatility and cytotoxicity and the low transformation rates obtained in most processes, difficult the commercialisation of biotransformation processes.

Production of Flavours and Fragrances

Monoterpenes are the most common fragrance and flavours, whilst higher terpenes have less olfactory importance because they are less volatile. Sesquiterpenes are relatively common in plants at small concentrations and are of some importance for the fragrance of cinnamon and ginger. Di- and triterpenes have a strong bitter taste and are rather exotic and rare as fragrances. Diterpenes are usually pharmacodynamically very active, conferring medicinal or poisonous properties.

Commercially available flavour compounds and fragrances are mainly produced nowadays via chemical synthesis or by extraction from plants and herbs. In chemical processes, the formation of undesired racemic mixtures, *i.e.*, mixtures that contain equal quantities of dextrorotatory and levorotatory enantiomers, is common. Chiral building blocks, pharmaceutical and agrochemical compounds and food additives are only considered as pure enantiomers when one enantiomer is present in excess of *ca.* 98%. The US Food and Drug Administration has declared that the biological effects of both enantiomers of a chiral compound must be studied because the non-therapeutic enantiomers may have unwanted side effects (Jirage and Martin, 1999). Besides, chirality is also important in fragrances and flavours as the perception of odour and flavour depends on the absolute conformation of the isomers. Extraction from plants has also several drawbacks, including the time needed for plant development which may depend on weather and soil conditions, the need for macro- and micronutrients, competition of weeds, necessity of pollinators, attacks by grasshoppers and leaf hoppers, etc. Furthermore, active compounds are usually secondary metabolites and are thus present in low quantities, making extraction and isolation difficult. If present in exotic plants, then the flavour products can be quite expensive. Microorganisms such as yeast, bacteria and fungi can be effectively used to produce fragrance and flavour compounds (de Carvalho and da Fonseca, 2006a, b).

Biotechnologically produced aroma compounds can be obtained by *de novo* synthesis, using simple cultivation medium, or by biotransformation/bioconversion of a substrate to the desired product. Although biotransformation and bioconversions are similar and the terms are often used interchangeable (Walker and Cox, 1995), there is a subtle difference between the two. Biotransformation refers to a specific modification of a defined compound to a defined product, structurally similar, by use of a biological catalyst (Lilly, 1994) whilst bioconversion refers to the catalytic activity of living organisms and thus the production of the major product from the precursor compound(s) may involve several biochemical steps.

De novo Synthesis

Since the early 1950s, fungal metabolites have been studied because of the wide diversity of chemical compounds and the great biochemical potential (Cole and Schweikert, 2003), but research has hardly focussed on volatile secondary metabolites until recent years.

Forty-eight volatile compounds (more than 80% contained 4-14 carbon atoms) were identified in the fruiting bodies of wild *Fistulina hepatica*, of which eleven were found to contribute significantly to its sweet and wild flowery flavour (Wu *et al.*, 2005). The sweet aroma was mainly attributed to phenylacetaldehyde, phenylacetic acid and bisabolol oxide B whilst the flowery fragrance was primarily attributed to phenylacetaldehyde, (*E*)-methyl cinnamate, linalool, bisabolol oxide B and (*E*)-2-methyl-2-butenoic acid. The latter two compounds were identified for the first time as native volatile secondary metabolites of fungi. In a study to assess the possibility of replacing *Penicillium camemberti* by *P. caseifulvum* as a fungal starter for the production of Brie and Camembert cheese, special attention was given to the capabilities of *P. caseifulvum* to produce volatile flavour compounds comparable to the usual strain (Larsen, 1998). A non-destructive diffuse sampling method was used to prevent further conversions during analysis. During the first sampling period, when growth became visible, both species showed strong lipolytic activity producing large amounts of odd-chain methyl ketones and small primary alcohols. In the second period, when heavy sporulation could be observed, unsaturated hydrocarbons and other terpenes *e.g.*, limonene, 3,7-dimethyl-1,3,7-octatriene, β-caryphyllene, other sesquiterpenes and diterpenes, were produced by both *P. caseifulvum* and *P. camemberti*. Mono- and diterpene production by *P. camemberti* on a milk fat based medium was reported in this study for the first time, while Larsen and Frisvad (1995) had already observed terpene production at the end of the growth phase of *P. vulpinum*.

Xu *et al.* (2005) studied the fungi responsible for the fermentation of Puer tea and the volatile compounds produced. Puer tea is a fermented Chinese tea that, unlike black tea, depends on microbial fermentation to develop a large number of complex volatile compounds necessary to form its particular flavour. During the identification process of the microorganism responsible for the fermentation, *Aspergillus niger* was the only fungus isolated. Since the green tea leaves, used as raw material, contained a moisture content of just 10%, the authors hypothesised that this fungus carries out the fermentation, as hardly bacteria or yeast could survive such low moisture content. During the fermentation process of Puer tea, aldehydes and ketones in the green tea decreased significantly whilst the amount of terpene alcohols *e.g.,* linalool and linalool oxides, methoxybenzene and its derivatives and indole, increased remarkably in the resulting tea. While indole give a flower like aroma, the monoterpene alcohols, particularly linalool oxide I, II and III and methoxy-benzene are responsible for the stale and mellow aroma.

Some of the research carried out on fungal metabolites also aimed at finding compounds with antimicrobial and cytotoxic activities. Rasser *et al.* (2002) reported the presence of several cytotoxic terpenoids in extracts of basidiomycete *Bovista* sp. 96042. Besides the already known illudane psathyrellon B and photoilludane armillol and the new secoprotoilludane, illudane and drimane sesquiterpenes, a novel hexacyclic illudane-illudane bis-sesquiterpene bovistol was found.

Rulepidanol and rulepidadienes A and B are three sesquiterpenes that were found for the first time in fruiting bodies of *Russula lepida* (Vidari *et al.*, 1998). When Vidari and coworkers compared the composition of the compounds reported in the literature in *Russula* mushrooms, they came to the conclusion that the most acrid and hot pungent tasting Russulaceae species are very similar to each other in chemical content of sesquiterpenoids.

Several sesquiterpenes and their oxygenated derivatives, sesquiterpenoids, have been isolated from fungi extracts. Among them are the following: a novel cytotoxic ascochlorin glycoside vertihemipterin A from the insect pathogenic fungus *Verticillium hemipterigenum* (Seephonkai *et al.*, 2004); (-)-5-demethoxyfumagillol, a potent angiogenesis inhibitor and fumagiringillin from *Aspergillus fumigatus* (Jiao *et al.*, 2004); three tyrosine kinase inhibitors from cultures of the fungus *Stachybotrys chortarum* (Vazquez *et al.*, 2004); paecilomycines A–C (Kikuchi *et al.*, 2004a), tenuipesine A (Kikuchi *et al.*, 2004b), spirotenuipesine A and spirotenuipesine B (Kikuchi *et al.*, 2004c), isolated from fruiting bodies of the fungus *Paecilomyces tenuipes* (this entomopathogenic fungus is used in folk medicine and health foods in China, Korea and Japan); lignoren isolated from the fungus *Trichoderma lignorum* (Berg *et al.*, 2004); two novel sesquiterpenoids, boletunone A and B, were present in the methanolic extract of *Boletus calopus* (Kim *et al.*, 2004); four illudane sesquiterpenes with cytotoxic and antibiotic activities from cultures of *Leptosphaeria maculans* (Reina *et al.*, 2004) and also a highly selective phytotoxin, depsilairdin (David *et al.*, 2004). Fraga (2005) wrote a review on natural sesquiterpenoids which includes fungal compounds.

Biotransformation and Bioconversion

Terpenoids are widely used in industry as flavour and fragrances, whilst terpene hydrocarbons are usually a sub-product of the extraction of essential oils because of their weak odour and chemical instability. As a result, terpenes such as limonene and α-pinene are readily available and are relatively inexpensive. Bacteria, yeast, fungi, plants and plant cell cultures are interesting biocatalysts for the production of high-value terpenoids from cheap terpene compounds (de Carvalho and da Fonseca, 2006b).

The increasing demand by consumers for natural and healthy products is favouring the use of biotechnology processes to produce terpenoids for the food, cosmetic and pharmaceutical industries. The difference in price between a natural compound and its synthetic version can be significant. For example, the price of vanillin, which is the major component of vanilla flavour extracted from the fermented pods of *Vanilla* orchids and the principal flavouring compound in the world, depends strongly on the source. The synthetic vanillin costs one-hundredth of the price of the natural product (Walton *et al.*, 2000). According to a report from the International Trade Centre (ITC, 2006), the world imports for vanilla beans reached 3,500 tonne valued at US$395 million in 2004. Typhoons and political unrest in Madagascar, where most of the world's vanilla is produced, are responsible for the high prices of natural vanilla. Biotechnological processes could be a sound alternative for the production of natural vanillin.

Several biotransformation and bioconversions of terpenes have been described using fungi cells or enzymes (Table 9.1). As mentioned previously, the enormous advantage of biotransformation of terpenes is the production of "natural" enantiomerically pure flavours and fragrances under mild reaction conditions.

The most abundant monocyclic monoterpenes in nature are the two enantiomers of limonene. The availability of around 50,000 tonnes of enantiopure *d*-limonene, which is the major component of citrus peel oil, at US$1-2 led to an increased interest by chemists and biologists (Duetz *et al.*, 2001). Several important and valued flavour and medicinal compounds present the same carbon structure, *e.g.*, carveol, carvone, perillyl alcohol, menthol. Limonene functions in its natural sources are mainly protective and include prevention of dehydration, antimicrobial activity and antioxidant properties. Bioconversions using microorganisms must thus overcome limonene toxicity. In Figure 9.1 some products obtained by biotransformation of limonene by fungi are depicted. Spoiled mandarins, tangerines and oranges may be a good source of fungal strains able to convert limonene.

Adams *et al.* (2003) reported the conversion of (*R*)-(+)-limonene to (*R*)-(+)-α-terpineol by *P. digitatum* in 8 h with a yield of 93% and an enantiomeric excess (*ee*) higher than 99%. Due to the low water solubility of limonene (0.150 mmol/L; Fichan *et al.*, 1999), a co-solvent, in particular ethanol, was efficient in increasing the solubility of limonene in the medium and thus in enhancing mass transfer and the bioconversion rate. Limonene inhibition and evaporation that occur at high concentrations were overcome by sequential addition of appropriate amounts of substrate.

Onken and Berger (1999) were able to overcome the toxic effect of limonene to fungi cells by adding the substrate through the gas phase. Further adaptation of the precultures to small amounts of substrate led to a 2-fold increase in the concentration of carveol and to a 3-4-fold increase in the concentration of carvone, reaching a total product concentration above 100 mg/L.

Table 9.1 Examples of biotransformations of terpenes carried out by fungi.

Publication	*Microorganism*	*Substrate*	*Product*
Adams et al., 2003	*Penicillium digitatum*	(*R*)-(+)-limonene	(*R*)-(+)-α-terpineol
Aleu *et al.*, 2001	*Botrytis sp.*	several	several
Chen and Reese, 2002	*Aspergillus niger*	stemodin stemodinone stemarin	hydroxylated products two analogues four new compounds
Demyttenaere and Kimpe, 2001	*Penicillium digitatum*	geraniol and nerol geranial and neral	6-methyl-5-hepten-2-one
Fraga *et al.*, 2001	*Mucor plumbeus*	ribenone	several
Kaspera *et al.*, 2005a	*Penicillium citrinum* *Gongronella butleri* *Penicillium digitatum* *Fusarium proliferatum* *Fusarium proliferatum* *Pleurotus sapidus*	(*R*)-(+)-limonene (*R*)-(+)-limonene (*R*)-(+)-limonene (*R*)-(+)-limonene (*S*)-(-)-limonene *trans*-(-)-carveol	*α*-terpineol perillyl alcohol *α*-terpineol *cis*-(+)-carveol *trans*-(-)-carveol (*R*)-(-)-carvone
Kaspera *et al.*, 2005b	*Chaetomium globosum*	valencene	nootkatone
Krings *et al.*, 2005	*Stemphylium botryosum*	γ-terpinene	*p*-mentha-1,4-dien-9-ol
Lindmark-Henriksson *et al.*, 2004	*Picea abies*	*β*-pinene (*R*)-(+)-limonene	*trans*-pinocarveol and minor products limonene-(1,2)-epoxide and minor products
Müller *et al.*, 2006	9 filamentous fungi	citral	citronellal
Onken and Berger, 1999	*Pleurotus sapidus*	limonene	*cis/trans*-carveol and carvone
Rongmin *et al.*, 2001	Laccases of *Botrytis cinerea* and *Trametes versicolor*	valencene	nootkatone
Tan and Day, 1998	*Penicillium digitatum*	(*R*)-(+)-limonene	*(R)-(+)-α-terpineol*

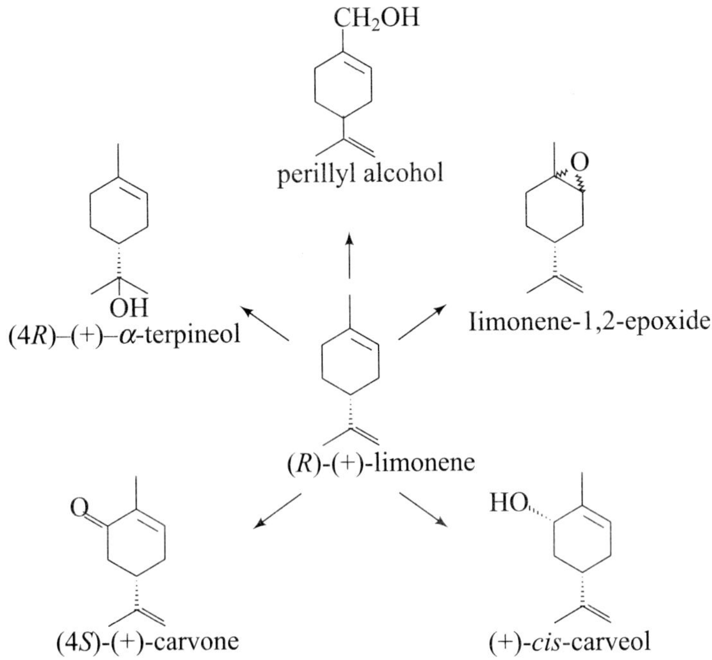

Figure 9.1 Products obtained by biotransformations of (*R*)-(+)-limonene with fungi.

Fungal mycelia of *P. digitatum*, immobilised in calcium alginate beads, produced 12.83 mg/(g_{beads}.day) of (4*R*)-(+)-α-terpineol at a concentration of dissolved oxygen of 50.0 μmol/L, in a batch system (Tan and Day, 1998). A 2.6-fold increase in productivity, compared to the control used, was achieved in a continuous reactor at an aeration rate of 0.3 L/min and a dilution rate of 0.014 h^{-1}. However, as referred by the authors, the production cost of α-terpineol using this bioconversion is above US$35. This was higher than the selling price of α-terpineol, which has a lilac odour and is one of the most used fragrance compounds.

P. citrinum also converted (*R*)-(+)-limonene to α-terpineol with a regioselectivity of 83% and a product yield of 80 mg/L, on a shake flask (Kaspera *et al.*, 2005a). On a laboratory scale fed-batch reactor, the same conversion was carried out with *P. digitatum*, yielding 506 mg/L. Other fungi converted limonene to different compounds (Table 9.1). *Pleurotus sapidus* converted limonene in *cis*- and *trans*-carveol and carvone (Onken and Berger, 1999) whilst *Fusarium proliferatum* converted (*R*)-(+)-limonene to *cis*-(+)-carveol and (*S*)-(-)-limonene to *trans*-(-)-carveol (Kaspera *et al.*, 2005a). *Piceas abies* oxidised (*R*)-(+)-limonene to limonene-(1,2)-epoxide and minor products (Lindmark-Henriksson *et al.*, 2004) and *Gongronella butleri* produced the valuable perillyl alcohol (Kaspera *et al.*, 2005a).

A. niger was used in hydroxylation reactions of terpenes (de Oliveira *et al.*, 1999) and according to Hashimoto *et al.* (2001), the bioconversion of terpenoid pharmaceuticals by *A. niger* and mammalian systems is similar. The same fungus was used by Chen and Reese (2002) for the bioconversion of terpenes from the plant *Stemodia maritima*. The bioconversions performed in 500 ml erlenmeyers containing 125 ml liquid medium, yielded hydroxylated products from stemodin and four novel compounds from stemarin, while stemodinone was converted into two analogues.

Sporulated surface cultures of *P. digitatum* transformed the pure terpene alcohols geraniol and nerol, a mixture of both, and a mixture of the corresponding aldehydes to 6-methyl-5-hepten-2-one (Demyttenaere and Kimpe, 2001). The mixture of aldehydes was converted faster than the alcohols. However, the yield of the biotransformation of the aldehydes was 76% whilst for alcohols was 83%. A spore suspension of *P. digitatum* was able to carry out the conversion of the mixture of alcohols to the same product, a maximum conversion of 43.7% being achieved after 10 days.

The distinct aroma of the so-called "noble" wine has been found to be the result of the biotransformation of terpenoids in the grapes into volatile compounds by the fungus *Botrytis cinerea*, which attacks the grape skin (Aleu and Collado, 2001). The *Botrytis* species as polyphage plant parasites produce toxins and wall-degrading enzymes and can also survive as saprophytes, decomposing plant tissue (Coley-Smith *et al.*, 1980). Laccases of *B. Cinerea* and *Trametes versicolor* can convert valencene to nootkatone, an important flavour compound of grapefruit (Rongmin *et al.*, 2001; Linke *et al.*, 2005). Laccases have been employed both in the removal of phenolic compounds, responsible for browning, astringency and bitterness of fruit juices, and as an analytical tool to detect and quantify phenols of fruits and vegetables.

Symbioses between fungi and bark beetles to overcome resistance mechanisms of trees have been described. To many fungal species, transmission of spores, which are distributed laterally or internally in the digestive tract of bark beetles while they construct breeding chambers and galleries in host trees, is vital. The odour emanated from fungi can be used to attract associated insects (Fäldt *et al.*, 1999). In several conifers, accumulation of monoterpenes at high concentrations is a defensive response to inoculation with fungi. Among them are α- and β-pinene, myrcene, terpinolene, sabinene, Δ-3-carene and limonene (Viiri *et al.*, 2001). Isolation of fungi able to resist and grow on trees in the presence of high concentrations of those compounds may be a good source of biocatalysts for terpene conversions.

Stemphylium botryosum converted γ-terpinene to *p*-mentha-1,4-dien-9-ol, which could be further oxidised to the aromatic alcohol *p*-cymene-9-ol with an enantiomeric excess of 74 and 70%, respectively (Krings *et al.*, 2005). These reactions occurred by enzymatic introduction of oxygen at non-activated carbons into a molecule with allylic and double-allylic positions, showing the advantage of biotransformations over chemical reactions.

Production of Chiral Building Blocks

In most cases, biotransformations are enantiospecific, resulting in chiral products that may be used as chiral building blocks. Furthermore, in the majority of the cases, there is no need to protect functional groups and it is possible to carry out biotransformations at centres that are chemically unreactive.

Sesqui- and diterpenes can be functionalised through microbial hydroxylations of terpenoid compounds. Hydroxylated drimanic compounds have been particularly studied because they present a wide range of biological activities and possible industrial applications. Hydroxylation at 3β- and/or 6α-positions were obtained from the hydroxylation of drimenol and related compounds by *M. plumbeus* and *Rhizopus arrhizus* (Aranda *et al.*, 1992). *M. plumbeus* also introduced a 2α-hydroxyl group in larixol and derived compounds (Herlem *et al.*, 1996). The resulting 2α-hydroxylated compounds can be used in the hemisynthesis of forskolin type compounds (forskolin has been used in the medical field for the treatment of allergies, cardiovascular diseases, glaucoma and recently has become a popular fat loss agent). Several fungal cultures were able to 3β-hydroxylate confertifolin, *A. niger* allowing the best yield at 80-90% (Aranda *et al.*, 2001). A further dehydration/oxidation reaction sequence resulted in a functionalisation transfer to position 1α, which is a difficult hydroxylation position.

Epoxidation of the vinyl group is the main reaction carried out by the fungus *Mucor plumbeus* on the diterpene ribenone and the hydroxylation was carried out preferentially at C1, C6 and C12 (Fraga *et al.*, 2001).

Shiomi and coworkers (1999) isolated acyl-CoA:cholesterol acyltransferase inhibitors (expected to retard atherosclerosis), acetylcholinesterase inhibitors (likely to retard Alzheimer's disease) and protein farnesyltransferase inhibitors (could have some role in cancer chemotheraphy) from fungi. They are all biosynthesised from terpene and polyketide, and may be called meroterpenoids.

The sesquiterpenes aristolane and nardosinane are rather rare in nature. They are present only in some terrestrial plants and marine organisms, but have also been isolated from the fruiting bodies of *Russula lepida* (Vidari *et al.*, 1998). Extracts of the basidiomycete *Bovista* sp. also revealed several terpenoids with cytotoxic activities, among which were the novel secoprotoilludane, illudane and drimane sesquiterpenes and an illudane-illudane bis-sesquiterpene bovistol (Rasser *et al.*, 2002). Basidiomycetes appear to be a good source of rare sesquiterpenes that may be used as bioactive compounds. Collybolides, which are an unusual class of sesquiterpenes, have only been isolated from the basidiomycete strains *Collybia maculata* and *C. peronata* (Castronovo *et al.*, 2001). Fraga (2005) wrote a review on natural sesquiterpenes that includes microbiological transformations.

Past and Future Perspectives

In the last decade, 18.2% of the work describing fungal production/transformation of terpenes was performed using agar plates. Another 18.2% of the studies were carried out in fermenters and stirred tank reactors. However, the large majority of the papers, 63.6% of all published work in this field, described work carried out in 100 ml–5L shaken flasks. Small-scale fermentations and biotransformations were performed throughout the last decade, whilst experiments with volumes above 1000 ml were carried out only in the last 5 years (Fig. 9.2).

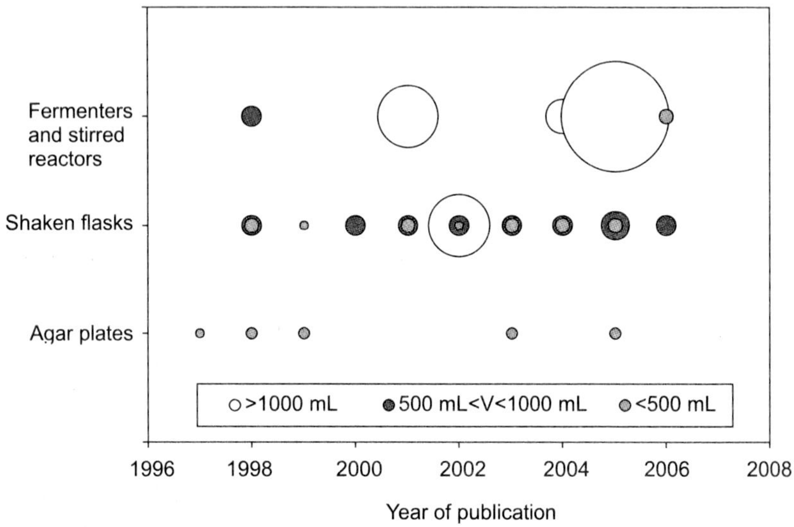

Figure 9.2 Type and volume of reactor used in the work described in published papers in 1996-2006. The volume of the reactor is represented by the diameter of the bubbles.

Petri dishes with agar media are good reactors for carrying out reactions involving fungi and terpenes, allowing a fast and reliable way of acquiring data (Oda and Ohta, 1997; Fischer *et al.*, 1999). Solid-state fermentation processes present several advantages over submerged cultures, such as high productivities and stability of products at low costs (Hölker and Lenz, 2005). Although they are difficult to scale-up, solid-state fermentations are used in several commercial processes at large scale (Chisti, 1999) and the food industry uses this type of fermentations with mixed cultures that are important to the flavour of the foods and beverages produced (Hölker and Lenz, 2005).

The commercial feasibility of a terpene biotransformation on large industrial scale depends on the response of the fungus to terpene toxicity (Sikkema et al., 1995; Schäfer *et al.*, 2004). One of the responses is the biotransformation of the toxic compound into a less toxic substance or its metabolism. The concentration of terpene in the reactor should thus be between the concentration needed to promote the biotransformation and the concentration responsible for loss of cell viability.

The cytotoxicity of terpenes and terpenoids together with low water solubility and high volatility, difficult their production and biotransformation by conventional processes. Biotechnologically produced flavours and fragrances are rare at the industrial scale. Nevertheless, fungi have been used since ancient times in solid-state fermentations in the production of food and beverages, mainly in the Far East Asia (Couto and Sanromán, 2006). One of the great advantages of these fermentations is that rich nutrient wastes (e.g. peanut shells) from agriculture and food industry may be used as cheap raw materials for the production of valuable compounds (Zorn *et al.*, 2005). However, since heterogeneous organic growth substrates are used, production of a single product is hardly achieved by solid-state fermentation, making downstream processing complicated and expensive. Furthermore, scale-up is difficult and on-line measurement and control of variables such as pH, temperature, moisture content, biomass and oxygen transfer still require the development of appropriate probes (Durand, 2003; Couto and Sanrománm, 2006).

The natural flavours produced by microorganisms should be at least five times cheaper than natural flavours extracted from plants to increase their importance and interest to the flavour industry (Nandakumara *et al.*, 2003). To improve process yields, the design of more appropriate bioreactors, in parallel to the screening of appropriate culture conditions for *de novo* synthesis of terpenes and/or good biotransformation media, must be achieved. The study of biosynthetic pathways involved in the production or biotransformation of terpenes in fungi could help, by genetic and metabolic engineering, the overexpression of the enzymes needed. However, processes using genetically modified organisms are difficult to implement due to consumer apprehension and lack of proper legislation (Nandakumara *et al.*, 2003). Nevertheless, the yields obtained with wild strains suggest that, with the suitable technology, economically viable production of terpenoid and terpene compounds will be achieved in a near future. Krings and Berger (1998) estimated the break-even price of a hypothetical flavour, assuming a production of 1g/L and an annual depreciation calculated for an amortisation period of 5 years with no discount rate. By increasing the production hypothetically from 1000 kg/year to 10 000 and 100 000 kg/year, the break-even price decreases from US$1240/kg to $300/kg and $202/kg, respectively.

CONCLUSION

Fungi are promising biocatalysts for the production of fragrances, flavours and building blocks of pharmaceutical compounds. Nevertheless, the product yields obtained so far by *de novo* synthesis and biotransformation processes are low and the flavour industry still uses mainly extraction processes from essential oils of plants and chemical routes. Further research is required to the application of biotransformation of terpenes at industrial scale. However, the benefit of producing natural substances in an increasingly health-aware market may help the development of profitable and sound biotechnological processes in a near future.

References

Adams, A., Demyttenaere, J.C.R., and De Kimpe, N. (2003). Biotransformation of (*R*)-(+)- and (*S*) ( ) limonene to α-terpineol, *Food Chemistry* 80: 525–534.

Aleu, J., and Collado, I.G. (2001). Biotransformations by *Botrytis* species, *Journal of Molecular Catalysis B: Enzymatic* 13(1-3): 77–93.

Aranda, G., Facon, I., Lallemand, J.Y., Leclaire, M., Azerad, R., Cortes, M., Lopez, J., and Ramirez, H. (1992). Microbiological hydroxylation in the drimane series, *Tetrahedron Letters* 33 (51): 7845–7848.

Aranda, G., Moreno, L., Cortés, M., Prangé, T., Maurs, M., and Azerad, R. (2001). A new example of 1α-hydroxylation of drimanic terpenes through combined microbial and chemical processes, *Tetrahedron* 57(28): 6051–6056.

Berg, A., Wangun, H.V.K., Nkengfack, A. E., and Schlegel, B. (2004). Lignoren, a new sesquiterpenoid metabolite from *Trichoderma lignorum* HKI 0257, *Journal of Basic Microbiology* 44 (4): 317–319.

Brown, L.M., Springer, J., and Bower, M. (1992). Chemical substitution for 1,1,1-trichloroethane and methanol in an industrial cleaning operation, *Journal of Hazardous Materials* 29(2): 179–188.

Bull, A.T., Bunch, A.W., and Robinson, G.K. (1999). Biocatalysts for clean industrial products and processes, *Current Opinion in Microbiology* 2(3): 246–251.

de Carvalho, C.C.C.R., and da Fonseca, M.M.R. (2006a). Carvone: Why and how should one bother to produce this terpene, *Food Chemistry* 95(3): 413–422.

de Carvalho, C.C.C.R., and da Fonseca, M.M.R. (2006b). Biotransformation of terpenes, *Biotechnology Advances* 24(2): 134–142.

Castronovo, F., Clericuzio, M., Toma, L., and Vidari, G. (2001). Fungal metabolites. Part 45: The sesquiterpenes of *Collybia maculata* and *Collybia peronata*, *Tetrahedron* 57(14): 2791–2798.

Chen, A.R.M., and Reese, P.B. (2002). Biotransformation of terpenes from *Stemodia maritima* by *Aspergillus niger* ATCC 9142. *Phytochemistry* 59(1): 57–62.

Chisti, Y. (1999). Solid substrate fermentations, enzyme production, food enrichment. In: Encyclopedia of Bioprocess Technology: Fermentation, Biocatalysis, and Bioseparation, vol. 5, Flickinger M.C. and Drew S.W. Eds, Wiley, New York, 2446–2462.

Cole, R.J., and Schweikert, M.A. (2003). Handbook of secondary fungal metabolites (Vol. I), Academic Press, New York, 1–988.

Coley-Smith, J.R., Verhoeff, K., and Jarvis, W.R. (1980). The Biology of *Botrytis*, Academic Press, London, 153–175.

Couto, S.R., and Sanromán, M.A. (2006). Application of solid-state fermentation to food industry–A review, *Journal of Food Engineering* 76(3): 291–302.

David, J.P., dos Santos, I.D., and David, J.M. (2004). A new sesquiterpene from the fruits of *Allophylus laevigatus*, *Fitoterapia* 75(7-8): 795–798.

Demyttenaere, J., and De Kimpe, N. (2001). Biotransformation of terpenes by fungi: Study of the pathways involved, *Journal of Molecular Catalysis B: Enzymatic* 11(4-6): 265–270.

de Oliveira, B.H., dos Santos, M.C., and Leal, P.C. (1999). Biotransformation of the diterpenoid isosteviol by *Aspergillus niger*, *Penicillium chrysogenum* and *Rhizopus arrhizus*, *Phytochemistry* 51(6): 737–741.

Duetz, W.A., Fjällman, A.H.M., Ren, S., Jourdat, C., and Witholt, B. (2001). Biotransformation of D-limonene to (+)-*trans*-carveol by toluene-grown *Rhodococcus opacus* PWD4 cells, *Applied and Environmental Microbiology* 67(6): 2829-2832.

Durand, A. (2003). Bioreactor designs for solid-state fermentation, *Biochemical Engineering Journal* 13(2-3): 113–125.

Fichan, I., Larroche, C., and Gros, J.B. (1999). Water solubility, vapor pressure, and activity coefficients of terpenes and terpenoids, *Journal of Chemical & Engineering Data* 44(1): 56–62.

Fäldt, J., Jonsell, M., Nordlander, G., and Borg-Karlson, A.K. (1999). Volatiles of bracket fungi *Fomitopsis pinicola* and *Fomes fomentarius* and their functions as insect attractants, *Journal of Chemical Ecology* 25(3): 567–590.

Fischer, G., Schwalbe, R., Möller, M., Ostrowski, R., and Dott, W. (1999). Species-specific production of microbial volatile organic compounds (MVOC) by airborne fungi from a compost facility, *Chemosphere* 39(5): 795–810.

Fraga, B.M. (2005). Natural sesquiterpenoids. *Natural Products Report* 22(4): 465-486.

Fraga, B.M., Hernández, M.G., González, P., López, M., and Suárez, S. (2001). Biotransformation of the diterpene ribenone by *Mucor plumbeus*, *Tetrahedron* 57(4): 761–770.

Hashimoto, T., Noma, Y., and Asakawa, Y. (2001). Biotransformation of terpenoids from the crude drugs and animal origin by microorganisms, *Heterocycles* 54(1): 529–559.

Herlem, D., Ouazzani, J., and Khuong-Huu, F. (1996). Chemistry of larixol I- Degradation of the side-chain and microbial hydroxylation, *Tetrahedron Letters* 37(8): 1241–1244.

Hölker, U, and Lenz, J. (2005). Solid-state fermentation — are there any biotechnological advantages?, *Current Opinion in Microbiology* 8(3): 301–306.

ITC (2006). World markets in the spice trade 2000-2004, International Trade Centre UNCTAD/WTO, Geneva, vi, 111p.

Jiao, W., Blunt, J.W., Cole, A.L.J. and Munro, M.H.G. (2004). Fumagiringillin, a new fumagillin derivative from a strain of the fungus *Aspergillus fumigatus*, *Journal of Natural Products* 67(8): 1434–1437.

Jirage, K.B., and Martin, C.R. (1999). New developments in membrane-based separations, *Trends in Biotechnology* 17(5): 197–200.

Kaspera, R., Krings, U., Pescheck, M., Sell, D., Schrader, J., and Berger, R.G. (2005a). Regio- and stereoselective fungal oxyfunctionalisation of limonenes. *Zeitschrift für Naturforschung* 60: 459–466.

Kaspera, R., Krings, U., Nanzad, T., and Berger, R.G. (2005b). Bioconversion of (+)-valencene in submerged cultures of the ascomycete *Chaetomium globosum*, *Applied Microbiology and Biotechnology* 67(4): 477–483.

Kikuchi, H., Miyagawa, Y., Sahashi, Y., Inatomi, S., Haganuma, A., Nakahata, N. and Oshima Y. (2004a). Novel trichothecanes, paecilomycine A, B, and C, isolated from entomopathogenic fungus, *Paecilomyces tenuipes*, *Tetrahedron Letters* 45(33): 6225–6228.

Kikuchi, H., Miyagawa, Y., Nakamura, K., Sahashi, Y., Inatomi, S., and Oshima, Y. (2004b). A novel carbon skeletal trichothecane, tenuipesine A, isolated from an entomopathogenic fungus, *Paecilomyces tenuipes*, *Organic Letters* 6(24): 4531–4533.

Kikuchi, H., Miyagawa, Y., Sahashi, Y., Inatomi, S., Haganuma, A., Nakahata, N., and Oshima, Y. (2004c). Novel spirocyclic trichothecanes, spirotenuipesine A and B, isolated from entomopathogenic fungus, *Paecilomyces tenuipes*, *Journal of Organic Chemistry* 69(2): 352–356.

Kim, W.G., Kim, J.W., Ryoo, I.J., Kim, J.P., Kim, Y.H., and Yoo, I.D. (2004). Boletunones A and B, highly functionalized novel sesquiterpenes from *Boletus calopus*, *Organic Letters* 6(5): 823–826.

Kirchner, EM. (1994). Environment, health concerns force shift in use of organic solvents. *Chemical and Engineering News* 72: 13–20.

Krings, U., and Berger, R.G. (1998). Biotechnological production of flavours and fragrances, *Applied Microbiology and Biotechnology* 49(1): 1–8.

Krings, U., Brauer, B., Kaspera, R., and Berger, R.G. (2005). Biotransformation of γ-terpinene using *Stemphylium botryosum* (Wallroth) yields *p*-mentha-1,4-dien-9-ol, a novel odorous monoterpenol, *Biocatalysis and Biotransformation* 23(6): 457–463.

Langenheim, JH. (1994). Higher plant terpenoids: a phytocentric overview of their ecological roles, *Journal of Chemical Ecology* 20(6): 1223–1280.

Lindmark-Henriksson, M., Isaksson, D., Vanìk, T., Valterová, I., Högberg, H.E., and Sjödin K., (2004). Transformation of terpenes using a *Picea abies* suspension culture, *Journal of Biotechnology* 107(2): 173–184.

Linke, D., Bouws, H., Peters, T., Nimtz, M., Berger, R.G., and Zorn, H. (2005). Laccases of *Pleurotus sapidus*: characterization and cloning, *Journal of Agricultural and Food Chemistry* 53(24): 9498–9505.

Larsen, T.O. (1998). Volatile flavour production by *Penicillium caseifulvum*, *International Dairy Journal* 8: 883–887.

Larsen, T.O., and Frisvad, J.C. (1995a). Comparison of different methods for collection of volatile chemical markers from fungi, *Journal of Microbiological Methods* 24(2): 135–144.

Lilly, M.D. (1994). Advances in biotransformation processes, *Chemical Engineering Science* 49(2), 151–159.

Müller, A., Hauer, B., and Rosche, B. (2006). Enzymatic reduction of the α, β-unsaturated carbon bond in citral, *Journal of Molecular Catalysis B: Enzymatic* 38(3-6): 126–130.

Nandakumara, R., Yoshimune, K., Wakayama, M., and Moriguchi, M. (2003). Microbial glutaminase: biochemistry, molecular approaches and applications in the food industry, *Journal of Molecular Catalysis B: Enzymatic* 23(2-6): 87–100.

Onken, J., Berger, R.G. (1999). Effects of *R*-(+)-limonene on submerged cultures of the terpene transforming basidiomycete *Pleurotus sapidus*, *Journal of Biotechnology* 69(2,3): 163–168.

Oda, S., and Ohta, H. (1997). Double coupling of acetyl coenzyme A production and microbial esterification with alcohol acetyltransferase in an interface bioreactor, *Journal of Fermentation and Bioengineering* 83(5): 423–428.

Rasser, F., Anke, T., and Sterner, O. (2002). Terpenoids from *Bovista* sp. 96042, *Tetrahedron* 58(39): 7785–7789.

Reina, M., Orihuela, J.C., González-Coloma, A., de Inés, C., de la Cruz, M., González del Val A., Torno, J.R., and Fraga, B.M. (2004). Four illudane sesquiterpenes from *Coprinopsis episcopalis*, *Phytochemistry* 65(4), 381–385.

Rongmin, H., Christenson, P.A., and Labuda, I.M. (2001). Prodution of natural flavors by laccase catalysis. European Patent EP 1 083 233 A1.

Schäfer, S., Schrader, J., and Sell, D. (2004). Oxygen uptake rate measurements to monitor the activity of terpene transforming fungi, *Process Biochemistry* 39(12): 2221–2228.

Seephonkai, P., Isaka, M., Kittakoop, P., Boonudomlap, U., and Thebtaranonth, Y. (2004). A novel ascochlorin glycoside from the insect pathogenic fungus *Verticillium hemipterigenum* BCC 2370, *The Journal of Antibiotics* 57(1), 10–16.

Shiomi, K., Tomoda, H., Otoguro, K., and Omura, S. (1999). Meroterpenoids with various biological activities produced by fungi, *Pure and Applied Chemistry* 71(6): 1059–1064.

Sikkema, J., Bont de, J.A.M., and Poolman, B. (1995). Mechanisms of membrane toxicity of hydrocarbons, *Microbiological Reviews* 59(2): 201–22.

Tan, Q., and Day, D.F. (1998a). Bioconversion of limonene to α-terpineol by immobilized *Penicillium digitatum*, *Applied Microbiology and Biotechnology* 49(1): 96–101.

Vazquez, M.J., Vega, A., Rivera-Sagredo, A., Jimenez-Alfaro, M.D., Diez, E., and Hueso-Rodríguez J.A. (2004). Novel sesquiterpenoids as tyrosine kinase inhibitors produced by *Stachybotrys chortarum*, *Tetrahedron* 60(10): 2379–2385.

Vidari, G., Che, Z., and Garlaschelli, L. (1998). New nardosinane and aristolane sesquiterpenes from the fruiting bodies of *Russula lepida*, *Tetrahedron Letters* 39(33): 6073–6076.

Viiri, H., Annila, E., Kitunen, V., and Niemelä, P. (2001). Induced responses in stilbenes and terpenes in fertilized Norway spruce after inoculation with blue-stain fungus, *Ceratocystis polonica*, *Trees* 15: 112–122.

Walker, J.M., and Cox, M. (1995). The language of biotechnology - A dictionary of terms (2nd Edition), Amer Chemical Society, USA, 300 p.

Walton, N.J., Narbad, A., Faulds, C.B., and Williamson, G. (2000). Novel approaches to the biosynthesis of vanillin, *Current Opinion in Biotechnology* 11(5):490–496.

Wu, S., Krings, U., Zorn, H., and Berger, R.G. (2005). Volatile compounds from the fruiting bodies of beefsteak fungus *Fistulina hepatica* (Schaeffer: Fr.) Fr., *Food Chemistry* 92(2): 221–226.

Xu, X., Yan, M., and Zhu, Y. (2005). Influence of fungal fermentation on the development of volatile compounds in the pure tea manufacturing process, *Engineering in Life Sciences* 5(4): 382–386.

Zorn, H., Peters, T., Nimtz, M., and Berger, R.G. (2005). The secretome of *Pleurotus sapidus*, *Proteomics* 5(18): 4832–4838.

10

Molecular and Physiological Aspects of Relevance in the Design of Industrial Strains of Lactose/Galactose—Metabolising Yeasts for Biotechnological Applications

Marta M. Rubio-Texeira

Whitehead Institute, Massachusetts Institute of Technology
9 Cambridge Center—653, Cambridge, MA 02142,
Tel: 00-1-617 258-5190; Fax: 00-1-617-258-7226; E-mail: rubio@wi.mit.edu

INTRODUCTION

For centuries before their discovery, numerous microorganisms were already playing an essential role in the manufacturing and processing of a variety of human foods, which are staple in our diet (e.g., bakery products, alcoholic beverages, and milk fermented products). The earliest reports in the history of the utilisation of yeasts in food manufacturing date back to 7000–4000 B.C. Evidence has been found that, by that time, fermented beverages such as beer were already being produced in places like ancient Sumeria and Egypt (De Winde, 2003). Manufacturing of dairy products such as yogurt, kefir, and cheese, which imply the use of microorganisms resident in milk, has also been carried out in Eurasia for millennia (Kurtzman and Fell, 1998; Rubio-Texeira, 2007). Although the presence of microscopic entities in our food and their actual requirement for the production of fermented products was far from known (let alone understood) until the period between 17th–19th centuries, their utilisation has, in fact, co-evolved with the development of agriculture and livestock technologies. In reality, as it has happened for certain plants and animals, a number of species of yeasts and other microorganisms were inadvertently selected some time back in our ancient past and furtherly adapted for growth, over thousands of years, in particular environmental conditions controlled by humans, in a process which we commonly know as "domestication" (Fay and Benavides, 2005).

Milk, one of such particular ecological niches, is a rich medium, able to sustain growth of a complex mixture of microorganisms. Among these, several species of yeasts from the genus *Kluyveromyces* have been found to play a major role in the process of fermentation of milk. Certain *Kluyveromyces* yeasts can live in milk because of their particular ability to metabolise milks' main sugar lactose. Recent efforts have been focused in transferring this trait to *Saccharomyces cerevisiae*, the alcohol-producer microorganism by excellence. The maximum goal for this industrial sector is to combine the best characteristics of both yeasts into one by means of the newest technologies of genetic and metabolic engineering that science has recently made available. Although several other microbes also have the ability to metabolise galactose and lactose, yeasts of *Kluyveromyces* and *Saccharomyces* genera are particularly convenient for the pharmaceutical and food industries because of their 'generally regarded as safe' (GRAS) status, officially approved not long ago by the United States Food and Drug Administration (Bonekamp and Oosterom, 1994). The versatility inherent to the use of these yeasts has definitely pushed forward their characterization, especially in the last 50 years, both at the molecular and the biotechnological levels. Given their importance for industrial biotechnology, the current chapter will now be focused on the review of the latest knowledge gathered about the most important physiological differences between these two yeasts and the most recent technologies that have been developed in order to take advantage of them.

Historical Origins of *Kluyveromyces* Research and Industrial Applicability of the Lactose-/Galactose-Consuming Trait

As opposed to the widely studied *S. cerevisiae*, and in spite of the obvious interest that *K. lactis* has long had for industry and biotechnology, attempts to research this latter yeast did not pick a real interest within the scientific community until the 1960s. Even then, such research was primarily motivated by the growing academic interest in studies of adaptive genetic and enzymatic regulation of sugar metabolism (Barnett, 2004; Barnett and Entian, 2005; Fukuhara, 2006). Pioneer studies in the genetic and molecular basis of lactose regulation in *Escherichia coli* were done at that time that would later constitute the basis for the development of molecular biology as a new methodological science. Following the achievements of the investigators in the *E. coli* system, a number of laboratories aimed to advance a further step by redirecting their interest to the search and characterization of a similarly inducible enzymatic system in lower eukaryotes. One of these groups was led by Harlyn O. Halvorson (University of Wisconsin, Madison), who sought advice from the taxonomist Lynferd J. Wickerham (U.S. Department of Agriculture, Peoria, Illinois). Wickerham recommended the use of *K. lactis*, the yeast that assimilated β-glucosides in an adaptive mode. Halvorson and colleagues obtained two different isolates from Peoria, NRRLY-1140 (CBS 2359 [Mat a] and Y-1118 (CBS 6315 [Mat a]) (see Table 10.1 for more details about the properties of different laboratory isolates used in *K. lactis* research). The 'paraconstitutive' (half-inducible, half-constitutive) nature of lactose regulation made it difficult at the time for this group to continue their research of this regulatory system, so after the pioneer studies of this group only a few laboratories continued to use *K. lactis*, mostly in comparative studies with the *S. cerevisiae* system in the field of mitochondrial genetics and biogenesis.

Table 10.1 *K. lactis* strains most commonly used for laboratory studies and their relative sensitivity to glucose repression.[1]

Strain	*Glucose transporter gene/s*	*Glucose repression*	*References/other aliases*
CBS 6315	*KHT1-KHT2*	Strong	NRRL Y-1118/ATCC 8563
JA6[2]	*KHT1-KHT2*	Strongly	ATCC 96263
CBS 2359	*RAG1*	Weakly	NRRL Y-1140/ATCC 8585
CBS 2360	*rag1⁻*	Non-sensitive	NRRL Y-1205/ATCC 8651

[1] Data from Breunig *et al.*, 2000, and Zachariae *et al.*, 1993.

[2] This strain carries a two-base pair alteration in the *KlGAL4* promoter (*LAC9-2*) that results in slightly reduced KlGal4p levels because of a reduced binding of the transactivator to its own promoter. This additional effect further enhances the sensitivity of this strain to glucose.

It was not until the early 1980s, the period in which the first studies in genetic engineering and biotechnology started to emerge, that a new interest in research into this yeast was reinitiated, this time because of the simultaneous influence of three unrelated topics: first, the discovery in *K. lactis* of the killer system based on linear DNA plasmids; second, the studies of regulated lactose metabolism; and third, the secretion of recombinant proteins. The discovery of the killer system (reviewed by Wésolowski-Louvel *et al.*, 1996) allowed the development of novel techniques for the genetic manipulation of *K. lactis*, which opened new perspectives for the use of molecular technologies to introduce and modify genes in yeasts other than the more conventional *S. cerevisiae*. At the same time, studies of this phenomenon revealed the fact that *K. lactis* was able to secrete a high-molecular weight killer toxin, which attracted the attention of laboratories looking for an efficient system to produce recombinant proteins in a secreted form. Moreover, an increasing understanding of the regulation of the lactose regulon of *K. lactis*, mostly derived from basic research carried out in the laboratories of R. C. Dickson (University of Kentucky, USA) and K.D. Breunig (Institut für Genetik, Martin-Luther-Universität Halle-Wittenberg, Halle, Germany), started to unveil the possibility of using this regulatory system in recombinant technologies for the production of heterologous proteins. Studies carried out by both groups revealed that the regulation of genes involved in the metabolisation of lactose in eukaryotes markedly differs from the bacterial lactose operon. *Kluyveromyces LAC/GAL* regulon showed a more complex mechanism that involved many genetically unlinked positive and negative regulatory genes. Simultaneous comparative analysis with the galactose regulon of the evolutionarily related yeast *S. cerevisiae*, also demonstrated that both genetic systems are very similarly structured but that the evolution of both yeasts in different environments (basically, fruits and its own fermentation products for *S. cerevisiae*, as oppossed to milk and dairy products for *K. lactis*) has led to remarkable differences in the way in which these genes are regulated by carbon sources (Johnston, 1987; Lohr *et al.*, 1995; Rubio-Texeira, 2005).

Even today, as we will have the opportunity to review in the next sections of this chapter, parallel studies of the lactose/galactose regulons of *K. lactis* and *S. cerevisiae* are still providing invaluable information to the scientific/industrial communities at various different levels. In the first place, molecular characterization of this genetic system has improved our general understanding of how interconnected gene regulatory networks respond to external stimuli, such as nutrients, and how cells make adaptations accordingly. The characterization of this system in yeasts has not only helped to elucidate the molecular basis of human genetic disorders directly related to the metabolism of lactose and galactose (e.g., lactose intolerance, and galactosemia; extensively reviewed in Swallow, 2003; Adam *et al.*, 2004), but has also helped to create therapies for their treatment, many of which are based on the utilisation of microbially-produced and purified β-galactosidases. β-galactosidases are also extensively used in the generation of galactooligosaccharides (GOS), which have been identified as essential dietary components for the proper development of intestinal flora, especially in young children (Szilagyi, 2002). Identification of the main regulators of this genetic circuit has also allowed the development of powerful molecular biology tools and techniques, such as the two-hybrid system, and the use of β-galactosidases as efficient genetic reporters to monitor gene expression (Fields and Song, 1989; Amberg *et al.*, 2005). The inducible character and on/off switching capability of the system in response to different carbon sources has provided an important genetic tool: the use of galactose-inducible promoters for homologous and heterologous gene expression (Romanos *et al.*, 1992; van Ooyen *et al.*, 2006).

The accumulation of all of this knowledge has also enabled the generation of multiple combinations of genetically modified versions of *K. lactis* and *S. cerevisiae* yeast strains with the aim to develop highly-efficient industrial systems that eliminate lactose (which, as we will explain below, is a major source of environmental pollution) from by-products resulting from the manufacturing of cheese and other dairy products (Moulin and Galzy, 1984; Mawson, 1994; Yang and Silva, 1995; Siso, 1996). One of the most obvious ways to couple lactose fermentation to an economically-rewarding methodology for dairy industries is by constructing yeasts that are able to generate products of commercial interest out of the process. Hence, the interest recently developed in adapting lactose-consuming yeasts for the synthesis of certain recombinant proteins of pharmaceutical and biotechnological interest as the main by-product of lactose metabolism (reviewed in Rubio-Texeira, 2006). This particular aspect is perhaps one of the main points in which a deeper molecular understanding of most, if not all, of the genetic components playing a role in the cellular pathways involved in the metabolisation of the carbon source for the production of heterologous proteins, will in the near future enable a more successful generation of efficient strains through metabolic engineering. In this respect, as we will see in further sections, this chapter not only discusses the most relevant technologies involving the use of lactose/galactose-metabolising yeasts but also proposes new alternatives which the latest research at the molecular level is now unveiling.

Divergent Evolution and Acquisition of Specific Traits by Yeasts of the Genus *Kluyveromyces* and *Saccharomyces*

The kingdom Fungi is conformed by six separated phyla (Wikipedia, the Free Encyclopedia, http://en.wikipedia.org). Each of these phyla consists of a high variety of eukaryotic organisms, adapted to multiple ecosystems. 'True yeasts', fungi that live in a predominantly unicellular life cycle, mostly belong to the phylum Ascomycota (fungi which form their spores in a sac-like *ascus*). This phylum is composed of three major subphyla: Pezyzomicotina (e.g., fungi that reproduce by fission rather than budding, including almost all the ascus fungi which have fruiting bodies visible to the naked eye, such as *Neurospora crassa*), Taphrinomycotina (diverse group considered, by molecular analysis, to be more primitive and which includes, among others, fission yeasts like *Schizosaccharomyces pombe*), and Saccharomycotina (which contains most of the 'true yeasts', formerly known as Hemiascomycetes phylum). *S. cerevisiae* and *K. lactis*, therefore, belong to this latter phylum. By the time this phylum was described as Hemiascomycetes it was already a highly diverse group of ascomycetes comprising most of the known yeast species, more than 1200 species (Kurtzman and Fell, 1998; Boekhout, 2005). The Saccharomycotina yeasts are believed to have diverged from a common ancestral fungus at least 400 million years ago.

Advances in molecular techniques have allowed siginificant progress in studies related to the origin of 'true yeasts'. The first completed genome sequence of a 'true yeast' was that of *S. cerevisiae*, which has been available since 1996 (Bassett *et al.*, 1996; Butler, 1996). The *S. cerevisiae* genome contains more than 30% of duplicated genes. This finding led to the, initially controversial and later confirmed, hypothesis that *S. cerevisiae* is a paleopolyploid, i.e. the entire genome of an ancestral yeast became duplicated at some point in evolution (whole-genome duplication –WGD-hypothesis; Wolfe and Shields, 1997) and that subsequent events of rearrangement, gene loss, and speciation of gene paralogues, forced by adaptation to novel ecological niches, gave birth to what is today known as the *Saccharomyces sensus stricto* and *sensus lato* species. Analysis of the complete genome sequences of Hemiascomycete species closely related to *S. cerevisiae* has dated the whole genome duplication after the emergence of *Kluyveromyces waltii* (Kellis *et al.*, 2004), and the filamentous yeast *Ashbya gossypii* (Dietrich *et al.*, 2004).

As a part of the Génolevure Project, a fungal genome sequencing initiative started by eight French research institutions, the complete genome of the haploid type of four other yeast species representing distant branches of Hemiscomycetes has very recently been sequenced (Dujon *et al.*, 2004). The selected species were: *Candida glabrata* (second causative agent of human candidiasis and phylogenetically more closely related to *S. cerevisiae* than to *C. albicans*) *K. lactis* (because of its interesting intermediate position within the phylogeny of Hemiascomycetes as well as its growing interest in biotechnology) *Debaryomyces hansenii* (halotolerant yeast related to *C. albicans* and other pathogenic yeasts, often found in fish and salted dairy products); and *Yarrowia lipolytica* (an alkane assimilating yeast, very distantly related to the rest of yeasts and with a number of common properties with filamentous fungi).

The availability of all these new genetic databases has been crucial for the latest advances in the identification of the mechanisms of eukaryotic genome evolution. These studies indicate that, the genomes of all of these yeasts are relatively small (9-20 Mb), containing around 6000 open reading frames (ORFs) similarly to *S. cerevisiae*. Within these genomes there are few introns and repetitive elements, intergenic regions are short, and only a few cases of horizontal transfer have been detected. In fact, they are highly compact genomes, containing a high gene density of about 1 gene per 2 kb (with the exception of *Y. lipolytica* which contains an average of 1 gene per 3 kb). More than 40% of protein families classified by the authors of the Génolevure project (2,014 families, representing 17,153 genes) are common to all five yeasts. The majority of these families contain a single protein from each species that are considered 'direct orthologues', and the remaining families contain at least one paralogous pair in at least one species. Approximately 98% of the universal Hemiascomycete protein families have homologues in other Ascomycetes and 92% have homologues in the Basidiomycetes, indicating that most of these proteins are largely conserved through evolution. However, the level of sequence identity between orthologous proteins ranges around 65% between *S. cerevisiae* and *C. glabrata*, 60-61% between *K. lactis* and either *S. cerevisiae* or *C. glabrata*, and even lower, 48-49% between the less related *Y. lipolytica* and the other three yeasts. This contrasts remarkably with the average protein sequence identity between mammals and fish (70%) surprisingly indicating that, within the Hemiascomycete phylum there is more divergence at the molecular level than in the entire phylum of Chordates, despite the similar morphologies and lifestyles of yeasts. The most recently supported hypothesis proposed by Dujon *et al.* (2004) to explain the evolution of the Hemiascomycetes phylum is shown in Fig. 10.1. According to this model several events of segmental duplication, formation of tandem repeats, and subsequent losses and rearrangements to keep a control in the genome sizes, have occurred at different stages in the evolution.

Studies before the sequencing of these four new genomes have already shown that *K. lactis* must have diverged from the *Saccharomyces* clade at least 150 million years ago, i.e., before the massive event of WGD that led to the appearance of *S. cerevisiae* and *C. glabrata*. A remarkable synteny (similarity in gene order) had previously been identified between the *K. lactis* and *S. cerevisiae* genomes (Ozier-Kalogeropoulos *et al.*, 1998; Keogh *et al.*, 1998). Approximately 59% of the adjacent pairs in *K. lactis* are also adjacent in *S. cerevisiae*. However, extensive gene duplications in the *K. lactis* genome are not found, despite the fact that some single-copy genes of *S. cerevisiae* have duplicates in *K. lactis*.

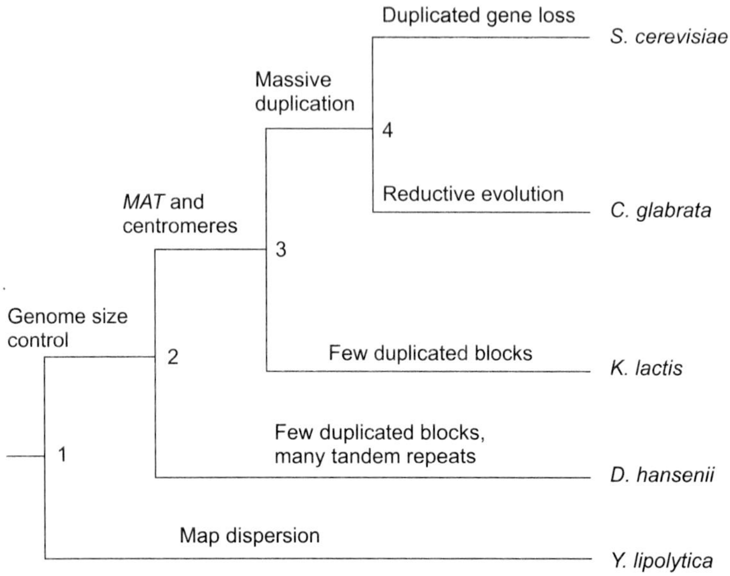

Figure 10.1 Tree topology representing the proposed major evolutionary events in hemiascomycete genomes, based on 25S rDNA sequence. This representation shows how an accidental whole-genome duplication event occurred in the common ancestor of *S. cerevisiae* and *C. glabrata* that separated the further evolution of these yeasts from that of *K. lactis*. This duplication event was followed by extensive gene loss of paralogous copies. Reproduced with permission from the reference Dujon *et al.*, 2004. (*Nature* MacMillan Magazines Ltd.).

In a study that identified 2,500 ORFs in *K. lactis*, 25 of them were found encoding proteins with no orthologues in *S. cerevisiae* (Bolotin-Fukuhara *et al.*, 2000). Most of these may represent enzymes involved in metabolic pathways allowing *K. lactis* to assimilate carbon and nitrogen sources that are not used by *S. cerevisiae*. Such functions may have evolved independently in *K. lactis* after the WGD event, or otherwise been the result of lost functions readapted to new activities through rearrangements, and/or of mutations leading to significant loss of sequence similarity with *S. cerevisiae*. In this respect, a recent study has shown that the common ancestor of modern yeasts seemed to have the entire GAL genetic pathway and that this pathway was progressively lost in several yeasts, such as *C. glabrata* or *K. waltii*, through rapid and irreversible gene inactivation events associated to the adaptation to new ecological niches (Hittinger *et al.*, 2004).

Another interesting study from Bolotin-Fukuhara and colleagues, consisted of the comparison of the repertoire of transcriptional regulators encoded by the *K. lactis* and *S. cerevisiae* genomes (Bussereau *et al.*, 2006). This analysis has shown that many duplicated or multiplied genes encoding transcriptional activators in *S.*

cerevisiae are reduced to one copy in *K. lactis*. However, when they looked at the number of proteins containing different DNA-binding motifs, this number was not reduced relative to *S. cerevisiae*. This may either mean that the transactivating domains have strongly diverged to adapt to new functions according to different evolutionary forces, or that the *K. lactis* genome simply has fewer transactivators because of its divergence from *S. cerevisiae* prior to the WGD event. In order to discern between both hypotheses, more experimental work will now be needed to characterize the function of all of these putative transactivators present in the *K. lactis* genome.

The enormous progress in molecular techniques applicable to taxonomic studies has, therefore, helped to unveil an important level of heterogeneity, not only between *K. lactis* and the relatively closely related *S. cerevisiae*, but also among isolates within this same species. When genetic studies of *Kluyveromyces lactis* (Dombrowski) van der Walt began, the species was incorrectly named *Saccharomyces lactis*, but later on *Kluyveromyces* species were separated from the *Saccharomyces* genus based on physiological criteria, like their petite-negative character, their inability to grow anaerobically, and their Crabtree (-) character (for a comparative review of all these properties, see the next section). Molecular typing techniques also led to the separation of *Kluyveromyces lactis* from *K. marxianus* (also known as *K. fragilis*) into a separate species (Vaughan-Martini *et al.*, 1993; Kurtzman and Robnett, 1998; for more information go to NEWT taxonomy browser, maintained by the Uniprot group, that integrates taxonomy data compiled in the NCBI database and data specific to the UniProt Knowledgebase: http://www.ebi.ac.uk/newt/display).

The currently most accepted subdivision of the species of *Kluyveromyces lactis* consists of two taxonomic varieties, *K. lactis* var. *drosophilarum* and *K. lactis* var. *lactis* (Lachance, 1998, 2007; Kurtzman, 2003). This subdivision is still primarily based on phenotypic and ecological distribution rather than in molecular analysis, but it helps to distinguish lactose-negative strains (the former, mainly isolated from plants and invertebrates) from the lactose-positive strains (the latter, isolated from dairy products). However, it is expected that molecular techniques will soon improve the phylogenetic analysis and lead to a more molecularly correct classification of the multiple isolates of the genus *Kluyveromyces* van der Walt amends. van der Walt still appearing in nature (Naumova *et al.*, 2004).

Main Physiological Differences between *Kluyveromyces lactis* and *Saccharomyces cerevisiae*

S. cerevisiae and *K. lactis* have different physiology in regards of their metabolisation of carbon sources. This is an interesting subject for the studies of comparative analysis because it reflects how evolution has caused similar regulatory networks to diverge for the adaptation of two related species to their respective ecological niches. Many overlapping regulatory circuits are simultaneously operating in cells that indeed have an effect in the way in which these two yeasts have evolved their different physiologies in terms of carbon utilisation. Because of spatial limitations, we will only review some of the most important physiological aspects that make these two yeasts so different. For more information on the matter, excellent reviews are being cited below that will further complete the readers overview.

Production of Ethanol Versus Generation of Biomass as Mechanisms Evolved in a Competition for Environmental Nutrients

The major physiological differences observed between these two yeasts are related to their preferential ability to catabolise a carbon source, either by fermentation or respiration (for a review, see Flores *et al.*, 2000). *S. cerevisiae* exhibits a predominantly fermentative growth in the presence of an extensive variety of sugars. Production of the relatively toxic compound ethanol is, in fact, an interesting evolutionary trait that allows yeasts to outcompete growth of other microorganisms within their particular environment. Since *S. cerevisiae* is particularly exposed to environments rich in sugars containing glucose and this monosaccharide is the most direct sugar intermediate entering glycolysis, it is not surprising to find that this yeast also has an extensive part of its carbon and energy metabolism regulated for the preferential use of glucose. *S. cerevisiae* is considered a Crabtree (+) yeast, meaning that in the presence of moderate to high concentrations of glucose, regardless of the oxygen levels, genes for the utilisation of alternative carbon sources, as well as genes involved in respiration, are extensively repressed so that the overall metabolism is deviated to the production ethanol (van Dijken *et al.*, 1993; Carlson, 1999). Recent studies have shown that a great deal of the so-called Crabtree effect, in other words, the onset of overflow metabolism toward ethanol, is also due to a limited capacity of the respiratory system to reoxidize all the NADH that is produced by the high glycolytic flux experienced by *S. cerevisiae* cells when growing on glucose (Vemuri *et al.*, 2007). NAD^+ is an essential coenzyme for the function of several enzymes of the glycolysis and TCA cycle, which plays a key role as carrier of electrons in the transfer of reduction potential. In order to obtain energy, enzymes must reduce it to NADH. When the rate of NADH produced by the glycolysis surpasses its rate of reoxidation to NAD^+, alternative mechanisms such as production of ethanol and glycerol must be turned on in order to keep this pathway working.

Production of ethanol is supposed as a temporary investment for yeasts that, in the short term, helps them to eliminate environmental competition and, in the long term, ensures them overall profit of the carbon nutrients that were present in the medium, through ultimate consumption of the ethanol produced, once the more preferred carbon sources are depleted. In spite of its convenience, the fermentative physiology of *S. cerevisiae* is more an exception than a rule in the world of yeasts species, which for the most part keep a much more aerobic lifestyle. Strictly aerobic yeasts like *Cryptococcus* or *Yarrowia* lack the ability to ferment any sugars. *K. lactis* occupies an intermediate position between strictly aerobic yeasts and the anaerobic facultative *S. cerevisiae*, being able to switch between both respiration and fermentation under determined conditions, although respiratory metabolism is always prevalent in this yeast. In fact, *K. lactis* is considered a petite (-) yeast. This means that it is unable to form viable respiratory-deficient mutants on treatment with DNA-targeting drugs that eliminate the mitochondrial DNA (Chen and Clark-Walker, 2000). *K. lactis* has therefore evolved in circumstances that favour oxidative growth by respiration. Fermentative growth has the disadvantage of producing less ATP, which therefore translates into a lower rate of biomass generation, as opposed to the more efficient respiratory pathway. An alternative way to outcompete rivals in nature that seems to have been adopted with almost equal success by *Kluyveromyces* yeasts, thus, consists of being able to grow and multiply much faster than these rivals so that the population can take full advantage of the nutrient resources before any slow-grower microorganisms have a chance to compete for their use. Precisely because of this characteristic, under certain conditions *S. cerevisiae* cultures may become contaminated and eventually overgrown by wild *Kluyveromyces* yeasts in industrial-scale processes (van Urk *et al.*, 1989). Given all the evidence, it is quite plausible that the fermentative physiology is a more sophisticated way of life that has been particularly mastered by the *Saccharomyces* yeasts branch, because this trait has been continuously favoured by the domestication and genetic selection processes experienced by these yeasts populations after millenia of evolutive development in human hands (i.e., by continuous growth in conditions of high cell density, in fermentors loaded with high concentrations of glucose-rich sugar mixtures, and in practical deprivation of O_2).

These are therefore, the main molecular factors being responsible for the relative sensitivity to glucose and predominance of either respiratory or fermentative metabolisms in yeasts.

Differences in Glycolytic Fluxes Determined by Sugar Transporters

In the first place, it has progressively been demonstrated that the presence and interplay of particular sets of sugar transporters (their relative amounts, regulation at the levels of expression and activity, and their relative affinities for the particular sugars) have an important contribution in the glycolytic flux rate, which in turn determines the deviation of the carbon catabolism either towards fermentation or towards respiration. One of the main causes of high glycolytic fluxes diverting metabolism towards fermentation is the saturability of the pyruvate dehydrogenase branch (Pronk *et al.*, 1996) Fig. 10.2. Low rates in the uptake and conversion of the particular sugar into a glycolytic intermediate determine, instead, low glycolytic fluxes. Such conditions do not saturate the pyruvate dehydrogenase branch and are, therefore, associated with predominance of the respiratory metabolism, as long as enough O_2, an active tricarboxylic acids (TCA) cycle, and oxidative phosphorylation pathways are available.

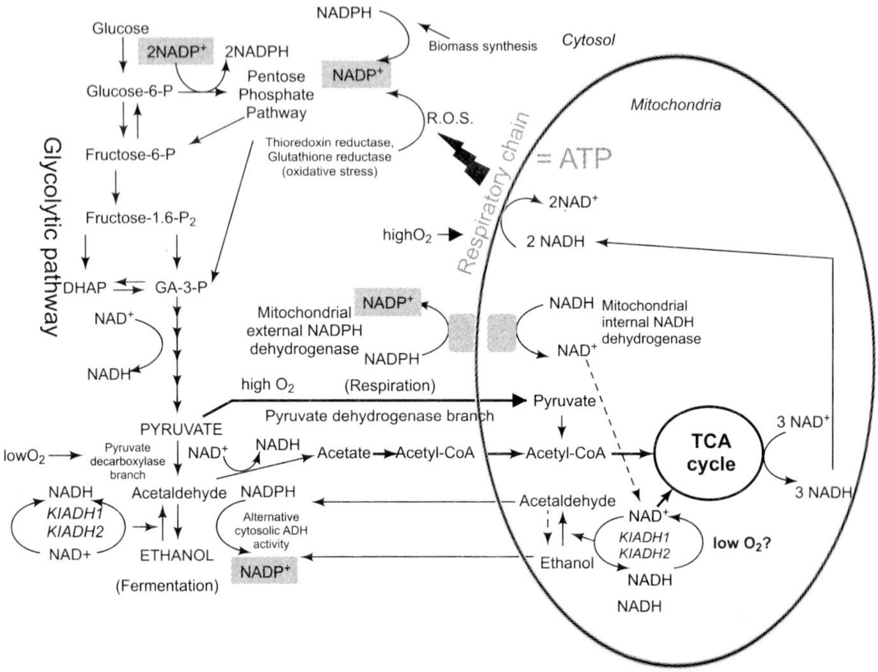

Figure 10.2 ***K. lactis* main sources of NADPH reducing power.** In *K. lactis*, a particularly active pentose phosphate pathway (PPP) bypasses different glycolytic steps. This allows glycolytic mutants to grow as long as mitochondrial respiration is active. The reason for this is that NADPH needs to be reoxidized to $NADP^+$ to keep an active pentose phosphate pathway. The major source of $NADP^+$ comes from the activity of mitochondrial dehydrogenases located in the outer face of the inner mitochondrial membrane. Their activity is coupled to the transference of electrons to the ubiquinone in the respiratory chain. Agents disrupting respiration after but not before this step, prevent the glycolytic mutants from growing because of an inability to regenerate sufficient levels of cytosolic $NADP^+$ for the pentose phosphate pathway. Biosynthetic pathways and the activity of certain oxidoreductases with a role in resistance to oxidative stress (in the elimination of 'reactive oxidative species'; ROS), constitute auxiliary sources of $NADP^+$. The pyruvate dehydrogenase branch (respiration) is more active than the pyruvate decarboxylase branch (fermentation), the latter being only significantly turned on in conditions of low O_2. Under these conditions, the activity of different sets of cytosolic and mitochondrial alcohol dehydrogenases (ADH) causes the shuttling of acetaldehyde and ethanol between the cytosol and the mitochondria, which has been proposed as a mechanism to help reoxidize NADH (to help restore the lowered levels of respiratory chain activity), and NADPH (to keep an active PPP).

A 'Kluyver effect' has been described as the inability of yeasts to grow on certain sugars in the absence of respiratory pathways (Fukuhara, 2003). When this phenomenon takes place, metabolisation of particular sugars is blocked by respiratory inhibitors and respiratory-deficient mutations, in spite of the presence of oxygen. Since it mostly relies on the fermentative pathway, *S. cerevisiae* is a very efficient anaerobic facultative yeast, able to ferment a variety of carbon sources in the absence of respiration/oxygen (Kluyver-negative). However, *K. lactis* is in general considered a Kluyver-positive yeast because of its inability to ferment and grow on several sugars, in the absence of an active respiratory pathway. An increasing amount of evidence has been gathered showing that the culprit of this phenomenon is, in fact, that differences in the rates of uptake and conversion to glycolytic intermediates of certain sugars determine different glycolytic fluxes on each of the two yeasts. As unequivocal proof, it has been shown that, in virtually almost any yeast, the Kluyver-positive/negative behaviour with respect to the utilisation of a particular sugar can be potentially switched merely through changes in the uptake rate of that sugar. For example, *S. cerevisiae* can naturally grow on galactose aerobically and anaerobically but this yeast can be converted into a Kluyver-positive yeast for the utilisation of galactose if the *GAL2* gene, encoding for the major galactose permease activity, is eliminated by mutation (Ulery *et al.*, 1991). The Kluyver-positive character of *K. lactis* on raffinose is also reversible, since *K. lactis* strains capable of growing on raffinose plus antimycin A can be obtained by overexpression, on this yeast, of the hexose transporter *HXT4* from *S. cerevisiae* (Goffrini *et al.*, 2002). (More aspects regarding the effect of certain sugar transporters in the glycolytic flux and fermentative versus respiratory metabolism, are detailed in further sections of this chapter.)

Differences at the Level of Pyruvate Metabolism

Many genes involved in the glycolytic pathway of *K. lactis*, were initially identified in a screen for mutations causing an inability to grow in high concentrations of glucose in the presence of the respiration inhibitor antimycin A. These mutants were classified as Rag –/+ (for 'resistance to antimycin A in glucose'; Ferrero *et al.*, 1981). Through this study it was then discovered that, in contrast with the high redundancy observed for this pathway in *S. cerevisiae*, many of the genes encoding for glycolytic functions are only present in one copy in *K. lactis*. But perhaps the most striking difference with *S. cerevisiae* is that, in spite of these genes being in just one copy, *K. lactis rag* null mutations in functions involved in earlier steps of glycolysis (e.g., in *RAG2*, the phosphoglucoisomerase gene) still allow normal growth of this yeast in glucose. This and other evidence (reviewed by Breunig and Steensma, 2003) have demonstrated that this happens because, in *K. lactis*, the first part of the glycolytic pathway can, in fact, be bypassed by the alternative pentose phosphate pathway (PPP) Fig. 10.2. In contrast to *S. cerevisiae*, *K. lactis* has a more active PPP so that the metabolisation of sugars is not primarily dependent on the glycolytic pathway (González-Siso *et al.*, 1996). However, the activity of the PPP is strongly dependent on the activity of the respiratory chain for the reoxidation of NADPH ($NADP^+$ is an another essential coenzyme with similar properties to NAD^+), such that the inability of a *rag2* mutant to grow on glucose in the presence of antimycin A is not caused by an inability to produce ethanol but because the NADPH formed in the PPP can not be reoxidized when the respiration is blocked. The reason for this is that NADPH reoxidation in *K. lactis* is mainly dependent upon the activity of mitochondrial external alternative dehydrogenases (Overkamp *et al.*, 2002). These enzymes, located in the external face of the inner mitochondrial membrane, oxidize cytosolic NADPH derived from the PPP by transference of electrons to the ubiquinone in the respiratory chain. Growth of the *rag2* mutant on glucose is therefore prevented by a blockage of the mitochondrial respiratory chain right after the ubiquinone step (e.g., by antimycin A), but not if the blockage takes place before the ubiquinone step (e.g., by malonate). That explains how this mutant can grow in the presence of malonate, even if respiration is blocked, at the expense of fermentation (Tarrio *et al.*, 2006). When alternative mechanisms for reoxidation of NADPH were investigated in *K. lactis*, an interesting finding was the fact that the high respiratory rate in *K. lactis rag2* mutant may increase the intracellular levels of ROS (reactive oxygen species like superoxide, hydrogen peroxide and hydroxyl radicals), which would induce the oxidative stress response as an alternative method of NADPH reoxidation (since this pathway includes the activation of enzymes that use NADPH to keep glutathione and thiredoxin reduced; Tarrio *et al.*, 2006). In favour of this possibility, an unrelated study has shown that in the laboratory *K. lactis* strain CBS2359, genes involved in detoxification and in the oxidative stress response are upregulated when the strain is grown in lactose (Suleau *et al.*, 2005). In summary, the lack of a redundancy in glycolytic genes may imply a necessity for *K. lactis* to keep high levels of activity of the alternative PPP as a back-up system, and this is probably one of the main reasons forcing this yeast to keep an oxidative metabolism.

K. lactis only initiates alcoholic fermentation in conditions of very low concentrations of O_2, such as when the population reaches high cell densities in the culture medium (Kiers *et al.*, 1998). Besides the facts mentioned above, there are additional factors that contribute to this different physiology. For instance, *K. lactis* has a pyruvate dehydrogenase branch that is twice as active and a pyruvate decarboxylase branch that is half as active as in *S. cerevisiae* (Zeeman *et al.*, 2000). The pyruvate decarboxylase gene (*KlPDCA*/*KlPDC1*/*RAG6*) in *K. lactis* is autoregulated, induced by glucose and repressed by ethanol (Destruelle *et al.*, 1999), but most importantly it is only strongly induced by low O_2 concentrations (Kiers *et al.*, 1998). Paradoxically, *K. lactis* has four alcohol dehydrogenase (*ADH*) genes. *KlADH1* and *KlADH2* are constitutive genes encoding two cytoplasmic alcohol dehydrogenases (Saliola *et al.*, 1990; Shain *et al.*, 1992), whereas *KlADH3* and *KlADH4* encode two mitochondrial activities. *KlADH4* is specifically induced by ethanol at the transcriptional level and is not sensitive to glucose repression (Mazzoni *et al.*, 1992) whereas *KlADH3* is induced by respiratory carbon sources except for ethanol (Saliola and Falcone, 1995). These two activities are therefore mutually exclusive. Ethanol produced or added to the medium activates

KlADH4 and represses *KlADH3*, whereas nonfermentable carbon sources other than ethanol activate *KlADH3* and not *KlADH4*. Possible reasons for *K. lactis* need to keep four genes encoding for ADH activities in spite of its predominant respiratory lifestyle are, firstly, that these activities help to reoxidize NADH in conditions of low O_2 concentrations in which the respiratory chain is not as active, and secondly, once the pyruvate decarboxylase gets activated by low O_2, ADH activities are also needed to prevent a build-up of toxic acetaldehyde within cells (Breunig and Steensma, 2003). Additional studies have indicated the existence of a poorly characterized cytosolic NADPH-dependent acetaldehyde reductase (ADH), which along with the mitochondrial ADH activities may generate a two-compartment transhydrogenase shuffling cycle for acetaldehyde and ethanol (Overkamp *et al.*, 2002). This could be an additional mechanism present in this yeast to reoxidize cytosolic NADPH (Fig. 10.2).

Additional factors are, nevertheless, accountable for the respiratory metabolism of *K. lactis*. For example, in *S. cerevisiae* the biosynthesis of sterols requires oxygen. Twelve molecules of molecular oxygen (O_2) are needed for the synthesis of one molecule of ergosterol (Rosenfeld and Beauvoit, 2003). Under anaerobiosis, *Saccharomyces* cells no longer synthesize sterols but instead import them. This sterol uptake is, therefore, essential under anaerobic conditions (Wilcox *et al.*, 2002). A recent study in which orthologues for genes that are essential for anaerobic growth in *S. cerevisiae* have been sought in the *K. lactis* genome, has revealed the apparent lack of 20 of these genes in *K. lactis* (Snoek and Steensma, 2006). Particularly, this study has shown that a system for the uptake of sterols seems to be missing in *K. lactis*, which has brought up the possibility of this being another among the most important reasons preventing *K. lactis* from growing in strictly anaerobic conditions.

Differences at the Level of Sensitivity to Glucose

K. lactis also has a significantly lower degree of glucose sensitivity, being classified as a rather Crabtree (–) yeast. Only certain laboratory strains (which very likely are the result of continued exposure to artificially increased levels of sugar in synthetic media) and a few wild isolates carrying specific sets of genes, which are discussed below, show a significant level of glucose repression (Table 10.1). Although sharing common elements with *S. cerevisiae*, the mechanisms underlying glucose repression have still evolved significant differences in *K. lactis*. Among these differences one fact that stands out is that the respiratory genes are not repressed by glucose as in *S. cerevisiae*. However, after an increased knowledge of all the factors influencing this phenomenon, it is not hard to imagine ways to convert *K. lactis* into a Crabtree (+) yeast. In fact, groups such as that of H. Y. de Steensma (Leiden University, the Netherlands) have succeeded in committing such a task just by playing with the levels of expression of acetylCoA synthetases and pyruvate dehydrogenases (Zeeman *et al.*, 2000; Fukuhara, 2002).

As raised in a previous section, the rate of glucose uptake plays a significant role in glucose repression by control of the glycolytic flux. *S. cerevisiae* has 20 genes encoding for glucose/hexose transporters (*HXT1* to *HXT17*, *GAL2*, *SNF3*, and *RGT2*; reviewed by Özcan and Johnston, 1999). None of these genes are essential because of their functional redundancy, which also allows this yeast to modulate the uptake rate in response to a whole range of environmental glucose concentrations through the combined expression of different sets of genes within this class. Overall, there is a low-affinity glucose uptake that appears constitutively expressed, and a high-affinity system that is glucose repressed (Bisson and Fraenkel, 1984). Low-affinity and high-affinity glucose transport result from the combined expression of several transporters rather than from individual transporting activities. The way in which these *HXT* genes are transcriptionally regulated in response to glucose is also consistent with their function either as low- or high-affinity transporters.

In *K. lactis* the situation looks quite different. First, the two systems (high- versus low-affinity) are encoded by two single genes: *RAG1*, encoding the low-affinity permease (Wésolowski-Louvel *et al.*, 1992a), and *HGT1*, encoding the high-affinity permease (Billard *et al.*, 1996). Second, in this yeast *HGT1* is constitutively expressed whereas *RAG1* is activated in the presence of high glucose concentrations (Chen *et al.*, 1992; Wésolowski-Louvel, 1992a). The *Rag*1 permease seems to be essential for fermentative growth, since a *rag1* mutant in the presence of antimycin A is unable to grow in high glucose concentrations (which usually force fermentative growth because of the requirement for a high substrate flow; Goffrini *et al.*, 1989; Wésolowski-Louvel, 1992b). Certain *K. lactis* isolates have the two closely linked and tandemly transcribed hexose transporter genes, *KHT1* and *KHT2*, in place of *RAG1* (Breunig, 1989). Comparison of their sequences shows that *RAG1* is almost identical to *KHT1*, and that *RAG1* very likely arose by recombination between *KHT1* and *KHT2*. Strains carrying the *KHT1* and *KHT2* genes show the highest level of sensitivity to glucose repression (Chen *et al.*, 1992; Weirich *et al.*, 1997). Conversely, natural isolates have also been found that are totally insensitive to glucose repression because they carry a defective *rag1* allele (Table 10.1); (Goffrini *et al.*, 1989; Suleau *et al.*, 2006).

Digging further into the regulatory mechanisms influencing the expression of hexose transporters in *S. cerevisiae*, Snf3 and Rgt2 are two proteins that, in spite of their similarity to the glucose carrier superfamily, do not transport glucose but instead serve as extracellular glucose sensors that generate an intracellular signal to promote the induction of *HXT* genes (Fig. 10.3). In the absence of glucose, transcription of the *HXT* genes is repressed at the nuclear level by Rgt1p. In conditions of low glucose, Snf3p is responsible for induction of the expression of several *HXT* genes, whereas in conditions of high glucose, Rgt2p is responsible for strongly inducing *HXT1*. In the absence of glucose, Mth1p and Std1p interact with the C-terminal tail of both Snf3p and Rgt2p. In sensing glucose, Snf3/Rgt2 sensors stimulate the phosphorylation of Mth1 and Std1 proteins by the casein kinases Yck1/2p, which result in their ubiquitination and subsequent degradation by the proteasome. This event triggers the release of Rgt1p from its repressor upstream binding sites, which leads to derepression of downstream *HXT* genes (for a more extensive review see Santangelo, 2006; Holsbeeks *et al.*, 2004).

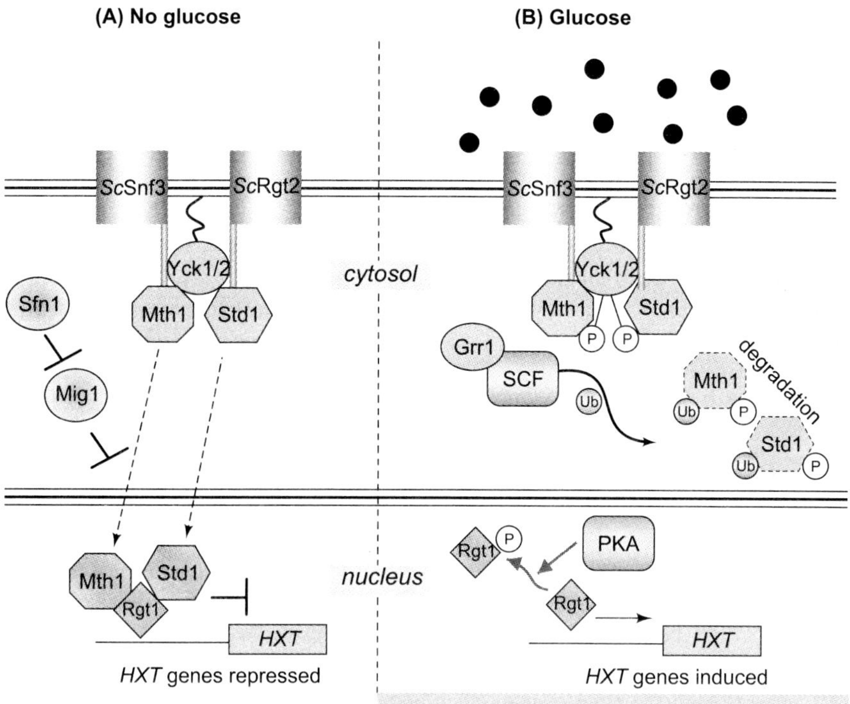

Figure 10.3 **Regulation of the hexose transporter genes (*HXT*) in *S. cerevisiae* through the glucose sensing/signaling Snf3/Rgt2 pathway.** Although some aspects of the pathway remain to be elucidated this seems to be the most accepted model to explain its functioning: A) in the absence of glucose, Mth1 and Std1 proteins enter the nucleus where they interact with the Rgt1p and the three proteins then act as co-repressors to inactivate the expression of *HXT* genes. In these conditions, this phenomenon is favored by Snf1p-dependent repression of Mig1p, which would otherwise repress *MTH1*. B) Snf3 and Rgt2 proteins sense the presence of glucose, promoting Mth1p and Std1p phosphorylation by Yck1/2, a plasma membrane lipid-anchored kinase activity. Snf3p and Rgt2p interact, through their cytosolic C-terminal tails, with the kinase and its substrates in such a way that a conformational change signaled through this contact region activates the phosphorylation event. Phosphorylated Mth1 and Std1 proteins are recognized by the SCF^{Grr1}-ubiquitin ligase complex, which ubiquitinates and targets them for proteolytic degradation. The lack of Mth1p and Std1p seems to switch Rgt1p into a transcriptional activator mode which then stimulates *HXT* gene expression. This switch requires phosphorylation of Rgt1p by PKA, which turns on Rgt1p intramolecular interaction, making it loose its ability to bind to the promoters in *HXT* genes.

Rgt1p regulation occurs through an intramolecular interaction between the N terminus and middle region of the protein, which has been suggested to inhibit the DNA-binding function of Rgt1p (Polish *et al.*, 2005). Such intramolecular interaction has recently been found to be dependent upon the activity of the Ras/cAMP-dependent protein kinase A (PKA), which is one of the central regulator pathways for nutrient sensing (Kim and Johnston, 2006). In the presence of high levels of glucose, PKA becomes activated and then regulates Rgt1p function through phosphorylation, which then induces Rgt1 removal from the *HXT* promoters by stimulating its intramolecular interaction. Rgt1 also controls the glycolytic flux by similarly controlling the expresion of the glycolytic hexokinase and regulator gene, *HXK2* (Palomino *et al.*, 2005). (More aspects of this particular regulation are discussed in the next section.)

Orthologues for many of these genes have also been found in *K. lactis*. For instance, *RAG4* encodes a glucose sensor that combines the functions of Snf3 (low) and Rgt2 (high) glucose sensors of *S. cerevisiae* (Betina *et al.*, 2001). *RAG8* encodes a casein kinase I-like protein, whose disruption abolishes transcription of *RAG1* (Chen *et al.*, 1992; Blaisonneau *et al.*, 1997). A *KlRGT1* gene has been identified that encodes a protein 31% identical to ScRgt1p (Rolland *et al.*, 2006; Özcan *et al.*, 1996), and a single orthologue gene for the *MTH1/STD1* paralogs is also present in the *K. lactis* genome (http://cbi.labri.fr/Genolevures/). Recent work has shown that, although KlRgt1 also acts as a repressor of *RAG1*, major differences exist in the regulation of this function between these two yeasts (Rolland *et al.*, 2006). KlRgt1p repressor function is controlled by Rag4- and Rag8-dependent phosphorylation in response to high glucose concentrations. When phosphorylated, it no longer represses *RAG1* expression. KlRgt1p only represses transcription of *RAG1* in the absence or in the presence of low levels of glucose. However, unlike ScRgt1p, KlRgt1p seems to bind to the *RAG1* promoter even when *RAG1* is derepressed by the presence of glucose. The fact that it does not leave the promoter under any condition studied, may be an indication that the *HXT* gene system is less easily derepressed than in *S. cerevisiae* and not totally derepressed even when in the presence of high concentrations of glucose. Moreover, it has been found that KlRgt1p, as ScRgt1p, also regulates the expression of the hexokinase gene *RAG5*, which suggests that the internal glycolytic flux is also regulated by the external glucose signal through direct control of the expression of this regulator/glycolytic enzyme. Wésolowski and colleagues explain these differences by taking into consideration that, in its normal environmental conditions (where no glucose or very low levels of it are present), the metabolism of *K. lactis* mostly relies on respiration because of a low glycolytic flux determined by the sole expression of the high-affinity glucose permease Hgt1p. The regulated low-affinity transporter Rag1p may be an alternative mechanism being turned on only in conditions in which *K. lactis* is suddenly facing a high-glucose environment, in which it needs to compete with other microorganisms for the use of this carbon source. Such situation would be an exceptional circumstance, rather than something commonly experienced by this yeast.

Comparative Overview of the Galactose Utilisation Pathway in *S. cerevisiae* and *K. lactis*

In the last two decades extensive information has been gathered about the functioning of this regulatory network that reveal both important similarities and differences between the galactose utilisation system of *S. cerevisiae* and *K. lactis*. (For a more detailed review the author recommends reading the following reviews: Johnston, 1987; Lohr *et al.*, 1995; Bash and Lohr, 2001; Rubio-Texeira, 2005; Traven *et al.*, 2006).

Catabolic/Structural Genes of the GAL Regulon

One of the most important differences relies on how *S. cerevisiae* and *K. lactis* encounter galactose in their particular environments. Galactose can be found as the free molecule or as part of particular disaccharides/polysaccharides so that both yeasts have evolved different mechanisms in order to release galactose monomers from them extracellular or intracellularly. *S. cerevisiae* may find galactose in the form of melibiose (6-*O*-α-D-galactopyranosyl-D-glucopyranose) and has developed a mechanism to extracellularly hydrolyze melibiose to glucose and galactose, through the action of a secretable α–galactosidase encoded by *MEL1* (Post-Beittenmiller *et al.*, 1984). Galactose is transported across the cell membrane by the combination of a facilitated diffusion mechanism and the action of the major galactose permease encoded by *GAL2* (Tschopp *et al.*, 1996). Gal2p has both high-affinity and low-affinity galactose transport mechanisms that have shown to be, in a poorly characterized way, dependent upon the presence of the galactokinase activity encoded by *GAL1* (Ramos *et al.*, 1989). In the absence of Gal2p, a certain level of galactose uptake may yet occur through the action of other hexose transporters, but these mutants grow very poorly with galactose as the only carbon source (Boles and Hollenberg, 1997).

K. lactis basically finds galactose as the free molecule or in the form of lactose (*O*-β-D-galactopyranosyl-(1-4)-β-D-glucose). The ability of *K. lactis* to assimilate lactose has been acquired, throughout evolution, by development of the genes *LAC12* and *LAC4*, respectively, which encode a lactose/galactose permease, and a cytosolic β-galactosidase that hydrolyzes lactose intracellularly (Sreekrishna and Dickson, 1985). For quite a while it was thought that *LAC12* encoded the only lactose/galactose transporter in *K. lactis*. However, in a recent work searching for multicopy suppressors of *K. lactis snf1* poor growth in galactose, a new hexose transporter encoded by *KHT3* was identified that has galactose transporter activity (Wiedemuth and Breunig, 2005). Multiple copies of *KHT3* restore growth of a *lac12* mutant strain in galactose but not in lactose. Another very recent report has also pointed out to the activity of Hgt1p not only as a glucose permease but also as an additional galactose transporter in *K. lactis* (Baruffini *et al.*, 2006). This study has shown that, in fact, Hgt1p acts as a high-affinity transporter of galactose, whereas Lac12p acts as a low-affinity transporter. *HGT1* expression is strongly induced by galactose and insensitive to glucose repression. The existence of this additional mechanism may now be accounted for the rapid adaptation to galactose observed in *K. lactis* when shifted from glucose to galactose medium.

The remaining structural genes of the GAL regulon encode the enzymatic components of the highly conserved Leloir pathway, involved in the cytosolic conversion of galactose into the glycolytic intermediate, glucose-6-phosphate (Leloir, 1971) (Fig. 10.4). Through the concerted action of these enzymes,

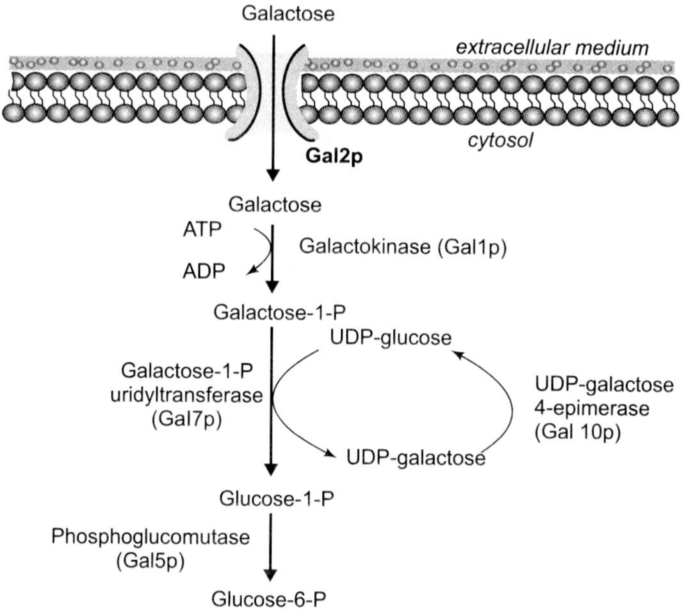

Figure 10.4 **The Leloir pathway in *S. cerevisiae*.** Galactose enters the cells through the galactose-specific permease, Gal2p. In the cytosol, the galactokinase Gal1p uses ATP to phosphorylate galactose to galactose-1-phosphate. The UDP-galactose 4-epimerase encoded by *GAL10* exchanges UDP-between galactose and glucose molecules, generating UDP-glucose that is used by the galactose-1-phosphate uridyltransferase Gal7p to convert galactose-1-phosphate into glucose-1-phosphate. The phosphoglucomutase, Gal 5p (in fact, present in two isozyme forms encoded by the *PGM1* and *PGM2* genes), transforms glucose-1-phosphate into the glycolytic intermediate, glucose-6-phosphate. The Leloir pathway is extensively conserved from bacteria to humans with only small differences in certain components of the pathway and/or the presence of alternative branches that bypass parts of the pathway and whose existence has yet to be shown in yeasts.

galactose is first phosphorylated to galactose-1-phosphate in an ATP-dependent manner by the galactokinase Gal1p (EC 2.7.1.6). The uridine diphosphoglucose 4-epimerase, Gal10p (EC 5.1.3.2), exchanges galactose with the glucose group in

UDP-glucose to create UDP-galactose, and modifies the stereochemistry of C4 in UDP-galactose, creating again UDP-glucose. Using the glucose generated in this step, galactose-1-phosphate is then converted into glucose-1-phosphate by the galactose-1-phosphate uridylyltransferase, Gal7p (EC 2.7.7.12). Glucose-1-phosphate is finally converted into glucose-6-phosphate through the action of the glycolytic enzyme phosphoglucomutase, Gal5 (EC 2.7.5.1). *S. cerevisiae* has, in fact, two phosphoglucomutase genes, *PGM1* and *PGM2*, from which *PGM2* encodes the major isozyme form (Oh and Hopper, 1990; Fu *et al.*, 1995).

The possible existence of alternative pathways for the utilisation of galactose in prokaryotes and fungi that could also be present on yeast has recently picked a new interest in this field. However, stronger evidence has yet to be collected that shows the relevance of these alternative pathways (Lai and Klapa, 2004).

Cis- and *Trans-* elements Regulating the GAL Regulon

In *S. cerevisiae* and *K. lactis*, fine-tuning regulation of the structural genes in the GAL regulon is achieved through clustering. The Leloir genes are clustered in one chromosome (chromosome II), with a tight linkage between *GAL7* and *GAL10*, and divergent transcription of *GAL1* and *GAL10* from an intergenic region (Lohr *et al.*, 1995). *K. lactis LAC4* and *LAC12* genes are similarly clustered and divergently transcribed from an unusually large intergenic region of 2.6 kb (Leonardo *et al.*, 1987; Gödecke *et al.*, 1991). Clustering allows co-regulation of these genes by means of common intergenic regulatory promoter sequences. As we shall see below, in spite of being located on separated locations and chromosomes, genes such as *GAL2* (in *S. cerevisiae*), *PGM2/GAL5*, and genes encoding for regulatory elements acting in *trans*, are controlled by variable numbers of the same type of positive and negative regulatory sequences present in their promoters. This variability in the combination of regulatory sequences allows specific levels of induction/repression for the different genes, which are crucial for the striking sensitivity that this regulatory circuitry has in response to carbon sources.

Induction of the *GAL* genes is determined by the interplay between three main GAL-specific regulatory proteins: a transcriptional activator, Gal4p; a repressor, Gal80p; and a ligand sensor/inducer, ScGal3p (or KlGal1, in *K. lactis*) that activates *GAL* gene expression through galactose- and ATP-dependent repression of Gal80p.

ScGAL4 and its *Kluveromyces* ortholog, *KlGAL4/LAC9*, encode the *GAL* gene-specific transcriptional activator, Gal4p. ScGal4p is a transcriptional activator with a Zn-Cys binuclear cluster-type DNA-binding domain, a linker domain, a dimerization domain and two acidic (AR1 and AR2) activation domains (Lohr *et al.*, 1995). In particular, the DNA-binding specificity is directly related to a 19-amino acid region that is located C-terminal from the Zn-Cys cluster domain (Reece and Ptashne, 1993). KlGal4p and ScGal4p can substitute for each other in their activating function although their overall similarity is low, being mostly restricted to the homology shown in the DNA-binding and the transactivating/repressor binding regions. As we will see in the next subsection, differences of regulation in *GAL4* function at the transcriptional and post-translational levels are also important in terms of the different behaviour of the GAL regulon displayed by *K. lactis* and *S. cerevisiae*.

The consensus Gal4p-binding site is the 17-mer sequence 5′-CGG-N11-CCG-3′ (most commonly known as 'galactose upstream activating sequence', or UAS_G). Gal4p binds to this sequence as a dimer (Marmorstein *et al.*, 1992). The Gal4p Zn-Cys domains contact the CGG elements directly, whereas the linker and dimerization domains interact with the phosphate backbone of the spacer residues in the 17-mer. Both the CGG residues and the length of the spacer region determine the strength with which Gal4p binds to the particular UAS_G (Liang *et al.*, 1996). Interestingly, Gal4p has also a certain ability to bind non-UAS_G elements *in vivo*, which indicates influence of the carbon source galactose in other less related pathways such as those involved in nitrogen utilisation (D'Alessio and Brandriss, 2000).

Gal4p binding to multiple adjacent UAS_G elements is cooperative and leads to synergistic activation of transcription (Giniger and Ptashne, 1988; Kang *et al.*, 1993). But, paradoxically, this fact also implies a tighter level of repression such that genes with multiple Gal4p binding sites are more tightly repressed and show lower levels of basal expression under non-inducing conditions than those regulated by a single Gal4p binding site (Bram *et al.*, 1986). It is possible to understand this paradox if we take into consideration the mode of action of the second most important transregulatory element of the GAL regulon, the repressor protein Gal80p.

GAL80 encodes a protein repressor that binds to Gal4p in order to counteract its activating function (Johnston *et al.*, 1987). Gal4 and Gal80 proteins bind stoichiometrically with high affinity (Chasman and Kornberg, 1990; Parthun and Jaehning, 1990; Schulz *et al.*, 1993). Recent studies using fluorescence resonance energy transfer (FRET) microscopy have confirmed that Gal4p and Gal80p remain associated even in the present of galactose (Bhaumik *et al.*, 2004). Through binding to a 28-amino acid region inside the second activation domain of Gal4p, Gal80p physically blocks interactions of the activation domain with the transcriptional machinery (Ma and Ptashne, 1987; Johnston *et al.*, 1987). For a complete inhibition, Gal80p dimerizes and interacts as such with the Gal4p dimer (Melcher and Xu, 2001). The extent of the repression increases with the number of DNA binding sites for Gal4p. This happens because adjacent sites allow the stabilization of transient interactions between adjacent Gal80p dimers, causing a complete shielding of the activation domains of Gal4p (Melcher and Xu, 2001).

KlGal80p has a high degree of homology with ScGal80p, with an overall identity of 60% and two blocks with over 80% of identical residues (Zenke *et al.*, 1993). As a difference with the Gal4p orthologs, in which separated domains can be recognized and remain functionally active once separated (e.g., in two-hybrid techniques), both Gal80p orthologs are particularly sensitive to single amino acid substitutions in different parts of the protein that can simultaneously eliminate functions such as binding to Gal4p, to Gal3/Gal1p, and/or homodimerisation

(Zenke *et al.*, 1999; Pilauri *et al.*, 2005; Melcher, 2005). These mutations do not cluster in particular regions but apparently affect the overall structural integrity of Gal80p rather than specific functions. KlGal80p three-dimensional structure has recently been solved (Thoden *et al.*, 2007). Interestingly, this study has shown that Gal80p is structurally homologous to glucose-fructose oxidoreductase, an enzyme in the sorbitol-gluconate pathway. This indicates that molecular scaffolds initially employed by enzymes of the sugar metabolism may have, at some point in evolution, been hijacked to function as transcriptional regulators.

The release from Gal80p inhibition depends on the competitive interaction with a third regulator protein, the ligand galactose sensor/inducer (Gal3 and Gal1 proteins in *S. cerevisiae*, or KlGal1p in *K. lactis*), which binds to Gal80p with 1:1 stoichiometry (Suzuki-Fujimoto *et al.*, 1996; Blank *et al.*, 1997; Yano and Fukasawa, 1997; Timson *et al.*, 2002). This third step of regulation is where the GAL regulons of *S.cerevisiae* and *K. lactis* seem to show the highest level of divergence. The development of *ScGAL1* and *ScGAL3* functions is one of the cases in which further evolution of *S. cerevisiae* has led to a case of initially redundant function with further partial loss of function and speciation of the corresponding paralogues. ScGal1p is bifunctional, presenting both galactokinase (conversion of galactose into galactose-1-P) and ligand sensing/inducing (ATP- and galactose-dependent binding to Gal80p and inhibition of its repressor function) activities (Bhat and Hopper, 1990; Meyer *et al.*, 1991). Similarly, *K. lactis GAL1* encodes a bifunctional galactokinase/sensor/inducer protein, which binds to KlGal80p, as Gal3p to ScGal80p, in an ATP- and galactose-dependent manner (Meyer *et al.*, 1991; Zenke *et al.*, 1996; Vollenbroich *et al.*, 1999; Menezes *et al.*, 2003). ScGal3p has a 73% identity and 92% similarity to the galactokinase enzyme ScGal1p at the amino acid level however it lacks the galactokinase activity (Bhat and Hopper, 1990; 1992). The insertion of two amino acids in a specific region suffices to restore galactokinase activity in Gal3p (Platt *et al.*, 2000). *S. cerevisiae* mutant strains lacking Gal3p suffer a substantially delayed induction of the *GAL* genes in galactose (phenomenon known as 'long term adaptation', reviewed in Bhat and Murthy, 2001). This delay is only temporary because Gal1p can replace Gal3p sensor/inducer function. However, ScGal1p ligand sensor activity is 40-fold less efficient, so that it is only able to replace Gal3p after its gene is fully induced and the protein is present in sufficient levels (Bhat and Hopper, 1991; Platt and Reece, 1998).

A particular meaning must be inherent to the evolutive separation of the sensor/inducer and galactokinase functions into separate paralogue genes. An interesting hypothesis to explain the replacement of Gal1p by Gal3p as a sensor/inducer in *S. cerevisiae* proposes that a simultaneous galactokinase activity would lead to a potentially toxic accumulation of galactose-1-phosphate at an early stage of induction in which the expression of the subsequent catabolic activities of the Leloir pathway has not yet been induced (Meyer *et al.*, 1991). Such functional redundancy and speciation has not been favoured in *K. lactis* because this yeast would not face a similar situation leading to accumulation of toxic levels of Gal-1-P. In this case, *LAC4*, required for the hydrolysis of lactose into glucose and galactose, is coinduced simultaneously to the rest of *GAL* genes (Zenke *et al.*, 1996). A low concentration of β-galactosidase at earlier stages of induction therefore prevents the intracellular release of high levels of galactose and the subsequent accumulation of a toxic excess of galactose-1-phosphate. Eventual full induction of the Leloir genes would further prevent accumulation of galactose-1-P because the precursor would immediately be converted to glucose-6-P.

Induction takes place when Gal3p (or KlGal1p) binds to its two allosteric effectors, first ATP and then galactose (Timson and Reece, 2002). Upon binding to these two ligands, a structural change happens in the sensor/inducer protein that facilitates its interaction with Gal80p. The structure of ScGal1p complexed with galactose and a nonhydrolizable ATP analogue has recently been solved (Thoden *et al.*, 2005). According to this and previous structural data, proteins from the galactokinase, homoserine kinase, mevalonate kinase and phosphomevalonate kinase (GHMP) family, contain a deep cleft located between the N-terminal and the C-terminal domains. The protein therefore has a bilobal shape that, upon binding to galactose and ATP, closes up creating a binding site for Gal80p. Gal80p binds to at least, two separate areas located on opposite sides of the active site, which is sandwiched between the N- and C-terminal domains. The sugar- and nucleotide-binding sites are also sandwiched between these two domains of the protein (Thoden *et al.*, 2005; Sellick and Reece, 2006). Substrate binding in ScGal1p apparently involves a transition from a flexible to a rigid structure. Recent mutational analysis indicates that Gal3p may exist in a conformation that has the ability to interact with Gal80p even in the absence of galactose, but that this interaction is weak and unable to sequester enough Gal80p to cause activation of the *GAL* genes, so that only the presence of galactose stabilizes the interaction increasing the half-life of the Gal3p-Gal80p complex (Lakshminarasimhan and Bhat, 2005).

S. cerevisiae and *K. lactis* seem to show additional differences in the way in which Gal3p/KlGal1p sequester Gal80p to release Gal4p from repression (Fig. 10.5). Some of the earliest models proposed to explain Gal3p mode of action involved the formation of a transcriptionally-active ternary complex Gal3-Gal80-Gal4p, and/or the possibility that Gal3p binding displaces Gal80p from the activation domain of Gal4p without promoting total dissociation (Platt and Reece, 1998; Sil *et al.*, 1999). This model implies that Gal3p is able to enter the nucleus. However, studies in *S. cerevisiae* later showed that Gal3p is exclusively cytosolic and that its interaction with Gal80p, therefore, occurs in the cytosol (Peng and Hopper, 2000; 2002). According to these data, a new model has been proposed in which Gal3p sequesters Gal80p monomers in the cytosol, which reduces Gal80p intranuclear levels impairing its self-association, in turn leading to a Gal4p gradual release from repression (Verma *et al.*, 2004; Pilauri *et al.*, 2005). Nevertheless, very recent data have shown that this mechanism does not seem to be the same as the one acting in *K. lactis*. The situation in this yeast points back instead, to the ternary complex model of interaction. In this case, KlGal1p has the capacity to enter the nucleus and interact with KlGal80 in this compartment and this interaction is exclusively nuclear (Anders *et al.*, 2006) (Fig. 10.5). Such data are in accordance with other related differences manifested by both yeasts at this

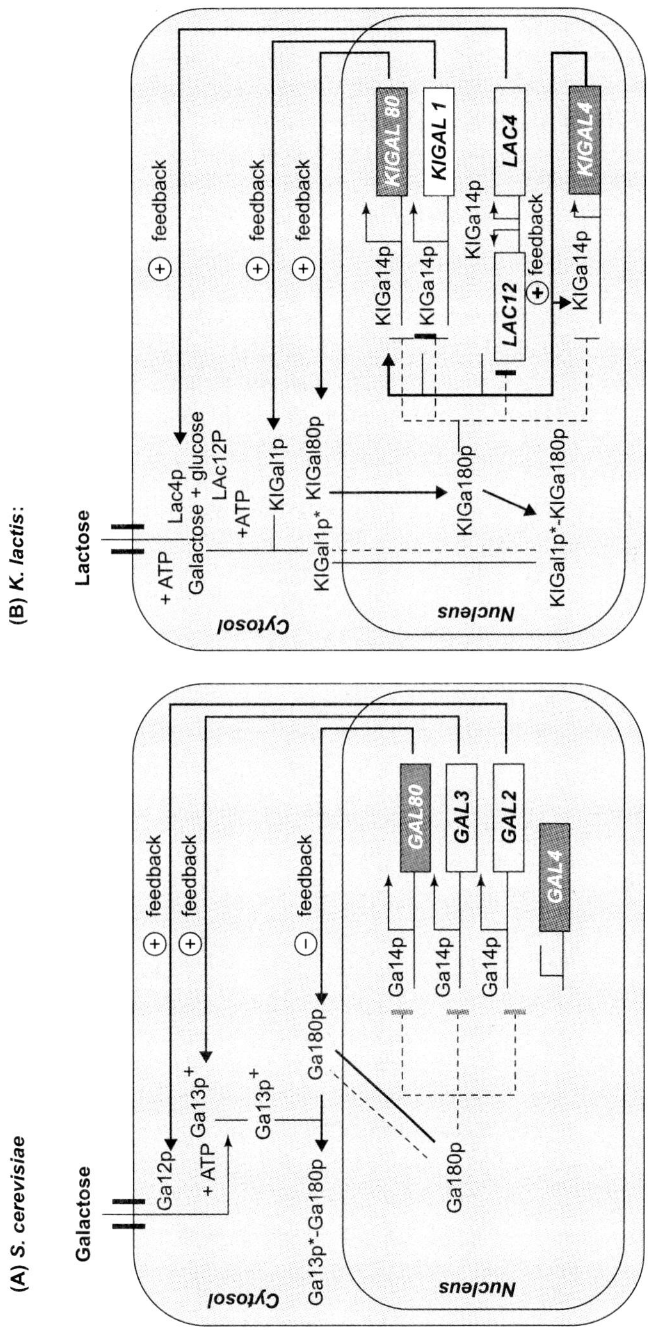

Figure 10.5 Simplified model showing the main differences between *S. cerevisiae* and *K. lactis* in the coordinated expression and interaction of the principal regulators acting in the induction of **the** *GAL* **genes.** A) In

(Cont.)

level of regulation. In spite of being 57% identical to KlGal1p, ScGal3p binds only poorly to KlGal80p. Conversely, *KlGAL1* can substitute for *GAL3* in *S. cerevisiae* but neither *GAL3* nor *ScGAL1* can complement the non-inducible phenotype of a *K. lactis gal1* deletion mutant unless the *KlGAL80* gene is also exchanged by *ScGAL80* (Zenke *et al.*, 1996). The resolution of the three-dimensional structure of KlGal80p has led the authors of this recent work to propose a more accurate model for the competition between KlGal4p dimers and KlGal1p monomers in which a mutually-exclusive interaction with distinct but overlapping parts of KlGal80p takes place (Thoden *et al.*, 2007).

An interesting hypothesis has, ultimately, emerged from all this recent evidence that speculates with the fact that by binding both to KlGal4p and/or KlGal1p, KlGal80p may not only regulate the GAL regulon by transcriptional repression but also by exerting a more direct control on the levels of galactokinase activity. In *S. cerevisiae*, the interaction between ScGal1p and ScGal80p is weak and unlikely to interfere with galactose metabolism (Platt and Reece, 1998). But in *K. lactis* the contribution of KlGal80 to regulation of the galactokinase by physical interaction within the nucleus might be more relevant. Galactokinase nuclear sequestration by KlGal80p could serve as a mechanism for the preferential use of glucose released from lactose hydrolysis, prior to the utilization of galactose (Anders *et al.*, 2006).

Induction of the GAL Regulon

As mentioned above, the number of regulatory sequences plays a determinant role in the relative levels of induction experienced by each of the *GAL* genes, which in turn affects the overall response of this network. A main difference between *S. cerevisiae* and *K. lactis* is the absence of *GAL4* promoter of an UAS_G,

---

(Cont.)

S. cerevisiae galactose enters the cells through the Gal2p permease. Galactose and ATP change the conformation of Gal3p, enhancing its binding to Gal80p. Since Gal3p is only cytosolic, the equilibrium of Gal80p molecules shifts towards the cytosol. This reduction in nuclear Gal80p concentration alleviates Gal4p from Gal80p-mediated repression, allowing it to activate *GAL* gene expression. Gal4p expression is constitutive and its levels are not influenced by the presence of galactose. However, all of the other *GAL* regulator genes respond to the presence of galactose, creating negative or positive overlapped feedback loops that modulate the overall response of the GAL network. B) In *K. lactis*, lactose enters the cells through the *LAC12* permease and is hydrolyzed to glucose and galactose by the β-galactosidase, Lac4p. The bifunctional galactokinase/sensor-inducer, KlGal1p, binds to ATP and galactose and is shuttled to the nucleus, which increases its chances to compete for interaction with nuclear KlGal80p. This phenomenon disrupts KlGal80p repressor interaction with the transcriptional activator, KlGal4p. Similar regulatory feedbacks as in *S. cerevisiae* are generated with the exception of KlGal4p, which is able to bind to its own promoter to positively regulate its own expression, generating an extra positive feedback loop. For more details, see the text.

in the S.cerevisiae which makes the expression of the activator constitutive and only indirectly regulated through Gal4p regulation of the expression of its repressor, Gal80p. *K. lactis GAL4* expression is instead autoregulated because of the presence of a weak UAS_G in its promoter (Zachariae and Breunig, 1993). This positive feedback loop is responsible for a two- to threefold higher level of Gal4p in *K.lactis* in non-inducing conditions, which in turn determines a higher level of basal expression for the *LAC/GAL* genes in this yeast. Also, two Gal4p-binding sites are located in the *KlGAL80* promoter, whereas only one is present in *ScGAL80* (Zenke *et al.*, 1993). That means a more stringent level of repression of the *KlGAL80* gene than the *ScGAL80* gene, which also helps to keep higher basal levels of *GAL* gene expression in *K. lactis*. Other *S. cerevisiae* genes with just one UAS_G, such as *MEL1*, and *GAL3*, also show relatively high levels of basal expression. In both yeasts, the tightest level of regulation occurs in the Leloir genes as well as in the genes encoding for lactose/galactose permeases (*GAL2* in *S. cerevisiae*, *LAC12* in *K. lactis*) and the β-galactosidase-encoding *LAC4* gene of *K. lactis*. The regulatory region of these genes contains multiple UAS_G and, therefore, they do not show particularly high levels of basal expression (Melcher and Xu, 2001). The lower the levels of basal expression the greater are the levels of induction experienced by the *GAL* genes in the presence of galactose (Lohr *et al.*, 1995). Genes from the Leloir pathway can reach ~1000-fold levels of increased expression in *S. cerevisiae*. *K. lactis* higher basal levels lead to a more modest level of induction of ~100-fold. However, overall, both species reach similar degrees of Gal protein expression. The only major difference is that, because of its different regulation of the regulatory *GAL* genes, *K. lactis* has permanently semi-induced levels of the *GAL* genes which probably have resulted from a more frequent exposure to lactose and galactose carbon sources throughout the evolution of this yeast.

Gal4p activates transcription by recruitment of coactivators and the general transcriptional machinery to promoter regions through its activation domain (Traven *et al.*, 2006). Gal4p physically interacts with TBP, TFIIB (transcriptional pre-initiation complex), Gal11 and Srb4 proteins (components of the Mediator complex; Lewis and Reinberg, 2003), Cdk8/Srb10 (cyclin kinase of the RNA polymerase II complex; Hirst *et al.*, 1999), SWI/SNF (chromatin remodeling complex; Mohrmann and Verrijzer, 2005), SAGA (Spt-Ada-Gcn5 hystone acetyltransferase complex; Timmers and Tora, 2005), and proteasome components such as Sug1 and Sug2 proteins (González *et al.*, 2002; Sulahian *et al.*, 2006). It has been shown that SAGA is the first complex that binds to the Gal4p activation domain and it does so through its subunit Tra1p (Bhaumik and Green, 2001; Bryant and Ptashne, 2003; Bhaumik *et al.*, 2004). All studies point at SAGA working as a scaffold that helps to assemble the pre-initiation complex at the promoter, rather than acting as a histone acetyl transferase (HAT) (Larschan and Winston, 2001; Bhaumik and Green, 2001; 2002). It is still controversial whether SAGA is also needed for assembly of the Mediator to the *GAL* promoters. The most likely is that the Mediator can directly bind to the activation domain of Gal4p but that the presence of SAGA helps to stabilize such interaction (Traven *et al.*, 2006). Mediator, TAFs, and RNA polymerase II are in turn, required for an efficient recruitment of SWI/SNF to the *GAL* promoters (Lemieux and Gaudreau, 2004). The role of SAGA as a scaffolding factor is now verified, since recent reports have shown that members of the SAGA complex and RNA export factors, such as Sac3p, determine the intranuclear association of active *GAL* genes to the nuclear pore (Cabal *et al.*, 2006; Taddei *et al.*, 2006; Luthra *et al.*, 2007). This finding confirms the 'gene gating' hypothesis: upon activation, genes get confined to the nuclear envelope in order to help the processing and export of their corresponding messenger RNAs (Fig. 10.6).

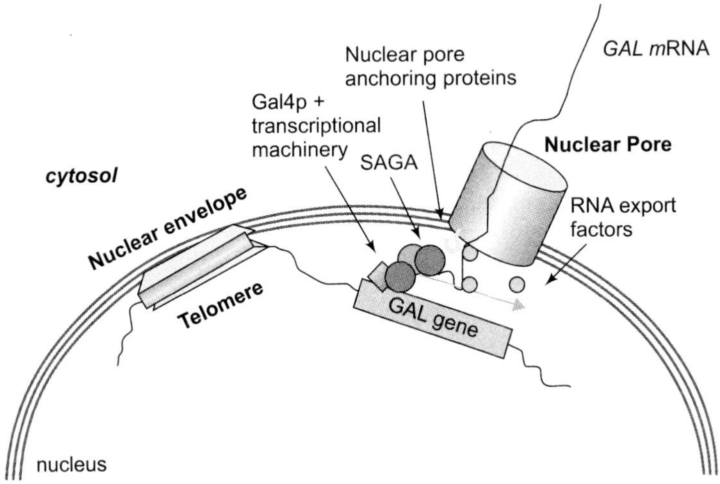

Figure 10.6 'Gating hypothesis' for the induction of the *GAL* genes. In the presence of galactose, members of the SAGA complex and interacting factors, such as messenger RNA export components, play a role in the association of *GAL* activated genes to the nuclear envelope to regions that are close to the nuclear pores. Furthermore, nuclear pore elements seem to also be involved, acting as anchors to physically link the activated *GAL* genes into these regions. In the absence of glucose, telomeric regions seem to play an additional role in facilitating the tethering of the *GAL* genes to the nuclear envelope prior to induction, such that when galactose is present, a rapid switch in telomeric chromatin may happen that then enhances the chances of activated genes to associate with nuclear pore regions. According to the 'gene gating' hypothesis, physical interaction and linking of activated genes with the nuclear envelope regions close to nuclear pores allows coupling of transcription to mRNA export into the cytosol.

It has now been known for quite a while that particular nucleosome positioning, influenced by binding of transcriptional regulators to certain regions within the *GAL* promoters, importantly determines their transcriptional activation state

(Bash and Lohr, 2001). For example, it has recently been found that deposition of the histone variant, Htz1p, typical from genes whose expression is frequently reprogrammed, is required for efficient Mediator recruitment to *GAL1* promoter and is dependent on Tup1p co-repressor (Gligoris *et al.*, 2007). The authors of this work propose that although being involved in the repression of *GAL* genes in the presence of glucose, through this action, Tup1p co-repressor marks the promoter for Mediator recruitment, therefore helping to poise the genes for a further transcriptional activation.

Chromatin remodeling complexes have also been shown to play an important role in contributing to epigenetic inheritance of transcriptional states. Transcriptional states produced by post-translational modification of nucleosomal histones can be passed from mother to daughter cells and be maintained through many generations, even when selective pressure is temporarily away. The discovery of this phenomenon has recently given birth to the new concept known as 'transcriptional memory' (Uffenbeck and Krebs, 2006; Levenson and Sweatt, 2006). In *S. cerevisiae*, the ATP-dependent chromatin remodeling enzyme, SWI/SNF, has an important role in keeping the rate of transcriptional induction of *GAL1* regulated by the prior expression state. It has recently been shown that the recruitment of SWI/SNF during the first round of *GAL1* transcription potentiates the reinduction of this gene following a period of glucose repression (Kundu *et al.*, 2007). A previous report has demonstrated that cells can also "remember" whether they were exposed to high or low concentrations of galactose in a separate way that is dependent upon the levels of cytosolic Gal3p (Acar *et al.*, 2005). Together, these and other recent studies have thus unveiled an interesting new level of regulation of genetic networks such as the GAL pathway.

Activation of the galactose promoters is not only dependent on the recruitment of all of these machineries but also on à series of post-translational modifications taking place on Gal4p. Gal4p phosphorylation is mediated by the cyclin-dependent protein kinases of the RNA Pol II holoenzyme (Hirst *et al.*, 1999; Rohde *et al.*, 2000). Gal4p undergoes a series of phosphorylations, which produce three distinct migratory forms (a, b, and c). Form c is the slowest migrating form and is exclusive of transcriptionally active Gal4p in the presence of galactose (Sadowski *et al.*, 1991). This form, in particular, corresponds to Gal4p phosphorylated at Ser 699 and Ser 837 (Sadowski *et al.*, 1991; Muratani *et al.*, 2005). Mutation of Ser 699 affects the induction of the *GAL* genes only in *GAL80*$^+$ cells (Sadowski *et al.*, 1996), which suggests that phosphor-Ser 699 may have a role in stabilizing a Gal80-Gal4 conformation that allows transcriptional activation. Alternatively it may serve as a landmark to target transcriptionally active Gal4p for proteolysis, which is an important process to allow transcriptional elongation once the activator has recruited the rest of the transcriptional machinery to the promoter (Muratani *et al.*, 2005). In this respect, recent reports have also shown a role for the ubiquitin degradation system in the turnover of active Gal4p. It has been demonstrated that Gal4p turnover is controlled through the action of two different F-box proteins, Grr1p and Dsg1p (Muratani *et al.*, 2005). Grr1p restricts the activity of Gal4p under non-inducing conditions, whereas Dsg1p acts positively on the expression of *GAL* genes by turning over the transcriptionally active Gal4c form. (Interestingly, Grr1 also plays a role in the downregulation of proteins involved in the glucose sensing/signaling Snf3/Rgt2 pathway; Santangelo, 2006.)

Induction of the *GAL* genes also involves recruitment by active Gal4p of a proteasomal subcomplex containing the six proteasomal ATPases, and Rpn1 and Rpn2 proteins, but lacking the 20S core and lid factors (González *et al.*, 2002; Sulahian *et al.*, 2006). This seems to be a common phenomenon that extends beyond the regulation of the *GAL* genes (Sikder *et al.*, 2006). It has been shown that the proteasomal ATPases play a role in SAGA recruitment to promoters during early stages of the transcription cycle (Lee *et al.*, 2005). They are also important to determine H2B histone monoubiquitination and H3 histone methylation, the latter mark being associated with recently transcribed genes (Ezhkova and Tansey, 2004). It is now demonstrated that these regulatory particles of the proteasome have, in fact, nonproteolytic activity that destabilises the transcriptional activator-DNA complex in an ATP-dependent fashion, and that this is a requirement for the promotion of transcriptional elongation (Nalley *et al.*, 2006). A very recent work has shown that transcriptional factors such as Gal4p are monoubiquitinated in their DNA binding domain and that this confers them with resistance to their 19S-mediated dissociation of the DNA (Ferdous *et al.*, 2007). This discovery now explains why certain E3 Ub ligases are important for the efficient function of some transcriptional activators (Salghetti *et al.*, 2001). The deletion of either *GRR1* or *DSG1* genes had no effect on Gal4p transcriptional activity so the E3 ligase responsible for this activating monoubiquitination of Gal4p has yet to be identified.

Main Factors Influencing Repression of the *GAL* Genes

Glucose sensing and signalling is a complex process that depends on the intricate and overlapping interplay of different pathways involved in central regulatory nutrient sensing and metabolism, such as TOR/MAP kinases (Wilson and Roach, 2002), and the Ras/cAMP/PKA pathway. Three main pathways seem to be in charge of transducing the glucose signal into the nucleus: the Ras-cAMP pathway (although some of its effects are mentioned here, this mechanism will not be discussed in detail because of spatial reasons, but for more information on this pathway can be found in Rolland *et al.*, 2002; Santangelo, 2006; Mirisola *et al.*, 2007); the Snf3/Rgt2/Rgt1 pathway, already discussed above (Holsbeeks *et al.*, 2004); and the Mig1-Hxk2/Snf1 glucose repression/derepression signalling pathway. This latter pathway is now quite characterized in *S. cerevisiae*, and deserves a special mention in this chapter because of its important role in shutting-off expression of the galactose genes (Fig. 10.7).

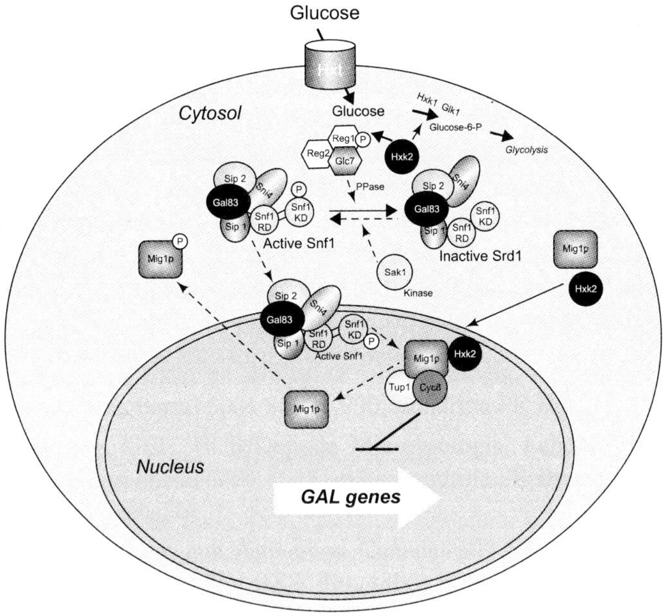

Figure 10.7 Simplified model of the main glucose repression pathway in yeast. In the presence of glucose, the hexokinase/regulator protein Hxk2 and the repressor DNA-binding protein Mig1p are recruited, along with the general co-repressor complex Tup1-Cyc8 into the *GAL* promoters which then inhibits their expression. The Snf1 complex composed of the protein kinase Snf1p and its regulators (Snf4p, Sip1p, Sip2p, and Gal83p) play a central role in catabolite derepression. In the presence of glucose, overlapping mechanisms keep the Snf1p kinase complex inactive and located at the cytosol. First, the interaction of the regulatory domain with the catalytic domain in Snf1p autoinhibits self-phosphorylation, which is a modification needed for its activation. Second, the Reg1/Glc7 phosphatase complex dephosphorylates any Snf1p that may remain active at the cytosol. Hxk2p is also involved in triggering repression of Snf1p activity, through the regulation of the phosphorylation state of Reg1p. When glucose levels decrease, these overlapping mechanisms are alleviated which permits binding of the Snf4p activating subunit to the regulatory domain of Snf1p, releasing the protein kinase from autoinhibition. Snf1 kinase phosphorylation is also reversed by the action of other kinases, as Sak1p. The subunit Gal83p determines active Snf1p nuclear localization, which then causes phosphorylation of Mig1p. This last event results in disruption of Mig1p interaction with Hxk2p and, therefore, release of Mig1p-mediated repression, by increasing the chances of Mig1p being exported back to the cytosol.

Mig1p is a Cys2-His2 zinc-finger protein with ability to bind specific upstream repressor sequences (URS_G) located in glucose-repressible promoters (Nehlin and Ronne, 1990; Cassart *et al.*, 1995). As with many other transcriptional activators and repressors, Mig1p activity is regulated through changes on its nuclear-cytosolic localisation. Mig1p nuclear concentration increases within minutes when cells are exposed to glucose and rapidly decreases upon glucose depletion (De Vit *et al.*, 1997). Mig1p does not inhibit transcription just by binding to its cognate sequence, but rather because it recruits the transcriptional co-repressor complex Cyc8-Tup1 to the promoters (Treitel and Carlson, 1995; Tzamarias and Struhl, 1995; Papamichos-Chronakis *et al.*, 2004). This repression is reversed by the Snf1 complex, a heterotrimeric kinase complex composed of a catalytic α-subunit (Snf1), a regulatory γ-subunit (Snf4), and three β-subunits controlling its subcellular localisation (Gal83, Sip1 and Sip2 proteins; Wilson and Roach, 2002). In particular, Gal83p plays a key role to determine Snf1 nuclear localisation in the absence of glucose and/or the presence of non-fermentable carbon sources (Vincent *et al.*, 2001). Upon derepression, all three subunits (α-, β-, and γ-) localize to the nuclear periphery, resulting in phosphorylation of perinuclear Mig1p (Treitel *et al.*, 1998; Sarma *et al.*, 2007). Derepression is then achieved because Mig1p phosphorylation in the perinuclear compartment favours its export back to the cytosol, mediated by the exportin Msn5p (De Vit and Johnston, 1999; Papamichos-Chronakis *et al.*, 2004). The recent finding that Mig1p requires perinuclear localisation for its repressor function has led to propose a 'reverse recruitment' model for gene regulation that is in accordance with the 'gene-gating' hypothesis (Sarma *et al.*, 2007). This theory proposes that transcriptional regulators may be tethered to nuclear pore complexes, recruiting the promoters of genes into the nuclear periphery in response to external signals (e.g., presence/absence of glucose), and controlling their expression by means of their own movement into or out of the perinuclear compartment.

The way in which cells initially sense glucose to trigger glucose repression through this signalling pathway is yet poorly understood, although it is known to involve the action of the main glycolytic hexokinase, Hxk2p (Santangelo, 2006). Both Snf1p and its antagonistic protein phosphatase, Glc7, seem to be receptors of this signal (Rández-Gil *et al.*, 1998; Mayordomo and Sanz, 2001). More light has been shed on the nature of this mechanism in the last few years. For example, it has recently been found that, in the presence of glucose, Hxk2p localises at the nucleus, where it interacts with Mig1p to form part of the repressor complex (Ahuatzi *et al.*, 2004). Ser311 of Mig1p is implicated on its interaction with Hxk2p and is also one of the four Snf1-dependent phosphorylation sites that govern nucleocytoplasmic transport of Mig1p (Ahuatzi *et al.*, 2007). Hxk2p also regulates the phosphorylation status of Glc7p phosphatase regulatory subunit, Reg1p (Sanz *et al.*, 2000). This induces Glc7p to negatively regulate Snf1 kinase activity. In low levels of glucose, a group of kinases (Sak1p, among them; Elbing *et al.*, 2006) counteract this effect, activating Snf1p-Gal83p so that the nuclear interaction of Hxk2p with Mig1p is outcompeted by Snf1 kinase, which can then reach and phosphorylate the critical Ser311 site on Mig1p, provoking its export back to the cytosol.

Further work showing the intricacies and intertwining of these three glucose repression pathways has recently demonstrated that the repression activity of

Rgt1p is regulated both by Snf1 and Tpk3 protein kinases (Palomino *et al.*, 2006). The cAMP-dependent protein kinase (PKA), which as mentioned above, is activated by extracellular glucose, is a heterotetramer composed of two regulatory subunits (Bcy1), and two catalytic subunits of three possible isoforms (Tpk1, Tpk2, and Tpk3; Rolland *et al.*, 2002). Tpk3 is responsible for Rgt1p hyperphosphorylation that promotes release of Rgt1p from the *HXK2* promoter. Snf1p is instead invoved in the modification of Rgt1p required for *HXK2* repression in the absence of glucose. As we have discussed in a previous section several parts of this pathway seem to work in a similar manner in *K. lactis*, although more characterization is still under way.

In *S. cerevisiae*, Mig1p binding to the URS_G in the *GAL3*, *GAL1*, and *GAL4* promoters reduces by 4-to 5-fold the levels of inducer, galactokinase, and transcriptional activator. This combined effect suffices to enhance Gal80p-mediated repression so that, overall, the *GAL* genes experience more than 1000-fold repression in the presence of glucose (Lohr *et al.*, 1995). In *K. lactis*, only one URS_G has been found in the *KlGAL1* promoter. This difference plus the self-activating circuit of *KlGAL4* expression are factors that importantly account for the relatively lower level of glucose sensitivity shown by yeast strains of this species (Dong and Dickson, 1997). Furthermore, in *S.cerevisiae* it has been shown that the glucose repression pathway operating through the Snf1 protein kinase and Mig1p also affects to the regulation of *HXT* expression. In the presence of glucose, Mig1p represses the expression of *MTH1*, which reinforces glucose-induced degradation of Mth1p and results in rapid glucose induction of *HXT* expression (Kaniak *et al.*, 2004; Kim *et al.*, 2006a, 2006b). Therefore, by boosting glucose transport, the system ensures a total shut-off of *GAL* and other genes for the metabolisation of alternative carbon sources.

Additional mechanisms at the level of post-transcriptional regulation are also playing an important role in the repression/derepression of the *GAL* genes. Gal proteins have, in general, a fast turnover, which ensures a rapid shut-off of the system once the genes are repressed (Johnston, 1987). Glucose triggers Gal2p permease ubiquitination, endocytosis, and targeting to the vacuole for degradation (Horak and Wolf, 1997). A recent report has also shown that plasma membrane sorting of Lac12p in *K. lactis* depends on active KlSnf1p, but through a process apparently independent from Mig1p, which suggests a more direct role of Snf1 at the protein sorting level (Wiedemuth and Breunig, 2005). Additional genes of yet uncharacterized function, such as *GAL6*, may also be regulating the *GAL* genes either at the level of mRNAs or other poorly understood post-transcriptional mechanisms (Zheng *et al.*, 1997; Zheng and Johnston, 1998).

Utilisation of the *GAL/LAC* Induction System and Genes in Biotechnology

The ability of certain microorganisms to metabolise lactose and/or galactose, and the inducibility of the genes involved in the utilization of these carbon sources, have repercussions in the pharmaceutical and food industries from different points of view. In the first place, the ability to metabolise lactose is applicable to solve a broad range of problems, from improvements in the quality of certain food products, or in the treatment of lactose intolerance, to the reduction of environmental pollution related to the elimination of lactose-containing cheese whey as a by-product from daily processing in worldwide dairy industries. Secondly, the inducibility of the *GAL* genes has proved to be an extremely useful system for the regulated over-expression in yeast of heterologous proteins of pharmaceutical and biomedical interest. As explained at the beginning of this chapter, the two main yeasts being currently used to take advantage of this property are *S. cerevisiae* and *K. lactis*, not only due to their quite extensive characterization but also because of their acknowledged GRAS status which makes them safe for utilization in food and biomedical-related industrial processes. In this section, we will first address the problem presented by the sugar lactose in food industry and how these microorganisms are used to solve different issues in this regard. Next, we will review some of the newest technologies developed in both yeasts to take advantage of the *GAL*-inducible promoter system. Finally, we will go through several new strategies of metabolic engineering that are now being developed for the genetic improvement of phenotypes of industrial interest.

Elimination of the Sugar Lactose for Different Biotechnological Purposes

The disaccharide lactose is the major sugar present in milk. Lactose has multiple industrial applications as a food ingredient, with the possibilities to be applied as a mild sweetener, flavour and dye carrier, emulsifier, and as positive effector in calcium absorption and the production of lactic acid in the intestine. It also has an important role as a supplement in infant formula since the contents of lactose in milk from animal origin is usually lower than in human milk (Rubio-Texeira, 2007). Lactose has also been extensively used in pharmaceutical industry for the coating of drug-containing pills and capsules (Yang and Silva, 1995; Ichikawa and Fukumori, 2007). However, lactose also has several undesirable effects. It has a lower solubility than other sugars, and a tendency to crystallize at low concentrations, which affects the texture of certain milk-derived foods, like ice cream (Marshall and Arbuckle, 1996). It also has lower sweetening power than sucrose, or its own hydrolytic product, galactose. In general, the hydrolysis of lactose improves the taste and quality of several dairy products. For instance, the generation of glucose and galactose mixtures, resulting from the hydrolysis of lactose in milk, significantly improves the ripening of cheese by specific microflora (Thomson and Gyuricsek, 1974). In fact, lactose/galactose-consuming yeasts like *K. lactis*, *K. marxianus*, and *S. cerevisiae* are frequently isolated from smear-ripened cheeses, where it has been shown that their action improves deacidification and the development of desirable volatile-sulphur-aromas (Kagkli *et al.*, 2006; Rubio-Texeira, 2007). The hydrolysis of lactose also enhances beneficial properties derived from the metabolism of other microorganisms such as Lactobacilli, reducing fermentation time and improving sweetness and viscosity in other dairy products like yogurt and kefir (O'Leary and Woychik, 1976; Hilgendorf, 1981). Lactose-hydrolyzed whey syrups also have better flavouring properties, becoming excellent supplements of protein and sugar in the manufacturing of food products such as dairy and bakery desserts, soft fruit drinks, livestock feeds, etc (Gekas and López-Leyva, 1985; Yang and Silva, 1995; Siso, 1996).

Apart from all of these effects of lactose in the quality of food, one of the most important problems that dairy industries face is the extensive percentage of human populations suffering from variable degrees of lactose intolerance, in many cases due to genetic deficiencies in the expression of lactase (Swallow, 2003; Adam *et al.*, 2004). Although with a lower incidence, lactose intolerance also affects some species of animals, particularly after weaning (Moulin and Galzy, 1984; Olchowy *et al.*, 1993). This poses a limit to the concentration of lactose that can be tolerated not only in human dairy-related foods, but also in dried concentrated whey-permeate for livestock feeds (which, as we will discuss below, has been considered among the alternative options for the elimination of this contaminant by-product).

The possibility to generate transgenic livestock expressing lactases to solve the general problem derived from human and animal lactose intolerance may soon turn into reality. Transgenic bovine chimeric offspring have already been produced from somatic cell-derived stem-like cells, opening new possibilities for the genetic manipulation of cattle in order to generate cows enabled for the production of milk with reduced levels of lactose (Cibelli *et al.*, 1998). In fact, to further prove the feasibility of this approach, Jost *et al.* (1999) have developed the first transgenic mice ectopically expressing rat intestinal lactase-phlorizin hydrolase in the mammary gland. The contents of lactose in the milk from these transgenic mice was reduced by 50–85% without affecting normal concentrations of fat and protein, and suckling mice nourished and developed normally from mothers producing low-lactose milk. Transgenic chickens expressing β-galactosidase in their intestine have also been generated (Mozdziak *et al.*, 2003). The acquisition of this new trait improves calcium absorption in birds favoring the use of cheese whey-derivates in poultry feeds. In the possible event that current restrictive legislations from different countries are overcome, this may turn out to be an interesting solution for the generalised problem of lactose-intolerance.

Lactose intolerance is not the only human genetic disorder related to galactose metabolism. Accumulation of galactose-1-phosphate, caused by the lack of functional galactose-1-phosphate uridyltransferase (E.C. 2.7.7.12; GALT), leads to galactosemia in humans (Leslie, 2003). If milk is not withdrawn from the affected infant, acute hepatotoxicity syndrome, *Escherichia coli* sepsis, and bleeding disorder eventually lead to death. Even under a galactose-restricted diet, patients still suffer from long-term complications such as ovarian failure, speech dyspraxia, ataxia, and mild mental retardation (Lai and Klapa, 2004). The exact mechanisms by which galactose-1-phosphate causes toxicity are still poorly understood, although a possible role of this molecule in the regulation of the inositol monophosphatase (IMPase), activity which is critical for the sustenance of neurotransmitter functions in the brain, has recently been discussed (Bhat, 2003).

Therefore, focusing on all of these problems, an interest has been raised in the last two decades for the development of technologies to isolate the enzyme responsible for the hydrolysis of lactose into glucose and galactose, β-galactosidase (E.C. 3.2.1.23), from microorganisms naturally possessing such activity (*K. lactis* among them). Enzymatic hydrolysis of lactose is more desirable in food technology than chemical hydrolysis because the latter requires drastic conditions of pH and temperature that ultimately reduce the quality of the dairy product after processing (Gekas and López-Leyva, 1985; Siso, 1996). Several bacteria, fungi, and yeast, are potential supplies of β-galactosidase. Differences in optimum pH, thermostability, and intracellular or secretable nature of the enzyme usually dictate the selection of the most suitable source for each particular biotechnological purpose. Although new potential sources for this enzymatic activity are continuously being found and analysed, the corroborated GRAS status of yeasts like *Kluyveromyces*, and of fungi like *Aspergillus niger* and *A. oryzae*, as stated previously, still place these among the favorite sources of β-galactosidase for food biotechnology and the pharmaceutical industry (Bonekamp and Oosterom, 1994). Yeast β-galactosidases have an important advantage over fungal lactases since certain physical properties, such as an optimal pH within the neutral range (6.9–7.5), makes them more suitable for the hydrolysis of lactose in dairy products, like milk and sweet cheese whey. The native intracellular *K. lactis* lactase, produced on an industrial scale and sold under the trade name Maxilact™, has already been on the market for decades and has a history of safe use (DSM Food Specialties, Delft, the Netherlands; Association of Manufacturers of Fermentation Enzyme Products, AMFEP, http://www.amfep.org). Neutralact®, is the DSM brand name of another commercially available lactase preparation, also produced from a *K. lactis* strain containing extra copies of the *LAC4* gene inserted in the rDNA locus (Coenen *et al.*, 2000).

Perhaps the most important problem to overcome in the utilisation of yeast lactases is their intracellular nature, which makes them less stable in solution, usually making it necessary to carefully control the physico-chemical parameters in their microenvironment in order to optimize their *in vitro* activity. Most of the methods developed involve the immobilization of the enzyme, or even whole cells, onto solid supports. The immobilisation usually implies insolubilisation of the enzyme that can be carried out through either chemical or physical binding. Chemical binding involves covalent binding, crosslinking, or co-polymerisation with a carrier. There is a great diversity in the type of carriers that can be used (i.e., cellulose, cotton cloth, nylon, acrylic copolymers, glyoxyl, glutaraldehyde, etc.). In physical binding, the enzyme is adsorbed or entrapped in gels, fibers, microcapsules, or membranes. The advantages and caveats of these different methods are extensively reviewed in Kilara and Shahani (1979); Gekas and López-Leyva (1985); and Clark (1994). These methods not only improve the stability of β-galactosidases but also reduce the competitive inhibition by galactose and non-competitive inhibition by glucose (Cabaille and Combes, 1995; Mateo *et al.*, 2004), at the same time as facilitating the recovery of the enzymatic activity for recyclatory use (e.g., *K. lactis* β-galactosidase immobilized in polyacrylamide gel beads showed significant stability over repeated runs; Ohmiya *et al.*, 1977). In terms of applying β-galactosidase preparations for the treatment of lactose intolerance, this enzyme has not only been used to hydrolyse the lactose present in dairy products like milk for consumption, but medical treatment methods have also been developed consisting of the inclusion of β-galactosidase entrapped in dried liposomes that can be ingested simultaneously with dairy meals (Kim *et al.*, 1999).

β-galactosidases also have an interesting ability to generate, under particular conditions, lactose derivatives by transgalactosylation. In this secondary reaction, the galactose moiety from a β-galactoside is transferred to an acceptor sugar molecule, which leads to the formation of galactooligosaccharides (GOS). GOS may then consist of a variable number of oligosaccharides bound through β-glycosidic linkages. Utilisation of combined techniques, such as immobilization of the isolated enzyme, use of high lactose concentrations, and microwave irradiation, has been reported as a method to improve the selectivity of the lactase reaction towards GOS production (Maugard *et al.*, 2003). In particular, the utilisation of mixtures of lactose and fructose allows the formation of lactulose (4-*O*-β-D-galactopyranosyl-D-fructose). GOS and lactulose have a very important use as prebiotics (Méndez and Olano, 1979; Sako *et al.*, 1999; Szilagyi, 2002). Prebiotics are considered as compounds that are able to go through the initial digestive barriers and reach the bacterial flora in the colon intact, where they are selectively metabolized by these benign Bifidobacteria and Lactobacilli, helping them in their proliferation and competition against colonization of the intestine by other pathogenic microorganisms. This beneficial property of prebiotics makes them important additives in infant formula and other dairy products (Yang and Silva, 1995). Lactulose is also used in pharmaceuticals as a mild laxative and in the treatment of hepatic encephalopathy, a disorder characterized by abnormalities in the metabolism of ammonia that are associated with liver failure and brain deficiencies (Hallmann, 2000; Dasarathy, 2003). Lactulose acidifies ammonia in the colon, which reduces its built-up and facilitates its excretion. A recent study has shown that *K. lactis* β-galactosidase is able to produce 5- to 10-fold more lactulose in a 2- to 7-times shorter period of time, than the β-galactosidases from *E. coli* and *A. oryzae* (Lee *et al.*, 2004).

As previously mentioned, the elimination of the lactose fraction from cheese-whey is another crucial factor for the economy of dairy industries. Several million tons of cheese-whey are produced annually worldwide and are, for the most part, released to the environment. Cheese-whey is an industrial by-product with a high biological oxygen demand (BOD) because of its high content (~5%) of the sugar lactose. When released into the environment, its high content of this sugar causes rapid bacterial proliferation in rivers and streams, placing whey on the list of major industrial contributors to environmental pollution. In several countries legislative procedures have been developed to restrict the amount of lactose-containing cheese-whey that can be discarded into the environment. However, most methods that successfully eliminate the lactose fraction from the cheese-whey prior to its elimination are economically unviable for many of the worldwide smaller dairy companies. Hence, the massive interest awakened in the last two decades to come up with a globally economic solution for the elimination of lactose from cheese-whey (Mawson, 1994).

Elimination of lactose from cheese-whey through fermentation carried out by microorganisms promises to be among the most economical solutions, since it not only allows a fast elimination of this sugar but it also has the potential for its simultaneous conversion into several products of higher commercial value. In some cases, anaerobic digestion of cheese-whey by methanogenic bacteria is used to produce biogas that can, in turn, be used as a source of energy to maintain the process, reducing its overall costs (Mawson, 1994; Ergüder *et al.*, 2001). In other cases, lactic acid bacteria (LAB) have been used because of their ability to produce bacteriocins, polysaccharides, and other potentially useful products (Flôres and Alegre, 2001; Briczinski and Roberts, 2002; Simova *et al.*, 2004; Nicolas *et al.*, 2004). Certain fungi produce alternative enzymes of industrial interest, such as the ligninolytic enzyme manganese peroxidase, by growing on cheese whey (Feijoo *et al.*, 1999). Despite the potential inherent in the use of all of these alternative microorganisms, most efforts are still being focused in improving the utilisation of the lactose-metabolising species of *Kluyveromyces* (Bonekamp and Oosterom, 1994; Maullu *et al.*, 1999; Ghaly and Kamal, 2004). The predominance of oxidative metabolism in these yeasts favours fast growth and rapid translation of the carbon source being metabolised, into biomass and other products of industrial interest.

One of the most immediate products of commercial interest that can be obtained from the industrial fermentation of lactose is indeed cell biomass. Yeast cell biomass recovered after the fermentative process is an excellent supplement for animal and human diets because of its content of ribonucleotides, poly- and oligosaccharides, and single cell protein (SCP) (Belem and Lee, 1998). SCP has multiple applications as a food supplement because it has interesting effects, such as acting as flavour carrier, thickener, stabilizer, etc. But perhaps, one of the most important uses of this microbial cell biomass is in dietary supplements, as probiotic (Fleet, 2007). Probiotics are microorganisms which, as prebiotic compounds, have the ability to survive the gastric barrier and reach the intestines where they help the immune system to compete and exclude pathogens and undesirable bacteria from adherence to the intestinal mucosa. Lactose-consuming microorganisms, such as lactic acid bacteria (LAB), and yeasts as *K. lactis* (and therefore, the consumption of dairy products in which these organisms are present) have beneficial effects on human health because of their probiotic condition. A recent study has shown that *K. lactis* has the ability to adhere to human enterocyte cells, showing resistance to bile acids and proteases, without eliciting allergenic or inflammatory responses (Kumura *et al.,* 2004). *S. cerevisiae* var. *boulardii* is also being studied in humans, for its use as probiotic in the treatment of serious intestinal disorders such as Crohn's disease (Fiocchi, 2006). The use of yeast biomass from *S. cerevisiae* and *K. lactis* produced by industrial fermentation processes is, in fact, displacing the traditional soya bean-based meals for livestock. In a recent study, it was shown that the replacement of soya bean meal with whey-derived yeast protein in the diets of reared piglets had positive effects in daily weight gain, and reduced the appearance of allergenicity and intestinal disorders (Spark *et al.*, 2005).

Heterologous and Homologous Gene Expression using the Galactose-inducible Genetic System

Another obvious way to profit from the microbial fermentation processes at a large scale is to couple the generation of biomass with the synthesis of alternative products of commercial interest. The increasing characterization of how regulatory networks dictate gene expression in yeasts like *S. cerevisiae* and *K. lactis* has allowed, in recent decades, to develop new technologies of recombinant DNA that enable the production, at industrial levels, of multiple heterologous proteins of pharmaceutical interest. The utilisation of inducible promoters to express heterologous proteins is, in particular cases, preferable over the use of constitutive promoters because it allows the separation of the growth phase from the production phase, which is particularly important when the protein being expressed is toxic to the cells (for a review about pioneer studies in heterologous gene expression using yeast, see Romanos *et al.*, 1992; Gellisen and Hollenberg, 1997). Galactose-inducible promoters are particularly attractive for this purpose, because this system allows the trial of different mixtures of carbon sources in the medium to optimize biomass growth yield and protein production rates.

While the use of a galactose-inducible system for heterologous protein production by *S. cerevisiae* has proved very successful at the laboratory level, one major problem for the use of this inducible system in industrial scale processes is the cost of the prime material for culture growth. Galactose is a relatively expensive carbon source in its purified form, so *S. cerevisiae* is instead usually grown in fermentation media consisting of complex mixtures of different fermentable sugars. These prime materials such as sugar cane molasses and grape contain high proportions of glucose, sucrose, and fructose. Beer wort and beet molasses are also rich in glucose, fructose, sucrose, maltose, and maltotriose. The initial concentration of all of these sugars is high enough to trigger undesirable responses in the cells, such as catabolite repression of respiration, and of the use of alternative carbon sources such as galactose. This is a caveat because, even in the case in which galactose was added as an inducer, catabolite repression would dampen the process not only by repressing *GAL*-inducible promoters but also because repression of respiration in favor of alcoholic fermentation results in lower biomass yields, hence, in a lower production of the heterologous product. Growth in these mixtures also triggers several stress pathways in yeast that are detrimental in terms of overall heterologous productivity (Verstrepen *et al.*, 2004). As we will see in the next subsection, genetic modifications of yeasts have been carried out in order to solve this problem, that allow the use of galactose as a non-metabolisable inducer. Lactose-containing cheese-whey is a substrate that for all the obvious reasons mentioned above has for quite a while been considered as an alternative prime material for the industrial growth of yeasts in fermentative processes. Whey, and in particular, whey permeate have the disadvantage of being low in organic nitrogen sources, but this problem is overcome by the addition of nitrogen supplements or through hydrolysation of whey proteins that reduce the need for more expensive supplementations (Yang and Silva, 1995).

As mentioned previously, *Kluyveromyces* industrial biomass obtained by growth in cheese-whey was primarily utilised in the 1960s as a supplement for food. However, the increasing characterization of the molecular mechanisms governing gene expression in *Kluyveromyces* yeasts has made this yeast an excellent alternative to *S. cerevisiae* for heterologous protein production. Soon after the development of new mechanisms for its genetic manipulation, heterologous protein production was successfully achieved using this yeast (reviewed by van Ooyen *et al.*, 2006). Furthermore, the current availability of the fully sequenced *Kluyveromyces* genome promises to soon unveil novel perspectives in this respect (Dujon *et al.*, 2004). *K. lactis LAC4* promoter was one of the first promoters employed to direct heterologous protein expression in this yeast and it has been used for industrial-scale expression of an extensive variety of heterologous proteins of pharmaceutical interest with high yields, such as in the case of the synthesis of bovine prochimosin (for more examples, see van Ooyen *et al.*, 2006). Other galactose-responsive promoters such as *KlGAL7* have also provided good yields at the fermentor level, such as in the production of human lysozyme (Maullu *et al.*, 1999). Even the utilisation of heterologous promoters such as *GAL*-inducible promoters from *S. cerevisiae* has also proved their validity in the generation of chimeric constructs to secrete recombinant proteins in *K. lactis* (Bergkamp *et al.*, 1992; González *et al.*, 1997). A lactose promoter variant ($P_{LAC4\text{-}PBI}$) has recently been constructed by deletion of sequences in the promoter ressembling the bacterial Pribnow box transcriptional element, with the aim to prevent bacterial expression of deleterious proteins that could affect intermediate cloning steps in the construction of *LAC4* promoter-based expression vectors. This variant exhibits negligible bacterial expression without affecting the expression levels normally attainable in *K. lactis* (Colussi and Taron, 2005). Basal/leaky expression from promoters such as *GAL1* in non-inducing conditions may also lead to undesirable phenotypes prior to induction in yeast. In order to fix this problem, a group has made advantage of the fact that *GAL1* promoter fused to the bacterial endonuclease gene encoding PvuII restriction enzyme is toxic for cells in noninducing conditions (Lewis *et al.*, 2005). Using this cytotoxic assay they PCR-mutagenized the *GAL1* regulatory region and, selecting for the loss of toxicity, they were able to isolate mutant promoters that retained high inducibility but exhibited reduced basal level of expression.

Apart from *Kluyveromyces* natural ability to use lactose as a carbon source, additional characteristics of this yeast makes it advantageous over the use of the more traditional *S. cerevisiae* for heterologous protein production. In terms of protein secretion, *Kluyveromyces* shows a stronger ability to secrete proteins, which is a desirable character that avoids the need to lyse cells at the end of the fermentation in order to recover the product. However, especially in the production of proteins for human therapeutic uses, problems such as yeast differential glycosylation and protein folding patterns still need to be overcome in order to express proteins as close as possible in their properties to their human native counterparts. These and other problems are now being tackled through the use of a novel emerging science: metabolic engineering.

Metabolic Engineering Strategies to Improve the Utilisation of Lactose/galactose Consuming Yeasts in Biotechnology

Metabolic engineering can be considered a new science that has evolved from the application of new technologies in molecular biology, biochemistry, and genetics, as well as the increasing molecular characterization of certain biological systems (precisely attained through the use of these technologies), in order to create and/or improve a particular phenotype in an organism by means of recombinant DNA technology (Stephanopoulos *et al.*, 1998). Metabolic engineering can follow two main different strategies. In 'constructive metabolic engineering', the metabolism in a wild-type cell is first characterized so that this knowledge can be later applied to execute genetic modifications that improve the expression of a relevant phenotype. In 'inverse metabolic engineering' a particular host is genetically modified with the aim to improve a determined property through transference onto that host of the genes known to be responsible for the desired phenotype in a different organism (Bailey *et al.*, 2002). In the latter case, full characterization of the host of interest is usually not required. However, in both cases changes in the overall cell physiology of the modified organism may be assessed to ensure that the introduction of the new trait does not negatively affect other properties previously present in the host.

Displacement of entire biosynthetic pathways into alternative host organisms is now possible by means of this new methodology. This possibility expands the potentials in the industrial utilisation of microbial fermentations not only for the generation of biomass or for the synthesis of proteins and peptides with pharmaceutical relevance, but also for the production of additional compounds and metabolites, which until now could only be produced at a high-scale level through methods of chemical organic synthesis. For instance, an artificial and fully self-sufficient mammalian biosynthetic pathway for the synthesis of hydrocortisone involving a total of 13 engineered genes has been assembled in yeast, that enables these new strains to produce hydrocortisone as the major steroid (up to a 70% of the total steroids) by growing on glucose or ethanol (Szczebara *et al.*, 2003). In a separate work, amorphadiene, a sesquiterpene precursor to the anti-malarial drug artemisinin, has been successfully produced from *S. cerevisiae* after the galactose-regulated overproduction of acetaldehyde dehydrogenase and introduction of a *Salmonella enterica* acetyl-CoA synthetase variant (Shiba *et al.*, 2007).

An example of inverse metabolic engineering that is of particular relevance to the present chapter, is precisely demonstrated by the multiple strategies proposed over the years to confer *S. cerevisiae* with the ability to metabolise lactose, in order to generate industrial strains of further commercial interest e.g., baker or brewer yeasts, able to grow in the cheap and extensively available substrate cheese whey. The earliest atempts to transfer the lactose-consuming trait onto *S. cerevisiae* consisted of rudimentary techniques such as the generation of *Kluyveromyces/ Saccharomyces* hybrid cells by protoplast fusion (Farahnak *et al.*, 1986), or the addition of purified β-galactosidase into the medium to produce a mixture of glucose and galactose, which could then be metabolized by *Saccharomyces* (Champagne and Goulet, 1988). Several genetic manipulations involving a cell lysis strategy for the extracellular hydrolysis of lactose were also explored. For instance, Porro *et al.* (1991) obtained *Saccharomyces* Lac^+ strains by simultaneous overproduction of *GAL4* and the *lacZ* gene from *E. coli*. Gal4p overproduction seemed to cause a certain level of lysis that allowed the release of the heterologous β-galactosidase onto the medium. Cell permeabilization with toluene or ethanol has been used in other cases to release intracellular β-galactosidase (Compagno *et al.*, 1993; Lee *et al.*, 2004). Other groups have developed thermosensitive autolytic mutants with a similar objective (Becerra *et al.*, 1997; 2004). These 'autolytic' strategies may be convenient when an intracellularly synthesized product needs to be recovered from the cells at the end of a fermentative process (Belem and Lee, 1998). The extracellular hydrolysis of lactose achieved by *S. cerevisiae* cells has also been attempted with relative success at the laboratory level, by means of heterologous expression of secretable versions of intracellular β-galactosidases accomplished through the fusion of these genes to *Saccharomyces* secretory signals, such as that of the *STA2* gene fused to *E. coli lacZ* (Porro *et al.*, 1992), or that of the alpha-factor pheromone gene to *Kluyveromyces LAC4* (Becerra *et al.*, 2002). However, the expression of fungal, naturally secretable β-galactosidases (e.g., from *A. niger*) has proved, so far, to be the most efficient method for the extracellular hydrolysis of lactose by *Saccharomyces* (Kumar *et al.*, 1992; Ramakrishnan and Hartley, 1993; Domingues *et al.*, 2002). Even so, the extracellular hydrolysis of lactose still faces a major problem when using *S. cerevisiae*-based cultures, which is that the glucose released from the hydrolysis of lactose reaches levels that are high enough to cause glucose repression and, therefore, impairment of the simultaneous use of galactose. This phenomenon causes diauxic growth, compromising the overall fermentation time, and biomass versus ethanol yield (Ramakrishnan and Hartley, 1993).

An alternative strategy involves the transfer of genes conferring *Saccharomyces* with a system for the permeabilization and intracellular hydrolysis of lactose. Lac^+ strains were obtained for the first time using this approach, by introduction in *S. cerevisiae* of the *LAC4* and *LAC12* genes from *K. lactis* through genetic transformation (Sreekrishna and Dickson, 1985). However, the endogenous *K. lactis* promoters were weakly expressed in *Saccharomyces*, which led to poor growth yields, and the type of plasmid used for the genetic transformation was also inadequate in terms of phenotypic stability. A significant improvement in the utilisation of these genes consisted of their expression from *CYC-GAL*, a *Saccharomyces*-derived, galactose-inducible hybrid promoter (Guarente *et al.*, 1982). Maintenance of high levels of expression and a mitotically stable phenotype were simultaneously achieved by coupling the genomic integration of this construct into noncoding regions of the ribosomal DNA (Rubio-Texeira *et al.*, 1998)

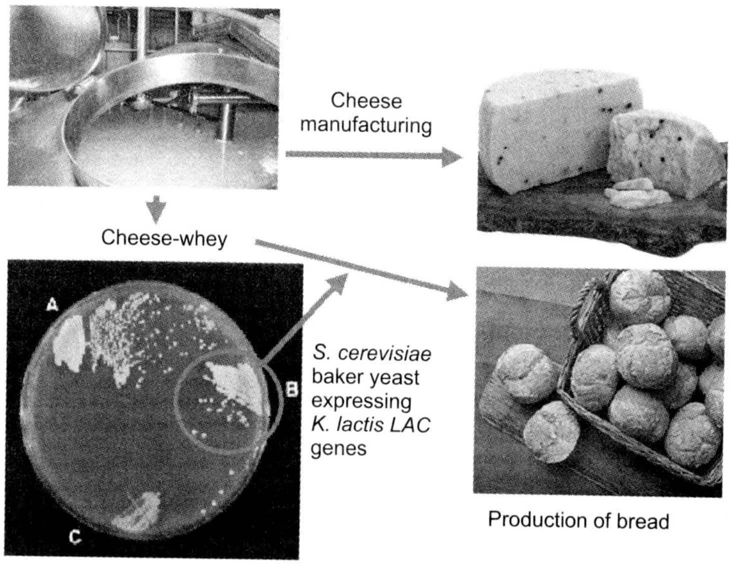

Figure 10.8 Strategy followed for the generation of a baker *S.cerevisiae* Lac$^+$ strain. A baker *S. cerevisiae* strain was genetically modified through insertion of the *LAC4* and *LAC12* genes from *K. lactis*, under the control of the galactose-inducible $P_{CYC\text{-}GAL}$ promoter, in the *RDN1* locus in order to obtain mitotically stable, high levels of expression of the heterologous *LAC* genes. The resulting strain was able to produce bread by growing on cheese whey (more details in Rubio-Texeira *et al.*, 1998; Adam *et al.*, 1999).

(Fig.10.8). Interestingly, the resulting strains displayed an enhanced oxidative metabolism that closely resembled that of *K. lactis*, generating high biomass yields. This effect can now be understood if we take into account the effect that the lactose transport rate very likely has on reducing the overall glycolytic flux of this engineered strain. This additional trait has therefore proved particularly useful to obtain high biomass yields and the further adaptation of this system into a baker's yeast has enabled the successful production of bread by strains able to grow on cheese-whey (Adam *et al.*, 1999). Similar approaches have, nevertheless, subsequently been explored in order to further improve the efficiency of the strategy (Jurascik *et al.*, 2006). Additional advantageous properties such as flocculation to facilitate the recovery of biomass at the end of the fermentative process have, for instance, been introduced into these lactose-consuming strains (Rubio-Texeira *et al.*, 2000; Domingues *et al.*, 2002).

But the possibilities to improve lactose bioconversion into products of higher commercial value do not stop here. In fact, the progressive characterization of the regulatory networks affecting the utilisation of different carbon sources in yeast, nowadays allows the researchers to make additional adjustments to optimize a particular phenotype just by changing the level of expression of the crucial functions involved in the regulation of a particular pathway. For example, significant alterations in the response of inducible promoter systems can be attained by genetic modification of particular interactions between regulatory components of a genetic network. For instance, *S. cerevisiae* cells simultaneously overexpressing three of the main *GAL*-gene regulators, Gal3p, Gal80p, and Gal4p, expressed 12-fold more β-galactosidase reporter from a galactose promoter, than the wild type cells (Sil *et al.*, 2000). This strategy did not seem to have any negative effects in the general physiology of the engineered strain. Its use is, then, particularly appealing, not only to improve homologous and heterologous protein expression driven by *GAL* promoters, but also because this strategy could also be adapted to lactic yeasts to perhaps enhance their already vigorous capability to consume lactose from cheese-whey.

The opposite strategy, knocking out the genes encoding the main repressor proteins of the *S. cerevisiae GAL* regulon, has also shown interesting results. In a recent study, the three repressors *GAL6*, *GAL80*, and *MIG1*, were simultaneously knocked-out (Ostergaard *et al.*, 2000). A 41% increased flux through the galactose utilization pathway was accomplished through the use of this strategy. Interestingly, this metabolic alteration does not yield higher levels of biomass but instead directs galactose utilisation towards ethanol production, which therefore makes this strategy potentially suitable to improve the performance of brewing and lactic yeasts in processes of alcoholic fermentation.

In a different work, it has also been shown that the expression of carboxypeptidase Y (CPY) from a *GAL10* promoter is enhanced in *GAL80* mutant cells growing in ethanol as a carbon source, which again indicates how an extensive part of the catabolite-repression effect exerted over galactose promoters containing multiple UAS_G can be prevented by the elimination of *GAL80* function (Shiba *et al.*, 2000). A further validation for the use of this type of approach to improve cells' behavior in determined environments lies in the fact that evolution was first in using such strategy to modify cellular fitness and adaptation in response to environmental variability. In a very recent study, a method was developed to map the genetic basis of naturally and experimentally evolved complex traits in order to understand the mechanisms of parallel evolutionary adaptation (Segrè *et al.*, 2006). This work showed that four *S. cerevisiae* yeast strains, independently adapted to a fluctuating glucose-galactose environment, ended up acquiring one or more missense mutations in *GAL80*. When transferred into the ancestral strain, the *gal80* mutations conferred the same fitness advantage that the evolved strains had shown in the transition from glucose to galactose. Other works involving bioinformatic analysis have also shown a connection between *GAL80* and the diauxic shift. Apparently, cultures of mutants lacking *GAL80* function experience elevated expression levels of the diauxic shift genes, which may improve their general fitness (Scott *et al.*, 2005).

One of the most interesting approaches recently developed for the utilisation of *GAL*-directed expression consists of the generation of a gratuitous induction system for galactose. The first attempts to create such a system were based on the combined utilisation of the *LAC4* promoter and *Kluyveromyces* strains deficient in the galactokinase function but not in the sensor/inducer function of KlGal1p. Deficiency on this enzymatic activity of the Leloir pathway allows galactose-driven induction of the *GAL* promoters but impairs the metabolisation of galactose (Hsieh and Da Silva, 2000; Panuwatsuk and Da Silva, 2003). This strategy has two important advantages. Firstly, it prevents the need to supplement the growth medium with high concentrations of the inducer galactose, which helps to reduce the overall cost of the fermentative process. Secondly, the combination of this technique with the elimination of repressors, or with enhanced expression of activators, could solve the problem of using cheap prime material mixtures, such as molasses, in industrial processes. The reason for this is that catabolite repression over the *GAL* genes may significantly be overcome through the use of this strategy. For example, the construction of a *S. cerevisiae* double mutant *gal1*Δ *hxk2*D has proved successful at a laboratory scale to express increased levels of secreted recombinant human serum albumin (HAS) protein from the *GAL10* promoter by growing cells in medium containing 2% glucose as carbon source (Kang *et al.*, 2005). The maximum levels of HSA protein and transcript in the *gal1*Δ strain were detected with 0.05% galactose, which is one-twentieth of the amount required for full induction in a wild type strain. A different approach consisted of the generation of a *S. cerevisiae* strain in which the *GAL1* gene was replaced by an extra copy of the *GAL4* gene to evaluate whether both gratuitous induction and regulated overexpression of the transactivator could further improve protein production from low- and multi-copy expression vectors containing the *GAL1* promoter fused to the structural gene for the green fluorescent protein, GFP (Stagoj *et al.*, 2006). This strategy increased up to 4.6-fold the levels of GFP expression. These levels were even higher than those obtained by the use of a *gal1*Δ mutant alone. Another strategy has recently been ideated as an alternative to the 'gratuitous inducer' methodology. In this case, a combinatorial library of engineered zinc finger proteins fused to the appropriate transcriptional domain to function as artificial transcriptional factors (ATFs) was generated, and screened for ATFs that are able to counteract glucose repression, and/or activate transcription from *GAL1* promoter in a galactose-independent manner (Park and Kim, 2006).

Another interesting target for metabolic engineering is the achievement of an improved regulated tuning of the galactose genetic switch, relying on the levels of expression of the galactose permease, Gal2p. A very recent study has shown that the removal of the positive feedback control loop regulating Gal2p expression either by deletion, controlled titration, or constitutive expression of *GAL2*, is sufficient to make the *GAL* promoters respond in a linear manner to changes in the levels of galactose (Hawkins and Smolke, 2006). Such behaviour of the network cannot be achieved by similarly changing the levels of other regulator proteins such as Gal80p or Gal3p, which once again points out the importance that the rate of the sugar transport has in the overall control of carbon metabolism. In this same study, it is also shown how Gal1p plays a dual role in the pathway. Removal of the catabolic activity of Gal1p increases the effective concentration of galactose in cells, which is expected to increase the response of the network at all galactose concentrations. However, the removal of the effect that Gal1p has in Gal2p transport activity (demonstrated by Ramos *et al.*, 1989) also eliminates the high affinity Gal2p transport mechanism, and this decreases the response of the network, particularly at lower galactose concentrations in which this transport mechanism is essential for efficient galactose uptake. In the absence of Gal2p, the effect of Gal1p in transport is no longer affecting the pathway but a galactokinase mutation in the presence of constitutively expressed Gal2p can further enhance the linearity of the response in a broader range of galactose concentrations precisely because of the overlapping effects of the Gal2p-regulating and catabolic roles of *GAL1*.

In yet another separate example of inverse genetic metabolic engineering it has recently been found through genome-wide transcriptomic analysis that the overexpression in *S. cerevisiae* of the major phosphoglucomutase gene, *PGM2*, increases the galactose uptake rate by a 70%, indicating that the enhancement of this activity has a positive effect in increasing the Leloir pathway flux because of a more rapid conversion of glucose-1-phosphate into glucose-6-phosphate (Bro *et al.*, 2005). Switches between the fermentative and respiratory metabolisms are also currently possible by means of metabolic engineering technology. For instance, a group has recently been able to alleviate the Crabtree effect in *S. cerevisiae* through the overexpression of heterologous NADH oxidases from *Streptococcus pneumoniae* and *Histoplasma capsulatum* (Vemuri *et al.*, 2007). The expression of these heterologous enzymes allowed *Saccharomyces* cells to overcome their deficient ability to reoxidize NADH in the presence of high glycolytic fluxes. Interestingly, one of the approaches used in the study led to increased activity of the TCA cycle. Other studies have shown that by potentiating the activity of the TCA cycle an overall increase in amino acid biosynthesis may be reached (Mathy *et al.*, 2006). This detail may be of particular importance if this type of strategy is ever coupled to heterologous protein production.

Similar techniques are also being applied to the progressively more characterized *K. lactis*. For instance, a mutation in *RAG5*, encoding the single hexokinase present in this yeast, which leads to a glucose non-repressible phenotype, resulted in a significant improvement in the ability of this yeast to secrete several heterologous proteins (Donnini *et al.*, 2004). The authors discuss the possibility that altered ratios in the levels of glucose-6-phosphate and glucose-1-phosphate may alter intracellular Ca^{2+} homeostasis, which in turn may affect endoplasmic reticulum and Golgi functions. This mutant also shows a slight defect in glycosylation. As mentioned previously, *K. lactis* has a higher secretory capacity compared to *S. cerevisiae*, which makes it a more appealing organism for heterologous protein secretion. The main problem to overcome with both yeasts relies in the glycosylation pattern, which is different from that of mammalian cells and can cause problems in the native activity, and potential immunogenicity when intended for therapeutic use. New metabolic engineering strategies are now being designed to 'humanize' the glycosylation pattern of heterologous proteins produced by yeast and fungi (reviewed by Gerngross, 2004). *K. lactis* is a more enticing microorganism than *S. cerevisiae* in terms of solving this particular aspect since in general it produces a lower level of mannosylation in secreted proteins. Mutations in genes encoding for mannosyltransferases showed reduced levels of N-glycosylation and improved secretory capability (Bartkeviciute and Sasnauskas, 2004; Uccelletti *et al.*, 2006). Further potential improvements in the secretion of heterologous proteins in both yeasts involve either the deletion or overexpression of functions involved in ER protein folding, proteolytic processing, and transport through the secretory pathway (Romanos *et al.*, 1992; van Ooyen *et al.*, 2006).

Future Perspectives

Functional genomics ('ome') technologies are already enabling a more successful design of metabolic engineering strategies (Fig. 10.9) Bro and Nielsen, 2004). Genome-wide screens using yeast deletion or gene-over expression libraries are now traditional methods to search for genes affecting a particular phenotype. Whole-genome DNA sequencing and hybridization of genomic DNA to whole-genome DNA microarrays allow the researchers to screen for polymorphisms and genotypic differences not only between different species but also within strains of the same species. The utilisation of these technologies is, for instance, very useful to identify the microorganisms responsible for different organoleptic characteristics of fermented food products, which in turn, increases our chances to further improve them (Diaz Ruiz and Wacher Rodarte, 2003). Transcriptional profiling, also through whole-genome microarray techniques, is very useful to detect changes at the mRNA level caused by the generation of particular mutations. For example, a high-throughput transcriptome analysis using a yeast model has been carried out to detect the *in vivo* targets of Gal-1-P toxicity, which as explained above, is a major cause of galactosemia (Slepak *et al.*, 2005). Proteomic analysis also enables the detection of changes in the pattern of expression and modifications experienced by certain proteins in response to environmental or genetic perturbations. For instance, by using DNA microarrays, quantitative proteomics, and databases of known physical interactions, a group was able to identify 997 messenger RNAs responding to 20 systematic perturbations of the yeast galactose-utilisation pathway, providing evidence that approximately 15 of 289 detected proteins are regulated post-transcriptionally, and identifying explicit physical interactions governing the cellular response to each perturbation (Ideker et al., 2001). Metabolite profiling (metabolome) and flux analysis (fluxomes), i.e. the analysis of the degree of activity of different cellular pathways and the metabolites produced by them, are also emerging technologies that will soon allow more complete comparisons of the overall changes occurring within the engineered microorganisms. As an example of the latter, monitoring the levels of cellular NADH proved to be a useful tool in order to determine the effect in the glycolytic flux that overexpressing bacterial NADH oxidases has in *S. cerevisiae* (Vemuri *et al.*, 2007). And these few examples are only the tip of the iceberg.

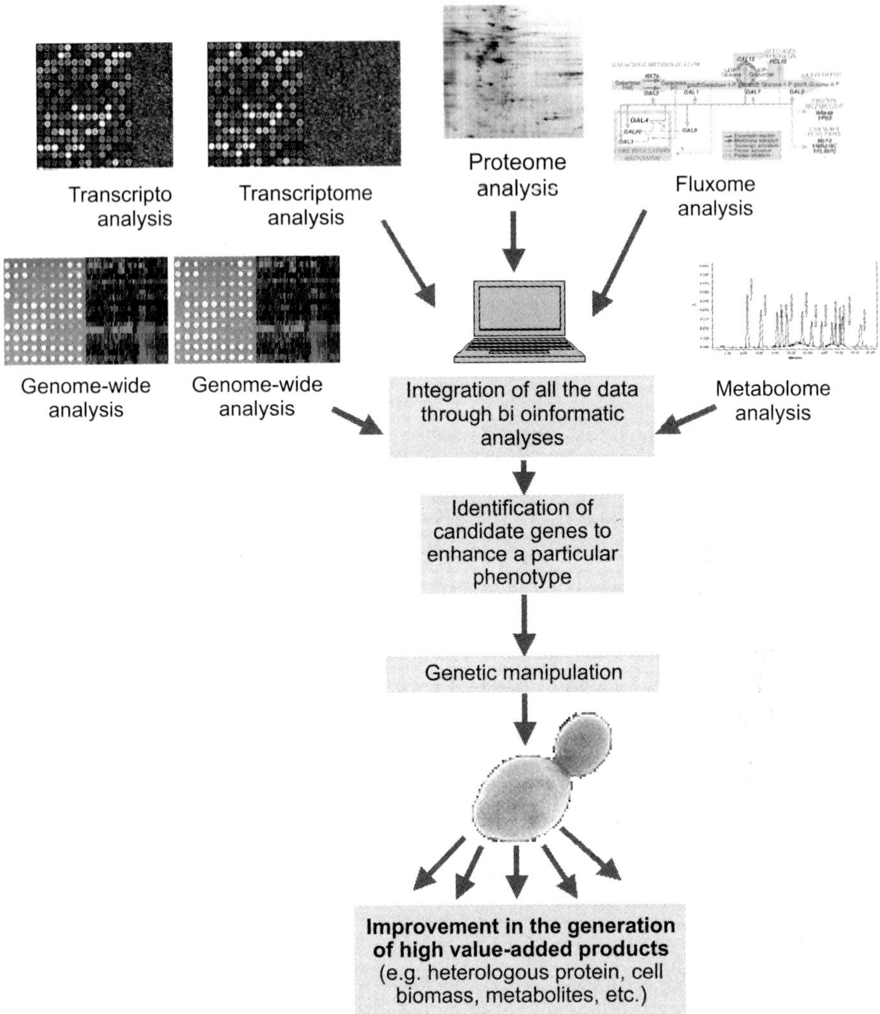

Figure 10.9 **Application of all the new 'ome' analyses in the design of metabolic engineering strategies to improve phenotypes with industrial interest.** The combination of comparative analyses of whole-genome sequences, and genome-wide genetic screens; transcriptional profiling by microarray (transcriptome); analysis of the whole protein contents through protein arrays, bidimensional gels, and mass spectrometry (proteomic analysis); analysis of the rate at which all cellular pathways work and co-interact (fluxome analysis); analysis of the relative levels of all of the metabolites present within cells (metabolome), and the gathering and processing of all of these data through bioinformatic technologies improves the design of metabolic engineering strategies to introduce and/or enhance industrially-desirable traits in yeasts and other microorganisms through genetic modification.

Summarizing, the availability of all of these new tools and the integration of the data generated through their use has enormous potential to let us reach a much more complete stage in the characterization of model cell systems that will soon allow us to switch from inverse metabolic engineering to constructive metabolic engineering for the design of genetically-improved 'superbug cell factories'.

References

Acar, M., Becskei, A. and van Oudenaarden, A. (2005). Enhancement of cellular memory by reducing stochastic transitions. *Nature* 435:228–232.

Adam A.C., Rubio-Texeira, M. and Polaina, J. (2004). Lactose: The milk sugar from a biotechnological perspective. *CRC Crit. Rev. Food Sci. Nutr.* 44:553–557.

Adam, A.C., Prieto, J.A., Rubio-Texeira, M. and Polaina, J. (1999). Construction of a lactose-assimilating strain of baker's yeast. *Yeast* 15:1299–1305.

Ahuatzi, D., Riera, A., Pelaez, R., Herrero, P. and Moreno, F. (2007). Hxk2 regulates the phosphorylation state of Mig1 and therefore its nucleocytoplasmic distribution. *J. Biol. Chem.* 282:4485–4493.

Ahuatzi, D., Herrero, P., de la Cera, T. and Moreno, F. (2004). The glucose-regulated nuclear localization of hexokinase 2 *in Saccharomyces cerevisiae* is Mig1-dependent. *J. Biol. Chem.* 279:14440–14446.

Amberg, D.C., Burke, D. and Strathern, J.N. (2005). Methods in yeast genetics: a Cold Spring Harbor Laboratory Course Manual. New York: Cold Spring Harbor Laboratory Press. pp 230.

Anders, A., Lilie, H., Franke, K., Kapp, L., Stelling, J., Gilles, E.D. and Breunig, K.D. (2006). The galactose switch in *Kluyveromyces lactis* depends on nuclear competition between GAL4 and GAL1 for GAL80 binding. *J. Biol. Chem.* 281:29337–29348.

Bailey, J.E., Sburlati, A., Hatzimanikatis, V., Lee, K., Renner, W.A. and Tsai, P.S. (2002). Inverse metabolic engineering: a strategy for directed genetic engineering of useful phenotypes. *Biotechnol. Bioeng.* 52:109–121.

Barnett, J.A. and Entian, K.D. (2005). A history of research on yeasts 9: regulation of sugar metabolism. *Yeast* 22:835–894.

Barnett, J.A. (2004). A history of research on yeasts 7: enzymic adaptation and regulation. *Yeast* 21:703–746.

Bartkeviciute, D. and Sasnauskas, K. (2004). Disruption of the *MNN10* gene enhances protein secretion in *Kluyveromyces lactis* and *Saccharomyces cerevisiae. FEMS Yeast Res.* 4:833–840.

Baruffini, E., Goffrini, P. Donnini, C. and Lodi, T. (2006). Galactose transport in *Kluyveromyces lactis*: major role of the glucose permease Hgt1. *FEMS Yeast Res.* 6:1235–1242.

Bash, R. and Lohr, D. (2001). Yeast chromatin structure and regulation of *GAL* gene expression. *Prog. Nucleic Acid Res. Mol. Biol.* 65:197–259.

Bassett, D.E.Jr., Basrai, M.A., Connelly, C., Hyland, K.M., Kitagawa, K., Mayer, M.L., Morrow, D.M., Page, A.M., Resto, V.A. Skibbens, R.V. and Hieter, P. (1996). Exploiting the complete yeast genome sequence. *Curr. Opin. Genet. Dev.* 6:763-766.

Becerra, M., Rodríguez-Belmonte, E., Cerdán, M.E. and González-Siso, M.I. (2004). Engineered autolytic yeast strains secreting *Kluyveromyces lactis* β-galactosidase for production of heterologous proteins in lactose media. *J. Biotechnol.* 109:131–137.

Becerra, M., Prado, S.D., Cerdán, M.E. and González-Siso, M.I. (2002). Metabolic engineering for direct lactose utilization by *Saccharomyces cerevisiae*. *Biotechnol. Lett.* 24:1391–1396.

Becerra, M., Cerdan, E. and Gonzalez-Siso, M.I. (1997). Heterologous *Kluyveromyces lactis* β-galactosidase production and release by *Saccharomyces cerevisiae* osmotic-remedial thermosensitive autolytic mutants. *Biochem. Biophys. Acta* 1335:235–241.

Belem, M.A. and Lee, B.H. (1998). Production of bioingredients from *Kluyveromyces marxianus* grown on whey: an alternative. *CRC Crit. Rev. Food Sci. Nutr.* 38:565–598.

Bergkamp, R.J., Kool, I.M., Geerse, R.H. and Planta, R.J. (1992). Multiple-copy integration of the alpha-galactosidase gene from *Cyamopsis tetragonoloba* into the ribosomal DNA of *Kluyveromyces lactis*. *Curr. Genet.* 21:365–370.

Betina, S., Goffrini, P., Ferrero, I. and Wésolowski-Louvel M. (2001). *RAG4* gene encodes a glucose sensor in *Kluyveromyces lactis*. *Genetics.* 158:541–548.

Bhat, P.J. (2003). Galactose-1-phosphate is a regulator of inositol monophosphate: a factor or a fiction? Med. Hypotheses 60:123–128.

Bhat, P.J. and Murthy, T.V. (2001). Transcriptional control of the *GAL/MEL* regulon of yeast *Saccharomyces cerevisiae*: mechanism of galactose-mediated signal transduction. *Mol. Microbiol.* 40:1059–1066.

Bhat, P.J. and Hopper, J.E. (1992). Overproduction of the GAL1 or GAL3 protein causes galactose-independent activation of the GAL4 protein: evidence for a new model of induction for the yeast *GAL/MEL* regulon. *Mol. Cell. Biol.* 12:2701–2707.

Bhat, P.J. and Hopper, J.E. (1991). The mechanism of inducer formation in gal3 mutants of the yeast galactose system is independent of normal galactose metabolism and mitochondrial respiratory function. *Genetics.* 128:233–239.

Bhat, P.J., Oh, D. and Hopper, J.E. (1990). Analysis of the GAL3 signal transduction pathway activating GAL4 protein-dependent transcription in *Saccharomyces cerevisiae*. *Genetics* 125:281–291.

Bhaumik, S.R., Raha, T., Aiello, D.P. and Green, M.R. (2004). *In vivo* target of a transcriptional activator revealed by fluorescence resonance energy transfer. *Genes Dev.* 18:333–343.

Bhaumik, S.R. and Green, M.R. (2002). Differential requirement of SAGA components for recruitment of TATA-box-binding protein to promoters *in vivo*. *Mol. Cell Biol.* 22:7365–7371.

Bhaumik, S.R. and Green, M.R. (2001). SAGA is an essential *in vivo* target of the yeast acidic activator Gal4p. *Genes Dev.* 15:1935–1945.

Billard, P., Menart, S., Blaisonneau, J., Bolotin-Fukuhara, M., Fukuhara, H. and Wésolowski-Louvel, M. (1996). Glucose uptake in *Kluyveromyces lactis*: role of the *HGT1* gene in glucose transport. *J. Bacteriol.* 178:5860–5866.

Bisson, L.F. and Fraenkel, D.G. (1984). Expression of kinase-dependent glucose uptake in *Saccharomyces cerevisiae*. *J. Bacteriol.* 159:1013–1017.

Blaisonneau, J., Fukuhara, H. and Wésolowski-Louvel M. (1997). The *Kluyveromyces lactis* equivalent of casein kinase I is required for the transcription of the gene encoding the low-affinity glucose permease. *Mol. Gen. Genet.* 253:469–477.

Blank, T.E., Woods, M.P., Lebo, C.M., Xin, P. and Hopper, J.E. (1997). Novel Gal3 proteins showing altered Gal80p binding cause constitutive transcription of Gal4p-activated genes in *Saccharomyces cerevisiae*. *Mol. Cell. Biol.* 17:2566–2575.

Boekhout, T. (2005). Biodiversity gut feeling for yeasts. *Nature.* 434: 449–451.

Boles, E. and Hollenberg, C. (1997). The molecular genetics of hexose transport in yeasts. *FEMS Microbiol. Rev.* 21:85–111.

Bolotin-Fukuhara, M., Toffano-Nioche, C., Artiguenave, F., Duchateau-Nguyen, G., Lemaire, M., Marmeisse, R., Montrocher, R., Robert, C., Termier, M., Wincker, P. and Wésolowski-Louvel, M. (2000). Genomic exploration of the hemiascomycetous yeasts:11. *Kluyveromyces lactis. FEBS Lett.* 487:66–70.

Bonekamp F.J. and Oosterom J. (1994). On the safety of *Kluyveromyces lactis*—a review. *Appl. Microbiol. Biotechnol.* 41:1–3.

Bram, R.J., Lue, N.F. and Kornberg, R.D. (1986). A *GAL* family of upstream activating sequences in yeast: roles in both induction and repression of transcription. *EMBO J.* 5:603–608.

Breunig, K.D. and Steensma, H.Y. (2003). *Kluyveromyces lactis*: genetics, physiology, and application. In: Topics in Current Genetics, Vol. 2. Functional Genetics of Industrial Yeasts. De Winde, J.H. Ed., Springer-Verlag, Berlin, pp 171–205.

Breunig, K.D. (1989) Glucose repression of *LAC* gene expression in yeast is mediated by the transcriptional activator LAC9. *Mol. Gen. Genet.* 216:422–427.

Briczinski, E.P. and Roberts, R.F. (2002). Production of exopolysaccharide-containing whey protein concentrate by fermentation of whey. *J. Dairy Sci.* 85:3189–3197.

Bro, C., Knudsen, S., Regenberg, B., Olsson, L. and Nielsen, J. (2005). Improvement of galactose uptake in *Saccharomyces cerevisiae* through overexpression of phosphoglucomutase: example of transcript analysis as a tool in inverse metabolic engineering. *Appl. Environ. Microbiol.* 71:6465–6472.

Bro, C. and Nielsen, J. (2004). Impact of 'ome' analyses on inverse metabolic engineering. *Metab. Eng.* 6:204–211.

Bryant, G.O. and Ptashne, M. (2003). Independent recruitment *in vivo* by Gal4 of two complexes required for transcription. *Mol. Cell* 11:1301–1309.

Bussereau, F., Casaregola, S., Lafay, J.F. and Bolotin-Fukuhara, M. (2006). The *Kluyveromyces lactis* repertoire of transcriptional regulators. *FEMS Yeast Res.* 6:325–335.

Butler, D. (1996). Interest ferments in yeast genome sequence. *Nature.* 380:660–661.

Cabaille, D. and Combes, D. (1995). Characterization of β-galactosidases from *Klyveromyces lactis. Biotechnol. Appl. Biochem.* 22:55–64.

Cabal, G.G., Genovesio, A., Rodriguez-Navarro, S., Zimmer, C., Gadal, O., Lesne, A., Buc, H., Feuerbach-Fournier, F., Olivo-Marin J.C., Hurt, E.C. and Nehrbass U. (2006). SAGA interacting factors confine sub-diffusion of transcribed genes to the nuclear envelope. *Nature.* 441:770–773.

Carlson, M. (1999). Glucose repression in yeast. *Curr. Opin. Microbiol.* 2:202-207.

Cassart, J.P., Georis, I., Ostling, J., Ronne, H. and Vandenhaute, J. (1995). The MIG1 repressor from *Kluyveromyces lactis*:cloning, sequencing and functional analysis in *Saccharomyces cerevisiae. FEBS Lett.* 371:191–194.

Champagne, C.P. and Goulet, J. (1988). Growth of baker's yeast (*Saccharomyces cerevisiae*) in lactose-hydrolyzed cheese-whey ultrafiltrate. *Can. Inst. Food Sci. Technol.* 21:545–548.

Chasman, D.I. and Kornberg, R.D. (1990). GAL4 protein: purification, association with GAL80 protein, and conserved domain structure. *Mol. Cell. Biol.* 10:2916–2923.

Chen, X.J. and Clark-Walker, G.D. (2000). The petite mutation in yeasts: 50 years on. *Int. Rev. Cytol.* 194:197–238.

Chen, X.J. Wésolowski-Louvel, M. and Fukuhara, H. (1992). Glucose transport in the yeast *Kluyveromyces lactis*. II. Transcriptional regulation of the glucose transporter gene *RAG1*. *Mol. Gen. Genet.* 233:97–105.

Cibelli, J.B., Stice, S.L., Golueke, P.J., Kane, J.J., Jerry, J., Blackwell, C., Ponce de León, F.A. and Robl, J.M. (1998). Transgenic bovine chimeric offspring produced from somatic cell-derived stem-like cells. *Nature Biotechnol.* 16:642–646.

Clark, D.S. (1994). Can immobilization be exploited to modify activity? *Trends in Biotechnol.* 12:439–449.

Coenen, T.M., Bertens, A.M., De Hoog, S.C. and Verspeek-Rip, C.M. (2000). Safety evaluation of a lactase enzyme preparation derived from *Kluyveromyces lactis*. *Food. Chem. Toxicol.* 38:671–677.

Colussi, P. and Taron, C.H. (2005). *Kluyveromyces lactis LAC4* promoter variants that lack function in bacteria but retain full function in yeast. *Appl. Environ. Microbiol.* 71:7092–7098.

Compagno, C., Tura, A., Ranzi, B.M. and Martegani, E. (1993). Bioconversion of lactose/ whey to fructose diphosphate with recombinant *Saccharomyces cerevisiae* cells. *Biotechnol. Bioeng.* 42:398–400.

D'Alessio, M. and Brandriss, M.C. (2000). Cross-pathway regulation in *Saccharomyces cerevisiae*: activation of the proline utilization pathway by Gal4 in *vivo*. *J. Bacteriol.* 182:3748–3753.

Dasarathy S. (2003). Role of gut bacteria in the therapy of hepatic encephalopathy with lactulose and antibiotics. *Indian J. Gastroenterol. Suppl.* 2:S50–53.

Destruelle, M., Menghini, R., Frontali, L. and Bianchi, M.M. (1999). Regulation of the expression of the *Kluyveromyces lactis PDC1* gene: carbon source-responsive elements and autoregulation. *Yeast* 15:361–370.

De Vit, M.J. and Johnston, M. (1999). The nuclear exportin Msn5 is required for nuclear export of the Mig1 glucose repressor of *Saccharomyces cerevisiae*. *Curr. Biol.* 9:1231–1241.

De Vit, M.J., Waddle, J.A. and Johnston, M. (1997). Regulated nuclear translocation of the Mig1 glucose repressor. *Mol. Cell. Biol.* 8:1603–1618.

De Winde, H. (2003). Functional genetics of industrial yeasts; of ancient skills and modern applications. In: Functional Genetics of Industrial Yeasts, De Winde J.H. Ed., Springer-Verlag, Berlin, pp 1-14.

Diaz Ruiz, G. and Wacher Rodarte, C. (2003). Methods for the study of microbial communities in fermented foods. *Revista Latinoamericana de Biologia* 45:30–40.

Dietrich, F.S., Voegeli, S., Brachat, S., Lerch, A., Gates, K., Steiner, S., Mohr, C., Pohlmann, R., Luedi, P., Choi, S., Wing, R.A., Flavier, A., Gaffney, T.D. and Philippsen, P. (2004). The *Ashbya gossypii* genome as a tool for mapping the ancient *Saccharomyces cerevisiae* genome. *Science.* 304:304–307.

Domingues, L., Teixeira, J.A., Penttilä, M. and Lima, N. (2002). Construction of a flocculent *Saccharomyces cerevisiae* strain secreting high levels of *Aspergillus niger* β-galactosidase. *Appl. Microbiol. Biotechnol.* 58:645–650.

Dong, J. and Dickson, R.C. (1997). Glucose represses the lactose-galactose regulon in *Kluyveromyces lactis* through a SNF1and MIG1-dependent pathway that modulates galactokinase (*GAL1*) gene expression. *Nucleic Acids Res.* 25:3657–3664.

Donnini, C., Farina, F., Neglia, B., Compagno, M.C., Uccelletti, D., Goffrini, P. and Palleschi, C. (2004). Improved production of heterologous proteins by a glucose repression-defective mutant of *Kluyveromyces lactis. Appl. Environ. Microbiol.* 70:2632–2638.

Dujon, B., *et al.* (2004). Genome evolution in yeasts. *Nature.* 430:35–44.

Elbing, K., McCartney, R.R. and Schmidt, M.C. (2006). Purification and characterization of the three Snf1-activating kinases of *Saccharomyces cerevisiae. Biochem. J.* 393:797–805.

Ergüder, T.H., Tezel, U., Güven, E. and Demirer, G.N. (2001). Anaerobic biotransformation and methane generation potential of cheese whey in a batch and UASB reactors. *Waste Manag*. 21:643-650.

Ezhkova, E. and Tansey, W.P. (2004). Proteasomal ATPases link ubiquitylation of histone H2B to methylation of histone H3. *Mol. Cell* 13:435–442.

Farahnak, F., Seki, T., Ryu, D.Y. and Ogrydziak, D. (1986). Construction of lactose-assimilating and high ethanol producing yeasts by protoplast fusion. *Appl. Environ. Microbiol.* 51:362–367.

Fay, J.C. and Benavides, J.A. (2005). Evidence for domesticated and wild populations of *Saccharomyces cerevisiae. PLOS Genetics* 1:66–71.

Feijoo, G., Moreira, M.T., Roca, E. and Lema, J.M. (1999). Use of cheese whey as a substrate to produce manganese peroxidase by *Bjerkandera* sp BOS55. *J. Ind. Microbiol. Biotechnol.* 23:86–90.

Ferdous, A., Sikder, D., Gillette, T., Nalley, K., Kodadek, T. and Johnston, S.A. (2007). The role of proteasomal ATPases and activator monoubiquytilation in regulating Gal4 binding to promoters. Genes & Dev. 21:112–123.

Ferrero, I., Viola, A.M. and Goffeau, A. (1981). Induction by glucose of an antimycin-insensitive, azide-sensitive respiration in the yeast *Kluyveromyces lactis. Antonie van Leeuweenhoek.* 47:11–24.

Fields, S. and Song, O. (1989). A novel genetic system to detect protein-protein interactions. *Nature*, 340:245–246.

Fiocchi, C. (2006). Probiotics in inflammatory bowel disease: yet another mechanism of action? *Gastroenterology* 131:2009–2012.

Fleet, G.H. (2007). Yeasts in food and beverages: impact on product quality and safety. *Curr. Opin. Biotechnol.* 18:170–175.

Flores, C.L., Rodriguez, C., Petit, T. and Gancedo, C. (2000). Carbohydrate and energy-yielding metabolism in non-conventional yeasts. *FEMS Microbiol. Rev.* 24:507–529.

Flôres, S.H. and Alegre, R.M.(2001). Nisin production from *Lactococcus lactis* A.T.C.C. 7692 using supplemented whey permeate. *Biotechnol. Appl. Biochem.* 34:103–107.

Fu, L., Bounelis, P., Dey, N., Browne, B.L., Marchase, R.B. and Bedwell, D.M. (1995). The posttranslational modification of phosphoglucomutase is regulated by galactose induction and glucose repression in *Saccharomyces cerevisiae. J. Bacteriol.* 177:3087–3094.

Fukuhara, H. (2006). *Kluyveromyces lactis*-a retrospective. *FEMS Yeast Res.* 6:323–324.

Fukuhara, H. (2003). The Kluyver effect revisited. *FEMS Yeast Res.* 3:327–331.

Fukuhara, H. (2002). *Kluyveromyces lactis* and the 14th workshop 'Biology of *Kluyveromyces*'. *FEMS Yeast Res.* 1:333–335.

Gekas, V. and López-Leyva, M. (1985). Hydrolysis of lactose: a literature review. *Proc. Biochem.* 20:2–12.

Gellisen, G. and Hollenberg, C.P. (1997). Application of yeasts in gene expression studies: a comparison of *Saccharomyces*, *Hansenula polymorpha*, and *Kluyveromyces lactis*-a review. *Gene* 190:87–97.

Gerngross, T.U. (2004). Advances in the production of human therapeutic proteins in yeasts and filamentous fungi. *Nature Biotechnol.* 22:1409–1414.

Ghaly, A.E. and Kamal, M.A. (2004). Submerged fermentation of acid cheese whey for protein production and pollution potential reduction. *Water Res.* 38:631-644.

Giniger, E. and Ptashne, M. (1988). Cooperative DNA binding of the yeast transcriptional activator GAL4. *Proc. Natl. Acad. Sci.* USA 85:382–386.

Gligoris, T., Thireos, G. and Tzamarias, D. (2007). The Tup1 co repressor directs Htz1 deposition at a specific promoter nucleosome marking the *GAL1* gene for rapid activation. *Mol. Cell. Biol. In Press.*

Gödecke, A., Zachariae, W., Arvanitidis, A. and Breunig, K.D. (1991). Coregulation of the *Kluyveromyces lactis* lactose permease and *β*-galactosidase genes is achieved by interaction of multiple LAC9 binding sites in a 2.6 kbp divergent promoter. *Nucleic Acids Res.* 19:5351–5358.

Goffrini, P., Ferrero, I. and Donnini, C. (2002). Respiration-dependent utilization of sugars in yeasts: a determinant role of sugar transporters. *J. Bacteriol.* 184:427–432.

Goffrini, P., Algeri, A.A., Donnini, C., Wésolowski-Louvel, M. and Ferrero, I. (1989). *RAG1* and *RAG2*: nuclear genes involved in the dependence/independence on mitochondrial respiratory function for growth on sugars. *Yeast.* 5:99–106.

González, F., Delahodde, A., Kodadek, T. and Johnston, S.A. (2002). Recruitment of a 19S proteasome subcomplex to an activated promoter. *Science*, 296:548–550.

Gonzalez, F.J., Montes, J., Martin, F., Lopez, M.C., Ferminan, E., Catalan, J., Galan, M.A. and Dominguez, A. (1997). Molecular cloning of *TvDAO1*, a gene encoding D-amino acid oxidase from *Trigonopsis variabilis* and its expression in *Saccharomyces cerevisiae* and *Kluyveromyces lactis*. *Yeast 13:*1399–1408.

Gonzalez-Siso, M.I., Freire-Picos, M.A. and Cerdan, M.E. (1996). Reoxidation of the NADPH produced by the pentose phosphate pathway is necessary for the utilization of glucose by *Kluyveromyces lactis rag2* mutants. *FEBS Lett.* 387:7–10.

Guarente, L., Yocum, R.R. and Gifford, P. (1982). A *GAL10-CYC1* hybrid yeast promoter identifies the GAL4 regulatory region as an upstream site. *Proc. Natl. Acad. Sci.* USA 79:7410–7414.

Hallmann, F. (2000). Toxicity of commonly used laxatives. *Med. Sci. Monit.* 6:618-628.

Hawkins, K.M. and Smolke, C.D. (2006). The regulatory role of the galactose permease and kinase in the induction response of the GAL network in *Saccharomyces cerevisiae. J. Biol. Chem.* 281:13485–13492.

Hilgendorf, M.J. (1981). Optimization of fungal lactase levels in yogurt manufacturing. *Cult. Dairy Prod. J.* 16:5–7.

Hirst, M., Kobor, M.S., Kuriakose, N., Greenblatt, J. and Sadowski, I. (1999). GAL4 is regulated by the RNA polymerase II holoenzyme-associated cyclin-dependent protein kinase SRB10/CDK8. *Mol. Cell.* 3:673–678.

Hittinger, C.T., Rokas, A. and Carroll, S.B. (2004). Parallel inactivation of multiple GAL pathway genes and ecological diversification in yeasts. *Proc. Natl. Acad. Sci.* USA 101:14144–14149.

Holsbeeks, I., Lagatie, O., Van Nuland, A., Van de Velde, S. and Thevelein, J.M. (2004). The eukaryotic plasma membrane as a nutrient-sensing device. *Trends Biochem. Sci.* 29:556–564.

Horak, J. and Wolf, D.H. (1997). Catabolite inactivation of the galactose transporter in the yeast *Saccharomyces cerevisiae*: ubiquitination, endocytosis, and degradation in the vacuole. *J. Bacteriol.* 179:1541–1549.

Hsieh, H.B. and Da Silva, N.A. (2000). Development of a *LAC4* promoter-based gratuitous induction system in *Kluyveromyces lactis*. *Biotechnol. Bioeng.* 67:408–416.

Ichikawa, H. and Fukumori, Y. (2007). Design of nanohydrogel-incorporated microcapsules for appropriate controlled-release of peptide drugs. *Yakugaku Zasshi* 127:813–823.

Ideker, T., Thorsson, V., Ranish, J.A., Christmas, R., Buhler, J., Eng, J.K., Bumgarner, R., Goodlett, D.R., Aebersold, R. and Hood, L. (2001). Integrated genomic and proteomic analysis of a systematically perturbed metabolic network. *Science* 292:929–934.

Johnston, M. (1987). A model fungal gene regulatory mechanism: the *GAL* genes of *Saccharomyces cerevisiae*. *Microbiol. Rev.* 51:458–476.

Johnston, S.A., Salmeron, J.M. Jr. and Dincher, S.S. (1987). Interaction of positive and negative regulatory proteins in the galactose regulon of yeast. *Cell* 50:143–146.

Jost, B., Vilotte, J.L., Duluc, I., Rodeau, J.L. and Freund, J.N. (1999). Production of low-lactose milk by ectopic expression of intestinal lactase in the mouse mammary gland. *Nature Biotechnol.* 17:160–164.

Jurascik, M., Guimaraes, P., Klein, J., Domingues, L., Teixeira, J. and Markos, J. (2006). Kinetics of lactose fermentation using by a recombinant *Saccharomyces cerevisiae* strain. *Biotechnol. Bioengin.* 94:1147–1154.

Kagkli, D.M., Tâche, R., Cogan, T.M., Hill, C., Casaregola, S. and Bonnarme, P. (2006). *Kluyveromyces lactis* and *Saccharomyces cerevisiae*, two potent deacidifying and volatile-sulphur-aroma-producing microorganisms of the cheese ecosystem. *Appl. Microbiol. Biotechnol.* 73:434–442.

Kang, H.A., Kang, W.K., Go, S.M., Rezaee, A., Krishna, S.H., Rhee, S.K. and Kim, J.Y. (2005). Characteristics of *Saccharomyces cerevisiae gal1*Δ and *gal1*Δ*hxk2*Δ mutants expressing recombinant proteins from the *GAL* promoter. *Biotechnol. Bioeng.* 89:619-629.

Kang, T., Martins, T. and Sadowski, I. (1993). Wild type GAL4 binds cooperatively to the *GAL1-10* UASG *in vitro*. *J. Biol. Chem.* 268:9629–9635.

Kaniak, A., Xue, Z., Macool, D., Kim, J.H. and Johnston, M. (2004). Integration of transcriptional and posttranslational regulation in a glucose signal transduction pathway in *Saccharomyces cerevisiae*. *Eukaryot. Cell* 3:221–231.

Kellis, M., Birren, B.W. and Lander, E.S. (2004). Proof and evolutionary analysis of ancient genome duplication in the yeast *Saccharomyces cerevisiae*. *Nature* 428:617–624.

Keogh, R.S., Seoighe, C. and Wolfe, K.H. (1998). Evolution of gene order and chromosome number in Saccharomyces, Kluyveromyces, and related fungi. *Yeast* 14:443–457.

Kiers, J., Zeeman, A.M., Luttik, M., Thiele, C., Castrillo, J.I., Steensma, H.Y., van Dijken, J.P. and Pronk, J.T. (1998). Regulation of alcoholic fermentation in batch and chemostat cultures of *Kluyveromyces lactis* CBS2359. *Yeast* 14:459–469.

Kilara, A. and Shahani, K.M. (1979). The use of immobilized enzymes in the food industry: a review. *CRC Crit. Rev. Food Sci. Nutr.* 12:161–198.

Kim, J.H. and Johnston, M. (2006a). Two glucose-sensing pathways converge on Rgt1 to regulate expression of glucose transporter genes in *Saccharomyces cerevisiae*. *J. Biol. Chem.* 281:26144–26149.

Kim, J.H., Brachet, V., Moriya, H. and Johnston, M. (2006b). Integration of transcriptional and posttranslational regulation in a glucose signal transduction pathway in *Saccharomyces cerevisiae*. *J. Biol. Chem.* 5:167–173.

Kim, C.K., Chung, H.S., Lee, M.K., Choi, L.N. and Kim, M.H. (1999). Development of dried liposomes containing beta-galactosidase for the digestion of lactose in milk. *Int. J. Pharm.* 183:185–193.

Kumar, V., Ramakrishnan, S., Teeri, T.T., Knowles, J.K. and Hartley, B.S. (1992). *Saccharomyces cerevisiae* cells secreting an *Aspergillus niger* β-galactosidase grow on whey permeate. *Biotechnology* (NY) 10:82–85.

Kumura, H., Tanoue, Y., Tsukahara, M., Tanaka, T. and Shimazazi, K. (2004). Screening of dairy yeast strains for probiotic applications. *J. Dairy Sci.* 87:4050–4056.

Kundu, S., Horn, P.J. and Peterson, C.L. (2007). SWI/SNF is required for transcriptional memory at the yeast *GAL* gene cluster. *Genes & Dev.* 21:997–1004.

Kurtzman, C.P. (2003). Phylogenetic circumscription of *Saccharomyces*, *Kluyveromyces* and other members of the Saccharomycetaceae, and the proposal of the new genera *Lachancea*, *Nakaseomyces*, *Naumovia*, *Vanderwaltozyma* and *Zigotorulospora*. *FEMS Yeast Res.* 4:233–245.

Kurtzman, C.P. and Fell, J.W. (1998). The yeasts, a taxonomic study, 3rd Ed., Elsevier Science Publishers, Amsterdam. 1082 pp.

Kurtzman, C.P. and Robnett, C. J. (1998). Identification and phylogeny of ascomycetous yeasts from analysis of nuclear large subunit (26S) ribosomal DNA partial sequences. *Antonie Van Leeuwenhoek* 73:331–371.

Lachance, M.A. (2007). Current status of *Kluyveromyces* systematics. FEMS Yeast Research. *In press.*

Lachance, M.A. (1998). *Kluyveromyces* van der Walt amend. van der Walt. In: The Yeasts, a Taxonomic Study. Kurtzman, C.P. and Fell, J.W., Eds., 4th Ed., Elsevier Science, Amsterdam, 227–247.

Lai, K., and Klapa, M.I. (2004). Alternative pathways of galactose assimilation: could inverse metabolic engineering provide an alternative to galactosemic patients? *Metab. Eng.* 6:239–244.

Lakshminarasimhan, A. and Bhat, P.J. (2005). Replacement of a conserved tyrosine by triptophan in Gal3p of *Saccharomyces cerevisiae* reduces constitutive activity: implications for signal transduction in the *GAL* regulon. *Mol. Gen. Genomics* 274:384–393.

Larschan, E. and Winston, F. (2001). The *S. cerevisiae* SAGA complex functions *in vivo* as a coactivator for transcriptional activation by Gal4. *Genes Dev.* 15:1946–1956.

Lee, D., Ezhkova, E., Li, B., Pattenden, S.G., Tansey, W.P. and Workman, J.L. (2005). The proteasome regulatory particle alters the SAGA coactivator to enhance its interactions with transcriptional activators. *Cell* 123:423–436.

Lee, Y.J., Kim, C.S. and Oh, D.K. (2004). Lactulose production by β-galactosidase in permeabilized cells of *Kluyveromyces lactis*. *Appl. Microbiol. Biotechnol.* 64:787–793.

Leloir, L.F. (1971). Two decades on research of the biosynthesis of saccharides. *Science.* 172: 1299–1303.

Lemieux, K. and Gaudreau, L. (2004). Targeting of Swi/Snf to the yeast *GAL1UASG* requires the Mediator, TAFIIs, and RNA polymerase II. *EMBO J.* 23:4040–4050.

Leonardo, J.M., Bhairi, S.M. and Dickson, R.C. (1987). Identification of Upstream Activator Sequences that regulate induction of the β-galactosidase gene in *Kluyveromyces lactis*. *Mol. Cell. Biol.* 7:4369–4376.

Leslie, N.D. (2003). Insights into the pathogenesis of galactosemia. *Ann. Rev. Nutr.* 23:59-80.

Levenson, J.M. and Sweatt, J.D. (2006). Epigenetic mechanisms: a common theme in vertebrate and invertebrate memory formation. *Cell. Mol. Life Sci.* 63:1009-1016.

Lewis, B.A. and Reinberg, D. (2003). The mediator coactivator complex: functional and physical roles in transcriptional regulation. *J. Cell Sci.* 116:3667–3675.

Lewis, L.K., Lobachev, K., Westmoreland, J.W., Karthikeyan, G., Williamson, K.M., Jordan and J.J., Resnick, M.A. (2005). Use of a restriction endonuclease toxicity assay to identify inducible *GAL1* promoter variants with reduced basal activity. *Gene* 363:183–192.

Liang, S.D., Marmorstein, R., Harrison, S.C. and Ptashne, M. (1996). DNA sequence preferences of GAL4 and PPR1: how a subset of Zn2 Cys6 binuclear cluster proteins recognizes DNA. *Mol. Cell. Biol.* 16:3773–3780.

Lohr, D., Venkov, P. and Zlatanova, J. (1995). Transcriptional regulation in the yeast *GAL* gene family: a complex genetic network. *FASEB J.* 9:777–787.

Luthra, R., Kerr, S.C., Harreman, M.T., Apponi, L.H., Fasken, M.B., Ramineni, S., Chaurasia, S., Valentini, S.R. and Corbett, A.H. (2007). Actively transcribed *GAL* genes can be physically linked to the nuclear pore by the SAGA chromatin modifying complex. *J. Biol. Chem.* 282:3042–3049.

Ma, J. and Ptashne, M. (1987). The carboxy-terminal 30 amino acids of GAL4 are recognized by GAL80. *Cell* 50:137–142.

Marmorstein, R., Carey, M. Ptashne, M. and Harrison, S.C. (1992). DNA recognition by GAL4: structure of a protein-DNA complex. *Nature* 356: 408–414.

Marshall, R.T. and Arbuckle, W.S. (1996). *Ice Cream*. 5th Ed. Chapman & Hall, New York, NY. 349 pp.

Mateo, C., Monti, R., Pessela, B.C., Fuentes, M., Torres, R., Guisan, J.M. and Fernández-Lafuente, R. (2004). Immobilization of lactase from *Kluyveromyces lactis* greatly reduces the inhibition promoted by glucose. Full hydrolysis of lactose in milk. *Biotechnol. Prog.* 20:1259–1262.

Mathy, G., Navet, R., Gerkens, P., Leprince, P., De Pauw, E., Sluse-Goffart, C.M. Sluse, F.E. and Douette, P. (2006). *Saccharomyces cerevisiae* mitoproteome plasticity in response to recombinant alternative ubiquinol oxidase. *J. Proteome Res.* 5:339–348.

Maugard, T., Gaunt, D., Legoy, M.D. and Besson, T. (2003). Microwave-assisted synthesis of galacto-oligosaccharides from lactose with immobilized β-galactosidase from *Kluyveromyces lactis*. *Biotechnol. Lett.* 25:623–629.

Maullu, C., Lampis, G., Desogus, A., Ingianni, A., Rossolini, G.A. and Pompei, R. (1999). High-level production of heterologous protein by engineered yeasts grown in cottage cheese whey. *Appl. Environ. Microb.* 65:2745–2747.

Mawson, A.J. (1994). Bioconversions for whey utilization and waste abatement. *Biores. Technol.* 47:195–203.

Mayordomo, I. and Sanz, P. (2001). Hexokinase PII: structural analysis and glucose signaling in the yeast *Saccharomyces cerevisiae*. *Yeast* 18:923–930.

Mazzoni, C., Saliola, M. and Falcone, C. (1992). Ethanol-induced and glucose-insensitive alcohol dehydrogenase activity in the yeast *Kluyveromyces lactis*. *Mol. Microbiol.* 6:2279–2286.

Melcher, K. (2005). Mutational hypersensitivity of a gene regulatory protein: *Saccharomyces cerevisiae* Gal80p. *Genetics* 171:469–476.

Melcher, K. and Xu, H.E. (2001). Gal80-Gal80 interaction on adjacent Gal4p binding sites is required for complete *GAL* gene repression. *EMBO J.* 20:841–851.

Méndez, A. and Olano, A. (1979). Lactulose: a review on some chemical properties and applications in infant nutrition and medicine. *Dairy Sci. Abstr.* 41:531–535.

Menezes, R.A., Amuel, C., Engels, R., Gengenbacher, U., Labahn, J. and Hollenberg, C.P. (2003). Sites for interaction between Gal80p and Gal1p in *Kluyveromyces lactis*: structural model of galactokinase based on homology to the GHMP protein family. *J. Mol. Biol.* 333:479–92.

Meyer, J., Walker-Jonah, A. and Hollenberg, C.P. (1991). Galactokinase encoded by *GAL1* is a bifunctional protein required for the induction of the *GAL* genes in *Kluyveromyces lactis* and is able to suppress the gal3 phenotype in *Saccharomyces cerevisiae*. *Mol. Cell. Biol.* 11:5454–5461.

Mirisola, M.G., Gallo, A. and De Leo, G. (2007). Ras-pathway has a dual role in yeast galactose metabolism. *FEBS Lett.* 581:2009–2016.

Mozdziak, P.E., Pophal, S., Borwornpinyo, S. and Petitte, J.N. (2003). Transgenic chickens expressing *β*-galactosidase hydrolyze lactose in the intestine. *J. Nutr.* 133:3076–3079.

Mohrmann, L. and Verrijzer, C.P. (2005). Composition and functional specificity of SWI2/SNF2 class chromatin remodeling complexes. *Biochim. Biophys. Acta* 1681:59–73.

Moulin, G. and Galzy, P. (1984). Whey, a potential substrate for biotechnology. *Biotechnol. Genet. Eng. Rev.* 1:347–373.

Muratani, M., Kung, C., Shokat, K.M. and Tansey, W.P. (2005). The F box protein Dsg1/Mdm30 is a transcriptional coactivator that stimulates Gal4 turnover and cotranslational mRNA processing. *Cell* 120:887–899.

Nalley, K., Johnston, S.A. and Kodadek, T. (2006). Proteolytic turnover of the Gal4 transcription factor is not required for function *in vivo*. *Nature* 442:1054–1057.

Naumova, E.S., Sukhotina, N.N. and Naumov, G.I. (2004). Molecular-genetic differentiation of the dairy yeast *Kluyveromyces lactis* and its closest wild relatives. *FEMS Yeast Res.* 5:263–269.

Nehlin, J.O. and Ronne, H. (1990). Yeast MIG1 repressor is related to the mammalian early growth response and Wilms' tumour finger proteins. *EMBO J.* 9:2891–2898.

Nicolas, G., Auger, I., Beaudoin, M., Hallé, F., Morency, E., LaPointe, G. and Lavoie, M.C. (2004). Improved methods for mutacin detection and production. *J. Microbiol. Methods.* 59:351–361.

Oh, D. and Hopper, J.E. (1990). Transcription of a yeast phosphoglucomutase isozyme gene is galactose inducible and glucose repressible. *Mol. Cell. Biol.* 10:1415–1422.

1. INTRODUCTION

Lichens are among nature's most elegantly coevolved symbiotic associations involving filamentous fungi (mycobiont) and microalgae (phycobiont) or cyanobacteria (cyanobiont) organized into complex tissues that bear little resemblance to the growth patterns of the individual partners (Hawksworth, 1988). This peculiar and exclusive association of the two symbionts makes lichens one of the most tolerant organisms on the planet, capable of inhabiting the subzero environment of the poles and alpine habitats on one extreme and the heat scorching deserts on the other extreme. Depending on their substrate and climatic conditions lichens have evolved, various growth forms, which can be mainly categorized in two types. The bigger, leafy and thread-like shrubby lichens (fruticose and foliose) belong to macrolichens while the crust or powdery crust forming smaller sized lichens belong to microlichens category. Beside these, there are some intermediate forms such as Leprose (granular or powdery), Placoid (closely attached to the substratum at centre and lobate or free at margin), Squamulose (minute lobes having dorsiventral differentiation, intermediate between crustose and foliose) and Dimorphic (single thallus having squamules as primary thallus bearing erect fruticose body as secondary thallus).

Lichens are among the most significant indicators of ecosystem functioning (Wolesley *et al.,* 1994), as they are very sensitive to habitat destruction and climate change. Ecosystem monitoring studies need to focus on a subset of lichen community as an indicator of ecosystem quality. The most commonly selected kinds of subsets of lichen species for monitoring are habitat subsets (e.g., Corticolous—on bark, Saxicolous—on rocks, Terricolous—on soil, Muscicolous—on moss, Ramicolous—on twigs and Omnicolous—on various substrates), and lichen guilds or morphological groups. Among the different lichen groups, epiphytic (corticolous) and soil lichens (terricolous) are found to be good indicator of ecosystem functioning (S. Will-Wolf *et al.,* 2002).

Soil lichens are known to be primary sources of nitrogen and carbon fixation in arid environments (Beymer & Klopatek, 1992; Evans & Belnap, 1999), vital soil stabilizers (Belnap & Gillette, 1998; Eldridge, 1998; Eldridge & Leys, 2003), and providers of habitat and food sources for other organisms (Zaady & Bouskila, 2002; Lalley *et al.,* 2006 a). The growth patterns of lichens, their distribution and varying morphological structures have been attributed to environmental factors such as elevation (Pirintsons *et al.*, 1995; Lalley *et al.,* 2006b), substrate aspect (John & Dale, 1990, Lalley *et al.,* 2006b), slope degree (Lalley *et al.,* 2006b), proximity to water (Lalley *et al.,* 2006b), exposure to radiation (Bjelland, 2003), and rainfall patterns or fog deposition (Eldridge & Tozer, 1997; Lalley *et al.,* 2006b), in addition to smaller scale factors such as substrate stability and chemistry (Ponzetti & McCunne, 2001; Belnap & Gillete, 1998; Lalley *et al.,* 2006 b) and association with other vegetation (Sheard, 1968).

India with only 2.4% share of global land surface, is a rich centre of lichen biodiversity, harbouring nearly 15% of total lichen species recorded globally (Negi & Upreti, 2000; Negi, 1999 and Negi & Gadgil, 1996). While there have been systematic studies on lichens for several decades, investigations of their community ecology have only recently begun (Awasthi, 2007; Negi, 2000 a,b,c; Negi and Upreti, 2000; Negi 1999; Negi and Gadgil, 1996; Awasthi, 1991; Upreti & Büdel, 1991; Awasthi, 1988). Though large-scale ecological studies of lichen communities give proper view of ecological preferences of lichen species, there is still a large scope of identifying a sensitive lichen functional group which can be treated as an indicator of ecosystem functioning in Indian habitats. Soil inhabiting lichens (Terricolous), are potential group of lichen species which can be calibrated for ecological studies, as their direct growth on the soil make them very sensitive to any change in the topographic and climatic environment. In the present study, groups of soil-inhabiting lichens are defined using ecological and morphological characters of the lichen species and the various habitat and climatic variables are analyzed for their effect on the determination of structure and distribution pattern of terricolous lichens in a temperate-alpine habitat of Garhwal Himalayas.

2. STUDY AREA

Chopta-Tungnath lies between 30°28′39″– 30°29′51″ N latitude and 79°12′9″ to 79°13′ 21″ E longitude in the Garhwal Himalayas of Rudraprayag district of Uttarakhand (Fig. 1). The area is famous for the holy shrine of Tungnath, situated at the top of the ridge dividing the waters of the Mandakini river (raising from Kedarnath) from those of the Alaknanda river (rising above Badrinath) and is third of the five and highest situated Panch Kedar temples—the supreme seat of Shaivite sect dedicated to Lord Shiva. The temple and associated alpine grassland lies about 2 km (1.2 mi) below the Chandrashila Peak (4,000 mt–13,123.4 ft) (Fig. 2). The road to Chopta is just below this ridge and hence provides the shortest bridle approach path for trekking to the temple from Chopta, over a short distance of about 4 km (2.5 mi). From the top of the Chandrashila peak, picturesque views of the Himalayan range comprising snow peaks of Nanda Devi, Panch Chuli, Banderpoonch, Kedarnath, Chaukhamba and Neelkanth on one side, and the Garhwal valley on the opposite side could be witnessed. The valley between Chopta and Tungnath temple has wooded hills with rich alpine meadows, oak woodlands, Rhododendron coppices and also agricultural fields.

The landscape of the area is typically mountainous with moderate to steep slopes (30° to 60°) and elevation rising from 2700 m– 4001 m above mean sea level. The topography of the area comprises ridges and exposed rocks with patches of flat alpine grassland. The pedology of the area is made up of crystalline and metamorphic weathering bedrock with sedimentary deposits formed during the Paleozoic. Soils in the area are of coarse texture, well drained and acidic (pH 4 to 5.5) (Sundriyal, 1992). Average annual precipitation in the area as recorded at Okhimath (30°30′N; 70°15′E, 2500 m) station, about 10 km west from Chopta, is 1888 × 5 ± 98 × 5 mm (SD) for 50 years of observations along with low to heavy snowfall during December to March (Negi, 2000). The maximum monthly temperature in the area varies from 19°C-37°C in the months of May to October, while the minimum temperature as low as –15°C in the alpine grasslands during the months of December to February. The higher plant vegetation of the study area is broadly classified as temperate mixed oak and coniferous forest through subalpine forest to alpine scrub or grassland along the altitude gradient (Gadgil and Meher-Homji 1990). The land use by the local human population is mainly confined to semipastoral mode of agriculture based mainly on livestock grazing, agriculture and fodder and fuel collection from the Oak-Rhododendron woodlands, besides some land around the Tungnath shrine is occupied and is used for religious tourism during the pilgrimage season. Thus, the anthropogenic pressure is mainly confined to lower altitudes of the landscape.

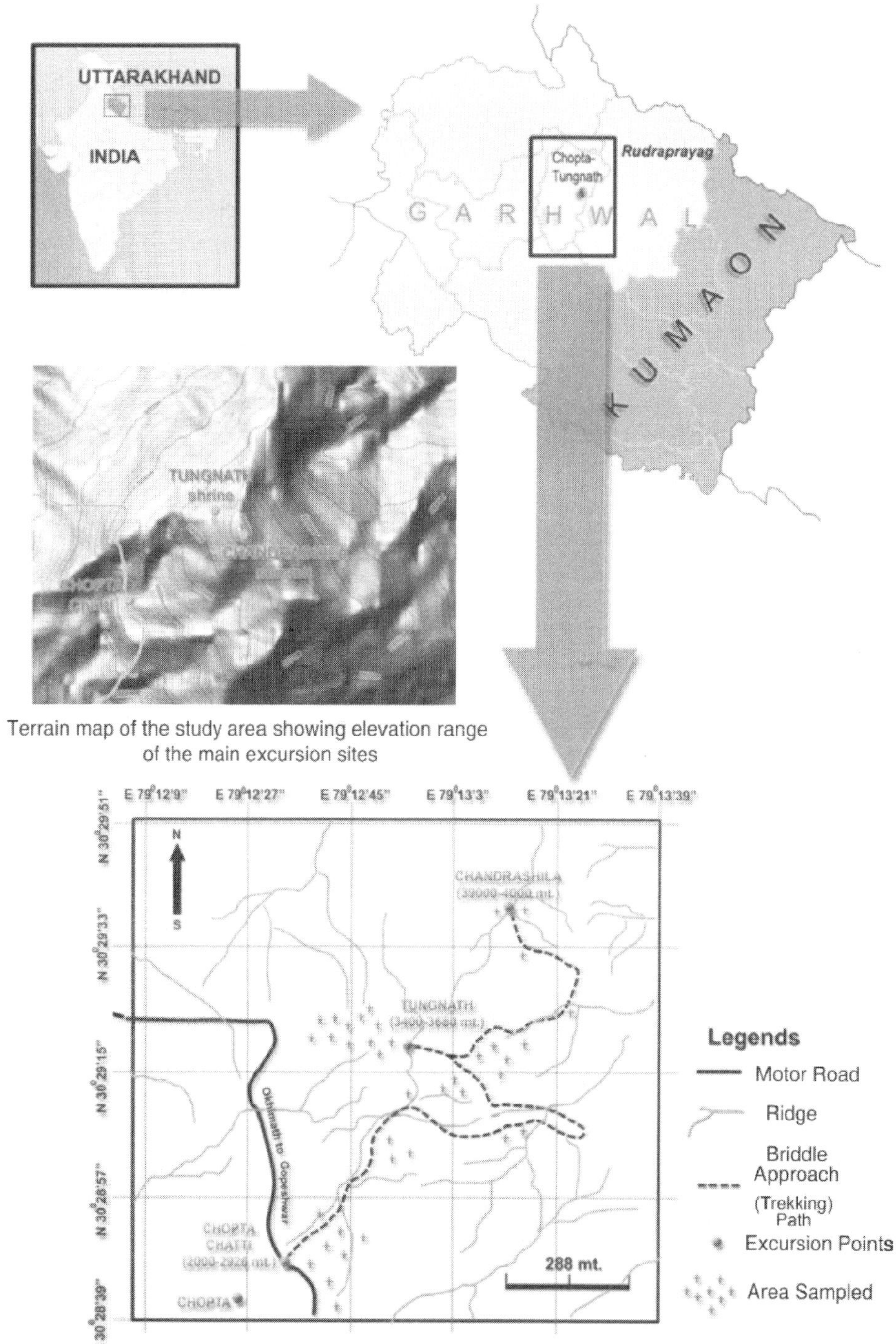

Fig. 1. Location map of the study area—Chopta-Tungnath tract, in Rudraprayag district of Uttarakhand, Garhwal Himalayas, India

Fig. 2. An overview of the Tungnath ridge showing the bridle path from Chopta to Tungnath and spatial orientation of some of the reference excursion points at the higher reaches (3400-4001 mt.) of study area.

3. FIELD METHODS AND DATA RECORDING

Sampling of terricolous lichens was done along the bridle approach path from Chopta to Tungnath and Tungnath to Chandrashila summit, employing stratified random sampling method (Greig-Smith, 1983: Krebs, 1989). The studied landscape was stratified into three macrohabitat types based on the dominant vegetation along the elevation gradient (Fig. 3, Table 1). These types are: (i) Oak-Rhododendron mixed forest (2700-3000 mt.) —with average altitude of 2862.5 ± 62.5 mt. the forests dominated by *Quercus semicarpifolia* forming the upper vegetational strata followed by *Rhododendron* spp. forming the lower story of vegetational strata; (ii) *Rhododendron* rich middle altitude subalpine forest (3000-3400 mt.) —with average altitude of 3261 ± 11.4 mt. dominated by *Rhododendron arboretum* and *Rhododendron campanulatum*, along with isolated stands of *Quercus semicapifolia* and few *Abies pindrow* and *Taxus baccata* trees; (iii) Higher altitude Alpine grassland, Bugyal (3400-4001 mt.) —with average altitude of 3706 ± 9.62 mt. grasslands vegetation dominated by herb species of *Anemone*, *Potentilla*, *Aster*, *Geranium*, *Meconopsis*, *Primula* and *Polemonium*, and scattered pockets of shrubs of *Rhododendron anthopogon* and *Juniperus* species (Negi, 2000c).

Terricolous lichens along with their substratum were randomly sampled, using soil core sampler (5 cm × 5 cm) (Pinokiyo *et al.,* 2008), along the bridle tract form Chopta to Tungnath and Tungnath to Chandrashila summit, covering the three macrohabitat types. The data of macroscale climatic variables–minimum-maximum temperature, elevation and relative humidity was taken for all samples in the three macrohabitats.

Ohmiya, K., Ohashi, H., Kobayashi, T. and Shimizu, S. (1977). Hydrolysis of lactose by immobilized microorganisms. *Appl. Environ. Microbiol.* 33:137-146.

Olchowy, T.W., Linnabary, R.D., Andrews, F.M. and Longshore, R.C. (1993). Lactose intolerance in a calf. *J. Vet. Intern. Med.* 7:12–15.

O'Leary, V.S. and Woychik, J.H. (1976). A comparison of some chemical properties of yogurts made from control and lactase-treated milks. *Food. Sci.* 41:791–793.

Ostergaard, S., Olsson, L., Johnston, M. and Nielsen J. (2000). Increasing galactose consumption by *Saccharomyces cerevisiae* through metabolic engineering of the *GAL* gene regulatory network. *Nature Biotechnol.* 18:1283–1286.

Overkamp, K.M., Bakker, B.M., Steensma, H.Y., van Dijken, J.P. and Pronk, J.T. (2002). Two mechanisms for oxidation of cytosolic NADPH by *Kluyveromyces lactis* mitochondria. *Yeast* 19:813–824.

Özcan, S. and Johnston, M. (1999). Function and regulation of yeast hexose transporters. *Microbiol. Mol. Biol. Rev.* 63:554–569.

Özcan, S., Leong, T. and Johnston, M. (1996). Rgt1p of *Saccharomyces cerevisiae*, a key regulator of glucose-induced genes, is both an activator and a repressor of transcription. *Mol. Cell. Biol.* 16:6419–6426.

Ozier-Kalogeropoulos, O., Malpertuy, A., Boyer, J., Tekaia, F. and Dujon, B. (1998). Random exploration of the *Kluveromyces lactis* genome and comparison with that of *Saccharomyces cerevisiae. Nucl. Acids Res.* 26:5511–5524.

Palomino, A., Herrero, P. and Moreno, F. (2006). Tpk3 and Snf1 protein kinases regulate Rgt1 association with *Saccharomyces cerevisiae HXK2* promoter. *Nucl. Acids Res.* 34:1427–1438.

Palomino, A., Herrero, P. and Moreno, F. (2005). Rgt1, a glucose sensing transcription factor, is required for transcriptional repression of the *HXK2* gene in *Saccharomyces cerevisiae. Biochem. J.* 388:697–703.

Panuwatsuk, W. and Da Silva, N.A. (2003). Application of a gratuitous induction system in *Kluyveromyces lactis* for the expression of intracellular and secreted proteins during fed-batch culture. *Biotechnol. Bioeng.* 81:712–718.

Papamichos-Chronakis, M., Gligoris, T. and Tzamarias, D. (2004). The Snf1 kinase controls glucose repression in yeast by modulating interactions between the Mig1 repressor and the Cyc8-Tup1 co-repressor. *EMBO Rep.* 5:368–372.

Park, K.S. and Kim, J.S. (2006). Engineering of *GAL1* promoter-driven expression system with artificial transcription factors. *Biochem. Biophys. Res. Commun.* 351:412–417.

Parthun, M.R. and Jaehning, J.A. (1990). Purification and characterization of the yeast transcriptional activator GAL4. *J. Biol. Chem.* 265:209–213.

Peng, G. and Hopper, J.E. (2002). Gene activation by interaction of an inhibitor with a cytoplasmic signaling protein. *Proc. Natl. Acad. Sci.* USA 99:8548–8553.

Peng, G. and Hopper, J.E. (2000). Evidence for Gal3p's cytoplasmic location and Gal80pís dual cytoplasmic-mechanisms for controlling Gal4p activity in *Saccharomyces cerevisiae. Mol. Cell. Biol.* 20:5140–5148.

Pilauri, V., Bewley, M., Diep, C.Q. and Hopper, J.E. (2005). Gal80 dimerization and the yeast *GAL* gene switch. *Genetics* 169:1903–1914.

Platt, A., Ross, H.C., Hankin, S. and Reece, R.J. (2000). Insertion of two amino acids into a transcriptional inducer converts it into a galactokinase. *Proc. Natl. Acad. Sci.* USA 97:3154–3159.

Platt, A. and Reece, R.J. (1998). The yeast galactose genetic switch is mediated by the formation of a Gal4p-Gal80p-Gal3p complex. *EMBO J.* 17:4086–4091.

Polish, J.A., Kim, J.H., and Johnston, M. (2005). How the Rgt1 transcription factor of *Saccharomyces cerevisiae* is regulated by glucose. *Genetics* 169:583–594.

Porro, D., Martegani, E., Ranzi, B.M. and Alberghina, L. (1992). Development of high cell density cultures of engineered *Saccharomyces cerevisiae* cells able to grow on lactose. *Biotechnol. Lett.* 14:1085–1088.

Porro, D., Martegani, E., Ranzi, B.M. and Alberghina, L. (1991). Lactose/whey utilization and ethanol production by transformed *Saccharomyces cerevisiae* cells. *Biotechnol. Bioeng.* 39:799–805.

Post-Beittenmiller, M.A., Hamilton, R.W. and Hopper, J.E. (1984). Regulation of basal and induced levels of the *MEL1* transcript in *Saccharomyces cerevisiae. Mol. Cell. Biol.* 4:1238–1245.

Pronk, J.T., de Steensma, H.Y. and van Dijken, J.P. (1996). Pyruvate metabolism in *Saccharomyces cerevisiae. Yeast* 12:1607–1633.

Ramakrishnan, S. and Hartley, B.S. (1993). Fermentation of lactose by yeast cells secreting recombinant fungal lactase. *Appl. Environ. Microbiol.* 59:4230–4235.

Ramos, J., Szkutnicka, K. and Cirillo, V.P. (1989). Characteristics of galactose transport in *Saccharomyces cerevisiae* cells and reconstituted lipid vesicles. *J. Bacteriol.* 171:3539–3544.

Rández-Gil, F., Sanz, P., Entian, K.D. and Prieto, J.A. (1998). Carbon source-dependent phosphorylation of hexokinase PII and its role in the glucose-signaling response in yeast. *Mol. Cell. Biol.* 18:2940–2948.

Reece, R.J. and Ptashne, M. (1993). Determinants of binding-site specificity among yeast C6 zinc cluster proteins. *Science* 261:909–911.

Rohde, J.R, Trinh, J. and Sadowski, I. (2000). Multiple signals regulate *GAL* transcription in yeast. *Mol. Cell Biol.* 20:3880–3886.

Rolland, S., Hnatova, M., Lemaire, M., Leal-Sanchez, J. and Wésolowski-Louvel M. (2006). Connection between the Rag4 glucose sensor and the KlRgt1 repressor in *Kluyveromyces lactis. Genetics* 174:617–626.

Rolland, F., Winderickx, J. and Thevelein, J.M. (2002). Glucose-sensing and signalling mechanisms in yeast. *FEMS Yeast Res.* 2:183–201.

Romanos M.A., Scorer C.A. and Clare J.J. (1992). Foreign gene expression in yeast: a review. *Yeast* 8:423–488.

Rosenfeld, E. and Beauvoit, B. (2003). Role of the non-respiratory pathways in the utilization of molecular oxygen by *Saccharomyces cerevisiae. Yeast* 20:1115–1144.

Rubio-Texeira, M. (2007). Milk: microbes' heaven making human's paradise. *Dynamic Biochemistry, Process Biotechnology, and Molecular Biology. In Press.*

Rubio-Texeira, M. (2006). Endless versatility in the biotechnological applications of *Kluyveromyces LAC* genes. *Biotechnol Adv.* 24:212–225.

Rubio-Texeira, M. (2005). A comparative analysis of the *GAL* genetic switch between not-so-distant cousins: *Saccharomyces cerevisiae* versus *Kluyveromyces lactis. FEMS Yeast Res.* 5:1115–1128.

Rubio-Texeira, M., Arévalo-Rodriguez, M., Lequerica, J.L. and Polaina J. (2000). Lactose utilization by strains expressing *Kluyveromyces lactis LAC* genes. *J. Biotechnol.* 84:97–106.

Rubio-Texeira, M., Castrillo, J.I., Adam, A.C., Ugalde, U.O. and Polaina, J. (1998). Highly efficient assimilation of lactose by a metabolically engineered strain of *Saccharomyces cerevisiae*. *Yeast* 14:827–837.

Sadowski, I., Costa, C. and Dhanawansa, R. (1996). Phosphorylation of Gal4p at a single C-terminal residue is necessary for galactose-inducible transcription. *Mol. Cell Biol.* 16:4879–4887.

Sadowski, I., Niedbala, D., Wood, K. and Ptashne, M. (1991). GAL4 is phosphorylated as a consequence of transcriptional activation. *Proc. Natl. Acad. Sci.* USA 88:10510–10514.

Sako, T., Matsumoto, K. and Tanaka, R. (1999). Recent progress on research and applications of non-digestible galacto-oligosaccharides. *Int. Dairy J.* 9:69–80.

Salghetti, S.E., Caudy, A.A., Chenoweth, J.G. and Tansey, W.P. (2001). Regulation of transcriptional activation domain function by ubiquitin. *Science* 293:1651–1653.

Saliola, M. and Falcone, C. (1995). Two mitochondrial alcohol dehydrogenase activities of *Kluyveromyces lactis* are differently expressed during respiration and fermentation. *Mol. Gen. Genet.* 249:665–672.

Saliola, M., Shuster, J.R. and Falcone, C. (1990). The alcohol dehydrogenase system in the yeast, *Kluyveromyces lactis*. *Yeast* 6:193–204.

Santangelo, G.M. (2006). Glucose signaling in *Saccharomyces cerevisiae*. *Microbiol. Mol. Biol. Rev.* 70:253–281.

Sanz, P., Alms, G.R., Haystead, T.A. and Carlson, M. (2000). Regulatory interactions between Reg1-Glc7 protein phosphatase and the Snf1 protein kinase. *Mol. Cell. Biol.* 20:1321–1328.

Sarma, N.J., Haley, T.M., Barbara, K.E., Buford, T.D., Willis, K.A. and Santangelo, G.M. (2007). Glucose-responsive regulators of gene expression in *Saccharomyces cerevisiae* function at the nuclear periphery via a reverse recruitment mechanism. *Genetics* 175:1127-1135.

Schulz, W.A., Ebling, B., Hasse, A., Zenke, F. and Breunig, K.D. (1993). Highly efficient transactivation by the yeast *Kluyveromyces lactis* transcription factor LAC9 and its inhibition by the negative regulator GAL80 in mammalian cells. *Biol. Chem. Hoppe Seyler* 374:313–318.

Scott, M.S., Perkins, T., Bunnell, S., Pepin, F., Thomas, D.Y. and Hallett, M. (2005). Identifying regulatory subnetworks for a set of genes. *Mol. Cell. Proteomics* 4:683–692.

Segrè, A.V., Murray, A.W., and Leu, J.Y. (2006). High-resolution mutation mapping reveals parallel experimental evolution in yeast. *PLOS Biol.* 4:1–14.

Sellick, C.A. and Reece, R.J. (2006). Contribution of amino acid side chains to sugar binding specificity in a galactokinase, Gal1p, and a transcriptional inducer, Gal3p. *J. Biol. Chem.* 281:17150–17155.

Shain, D.H., Salvadore, C. and Denis, C.L. (1992). Evolution of the alcohol dehydrogenase (*ADH*) genes in yeast: characterization of a fourth *ADH* in *Kluyveromyces lactis*. *Mol. Gen. Genet.* 232:479–488.

Shiba, Y., Paradise, E.M., Kirby, J., Ro, D.K. and Keasling, J.D. (2007). Engineering of the pyruvate dehydrogenase bypass in *Saccharomyces cerevisiae* for high-level production of isoprenoids. *Metab. Eng.* 9:160–168.

Shiba, Y., Ono, C., Fukui, F. and Yoshikawa, H. (2000). Effect of ethanol on the production of carboxypeptidase Y using the *GAL10* promoter in a *Saccharomyces cerevisiae gal80* mutant. *J. Biosci. Bioeng.* 89:426–430.

Sikder, D., Johnston, S.A. and Kodadek, T. (2006). Widespread, but non-identical, association of proteasomal 19S and 20S proteins with yeast chromatin. *J. Biol. Chem.* 281:27346–27355.

Sil, A.K., Xin, P. and Hopper, J.E. (2000). Vectors allowing amplified expression of the *Saccharomyces cerevisiae* Gal3p-Gal80p-Gal4p transcription switch: applications to galactose-regulated high-level production of proteins. *Protein Expr. Purif.* 18:202–212.

Sil, A.K., Alam, S., Xin, P., Ma, L., Morgan, M., Lebo, C.M., Woods, M.P. and Hopper, J.E. (1999). Gal4p transcription switch of yeast: Gal3p destabilizes the Gal80-Gal4p complex in response to galactose and ATP. *Mol. Cell. Biol.* 19:7828–7840.

Simova, E.D., Frengova, G.I. and Beshkova, D.M. (2004). Exopolysaccharides produced by mixed culture of yeast *Rhodotorula rubra* GED10 and yogurt bacteria (*Streptococcus thermophilus* 13a + *Lactobacillus bulgaricus* 2-11). *J. Appl. Microbiol.* 97:512–519.

Siso, M.I.G. (1996). The biotechnological utilization of cheese whey: a review. *Biores. Technol.* 57:1–11.

Slepak, T., Tang, M., Addo, F. and Lai, K. (2005). Intracellular galactose-1-phosphate accumulation leads to environmental stress response in yeast model. *Mol. Genet. Metab.* 86:360–371.

Snoek, I.S. and Steensma, H.Y. (2006). Why does *Kluyveromyces lactis* not grow under anaerobic conditions? Comparison of essential anaerobic genes of *Saccharomyces cerevisiae* with the *Kluyveromyces lactis* genome. *FEMS Yeast Res.* 6:393–403.

Spark, M., Paschertz, H. and Kamphues, J. (2005). Yeast (different sources and levels) as protein source in diets of reared piglets: effects on protein digestibility and N-metabolism. *J. Anim. Physiol. Anim. Nutr. (Berl.).* 89:184–188.

Sreekrishna, K. and Dickson, R.C. (1985). Construction of strains of *Saccharomyces cerevisiae* that grow on lactose. *Proc. Natl. Acad. Sci.* USA. 82:7909–7913.

Stagoj, M.N., Comino, A. and Komel, R. (2006). A novel *GAL* recombinant yeast strain for enhanced protein production. *Biomol. Eng.* 23:195–199.

Stephanopoulos, G., Aristodou, A. and Nielsen, J. (1998). Metabolic engineering. Academic Press, Inc., San Diego, CA. pp. 750.

Sulahian, R., Johnston, S.A. and Kodadek, T. (2006). The proteasomal ATPase complex is required for stress-induced transcription in yeast. *Nucleic Acids. Res.* 34:1351–1357.

Suleau, A., Gourdon, P., Reitz-Ausseur, J. and Casaregola, S. (2006). Transcriptomic analysis of extensive changes in metabolic regulation in *Kluyveromyces lactis* strains. *Eukaryotic Cell* 5:1360–1370.

Suleau, A., Jacques, N., Reitz-Ausseur, J. and Casaregola, S. (2005). Intraspecific gene expression variability in the yeast *Kluyveromyces lactis* revealed by micro-array analysis. *FEMS Yeast Res.* 5:595–604.

Suzuki-Fujimoto, T., Fukuma, M., Yano, K.I., Sakurai, H., Vonika, A., Johnston, S.A. and Fukasawa, T. (1996). Analysis of the galactose signal transduction pathway in *Saccharomyces cerevisiae*: interaction between Gal3p and Gal80p. *Mol. Cell. Biol.* 16:2504–2508.

Swallow, D.M. (2003). Genetics of lactose persistence and lactose intolerance. *Annu. Rev. Genet.* 37:197–219.

Szczebara, F.M., Chandelier, C., Villeret, C., Masurel, A., Bourot, S., Duport, C., Blanchard, S., Groisillier, A., Testet, E., Costaglioli, P., Cauet, G., Degryse, E., Balbuena, D., Winter, J., Achstetter, T., Spagnoli, R., Pompon, D. and Dumas, B. (2003). Total biosynthesis of hydrocortisone from a simple carbon source in yeast. *Nature Biotechnol.* 21:43–149.

Szilagyi A. (2002). Review article: lactose—a potential prebiotic. *Alimentary Pharmacology and Therapeutics* 16:1591–1602.

Taddei, A., Van Houwe, G., Hediger, F., Kalck, V., Cubizolles, F., Schober, H. and Gasser, S.M. (2006). Nuclear pore association confers optimal expression levels for an inducible yeast gene. *Nature* 441: 774–778.

Tarrio, N., Becerra, M., Cerdan, M.E. and Gonzalez-Siso, M. (2006). Reoxidation of cytosolic NADPH in *Kluyveromyces lactis*. *FEMS Yeast Res.* 6:371–380.

Thoden, J.B., Sellick, C.A., Reece, R.J. and Holden, H.M. (2007). Understanding a transcriptional paradigm at the molecular level. The structure of yeast Gal80p. *J. Biol. Chem.* 282:1534–1538.

Thoden, J.B., Sellick, C.A., Timson, D.J., Reece, R.J. and Holden, H.M. (2005). Molecular structure of *Saccharomyces cerevisiae* Gal1p, a bifunctional galactokinase and transcriptional inducer. *J. Biol. Chem.* 280:36905–36911.

Thomson, M.P. and Gyuricsek, D.M. (1974). Manufacture of Cheddar cheese from hydrolyzed lactose milk. *J. Dairy Sci.* 57:598–602.

Timmers, H.T. and Tora, L. (2005). SAGA unveiled. *Trends Biochem. Sci.* 30:7–10.

Timson, D.J. and Reece, R.J. (2002). Kinetic analysis of yeast galactokinase: implications for transcriptional activation of the *GAL* genes. *Biochimie* 84:265–272.

Timson, D.J., Ross, H.C. and Reece, R.J. (2002). Gal3p and Gal1p interact with the transcriptional repressor Gal80p to form a complex of 1:1 stoichiometry. *Biochem. J.* 363:515–520.

Traven, A., Jelicic, B. and Sopta, M. (2006). Yeast Gal4: a transcriptional paradigm revisited. *EMBO Rep.* 7:496–499.

Treitel, M.A., Kuchin, S. and Carlson, M. (1998). Snf1 protein kinase regulates phosphorylation of the Mig1 repressor in *Saccharomyces cerevisiae*. *Mol. Cell Biol.* 18:6273–6280.

Treitel, M.A. and Carlson, M. (1995). Repression by SSN6-TUP1 is directed by MIG1, a repressor/activator protein. *Proc. Natl. Acad. Sci.* USA 92:3132–3136.

Tschopp, J.F., Emr, S.D., Field, C. and Scheckman, R. (1986). *GAL2* codes for a membrane-bound subunit of the galactose permease in *Saccharomyces cerevisiae*. *J. Bacteriol.* 166:313–318.

Tzamarias, D. and Struhl, K. (1995). Distinct TPR motifs of Cyc8 are involved in recruiting the Cyc8-Tup1 corepressor complex to differentially regulated promoters. *Genes Dev.* 9:821–831.

Uccelletti, D., Farina, F., Rufini, S., Magnelli, P., Abeijon, C. and Palleschi, C. (2006). The *Kluyveromyces lactis* α 1,6-mannosyltransferase KlOch1p is required for cell wall organization and proper functioning of the secretory pathway. *FEMS Yeast Res.* 6:449–457.

Uffenbeck, S.R. and Krebs, J.E. (2006). The role of chromatin structure in regulating stress-induced transcription in *Saccharomyces cerevisiae*. *Biochem. Cell Biol.* 84:477–489.

Ulery, T.L., Mangus, D.A. and Jaehning, J.A. (1991). The yeast *IMP1* gene is allelic to *GAL2*. *Mol. Gen. Genet.* 230:129–135.

van Dijken, J.P., Weusthuis, R.A. and Pronk, J.T. (1993). Kinetics of growth and sugar consumption in yeasts. *Antonie Van Leeuwenhoek* 63:343–352.

van Ooyen, A.J., Dekker, P., Huang, M., Olsthoorn, M.M., Jacobs, D.I., Colussi, P.A. and Taron, C.H. (2006). Heterologous protein production in the yeast *Kluyveromyces lactis*. *FEMS Yeast Res.* 6:381–392.

van Urk, H., Postma, E., Scheffers, W.A. and van Dijken, J.P. (1989). Glucose transport in Crabtree-positive and Crabtree-negative yeasts. *J. Gen. Microbiol.* 135:2399–2406.

Vaughan-Martini, A., Martini, A. and Cardinali, G. (1993). Electrophoretic karyotyping as a taxonomic tool in the genus *Saccharomyces*. *Antonie Van Leeuwenhoek* 63:145–156.

Vemuri, G.N., Eiteman, M.A., McEwen, J.E., Olsson, L. and Nielsen, J. (2007). Increasing NADH oxidation reduces overflow metabolism in *Saccharomyces cerevisiae*. *Proc. Natl. Acad. Sci. USA* 104:2402–2407.

Verma, M., Bhat., P.J., Bhartiya, S. and Venkatesh, K.V. (2004). A steady state modelling approach to validate an *in vivo* mechanism of the GAL regulatory network in *Saccharomyces cerevisiae*. *Eur. J. Biochem.* 271:4064–4074.

Verstrepen, K.J., Iserentant, D., Malcorps, P., Derdelinckx, G., van Dijck, P., Winderickx, J., Pretorius, I.S., Thevelein, J.M. and Delvaux, F.R. (2004). Glucose and sucrose: hazardous fast-food for industrial yeast? *Trends Biotechnol.* 22:531–537.

Vincent, O., Townley, R., Kuchin, S. and Carlson, M. (2001). Subcellular localization of the Snf1 kinase is regulated by specific beta subunits and a novel glucose sensing mechanism. *Genes Dev.* 15:1104–1114.

Vollenbroich, V., Meyer, J., Engels, R., Cardinali, G., Menezes, R.A. and Hollenberg, C.P. (1999). Galactose induction in yeast involves association of Gal80p with Gal1p or Gal3p. *Mol. Gen. Genet.* 261:495–507.

Weirich, J., Goffrini, P., Kuger, P., Ferrero, I. and Breunig, K.D.(1997). Influence of mutations in hexose-transporter genes on glucose repression in *Kluyveromyces lactis*. *European J. Biochem.* 249:248–257.

Wésolowski-Louvel M., Breunig K.D. and Fukuhara H. (1996). *Kluyveromyces lactis*. In: Nonconventional Yeasts in Biotechnology. A Handbook. Wolf K., Ed., Springer-Verlag, Berlin, pp. 139–201.

Wésolowski-Louvel M., Goffrini, P., Ferrero, I. and Fukuhara, H. (1992a). Glucose transport in the yeast *Kluyveromyces lactis*. I. Properties of an inducible low-affinity glucose transporter gene. *Mol. Gen. Genet.* 233:89–96.

Wésolowski-Louvel M., Prior, C., Bornecque, D. and Fukuhara, H. (1992b). Rag- mutations involved in glucose metabolism in yeast: isolation and characterization. *Yeast* 8:711–719.

Wiedemuth, C. and Breunig, K.D. (2005). Role of Snf1p in regulation of intracellular sorting of the lactose and galactose transporter Lac12p in *Kluyveromyces lactis*. *Eukaryotic Cell* 4:716–721.

Wilcox, L.J., Balderes, D.A., Whatron, B., Rao, G. and Sturley, S.L. (2002). Transcriptional profiling identifies two members of the ATP-binding cassette transporter superfamily required for sterol uptake in yeast. *J. Biol. Chem.* 277:32466–32472.

Wilson, W.A. and Roach, P.J. (2002). Nutrient-regulated protein kinases in budding yeast. *Cell* 111:155–158.

Wolfe, K.H. and Shields, D.C. (1997). Molecular evidence for an ancient duplication of the entire yeast genome. *Nature* 387:708–713.

Yang, S.T. and Silva, E.M. (1995). Novel products and new technologies for use of a familiar carbohydrate, milk lactose. *J. Dairy Sci.* 78:2541–2562.

Yano, K.I. and Fukasawa, T. (1997). Galactose-dependent reversible interaction of Gal3p with Gal80p in the induction pathway of Gal4p-activated genes of *Saccharomyces cerevisiae. Proc. Natl. Acad. Sci.* USA 94:1721–1726.

Zachariae W. and Breunig KD. (1993). Expression of the transcriptional activator LAC9 (KlGAL4) in *Kluyveromyces lactis* is controlled by autoregulation. *Mol. Cell. Biol.* 13:3058–66.

Zeeman, A.M. and Steensma, H.Y. (2003). The acetyl co-enzyme A synthetase genes of *Kluyveromyces lactis. Yeast* 20:13–23.

Zeeman, A.M. Kuyper, M., Pronk, J.T., van Dijken, J.P. and Steensma, H.Y. (2000). Regulation of pyruvate metabolism in chemostat cultures of *Kluyveromyces lactis* CBS 2359. *Yeast* 16:611–620.

Zenke, F.T., Kapp, L. and Breunig, K.D. (1999). Regulated phosphorylation of the Gal4p inhibitor Gal80p of *Kluyveromyces lactis* revealed by mutational analysis. *Biol. Chem.* 380:419–430.

Zenke, F.T., Engels, R., Vollenbroich, V., Meyer, J., Hollenberg, C.P. and Breunig, K.D. (1996). Activation of Gal4p by galactose-dependent interaction of galactokinase and Gal80p. *Science* 272:1662–1665.

Zenke, F.T., Zachariae, W., Lunkes, A. and Breunig, K.D. (1993). Gal80 proteins of *Kluyveromyces lactis* and *Saccharomyces cerevisiae* are highly conserved but contribute differently to glucose repression of the galactose regulon. *Mol. Cell. Biol.* 13:7566–7576.

Zheng, W. and Johnston, S.A. (1998). The nucleic acid binding activity of bleomycin hydrolase is involved in bleomycin detoxification. *Mol. Cell Biol.* 18:3580–3585.

Zheng, W., Xu, H.E. and Johnston, S.A. (1997). The cysteine-peptidase bleomycin hydrolase is a member of the galactose regulon in yeast. *J. Biol. Chem.* 272:30350–30355.

11

Current Advances in Citric Acid Production by Fungi

Maria Papagianni

Department of Hygiene and Technology of Food of Animal Origin, School of Veterinary Medicine, Aristotle University of Thessaloniki, Greece. E-mail: mp2000@vet.auth.gr. Tel: +30-2310-999804. Fax: +30-2310-999829.

INTRODUCTION

Citric acid (2-hydroxy-propane-1,2,3-tricarboxylic acid) is one of the most commonly used acids in the food and pharmaceutical industries. It was first isolated by Scheele in 1784 as a constituent of citrus fruits (Scheele, 1784). However, since the discovery by Krebs of the TCA cycle, within which citric acid is a key intermediate, its ubiquitous occurrence in almost all living systems has been established. Due to its pleasant taste, low toxicity and ease of assimilation interest in industrial production arose very early. Currie in 1917, laid down the basis for the industrial production of citric acid with the discovery that *Aspergillus niger* can grow abundantly in a nutrient medium with an initial pH of 2.5-3.5, while doing so it excretes large amounts of citric acid (Mattey and Kristiansen, 1999). Today, some 400 000 metric tonnes per year are produced by industrial fermentation (Arora, 2004).

The biochemical pathways related to citric acid accumulation and their regulation have been the subject of numerous investigations. The excretion of many metabolic products by filamentous fungi is due to the excess of carbon and a certain "disorganization" of metabolism which is often exacerbated by the lack of trace elements in the nutrient media (Wolschek and Kubicek, 1999). The nature of the growth limiting substrate is critical and it determines the metabolic flow pattern resulting in a particular type of "overflow" product dependent on the microorganism used.

MICROORGANISMS PRODUCING CITRIC ACID

A large number of fungi have the ability to produce citric acid, for example: *Aspergillus niger*, *A. clavatus*, *Penicillium luteum*, *P. citrinum*, *Mucor piriformis*, *Candida lipolytica*, and others (Jong *et al.*, 1994). The industry mainly employs selected strains of *Aspergillus niger*, a filamentous fungus, which is capable of converting as much as 90°/w of the carbohydrate supplied to citric acid (Wayman and Mattey, 2000). Citric acid is also produced by yeasts, mainly species of *Candida*, especially *C. lipolytica*. An increasing proportion of citric acid production is now moving over to *Candida* yeasts, which have a better tolerance for the trace metals commonly found in low-cost complex media. *Candida* strains are also tolerant of higher initial sugar concentrations and have faster conversion rates, but neither the yield nor the final concentration of citric acid are as high as those achieved with *A. niger* (Mattey, 1999).

THE CITRIC ACID FERMENTATION BY *ASPERGILLUS NIGER*

Industrial strains of citric acid producing *A. niger* are among the most secretly kept microorganisms in biotechnology. These might have been isolated from soils or plant materials and further improved by mutation and selection. The strain improvement strategy or its significance to the industrial know-how is also not revealed. However, mutant isolation procedures have been reported by academics and include among others tolerance against high sugar concentrations, low pH, and respiratory chain inhibitors.

Raw Materials

Beet molasses, cane molasses, corn starch, corn syrup, glucose, sucrose, wheat bran or sweet potato wastes can be used as the major raw material. Metal ions, such as iron, copper, zinc, etc. have an important influence on the fermentation. Since the concentration of trace elements affects citric acid production, these may have to be removed from the raw materials by chemical methods or using ion exchange resins (Atkinson and Mavituna, 1991). Table 11.1 gives some data for the production of citric acid from industrial raw materials.

Table 11.1 Production of citric acid by the submerged process (Prescott and Dunn, 1959).

Carbohydrate source	*Reagent used*	*Yield (%)*	*Fermentation*
Beet molasses	Potassium ferrocyanide	max 64[a]	8-9 days
	Potassium ferrocyanide	max 72[a]	70 hours
	Potassium ferrocyanide	68.5[a]	116 hours
	Potassium ferrocyanide	average 64[a]	80-116 hours
Cane molasses	Methanol	63.3[b]	9 days
	Bone char + Suchar CPS + amberlites	61.9[b]	12 days

(*Contd.*)

Carbohydrate source	*Reagent used*	*Yield (%)*	*Fermentation*
Corn starch	Methanol	71.5 [b]	10 days
Corn syrup	Morpholine	79.9	8.67 days
Glucose	Methanol	71.2 [b]	7 days
Sucrose	None	average 72	9 days

[a]Based on available sugar
[b]Based on sugar consumed

Process Outline

Most industrial citric acid using *A. niger* is produced by submerged fermentation in large, aerated, stainless steel fermentors of a stirred tank or airlift type. However, a 20% of world production is still obtained by solid-state fermentation in aluminium or stainless still trays. Solid-state fermentation is being replaced by submerged fermentation, which is more efficient and amenable to automation and control. A comparison of the two processes is given in Table 11.2.

Table 11.2 Comparison of solid-state and submerged citric acid fermentation processes

Criterion	*Solid-state process*	*Submerged process*
Productivity	0.85-1.0 $kgm^{-3}d^{-1}$	16.5-18.0 $kgm^{-3}d^{-1}$
Yield [a]		
Fermentation	65-75%	87-92%
Whole plant	58-70%	81-88%
Energy requirements		
Electricity	1.400-1.600 kWh $tonne^{-1}$	2.500-2.800 kWh $tonne^{-1}$
Steam	15-20 tonne $tonne^{-1}$	9.7-10.2 tonne $tonne^{-1}$
Space requirements[b]		
Fermentation section	370 $m^2tonne^{-1}d^{-1}$	~8 $m^2tonne^{-1}d^{-1}$

[a] To fermentable disaccharide
[b] For production of citric acid monohydrate

In the solid-state fermentation process, the inoculated broth is placed in shallow pans of high-purity aluminium or stainless steel. Humidified air is blown over the pans for five to six days, after which time is replaced by dry air. Spores' germination occurs within 24 hours and the mycelium covers the surface of the broth. The initial pH of 4.0 to 6.0, drops rapidly with germination to pH 1.5 to 2.0. Little citric acid is produced during growth. The initial sugar concentration of 20 to 25% is reduced to 1-3% eight to nine days from inoculation. The yield of citric acid on the accumulated carbohydrate is in the range of 80-85%.

In the submerged fermentation process, a spore inoculum is usually used. For decationized solutions, No. 316 stainless steel is used to avoid corrosion. The initial pH of about 4.0 falls to 1.5 to 2.0 with the uptake of ammonium ions. Air is sparged at a rate of 0.5 to 1.5 vvm. Table 11.3 gives an outline of the conditions required for the production of citric acid by the submerged process.

Table 11.3 Conditions for the citric acid production of in submerged fermentation

Parameter	*Conditions*
pH	1.5-2.8
pH>2.8	Oxalic acid formation, danger of contamination, low yields, low mycelium growth rate
pH<1.5	Low growth rate, abnormal morphologies
Temperature	28-33 °C
T>33°C	Oxalic acid production
T<28 °C	Low productivity
Fermentation duration	Depending on the sugar concentration in the raw material. Generally, within 6-15 days
Trace elements	Of special importance: manganese; also important: zinc, iron, copper, heavy metals and alkaline metals

Product Recovery

At the end of the process, lasting 6-15 days, the culture broth is filtered to remove mycelium. A polishing filtration may be necessary to remove residual antifoam, mycelia or oxalate (Atkinson and Mavituna, 1991). Calcium citrate is precipitated by addition of calcium hydroxide. Calcium citrate is then filtered off and the filter cake is transferred to a tank, where it is treated with sulphuric acid to precipitate calcium sulphate. The filtrate containing citric acid is then purified by passing over activated carbon, demineralized by ion exchangers and finally the purified solution is evaporated yielding crystals of citric acid which are removed by centrifugation.

The Commercial Market of Citric Acid

Citric acid is marketed as an anhydrous crystalline chemical which can be the crystalline monohydrate or the crystalline sodium salt. Anhydrous and monohydrate crystals may be granular (15-50 mesh), fine or powdery (35-100 mesh for anhydrous acid and 60-200 mesh for the monohydrate). Crystals are supplied in bags (moisture-proof) of drums. Storage should be at 21°C and 50-70% relative humidity. Caking may occur if stored at high temperature and humidity.

Citric acid was originally crystallized from lemons in 1784 and this remained the source for some time, with Italy supplying the bulk of the fruit. In 1923 Pfizer started production in the U.S.A. using *A. niger* growing in surface culture on a sucrose-salts medium. Several processes for synthetic production have been described but none is able to compete the product of *A. niger* which accounts for almost 99% of production, the remainder being derived from lemons and pineapple wastes. After the monopoly of the Italian government was broken, price reduction was followed by substantial increases in production, which rose from 5 000 metric tones in 1929 to the current level of almost 400 000 tonnes per annum. In the West, a few companies dominate production and there is a substantial excess capacity.

Citric acid production and use are profitable from an economic perspective. Profit margins are small, and there is keen competition among suppliers. In the developed world, manpower requirements for production have been reduced to minimal levels and a very large proportion of the bulk sale price of citric acid is due to the cost and treatment of raw materials. Therefore, new producers are unlikely to enter the market except in countries where substantial quantities of very cheap feedstock are available locally. In developing countries, the relatively high cost of importing citric acid has driven the development of alternative production methods. These methods typically use small fermentation vessels, are quite labor-intensive and use locally sourced raw materials (for example, by-products of food processing, such as coconut husks and pineapple fruit pulp).

Citric acid forms a wide range of metallic salts, including complexes with divalent metal ions (e.g. Fe, Cu, Mg, Mn, Ca, etc.) It is this property that has lead to its widespread use as a sequestering agent and as an antioxidant (by reducing the opportunity for any catalytic effect of metal ions on the oxidation of fats and oils). Citric acid is used mainly as a flavouring agent in the production of food and in soft drinks. It is an efficient metal chelator and this property leads to a wide variety of industrial, pharmaceutical and food uses. It is used as a cleaning agent for metals, to keep metals in solutions for electroplating and in agriculture, to control the density of clay suspensions by regulating swelling and as a replacement for polyphosphates in the detergent industry. Also, citric acid acts as a blood anti-coagulant. Its acidity and non-toxicity are of obvious importance to the food and pharmaceutical industries. Citric acid and sodium citrate are used in conjunction as acidity regulators, and the pleasant flavour of the compound adds value to many products that are to be consumed. Table 11.4 summarizes some of the main commercial uses for citric acid.

Table 11.4 Commercial uses of citric acid.

Commercial use of citric acid	*Properties*	*Market Share**
Food industry		
Beverages, jellies, jam	Acidulant / flavouring	About 75%
fats and oils, frozen food	Antioxidant	
Pharmaceutical industry		
Effervescent products	Acid / flavouring	
Vitamin preparations	Antioxidant	
Iron preparations	Salt formation	
Anticoagulants	Sequestering / buffering	About 10 %
Cosmetics	Buffering / antioxidant	
Other Industry		
Photographic	Buffering	
Detergents	Buffering / sequestering	About 15%
Cleaning (of metals)	Sequestering	
Primer / polymer production	Sequestering	

* Moses V, Cape RE, 1994.

BIOCHEMICAL MECHANISMS

The biochemical pathways related to citric acid accumulation and their regulation have been the subject of numerous investigations. The excretion of many metabolic products by filamentous fungi is due to the excess of carbon and a certain "disorganization" of the metabolism which is often exacerbated by lack of trace elements in the nutrient media (Kubicek and Röhr, 1989). The nature of the growth limiting substrate is critical and it determines the metabolic flow pattern resulting in a particular type of "overflow" product dependent on the microorganism used.

The metabolic pathway for citric acid production by *A. niger* involves the breakdown of carbon compounds via glycolysis (Fig.11.1), formation of oxaloacetate from pyruvate and CO_2 by the pyruvate carboxylase reaction, and then formation of citric acid by the ticarboxylic acid cycle (TCA) cycle. The key regulatory enzymes for citric acid production are: phosphofructokinase, pyruvate carboxylase, citrate synthase, and α-ketoglutarate dehydrogenase. Aconitase, which degrades citric acid within the TCA cycle, is not a regulatory enzyme and citric acid accumulation is not due to inhibition of aconitase activity (Kubicek and Röhr, 1985). The regulatory patterns for the citric acid pathway are summarized in Fig.11.2.

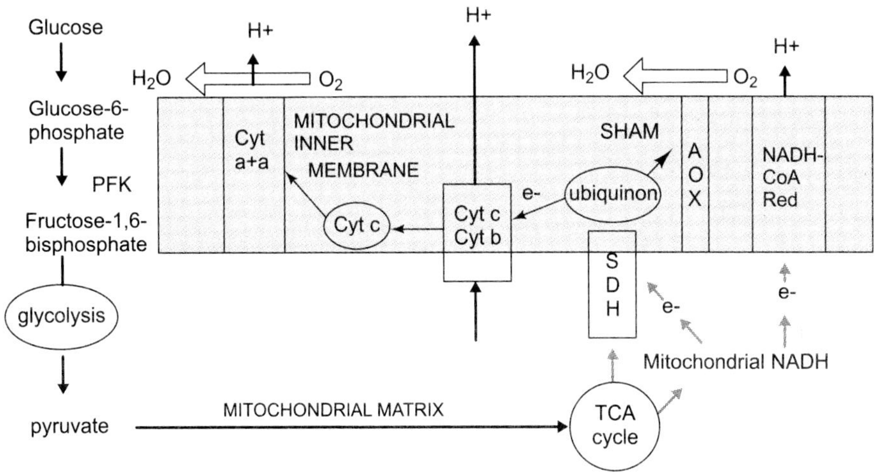

Figure 11.1 Organization of glycolysis and the respiratory chain in the mitochondrial membrane of A. niger.

Phosphofructokinase (PFK) is the central enzyme in citric acid accumulation. It is regulated by citric acid since the accumulation of citric acid exerts a negative feedback on the enzyme (Habison *et al.*, 1983). This effect is, however, counteracted by elevated levels of intracellular ammonium ions. This situation takes place when manganese is limiting. Increased NH_4^+ and protein turnover/amino acid pools have been associated with manganese deficiency in overproducing

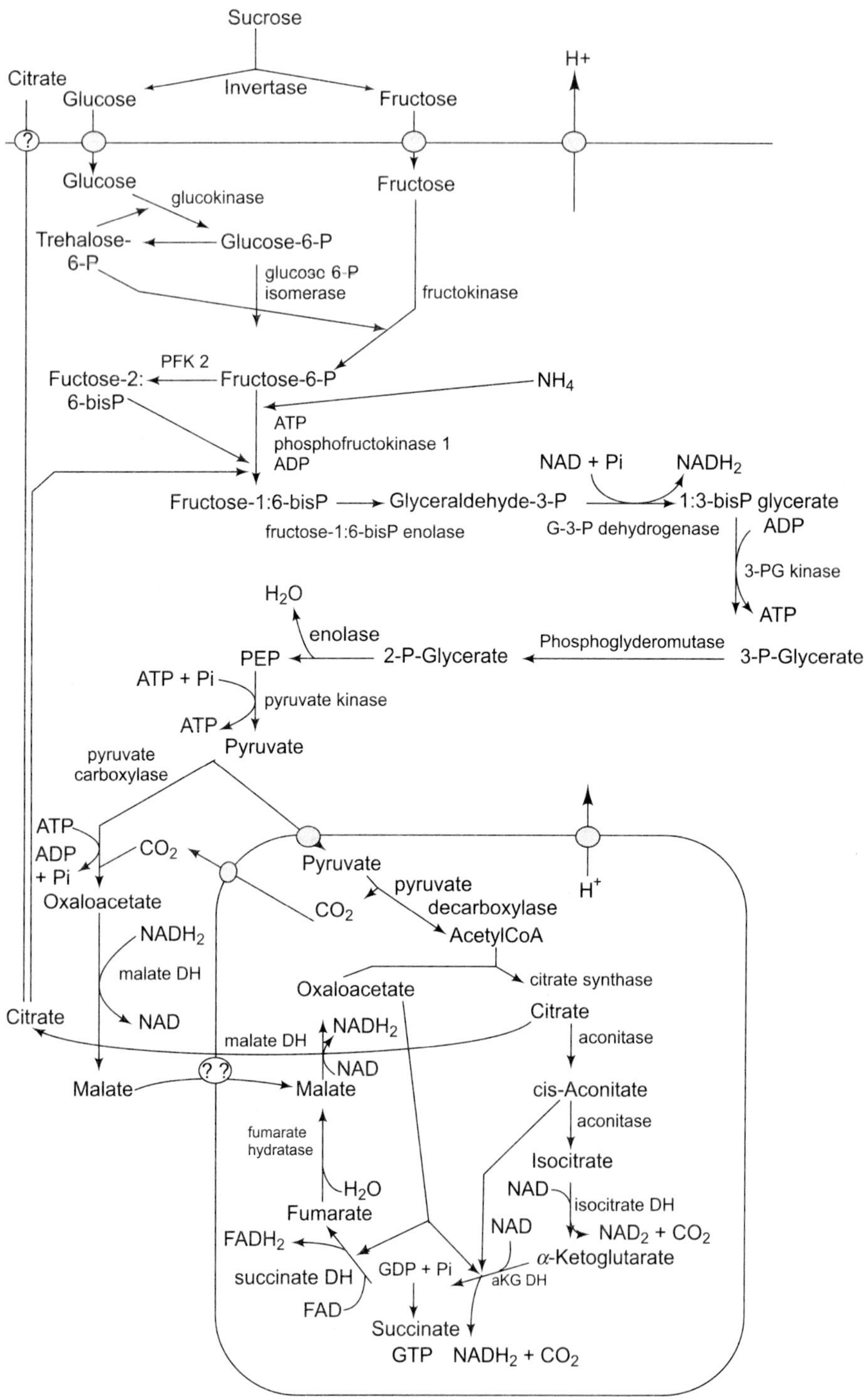

Figure 11.2 A summary of the regulatory patterns for the citric acid pathway.

strains of *A. niger*. Phosphofructokinase is also involved in the regulatory mechanism related to the formation of fructose-2,6-biphosphate (FBP) (Kubicek-Pranz *et al.*, 1990), which is formed by the 6-phosphofructo-2-kinase (PFK 2). FBP signals the presence of glucose and appears to play a critical role in regulating the main PFK activity. Increased levels of FBP are observed during the induction of acidogenesis by high sugar concentrations in fermentation media in *A. niger* fermentations. Such high sugar levels are employed in industrial fermentation media. It is now known that FBP activates PFK by lowering its K_m and it strongly inhibits gluconeogenesis. It also antagonizes citrate inhibition of PFK (Arts *et al.*, 1987).

Pyruvate carboxylase is the key enzyme that replenishes intermediates, supplying oxaloacetate to the TCA cycle. It forms oxaloacetate from pyruvate and CO_2, derived from oxidative decarboxylation of pyruvate (Kubicek and Röhr, 1989). This enzyme is induced by high sugar concentrations, like those necessary for acidogenesis (Hossain *et al.*, 1984). It is a cytosolic enzyme and thus able to act in conjunction with malate dehydrogenase to generate malate, the counterion of citrate efflux from the mitochondria. Therefore, under conditions of high sugar levels, pyruvate carboxylase plays a central role in citric acid accumulation by generating cytosolic malate.

The other two key regulatory enzymes are citrate synthase and α-ketoglutarate dehydrogenase. Citrate synthase catalyses the reversible condensation reaction between oxaloacetate and acetylCoA:

$$\text{acetylCoA} + \text{oxaloacetate} \leftrightarrow \text{citrate}^{3-} + \text{H}^+ + \text{CoA-SH}$$

The equilibrium of the reaction is very much in favor of the production of citrate because of the thioester hydrolysis that occurs as part of the reaction. The rate of the reaction is controlled *in vivo* by the availability of the substrates and not by any inhibitors. Oxaloacetate has a strong effect upon the reaction rate because the concentration of oxaloacetate has been shown to affect the K_m of the enzyme for acetylCoA (Kubicek and Röhr, 1980). This indicates that the reaction follows an ordered mechanism, with oxaloacetate bound to the enzyme first with acetylCoA being bound subsequently. ATP and coenzyme A inhibit the enzyme *in vitro*, both competing with acetylCoA for the active site. There is no or little inhibition at the concentrations of these metabolites seen *in vivo*. Elevated levels of oxaloacetate during citric acid accumulation block citric acid catabolism and promote accumulation, besides enhancing the rate of oxaloacetate conversion to citric acid. *A. niger* citrate synthase has been shown to be mainly regulated by oxaloacetate (Kubicek and Röhr, 1980), while the inhibitory effect of oxaloacetate on α-ketoglutarate dehydrogenase raises NADH/NADPH levels, which in turn lower the activity of isocitrate dehydrogenase and further promote citric acid overproduction (Meixner-Monori *et al.*, 1986).

A critical role on the citric acid fermentation process by *A. niger* has been attributed to manganese ions, the influence of which on protein synthesis, was considered to be of major importance since cycloheximide, an inhibitor of *de novo* protein synthesis, was found to antagonize the effect of manganese addition

(Wolschek and Kubicek, 1999). The manganese effect has attracted considerable attention since manganese deficiency is essential for citric acid overproduction (Kubicek and Röhr, 1977; Kisser *et al.*, 1980; Netic *et al.*, 1997; Papagianni *et al.*, 1999a). Cellular anabolism of *A. niger* is impaired under manganese deficiency and/or nitrogen and phosphate limitation. According to Habison *et al.* (1983) and Röhr and Kubicek (1981), the protein breakdown under manganese deficiency results in a high intracellular NH_4^+ concentration (the "ammonium pool"), which causes inhibition of the enzyme phosphofructokinase, an essential enzyme in the conversion of glucose and fructose to pyruvate, leading to a flux through glycolysis and the formation of citric acid. The high glucose and NH_4^+ concentrations on the other hand, strongly repress the formation of 2-oxoglutarate dehydrogenase and thus inhibit the catabolism of citric acid within the tricarboxylic acid cycle (Röhr and Kubicek, 1981). Manganese also affects cell wall biosynthesis, composition and morphology, changes lipid biosynthesis and alters membrane phospholipid composition.

However, recent results from studies on the early stages of citric acid accumulation by *A. niger* reported by Papagianni *et al.* (2005) come into direct contradiction with the early theory of the ammonium pool inside the cell, regarded as responsible for inhibition of the enzyme phosphofructokinase. Investigating the fate and role of ammonium ions in the early phase of the citric acid fermentation process, Papagianni *et al.* (2005) came to the conclusion that ammonium ions are not simply deposited inside the cell to make an ammonium pool but they enter the cell to combine with glucose and form glucosamine, which is immediately released in the fermentation broth. The slightly acidic intracellular pH, the extremely low ammonium ion concentrations inside the cell (about 1% of the external concentration), and the synthesis and release of glucosamine described in that case show that the inhibition of phosphofructokinase is certainly not due to increased concentrations of ammonium ions. The long-holding relationship between high glucose and ammonium ion concentrations from one side and the enzymes phosphofructokinase, 2-oxoglutarate dehydrogenase, and the synthase of glucosamine from the other side within the TCA cycle certainly needs further investigation.

CONDITIONS OF CITRIC ACID PRODUCTION FROM CARBOHYDRATES BY *A. NIGER*

The basic ingredients for successful citric acid fermentations have been known for quite long time. Media introduced by Shu and Johnson (1948a, 1948b) are still used for research with little or no modification. The conditions consist of certain nutrients being present in excess (sugar, protons, oxygen), others being limiting (nitrogen, phosphate) and some otherwise common feed components must be kept below defined limits (trace metals, especially manganese).

Carbon Source

The effect of carbohydrate concentration on citric acid yield in submerged fermentation has been studied by Xu *et al.* (1989), Honecker *et al.* (1989), and Kubicek-Pranz *et al.* (1990). The sugars maltose, sucrose, glucose, mannose and fructose were tested (Xu *et al.*, 1989) and highest yields were observed at sugar concentration of 10% w/v, with the exception of glucose where 7.5% gave the best results. No citric acid was produced on media containing less than 2.5% sugar. They also noted that an increase in sugar concentration from 1% to 14% increases the lag time of growth from 12 to 18 hours and decreases the growth rate by around 20%. Hossain *et al.* (1984) suggested that high concentrations of appropriate carbon sources lead to repression of α-keto-glutarate dehydrogenase, hence explaining the effect of the sugar concentration and source in terms of enzyme repression. The influence of glucose concentration on growth and citric acid production by *A. niger* in batch and fed-batch cultures, in which the concentration of glucose was maintained stable during the process (glucostat), was studied by Papagianni (1995) and Papagianni *et al.* (1999c). The level of glucose had a marked effect on specific glucose uptake rates by *A. niger* (Fig.11.3) and on citric acid production rates; the specific rate of citric acid production increased with increasing initial glucose concentration (batch culture) or glucose levels (glucostat culture), and in both culture methods, the specific growth rate increased with decreasing glucose concentration for the first 48 hours of fermentation. Fungal morphology was also affected by the sugar level and the size of mycelial clumps was reduced at low glucose levels while their shape was also affected as the ratio between the perimeter of the clump and the perimeter of its

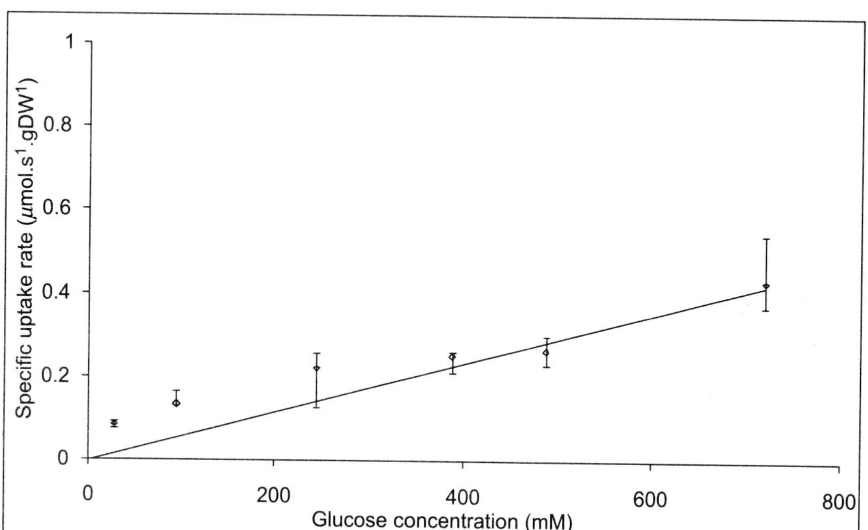

Figure 11.3 Comparison of specific uptake rate by *A. niger* in the citric acid fermentation and glucose concentration in glucostat experiments

core increased with decreasing glucose concentration. The glucose levels in the medium were shown to lead a significant non-catalyzed entry into the mycelium, which accounted for the citric acid productivity and growth rate changes (Wayman and Mattey, 2000; Papagianni and Mattey, 2004).

The carbon source used in industrial fermentations is typically beet molasses, although cane molasses, fruit pulp, polysaccharides and sugars are used if local conditions permit their economic use. Such low-grade carbon sources are waste products and generally contaminated with high levels of cations from previous industrial processes. These are typically precipitated out with potassium ferrocyanide or removed with cation exchange resins, although activated carbon, bone char and bauxite have all been proposed for this task (Prescott and Dunn, 1959). The complexity of these methods for small-scale laboratory work is such that much published research work has been carried out with refined sugars, generally glucose or sucrose.

Culture pH

The maintenance of a low pH during fermentation is vital for a good yield of citric acid. The pH should fall below 2.0 within a few hours of the initiation of the process otherwise yields are reduced (Mattey, 1992). pH values of more than 2.8 lead to formation of oxalic acid, decreased citric acid production and make the process susceptible to contamination. Increasing the pH to 4.5 during the production phase reduces the final yield of citric acid up to 80% (Papagianni, 1995). A pH below 1.5 leads to reduced growth rates and abnormal morphologies.

Information concerning the effect of broth pH upon the morphology of citric acid producing *A. niger* is very limited. Papagianni (1995) and Papagianni *et al.* (1999a) studied the effect of pH on morphology and citric acid production by *A. niger* MP1. Fermentations were carried out in a 10 L stirred tank biorector, at 500 rpm, 1 vvm, and uncontrolled pH, resulting in a final pH of 1.6, and pH controlled at 2.1 and 3.0. Citric acid production was the highest when pH was maintained at 2.1, with 122 g/L at the end of the batch at 168 hours of fermentation, compared to 75 g/L at pH 3.0, and 65 g/L at pH 1.6. Biomass levels were slightly increased with increasing pH: 5.65 g/L at pH 1.6, 6.20 g/L at pH 2.1 and 7.40 g/L at pH 3. The highest values of specific citric acid production rates were obtained at pH 2.1, with a maximum value of 0.35 h^{-1}, while almost 0.18 h^{-1} in the other two pH. Morphological parameters, such as the mean perimeter of clumps (P1, μm), mean perimeter of cores of clumps (P2, μm), mean length of filaments (l, μm) and mean diameter of filaments (d, μm), were measured throughout runs by means of digital image analysis. Results are shown in Table 11.5. P1, P2, and l increased with increasing pH, whilst there was no unidirectional response for the parameter d. In addition to the small size of aggregated mycelium at pH 1.6, there was an unusual high number of swollen hyphae in the mycelium. These experiments indicated also that productivity and morphology are closely linked.

Table 11.5 The influence of pH on selected morphology parameters of citric acid producing *A. niger* mycelium grown in a conventional STR (samples taken at 168 hours of fermentation)

pH	*P1 (μm)*	*P2 (μm)*	*l (μm)*	*d (μm)*
1.6	376	202	44	2.6
2.1	850	370	70	2.7
3.0	915	402	96	2.5

Aeration and Agitation

Aeration has a critical effect on the submerged process of citric acid fermentation (Clark and Lentz, 1961; Dawson *et al.*, 1988a). It has been shown that increasing the dissolved oxygen concentration in a fermenter by introducing pure oxygen instead of air reduces the fermentation time, while a short interaction of the oxygen supply irreversibly affects the rate of acid production (Kubicek *et al.*, 1980; Dawson *et al.*, 1988b). In submerged fermentation, agitation is important for adequate mixing, mass transfer and heat transfer. In an aerobic fermentation, mixing is required to ensure sufficient oxygen transfer throughout the reactor vessel. Agitation also creates mechanical (shear) forces that affect microorganisms in various ways, causing for example morphological changes, variation in their growth and product formation and also causing damage in the cell structure (Papagianni *et al.*, 1999b). It is well-known today that under increased stirrer speeds in a stirred tank bioreactor, filamentous fungi grow as thick, short and highly branched filaments which can form permanent aggregates in the form of mycelial pellets or clumps. The particular agitation induced morphological type has been described many times as the type of the acidogenic mycelium of *A. niger*, as shown in Fig. 11.4 (Papagianni, 2003). Studies by Papagianni *et al.* (1994, 1998, 2003) with two different reactor types, a conventional stirred tank and a tubular loop, showed that morphology and productivity are closely linked in citric acid producing *A. niger* and the amount of product and a sensitive morphological parameter, e.g. the perimeter of aggregates, is a function of either the stirrer speed (stirred tank reactor) or the circulation time (loop reactor).

Nitrogen and Phosphate Limitation

In defined media, nitrogen is supplied as ammonium sulphate or nitrate. The advantage of ammonium salts is that they decrease the pH as they are consumed which is a prerequisite of citric acid fermentation (Mattey, 1992).

Reports concerning the limitation of the above elements have been contradictory. According to Shu and Johnson (1948a,b), phosphate does not have to be limiting. However, when trace metal levels are not limiting, additional phosphate results in side reactions and increased growth (Papagianni *et al.*, 1999a). Kubicek and Röhr (1977) showed that citric acid accumulated whenever phosphate was limited even when nitrogen was not available. In contrast, Kristiansen and Sinclair (1979) using continuous culture concluded that nitrogen limitation was essential for citric acid production.

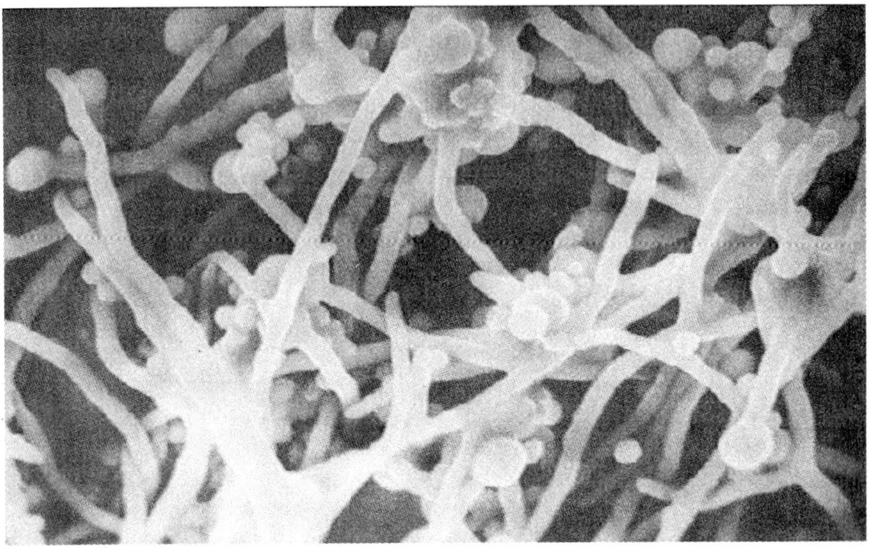

Figure 11.4 The acidogenic mycelium of *A. niger* (Scanning electron microscopy).

Trace Elements

Several essential trace metals are necessary for the growth of *A. niger*. In submerged citric acid fermentation, however, some trace elements have to be limited. These are Zn, Cu, Fe, Mn, heavy metals and alkaline metals (Kubicek and Röhr, 1989). Shu and Johnson (1948a,b) found optimal levels of Zn and Fe at 0.3 and 1.3 ppm, respectively.

Addition of manganese in media in levels as low as 3μg/l was shown to reduce severely the yield of citric acid under otherwise optimal conditions (Clark *et al.*, 1966). Mattey and Bowes (1978) reported that addition of 10 mg/l Mn^{2+} resulted in halving the citric acid accumulation. Investigations of Clark *et al.* (1961, 1966) and Kisser *et al.* (1980) confirmed the key regulatory nature of manganese ions. Papagianni (1995) and Papagianni *et al.* (1999a) studied the morphology of *A. niger* in the citric acid fermentation under various manganese concentration levels. The swollen cells associated with overproduction of citric acid by *A. niger* in manganese deficient media were absent when manganese was included in levels as low as 1μg/l, while the mycelium lost its tendency to agglomerate and form clumps. There are many reports on the role of manganese ions in the citric acid fermentation by *A. niger*. The influence of manganese ions on protein synthesis was considered to be of major importance since cycloheximide, an inhibitor of *de novo* protein synthesis, was found to antagonize the effect of manganese addition (Kisser *et al.*, 1980).

Ferrous and zinc have an important role in fermentation and their levels have been related to the diversion of carbon between biomass and citric acid. Copper has been found to act through growth mechanisms rather than those leading to acid production (Mattey, 1992). The importance and the interdependence of trace metals and other medium constituents have been demonstrated in a number of reports. In particular Trumphy and Millis (1963) found a strong interdependence between Fe and Zn, Fe and nitrogen as well as nitrogen and phosphate.

FUNGAL MORPHOLOGY

In submerged culture the morphology of filamentous microorganisms varies between two extreme forms, pellets and free filaments, depending on culture conditions and the genotype of the strain. According to many reports, mycelial morphology is crucial to the process of fermentation, not only in relation to the shape of the hyphae themselves and the aggregation into microscopic clumps (micro-morphology), but also in the pelleted form of growth (macro-morphology). In all cases reported, the mycelium of acidogenic *Aspergillus niger* was found to confirm to the morphological type described by Snell and Schweiger (1951): short, swollen filaments with swollen tips (Fig. 11.4). The pellets should be small with a hard, smooth surface. This is brought about by adjustment of agitation and aeration, pH adjustment, concentration of important trace metals, inoculum type and level (Papagianni, 1995).

Since the morphological form can strongly influence the productivity in a fungal fermentation, research on various aspects of morphological development has attracted the interest of academia and the industry and attempts have been made to induce a particular form of growth and to relate morphology to product synthesis. The influence of fungal morphology on productivity in the citric acid fermentation by *A. niger* has been the subject of a large number of reports. Initial investigations relied on manual measurements from photographs and little quantitative work was presented until the 1980s. Later, the application of image analysis permitted the extraction of quantitative information on fungal morphology (Tucker *et al.*, 1992).

Most of the factors that affect the citric acid fermentation by *A. niger* (as discussed in section 5) have been shown to influence the morphology of the fungus in submerged culture and have been discussed so far, e.g. agitation, broth pH, manganese. Others, like aeration, seem to have no direct influence on fungal morphology. While oxygen acts as a direct regulator of citric acid accumulation, reports on the effect of dissolved oxygen tension on the macro-morphology of *A. niger* suggest that no direct relationship exists between the two. Gomez *et al.* (1988), in their work on citric acid production by *A. niger*, found that no difference in morphology for pellets or free filaments could be ascribed to dissolved oxygen levels although production of citric acid was enhanced, particularly from pellets, by increasing the dissolved oxygen at different fermentation stages.

Among the factors that determine morphology and the overall course of fungal fermentations, the type (spore or vegetative) and amount of the inoculum is of prime importance. Early attempts have been made to standardize inocula for citric acid production in submerged culture (Martin and Waters, 1952; Steel *et al.*, 1955). Van Suijdam *et al.* (1980) reported that *A. niger* pellets would only form at inoculum levels below 10^{11} spores per m^3. The effect of inoculum on fungal morphology in submerged culture was assessed for long time mainly by the presence or absence of pellets and their characteristics. During the 1990s, however, image analysis applications permitted a more systematic approach and the effect of inoculum level was for the first time quantified for the penicillin producer *Penicillium chrysogenum* by Tucker and Thomas (1992).

Although, in terms of bulk production, citric acid is widely regarded as the most important of the organic acids produced by microbiological methods, the inoculum effect within the particular fermentation has been a rather neglected area for long time. In a recent work Papagianni and Mattey (2005), studied the morphological development of *A. niger* in submerged citric acid fermentations as a function of the spore inoculum level. A series of batch fermentations were carried out which were inoculated directly with spores in concentrations ranging from 10^4 to 10^9 spores per ml. The particular morphological features, evaluated by digital image analysis, were classified using an artificial neural network (ANN). The input layer consisted of the morphological features and cultivation time, while the output layer consisted of four object types: globular and elongated pellets, clumps and free mycelial trees. The significance of the particular morphological features and their combination was determined by cluster analysis. Cell volume fraction analysis for the various inoculum levels tested revealed that by raising the spore inoculum level from 10^4 to 10^9 spores per ml, a clear transition from pelleted to dispersed forms occur. Image analysis produces a large number of object features, which vary widely during fermentation. These features however, do not yield by themselves sufficient information about the morphological state. Their combination, including the fermentation time, by means of an artificial neural network, permits their classification into general object types. Cluster analysis allows the evaluation of the significance of the object features and their optimal combination for the entire fermentation type. The methods used in that study, allowed for the detailed quantification of the transition between the two extreme morphological forms.

FUNGAL GENETICS AND STRAIN IMPROVEMENT

Industrial citric acid fermentation relies on improved strains of *A. niger*, chosen for their increased productivity and adaptability to industrial bioreactors. These microorganisms are derived by mutagenesis and screening, or selection for specific biochemical features. Traditional genetic approaches however, have been hampered by the absence of a sexual cycle. The first studies on protoplast fusion in citric acid producer *A. niger* were reported by Kirimura *et al.* (1986, 1988), where they succeeded in producing fusion strains with increased productivities compared to parental strains in solid-state fermentation. Most recent advances suggest that we are not far from constructing recombinant strains suitable for industrial citric acid production. Several genes involved in biosynthetic pathways have been cloned and transformants have been obtained that show 20- to 30-fold overexpression of the enzymes (Kusters-van Someren *et al.*, 1991; Ruijter *et al.*, 1997; Mesojednik and Legiša, 2005). The introduction of key engineered genes will enlighten many aspects related to the mechanisms of enzyme expression and citric acid overproduction that still remain unclear.

CONCLUSION

The citric acid fermentation is one of the best-studied processes and one of the rare examples of industrial fermentation technology where academic findings have worked in tandem with industrial know-how to produce a very efficient process. Currently, the world production of citric acid is about 400 ktonnes per annum and most of it is made by *Aspergillus niger*. Certainly work will continue on new methods for citric acid production, strain improvement and metabolic pathway engineering.

References

Arora, D.K. (2004). Handbook of Fungal Biotechnology. Mycology, Vol. 20. New York: Marcel Dekker.

Arts, E., Kubicek, C.P. and Röhr, M. (1987). Regulation of phosphofructokinase from *Aspergillus niger*. Effect of fructose 2,6-bisphosphate on the action citrate, ammonium ions, and AMP. *J. Gen. Microbiol.* 133:1195–1200.

Arora D.K. (2004). Handbook of Fungal Biotechnology. Mycology, Vol. 20. New York: Marcel Dekker; 353–367.

Arts E., Kubicek C.P., and Röhr M. (1987). Regulation of phosphofructokinase from *Aspergillus niger*. Effect of fructose 2,6-bisphosphate on the action citrate, ammonium ions, and AMP. *J. Gen. Microbiol.* 133:1195–1200.

Atkinson B, and Mavituna F. (1991). Biochemical Engineering and Biotechnology Handbook. New York: Stockton Press; 1173–1174.

Clark D.S. and Lentz C.P. (1961). Submerged citric acid fermentation on beet molasses: effect of pressure and recirculation of oxygen. *Can. J. Microbiol.* 7: 447–452.

Clark D.S., Ito K, and Horitsu H. (1966). Effect of manganese and other heavy metals on submerged citric acid fermentation of molasses. *Biotechnol. Bioeng.* 8:465–471.

Currie J.N. (1917). The citric acid fermentation of *Aspergillus niger*. *J. Biol. Chem.* 31:5.

Dawson M.W., Maddox IS and Brooks J.D. (1988a). Effect of interruptions to the air supply on citric acid production by *Aspergillus niger*. *Enzyme Microb.Technol.* 8:37–40.

Dawson M.W., Maddox IS, Boag IF andBrooks J.D. (1988b). Application of fed-batch culture to citric acid production by *Aspergillus niger*: the effects of dilution rate and dissolved oxygen tension. *Biotechnol. Bioeng.* 32: 220–226.

Gomez R., Schnabel I. and Garrido J. (1988). Pellet growth and citric acid yield of *Aspergillus niger* 110. *Enzyme Microb. Technol.* 10: 188–191.

Habison A, Kubicek CP and Röhr M. (1983). Partial purification and regulatory properties of phosphofructokinase from *Aspergillus niger*. *Biochem. J.* 209:669–676.

Honecker S., Bisping B., Yang Z., Rehm H.J. (1989). Influence of sucrose concentration and phosphate limitation on citric acid production by immobilized cells of *Aspergillus niger. Appl. Microbiol. Biotechnol* 31:17–24.

Hossain M., Brooks J.D., Maddox I.S. (1984). The effect of the sugar source on citric acid production by *Aspergillus niger*. *Appl. Microbiol. Biotechnol.* 19:393–397.

Jong C.S., Birmingham JM, and Ma C. (1994). ATCC names of industrial fungi. Maryland USA: American Type Culture Collections.

Kirimura K., Yaguchi T, and Usami S. (1986). Intraspecific protoplast fusion of citric acid producing strains of *Aspergillus niger. J. Ferm. Technol.* 64:473–479.

Kirimura K., Nakajima I., Lee SP, Kawabe S, and Usami S. (1988). Citric acid production by diploid strains of *Aspergillus niger* obtained by protoplast fusion. *Appl. Microbiol. Biotechnol.*27:504–506.

Kisser M., Kubicek C.P., and Röhr M. (1980). Influence of manganese and cell wall composition of *Aspergillus niger* during citric acid fermentation. Arch. Microbiol. 128:26–33.

Kristiansen B., and Sinclair C.G. (1979) Production of citric acid in continuous culture. *Biotechnol. Bioeng.* 21:297–315.

Kubicek C.P., and Röhr M. (1977). Influence of manganese on enzyme synthesis and citric acid accumulation by *Aspergillus niger. Eur. J. Appl. Microbiol.* 4:167–173.

Kubicek C.P., and Röhr M. (1980). Regulation of citrate synthase from citric acid accumulating fungus *Aspergillus niger. Biochim. Biophys. Acta* 615:449–454.

Kubicek C.P., and Röhr M. (1985). Aconitase and citric acid fermentation by *Aspergillus niger. Appl. Environ. Microbiol.* 50:1336–1338.

Kubicek C.P., and Röhr M. (1989). Citric acid fermentation. CRC *Crit. Rev. Biotechnol.* 4: 331–373.

Kubicek C.P., Zehentgruber O., El-Kalak H, and Röhr M. (1980). Regulation of citric acid production by oxygen: effects of dissolved oxygen tension on adenylate levels and respiration in *Aspergillus niger. Eur. J. Appl. Microbiol. Biotechnol.* 9:101–116.

Kubicek-Pranz E.M., Mozelt M., Röhr M., and Kubicek C.P. (1990). Changes in the concentration of fructose 2,6-bisphosphate in *Aspergillus niger* during stimulation of acidogenesis by elevated sucrose concentration. *Biochim. Biophys. Acta* 1033:250–255.

Kusters-van Someren M.A., Harmsen J.A., Kester H.C., and Visser J. (1991). Structure of the *Aspergillus niger pel*A gene and its expression in *Aspergillus niger* and *Aspergillus nidulans. Curr. Genet.* 20:293–299.

Martin S.M., and Waters W.R. (1952). Production of citric acid by submerged fermentation. *Ind. Eng. Chem.* 44: 2229–2233.

Mattey M. (1992).The production of organic acids. CRC *Crit. Rev. Biotechnol.* 12:87–132.

Mattey, M. (1999). Biochemistry of citric acid production by yeasts. In: Kristiansen B, Mattey M, Linden J, editors. Citric acid biotechnology. London: Taylor and Francis; pp. 33–54.

Mattey M., and Bowes I. (1978). Citrate regulation of NADP$^+$-specific isocitrate dehydrogenase of *Aspergillus niger. Biochem. Soc. Trans.* 6:1224–1226.

Mattey M., and Kristiansen B. (1999). A brief introduction to citric acid biotechnology. In: Kristiansen B., Mattey M., Linden J., editors. Citric acid biotechnology. London: Taylor and Francis; pp. 1–9.

Meixner-Monori B., Kubicek C.P., Harrer W., Schreferl G., and Röhr M. (1986) NADP-specific isocitrate dehydrogenase from the citric acid accumulating fungus *Aspergillus niger. Biochem. J.* 236:549–557.

Mesojednik S., and Legiša M. (2005). Posttranslational modification of 6-phosphofructo-1-kinase in *Aspergillus niger. Appl. Environ. Microbiol.*; 71:1425–32.

Moses V., and Cape R.E. (1994). Biotechnology: The science and the business. Chur, Switzerland: Harwood Academic Publishers; Provide pp. 284-288page range

Netik A., Torres N.V., Riol J.M., and Kubicek C.P. (1997). Uptake and export of citric acid by *Aspergillus niger* is reciprocally regulated by manganese ions, *Biochim. Biophys. Acta-Biomembranes* 1326:287–294.

Papagianni M. Morphology and citric acid production of *Aspergillus niger* in submerged culture. Ph.D. Thesis, University of Strathclyde, Glasgow, UK; Provide page rangel refer to the thesis 1995.

Papagianni M. (2003). Fungal morphology and metabolite production in submerged mycelial processes. *Biotechnol. Adv.* 22:189–259.

Papagianni M., and Mattey M. (2004). Modeling the mechanisms of glucose transport through the cell membrane of *Aspergillus niger* in submerged citric acid fermentation processes. Biochem. Eng. J. 20:7–12.

Papagianni M., and Mattey M. (2006). Morphological development of *Aspergillus niger* in submerged citric acid fermentation as a function of the spore inoculum level. Application of neural network and cluster analysis for characterization of fungal morphology. *Microbial Cell Factories* 5:3.

Papagianni M., Mattey M., Berovic M., and Kristiansen B. (1999a). *Aspergillus niger* morphology and citric acid production in submerged batch fermentation: Effects of culture pH, phosphate and manganese levels. *Food Technol. Biotechnol.*37:165–171.

Papagianni M., Mattey M., and Kristiansen B. (1994). Morphology and citric acid production of *Aspergillus niger* PM1. *Biotechnol. Lett.* 16:929–934.

Papagianni M., Mattey M., and Kristiansen B. (1998). Citric acid production and morphology of *Aspergillus niger* as functions of the mixing intensity in a stirred tank and a tubular loop bioreactor. *Biochem. Eng. J.* 2:197–205.

Papagianni M., Mattey M., and Kristiansen B. (1999b). Hyphal vacuolation and fragmentation in batch and fed-batch culture of *Aspergillus niger* and its relation to citric acid production. *Process Biochem.* 35:359–366.

Papagianni M., Mattey M., and Kristiansen B. (1999c). The influence of glucose concentration on citric acid production and morphology of *Aspergillus niger* in batch and fed-batch culture. *Enzyme Microb. Technol.* 25:710–717.

Papagianni M., Mattey M., and Kristiansen B. (2003). Design of a tubular loop bioreactor for scale-up and scale-down of fermentation processes. *Biotechnol. Prog.* 19: 1498–1504.

Papagianni M., Wayman F.M., and Mattey M. (2005). Fate and role of ammonium ions during fermentation of citric acid by *Aspergillus niger. Appl. Environ. Microbiol.* 71:7178–7186.

Prescott S., and Dunn C.G. (1959). The citric acid fermentation. In: Industrial Microbiology, 3rd edition, London: McGraw Hill; pp. 533–577.

Röhr, M., and Kubicek C.P. (1981). Regulatory aspects of citric acid fermentation by *Aspergillus niger*. Process Biochem. 16:34–37.

Röhr M., Zehentgruber O., and Kubicek C.P. (1981). Kinetics of Biomass formation and citric acid production by *Aspergillus niger* on pilot plant scale. *Biotechnol. Bioeng.* 23:2433–2445.

Ruijter G.J., Panneman H, and Visser J. (1997). Overexpression of phosphofructokinase and pyruvate kinase in citric acid-producing *Aspergillus niger. Biochim. Biophys. Acta* 1334:317–26.

Scheele C. (1984). Crells Ann. Sammtliche Physische and Chemische Werke. Hermbstadt, Berlin. 2:1.

Shu P., and Johnson M.J. (1948a). Citric acid production by submerged fermentation with *Aspergillus niger*. Ind. Eng. Chem. 40:1202–1205.

Shu P., and Johnson M.J. (1948b). The interdependence of medium constituents in citric acid production by submerged fermentation. *J. Bacteriol.*54:161–167.

Snell R.L., and Schweiger L.B. (1951). Citric acid by fermentation. British Patent No 653 808. *Chemical Abstracts* 45:8719a.

Steel R., Lentz C.P., and Martin S.M. (1954). A standard inoculum for citric acid production in submerged culture. *Can. J. Microbiol.* 1, 150–157

Trumphy B.H., and Millis N.F. (1963). Nutritional requirements of an *Aspergillus niger* mutant for citric acid production. *J. Gen. Microbiol.* 30:381–393.

Tucker, K.G., and C.R. Thomas. (1992). Mycelial morphology: The effect of spore inoculum level. *Biotechnol. Lett.* 14: 1071–1074.

Tucker, K.G., T. Kelly, P. Delgrazia, and C.R. Thomas. (1992). Fully automatic measurement of mycelial morphology by image analysis. *Biotechnol. Prog.* 8: 353–359.

van Suijdam J.C., Kossen NWF, and Paul P.G. (1980). An inoculum technique for the production of fungal pellets. *Eur. J. Appl. Microbiol. Biotechnol.* 10: 211–221.

Wayman F.M., and Mattey M. (2000). Simple diffusion is the primary mechanism for glucose uptake during the production phase of the *Aspergillus niger* citric acid process. *Biotechnol. Bioeng.* 67:451–456.

Wolschek M.F., and Kubicek C.P.(1999). Biochemistry of citric acid accumulation by *Aspergillus niger*. In Citric acid biotechnology, Kristiansen B., Mattey M., Linden J., editors. London: Taylor and Francis; pp. 11–33.

Xu B.D., Madrit C., Röhr M., and Kubicek C.P. (1989). The influence of type and concentration of the carbon source on production of citric acid by *Aspergillus niger*. *Appl. Microbiol. Biotechnol.* 30, 553–558.

12

Arbuscular Mycorrhizal Fungi Used as Biofertilizers in Revegetation Programmes

Waldemar Zangaro[1], Marco A. Nogueira[2] and Galdino Andrade[2*]

[1]Universidade Estadual de Londrina - UEL. CCB/Dept.de Biologia Animal e Vegetal.
[2]UEL. CCB/Dept.de Microbiologia, Laboratório de Ecologia Microbiana.
Caixa Postal 6001 86051-990
Londrina PR Brazil *Corresponding Author. E-mail: andradeg@uel.br

INTRODUCTION

The soil microbial community is composed of several species that act on the biogeochemical cycles, which are important for soil fertility and plant growth (Andrade, 2004). Among the soil microorganisms, the external net of arbuscular mycorrhizal (AM) fungi hyphae play an important role in nutrient uptake, especially those of low mobility in the soil like phosphorus (P). The external hyphae work as an extention of the host's roots, increasing its effectiveness on soil exploration (Gilroy and Jones, 2000). For example, the soil P can be reached beyond the depletion zone around the root surface (Miller and Jastrow, 1994; Jakobsen, 1995). The low-diameter hyphae are more able to uptake P from small sites that can not be accessed by the root hairs (Jakobsen, 1995).

Arbuscular mycorrhizas make symbiosis with many plant species, including herbaceous and woody (Janos, 1983; Sanders *et al.*, 1996; Zangaro *et al.*, 2003). This symbiosis play important role in plant nutrition, growth and survival (Sylvia and Williams, 1992; Janos, 1996), besides increasing the photosynthetic rate and roots longevity (Linderman, 1988; Comas *et al.*, 2002).

In this chapter, some experimental results of AM fungi inoculation in 80 native woody species from different ecological groups will be presented. Approaches related to the AM fungi relationship with woody native species from the Rio Tibagi basin (Paraná State, Brazil) will be discussed, especially in relation to the woody species development, the vegetal succession and the importante of AM fungi on the reclamation of degraded areas by its use as biofertilizers.

ARBUSCULAR MYCORRHIZAL FUNGI IN NATIVE WOODY SPECIES

Eighty woody species native from the Rio Tibagi basin (Paraná, Brazil) were tested in relation to its initial growth response to inoculation with AM fungi. The soil for growth was obtained by mixing the subsoil of an oxisoil (85% w/w) with sand (15% w/w) and disinfected with methyl bromide. Some chemical characteristics of the mix was pH 5.1; C (1,9 g $100g^{-1}$); P (1,8 mg kg^{-1}), Al (0,08 meq 100 ml^{-1}), Ca (5,47 meq 100 ml^{-1}), Mg (2,46 meq 100 ml^{-1}), and K (0,61 mcq 100 ml^{-1}). The microbial community, except AM fungi, was restablished with 100 ml of a filtrate obtained from the original soil. Each pot containing 2 kg of soil received ca. 350 spores of indigenous AM fungi or left uninoculated. Ten replications per woody species were performed for each woody species and mycorrhizal status. The plantlets were pre-germinated in washed sand and transferred to the pots 6 to 10 days after emergence. The pots were distributed randomly in the greenhouse and the soil was kept at field capacity during the experimental period.

The mycotrophic status and successional group of each species are presented in Table 12.1. Among the successional groups, 16 species were pioneers, 20 early secondary, 29 late secondary and 15 climaces. According to the response to AM fungi and the period necessary to reach the maximum development in the greenhouse, the plantlets of pioneer species were kept in the greenhouse for 16.8 weeks, the early secondary for 19.3 weeks, the late secondary for 21.2 weeks and the climaces for 25.8 weeks.

Table 12.1 Native woody species, ecological successional groups and mycotrophic status. *In relation of the root colonization of seedlings grown in greenhouse: high (> 50%), middle (49 - 20%) and low (19 - 1%). **Seedling from the field.

Pioneer	*Micotrophy**
Aegiphila sellowiana Cham. (Verbenaceae)**	High
Cecropia glaziovii Snethl. (Cecropiaceae)**	High
Cecropia pachystachya Trec. (Cecropiaceae)**	High
Cestrum intermedium Semdth. (Solanaceae)**	High
Citharexylum myrianthum Cham. (Verbenaceae)	High
Croton floribundus Spreng. (Euphorbiaceae)**	High
Croton urucurana Baill. (Euphorbiaceae)	High
Mimosa scabrella Benth. (Mimosaceae)	High
Piptocarpha axillaris (Less.) Baker (Asteraceae)	High
Schinus terebinthifolius Raddi (Anacardiaceae)**	High
Senna macranthera (Collad.) H. S. Irwim & Barreby (Caesalpinaceae)**	High
Solanum argenteum Dun. (Solanaceae)**	High
Tabernaemontana australis (M. Arg.) Miers (Apocynaceae)**	High
Trema micrantha (L.) Blum. (Ulmaceae)**	High
Xylosma ciliatifolium (Clos) Eichler (Flacourtiaceae)**	High

(Contd.)

Table 12.1 (*Contd.*)

Pioneer	*Micotrophy**
Xylosma pseudosalzmannii Sleumer (Flacourtiaceae)	High
Early secondary	
Albizia hasslerii (Chodat) Burkart (Mimosaceae)	High
Anadenanthera colubrina (Vell.) Brenan (Mimosaceae)	High
Anadenanthera macrocarpa (Benth.) Brenan (Mimosaceae)	Middle
Bastardiopsis densiflora (Hook. and Arn.) Hassl. (Malvaceae)	High
Bauhinia forficata Link. (Caesalpinaceae)	High
Casearia sylvestris Sw. (Flacourtiaceae)**	High
Colubrina glandulosa Perk. (Rhamnaceae)**	High
Cordia trichotoma (Vell.) Arrab. Ex Steud. (Boraginaceae)	High
Eugenia uniflora L. (Myrtaceae)**	High
Guazuma ulmifolia Lam. (Sterculiaceae)**	High
Heliocarpus americanus L. (Tiliaceae)**	High
Lafoensia pacari A. St. Hil. (Lythraceae)	High
Lonchocarpus campestris Mart. & Benth. (Fabaceae)	High
Luehea candicans Mart. Zucc. (Tiliaceae)**	High
Luehea divaricata Mart. (Tiliaceae)	High
Parapiptadenia rigida (Benth.) Brenan (Mimosaceae)**	High
Poecilanthe parviflora Benth. (Fabaceae)	High
Pseudobombax grandiflorum (Cav.) A. Robyns (Bombacaceae)	High
Sebastiania commersoniana (Baill.) Smith & Downs (Euphorbiaceae)**	High
Tabebuia chysotricha (Mart. Ex DC.) Standl. (Bignoniaceae)	High
Late secondary	
Acacia polyphylla DC. (Mimosaceae)	Middle
Astronium graveolens Jacq. (Anacardiaceae)**	Middle
Bougainvillea spectabilis Willd. (Nyctaginaceae)**	Low
Campomanesia xanthocarpa O. Berg (Myrtaceae)**	High
Cedrela fissilis Vell. (Meliaceae)**	Middle
Centrolobium tomentosum Guill. Ex Benth. (Fabaceae)**	Middle
Chorisia speciosa St. Hil. (Bombacaceae)**	Low
Cordia ecalyculata Vell. (Boraginaceae)	Middle
Enterolobium contorfisiliquum (Vell.) Morong (Mimosaceae)	Low
Ficus guaranítica Schodat (Moraceae)**	Low
Genipa americana L. (Rubiaceae)	Low
Ingá sessilis (Vell.) Mart. (Mimosaceae)	Low
Ingá striata Benth. (Mimosaceae)**	Low
Jacaranda mimosaefolia D. Don (Bignoniaceae)	Low
Jacaranda puberula Cham. (Bignoniaceae)**	Low
Jacaratia spinosa (Aubl.) A. DC. (Caricaceae)**	Low
Lithraea molleoides (Vell.) Engl. (Anacardiaceae)**	Low
Lonchocarpus muehlbergianus Hassl. (Fabaceae)**	Low
Machaerium minutiflorum Tul. (Fabaceae)	Low

(*Contd.*)

Table 12.1 (*Contd.*)

Pioneer	*Micotrophy**
Ocotea puberola (Reich.) Ness (Lauraceae)**	Middle
Peltophorum dubium (Spreng.) Taub. (Caesalpinaceae)**	Absent
Prunus sellowii Hoehne (Rosaceae)**	Low
Pterogyne nitens Tul. (Caesalpinaceae)**	Low
Ruprechtia laxiflora Meisn. (Polygonaceae)**	Low
Strichinus brasiliensis (Spreng.) Mart. (Styracaceae)**	Low
Syagrus romanzoffiana (Cham.) Glassm. (Arecaceae)**	Low
Tabebuia roseo-alba (Ridl.) Sand. (Bignoniaceae)	High
Vitex montevidensis Cham. (Verbenaceae)**	High
Zeyheria tuberculosa (Vell.) Bureau (Bignoniaceae)	Absent
Climax	
Actinostemon concolor (Spreng.) M. Arg. (Euphorbiaceae)**	Low
Aspidosperma polyneuron M. Arg. (Apocynaceae)**	Absent
Cariniana estrellensis (Raddi) Kuntze (Lecythidaceae)	Low
Copaifera langsdorffii Desf. (Caesalpinaceae)	Low
Euterpe edulis Mart. (Arecaceae)**	Low
Guarea kunthiana A. Juss. (Meliaceae)**	Low
Holocalyx balansae Micheli (Fabaceae)**	Low
Hymenaea courbaril L. (Caesalpinaceae)	Low
Ocotea indecora Schost. (Lauraceae)**	Low
Ormosia arborea (Vell.) Harms (Mimosaceae)	Low
Plinia rivularis (Cambess.) Rotman (Myrtaceae)**	Low
Sorocea bonplandii (Baill.) Burg. Lanj. & Bcer (Moraceae)**	Low
Trichilia casaretti C. DC. (Meliaceae)	Low
Trichilia claussenii C. DC. (Meliaceae)**	Low
Trichilia elegans A. Juss. (Meliaceae)**	Low

The ratio between the growth rate of mycorrizal and non-mycorrhizal plants (in height basis) decreased along the successional groups (Fig. 12.1), as also observed for the shoot and root dry biomass (Fig. 12.2). On the other hand, the ratio between the root to shoot ratio of mycorrhizal and non-mycorrhizal plants increased along the ecological groups (Fig. 12.3), indicating that AM fungi affect differently the C allocation between roots and shoots along the different successional groups. Mycorrhiza increased the C allocation to the shoots in the pioneer and early secondary species, resulting in lower root to shoot ratio. The late secondary and climax species showed low responsiveness to mycorrhiza, indicating that the symbiosis did not improve the vegetative growth of these species.

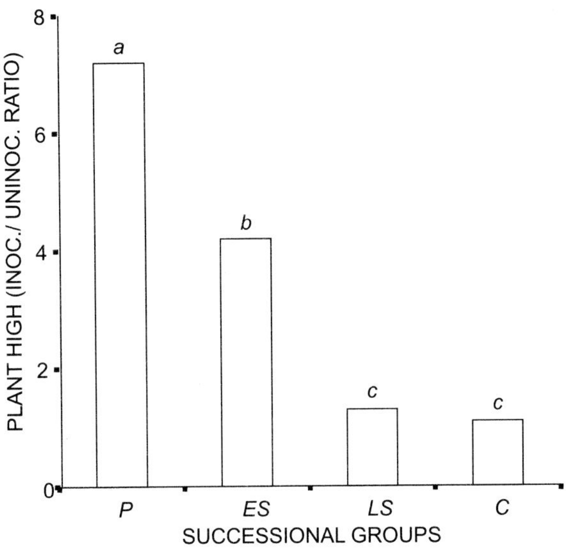

Figure 12.1 Plant height (inoculated/uninoculated ratio) of native woody species belonging to different successional stages. Means followed by same letter are not different by Tukey-Kramer HSD test at 0.05 level.

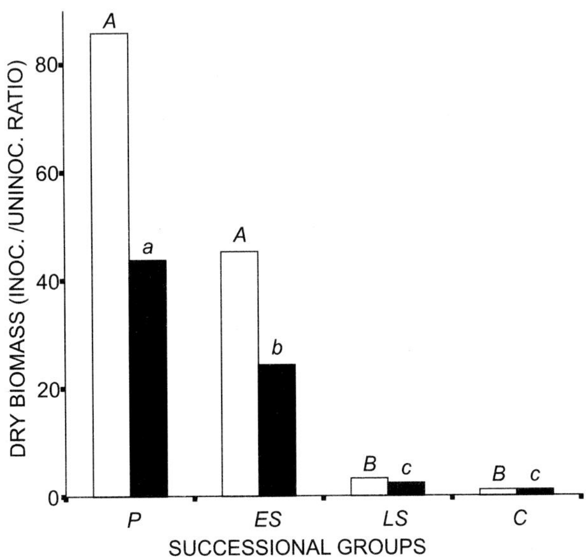

Figure 12.2 Dry biomass (inoculated/uninoculated ratio) of shoot (□) and root (■) of native woody species belonging to different successional stages. Means followed by same letter, capital for shoot and small for root, are not different by Tukey-Kramer HSD test at 0.05 level.

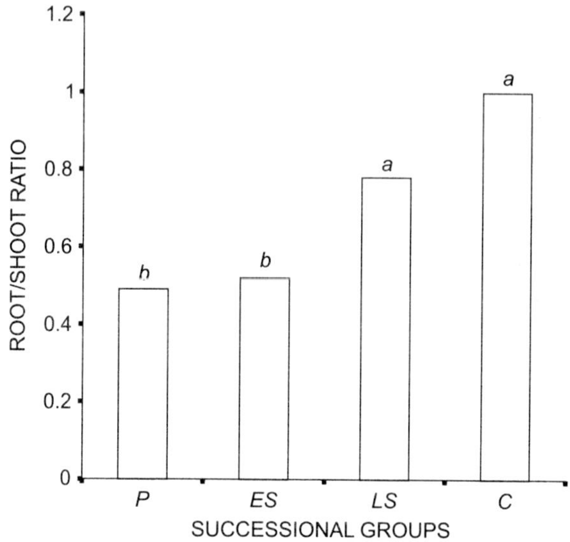

Figure 12.3 Root/shoot ratio (inoculated/uninoculated ratio) of native woody species belonging to different successional stages. Means followed by same letter are not different by Tukey-Kramer HSD test at 0.05 level.

The plant response to AM fungi, based on the shoot dry biomass of mycorrhizal and non-mycorrhizal plantas, was calculated according to Plenchette *et al.* (1983). The response to AM fungi decreased with the advancement of the successional groups and was very contrasting from the first to the last successional group (Fig. 12.4). Such behaviour emphasizes the importance of AM fungi

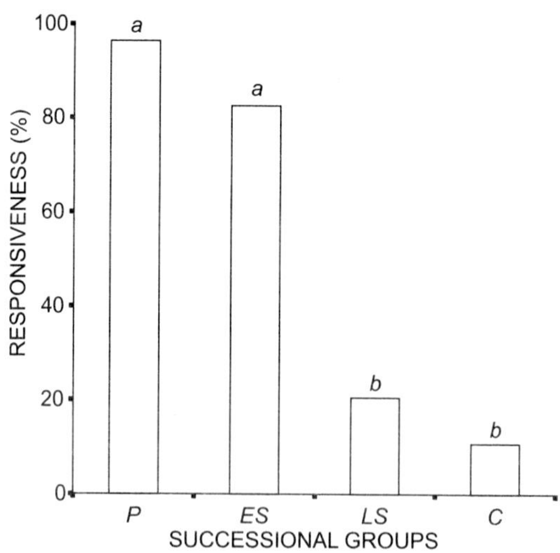

Figure 12.4 Plant responsiveness to AM fungi of native woody species belonging to different successional stages. Means followed by same letter are not different by Tukey-Kramer HSD test at 0.05 level.

in the initial growth of all pioneer and early secondary species. All late secondary, except *Campomanesia xanthocarpa*, *Tabebuia roseo-alba* and *Vitex montevidensis* and all climaces species showed low responsiveness to AM fungi, even grown in chemically poor soil. These results corroborate previous findings of Zangaro *et al.* (2000; 2003; 2005) for species from the Rio Tibagi basin and others from Siqueira *et al.* (1998) for species from South-easthern Brazil.

Minerals Content in Plants

When non-inoculated with AM fungi, the pioneer and early secondary species showed lower P, Ca and K concentration in leaves than the species from the late secondary and climaces species (Table 12.2). Conversely, when the pioneer and early secondary species were inoculated with AM fungi, the concentration of these nutrients in leaves increased strongly in relation to the non-inoculated plants. For the late secondary and climaces species there was no increase of nutrient concentration due to mycorrhiza.

Table 12.2 Leaf nutrient concentration of the native woody species belonging to different ecological groups, inoculated and uninoculated with arbuscular mycorrhizal fungi. Nutrient concentration ratio between inoculated/uninoculated NCR[1].

Successional groups	Uninoculated	Inoculated	NCR[1]
P (mg 100 mg^{-1})			
Pioneer	0.07 bB	0.14 aA	2.0
Early secondary	0.09 bB	0.14 aA	1.56
Late secondary	0.14 aA	0.15 aA	1.07
Climax	0.15 aA	0.15 aA	1.0
Ca (mg 100 mg^{-1})			
Pioneer	0.32 bB	0.59 aA	1.84
Early secondary	0.37 bB	0.52 aA	1.41
Late secondary	0.54 aA	0.56 aA	1.04
Climax	0.65 aA	0.66 aA	1.02
K (mg 100 mg^{-1})			
Pioneer	0.72 bB	1.54 aA	2.14
Early secondary	0.82 bB	1.43 aA	1.74
Late secondary	1.13 aA	1.15 aB	1.02
Climax	1.11 aA	1.17 aB	1.05

Means followed by the same latter (smal in row and capital in column) are not different by Tukey-Kramer HSD test at 5%.

The mycorrhizal symbiosis generally improves some nutrients uptake that contributes for plant growth (Merryweather and Fitter, 1996; Smith and Read, 1997). In the uninoculated treatment, concentrations of P, Ca and K in the late secondary and climaces were, respectively, 1.8, 1.7 and 1.4 times greater than in pioneers and early secondary species. This suggests that the pioneer and early secondary species are less able to acquire nutrients from a low-fertility soil in the absence of

AM fungi, as compared to later successional species. However, in mycorrhizal symbiosis the pioneer and early secondary species had an improvement in nutrient uptake capacity. In the case of P and Ca the concentrations in pioneers and early secondary were similar to the late secondary and climaces. For K the increase in uptake was 1.28 times greater than the late secondary and climaces. This observation strengthens the importance of AM fungi in helping the pioneer and early secondary species in taking up nutrients from low-fertility soils. The nutrients accumulation in the plant shoots was calculated based on the nutrient concentration and plant biomass (Table 12.3). The mycorrhizal symbiosis increased the uptake and accumulaion of nutrients only in the pioneer and early secondary plants. In relation to acumulation of P and Ca, the pioneer, early secondary and late secondary plants accumulated similar amounts of nutrients, but surpassed the climaces plants. The most K accumulation was found in the pioneer plants. Again, these results reinforce the importance of AM fungi for plant of the initial succession stages in relation to mineral nutrition, in comparison to plants from later successional stages.

Table 12.3 Leaf nutrient content of the native woody species belonging to different ecological groups, inoculated and uninoculated with arbuscular mycorrhizal fungi. Accumulation fator: ratio between inoculated/uninoculated. Nutrient concentration ratio between inoculated/uninoculated NCR^2.

Successional groups	Uninoculated	Inoculated	NCR^2
P (mg per leaf)			
Pioneer	0.39 bB	5.16 aA	13.23
Early secondary	0.56 bB	4.29 aA	7.66
Late secondary	2.42 aA	3.29 aA	1.36
Climax	0.99 aB	1.12 aB	1.13
Ca (mg per leaf)			
Pioneer	1.79 bB	19.45 aA	10.87
Early secondary	3.86 bB	15.69 aA	4.06
Late secondary	6.86 aA	14.56 aA	2.12
Climax	4.79 aB	5.61 aB	1.17
K (mg per leaf)			
Pioneer	4.62 bB	53.63 aA	11.61
Early secondary	10.12 bB	39.36 aB	3.89
Late secondary	17.21 aA	20.75 aB	1.21
Climax	8.11 aB	9.32 aC	1.15

Means followed by the same latter (smal in row and capital in column) are not different by Tukey-Kramer HSD test at 5%.

The high nutrient accumulation among the earlier successional species suggests its high external demand for nutrients. Seeing that these plantlets grow up generally in low-fertility soils, the dependence of pioneer and early secondary plants on AM fungi is highlighted.

Most of the late secondary and climaces species had no advantage with the symbiosis. Its growth and biomass production were small independently of AM fungi. In this case, the low soil fertility may have limited the plant growth. Nevertheless, these species generaly develop well in soils with moderate to high fertility, such as the case of the mature forest at the Mata dos Godoy State Park in Londrina, PR, Southern Brazil.

Mycorrhizal Root Colonization

All the above mentioned plants were assessed for mycorrhizal root colonizatio. In addition, the mycorrhizal root colonization was also assessed in plantlets from 49 woody native species grown naturally in the understorey of the deciduous seasonal forest at the Mata dos Godoy State Park (23° 27'S and 51° 15'W) and in the understorey of an bordering secondary forest dominated by pioneer and early secondary species.

Either the species grown in the greenhouse or those collected from the field showed decrease on mycorrhizal root colonization along the successional groups (Fig. 12.5). The mycorrhizal root colonization in pioneers and early secondary grown in greenhouse was four times higher than the root colonization in late secondary and climaces species. For the plants collected from field, this proportion was 9 fold higher, considering the same successional groups. The lower level of mycorrhizal root colonization under natural conditions may be result of competition and predation of mycorrhizal hyphae by soil organisms, contributing to reduce the root colonization (Miller and Jastrow, 1994; Barea and Jeffries, 1995).

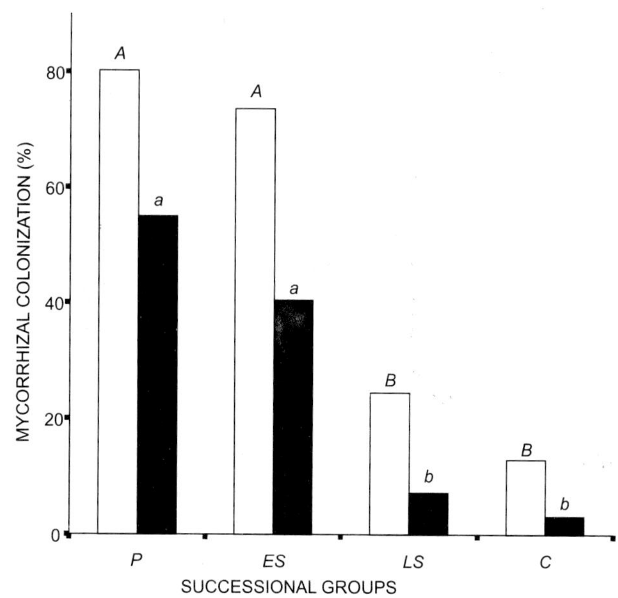

Figure 12.5 Root mycorrhizal colonization in greenhouse (□) and in field (■) of native woody species belonging to different successional stages. Means followed by same letter, capital for greenhouse and small for field, are not different by Tukey-Kramer HSD test at 0.05 level.

The percentage of mycorrhizal root colonization in greenhouse correlated positively with the root colonization in the species collected under natural conditions (Fig.12.6). This observation suggests that the root colonization found in greenhouse is secure for interpreting the mycorrhizal colonization in natural conditions. In addition, the level of mycorrhizal root colonization in the field is an indicative of potential inoculum (Gange *et al.*, 1993).

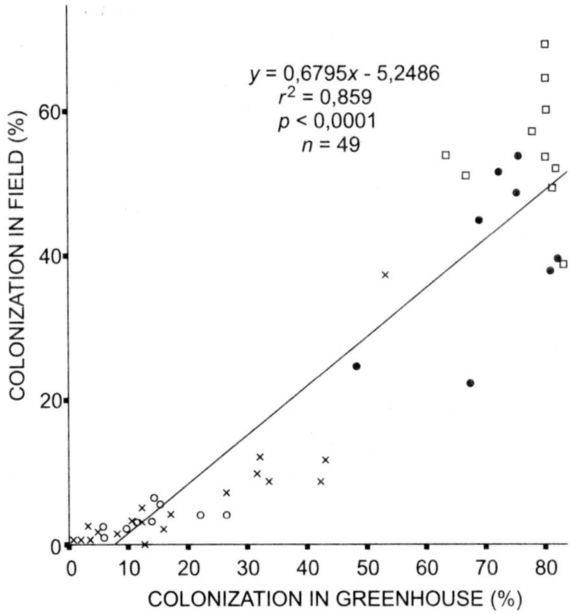

Figure 12.6 Relationship between mycorrhizal colonization in greenhouse with mycorrhizal colonization in the field of native woody species belonging to different successional stages: pioneer (•), early secondary (□), late-secondary (×), and climax (o).

The plant responsiveness to AM fungi correlated positively with mycorrhizal root colonization along the successional groups in both greenhouse (Fig 12.7) and field conditions (Fig. 12.8). The seed fresh biomass of the woody species increased along the successional groups (Fig. 12.9), but, conversely to plant responsiveness correlated inversely with mycorrhizal responsiveness (r^2 = 0.76). Inverse correlation was also observed between seed fresh biomass and root colonization either in greenhouse (r^2 = 0.80) or in natural conditions (r^2 = 0.72).

The inverse correlation between seed fresh biomass with plant responsiveness and mycorrhizal colonization is attributed to the fact that species belonging to the final succession stages have larger seeds containing more reserves than the species at the beginning of the succession. Thus, plants with larger seeds containing more reserves would not be so dependent on AM fungi for the initial growth,

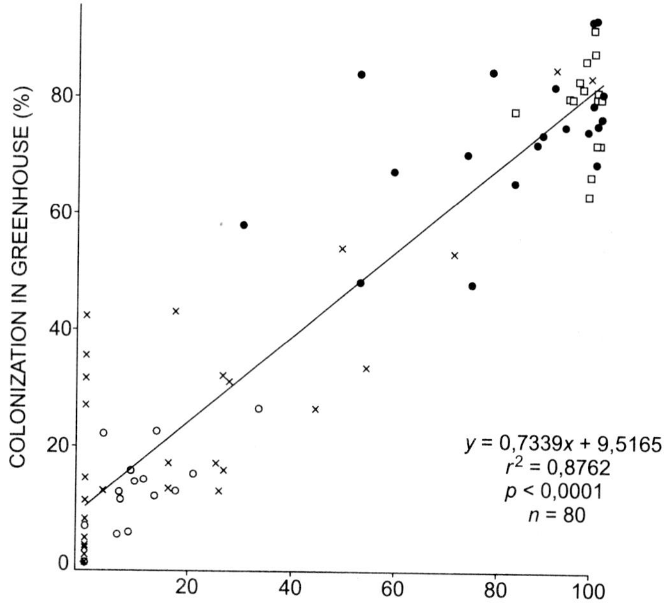

Figure 12.7 Relationship between plant responsiveness to AM fungi with mycorrhizal root colonization in greenhouse of native woody species belonging to different successional stages: pioneer (•), early secondary (□), late-secondary and climax (o).

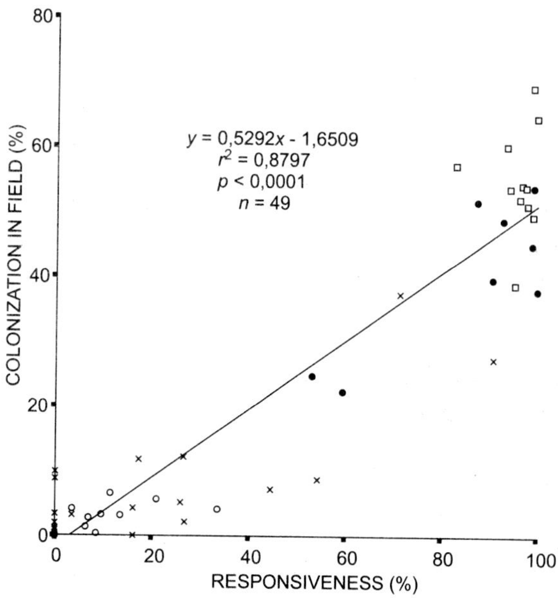

Figure 12.8 Relationship between plant responsiveness to AM fungi with mycorrhizal root colonization in field of native woody species belonging to different successional stages: pioneer (•), early secondary (□), late-secondary and climax (o).

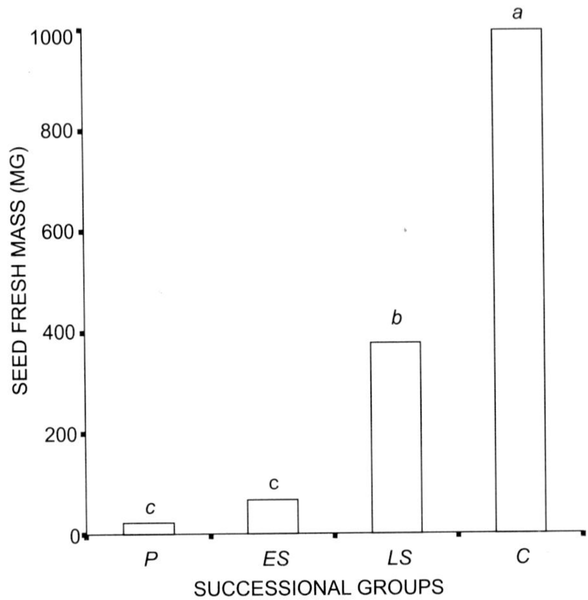

Figure 12.9 Seed fresh mass of native woody species belonging to different successional stages. Means followed by same letter are not different by Tukey-Kramer HSD test at 0.05 level.

resulting in lower root colonization levels. On the other hand, species from the initial succession stages have small seeds, consequently with low nutritional reserves that probably are not enough to bear the initial plant growth for long periods. Most of the pioneer and early secondary species that were not inoculated with AM fungi in greenhouse conditions stopped growing after 3–5 months and presented most of the leaves in senescence. This suggests that the woody species that initiate the vegetal succession can only establish and grow in low-fertility soils when in symbiosis with AM fungi. Consequently, the fast growth of pioneer and early secondary species in degraded areas can only be reached when associated with AM fungi, due to its role in improving the plant capacity to uptake nutrients from soil.

INOCULUM POTENTIAL OF ARBUSCULAR MYCORRHIZAL FUNGI IN SOIL

The present study also assessed the potential inoculum of AM fungi in the soils from the mature forest (Mata dos Godoy State Park), in a gap in the mature forest and in the bordering secondary forest. The natural soils were diluted (0, 10, 20, 30, 40, 50, 60, 80, 90 and 100% w/w) with a mix of subsoil and sand (4:1), previously fumigated with methyl bromide. The mixture of natural soil and fumigated soil (1 kg) was set into pots and sowed with *Cecropia pachystachya* Trec. (Cecropiaceae) as trap plant, in three replications for each soil diluition. Mycorrhizal root colonization was assessed with 90 days after plant growth.

The potential inoculum of AM fungi in the soil of the secondary forest was aproximately five times greater than in the mature forest (Fig. 12.10). The secondary forest had large amounts of AM fungi propagules and the mycorrhizal colonization decreased after 50% dilution. At this dilution level, the mycorrhizal colonization fell by 19%, lower than the soil dilution rate. In the gap area, the colonization level was 14% in the original soil and fell by 73% at 50% of soil dilution rate. In the mature forest, the amount of propagules in the original soil was lower than in the gap and the mycorrhizal colonization was 11%. At 50% of soil dilution rate, the root colonization also fell by 73%.

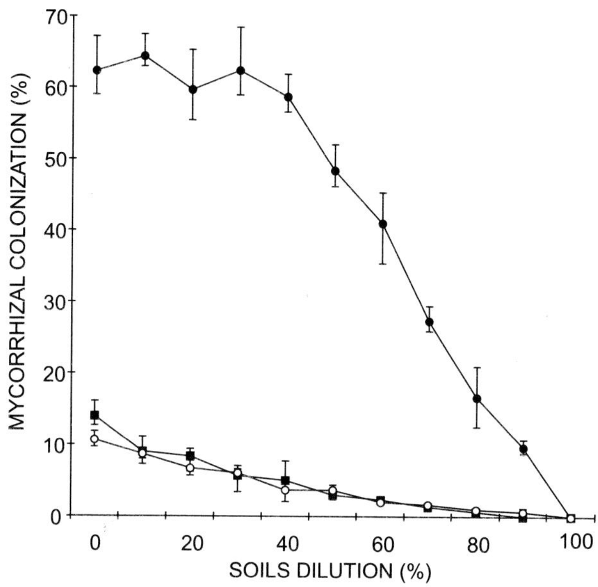

Figure 12.10 Soils dilution effect from the areas of early succession (•), gap (▪) and mature forest (o), on root mycorrhizal colonization of *Cecropia pachystachya*. Vertical bars show the ranges.

The higher colonization ratio in the plants with the soil from the secondary forest, dominated by plants from the initial successional groups in relation to the gap and the mature forest emphasizes the lower inoculum potential in the latter ones. These results suggest that woody pioneer and early secondary plants may be able to multiply the AM fungi in large amounts, allowing high inoculum potential for their offsprings. On the other hand, in the mature forest, the inoculum potential was much lower as compared to the secondary forest at the beginning of vegetal succession. These results indicate that the late secondary and climaces species that compose the most part of the vegetation in a mature forest have lower levels of micotrophy. Consequently, the potential of AM fungi inoculum is low in such situations.

ARBUSCULAR MYCORRHIZAL FUNGI AND SUCCESSION

In the forestal ecosystems, the litter is the main source of organic matter and nutrients and controls most of the functional processes in the soil. Such ecosystems have large variability in relation to the nutrient availability, ranging from oligotrophics (infertile soils) to eutrophics (fertile soils) (Vitousek, 1984; Power *et al.*, 2005). Different ecosystems have different patterns of nutrient cycling. In the oligotrophic ones, such as in several Amazonian ecosystems, plants have mechanisms to conserve nutrients and avoid its losses. In these cases, one of the mechanisms is the nutrient translocation before the leaves fall, what is in agreement with the low concentrations of nutrients found in the soil and litter. On the other hand, plants in chemically richer soils allow greater return of nutrients to the soil through the litter to be later recycled in soil (Vitousek and Sanford, 1986; Pagano, 1989; Santos, 1989).

Kingle and Rodrigues (1968) found 152 kg ha^{-1} $year^{-1}$ of macronutrients transferred to the soil via litter in an Amazonian dry land forest. In a tropical forest in Costa Rica, Heaney and Proctor (1989) found 219 kg ha^{-1} $year^{-1}$ of macronutrients transferred to the soil in the same way. Janos (1980a, b) working with plants of native woody species from forests on low-fertility soils in Costa Rica (Asbjornsen and Montagnini, 1994; Johnson and Wedin, 1997), suggested that AM fungi play important role in the uptake of nutrients for the mature forest.

When studying nutrients recycling in deciduous mesophile forests on eutrophic soils at São Paulo State, Brazil, the amounts of nutrients returned to the soil by means of the litter mineralization was higher: 446 kg ha^{-1} $year^{-1}$ at Rio Claro (Pagano, 1989), 427 kg ha^{-1} $year^{-1}$ at Anhambi (Schlittler *et al.*, 1993) and 615 kg ha^{-1} $year^{-1}$ at Araras (Diniz, 1987). In a caducipholy tropical forest at Espírito Santo, Brazil, Santos (1989) found 512 kg ha^{-1} $year^{-1}$ of macronutrients recycled from litter. All above-mentioned authors concluded that the plants from these forests have no mechanisms to retain nutrients in the biomass, which are returned to the soil in the decaying leaves, stems, branches, barks and flowers and are further recycled by mineralization.

More productive forests, developing on fertile soils, show higher concentrations of P, K and Ca in their biomasses, while forests developing on poorer soils are less productive and have less nutrients to be recycled (Vitousek and Sanford, 1986). Thus, in environments with contrasting levels of soil fertility, the nutrient recycling also seems to be distinct. In low-fertility soils, the AM fungi are expected to play an important role on mature forest nutrition, such as in Amazonian and Costa Rica soils.

In forests on eutrophic soils, such as the stational decidue forest at the State Park Mata dos Godoy, the mechanisms to conserve the nutrients in the biomass before the fall of the vegetal constituents are less active, what turn the litter richer in nutrients and consequently more nutrients are released to the soil solution after mineralization. As plant responses to AM fungi decrease with increasing soil fertility, especially P availability (Graham and Eissenstat, 1994), the importance of

AM fungi on plant nutrition in the mature forest seems to be lower than for plants from the initial succession stages.

The early stages of vegetal succession is marked by dominance of pioneer species, which have short lifetime, fast growth, intolerance to shadowing, small fruits and seeds that are dispersed to long distances, high photosynthetic rates and high capacity for uptake and accumulation of nutrients (Ferretti *et al.*, 1995; Guariguata and Ostertag, 2001). These characteristics are typical of highly mycotrophic hosts (Siqueira *et al.*, 1998; Zangaro *et al.*, 2003, 2005), due to its high need for nutrients, generally found in low concentrations in degraded soils. These factors make the arbuscular mycorrhizas important partners for pioneers and early secondary species in helping them to establish in low-fertility areas. In later stages of the vegetative succession, the level of soil fertility is generally increased. This occurs due to the mineralization of the litter initially produced by the early successional species, whose biomasses have larger nutrient accumulation rates (Lugo and Helmer, 2004). For instance, the soil from mature forest at the State Park Mata dos Godoy contains six times more phosphorus than a degraded neighbour soil under starting vegetal succession, in which only pioneer species are found (Table 12.4). The higher availability of nutrients in the soil of the mature forest, in addition to the adaptation of the plantlets to shadowing, slow growth rates and lower requirement for nutrients for these species, may have been crucial for the low mycorrhizal colonization and the poor response of the late sucessional species to AM fungi. Moreover, the lower Al concentration found in the 0–20 cm depth soil under mature forest make that soil horizon more feasible to be explored by fine roots (Jordan and Herrera, 1981; Cooperband *et al.*, 1994).

Table 12.4 Nutrient concentration of the soils from mature forest and early successional area*.

	$g\ 100\ g^{-1}$	$mg\ kg^{-1}$			*Meq* $100\ mL^{-1}$		
	pH	C	P	Al	Ca	Mg	K
Mature forest	4.9 a	4.2 a	9.96 a	0.04 b	11.7 a	3.61 a	0.52 b
Early succession	5.2 a	1.8 b	1.89 b	0.09 a	6.89 b	2.77 b	0.73 a

Means of 10 soil samples from 0–10 cm of deep. Means followed by the same latter are not different by Tukey test 5%.

The low light intensity may be negative for mycorrhizal plants (Gehring, 2003). The low rate of C acquisition by photosynthesis limits the growth rate, and thus the nutritional plant demand is lower. In this case, the fungus consists of an extra C sink to the host, that does not need nutrients in high demand (Hodge, 2004). Thus, there is a slight balance between the nutritional benefits and the C costs to keep the mycorrhizal fungus (Smith and Read, 1997). It is known that the AM fungi may demand up to 20% of the total C fixed photosynthetically by plant (Lynch and Ho, 2005).

Although arbuscular mycorrhizas demand for C from the host plant, the CO_2 fixation is improved in mycorrhizal plants (Smith and Read, 1997). This statement seems not to fit to the mature forest because the plantlets that take part in the later successional stages are under shadowing. These plants remain for long time in the photosynthetic compensation point and do not produce enough carbohydrates to support the mycorrhizal growth in its roots. This may be the reason that Power *et al.* (2005) found very low amounts of AM fungal hyphae in four mature forests in the tropical region in Central and South America. On the other hand, the pioneer plantlets, which grow in environments under high light incidence, have fast growth and fix greater amounts of C as carbohydrates. These plants are thus more able to support mycorrhizal growth in their roots, that by it turn help the host to adapt in low-fertility soils.

The AM fungi may play an important role in the vegetal succession. The result pointed out showed that there is a relationship between the potential of inoculum found in the field and the occurrence of woody species in adult stage in different habitats. The pioneers and early secondary species show high mycorrhizal colonization levels and respond to this colonization. Conversely, the late secondary and climaces species show low mycorrhizal root colonization and low response to AM fungi colonization. While the first species usually colonizes natural areas with high mycorrhizal inoculum potential, the latter colonizes the understory of mature forests where the mycorrhizal inoculum potential is low. Thus, a gradient of mycorrhizal colonization and responsiveness to AM fungi may be expected in relation to the preferential plant habitat.

ARBUSCULAR MYCORRHIZAL FUNGI AND REVEGETATION PROGRAMS

The soil and water degradation are some of the consequences of unsuitable soil use and management. Erosion is one of the main causes of soil degradation, mainly in tropical and subtropical region. In this case, the superficial horizon is generally lost, decreasing the soil fertility due to losses of nutrients and organic matter. Besides the physical and chemical degradation, the biological degradation in such areas also restricts the plant reestablishment.

Areas of tropical forest has been deforested and substituted by crops and pastures, which contributes to decrease the soil fertility. In many cases, these areas are abandoned and are recolonized by the native vegetation, which allow for a certain degree of soil fertility restoration. Thus, the remainings of native vegetation are important as germoplasm reservoir for plants and AM fungi.

In areas where the the natural succession cannot occur and establish naturally due to long distances from the propagules sources, one alternative is the artificial reforestation either with native or exotic species. In this case, the success of the reforestation depend on the capacity of the seedling in taking water and nutrients to resist and survive in the field (Perry *et al.*, 1987). In the case of degraded areas, revegetation is generally costly because the production and installation of seedlings is generally expensive. This cost is still increased due to the need for chemical fertilizers in low-natural fertility degraded soils. Nevertheless, some alternatives may contribute to reduce costs in revegetation programs. The use of indigenous leguminous woody species is less impacting to the environment to be revegetated and contributes to recover and conserve the biodiversity. The soil biological diversity and activity is also stimulated, which contributes to improve the nutrient cycling and enable the establishment of other vegetal species.

The seedling survival after transplanting to the field is generally increased when they are inoculated with AM fungi (Siqueira *et al.*, 1998). This fact is atributed to an increased capacity for water and nutrients uptake, improved root longevity and resistance against pathogens (Smith and Read, 1997).

The AM fungi inoculum potential in the soil affects the establishment, growth and competition among plant species and may affect the reforestation in degraded areas (Asbjornsen and Montagnini, 1994; Zangaro *et al.*, 2000). Some soil factors such as aluminium toxicity, acidity, high soil temperatures may impair the restoration of plant community and AM fungi, especially in tropical areas. Once the vegetal community is reestablished, the increased root density and organic matter in the soil are increased and favours the AM fungi proliferation (Janos, 1996).

For the success of the restablishment of a diverse and self-sustainable ecosystem, an adequate inoculum potential of AM fungi is desirable (Jasper *et al.*, pp 1992). In areas with poor inoculum potential, the use of appropriate host plants is the key for the reestablishment of a diverse AM fungi population. In the case of areas deprived of AM fungi inoculum, its reintroduction may be done by means of plantlets previously inoculated in nursery (Janos, 1996). Thus, the knowledge on the degree of micotrophy among the species that will be employed in the reforestation is very important for the reforestation success.

Importance of Arbuscular Mycorrhizal Fungi for Recovering Degraded Lands

The elimination of large areas of native vegetation sometimes results in environmental degradation, including soil and water (Caravaca *et al.*, 2003; Thrall *et al.*, 2005). In some cases, the native vegetation must be reestablished in order to revert or stop the degradation process.

Results shown above reinforce the importance of mycorrhizas for the success of early successional species, mainly pioneers and early secondary ones. The pioneer species are responsible for the initial forest structuration (Montagnini and Sancho, 1994; Guariguata and Ostertag, 2001). Establishment of these plants results in the improvement of the environment to be colonized by later successional species, such as less water loss, improvement of soil structure and aeration, less occurrence of pathogens (Sylvia and Williams, 1992). These improvements result in higher photosynthetic rates, root longevity (Linderman, 1988; Comas *et al.*, 2002), root density, organic matter, factors that favour the AM fungi proliferation (Janos, 1996). The benefits from the mycorrhizal symbiosis, especially nutritional (Barea and Jaffries, 1995) may favour direct and indirectly the establishment of species of later successional stages.

The vegetal succession in degraded areas is unpredictable (Jordan, 1991), mainly due to the absence of a bank of seeds and to the low soil fertility in such conditions. In these cases, it is necessary to make the artificial reforestation with indigenous species or with adequate exotic species.

During the reforestation, there are two essential components. The soil and its physical, chemical and biological properties and the selected plant species to be used, whose success in establishing will depend on its capacity on taking resources from soil for its development. Thus, the basic unity for artificial reforestation is the individual tree, whose environmental and nutritional requirements must be known (Carneiro *et al.*, 1996).

The characteristics of woody early successional species, such as intolerance to shadowing, high capacity for uptake and accumulation of nutrients and small and abundant seeds, fit as a good partner for AM fungi (Zangaro *et al.*, 2000, 2003). The high mycotrophism of pioneer and early secondary species makes the mycorrhizal symbiosis one of the main tools for installation and start the vegetal succession. This reinforces the importace of AM fungi inoculation in plantlets to be used in reforestation programs (Zangaro *et al.*, 2000).

Spores of AM fungi are generally found in large amounts in pastures, areas that may be useful for reforestation programs (Johnson and Wedin, 1997). Fischer *et al.* (1994) found 62.8 spores of AM fungi in 100 g^{-1} of soil in pastures of Costa Rica, while Sieverding (1991) found 103 spores in pastures in tropical region. Zangaro (unpublished data) found 567 and 623 spores of AM fungi in 100 g of soil from pastures under *Brachiaria* sp. and *Paspalum* sp., respectively. In the southern region of Brazil, the forest fragments are generally surrounded by pastures or monocrops that are frequently colonized by pioneer and early secondary species originated from these fragments (Zangaro, personal obsevation). This suggests that the AM fungi propagules in these soils are enough to promote the establishment, growth and survival of such plants. In case of degraded areas with low AM fungi propagules, the plantlets to be used in the reforestation should be inoculated in the nursery. Nevertheless, the plant response to mycorrhizal inoculation must be known previously. The use of plants that respond to AM fungi inoculation is a tool for the success of the establishment of plants to start the arboreal succession in programmes destinated to rehabilitation of degraded areas. Once the pioneer plants are well established, they will favour the microbial, faunal and floral biodiversity and help the ecosystem in reestablishment to be functional.

Arbuscular Mycorrhizas as Biofertilizers

The use of some species of woody indigenous trees for reforestation is limited by its low growth rate. Conversely, other species are fast growing, but its success in establishing degraded areas depend on an adequate nutrient supply (Marques *et al.*, 2001). Thus, the needs for nutrients usually increase the costs for reforestation (Franco and Faria, 1997).

Nitrogen (N) and phosphorus (P) are the most required plant nutrients, and are generally limiting in degraded soils. N is easily lost from soil due to its dynamics, while P have low mobility in soil and is fixed in clay and oxidic surfaces. These properties make necessary the use of fertilizers containing N and P for revegetation of degraded areas, which increases the revegetation costs. In addition, the excessive or inadequate use of mineral nutrients may result in environmental pollution, mainly when these nutrients reach the underground water, streams or lakes.

The role of microorganisms as biofertilizers is useful to reduce the environmental and economical costs. When applied to soil or plants, they help their growth mainly by improving nutrients availability. Among the microorganisms with great potential as biofertilizers are the diazotrophic bacteria and the AM fungi. The double inoculation with such microorganisms improves the hosts' ability in taking up nutrients, improving its growth and survival. All these benefits decrease the environmental and economical costs related to the chemical fertilizers.

In relation to the tripartite symbiosis (host plant, AM fungi and diazotrophic bacteria), most of the genera from the Mimosaceae family have capacity to form nodules with N-fixing bacteria. In addition, most of plants are also able in forming symbiosis with AM fungi, which may improve the biological nitrogen fixation (BNF), as a result of an improvement of the host P nutritional status. By it turn, the AM fungi root colonization may also be stimulated by diazotrophic bacteria. In this case, the inoculation of the host plant with AM fungi and diazotrophic bacteria may confer additional advantages for all partners in this tripartite symbiosis.

Biofertilizers are defined as substances containing living microorganisms that are able to colonize the rhizosphere or plant tissues and promotes its growth (Vessey, 2003). This growth improvement is generally consequence of improved nutrient supply or availability. Under this view, the use of biofertilizers for woody native species is promising for decreasing costs related to plantlets production and establishment in degraded areas. Nevertheless, the use of microorganisms depends on its ability in establishing and benefits the host plant.

According to Barea *et al.* (2002), the establishment of mycorrhizal symbiosis affects the plant physiologically, mainly in relation to C allocation. The extraradical mycorrhizal mycelium grows throughout the soil and increases the surface area for nutient acquisition such as P, Ca, Zn and even N (Hawkins and George, 2001; Neumann and George, 2005). The external mycorrhizal hyphae can be considered a way that shortens the distance between the above-mentioned nutrients to the plant (Zangaro *et al.*, 2005). Such structures can reach up to 11 cm far from the root surface (Li *et al.*, 1991), and increase the plant potential in obtaining water and nutrients. In addition, according to Neumann and George (2005), the external hyphae contribute to soil aggregate stability, by means of a typical protein termed glomalin (Caravaca *et al.*, 2005). The external hyphae also protect the host plant against excessive heavy metal in the environment and in some situations serve as inoculum source for root colonization.

The knowledge on the ability of plants to form symbiosis with AM fungi is important for the success of reforestation in degraded areas due to the low natural soil fertility and the low mycorrhizal inoculum potential in such areas (Zangaro *et al.*, 2003).

Mycorrhizal plants usually show increase on shoot and root dry weights and higher P concentration in tissues (Zangaro *et al.*, 2003). According to Carneiro *et al.* (1996), apart from the effects on initial growth and seedling quality, there are evidences that mycorrhiza affects the further stages of vegetal succession and the structure of vegetal communities. Thus, AM fungi are important for artificial re-forestation, especially in low-fertility soils, where mycorrhiza may reduce the host nutritional requirements, decrease the implantation costs and increase the seedlings survival in the ecosystem in restoration.

The plant responses to mycorrhiza depend on its degree of mycotrophy. According to Janos (1996), facultative mycotrophs may have high dependence on mycorrhiza in low-fertility soils; the obligate ones grow without mycorrhiza only when supplied with fertilizers, while the non-micotrophs do not form mycorrhiza. As shown before, the degree of mycotrophism along the successional species is the highest in the pioneer and early secondary group. Thus, the mycorrhizal symbiosis is one of the most important tool for installation of such plants in degraded, low-fertility soils.

The selection for AM fungi efficient in promoting plant growth is essential in inoculation programmes, once there are different levels of compatibility between AM fungi and host plants (Roldán *et al.*, 1992; Caravaca *et al.*, 2003; Lovato *et al.*, 2006). Even being able to colonize the host plant, the AM fungi species are not able to promote plant growth. This lack of response may be related to the required period for the full establishment of the symbiosis, mainly in annual crops (Abbott and Robson, 1981). For example, AM fungi from *Gigaspora* genus are not so efficient in promoting growth in annual crops (Nogueira and Cardoso, 2000), but are efficient in promoting growth of perennial ones (Siqueira *et al.* 1993; Antunes *et al.*, 1998). In some cases of non-efficient symbiosis, although AM fungi colonizes plant root extensively, they do not produce external mycelium enough to help the plant to explore the surrounding soil (Nogueira and Cardoso, 2000; Lovato *et al.*, 2006).

Synergism between AM Fungi and Rhizobacteria in Woody Leguminous Trees

Plants able to form symbiosis with diazotrophic bacteria and AM fungi present great potential to be used in revegetation programmes (Franco and Faria, 1997). In this case, plants have access to two of the most required nutrients, i.e. N by diazotrophic bacteria and P by mycorrhiza. Thus, mycorrhizal plantlets show greater survival indexes after transplanting and also higher initial growth (Siqueira *et al.*, 1993; Siqueira *et al.*, 1998; Marques *et al.*, 2001; Lovato *et al.*, 2006). Moreover, the BNF, either symbiotic or associative, is benefited in mycorrhizal plants. On the other hand, some bacteria may also stimulate the mycorrhizal colonization (Toro *et al.*, 1997). Results from Lovato *et al.* (2006) indicated that leguminous, early and late secondary, were responsive to certain AM fungi isolates, but the responsiveness was still stimulated when plants were also inoculated with diazotrophic bacteria (*Rhizobium* sp.) (Fig. 12.11). The four woody leguminous species benefited mainly from two mycorrhizal species. Plant response to rhizobacteria was observed only after AM fungi stimulated plant growth. Even the non-noduliferous species, *P. dubium*, had its height stimulated when associated to *G. clarum* and a rhizobacteria from the genus *Burkholderia*. Similarly, Burity *et al.* (2000) observed stimulus on *Mimosa caesalpinifolia* height when associated to *Rhizobium* and arbuscular mycorrhiza, emphasizing the importance of microbial symbiosis on plant growth.

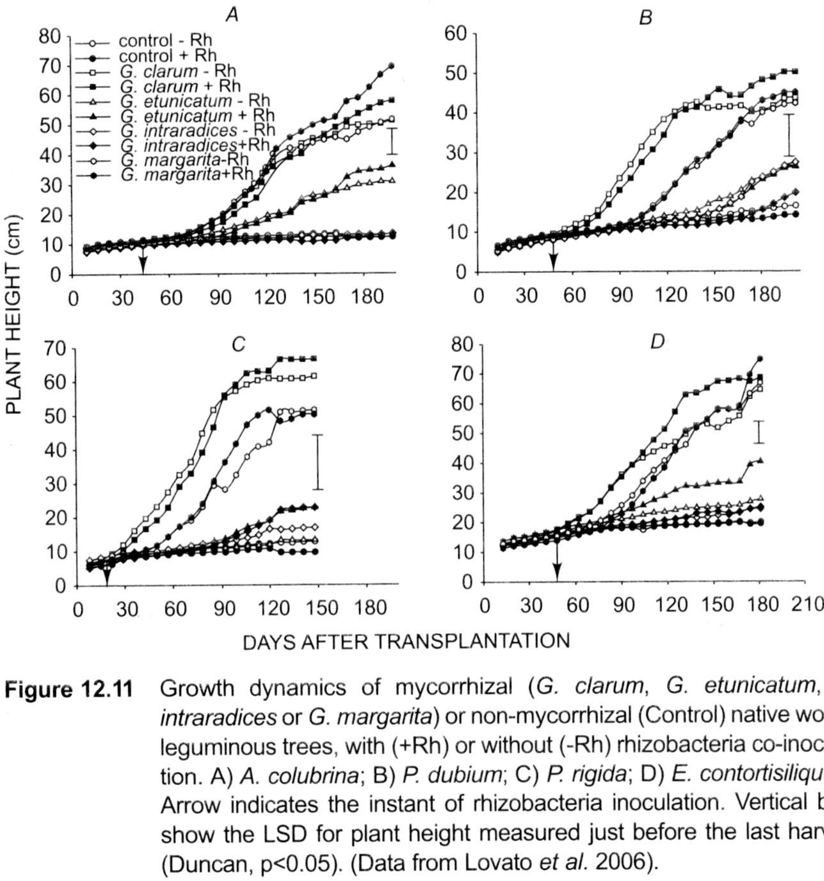

Figure 12.11 Growth dynamics of mycorrhizal (*G. clarum*, *G. etunicatum*, *G. intraradices* or *G. margarita*) or non-mycorrhizal (Control) native woody leguminous trees, with (+Rh) or without (-Rh) rhizobacteria co-inoculation. A) *A. colubrina*; B) *P. dubium*; C) *P. rigida*; D) *E. contortisiliquum*. Arrow indicates the instant of rhizobacteria inoculation. Vertical bars show the LSD for plant height measured just before the last harvest (Duncan, $p<0.05$). (Data from Lovato *et al.* 2006).

Plant growth promoting rhizobacteria (PGPR) and AMF may interact synergistically on plant growth, as a result of improved nutrient acquisition or production of phytohormones (Artursson *et al.*, 2006). Inoculation with *G. clarum* and *G. margarita* also stimulated significantly the shoot and root biomasses and total root length (Lovato *et al.*, 2006). The inoculation with *G. etunicatum* resulted in lower response, while *G. intraradices* usually did not affect plant growth. AMF are generally not specific for host colonization, but different AM fungi species or even isolates have different effects onto host plant (Barea *et al.*, 2002). These results strengthen the need for screening an efficient mycorrhizal symbiosis for a successful plant growth in a low-fertility soil (Caravaca *et al.*, 2003).

Besides the benefits from the BNF, the AM fungi species associated to the host plant may change the plant response to diazotrophic bacteria in the tripartite symbiosis. In general, the plant benefits from the BNF are generally greater when also in symbiosis with AM fungi (Xavier and Germida, 2002; Lovato *et al.*, 2006). The sum of the benefits from each symbiont isolated is lower than the benefits reached when both are together. This improved benefit is the result of a synergism between AM fungi and diazotrophic bacteria (Barea *et al.*, 2002).

In general, the benefits for plants in the triple interaction are: faster growth, improved N and P nutrition, diseases control, drought resistance, phosphate solubilization, increase in nodulation and higher mycorrhizal colonization (Burity *et al.*, 2000; Marques *et al.*, 2001; Xavier and Germida, 2003; Jia *et al.* 2004). The enhancement of BNF in mycorrhizal plants is attributed to a higher P uptake capacity in mycorrhizal plants, which allows for a better energy supply to the BNF as ATP (Artursson *et al.*, 2006). In addition, mycorrhizal plants also show higher levels of nodulation. By it turn, the rhizobacteria also benefit the AM fungi (Johansson, 2004; Lovato *et al.* 2006). Previous findings have shown that not only plants benefits from BNF, but also the AM fungi (Minerdi *et al.*, 2001; Lovato *et al.*, 2006) (Fig. 12.12). Mycorrhizal root colonization varied according to AM fungi species and rhizobacteria, which stimulated mycorrhizal colonization in some cases, and decreased in other (Lovato *et al.*, 2006). Burity *et al.* (2000) also observed increase in root colonization in *Rhizobium*-inoculated plants. Root hair exudates substances derived from photoassimilates, generally amino acids, organic acids, sugars, mucilage, proteins and flavonoids. The latter may attract fungal hyphae towards the root surface, increasing root colonization. Any factor affecting plant physiology will probably also affect the mycorrhizal fungi associated. Indeed, *G. clarum* colonization was stimulated in three hosts, while *G. margarita* was stimulated only in *A. colubrina* and reduced in the other hosts. In the host *P. rigida*, both AM fungi had the root colonization decreased in the presence or rhizobacteria. As observed from these data, the stimulation or suppression of mycorrhizal root colonization depends on the interaction among rhizobacteria-host-AM fungus.

According to Barea *et al.* (2002), some bacteria, so-called helper, improve the mycorrhiza formation and functioning by stimulating the fungal mycelial growth and producing substances that improve the root permeability that make easier the mycorrhizal colonization. As shown before, a previous screening should be done before these two microorganisms are co-inoculated. In the effective interactions, plants are stimulated in P and N nutrition, which reflects in improved plant growth and better chance to survive in the environment, especially in low-fertility soils (Artursson *et al.*, 2006).

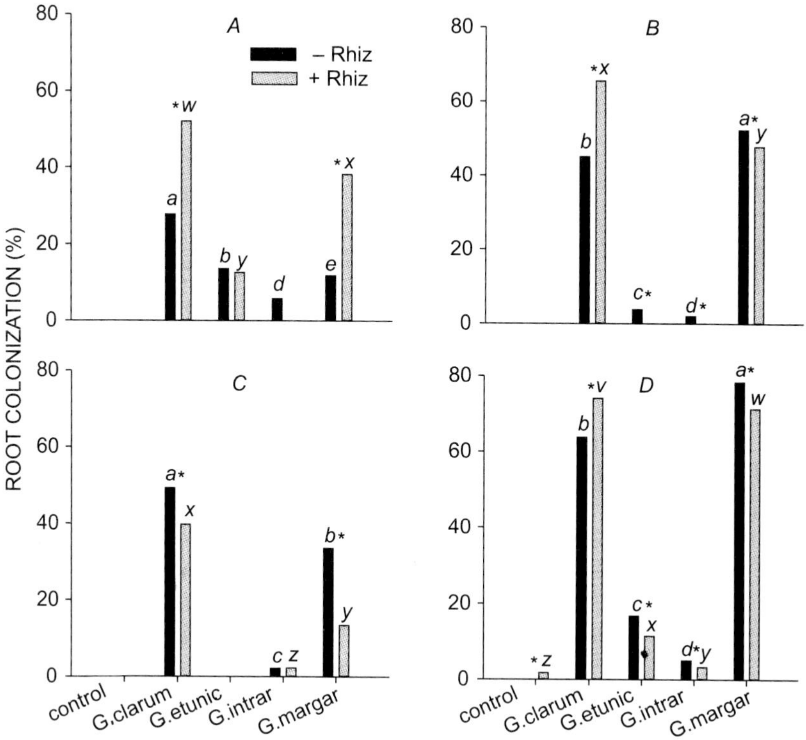

Figure 12.12 Per cent root colonization of native woody leguminous trees inoculated (*G. clarum*, *G. etunicatum*, *G. intraradices* or *G. margarita*) or not (Control) with arbuscular mycorrhizal fungi, with (+Rhiz) or without (-Rhiz) rhizobacteria. A) *A. colubrina*; B) *P. dubium*; C) *P. rigida*; D) *E. contortisiliquum*. Means sharing same letter, in each leguminous tree, do not differ each other (Duncan, $p<0.05$). a, b, c, d and e: compare arbuscular mycorrhizal fungi without rhizobacteria; v, w, x, y, z, compare arbuscular mycorrhizal fungi in the co-inoculation with rhizobacteria. * Indicate significant effect or rhizobacteria in each AMF treatment (Duncan, $p<0.05$). (Data from Lovato *et al.* 2006).

CONCLUSION

The role of AM fungi in helping the plant establishment in low-fertility soils is an important tool for reclamation of degraded environments. Besides decreasing the installation costs by lowering the fertilizer needs, also increases the chances of pioneers and early-secondary plants establishment, which creates an adequate environment for the vegetal succession. When the soil of the environment to be recuperated has insufficient inoculum potential, the artificial inoculation during the nursery plantlet production is a good alternative to improve the success of revegetation strategies. Besides increasing the plantlets establishment in the area, also contributes to increase the potential of AM fungi inoculum that will help thc establishment of mycotrophic plants that will naturally colonize the area during the later successional stages. An alternative that seems to be very feasible is the use of noduliferous leguminous species, which in addition to benefit from the mycorrhizal symbiosis, also benefit from the nitrogen-fixing bacteria. The search for the most effective tripartite interaction in the case of leguminous species will help the vegetal establishment in low-fertility soils and enable for a feasible environment for species from further stages of vegetal succession.

References

Abbott, L.K. and Robson, A.D. (1981). Infectivity and effectiveness of vesicular arbuscular mycorrhizal fungi: effect of inoculation type, *Australian Journal of Agricultural Research* 32:631–639.

Andrade, G. (2004). Role of functional groups of microorganisms on the rhizosphere microcosm dynamics. In: (Eds.) Ajit Varma, Lynette Abbott, Dietrich Werner, Rudiger Hampp. Plant surface microbiology. Springer-Verlag, Berlim. 51–69.

Antunes, V., Silveira A.P. and Cardoso, E.J.B.N. (1998). Interação entre diferentes tipos de solo e fungos micorrízicos vesículo-arbusculares na produção de mudas de café (*Coffea arabica*, L.), *Turrialba* 38:117–122.

Artursson, V., Finlay R.D. and Jansson, J.K. (2006). Interactions between arbuscular mycorrhizal fungi and bacteria and their potential for stimulating plant growth, *Environmental Microbiology* 8: 1–10.

Asbjornsen, H. and Montagnini, F. (1994). Vesicular-arbuscular mycorrhizal inoculum potential affects the growth of *Stryphnodendron microstachyum* seedlings in a Costa Rican human tropical lowland, *Mycorrhiza* 5:45–51.

Barea, J.M. and Jeffries, P. (1995). Arbuscular mycorrhizas in sustainable soil-plant systems. In: Mycorrhiza: Structure, Function, Molecular Biology and Biotechnology, Varma, A. and Hock, B. Eds., Springer - Verlag, Berlin Heidelberg. 521–560.

Barea, J.M., Azcón R. and Azcón-Aguilar, C. (2002). Mycorrhizosphere interactions to improve plant fitness and soil quality, *Antonie van Leeuwenhoek* 81:343–351.

Burity, H.A., Lyra, M.C.C.P., Souza, E.S., Mergulhão, A.C.E.S. and Silva, M.L.R.B. (2000). Effectiveness of inoculation with arbuscular mycorrhizal fungi and *Rhizobium* sp. on *Mimosa caesalpinifolia* seedlings, under different phosphorus levels, *Pesquisa Agropecuária Brasileira* 35:801–807.

Caravaca, F., Barea, J.M., Palenzuela, J., Figueroa, D., Alguacil, M.M. and Roldán, A. (2003). Establishment of shrub species in a degraded semiarid site after inoculation with native or allochthonous arbuscular mycorrhizal fungi, *Appied Soil Ecology* 22:103–111.

Caravaca, F., Alguacil, M.M., Barea, J.M. and Roldán, A. (2005). Survival of inocula and native AM fungi species associated with shrubs in a degraded Mediterranean ecosystem, *Soil Biology and Biochemistry* 37:227–233.

Carneiro, M.A.C., Siqueira, J.O., Davide, A.C., Gomes, L.J., Curi, N. and Vale, F.R. (1996). Mycorrhizal fungi and superphosphate on growth of tropical woody species, *Scientia Forestalis* 50:21–36.

Cooperband, L.R., Boerner, R.E.J. and Logan, T.J. (1994). Humid tropical leguminous tree and pasture grass responsiveness to vesicular-arbuscular mycorrhizal infection, *Mycorrhiza* 4:233–239.

Comas, L.H., Bouma, T.J. and Eissenstat, D.M. (2002). Linking root traits to potential growth rate in six temperate tree species, *Oecologia* 132:34–43.

Diniz, S. (1987). Ciclagem de nutrientes associados aos processos de produção e decomposição do folhedo em um ecossistema de mata mesófila semidecídua, no município de Araras - São Paulo. Dissertação de mestrado. Universidade Estadual Paulista. Rio Claro, SP.

Ferretti, A.R., Kageyama, P.Y., Arbocz, G.F., Santos, J.D., Barros, M.I.A., Lorza, R.F. and Oliveira, C. (1995). Classificação das espécies arbóreas em grupos ecológicos para revegetação com nativas no Estado de São Paulo, *Florestar Estatístico* 3:73–77.

Fischer, C.R., Janos, D.P., Perry, D.A., Linderman, R.G. and Sollins, P. (1994). Mycorrhiza inoculum potentials in tropical secondary succession, *Biotropica* 26:369–377.

Franco, A.A. and Faria, S.M. (1997). The contribution of N_2-fixing tree legumes to land reclamation and sustainability in the tropics, *Soil Biology and Biochemistry* 29:897–903.

Gange, A.C., Brown, V.K. and Sinclair, G.S. (1993). Vesicular-arbuscular mycorrhizal fungi: a determinant of plant community structure in early succession, *Functional Ecology* 7:616–622.

Gehring, C.A. (2003). Growth responses to arbuscular mycorrhizae by rain forest seedlings vary with light intensity and tree species, *Plant Ecology* 167:127–139.

Gilroy, S. and Jones, D.L. (2000). Through form to function: root hair development and nutrient take up, *Trends in Plant Science* 5:56–60.

Graham, J.H. and Eissenstat, D.M. (1994). Host genotype and the formation and function of VA mycorrhizae, *Plant and Soil* 159:179–185.

Guariguata, M.R. and Ostertag, R. (2001). Neotropical secondary forest sucession: changes in structural and functional characteristics, *Forest Ecology and Management* 148:185–206.

Hawkins, H.J. and George, E. (2001). Reduced N-15-nitrogen transport through arbuscular mycorrhizal hyphae to *Triticum aestivum* L. supplied with ammonium vs. nitrate nutrition, *Annals of Botany* 87:303–311.

Heaney, A. and Proctor, J. (1989). Chemical elements in litter in forests on Volcan Barva, Costa Rica. In: Mineral Nutrients in Tropical Forest and Savanna Ecosystems, Proctor, J. Ed., Blackwell Sci. Publi., Oxford, 255–271.

Hodge, A. (2004). The plastic plant: root responses to heterogeneous supplies of nutrients, *New Phytologist* 162:9–24.

Jakobsen, I. (1995). Transport of phosphorus and carbon in VA Mycorrhizas. In: Mycorrhiza: Structure, Function, Molecular Biology and Biotechnology. Varma, A. and Hock, B. Eds., Springer-Verlag, Berlin Heidelberg, 297–324.

Janos, D.P. (1980a). Vesicular-arbuscular mycorrhizae affect lowland tropical rainforest plant growth, *Ecology* 61:151–162.

Janos, D.P. (1980b). Mycorrhizae influence tropical succession, *Biotropica* 12:56–64.

Janos, D.P. (1983). Tropical mycorrhizas, nutrient cycles and plant growth. In: Tropical Rain Forest: Ecology and Management. Sutton, S.L., Whitmore, T.C. and Chadwick, A.C. Eds., Blackwell Scientific Publications, Oxford, 327–345.

Janos, D.P. (1996). Mycorrhizas, succession, and the rehabilitation of deforested lands in the humid tropics. In: Fungi and Environmental Change. Frankland, J.C., Magan, N. and Gadd, G.M. Eds., British Mycological Society Symposium Vol. 20. Cambridge University Press, Cambridge, 129–162.

Jasper, D.A., Abbott, L.K. and Robson, A.D. (1992). Soil disturbance in native ecosystems—the decline and recovery of infectivity of VA mycorrhizal fungi. In: Mycorrhizas in Ecosystems. Read D.J., Lewis, D.H., Fitter, A.H. and Alexander, I.J. Eds., Wallingford: CAB International, 151–155.

Jia, Y., Gray, V.M. and Straker, C.J. (2004) The influence of *Rhizobium* and arbuscular mycorrhizal fungi on nitrogen and phosphorus accumulation by *Vicia faba*, *Annals of Botany* 94:251–258.

Johansson, J., Paul, L.R. and Finlay, R.D. (2004). Microbial interactions in the mycorrhizosphere and their significance for sustainable agriculture, *FEMS Microbiology Ecology* 48:1–13.

Johnson, N.C. and Wedin, D.A. (1997). Soil carbon, nutrients, and mycorrhizae during conversion of dry tropical forest to grassland, *Ecological Applications* 7:171–182.

Jordan, C.F. and Herrera, R. (1981). Tropical rainforests: are nutrients really critical?, *American Naturalist* 117:167–180.

Jordan, C.F. (1991). Nutrient cycling processes and tropical forest management. In: Rain forest regeneration and management, Gomez-Pompa, A., Whitmore, T.C., Hadley, M. Eds., The Parthenon Publishing Group, Paris, 159–179.

Klinge, H. and Rodrigues, W.A. (1968). Litter production in an area of Amazoniam terra firme forest, *Amazoniana* 1:287–310.

Linderman, R.G. (1988). VA (vesicular-arbuscular) mycorrhizal simbiosis, *Atlas Sci. Anim. Plant Sci.*, 1:183–188.

Li, X.L., Gegorge, E. and Marshner, H. (1991). Extension of the phosphorus depletion zone in VA-mycorrhizal white clover in a calcareous soil, *Plant and Soil* 136:41–48.

Lovato, G.M., Nogueira, M.A., Knob, A., Lima, D.S, Spago, F.R., Murate, L.S., Myiauchi, M.Y.H., Zangaro, W., Colozzi-Filho, A. and Andrade, G. (2006). Interactions between arbuscular mycorrhiza and rhizobacteria on native woody leguminous trees useful in reforestation, *Applied Soil Ecology* (Submitted).

Lugo, A.E. and Helmer, E. (2004). Emerging forest on abandoned land: Puerto Rico's new forests, *Forest Ecology and Management* 190:145–161.

Lynch, J.P. and HO, M.D. (2005). Rhizoeconomics: carbon costs of phosphorus acquisition, *Plant and Soil* 269:45–56.

Marques, M.S., Pagano, M. and Scotti, M.R.M.M.L. (2001). Dual inoculation of a woody legume (*Centrolobium tomentosum*) with rhizobia and mycorrhizal fungi in south-eastern Brazil, *Agroforestry Systems* 52,107–117.

Merryweather, J. and Fitter, A. (1996). Phosphorus nutrition of an obligately mycorrhizal plant treated with the fungicide benomyl in the field, *New Phytologist* 132:307–311.

Miller, R.M. and Jastrow, J.D. (1994). Vesicular-arbuscular mycorrhizae and biogeochimical cycling. In: Mycorrhizae and Plant Health. Pfleger, F.L and Linderman, R.G. Eds., APS Press, Minnesota, 189–212.

Minerdi, D., Fani, R., Gallo, R., Boarino, A. and Bonfante, P. (2001). Nitrogen fixation genes in an endosymbiotic *Burkholderia* strain, *Applied and Environmental Microbiology* 67:725–732.

Montagnini, F. and Sancho, F. (1994). Nutrient budgets of young plantations with native trees: strategies for sustained management. In: Forest Resources and Wood-based Biomass Energy as Rural Development Assets. Bentley, W. and Gowen, M. Eds., Oxford and IBH Publishing, New Delhi, 213–233.

Neumann, E. and George, E. (2005). Extraction of extraradical arbuscular mycorrhizal mycelium from compartments filled with soil and glass beads, *Mycorrhiza* 15:533–537.

Nogueira, M.A. and Cardoso, E.J.B.N. (2000). Colonização radicular e produção de micélio externo por duas espécies de fungos micorrízicos arbusculares em soja, *Revista Brasileira de Ciência do Solo* 24:329–338.

Pagano, S.N. (1989). Nutrientes minerais do folhedo produzido em mata mesófila semidecídua no município de Rio Claro, SP., *Revista Brasileira de Biologia* 49:641–647.

Perry, D.L., Molina, R. and Amaranthus, M.P. (1987). Mycorrhizae, mycorrhizospheres, and reforestation: current knowledge and research needs, *Canadian Journal of Forest Research* 17: 929–940.

Plenchette, C., Fortin, J.A. and Furlan, V. (1983). Growth response of several plants species to mycorrhiza in a soil of moderate P fertility. I. Mycorrhizal dependency under field conditions, *Plant and Soil* 70:191–209.

Powers, J.S., Treseder, K.K. and Lerdau, M.T. (2005). Fine roots, arbuscular mycorrhizal hyphae and soil nutrients in four neotropical rain forests: patterns across large geographic distance, *New Phytologist* 165:913–921.

Roldán, Á., Díaz, G. and Albaladejo, J. (1992). Effect of VAM-fungal inoculation on growth and phosphorus uptake of two Hedysarum species in a Xeric Torrorthent soil from southeast Spain, *Arid Soil Reserch Rehabilitation* 6:33–39.

Sanders, I.R., Clapp, J.P. and Wiemken, A. (1996). The genetic diversity of arbuscular mycorrhizal fungi in natural ecosystems—a key to understanding the ecology and functioning of the mycorrhizal symbiosis, *New Phytologist* 133:123–134.

Santos, V.D. (1989). Ciclagem de nutrientes minerais em mata tropical subcaducifolia dos planaltos do Paraná (Parque Estadual Vila Rica do Espirito Santo-Fênix-PR). Tese de doutorado, Universidade Federal de São Carlos, São Carlos.

Schlitter, F.H.M., de Marinis, G. and Cesar, O. (1993). Transferência de macronutrientes pela serapilheira produzida na floresta do Morro do Diabo (região do pontal do paranapanema, estado de São Paulo), *Arquivos de Biologia e Tecnologia* 36:597–611.

Sieverding, E. (1991). Vesicular-Arbuscular Mycorrhiza Management in Tropical Agrosystems. Deutsche Gesellschaft fur Technissche Zusammenarbeit (GTZ) GmbH. Eschbon.371 p.

Siqueira, J.O., Colozzi-Filho, A., Saggin-Júnior, O.J., Guimarães, P.T.G. and Oliveira, E. (1993). Crescimento de mudas e produção do cafeeiro sob influência de fungos micorrízicos e superfosfato, *Revista Brasileira de Ciência do Solo* 17:53–60.

Siqueira, J.O., Carneiro, M.A.C., Curi, N., Rosado, S.C.S. and Davide, A.C. (1998). Mycorrhizal colonization and mycotrophic growth of native woody species as related to successional groups in South-eastern Brazil, *Forest Ecology Managment* 107:241–252.

Siqueira, J.O., Curi, N., Vale, F.R., Ferreira, M.M. and Moreira, F.M.S. (1995). Aspectos de solos, nutrição vegetal e microbiologia na implantação de matas ciliares. Companhia Energética de Minas Gerais, Universidade Federal de Lavras. 54 p.

Smith, S.E. and Read, D.J. (1997). Mycorrhizal Symbiosis, Academic Pess, London, 605 p.

Sylvia, D.M. and Williams, S.E. (1992). Vesicular-arbuscular mycorrhizae and environmental stress. In: Mycorrhizae in Sustainable Agriculture. Bethlenfalvay, G.J. and Linderman, R.G. Eds., *ASA Special Publication* 54:101–124.

Thrall, P.H., Millsom, D.A., Jeavons, A.C., Waayers, M., Harvey, G.R., Bagnall, D.J. and Brockwell, J. (2005). Seed inoculation with effective root-nodule bacteria enhances revegetation success, *Journal of Applied Ecology* 42:740–751.

Toro, M., Azcón, R. and Barea, J.M. (1997). Improvement of arbuscular mycorrhiza development by inoculation of soil with phosphate-solubilizing rhizobacteria to improve rock phosphate bioavailability (32P) and nutrient cycling, *Applied and Environmental Microbiology* 63:4408–4412.

Vessey, J.K. (2003). Plant growth promoting rhizobacteria as biofertilizers. *Plant and Soil* 255:571–586.

Vitousek, P.M. and Sanford Jr, R.L. (1986). Nutrient cycling in moist tropical forest, *Ann. Rev. Ecol. Syst.* 17:137–167.

Vitousek, P.M. (1984). Litterfall, nutrient cycling, and nutrient limitation in tropical forests, *Ecology* 65:285–298.

Xavier, L.J.C. and Germida, J.J. (2002) Response of lentil under controlled conditions to co-inoculation with arbuscular mycorrhizal fungi and rhizobia varying in efficacy, *Soil Biology and Biochemistry* 34:181–188.

Xavier, L.J.C. and Germida, J.J. (2003) Selective interactions between arbuscular mycorrhizal fungi and *Rhizobium leguminosarum* bv. viceae enhance pea yield and nutrition. *Biology and Fertility of Soils* 37:261–267.

Zangaro, W., Bononi, V.L.R. and Trufen, S.B. (2000). Mycorrhizal dependency, inoculum potential and habitat preference of native woody species in South Brazil, *Journal of Tropical Ecology* 16: 603–622.

Zangaro, W., Nishidate, F.R., Camargo, F.R.S., Romagnoli, G.G. and Vandresen, J. (2005). Relationships among arbuscular mycorrhizas, root morphology and seedling growth of tropical native woody species in southern Brazil, *Journal of Tropical Ecology* 21:529–540.

Zangaro, W., Nisizaki, S.M.A., Domingos, J.C.B. and Nakano, E.M. (2003). Mycorrhizal response and successional status in 80 woody species from South Brazil, *Journal of Tropical Ecology* 19: 315–324.

13

Ectomycorrhizal Fungi and its Application in Forest Nursery

Miguel Marin

Department of Physiology, Genetics and Microbiology, University of Alicante
Carretera San Vicente del Raspeig s/n, E-03690 San Vicente del Raspeig (Alicante), Spain
E-mail: mmarinz@ua.es

INTRODUCTION

Fungi are eukaryotic organisms that form the major part of microbial biomass in boreal, temperate and Mediterranean forest soils. Fungi are divided into five major groups: Deuteromycetes (Fungi imperfecti), Oomycetes, Zygomycetes, Ascomycetes and Basidiomycetes. The function, reproduction and ecological function of the species belonging to different taxonomic groups, outside a limited number of 'model' species, are poorly understood. In 1995, approx. 72,000 fungal species were known and it has been estimated that there may be about 1.6 million fungal species (Carlile *et al.*, 2001). However, functional groups such as saprotrophs (wood- and litter-decomposing fungi), parasitic, pathogenic and symbiotic (e.g., mycorrhizal) fungi are all known to exist in the soil environment (Trappe and Luoma, 1992).

Wood and litter-decomposing fungi are saprotrophic and thus obtain their energy from dead organisms. Wood-decomposing fungi are specialized in using lignin or cellulose compounds as their energy source. Wood-decomposers are generally divided into white rot (degrade lignin), brown rot (degrade cellulose) and soft rot fungi (degrade cellulose causing wood to lose its mechanical strength) (Dix and Webster, 1995). These organisms are mainly responsible for initiating the degradation process of woody material, which eventually ends in its complete mineralization.

Litter-decomposing fungi have similar abilities as wood-rotting fungi, but they live in an ecologically more heterogenic environment, the litter layer (Dix and Webster, 1995). As such, these fungi as a group are capable of degrading a wider range of compounds and they need to be more competitive in the soil environment that is also populated by many other organisms.

In contrast to saprotrophic fungi, mycorrhizal fungi generally tend to obtain a major part of their energy resources directly from the host plant (Harley and

Smith, 1983). Mycorrhizal fungi form a symbiotic relationship with many plant species. The earliest evidences of mycorrhizal associations are from the early Devonian era, approx. 400 million years ago, when the first land plants were developing. It has been estimated that this ancient form of symbiosis can be found in approximate 80% of land plants (Brundrett, 2002). Several different types of mycorrhizal associations exist (Harley and Smith, 1983). The main division is based on the type of cellular contact: in endomycorrhizas the fungal structures are formed within plant cells (intracellular), while in ectomycorrhizas, the fungi are restricted to the intercellular space in cortical and epidermal layers of fine roots and do not puncture the cell membrane. There are several different kinds of endomycorrhizas: VA (vesicular-arbuscular, often called only arbuscular), ectendo-, arbutoid, monotropoid, ericoid and orchid mycorrhizas (Harley and Smith, 1983). Vesicular-arbuscular mycorrhizae are formed by several hundred species in the family Endogonaceae, subdivision Zygomycotina; and, ectomycorrhizae are formed by several thousand fungal species in the subdivisions Basiomycotina and Ascomycotina.

The majority of plants are VA mycorrhizal (they are very important in agriculture like biofertilizers, see Marin, 2005), but a large number of trees and shrubs are ectomycorrhizal, including species in the Pinaceae, Fagaceae, Betulaceae, Salicaceae, and Tiliaceae as well as many members of the Rosaceae, Leguminoseae, Ericaceae, Juglandaceae, and other families (Brundrett, 2002). Some genera of forest tree species, e.g., poplars, eucalyptus and alders, form both types of mycorrhizae. The specificity for their host plants varies, but in the boreal forest zone tree species (Pinaceae) generally form ectomycorrhizal (sometimes ectendomycorrhizal) symbioses while shrubs are commonly ericoid mycorrhizal, and herbs and grasses are VA mycorrhizal (Harley and Smith, 1983; Smith and Read, 1997). In the mixed conifer/hardwood forests that characterize Mediterranean climates, both conifers and hardwoods are mostly ectomycorrhizal, but in the temperate mixed-species forests with summer rains the two mycorrhizal groups tend to be evenly represented.

ECTOMYCORRHIZAS

Symbiosis in which both the partners get benefit from the association is called mutualism (Begon *et al.*, 1996). Ectomycorrhiza is the composite organ formed by the host plant roots and ectomycorrhizal fungi (Harley and Smith, 1983), and these can influence the host plant in numerous ways. Generally, in mycorrhizal symbiosis, the fungal symbiont (mycobiont) gains energy from the host plant in the form of photosynthetically fixed carbon compounds while the plant receives nutritional benefits, especially phosphorus and nitrogen uptake (Harley and Smith, 1983). Mycorrhizae confer other benefits to their plant hosts too: they can take up water and increase drought resistance of plant hosts (Garbaye, 2000); some mycorrhizal fungi can also detoxify certain soil toxins or enable seedlings to resist high soil temperatures or extreme acidity (Burton *et al.*, 2002; Godbold *et al.*, 2003; Pendall *et al.*, 2004); and, it is practical importance to nursery management, some mycorrhizal fungi can protect roots against certain pathogens protection against competitive plant pathogens (Marx, 1969; Smith and Read, 1997; Sen, 2001).

ECTOMYCORRHIZAL FUNGI

Decades ago, it was demonstrated that ectomycorrhizal fungi (EMF) improve plant nutrition by enhancing nitrogen, phosphorus and potassium uptake (Harley and Smith, 1983; Read *et al.*, 1989). Although the possibility of ectomycorrhizal involvement in nitrogen uptake was proposed by Frank (1885), until the last 20 years further investigations focussed on the capture of phosphorus (Smith and Read, 1997). Low availability of nitrogen is a characteristic feature of many ecosystems dominated by ectomycorrhizal symbioses, such as boreal forests (Tamm, 1991). Inorganic nitrogen can be taken up directly by plants, but ectomycorrhizas have been shown to improve N uptake (Rygiewicz *et al.*, 1984; Martin *et al.*, 2001). Since saprotrophic fungi and bacteria readily utilise inorganic nitrogen, these forms are competent in soil. EMF have a competitive advantage in harvesting these nutrients compared to saprotrophic organisms, because they have a direct access to plant photosynthates and do not therefore need to compete as much for energy.

Nutrients, particularly N and P, are taken up by ectomycorrhizal mycelium through active absorption and specific cell membrane transporters following degradation by exoenzymes (e.g., phosphatases) (Smith and Read, 1997). Nutrient exchange between fungus and plant occurs using cell-to-cell contact in ectomycorrhizas, especially in the Hartig net (it is formed by the fungal growth between plant cortical cells and it provides a large surface of contact between the two symbionts) where fungal mycelium occupies the intercellular space between cortical cells of the colonized fine root (Harley and Smith, 1983).

Fluctuations in the availability of water commonly occur in natural environments and drought periods can cause considerable stress symptoms for trees (Garbaye, 2000). Non-mycorrhizal root tips, which are covered with 'slimy' mucilage, offer only limited buffering against drought. In contrast, mycorrhizal root tips, which are typically unsheathed by a hyphal structure called the mantle or fungal sheath (it defines the layer of fungal tissue that develops on the surface of the mycorrhizal root and it varies greatly between fungal species both in forms and thickness) and which often are connected to extensive network of extraradical or extramatrical mycelium (it extends from the mantle into the surrounding soil forming a connection between the mantle and soil), with or without rhizomorphs, (they are linear aggregates of hyphae which specialize in the uptake and transport of water and nutrients) are much less sensitive to dry soil conditions. Duddridge *et al.* (1980) showed for the first time that fungal mycelia could take up and transport water from as far as 8 cm from the root. Such distances, developing from each individual ectomycorrhizal root tip, increase the plant's access to soil water resources significantly. There are species-specific differences in mycorrhizal fungal influence on a plant's tolerance to water stress and therefore maintaining fungal biodiversity indirectly guarantees better water efficiency at the ecosystem level (Garbaye, 2000).

Ectomycorrhizas also protect host plant roots against root pathogens and root herbivorous soil microfauna. Mycorrhizal fungi reduce the likelihood of pathogens invading the root through competition (physical barrier formation) support of associated antagonistic microorganisms unfavourable for pathogens or secretion of toxic compounds (Harley and Smith, 1983). Improved general fitness of mycorrhizal plants also decreases the risk of pathogenic attack.

Increased shoot biomass of mycorrhizal versus non-mycorrhizal seedlings has been demonstrated over the last 100 years, starting at the time when the symbiosis and its major functions were discovered (Harley and Smith, 1983). Specific ectomycorrhizal species and strains can give competitive advantages (Le Tacon *et al.*, 1992) and in certain cases improved shoot growth can positively influence seedling survival and later competitive success. This phenomenon is the basis of inoculating forest tree seedlings to be used in forest regeneration (Landis *et al.*, 1990), since it is often critical that the seedlings are not suppressed in the first years by the ground vegetation in the reforestation site.

MYCORRHIZOSPHERE

The growth pattern of fungal hyphae is characteristically apical with subapical branching (Carlile *et al.*, 2001). The branching habit of individual hyphae in the mycelium results locally in efficient harvesting of nutrients but mycelial strands or rhizomorphs are needed if new sources of nutrients are to be reached. In unfavourable conditions, sclerotia can be produced by some species (Carlile *et al.*, 2001).

Since the hyphal network colonises large volumes of soil, and since plant photosynthates flow into this hyphal network, it is obvious that the presence of mycorrhizal extramatrical mycelium in soil alters its environmental conditions compared to non-colonised, 'bulk soil'. The term mycorrhizosphere (Linderman, 1988) has been used to describe the soil volume influenced by the mycorrhizal mycelium. Thus, tree root associated mycorrhizal fungi increase greatly the volumes of soil that can be effectively exploited by the host tree (Smith and Read, 1997). This critical feature forms the basis for the observed positive influences gained through mycorrhizal symbiotic association between fungi and plants. Access to remote and recalcitrant nutrient sources and water with the help of fungi is essential for the host trees.

Fluctuations in soil fungal biomass occur seasonally (Berg *et al.*, 1998). The mycelial growth pattern in soil was proposed (Shearer, 1995) and later shown (Lindahl, 2001) to be dynamic, involving competition between different types of fungi and different fungal species. Environmental and biological factors (e.g., host plant diversity and competition) influence fungal community dynamics, but reciprocal fungal competitive patterns can also affect above ground host plant diversity and productivity (van der Heijden *et al.*, 1998).

Many soil animals graze mycorrhizal mycelium, which can have either positive (Setälä, 1995) or neutral/negative impact on tree growth performance (Ek *et al.*, 1994). Mycelium also supports bacteria and location-specific bacteria have been identified in mycorrhizospheres (Nurmiaho-Lassila *et al.*, 1997; Timonen *et al.*, 1998; Bertaux *et al.*, 2003). Further, differences in bacterial communities extracted from bulk soil, rhizosphere and mycorrhizosphere soil compartments have been detected in different podzolic forest and oil-contaminated soils (Timonen *et al.*, 1998; Heinonsalo *et al.*, 2000). Bacteria have been shown to specifically support mycorrhiza formation (Garbaye and Bowen, 1987; Meyer and Linderman, 1986) and a theory for the mode of action of mycorrhiza helper bacteria (MHB) has been presented by Garbaye (1994). Nitrogen-fixing genes have recently been detected in endosymbiont bacterial strain of mycorrhizal fungus (Minerdi *et al.*, 2001) suggesting mycorrhizosphere-associated nitrogen fixation in soil. In conclusion, it is evident that fungal mycorrhizosphere is a particular, extensive habitat supporting a variety of bacterial types and having specific functions of mostly unknown ecological importance.

ECTOMYCORRHIZAL FUNGAL DISPERSAL

EMF can disperse and form new mycorrhizas either via living hyphal connections, special resting structures like sclerotia (they are resting bodies, which can persist for years, formed by the fungal mycelium (Miller *et al.*, 1994)), or spores (Brundrett, 1991). Living hyphae, which form part of the plant-supported hyphal network, is probably the most common source of inoculum in forest conditions, where mycorrhizal roots are already available. Dahlberg and Stenlid (1995) also suggested that in the presence of an existing mycorrhizal community in an established forest, mycelial colonization probably dominant over spore colonization. Under conditions where the living hyphae are not available, e.g., after clear-cut logging, the major sources of inoculum for new tree seedlings may be sclerotia and spores. If the regeneration happens soon after logging (e.g., within a few years), the remaining stumps and dying roots have been shown to support living hyphae of EMF (Hagerman *et al.*, 1999) and new seedlings can achieve mycorrhizas that way.

Spores and sclerotia can survive in soil for a very long time and can be activated by suitable conditions e.g., increased soil moisture, heat shock after forest fire or stimulatory compounds in root exudates. Spore activation is followed by spore germination (Carlile *et al.*, 2001). The emanating primary monokaryotic mycelium of a basidiospore is already capable of forming mycorrhizae (Kope and Fortin, 1990), but normally mycorrhiza formation is generated by secondary heterokaryotic mycelium that arise from the mating of two sexually compatible primary mycelium. Little is known about forms of mycelium responsible of mycorrhiza formation by Ascomycetes, which are numerous in ectomycorrhizal communities (Tedersoo *et al.*, 2003). Some fungal species are known to spread successfully via spores while others form mycorrhizas preferentially via living connections or sclerotia (Brundrett, 1991; Dahlberg, 1997).

The succession of EMF is mainly connected to the dispersal ability of different fungi. The hypothesis of early- and late-stage fungi (Mason *et al.*, 1983), originating from a study of Mason *et al.* (1982) carried out in former agricultural soil, suggested that some species colonise the host tree roots earlier, while some others can form the symbiosis only in later growth stage of the trees. This hypothesis is no longer widely supported in the scientific community, but the principle of certain successive traits of EMF is reasonable. Nara *et al.* (2003a,b) studied primary succession of EMF in pristine soil developed by volcanic activities having no earlier mycorrhizal history. They found a clear trend of succession that was detected both in mycorrhizas in roots and in the sporocarp production. The conditions in a secondary successional site, where ectomycorrhizal inoculum from previous forest exists, are very different from these primary sites and thus the mechanisms of succession remain poorly defined. However, understanding the functional differences between different 'stage' fungi and the reasons for changes in ectomycorrhizal community structure due to manmade disturbances is an important and a demanding future task.

ECTOMYCORRHIZAL BIODIVERSITY

More than 1000 fungi are estimated to form ectomycorrhizas in the Scandinavian forests (Knudsen and Hansen, 1991), a number that greatly exceeds the number of ectomycorrhizal hosts in the region. Since many ectomycorrhizal fungal species do not form epigeous (above-ground) sporocarps but inconspicuous or hypogeous (below-ground) ones, it is probable that the actual species richness is much higher. Thus, the possible species combinations and intra- or inter-specific interactions in these highly diverse communities are enormous and have been a subject of increasing interest in the last 20 years (Dahlberg, 2001). The community studies of ectomycorrhizas normally aim either at simply describing the species present under specific conditions or at observing the effects of natural or anthropogenic disturbances on the community structure.

Many ectomycorrhizal community studies have focussed on mycorrhizas in the humus horizon. Although the fine roots of some tree species, e.g., Norway spruce, are more superficially distributed than others (Mikola *et al.*, 1966), the importance of mycorrhizal fungal communities in the mineral horizons should not to be neglected (Taylor, 2002). Only in recent years some community-level studies in world forests using modern molecular methodology have highlighted the importance of these deeper soil horizons (Dickie *et al.*, 2002; Landeweert *et al.*, 2003).

Ectomycorrhizal fungal identification has traditionally been based on observations on the morphological characters of mycorrhizal root tip, morphotyping (Agerer, 1987–2002). If morphotyping is carried out with care, good species identification is obtained (Sakakibara *et al.*, 2002). However, detailed morphotyping involving the use of microscopy requires long experience and training and in many cases in published literature, only gross morphotyping using easily detectable characters has been used. In those cases, morphotyping does not separate similar looking species and even the detailed analysis cannot uncover strain-level differences. Due to those deficiencies, the use of molecular methods has been increasing. The use of the polymerase chain reaction (PCR) has dramatically increased our ability to document the diversity of EMF primarily because fungi can be identified directly from vegetative structures, and genetic data is obtained rapidly enough to allow relatively large numbers of samples to be processed. Analyzing the ITS-region of the ribosomal RNA gene, which has multiple copies in chromosomal DNA, using PCR reactions and specific primers (see for methodological details in reviews by Dahlberg (2001) and Horton and Bruns (2001)), it has been possible to identify fungal species and strains from even single ectomycorrhizal root tip samples (Gardes *et al.*, 1991; Dahlberg *et al.*, 1997; Eberhardt *et al.*, 2000; Landeweert *et al.*, 2003). In most cases, a combined approach employs morphological sorting of EM root tips with molecular identification using restriction fragment length polymorphism (RFLP) analysis of the ITS region of the nuclear rRNA gene repeat (Sakakibara *et al.*, 2002). The range of molecular biological methods available is wide and new methods are being actively developed (Iotti and Zambonelli, 2006; Mello *et al.*, 2006). In the future, EMF molecular data will be computerized in Internet and by means of consultation it will know at any moment whereupon fungus we worked. At the moment, there are already some works in this sense (Koljalg *et al.*, 2005; Nilsson *et al.*, 2005).

ECTOMYCORRHIZAL INOCULATION

Controlled mycorrhization (Le Tacon *et al.*, 1997) has been used in some countries with the aims of introducing growth-promoting ectomycorrhizal fungi into nurseries, and modifying seedling production procedures, respectively, in order to achieve improved growth responses and rapid post-transplantation adaptation in the reforestation site (Kropp and Langlois, 1990). In North America, the research and practical applications have been established for decades (Landis *et al.*, 1990), whilst in Europe the procedure has been in commercial use at least in France since early 1990s (Le Tacon *et al.*, 1997).

The potential advantages of controlled mycorrhization in the nursery are not only the positive growth response of the seedlings, but also:

- Alleviation of nutrient stress. Under deficiency conditions, EMF can increase nutrient uptake (Boyd *et al.*, 1986; Jones *et al.*, 1991). Under toxic conditions as with heavy metals, mycorrhizal fungi pull metals out of the soil and sequester them (Jones and Hutchinson, 1986).
- Enhancement of rooting and reduction of transplant shock. EMF stimulate root production and dramatically increase the volume of soil the plant can explore. This is especially important on disturbed sites, where nutrients and water availability might otherwise inhibit plant growth (Teste *et al.*, 2004).

- Alleviation of drought stress. The increased root volume is able to take up more water. The mycorrhizal fungi also enhance the host's osmotic adjusting capabilities, allowing some plants to continue to extract water from soils as they become drier (Garbaye, 2000).
- Stabilization and aggregation of soil. Mycorrhizal fungi encircle soil particles and glue larger soil particles together into aggregates. This increases air and water movement though the soil while reducing erodibility (Borchers and Perry, 1992).
- Reduction in root disease. Mycorrhizae directly and/or indirectly: antagonize disease organisms, increase the number of biocontrol agents around the roots, occupy potential infection sites on the root, and increase host plant vigour to the extent that it can survive disease (Marx, 1969; Perrin and Garbaye, 1983).
- Enhancement of nutrient transfer between plants. This is especially important when nitrogen fixing and non-nitrogen fixing species are planted together (Simard *et al.*, 1997).
- Enhancement of beneficial interactions with other microbes (Garbaye, 1994). Mycorrhizal fungi will increase the nitrogen made available to the plant by both symbiotic and free-living bacteria. It will also increase phosphate uptake to the plant by phosphate solubilizing bacteria. And it will support biocontrol agents that are antagonistic to pathogenic organisms.

Seedling performance in the field after ectomycorrhizal inoculation has been variable in field studies. In some cases, six years after outplanting an increase in tree volume of approx. 90% has been observed, while in other conditions inoculated seedlings performed worse in the field compared to control ones (Le Tacon *et al.*, 1992). Variable environmental growth conditions and differences in local mycoflora obviously explain part of the variation.

Other cause can be the concept of 'early stage' and 'late-stage', explained before, applied to EMF inoculation in nursery. As it has been commented, these were adopted to differentiate between fungi found on young roots from EMF that colonized on older root systems. The persistence of early succession fungi on newly formed root systems resulted in distinct spatial patterning of EMF (Dighton and Mason, 1985; Dighton *et al.*, 1986; Mason *et al.*, 1987). The colonization and establishment of early- and late-stage EMF have been linked to physiological interactions between their hosts and environment and their reproductive strategies. Early stage EMF appear to demand small amounts of carbon from their hosts and require low concentrations of nitrogen and phosphorus, whereas late-stage fungi require greater amounts of carbon, nitrogen and phosphorus (Deacon and Fleming, 1992; Colpaert *et al.*, 1996). Colonization by early or late-stage EMF appears to be dependent on their reproductive mode (e.g., spores or mycelia), age of the host and substrate conditions. Early stage EMF establish readily by spores or mycelia in laboratory inoculation experiments, whereas late-stage EMF establish primarily from dikaryotic mycelia (Deacon and Fleming, 1992).

Consequently, for early-stage EMF, a high number of genotypes would be expected as a result of reproduction primarily by spores, whereas the formation of larger clones would be predicted if reproduction occurred primarily by mycelial expansion in late-stage EMF (Dahlberg and Stenlid, 1995; Dahlberg, 1997). Most studies of early stage EMF confirm these expectations (de la Bastide *et al.*, 1994; Henrion *et al.*, 1994; Gryta *et al.*, 1997; Guidot *et al.*, 1999).

A limited number of the formers known as early-stages species have been successfully grown *in vitro*. The candidates should be searched out in the following genera: *Hebeloma, Laccaria, Paxillus, Tuber, Scleroderma, Pisolithus, Suillus, Thelephora, Cenococcum, Rhizopogon,* etc. The nursery decided to develop the large-scale production of mycorrhizal seedlings with selected fungal strain to improve survival and early growth performances in the field. Some mycorrhizal associations are formed naturally during the first growing season at any nursery and are typical of growth conditions. The fungal species involved depends on environmental conditions, cultural practices and host species cultivated. Most of them are adapted to high fertility conditions and intensive cultivation and disappear quickly once saplings are outplanted in the fields. Such associations are considered as useless either at the nursery or in the fields.

Reports of the longevity and persistence of a genotype at a particular locale are not consistent. In independent studies of the early stage fungus, *Hebeloma cylindrosporum*, genets remained at a particular locale after many years (Henrion *et al.*, 1994), in contrast to genets that were not found in successive samplings of the same host tree (Gryta *et al.*, 1997). In early stage EMF, *Laccaria bicolor* and *L. laccata*, only a few of the genets mapped persisted over a sampling period of 3 years (Selosse *et al.*, 2000). Ectomycorrhizal root tips are of short duration, with a turnover time of 3 months to 3 years (Fogel and Hunt, 1983; Harley and Smith, 1983) therefore persistence over longer periods requires repeated colonization events at the root mycelium/spore interface or by persistent mycelia.

Published reports of the size and number of genets for late-stage fungi are also rare. *Russula* spp. are good candidates for examination of genetic structure of late-stage EMF since they often comprise the majority of sporocarps found in mature stands of temperate forests (Richardson, 1970; Harvey *et al.*, 1976; Watling, 1984; Bills *et al.*, 1986; Mason *et al.*, 1987; Kernaghan *et al.*, 1997), are generally absent or not abundant in early regeneration of stands (Keizer and Arnolds, 1994; Bradbury, 1997; Bradbury *et al.*, 1998; Kranabetter and Wylie, 1998) and rarely form mycorrhizae from spore inoculum in laboratory experiments (Ali and Jackson, 1988).

More research is required to understand the reasons for the variability, and to improve the inoculation success in the field. By these reasons, it is important that the planting of seedlings that have been inoculated with specific mycorrhizal strains, typically of exotic origin, into nature should not cause any negative impacts on the local, indigenous microflora. Ecological knowledge on the interactions of existing community and introduced strains should therefore be improved to prevent the risk of adverse effects. Understanding the survival and competitive abilities of selected strains under field conditions is of high priority in the risk management of controlled mycorrhization. The potential problems are in many ways analogous to those in studies on field release of genetically modified micro-organisms.

WHY IS IT NEEDED IN NURSERY?

The development of controlled mycorrhization in nursery plants arises before situations in which the survival of the plant is severely difficult to short or half term and/or in the possible economic advantage of the capacity of the mycorrhizal association when producing vegetal or fungal material. It cannot be forgotten that in a symbiosis the union of two species benefited mutually, and by this reason, in the artificial mycorrhization both the plant and the fungus must be observed. Here it is where the greater failures and deceptions found have taken place in the last years, as much in the field of agriculture like in the forest one. It is not sufficient to make a mycorrhization and find benefits in nursery phase, but when the plant is transplant to the field, the mycorrhization advantages must remain.

A plant cultivated in nursery in substratum artificial and provided of water and nutrients, probably will not be mycorrhized, or it will do with some EMF species of difficult survival in the field, which can carry to it inconvenient for its good establishment. Similar can happen when artificial mycorrhizations are made in nursery with no adapted EMF, because in this case the local EMF species would move to the inoculated EMF which they were implanted in the nursery. At the contrary, positive effects in survival and early growth can be expected at outplanting sites if the fungal strains selected for adaptation to the field conditions. Consequently, controlled mycorrhizal seedlings represent undoubtedly a great progress for reforestation

A program of inoculation in nursery must look for effective fungi and for it must know clearly the objectives. These can be all or a combination of the following ones:

- reduction of the percentage of dead plants in nursery
- increase of the caliber and size of the stem
- adapted development of the radical system
- protection against pathogens
- fast colonization
- increase of the survival in field

In addition, it would be necessary to add to these objectives the mycorrhization of forest plants with edible EMF, that once outplanting and after several years of growth, fruit bodies will be harvested with an additional economic interest for the new forest.

THE INOCULUM

The inoculum suitable for any type of mycorrhization always would be that formed by indigenous the EMF. But in the situations where the inoculum of native EMF is low or inefficient (for example, left agricultural grounds), or the actual culture in container, providing the inoculum of the appropriate EMF to the plant will be necessary for which the mycorrhization takes place and, therefore, the benefit to the plant. In a program of controlled mycorrhization, the inoculum would have the capacity to be produced on long scale, and to allow controlling its parameters of production to a low volume and cost. This inoculum must be applied in the nursery phase so that an acceptable amount of the EMF is distributed between each one of the plants before their transplant to field. The inoculation usually is made, for its greater effectiveness, when seedlings are emerging. It will not be made in the final phase of nursery or at the moment of the plantation in field due to the required time and the necessity of a greater amount of inoculum, not benefiting seedlings weeks after of the ectomycorrhiza formation (Marx *et al.*, 1991). The first course in any program of controlled inoculation is to obtain the fungal stock that is infective, able to be associated and grow effectively in the plant roots, and effective, able to benefit the development and stress tolerance from the host plant. The ectomycorrhiza formation of the fungus must be at least superior to 50% of the active ends by the root to assure the positive effects of the inoculation (Marx and Cordell, 1987). The EMF isolated varies much in their properties, reason why it is always important to select the stocks that are developed satisfactorily. The inocula of these species can be constituted by fungal mycelia (inoculum vegetative), basidiospores, or even of less specific inocula, like soils from zones with fungal mycelium or fruit bodies, in this case with the consequent loss of homogeneity of the inoculum.

Types of Inoculum

The main types of inoculum are: passive inoculation, soil and humus, mycorrhizal seedlings and roots, basidiospores, and pure mycelium. The first three were the used methods of inoculation in the first mycorrhizations in forest nurseries, and although they are even used, little by little are being left in order to adapt to the modern production systems of forest plant in pot. At the moment, they are applied to the basidospores and pure inoculum inoculation that allows obtaining a controlled mycorrhization.

Passive Inoculation

Nurseries established in forest zones or surrounded by ectomycorrhizal hosts usually produce seedlings with abundant and diverse ectomycorrhizae (Trappe, 1977). Spontaneous inoculation of containerised seedlings growing on mycorrhiza-free potting mix probably occurs mostly by airborne spores. In general, all seedlings will develop mycorrhizae early in the first growing season. Even if pots are fumigated, regular and prolific fruiting of sporocarps in neighbouring forests provides abundant spore inoculum, as does fruiting of fruit bodies in established nursery beds. Under these conditions, whenever spores enter the substrate, ectomycorrhizae can begin developing with the first production of feeder roots 6 to 10 weeks after germination. If fruiting populations of EMF have built up in greenhouses or cold frames, spore inoculation may be pervasive.

Natural sources of inoculum may prove to be reliable for some nurseries. Each nursery must evaluate the risks in light of its own situation and experience. To do so requires careful observation of mycorrhizae and careful record keeping on a continuing basis. Because mycorrhizal deficiency has frequently been misdiagnosed, few nurseries have adequate data to assess the reliability of natural sources of inoculum. Even if natural inoculation is acceptably reliable for a nursery, it preludes deliberate selection of inoculum for best seedling performance either in the nursery or after outplanting (El Karkouri *et al.*, 2005).

Soil and Humus

Incorporation of propagule-bearing soil and humus into planting holes was practiced in the mid-19th century truffle culture and has since been a successful method of mycorrhizal inoculation of nurseries around the world. The practices followed to date suffer from the same deficiency as spontaneous inoculation little or no success in the selection of the most desirable species of EMF (Selosse *et al.*, 1999). Once a desirable fungus is established, however, nursery beds can be managed to provide a continuing source of inoculum. When a portion of the soil is removed for use as inoculum for production beds, new inoculum grown in peat can be added and the soil planted again with the host tree needed to maintain the fungus. The major drawback is the logistics of collecting and transporting the large quantities of soil needed. Unfortunately, weed seeds, rhizomes, and potential pathogens may be introduced along with the beneficial fungi.

Mycorrhizal Seedlings and Roots

Transplantation of mycorrhizal seedlings into nursery beds has proven an effective means of inoculation. Theoretically, the mycorrhization takes place by proximity between the roots, allowing the introduction of the EMF wished by the roots of the mycorrhized plant to the near plant roots.

Freshly excised mycorrhizal root systems is the simplest method of inoculation in nursery when the wished species of EMF is not had, and this roots may come from forest directly or the other nursery. This system is used from the beginning of the mycorrhization practices (Mikola, 1970). The voluminous and expensive transport of the mycorrhized roots, and the intensive work in nursery are some of the problems of this kind of inoculum. Moreover, this method evidently does not allow controlling the EMF that it inoculates to the plant, and pathogenic agents may be introduced (nematodes, fungi, etc.) or not wished EMF.

Basidiospores

In the majority of species, the spores fall naturally from fruit bodies under the influence of gravity and so deposits of spores can usually be obtained at room temperature overnight on a clean surface placed beneath portions of fruit bodies. A bowl or beaker is inverted over the fruit bodies to protect the falling spores from droughts. Ideally some spores should be collected on a glass microscope slide because the dimensions and other morphological features of mature spores are useful in identification. To minimize the chances of isolating mycoparasitic fungi, spore morphology should be checked against published descriptions. Glass slide sandwiches, cellophane sheets and aluminium foil are all useful for keeping spore deposits clean when in transit and before they are used to isolate germling mycelia. When consulting some taxonomic keys, however, it must be borne in mind that spore colour is described as seen against a background of white paper (Moser, 1983), which may differ slightly but significantly from that recorded for spores on glass.

The inoculum is formed by a spore's suspension in distilled water sterile. The spore concentration is adjustable voluntarily diluting more or less the inoculum. The time of maintenance of the suspension varies according to the author and the species, but they only offer viability guarantee during few months (Torres and Honrubia, 1994). Marx *et al.* (1995) review different methods of producing a spore slurry inoculum. Dry spores can also be used, although utilizing a carrier such as hydro-mulch to apply the spore inoculum or pelletizing spores is suggested to increase its effectiveness (Marx and Bell, 1985).

An advantage of sporal inoculation is that since the fruiting body is the source of spores it is relatively easy to determine the species of fungus. However, basiodiospore germination has long been a problem in experimental mycology, and in some EMF this problem has been observed (*Amanita*, *Russula*, *Tuber*) (Fries, 1978). This has been interpreted in various ways, e.g., as an expression of dormancy, as a need for a germination-inducing factor, or as the effect of an endo- or exogenous inhibitor. Moreover, production of spore inoculum is often limited by the number of viable spores that are produced on the gills of most EMF. Another drawback, when spore inoculum is utilized is that mycorrhization of plants may not be feasible in pot studies because of slow spore germination and slow development of the mycelium complicated by on-going competition with other soil organisms, and conditions unfavourable for spore germination. In addition, by own nature, sporal inocula always have a high genetic variability (Kuek *et al.*, 1992).

Rincón *et al.* (2005) observed that spore inoculum of *Rhipogon*, *Pisolithus* and *Scleroderma* in *P. pinea* seedlings, produced in container, was effective to obtain mycorrhizal plant. Also, the high quantity of spores that can be obtained from one basidioma and the ease of application of this type of inoculum makes of it a method particularly well suited for large-scale inoculation of containerized seedlings in commercial nurseries. At last, a good review about sporal inoculation can be read in Brundrett *et al.* (2005) on the inoculation of different EMF on *Eucalyptus* seedlings.

Pure Mycelial Culture

Mycelium inoculation of pure cultures or vegetative inoculation is possibly most effective and safe of whichever systems exist, but at the same time it is the most expensive and complicated handling. Its application demands knowledge and facilities that is why its generalization is difficult in many nurseries of the world. It is possible to obtain mycelium from several sources, principally from fruit bodies, but also, from mycorrhizal roots (Bending and Read, 1995) or *in vitro* germination of basidiospores, host plants release into the rhizosphere critical metabolites that are able to trigger basidiospore germination (Fries *et al.*, 1987). These two last two methods are used when we do not have pure mycelium of the wished EMF and we want to obtain it. One advantage of using mycelium gained from fruiting bodies lies in the ease of identifying the fungal species, compared to other sources of fungal inoculum. On the other hand, use of fruit bodies excludes fungal species that do not readily produce these sexual reproductive structures, or that do not fruit regularly, but might be present in natural settings; these species could be critical for plant survival and fitness in that system, especially in impacted soils.

Fruit bodies of EMF collected in the field can be used to start pure cultures agar. The most successful method involves aseptically removing a small portion of tissue from the fruit body with forceps and placing it on an agar slope in sterile tubes. The fruit body should be carefully broken apart to reveal tissue that has not been exposed to the air or contaminated by insects. For mushrooms, tissue from apex of the stem or in the cap itself is often the most successful for obtaining cultures. For puffballs and truffles, the immature internal fertile tissue (before it becomes powdery at maturity) is suitable. Generally, the younger the fruit body the greater is the chance of success, as the tissue of fruit bodies tends to degenerate and become more susceptible to bacterial and other contaminations with age. Tissue from fruit bodies is used to culture the fungus, however, not all species grow in captivity. Some fungal genera are particularly difficult to isolate into pure culture from fruit body tissue (e.g., *Russula*, *Tuber*) while others are more readily amenable to culturing (e.g., *Pisolithus*, *Scleroderma*, *Laccaria*).

The production of a pure mycelium of individual fungal species from fruit bodies requires tissue culturing the fungus and growing it under sterile conditions. The survival and growth of the fungal mycelium on Petri dishes enriched with nutrient solution is critical for providing a supply of living fungal tissue for the development of inoculum. Modified Melin Norkans (MMN) nutrient media developed by Marx (1969) has been shown to be an effective nutrient medium for most EMF. Earlier studies have examined the nutritional requirements of EMF in more detail (Trappe, 1977). Different ecological and physiological factors affect the growth of mycorrhizal fungi and also mycorrhiza synthesis and function in nature (Bowen, 1994). Physical characteristics and nutrients of soils, especially carbon and nitrogen sources, greatly influence the establishment and growth of ectomycorrhizal fungi in the field and also in controlled conditions in the laboratory (France and Reid, 1983; Lilleskov *et al.*, 2002). Different authors have analysed the influence of nitrogen (Finlay *et al.*, 1992; Baar *et al.*, 1997; Anderson *et al.*, 2001; Sangtiean and Schmidt, 2002; Sawyer *et al.*, 2003a; Hobbie *et al.*, 2004; Daza *et al.*, 2006), carbon (Table 13.1) (Palmer and Hacskaylo, 1970; Marin, 1999; Ceccaroli *et al.*, 2001; Manjon *et al.*, 2006), phosphorus (Sawyer *et al.*, 2003b), temperature (Hacskaylo *et al.*, 1965; Sánchez *et al.*, 2001), pH (Hung and Trappe, 1983; Sundari and Adholeya, 2003) and water stress (Machado and Bragança, 1994) on the growth of ectomycorrhizal fungi.

Table 13.1 Produced biomass (mg) of several EMF after growing in liquid medium BAF* with different carbohydrates during 30 days. (Significant biomass differences in each fungus are indicated with different letter, $P \leq 0.05$).

	Glucose	*Sucrose*	*Sucrose + Glucose$_i$*	*Glucose$_i$*	*Without carbohydrate*
Lactarius deliciosus	440.0^{a}	195.6^{b}	278.8^{b}	74.6^{c}	57.4^{c}
Lactarius sanguifluus	265.8^{a}	322.9^{a}	189.0^{ab}	58.4^{b}	46.3^{b}
Lactarius chrysorrheus	167.8^{ab}	161.4^{ab}	119.7^{b}	82.3^{c}	45.6^{c}
Lactarius hepaticus	278.6^{a}	320.7^{a}	272.2^{a}	73.8^{b}	76.6^{b}
Boletus edulis	316.9^{a}	235.6^{a}	248.4^{a}	75.0^{b}	65.9^{b}
Pisolithus arhizus	73.1^{ab}	149.3^{a}	83.3^{ab}	29.4^{b}	23.1^{b}

*Oort (1981)

The cultured mycelium, applied in a liquid (Cline *et al.*, 1982; Buschena *et al.*, 1992), or in sterilized soil (Riffle and Maronek, 1982; Marx *et al.*, 1991) is typically utilized in experiments with EMF because of its ability to effectively infect host plants in pot studies compared to other types of inoculum. In addition, individual species of EMF can be studied for their effects. Brundrett *et al.* (1996) includes a good review of the procedures and steps involved in producing vegetative inoculum.

In addition to production of an actively growing culture of a fungus, successful inoculation requires that the fungus remain viable long enough for the plant roots to come into contact with the mycelium for mycorrhization to occur (Trappe, 1977). EMF that are unable to survive and grow in Petri dish media or in soil substrate can be excluded from further evaluation using this method. A liquid type inoculum typically consists of mycelial plugs of the fungi initially placed in a flask of nutrient solution (without agar) maintained under sterile conditions for subsequent growth (Cline *et al.*, 1982). Blending can be considered for distributing the liquid inoculum to the host plant, but it may reduce viability of the inoculum by disrupting hyphal growth. Conversely, other studies suggest blending, or mixing of inoculum can increase the number of fungal propagules, providing more points of contact for hyphal growth, thereby increasing the overall effectiveness of the inoculum in forming mycorrhizae for some species of fungi (Molina and Trappe, 1982). Based on experimental findings it appears that different species of ectomycorrhizal fungi have varying levels of sensitivity to blending or mixing of hyphae (Boyle *et al.*, 1987). Liquid inoculum is also susceptible to contamination by other organisms.

Development of a soil type inoculum is another possibility for mycorrhization. A soil inoculum is a mixture of liquid nutrient media and soil, colonized by the mycelium of ectomycorrhizal fungi under sterile conditions. Ectomycorrhizal fungi colonize the sterile soil to a point where the inoculum can be applied in non-sterile conditions. The goal is for the desired fungus to survive long enough to form mycorrhizae with the host plant. The slow growth of ectomycorrhizal fungi in comparison to pathogenic or saprophytic fungi is often a limiting factor in the formation of mycorrhizae in greenhouse studies. Soil inoculum, specifically when vermiculite is included in the mix, provides mycorrhizal fungi an important niche where it has been suggested that they are protected from environmental stress and microbial competition (Marx *et al.*, 1991). Marx and Bryan (1975), described a method for producing soil inoculum that is continued to be the basis for producing viable inoculum today (Brunner, 2001).

The correct combination of media properties, such as pH, temperature, water holding capacity and niche habitat, are critical considerations for fungal survival and growth in soil inoculum. Mason (1980), using a soil inoculum under sterile conditions, found that the different nutritional needs of various ectomycorrhizal fungi meant that certain nutrient media combinations or concentrations might be precluded from use for certain species.

In vitro studies involving mycorrhizal fungi typically use a nutrient solution of culture medium under sterile conditions in tubes or flasks often with peat and vermiculite (Molina and Palmer, 1982), which encourages fungal growth. However, under non-sterile conditions, the use of culture media can encourage the colonization of contaminant fungi, which out-compete mycorrhizal fungi in high nutrient environments. Brunner (2001) reported high levels of mycorrhization using a diluted mixture of MMN applied to seedlings of Norway spruce (*Picea abies*), which reduces contamination potential.

Seedling Inoculation

The methods for use of mycorrhizal fungi with trees for pot studies, for outplanting in field studies, and for practical reclamation purposes has primarily focused on conifer trees, such as pine, and use of a commercial inoculum such as *Pisolithus arhizus* (Marx *et al.*, 1991). However, inoculation techniques adapted to conventional nursery tasks for other tree species and EMF have been reported and, in some cases, developed at a pre-commercial level (Marx *et al.*, 1982; Last *et al.*, 1984; Marx *et al.*, 1984; Hung and Trappe, 1987). Subsequent field-performance testing of inoculated versus non-inoculated plants has covered many plant-fungus combinations and field situations (Marx, 1980; Le Tacon *et al.*, 1992; Jackson *et al.*, 1995; Castellano, 1996; Pera *et al.*, 1999). Comprehensive data, however, are difficult to obtain since experimental conditions are very variable and most research is concentrated in small geographical regions and with few fungus-plant combinations (Castellano, 1996). Among the situations where mycorrhizal inoculation seems to be effective to promote plant survival and growth after outplanting are the environmentally stressful sites (Perry *et al.*, 1989; Marx *et al.*, 1991; Castellano, 1996).

The method of inoculation, since we have seen before, is also important for successful mycorrhization of plants in the nursery. One of them, i.e., "banding" has shown to be an effective method for inoculation of plants (Riffle and Maronek, 1982). Placing a layer, or "band" of soil inoculum just below developing root systems, is expected to bring the fungus in contact with the roots as they grow through the band (Marx and Kennedy, 1982). The goal is to infect roots with the mycorrhizal fungus and ultimately mycorrhizae are formed. Root system development is an important consideration in mycorrhizal studies because roots that develop too quickly, especially in the nursery, might grow beyond the source of inoculum in the container, thus reducing mycorrhization. This is a particular problem with faster growing species such as aspen (*Populus*), birch (*Betula*), and oak (*Quercus*).

Other method, a liquid inoculum, can be applied to the soil surface of plants, or injected into the soil before or after plants are added (Marx *et al.*, 1991; Marin, 1999). Gagnon *et al.* (1995) produced a liquid inoculum in fermenters that were gently agitated, and applied the inoculum by mixing it with soil just before seeding, with good results. Similarly, mycelial plugs and basidiospores can be applied as inoculum in various ways. Plugs or spores can be added directly to the surface of the soil, or injected into it at various levels, all of which will affect survival of the fungus, its speed of growth, and ultimately mycorrhization levels.

Finally, the alginate-entrapped mycelial inoculum method has been used widely. An alginate inoculum can be easily controlled and mechanized (see Table 13.2) and has been effectively used with some fungal species such as *Laccaria bicolor* and *Hebeloma* spp. (Le Tacon *et al.*, 1983; Mauperin *et al.*, 1987; Mortier *et al.*, 1988). Also, alginate inoculum has been used to successfully inoculate *Pseudotsuga menziesii* with entrapped mycelium of L. *bicolor* plus spores of *Rhizopogon subareolatus* in dual inoculations (Parladé *et al.*, 1999). However, attempts to inoculate plants with *Pisolithus arhizus* entrapped in alginate consistently failed (Pera *et al.*, 1994). Parladé *et al.* (2004) showed the possibility of using alginate method mycorrhizations with *Lactarius*, an EMF with great commercial interest. A disadvantage of this method is the lack of success with the alginate-entrapped mycelial inoculum since viability of the mycelium entrapped in the beads was proven to be unaffected (Table 13.3). Further studies, which consider physical and chemical characteristics of the alginate inoculum and their influence on the colonization ability of the fungus under greenhouse conditions should be undertaken.

Nursery Conditions

The type and quality of the nursery substrate also conditioned seedling performance. A factor that can affect mycorrhization in the nursery is the level of fertilization (Dixon *et al.*, 1985; Gagnon *et al.*, 1995). While it may be necessary to fertilize plants in pots in nursery, conditions to sustain growth, the high fertilization levels typically used in greenhouses can preclude mycorrhization. Fertilization levels that allow for mycorrhization need to be determined for each plant species. And although peat:vermiculite and peat:bark are standard substrates in container-grown seedling production, the variability of peat and, especially, of composted bark is a major drawback to obtain homogeneous quality.

Table 13.2 Polymerization of alginate with entrapped mycelium of *Lactarius deliciosus* and fungal growth.

% alginate	*g/l $CaCl_2 \times 2H_2O$*	*Polymerization*	*Growth*
0.7	10	+	–
0.5	10	+	–
0.7	5	+	–
0.5	5	+	–
0.7	2	+	+
0.5	2	+	+
0.7	1	+	+
0.5	1	+	+
0.7	0.5	–	+
0.5	0.5	–	+

Table 13.3 Growth percentage of alginate-entrapped mycelium of several EMF stored during different times and liquid substrates. ng: not growing.

Time	*Lactarius deliciosus*		*Lactarius sanguifluus*		*Amanita caesarea*		*Pisolithus arhizus*	
(months)	*Medium*	*Water*	*Medium*	*Water*	*Medium*	*Water*	*Medium*	*Water*
0	100	100	100	100	100	100	100	100
1	100	100	90	60	100	100	83.5	29
2	100	100	80	60	100	100	60	21.7
3	100	100	70	60	100	100	65	20
6	100	100	50	60	100	100	66.7	20
9	100	100	37.5	60	100	100	60	20
12	100	100	37.5	60	100	100	ng	ng
15	100	100	ng	ng	ng	ng	ng	ng

Some works have studied the effect of the pesticides on EMF pure mycelium (Marin, 1999; Laatikainen and Heinonen-Tanski, 2002; Diaz *et al.*, 2003) and in nursery (Trappe *et al.*, 1984; O'Neill and Mitchell, 2000; Kernaghan *et al.*, 2003). However, no matter how much is revealed by research about effects of pesticides on mycorrhizal fungi and mycorrhiza formation, it is important to realize that most pesticides used in nurseries are synthetic compounds that organisms have never encountered before. Moreover, a given chemical will not necessarily produce the same responses in all species of fungi or hosts or in all nurseries. Hence, first use of a chemical in a nursery should always be in trials of limited scope that include evaluation of its effects on mycorrhiza development.

At last, the mycorrhization of forest plants in a commercial nursery, it is advisable first developed a laboratory specially designed for mycorrhizal studies. Staffs specially trained is in charge of all steps of controlled mycorrhization programme which must comply the following principles or recommendations:

- Select plant and EMF, grow both partners for site adaptation.
- Formulate inoculum adapted.

- Observe and record the effects of new or experimental management practices on mycorrhiza development as well as on other seedling characteristics.
- Be alert to and avoid practices that cause mycorrhiza deficiency.
- Bring partners together in appropriate conditions that favour their association.
- Monitor closely growth parameters and environmental conditions.
- Identify mycorrhizal association, control, approve and register mycorrhizal plants.
- Remain alert to research developments on mycorrhizal inoculation of nursery seedlings with EMF proven effective and beneficial.

Future Perspectives

New nursery production systems of forest plants are being developed that emphasize the use of slow-release fertilizers, minimize the use of pesticides and soluble herbicides, and more efficiently utilize water. However, these controlled conditions are difficult to obtain in industrial nursery and to integrate into the commercial production process a practical EMF inoculation method can be complicated. For this, it is critical to know the threshold colonization level required to ensure fungal persistence, for the economic viability of producing mycorrhizal plants for commercial purposes.

A first step is the knowledge of *in vitro* growth requirements of EMF species towards inoculum production, principally to the production infective mycelia. Also, further research is needed to optimize the effective inoculation methods, with dosage experiments, to obtain maximized root colonization, especially if a late-stage ectomycorrhizal fungus (e.g., *Lactarius*) is to be introduced. Injection of mycelial slurries obtained from blended fungal colonies grown in liquid media or alginate-entrapped EMF mycelia could be promising methods to produce mycorrhizal seedlings (Table 13.4) since the production of enough mycelia and control of contaminants can be easily achieved. On the other hand, spore-based inoculum was easy to apply and inexpensive relative to other tree seedling production costs, though the success rate of inoculation varied considerably between genera and functional groups of fungi and was influenced by growing conditions.

Molecular identification methods provide a novel tool for future studies on EMF nursery dynamics. These molecular methods allow identification of EM mycelium in roots and substrate and therefore permit a unique view of the EMF in forest plant mycorrhization, independent of EMF root tip occurrence. However, there is no doubt that the recent development of molecular methods coupled with the continued reduction in running costs, will allow researchers and commercial nurseries to produce large data sets which can be used in the future to better understand nursery mycorrhizal process.

The future is that the commercial nurseries are able to offer the mycorrhizal plant with the wished EMF, with species as *Lactarius*, *Tuber*, etc (they have high interest in Mediterranean area and high value for ours degraded forests and countryside), for field or nursery applications to increase the fruit body production and ecological benefits to trees.

Table 13.4 Mycorrhizal percentage of *Pinus pinaster* seedling roots after inoculation with liquid growing mycelium of *Lactarius deliciosus* and *Pisolithus arhizus*.

Inoculum (mg/plant)	*% mycorrhization L. deliciosus*	*% mycorrhization P. arhizus*
0.5	16	21
1	39	56
2.5	70	84
5	78	82

CONCLUSION

A lot of experimental studies reported beneficial effects of EMF inoculation in forest plants all over the world. Very rare papers mentioned negative effects or no effect but in most case the soil conditions were not convenient to the fungal partner. Consequently controlled mycorrhizal seedlings represent undoubtedly a great progress for reforestation. Production of seedlings for forestation, reclamation and sometimes reforestation needs not only favourable shoot and root sizes but also well-developed mycorrhizae. However, optimized protocols and skills in nursery practises are needed to obtain successful results.

The ultimate objective of ectomycorrhizal fungal inoculation is higher quality of nursery stock and improved field performance by the seedlings, meaning increased survival, growth and probably wood quality. Inoculated fungi must be multi-stage mycorrhizal fungi and should be able to compete with native mycorrhizal fungi in the nursery and in the field. Consequently, more research is required to match appropriate fungi to inoculation rates and growing conditions. A cost-benefit analysis of mycorrhizal inoculation in the nursery will require subsequent assessment of impacts on fungal diversity and inoculum levels in the field.

Regardless of inoculation method, the use of "native" ectomycorrhizal fungi is critical because it is assumed that these fungi are ecologically adapted to the site characteristics and will be able to persist on the roots of the host plant, which circumvents the problem with commercial inocula consisting of exotic strains or species of mycorrhizal fungi that are not adapted to the environmental conditions on site.

Although satisfactory mycorrhiza formation in the nursery has been obtained, new research will be able to characterize many of the physiological events (e.g., changes in chemical compositions, proteins, enzymes, hormonal effects, etc.) that may occur at the cellular level during the inoculation process. This realm requires further investigation. Understanding these processes may improve the potential use of ectomycorrhizal inoculation in commercial nurseries for successful reforestation.

REFERENCES

Ali, N.A. and Jackson, R.M. (1988). Effects of plant roots and their exudates on germination of basidiospores of ectomycorrhizal fungi, *Transactions of the British Mycological Society* 91: 253–260.

Agerer, R. (1987-2002). Colour atlas of ectomycorrhizae. 1st–12th, Einhorn-Verlag, Schwäbisch Gmünd.

Anderson, I.C., Chambers, S.M. and Cairney, J.W.G. (2001). Variation in nitrogen source utilisation by *Pisolithus isolates* maintained in axenic culture, *Mycorrhiza* 11: 53–56.

Baar, J., Comini, B., Elferink, M.O. and Kuyper, T.W. (1997). Performance of four ectomycorrhizal fungi on organic and inorganic nitrogen sources, *Mycological Research* 101: 523–529.

Begon, M., Harper, J.L. and Townsend, C.R. (1996). Ecology. Individuals, populations and communities, 3rd Edn. Blackwell Science Ltd, Oxford.

de la Bastide, P.Y., Kropp, B.R. and Piché, Y. (1994). Spatial distribution and temporal persistence of discrete genotypes of the ectomycorrhizal basidiomycete *Laccaria bicolor* (Maire) Orton, *NewPhytologist* 127: 547–556.

Bending, G.D. and Read, D.J. (1995). The structure and function of the vegetative mycelium of ectomycorrhizal plants. 6. Activities of nutrient mobilizing enzymes in birch litter colonized by *Paxillus involutus* (Fr.) Fr, *NewPhytologist* 130: 411–417.

Berg, M.P., Kniese, J.P., Bedaux, J.J.M. and Verhoef, H.A. (1998). Dynamics and stratification of functional groups of micro- and mesoarthropods in the organic layer of a Scots pine forest, *Biology and Fertility of Soils* 26: 268–284.

Bertaux, J., Schmid, M., Prevost-Boure, N.C., Churin, J.L., Hartmann, A., Garbaye, J. and Frey-Klett, P. (2003). *In situ* identification of intracellular bacteria related to *Paenibacillus* spp. in the mycelium of the ectomycorrhizal fungus *Laccatia bicolor* S238N, *Applied and Environmental Microbiology* 69: 4243–4248.

Bills, G.F., Holtzman, G.I. and Miller, O.K. (1986). Comparison of ectomycorrhizal basidiomycete communities in red spruce versus northern hardwood forests of West Virginia, *Canadian Journal of Botany* 64: 760–768.

Borchers, J.G. and Perry, D.A. (1992). The influence of soil texture and aggregation on carbon and nitrogen dynamics in southwest Oregon forests and clear cuts, *Canadian Journal of Forest Research* 22: 298–305.

Boyd, R., Furbank, R.T. and Read, D.J. (1986). Ectomycorrhiza and the water relations of trees. *In*: Physiological and genetical aspects of mycorrhizae, Proceedings, 1st European symposium on mycorrhizae, Gianinazzi-Pearson V. and Gianinazzi S. Eds., INRA, Paris, 689–693.

Boyle, C.C., Robertson, W.J. and Salanius, P.O. (1987). Use of mycelial slurries of mycorrhizal fungi as inoculum for commercial tree seedling nurseries, *Canadian Journal of Forest Research* 17: 1480–1486.

Bowen, G.D. (1994). The ecology of ectomycorrhiza formation and functioning. *In*: Management of mycorrhizas in agriculture, Robson A.D., Abbott L.K. and Malajczuk N. Eds., Horticulture and forestry, Kluwer, The Netherlands, 61–68.

Bradbury, S.M. (1997). Ectomycorrhizas of lodgepole pine (*Pinus contorta*) seedlings originating from seed in South-western Alberta cut blocks, *Canadian Journal of Botany* 67: 213–217.

Bradbury, S.M., Danielson, R.M. and Visser, S. (1998). Ectomycorrhizas of regenerating stands of lodgepole pine (*Pinus contorta*), *Canadian Journal of Botany* 76: 218–227.

Brundrett, M. (1991). Mycorrhizas in natural ecosystems, *Advances in Ecological Research* 21: 171–313.

Brundrett, M. (2002). Coevolution of roots and mycorrhizas of land plants, *New Phytologist* 154: 275–304.

Brundrett, M., Bougher, N., Dell, B., Grove, T. and Malajczuk, N. (1996). Working with mycorrhizas in forestry and agriculture, ACAIR, Canberra, Australia.

Brundrett, M., Malajczuk, N., Mingqin, G., Daping, X., Snelling, S. and Dell, B. (2005). Nursery inoculation of *Eucalyptus* seedlings in Western Australia and Southern China using spores and mycelial inoculum of diverse ectomycorrhizal fungi from different climatic regions, *Forest Ecology and Management* 209: 193–205.

Brunner, I. (2001). Increasing the nitrogen load reduces ectomycorrhization and alters element concentration in Norway spruce seedlings, *Environmental Pollution* 114: 223–233.

Buschena, C.A., Doudrick, L. and Anderson, N.A. (1992). Persistence of *Laccaria* spp. as ectomycorrhizal symbionts of container-grown black spruce, *Canadian Journal of Forest Research* 22: 1883–1887.

Burton, A.J., Pregitzer, K.S., Ruess, R.W., Hendrik, R.L. and Allen, M.F. (2002). Root respiration in North American forests: effects of nitrogen concentration and temperature across biomes, *Oecologia* 131: 559–568.

Castellano, M.A. (1996). Outplanting performance of mycorrhizal inoculated seedlings, *In*: Concepts in mycorrhizal research, Mukergi K.G. Ed., Kluwer Academic Publishers, Dordrecht, 223–301.

Carlile, M.J., Watkinson, S.C. and Gooday, G.W. (2001). The Fungi. 2nd Edn., Academic Press, London.

Ceccaroli, P., Saltarelli, R., Cesari, P., Zambonelli, A, and Stocchi, V. (2001). Effects of different carboyhydrate sources on the growth of *Tuber borchii* Vittad. mycelium strains in pure culture, *Molecular and Cellular Biochemistry* 218: 65–70.

Cline, M.L. and Reid, C.P.P. (1982). Seed source and mycorrhizal fungus effects on growth of containerized *Pinus contorta* and *Pinus ponderosa* seedlings, *Forest Science* 28: 237–250.

Colpaert, J.V., Van Laere, A. and van Assche, J.A. (1996). Carbon and nitrogen allocation in ectomycorrhizal and non-mycorrhizal *Pinus sylvestris* L. seedlings, *Tree Physiology* 16: 787–793.

Dahlberg, A. (1997). Population ecology of *Suillus variegatus* in old Swedish Scots pine forests, *Mycological Research* 101: 47–54.

Dahlberg, A. (2001). Community ecology of ectomycorrhizal fungi: an advancing interdisciplinary field, *New Phytologist* 150: 555–562.

Dahlberg, A., Jonsson, L. and Nylund, J.-E. (1997). Species diversity and distribution of biomass above, and below, ground among ectomycorrhizal fungi in an old-growth Norway spruce forest in south Sweden, *Canadian Journal of Botany* 75: 1323–1335.

Dahlberg, A. and Stenlid, J. (1995). Spatiotemporal patterns in ectomycorrhizal populations, *Canadian Journal of Botany* 73: S1222–S1230.

Daza, A., Manjón, J.L., Camacho, M., Romero de la Osa, L., Aguilar A. and Santamaría C. (2006). Effect of carbon and nitrogen sources, pH and temperature on *in vitro* culture of several isolates of *Amanita caesarea* (Scop.:Fr.) Pers., *Mycorrhiza* 16: 133–136.

Deacon, J.W. and Fleming, L.V. (1992). Interactions of ectomycorrhizal fungi. *In*: Mycorrhizal functioning, an integrative plant-fungal process, Allen M.F. Ed., Chapman & Hall, New York, USA, 249–300.

Diaz, G., Carrillo, C. and Honrubia, M. (2003). Differential responses of ectomycorrhizal fungi to pesticides in pure culture, *Cryptogamie Mycologie* 24: 199–211.

Dickie, I.A., Xu, B. and Koide, R.T. (2002). Vertical niche differentiation of ectomycorrhizal hyphae in soil as shown by T-RFLP analysis, *New Phytologist* 156: 527–535.

Dighton, J. and Mason, P.A. (1985). Mycorrhizal dynamics during forest tree development. *In*: Developmental biology of higher fungi, British Mycological Society Symposium 10, Moore D., Casselton L.A., Wood D.A. and Frankland J.C. Eds., Cambridge University Press Cambridge, USA, 117–139.

Dighton, J., Poskitt, J.M. and Howard, D.M. (1986). Changes in occurrence of basidiomycete fruit bodies during forest stand development: with specific reference to mycorrhizal species, *Transactions of the British Mycological Society* 87: 163–171.

Dix, N.J. and Webster, J. (1995). Fungal Ecology, Chapman & Hill, London.

Dixon, R.K., Behrns, G.T., Garrett, H.E., Cox, G.S. and Sander, I.L. (1985). Synthesis of ectomycorrhizae on container-grown oak seedlings, *Southern Journal of Applied Forestry* 9: 95–99.

Duddridge, J.A., Malibari, A. and Read, D.J. (1980). Structure and function of mycorrhizal rhizomorphs with special reference to their role in water transport, *Nature* 287: 834–836.

Eberhardt, U., Oberwinkler, F., Verbeken, A., Rinaldi, A.C., Pacioni, G. and Comandini, O. (2000). *Lactarius* ectomycorrhizae on *Abies alba*: morphological description, molecular characterization, and taxonomic remarks, *Mycologia* 92: 860–873.

Ek, H., Sjögren, M., Arnebrandt, K. and Söderström, B. (1994). Extramatrical mycelial growth, biomass allocation and nitrogen uptake in ectomycorrhizal systems in response to collembolan grazing, *Applied Soil Ecology* 1: 155–169.

Finlay, R.D., Frostegård, Å. and Sonnerfeldt, A.M. (1992). Utilization of organic and inorganic nitrogen sources by ectomycorrhizal fungi in pure culture and in symbiosis with *Pinus contorta* Dougl. Ex Loud, *New Phytologist* 120: 105–115.

Fogel, R. and Hunt, G. (1983). Contribution of mycorrhizae and soil fungi to nutrient cycling in a Douglas-fir ecosystem, *Canadian Journal of Forest Research* 13: 219–232.

France, R.C and Reid, C.P.P. (1983). Interactions of nitrogen and carbon in the physiology of ectomycorrhizae, *Canadian Journal of Botany* 61: 964–984.

Frank, A.B. (1885). Über die Wurzelsymbiose beruhende Ernährung gewisser Bäume durch unterirdische Pilze (On the nutrition of certain trees based on root symbiosis of underground fungi), Ber. Deutsche Botanische Gesellschaft 3: 128–145.

Fries, N. (1978). Basidiospore germination in some mycorrhizal forming Hymenomycetes, *Transactions of the British Mycological Society* 70: 319–324.

Fries, N., Serck-Hanssen, K., Häll Dimberg, L. and Theander, O. (1987). Abietic acid, an activator of basidiospore germination in ectomycorrhizal species of the genus *Suillus* (Boletaceae), *Experimental Mycology* 11: 360–363.

Gagnon, J., Langlois, C.G., Bouchard, D. and Le Tacon, F. (1995). Growth and ectomycorrhizal formation of container-grown Douglas-fir seedlings inoculated with

Laccaria bicolor under four levels of nitrogen fertilization, *Canadian Journal of Forest Research* 25: 1953–1961.

Garbaye, J. (1994). Helper bacteria—a new dimension to the mycorrhizal symbiosis, *New Phytologist* 128: 197–210.

Garbaye, J. (2000). The role of ectomycorrhizal symbiosis in the resistance of forests to water stress, *Outlook on Agriculture* 29: 63–69.

Garbaye, J. and Bowen, G.D. (1987). Effect of different microflora on the success of ectomycorrhizal inoculation of *Pinus radiata*, *Canadian Journal of Forest Research* 17: 941–943.

Gardes, M., White, T.J., Fortin, J.A., Bruns, T.D. and Taylor, J.W. (1991). Identification of indigenous and introduced symbiotic in ectomycorrhizae by amplification of the nuclear and mitochondrial ribosomal DNA, *Canadian Journal of Botany* 69: 180-190.

Godbold, D.L., Fritz, H.W., Jentschke, G., Meesenburg, H. and Rademacher, P. (2003). Root turnover and root necromass accumulation of Norway spruce (*Picea abies*) are affected by soil acidity, *Tree Physiology* 23: 915–921.

Gryta, H., Debaud, J.C., Effosse, A., Gay, G. and Marmeisse, R. (1997). Fine scale structure of populations of the ectomycorrhizal fungus *Hebeloma cylindrosporum* in coastal sand dune forest ecosystems, *Molecular Ecology* 6: 353–364.

Guidot, A., Lumini, E., Debaud, J.C. and Marmeisse, R. (1999). The nuclear ribosomal DNA intergenic spacer as a target sequence to study intraspecific diversity of the ectomycorrhizal basidiomycete *Hebeloma cylindrosporum* directly on *Pinus* root systems, *Applied and Environmental Microbiology* 65: 903–909.

Hacskaylo, E., Palmer, J.G. and Bozo, J.A. (1965). Effect of temperature on growth and respiration of ectotrophic mycorrhizal fungi, *Mycologia* 57: 748–756.

Hagerman, S.M., Jones, M.D., Bradfield, G.E., Gillespie, M. and Durall, D.M. (1999). Effects of clear cut logging on the diversity and persistence of ectomycorrhizae at a subalpine forest, *Canadian Journal of Forest Research* 29: 124–134.

Harley, J.L. and Smith, S.E. (1983). *Mycorrhizal Simbiosis*, Academic Press Inc, London.

Harvey, A.E., Larsen, M.J. and Jurgensen, M.F. (1976). Distribution of ectomycorrhizae in a mature Douglas fir-larch forest soil in western Montana, *Forest Science* 22: 393–398.

van der Heijden, M.G.A., Klironomos, J.N., Ursic, M., Moutoglis, P., Streitwolf-Engel, R., Boller, T., Wiemken, A. and Sanders, I.R. (1998). Mycorrhizal fungal diversity determines plant biodiversity, ecosystem variability and productivity, *Nature* 396: 69–72.

Heinonsalo, J., Jorgensen, K.S., Haahtela, K. and Sen, R. (2000). Effects of *Pinus sylvestris* root growth and mycorrhizosphere development on bacterial carbon source utilization and hydrocarbon oxidation in forest and petroleum-contaminated soils, *Canadian Journal of Microbiology* 46: 451–464.

Henrion, B., Di Battista, C. and Bouchard, D. (1994). Monitoring the persistence of *Laccaria bicolor* as an ectomycorrhizal symbionts of nursery-grown Douglas fir by PCR of the rDNA intergenic spacer, *Molecular Ecology* 3: 571–580.

Hobbie, E.A., Sanchez, F.S, and Rygiewicz, P.T. (2004). Carbon use, nitrogen use, and isotopic fractionation of ectomycorrhizal and saprotrophic fungi in natural abundance and C-13-labelled cultures, *Mycological Research* 108: 725–736.

Horton, T.R. and Bruns, T.D. (2001). The molecular revolution in ectomycorrhizal ecology: peeking into the black-box, *Molecular Ecology* 10: 1855–1871.

Hung, L. and Trappe, J.M. (1983). Growth variation between and within species of ectomycorrhizal fungi in response to pH *in vitro*, *Mycologia* 75: 234–241.

Hung, L. and Trappe, J.M. (1987). Ectomycorrhizal inoculation of Douglas-fir transplanted container seedlings with commercially produced inoculum, *New Forests* 1: 141–152.

Iotti, M. and Zambonelli, A. (2006). A quick and precise technique for identifying ectomycorrhizas by PCR, *Mycological Research* 110: 60–65.

Jackson, R.M., Walker, C., Luff , S. and McEvoy, C. (1995). Inoculation and field testing of Sitka spruce and Douglas-fir with ectomycorrhizal fungi in the United Kingdom, *Mycorrhiza* 5: 165–173.

Jones, M.D., Durall, D.M. and Tinker, P.B. (1991). Fluxes of carbon and phosphorus between symbionts in willow ectomycorrhizas and their changes with time, *New Phytologist* 118: 99–106.

Jones, M.D. and Hutchinson, T.C. (1986). The effect of mycorrhizal infection on the response of *Betula papyerifera* to nickel and copper, *New Phytologist* 102: 429–442.

El Karkouri, K., Martin, F., Douzery, J.P.E. and Mousain, D. (2005). Diversity of ectomycorrhizal fungi naturally established on containerised *Pinus* seedlings in nursery conditions, *Microbiological Research* 160: 47–52.

Keizer, P.J. and Arnolds, E. (1994). Succession of ectomycorrhizal fungi in roadside verges planted with common oak (*Quercus robur* L.) in Drenthe, The Netherlands, *Mycorrhiza* 4: 147–159.

Kernaghan, G., Currah, R.S. and Bayer, R.J. (1997). Russulaceous ectomycorrhizae of *Abies lasiocarpa* and *Picea engelmannii*, *Canadian Journal of Botany* 75: 1843–1850.

Kernaghan, G., Sigler, L. and Khasa, D. (2003). Mycorrhizal and root endophytic fungi of containerized *Picea glauca* seedlings assessed by rDNA sequence analysis, *Microbial Ecology* 45: 128–136.

Knudsen, H. and Hansen, L. (1991). New taxa and combinations in the Agaricales, Boletales and Polyporales, *Nordic Journal of Botany* 11: 477–481.

Koljalg, U., Larsson, K.H., Abarenkov, K., Nilsson, R.H., Alexander, I.J., Eberhardt, U., Erland, S., Hoiland, K., Kjoller, R., Larsson, E., Pennanen, T., Sen, R., Taylor, A.F.S., Tedersoo, L., Vralstad, T. and Ursing, B.M. (2005). UNITE: a database providing web-based methods for the molecular identification of ectomycorrhizal fungi, *New Phytologist* 166: 1063–1068.

Kope, H.H. and Fortin, J.A. (1990). Germination and comparative morphology of basidiospores of *Pisolithus arhizus*, *Mycologia* 82: 350–357.

Kranabetter, J.M. and Wylie, T. (1998). Ectomycorrhizal community structure across forest openings on naturally regenerated western hemlock seedlings, *Canadian Journal of Botany* 76: 189–196.

Kropp, B.R. and Langlois, C.G. (1990). Ectomycorrhizae in reforestation, *Canadian Journal of Forest Research* 20: 438–451.

Kuek, C., Tommerup, I.C. and Malajczuk, N. (1992). Hydrogel bead inocula for the production of ectomycorrhizal eucalyptus for plantations, *Mycological Research* 96: 273–277.

Laatikainen, T. and Heinonen-Tanski, H. (2002). Mycorrhizal growth in pure cultures in the presence of pesticides, *Microbiological Research* 157: 127–137.

Landeweert, R., Leeflang, P., Kuyper, T.W., Hoffland, E., Rosling, A., Wernars, K. and Smit, E. (2003). Molecular identification of ectomycorrhizal mycelium in soil horizons, *Applied and Environmental Microbiology* 69: 327–333.

Landis, T.D., Tinus, R.W., McDonald, S.E. and Barnett, J.P. (1990). The biological component: nursery pests and mycorrhizae, Vol. 5, The container tree nursery manual, Agricultural Handbook 674. US Department of Agriculture, Forest Service, Washington D.C., USA.

Last, F.T., Mason, P.A., Wilson, J. (1984). Controlled inoculation of Sitka spruce with sheating (ecto-) mycorrhizal fungi—a commercial experience in 1982, *Scottish Forestry* 38: 75–77.

Le Tacon, F., Alvarez, I.F., Bouchard, D., Henrion, B., Jackson, R.M., Luff, S., Parlade, J.I., Pera, J., Stenström, E., Villeneuve, N. and Walker, C. (1992). Variations in field response of forest trees to nursery ectomycorrhizal inoculation in Europe. In: Mycorrhizas in ecosystems, Read D.J., Lewis D.H., Fitter A.H. and Alexander I.J. Eds., CAB International, Wallingford, 119–132.

Le Tacon, F., Jung, G., Michelot, P. and Mugnier, M. (1983). Efficacité en pépinière forestière d'un inoculum de champignon ectomycorhizien produit en fermenteur et inclus dans une matrice de polymères, *Annales des Sciences Forestieres* 40: 165–176.

Le Tacon, F., Mousan, D., Garbaye, J., Bouchard, D., Churin, J.-L., Argillier, C., Amirault, J.-M. and Généré, R. (1997). Mycorrhizes, pépinières et plantations forestières en France (Summary in English), *Revue Forestière Française* 49: 131–154.

Lindahl, B. (2001). Nutrient cycling in boreal forest—a mycological perspective. Acta Universitatis Agriculturae Sueciae, Silvestria 214. SLU Service, Uppsala, Sweden.

Linderman, R.G. (1988). Mycorrhizal interactions with the rhizosphere microflora - the mycorrhizosphere effect, *Phytopathology* 78: 366–371.

Lilleskov, E.A., Hobbie, E.A. and Fahey, T.J. (2002). Ectomycorrhizal fungal taxa differing in response to nitrogen deposition also differ in pure culture organic nitrogen use and natural abundance of nitrogen isotopes, *New Phytologist* 154: 219–231.

Machado, H. and Bragança, H. (1994). *In vitro* study of ectomycorrhiza formation under drought stress conditions. *In*: Mycorrhizas in integrated systems. From genes to plant development, Azcón-Aguilar C. and Barea J.M. Eds., Proceedings of the 4th European symposium on mycorrhizas, Granada, 455–458.

Manjon, J.L., Camacho, M., de la Osa L.R., Aguilar A. and Santamaria C. (2006). Effect of carbon and nitrogen sources, pH and temperature on *in vitro* culture of several isolates of *Amanita caesarea* (Scop.: Fr.) Pers., *Mycorrhiza* 16: 133–136.

Marin, M. (1999). Optimization of mycelial growth and molecular characterization of ectomycorrhizal fungi: study in *Lactarius deliciosus*, Ph.D. thesis. University of Valencia, Valencia, Spain.

Marin, M. (2005). Arbuscular mycorrhizal inoculation in nursery practice. *In*: Handbook of microbial biofertilizers, Rai M.K. Ed., The Haworth Press, New York, 289–324.

Martin, F., Cliquet, J.-B. and Stewart, G. (2001). Nitrogen acquisition and assimilation in mycorrhizal symbioses. *In*: The Assimilation of Nitrogen by Plants, Lea P.J. and Morot-Gaudry J.F. Eds., Springer-Verlag, Berlin,147–166.

Marx, D.H. (1969). Influence of ectotrophic mycorrhizal fungi on resistance of pine roots to pathogenic infections. I. Antagonism of mycorrhizal fungi to root pathogenic fungi and soil bacteria, *Phytopathology* 59: 153–163.

Marx, D.H. (1980). Ectomycorrhizal fungus inoculations: a tool for improving forestation practices, *In*: Tropical mycorrhiza research, Mikola P. Ed., Oxford University Press, New York, 13–71.

Marx, D.H. and Bell, W. (1985). Formation of *Pisolithus* Ectomycorrhizae on Loblolly Pine Seedlings with Spore Pellet inoculum applied at different times, USDA Forest Service, Research Paper, SE-249.

Marx, D.H., Berry, C.R. and Kormanik, P.P. (1995). Application of municipal sewage sludge to forest and degraded land, American Society of Agronomy, Crop Science Society of America, Soil Science Society of America, ASA Special Publication 58: 275-295.

Marx, D.H. and Bryan, W.C. (1975). Growth and ectomycorrhizal development of loblolly pine seedlings in fumigated soil infested with the fungal symbiont *Pisolithus tinctorius*, *Forest Science* 21: 245–254.

Marx, D.H. and Cordell, C.E. (1987). Ecology and management of ectomycorrhizal fungi in regenerating forest in the Eastern United States. *In*: Mycorrhizae in the Next Decade: practical applications and Research Priorities. Seven North American Conference on Mycorrhizae, Sylvia D.M., Hung L.L. and. Graham J.H Eds., Institute of Food and Agricultural Sciences, University of Florida, Gainesville, 78–98.

Marx, D.H., Cordell, C.E., Kenney, D.S., Mexal, J.G., Artman, J.D., Riffle, J.W. and Molina, R.J. (1984). Commercial vegetative inoculum of *Pisolithus tinctorius* and inoculation techniques for development of ectomycorrhizae on bare-root tree seedlings, *Forest Science Monograph* 25.

Marx, D.H. and Kenney, D.S. (1982). Production of Ectomycorrhizal fungus inoculum. *In*: Methods and Principles of Mycorrhizal Research, Schenck N.C. Ed., APS Press, Minnesota, 131–146.

Marx, D.H., Ruehle, J.L. and Cordell, C.E. (1991). Methods for studying nursery and field response of trees to specific ectomycorrhiza, *In*: Methods in Microbiology, Norris J.R., Read D.J.and Varma A.K. Eds. , Academic Press, London, 383–411.

Marx, D.H., Ruehle, J.L., Kenney, D.S., Cordell, C.L., Riffle, J.W., Molina, R.J., Pawuk, W.H., Navratil, S., Tinus, R.W. and Goodwin, O.C. (1982). Commercial vegetative inoculum of *Pisolithus tinctorius* and inoculation techniques for development of ectomycorrhizae on container-grown tree seedlings, *Forest Science* 28: 373–400.

Mason, P.A. (1980). Aseptic synthesis of sheathing (ecto-) mycorrhizas. *In*: Tissue Culture Methods for Plant Pathologists,. Ingram D.S. and Helgeson J.P. Eds., Blackwell Scientific Publications, Oxford, 173–178.

Mason, P.A., Last, F.T., Pelham, J. and Ingleby, K. (1982). Ecology of some fungi associated with an aging stand of birches (*Betula pendula* and *Betula pubescens*), *Forest Ecology and Management* 4: 19–39.

Mason, P.A., Last, F.T., Wilson, J., Deacon, J.W., Fleming, L.V. and Fox, F.M. (1987). Fruiting and succession of ectomycorrhizal fungi. *In*: Fungal infection of plants, Pegg G.F. and Ayres P.G. Eds., Cambridge University Press, New York, USA, 253–268.

Mason, P.A., Wilson, J., Last, F.T. and Walker, C. (1983). The concept of succession in relation to the spread of sheathing mycorrhizal fungi on inoculated tree seedlings growing in unsterile soils, *Plant and Soil* 71: 247–256.

Mauperin, C., Mortier, F., Garbaye, J., Le Tacon, F. and Carr, G. (1987). Viability of an ectomycorrhizal inoculum produced in a liquid medium and entrapped in a calcium alginate gel, Canadian Journal of Botany 64: 848–852.

Mello, A., Ghignone, S., Vizzini, A., Sechi, C., Ruiu, P. and Bonfante, P. (2006). ITS primers for the identification of marketable boletes, *Journal of Biotechnology* 121: 318–329.

Meyer, J.R. and Linderman, R.G. (1986). Selective influence on populations of rhizosphere or rhizoplane bacteria and actinomycetes by mycorrhizas formed by *Glomus fasciculatum*, *Soil Biology and Biochemistry* 18: 191–196.

Mikola, P., Hahl, J. and Torniainen, E. (1966). Vertical distribution of mycorrhizae in pine forests with spruce undergrowth, *Annales Botanici Fennici* 3: 406–409.

Mikola, P. (1970). Mycorrhizal inoculation in afforestation, *International Review of Forest Research* 3: 123–196.

Miller, S.L., Torres, P. and McClean, T.M. (1994). Persistence of basidiospores and sclerotia of ectomycorrhizal fungi and *Morchella* in soil, *Mycologia* 86: 89–95.

Minerdi, D., Fani, R., Gallo, R., Boarino, A. and Bonfante P., (2001). Nitrogen fixation genes in an endosymbiotic *Burkholderia* strain, *Applied and Environmental Microbiology* 67: 725–732.

Molina, R., and Palmer, J. (1982). Isolation, maintenance, and pure culture manipulation of ectomycorrhizal fungi. *In*: Methods and principles of mycorrhizal research, Schenk N.C. Ed., American Phytopathological Society, St. Paul, Minnesota, USA.

Molina, R. and Trappe, J.M. (1982). Patterns of ectomycorrhizal host specificity and potential among Pacific Northwest conifers and fungi, *Forest Science* 28: 423–458.

Mortier, F., Le Tacon, F. and Garbaye, J. (1988). Effect of inoculum type and inoculum dose on ectomycorrhizal development, root necrosis and growth of Douglas-fir seedlings inoculated with *Laccaria laccata* in a nursery, *Annales des Sciences Forestieres* 45: 301–310.

Moser, M. (1983). Keys to Agarics and Boleti, Roger Phillips, London.

Nara, K., Nakaya, H. and Hogetsu, T. (2003a). Ectomycorrhizal sporocarp succession and production during early primary succession on Mount Fuji, *New Phytologist* 158: 193–206.

Nara, K., Nakaya, H., Wu, B.Y., Zhou, Z.H. and Hogetsu, T. (2003b). Underground primary succession of ectomycorrhizal fungi in a volcanic desert on Mount Fuji, *New Phytologist* 159: 743–756.

Nilsson, R.H., Kristiansson, E., Ryberg, M. and Larsson, K.H. (2005). Approaching the taxonomic affiliation of unidentified sequences in public databases—an example from the mycorrhizal fungi, BMC Bioinformatics 6: 178.

Nurmiaho-Lassila, E.L., Timonen, S., Haahtela, K. and Sen, R. (1997). Bacterial colonization patterns of intact *Pinus sylvestris* mycorrhizospheres in dry pine forest soil: an electron microscopy study, *Canadian Journal of Microbiology* 43: 1017–1035.

O'Neill, J.J.M. and Mitchell, D.T. (2000). Effects of benomyl and captan on growth and mycorrhizal colonization of Sitka-spruce (*Picea sitchensis*) and ash (*Fraxinus excelsior*) in Irish nursery soil, *Forest Pathology* 30: 165–174.

Oort, A.J.P. (1981). Nutritional requirements of *Lactarius* species, and cultural characters in relation to taxonomy, North Holland, Amsterdam.

Palmer, J.G. and Hacskaylo, E. (1970). Ectomycorrhizal fungi in pure culture. I. Growth on single carbon sources, *Physiologica Plantarum* 23: 1187–1197.

Parladé, J., Álvarez, I.F. and Pera, J. (1999). Coinoculation of containerised Douglas-fir (*Pseudotsuga menziesii*) and maritime pine (*Pinus pinaster*) seedlings with the ectomycorrhizal fungi *Laccaria bicolor* and *Rhizopogon* spp., *Mycorrhiza* 8: 189–195.

Parladé, J., Pera, J. and Luque, J. (2004). Evaluation of mycelial inocula of edible *Lactarius* species for the production of *Pinus pinaster* and *P. sylvestris* mycorrhizal seedlings under greenhouse conditions, *Mycorrhiza* 14: 171–176.

Pendall, E., Bridgham, S., Hanson, P.J., Hungate, B., Kicklighter, D.W., Johnson, D.W., Law, B.E., Luo, Y.Q., Megonigal, J.P., Olsrud, M., Ryan, M.G. and Wan, S.Q. (2004).

Below-ground process responses to elevated CO_2 and temperature: a discussion of observations, measurement methods, and models, *New Phytologist* 162: 311–322.

Pera, J., Álvarez, I.F., Rincón, A., Parladé, J. (1999). Field performance in Northern Spain of Douglas-fir seedlings inoculated with ectomycorrhizal fungi, *Mycorrhiza* 9: 77–84.

Pera, J., Parladé, J. and Álvarez, I.F. (1994). Eficacia del tipo de inóculo de *Pisolithus tinctorius* en la formación de micorrizas en *Pinus pinaster* y *Pseudotsuga menziesii*, *Investigaciones Agrarias y Sistemas de Recursos Forestales* 3: 19–29.

Perrin, R. and Garbaye, J. (1983). Influence of ectomycorrhizae on infectivity of *Pythium*-infested soils and substrates. In: Tree root systems and their mycorrhizas, Atkinson D. ED, Nijhoff/Junk, The Hague, 345–351.

Perry, D.A., Margolis, H., Choquette, C., Molina, R. and Trappe, J.M. (1989). Ectomycorrhizal mediation of competition between coniferous tree species, *New Phytologist* 112: 501–511.

Perry, D.A., Molina, R. and Amaranthus, M.P. (1987). Mycorrhizae, mycorrhizospheres, and reforestation: current knowledge and research needs, *Canadian Journal of Forest Reseach* 17: 929–940.

Read, D.J., Leake, J.R. and Langdale, A.R. (1989). The nitrogen nutrition of mycorrhizal fungi and their host plants. *In*: Nitrogen phosphorus and sulfur utilization by fungi, Boddy L., Marchant R. and Read D.J. Eds., Cambridge University Press, New York, 181–204.

Richardson, M.J. (1970). Studies on *Russula emetica* and other agarics in a Scots pine plantation, *Transactions of the British Mycological Society* 55: 217–229.

Riffle, J.W. and Maronek, D.M. (1982). Ectomycorrhizal inoculation procedures for greenhouse and nursery studies. *In*: Methods and Principles of Mycorrhizal Research, Schneck N.C. ED., APS Press. St. Paul, Minnesota, 147–155.

Rincón, A., Parladé, J. and Pera, J. (2005). Effects of ectomycorrhizal inoculation and the type of substrate on mycorrhization, growth and nutrition of containerised *Pinus pinea* L. seedlings produced in a commercial nursery, *Annals of Forest Science* 62: 817–822.

Rygiewicz, P.T., Bledsoe, C.S. and Zasoski, RJ (1984). Effects of ectomycorrhizae and solution pH on [N-15] ammonium uptake by coniferous seedlings, *Canadian Journal of Forest Research* 14: 885–892.

Sakakibara, S.M., Jones, M.D., Gillespie, M., Hagerman, S.M., Forrest, M.E., Simard, S.W. and Durall, D.M. (2002). A comparison of ectomycorrhiza identification based on morphotyping and PCR-RFLP analysis, *Mycological Research* 106: 868–878.

Sánchez, F., Honrubia, M. and Torres, P. (2001). Effects of pH, water stress and temperature on *in vitro* cultures of ectomycorrhizal fungi from Mediterranean forests, *Cryptogamie Mycologie* 22: 243–258.

Sangtiean, T. and Schmidt, S. (2002). Growth of subtropical ECM fungi with different nitrogen sources using a new floating culture technique, *Mycological Research* 106: 74–85.

Sawyer, N.A., Chambers, S.M. and Cairney, J.W.G. (2003a). Utilisation of inorganic and organic nitrogen sources by *Amanita* species native to temperate Eastern Australia, *Mycological Research* 107: 413–420.

Sawyer, N.A., Chambers, S.M. and Cairney, J.W.G. (2003b). Utilisation of inorganic and organic phosphorus sources by isolates of *Amanita muscaria* and *Amanita* species native to temperate Eastern Australia, *Australian Journal of Botany* 51: 151–158.

Selosse, E.M.A., Bouchard, D., Martin, F. and Le Tacon, F. (2000). Effect of *Laccaria bicolor* strains inoculated on Douglas-fir (*Pseudotsuga menziesii*) several years after nursery inoculation, *Canadian Journal of Forest Research* 30: 360–371.

Selosse, M.A., Martin, F., Bouchard, D. and Le Tacon, F. (1999). Structure and dynamics of experimentally introduced and naturally occurring *Laccaria* sp. discrete genotypes in a Douglas fir plantation, *Applied Environmental Microbiology* 65: 2006–2014.

Sen, R. (2001). Multitrophic interactions between a *Rhizoctonia* sp and mycorrhizal fungi affect Scots pine seedling performance in nursery soil, *New Phytologist* 152: 543–553.

Setälä, H. (1995). Growth of birch and pine-seedlings in relation to grazing by soil fauna on ectomycorrhizal fungi, *Ecology* 76: 1844–1851.

Shearer, C.A. (1995). Fungal competition, *Canadian Journal of Botany* 73: S1259-S1264.

Simard, S.W., Perry, D.A., Jones, M.D., Myrold, D.D., Durall, D.M. and Molina, R. (1997). Net transfer of carbon between tree species with shared ectomycorrhizal fungi, *Nature* 388: 579–582.

Smith, S.E. and Read, D.J. (1997). Mycorrhizal Symbiosis. 2nd ed. Academic Press, London.

Sundari, S.K. and Adholeya, A. (2003). Growth profile of ectomycorrhizal fungal mycelium: emphasis on substrate pH influence, *Antonie van Leeuwenhoek International Journal of General and Molecular Microbiology* 83: 209–214.

Tamm, C.O. (1991).: Nitrogen in terrestrial ecosystems. Questions of productivity, vegetational changes, and ecosystem stability. *In*: Ecological Studies 81, Springer-Verlag, London, 1–115.

Taylor, A.F.S. (2002). Fungal diversity in ectomycorrhizal communities: sampling effort and species detection, *Plant and Soil* 244: 19–28.

Tedersoo, L., Kõljalg, U., Hallenberg, N. and Larsson, K.H. (2003). Fine scale distribution of ectomycorrhizal fungi and roots across substrate layers including coarse woody debris in a mixed forest, *New Phytologist* 159: 153–165.

Teste, F.P., Schmidt, M.G., Berch, S.M., Bulmer, C. and Egger, K.N. (2004). Effects of ectomycorrhizal inoculants on survival and growth of interior Douglas-fir seedlings on reforestation sites and partially rehabilitated landings, *Canadian Journal of Forest Research* 34: 2074–2088.

Timonen, S., Jorgensen, K.S., Haahtela, K. and Sen, R. (1998). Bacterial community structure at defined locations of *Pinus sylvestris-Suillus bovinus* and *Pinus sylvestris-Paxillus involutus* mycorrhizospheres in dry pine forest humus and nursery peat, *Canadian Journal of Microbiology* 44: 499–513.

Torres, P. and Honrubia, M. (1994). Inoculation of containerized *Pinus halepensis* (Miller) seedlings with basidiospores of *Pisolithus arthizus* (Pers) Rauschert, *Rhizopogon roseolus* (Corda) and *Suillus collinitus* (Fr) O Kuntze, *Annals of Forest Science* 51: 1–8.

Trappe J.M. (1977). Selection of fungi for ectomycorrhizal inoculation in nurseries, *Annual Review of Phytopathology* 15: 203–222.

Trappe, J.M. and Luoma, D. (1992). The ties that bind: fungi in ecosystems, *In*: The fungal community-Its organization and role in the ecosystem, Carroll G.C. and Wicklow D.T. Eds. Marcel Dekker, New York, 17–27.

Trappe, J.M., Molina, R. and Castellano, M. (1984). Reactions of mycorrhizal fungi and mycorrhiza formation to pesticides, *Annual Review of Phytopathology* 22: 331–35.

Watling, R. (1984). Larger fungi around Kindrogan, Perthshire, *Transactions of the British Mycological Society* 44: 237–259.

14

Bioproduction and Significance of Fungal Vitamins

Hanna Dahm and Agnieszka Ciesielska

Department of Microbiology, Institute of General and Molecular Biology,
Nicolaus Copernicus University,
Gagarina 9, 87-100 Toruñ, Poland;
E-mail dahm@biol.uni.torun.pl

INTRODUCTION

Vitamins are heterogenous compounds. They are all organic substances, which are essential for a healthy development of humans and animals. Vitamins have a catalytic role in the organisms in enabling synthesis, conversion and degradation of macromolecules such as proteins, lipids, carbohydrates, nucleic acids (Vandamme, 1989).

The physiological function of water-soluble vitamins is well known. They are part of coenzymes and thus responsible for specific biochemical reactions (Abico, 1975; Iwashima, 1980). If vitamins cannot find protein partners, they either get stored in the body (fat-soluble vitamins) or flushed out in the urine (water-soluble vitamins). Water-soluble vitamins leave the body quickly with the exception of vitamin B_{12}. This vitamin is stored in the liver, which is enough for 3 to 5 years.

The function of fat-soluble vitamins is not known in all cases; their involvement can often be traced down to specific biochemical processes.

Besides their function as coenzymes most of the vitamins have been shown to possess various other biofunctions.

In addition to their nutritional and physiological importance, vitamins have also found large-scale technical applications as antioxidants (vitamins C and E), as acidulant (vitamin C) in the food, feed, chemical, cosmetic and pharmaceutical industry (Diplock, 1985; Vandamme, 1989).

Derived concentrates or extracts from vitamin-rich natural sources find little use in the food or pharmaceutical industry. Vandamme (1989) thinks that some of the reasons are: (1) the average daily intake of vitamins does not always seem to be supplied solely via these natural products; (2) the level of the natural source of vitamins is usually relatively low; (3) vitamins are labile molecules during the process of preservation, storage or preparation of food and are generally very sensitive to pH, heat, light, oxygen. These factors have led to the industrial manufacturing of most vitamins.

Several vitamins are produced only or mainly chemically (vitamins A, D_3 H, E, K, B_1) although microbiological methods exist or emerge; other are produced by fermentation (B_6, B_{12}, D_2); for others both chemical and microbial processes are run industrially (vitamin B_{12}), and vitamin C is produced via a combination of chemical and microbiological methods (Yamada *et al.*, 1971; Vandamme, 1989).

FAT-SOLUBLE VITAMINS

Vitamin A and Carotenoids

Vitamin A functions in vision, cell differentiation, and pattern formation during embryogenesis. It influences many physiologic processes, including growth, reproduction and immune response (Barua *et al.*, 2000).

The animal products usually provide preformed vitamin A and vegetables and fruits provitamin A carotenoids. Carotenoids can be oxidatively converted into retinal (aldehyde) in mammals primarily by central cleavage. Retinyl esters, the major dietary form of vitamin A from animal products are hydrolyzed in the intestine in the presence of pancreatic esterases and conjugated bile salts. Retinal produced from carotenoid is reduced to retinol which is estrified and corporated into chylomicra.

Within cells, vitamin A is bound to specific proteins, termed cellular retinoid-binding proteins (Barua *et al.*, 2000). The most prominent cellular manifestation of vitamin A deficiency is process of "keranization". Low concentrations of retinoic acid rapidly induce transglutaminase in macrophages, granulocytic differentiation in promyelocytic leukemia cells and the differentiation of embryonic teratocarcinoma cells (Barua *et al.*, 2000).

Several species of the fungi are potential sources of carotenoids (provitamin A), and a large scale has been based on *Blakeslea trispora*. The yearly production purified β-carotene in the only one country may amount 200 kg.

β-carotene predominates in the *Mucorales*, *Blekeslea trispora, Phycomyces blekesleeanus*, *Choanephora cucurbitarum*; in the yeast *Rhodotorula aurea, R. rubra, R. minuta, Neurospora crassa*; in *Aspergillus giganteus* and various species of *Penicillium* as well as in *Fusarium aquaeductum, Gibberella fujikuroi, Verticillium agaricinum*.

β-carotene and lycopene are found in different strains of the *Ustilago violacea* and canthaxanthin in the mushroom *Cantharellus cinnabarinus* (Cerda-Olmedo, 1989). Han *et al.* (2004) detected vitamin A in the culture of eleven species of genus *Paecilomyces* (closely related to *Penicillium*). Most species of *Paecilomyces* are common soil-borne saprobes, a section in this genus is known primarily as parasites of insect.

Most of the fungal carotenoids are also found in plants, but a few carotenoids torularhodin, torulene, neurosporaxanthin, plectaniaxanthin, β-γ carotene, are exclusive of certain fungi (Andrewes and Starr, 1976; Cerda-Olmedo, 1989).

Table 14.1 Examples of photocarotenogenesis in some fungi (Cerdá-Olmedo, 1989; modified).

Name of fungus	*Coloured carotenoids in illuminated cultures (μg/g)[a] main component*	*Reference*
Aspergillus giganteus	170 β-carotene	El-Jack *et al.*, 1988
Fusarium aquaeductuum	213 neurosporaxanthin	Bindl *et al.*, 1970
Gibberella fujikuroi	159 neurosporaxanthin	Avalos & Cerdá-Olmedo, 1987
Neurospora crassa	152 neurosporaxanthin	Harding *et al.*, 1969
Phycomyces blakesleeanus	550 β-carotene	Bergman *et al.*, 1973
Rhodosporidium diobovatum	700 β-carotene	Vasivnyuk & Getman, 1984
Rhodotorula minuta	150 torulene	Tada & Shiroishi, 1982
Verticillium agaricinum	445 torulene	Valadon & Mummery, 1973

[a] The numerical values are the total concentrations of the carotenoids detected in illuminated cultures, in μg/g dry mycelial weight.

In spite of many fungi attracted the attention of carotenoid chemists because of their bright colours, but the strains used in the laboratories may not be optimal for carotenoid production. The carotenoids concentration are usually too low for commercial application. The accumulation of carotenoids in the fungi is influenced by many environmental factors.

Blue illumination stimulates the production of carotenoids in many fungi, bacteria and plants. However, photocarotenogenesis does not occur in *Blekeslea trispora* (Sutter, 1970). *Neurospora crassa* offers magnificent prospects for all sorts of genetic manipulations for carotenoid biosynthesis. Extensive UV mutagenesis provide to obtain dark pigmanted mutants accumulating of a higher proportion of final products of the carotenoid pathway (Avalos *et al.*, 2001).

The accumulation of carotenoids is influenced by the culture condition of fungi but only to a minor extent. The nature of the carbon and nitrogen source affects carotenoids formation in fungi in unpredictable manner. In general, a high carbon to nitrogen ratio promotes carotenogenesis. A medium with a neutral to alkaline pH favors β-carotene production by *Blekeslea*. The best temperatures for carotenogenesis in *Blekeslea trispora* and *Phycomyces blakesleeanus* is about 26°C. Acetate, leucine and other amino acids increase the β-carotene content of *Phycomycetes* (can be converted directly to hydroxymethylglutaral-coenzyme A (Lilly *et al.*, 1960; Cerda-Olmedo, 1987).

The accumulation of β-carotene in the *Mucorales* is stimulated by β-ionone and retinol. However, such compounds are kept away from industrial fermentors by their high prices. In the production of β-carotene may be replaced by citrus pulps and by limone and citrus oils (Cerda-Olmedo, 1989).

Although over 100 chemical activators of carotenogenesis are known (many aromatic compounds, e.g., cinnamic alcohol, eugenol, safrol, 4 methoxyphenol, etc.) most are relatively ineffective, too expensive or too toxic for practical application (Cerda-Olmedo and Hüttermann, 1986).

Vitamin D

Among all vitamins, vitamin D is outstanding, because it the only vitamin with a steroid structure. Recently, vitamin D is considered to be a hormone Vitamin D and its metabolites are a group of 9, 10 secosteroides. The ring of the steroid nucleus is rotated about the C 6 position.

The conjugated cis-triene system gives rise to the characteristic UV absorption spectrum of vitamin D. The D_3 series has a side chain derived from cholesterol; the D_2 series has a side chain derived from ergosterol containing an additional C 22 (23) double bond and C 24 methyl group.

Ergosterol is the main sterol in yeasts and mycelial fungi. Ergosterol is not the only provitamin suitable for antirickets treatment. 7-Dehydrocholesterol as produced in the mammalian body may also be transformed into vitamin D (D_3). The human body produces its vitamin D from 7 hydrocholesterol upon exposure to sunlight. Vitamins D_2 and D_3 are derived by photoirradiation from their respective 5,7 diene sterol precursors: 7 dehydrosterol or ergosterol (Jones and Makin, 2000).

Vitamin D_3 can be synthesized in the skin. Once the previtamin D is formed in the upper layer of the skin as the result of the reaction of ultraviolet light and in the lower layers is slowly transformed into vitamin D_3. Vitamin D from skin does not circulate for long in the bloodstream, but is immediately taken up by adipose tissue or liver for storage or activation (Jones and Makin, 2000). The first step of activation of vitamin D_3 in the liver is 25-hydroxylation carried out by cytochrome P 450. It is bifunctional cytochrome which in addition to 25-hydroxylating vitamin D_3 also carries out the side-chain hydroxylation of intermediates involved in bile acid biosynthesis. 25-hydroxyvitamin D_3 is converted to the active form of vitamin D known as calcitrol or 1α 25 dihydroxyvitamin D_3. The second step of activation, 1α hydroxylation occurs in the kidney.

The analysis of vitamin D and its metabolites is worthy of attentions for many reasons: etiology, pathogenesis and treatment of diseases involving disturbances of calcium and phosphorus metabolism. Vitamin D stimulates DNA to produce transport proteins, which bind calcium and phosphorus increasing intestinal absorption of these minerals.

Ergosterol is a fat-soluble vitamin with steroid structure and differs from cholesterol in additional carbon (methyl group) at C-24 in the β-position, as well as additional double bonds between carbons 7:8 and 22:23 (Margalith, 1989).

The sterol was first isolated from sclerotia of *Claviceps purpurea* (rye ergot), but yeasts have been the object of many workers interested in the biotechnology of ergosterol and its biosynthesis. Bills and coworkers (1930) as the first detected ergosterol in cells of yeasts. They examinated 29 strains of 13 species of 5 genera: *Endomyces, Nadsonia, Mycoderma, Saccharomyces and Zygo-saccharomyces. Saccharomyces carlsbergensis* (strain ATCC 2345) had the highest ergosterol level (2–4%). Usdin and Burell (1952) found 2–7% ergosterol in the cells of *Rhodotorula gracilis.* A study for ergosterol in yeasts was carried out by Dulaney *et al.* (1954). They examined strains of the following genera: *Candida, Diploascus, Hansenula, Nematospora, Saccharomyces, Torulopsis, Cryptococcus, Endomyces, Kloeckera, Pichia, Sporobolomyces, Debaryomyces, Endomycopsis, Nadsonia, Rhodotorula, Schizosaccharomyces, Taphrina.* The best organism was *Saccharomyces cerevisiae* for which 3–9% of ergosterol was reported. In the cultures remaining fungi levels of ergosterol never exceeded 0–4%.

A method for the production of high sterol yeasts (10% dry weight) was patented by Dulaney in 1957 (Margalith, 1989).

Table 14.2 Some patents for the production of ergosterol (Margalith, 1989; modified).

Name	*Country*	*Patent No.*	*Year*	*Organism*
Ergosterol containing yeasts	USA	2 817 624	1957	*S. cerevisiae* NRRL strains
Ergosterol from yeasts	Britain	801 390	1958	
Extraction of sterols	USA	2 837 540	1958	
Production of ergosterol by yeasts from hydrocarbons	Japan	49 792 802	1971	*C. tropicalis* (pk 233)
Ergosterol production from petroleum	Japan	73 92 486	1973	*C. petrophilum* (ATCC 20226)
Fermentative production of sterols	Japan	78 86 092	1978	*Aureobasidium pullulans* (FERM + P3496)
Cultivation of yeasts producing ergosterol	Czechoslovakia	196 950	1981	*S. cerevisiae*
Production of ergosterol	Japan	81 58 496	1981	*S. uvarum* (UFO 1264)

Sterol production by yeasts is affected by environmental conditions. High concentration of carbohydrates stimulate the production of ergosterol, while nitrogen-rich media usually decrease levels of sterol biosynthesis (Dulaney *et al.*, 1954; Gal'tsowa *et al.*,1959).

Ethanol can serve as a carbon source for sterol production by yeasts. In the past, enhanced sterol production consisted of exposing yeast culture to vapors of ethanol. By this method a sterol level of ca. 10% could be obtained (Margalith, 1989).

However, according to results of many workers, the most important environmental factor for yeasts growth and sterol production is availability of oxygen. The sterols cannot be synthesized under anaerobic conditions. Growth yeasts under semianaerobic conditions yielded ca. 1/10 of sterols in comparison to fully aerated yeasts cells (Klein, 1955; Margalith, 1989).

Several filamentous fungi *Trichoderma, Cephalosporium* and *Fusarium* are also potential producers of ergosterol, however they are lower producers compared to the *Saccharomyces cerevisiae.*

Sterols belong to the lipid fraction of fungal cells and for that reason both sterols and fatty acids employ acetate as precursors. Studies showed that acetate, both the methyl and carboxyl carbons serves as a source for all the carbons of ergosterol (Schechter *et al.*, 1970).

Both ergosterol and 7-dehydrocholesterol can be transformed into previtamin D by UV irradiation; low temperatures are maintained to prevent the formation of side products.

Vitamin E

Evans and Bishop in 1922 discovered a new nutritional factor, initially termed as "factor x", but later named as vitamin E and tocopherol, that was essential for a normal development of animals, especially in small mammals and birds.

The compounds with vitamin E activity were subsequently isolated and named α, β, and γ-tocopherol.

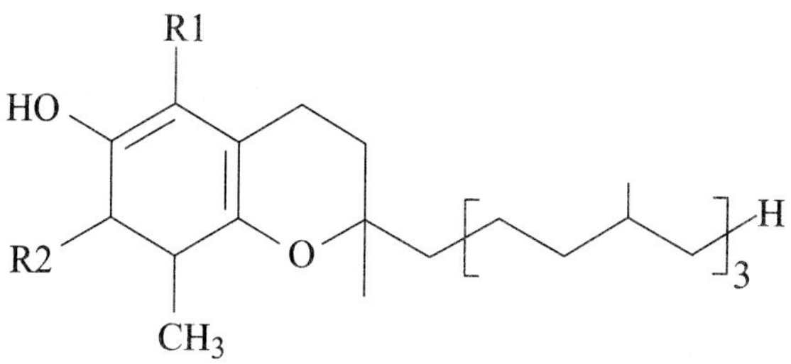

R1	*R2*	
CH_3	CH_3	α-Tocopherol
CH_3	H	β-Tocopherol
H	CH_3	γ-Tocopherol
H	H	δ-Tocopherol

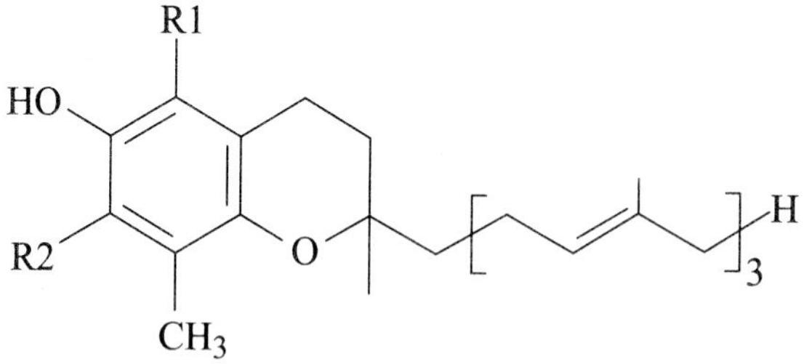

R1	*R2*	
CH_3	CH_3	α-Tocopherol
CH_3	H	β-Tocopherol
H	CH_3	γ-Tocopherol
H	H	δ-Tocopherol

Figure 14.1 Chemical structure of vitamin E group (Tani, 1989).

Now, vitamin E is a collective term for tocopherols and tocotrienols. Their basic structural element is tocol or 2-methyl-2-(4′,8′,12′-trimethyl-tridecyl)-6-chromanol (Nelis *at al.*, 2000).

Tocopherols and tocotrienols are biosynthesized only by plants. The richest sources of vitamin E are plant oils such as sunflower and wheat germ oil. Animals and humans are unable to biosynthesize vitamin E.

Vitamin E has antioxidant properties, occurs with polyunsaturated fatty acids protects polyunsaturated fatty acids from peroxidation and thus aids in the maintenance of the integrity of cellular membrane.

The biological antioxidant action of vitamin E forms an integral part of larger multicomponent cellular defense system including superoxide dismutase, catalase, glutathione peroxidase and selenium. Deficiency of vitamin E results aging, cancer, Parkinson's disease, cataract, artherosclerosis and cardiovascular diseases (Combs, 1992; Packer and Fuchs, 1993).

Diplock *et al.* (1961) and, Hughes and Tove (1982) studied the production ability of tocopherol by 162 strains of 26 genera of yeasts, 74 strains of 15 genera of *Phycomycetes, Ascomycetes* and *Fungi imperfecti*, 41 strains from 12 genera of *Basidiomycetes*, 4 strains of algae belonging to genera of *Euglena, Chlorella and Chlamydomonas.*

Tocopherols were detected in cells of yeasts, but no such substances were detected in cells and culture filtrates of molds and *Basidiomycetes* under cultures and detection conditions.

However, Han *et al.* (2004) detected vitamin E in the culture of eleven species of the genus *Paecilomyces* (closely related to *Penicillium*).

Vitamin K

In 1930, Henrik Dam (1935) described a compound named Koagulation Vitamin (VK), which prevented hemorrhagic disease in chickens. Though the role that vitamin K plays in blood coagulation seems more or less elucidated, but many questions remain unclearly, for example, role VK in bone metabolism (Hodges *et al.*, 1993; Vermeer *et al.*, 1995). According to Kransinski and Russell (1985), vitamin K converts glucose into glycogen for liver storage and aids in the transport of calcium.

Vitamin K (phylloquinone) is a common name for a group of molecules with the same 2-methyl-1,4 naphtoquinone structure, with differences in the side chain on position 3 of the naphtoquinone (Fauler *et al.*, 2000).

The number of carbon atoms in this side chain is used to characterize the different molecules. $VK_{1(20)}$ means 20 carbon atoms in the side; whereas MK_n (menaquinone, VK_2) is a synonym for the number of n-isoprene units.

Vitamin K_1 (phylloquinone) is insoluble in water, but slightly soluble in ethanol and can be easily dissolved in ether hexane or chloroform. It is destroyed in a strong alkaline medium, but is stable in a slightly acidic medium and under oxidizing conditions.

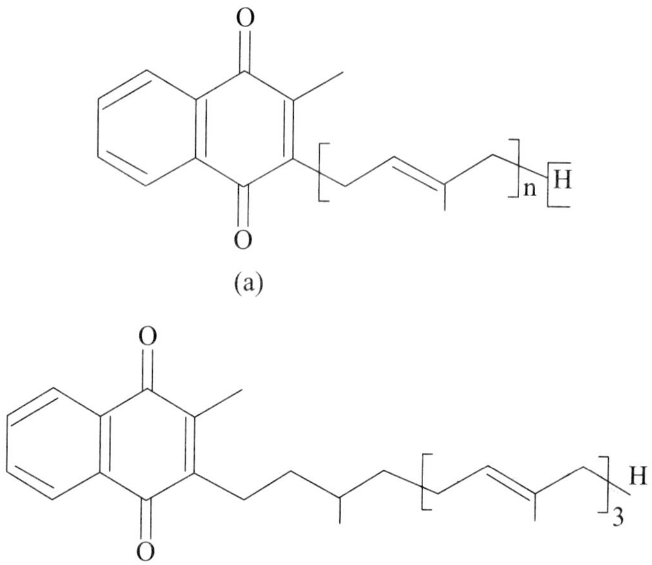

(a)

(b)

Figure 14.2 Chemical structure of vitamin K. (a) Menaquinone-n (MK-n); (b) phylloquinone (Tani, 1989).

Menaquinone (K_2) has a physical and chemical properties similar to VK_1, with the exception of the lipophilic character. MK_2 vitamin is formed by intestinal bacteria (e.g., *Escherichia coli, Staphylococcus aureus)* (Conly *et al.*, 1994; Fauler *et al.*, 2000).

The occurrence of isoprenoid quinones was studied in about 900 microbial strains (Tani *et al.*, 1984). The quinone system of molds was tested in 56 strains of 17 genera of *Phycomycetes, Ascomycetes* and *Fungi Imperfecti.* All of them lacked menaquinones (MK); their quinone system was coenzyme Q (CoQ). The same result was obtained with 88 strains of yeasts and with 28 strains of *Basidiomycetes* of 14 genera.

However, MK was found to be widely distributed in *Procaryota. Flavobacterium* species showed a higher content of MK than other bacteria.

All actinomycetes tested (40 strains) contained only MK.

Water-soluble Vitamins

Vitamin C (ascorbic acid)

Ascorbic acid (2,3-endiol-L-gulonic acid-γ-lactone) is biosynthesized in all chlorophyll-containing plants and in the liver and kidney of most mammals and birds.

In plants, two major biosynthetic pathways have been presented: glucose-glucuronic-gulonic path and the galactose-galacturonate-galactonolactone path (Nyyssönen *et al.*, 2000).

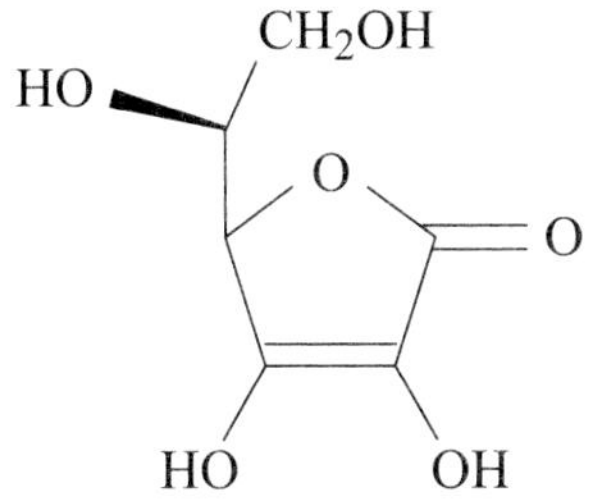

Figure 14.3 L-ascorbic acid (Delić *et al.*, 1989).

In most animals ascorbic aid is formed from D-glucose. The ability to synthesize vitamin C is lacking in the insects, invertebrates, most fish and humans.

L-ascorbic acid shows diverse biological activity. It is required for many hydrolase enzymes. Ascorbic acid is needed for conversion of tyrosine to the neurotransmitter dopamine and hydroxylation to adrenaline and noradrenaline and for hydroxylation of steroid hormones. Ascorbic acid is also known to participate in hydroxylation of aromatic compounds and carcinogens via microsomal mono-oxygenase system (Gaby *et al.*, 1991; Linder, 1991; Rose and Bode, 1992).

The L-ascorbic acid which is used as a supplement in tablets as an antioxidant is produced by the synthesis from D-glucose to D-sorbitol followed by oxidation of D-sorbitol to L-sorbose by *Acetobacter suboxydans*; the production of diacetone L-sorbose and its oxidation with permanganate in alkaline solution to the carbonic acid; hydrolytic cleavage of acetone to form 2-oxo-L-gulonic acid; and enolization to ascorbic acid. It is classic Reichstein-Grüssnert synthesis, most important for biotechnological processes (Nyyssõnen *et al.*, 2000).

Since fruits and vegetables are the main natural sources of L-ascorbic acid, biosynthetic pathways in plants were extensive studied. It has been shown, that L-ascorbic acid is a product of hexose phosphate metabolism including conversion of D-glucose or D-galactose to L-ascorbic acid.

Biosynthesis of L-ascorbic acid in animals is somewhat simpler, it is synthesized in the liver via the glucuronic acid pathway.

The first biooxidation of sorbitol to sorbose by the microorganisms *Bacterium xylinum (Acetobacter xylinum)* was described by Bertrand (1896). Since different genera of bacteria were screened for their ability to oxidize sorbitol and L-sorbose, but there are only a few genera of microorganisms capable of oxidizing sorbitol to 2-keto-L-gulonic acid: *Acetobacter sp., Pseudomonas sp.* (Obata *et al.*, 1975), *Gluconobacter melanogenus* (Okazaki *et al.*, 1968, 1969), *Gluconobacter oxydans* (Yin *et al.*, 1980).

Other groups of microorganisms were found to be able to produce L-ascorbic acid from different sugar substrates but concentrations of substrates and yields of L-ascorbic acid are very low.

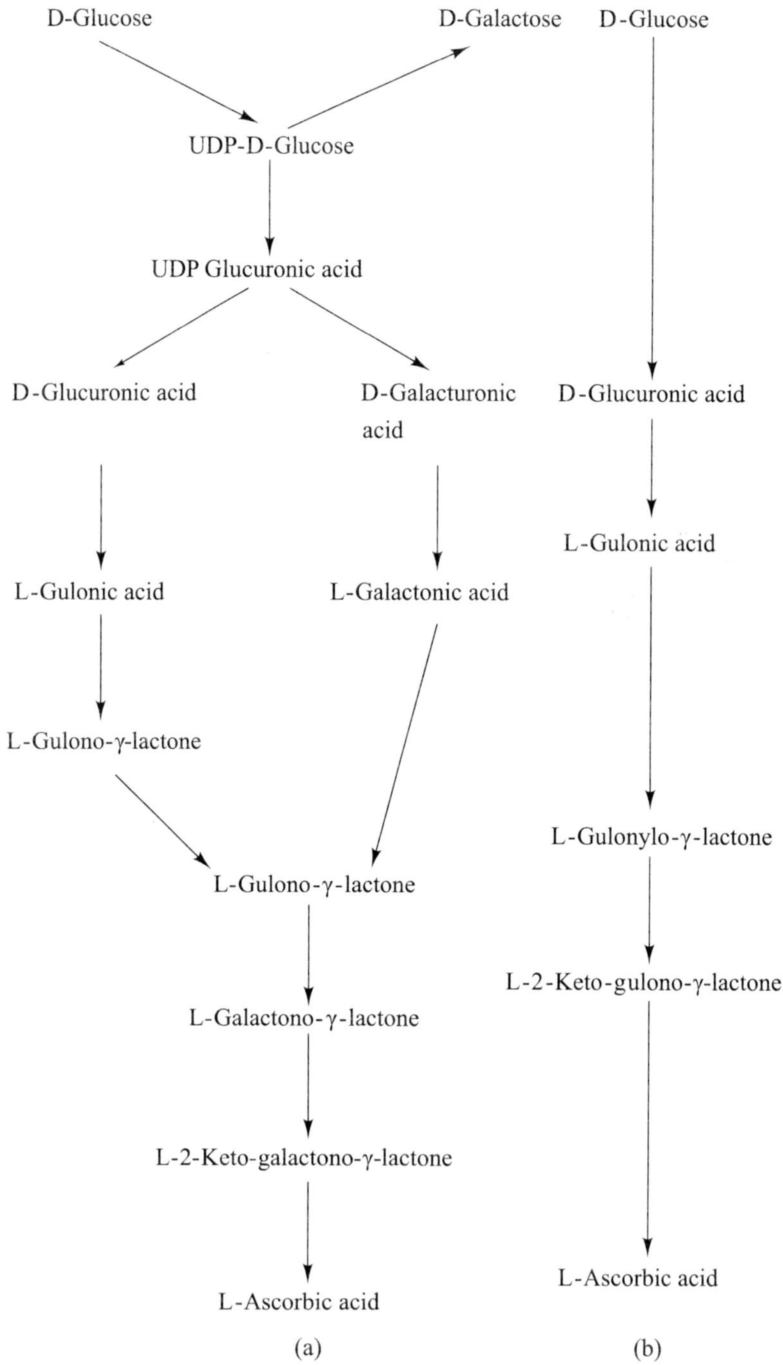

Figure 14.4 Proposed routes of L-ascorbic acid biosynthesis in plants (a) and animals (b), involving inversion of configuration; UDP – uridine diphosphate (Delić *et al.*, 1989).

Geiger-Huber and Galli (1945) found that *Aspergillus niger* was able to synthesize L-ascorbic acid (1.5 mg/100 ml) after 6-8 days in the medium containing 15% sucrose. Sucrose was also a good substrate for L-ascorbic biosynthesis with *Aspergillus flavus* (Ramakrishnan and Desai, 1956) and glucose for *A. niger* (Sastry and Sarma, 1957). Authors proposed the following metabolic pathway: D-glucose – D-glucuronic acid - D-glucono-lactone - 2-keto-L-gulonolactone - L-ascorbic acid.

Roland *et al.* (1985, 1986) detected that strains of *Candida norvegensis* were able to produce L-asorbic acid in the medium with L-galactonic substrates such as L-galactonic acid, L-galactonic acid esters or L-galactono-γlactone.

According to these results, it seems likely that biosynthetic pathways in some yeasts and molds are similar to plants.

Thiamine (Vitamin B_1)

Thiamine monophosphate (TMP), thiamine pyrophosphate (TPP) and thiamine triphosphate (TTP) are commonly found in organisms. However, 80% to 90% of the total thiamine content in cells is TPP, the coenzyme form of thiamine (Egi *et al.*, 1986).

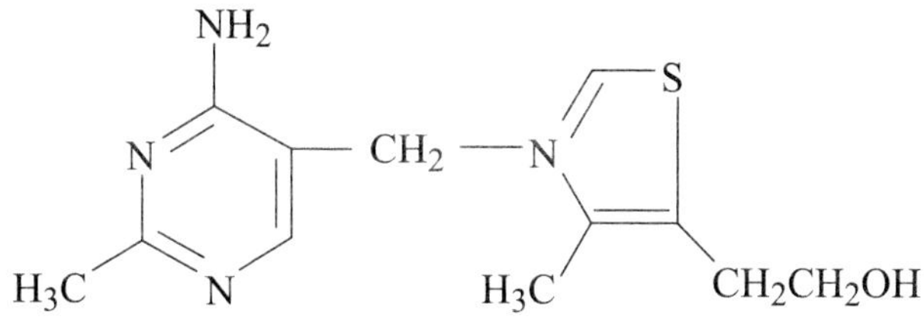

Figure 14.5 Structure of thiamine (Kawasaki and Egi, 2000).

Thiamine is metabolized to TPP by thiamine pyrophosphokinase in animal cells (red and white blood cells), in plants, yeast and bacteria (Sanemori and Kawasaki, 1980). A mutant of *Saccharomyces cerevisiae* excreting thiamin (200 mgL^{-1}) has been isolated (Haj-Ahmad *et al.*, 1992).

Thiamine has to be converted to TPP before exerting its physiological function as a coenzyme in cells. Thiamine pyrophosphate functions as a coenzyme for several important enzymes in the carbohydrate and amino acid metabolism (pyruvate dehydrogenase, oxoglutarate dehydrogenase, transketolase, α-acid dehydrogenase, pyruvate decarboxylase, carboxylate carboligase (Krampitz, 1969; Kawasaki and Egi, 2000).

Thiamine is known to affect nerve functions as an antipolyneuritis factor (emotional stabilizer), effects on nerve conductance or neurotransmission (Eder and Dunant, 1980). Vitamin B_1 is responsible for enzyme reactions, including the release of energy from glucose. Thiamine as well as other vitamins, are produced by microorganisms in extremely small amounts and it is not formed in great excess over the organism's own requirement.

Akagi and Kumaoka (1963) detected that production of thiamine by growing culture of *Escherichia coli* (ATCC 9637) was 1.6 higher than the amount of thiamine in cells (21-23 ng/mg dry weight).

Dahm *et al.* (1989) studied production of B-group vitamins by some soil-borne fungi (*Macrophomina phaseolina, Fusarium solani, F. oxysporum and Rhizoctonia solani*). Thiamine was detected in the culture medium of *F. solani* (2.7 μg L^{-1}), *F. oxysporum* (10.2 μg L^{-1}) and *R. solani* (8 μg L^{-1}).

Studies were carried out on B-group production by mycorrhizal fungi (*Suillus bovinus, Hebeloma crustuliniforme, H. mesophaeum, Cenococcum graniforme, Pisolithus tinctorius*) grown *in vitro* at different pH values. Thiamine was synthesized by the fungi in minute amounts (0–2 μg L^{-1}), mainly in the acidic medium (Strzelczyk *et al.*, 1991).

Silhankova (1985) isolated thiamine-excreting mutants of yeast, *Saccharomyces cerevisiae* and *S. carlsbergensis* which excreted upto 920 μg thiamine L^{-1} of the culture medium. Mutants were obtained by employing several rounds of UV mutagenesis and selection.

Many results seem to indicate that biosynthetic pathways of thiamine in microorganisms differ between bacteria, yeast and other fungi (Kumaoka and Brown, 1967; Estramareix and Therisod, 1984; Grue-Sørensen *et al.*, 1986).

Riboflavin (Vitamin B_2)

Riboflavin, 7,8 dimethyl-10-(1′-D-ribityl) isoalloxazine (RF), a yellow-green, light-sensitive compound is widely distributed in animals, plants and microorganisms.

Biologically, the two coenzyme forms, flavin mononucleotide (FMN) and flavin adenine dinucleotide (FAD), are predominant. These terms are formally incorrect, because FMN is no nucleotide and FAD is no dinucleotide, however, these names are accepted (Nielsen, 2000).

Based on their unique structure (benzene, pirimidine, pyrazine ring), the reactivity of flavin compounds in biochemical reactions is very complex. Flavin coenzymes are most versatile catalysts in biological redox systems; can act as a redox switch between two electron donors (NADH, succinate) and one electron acceptors (iron-sulphur proteins). Vitamin B_2 supplies oxygen to the body tissues. Neutral aqueous solutions of riboflavin have a greenish-yellow colour.

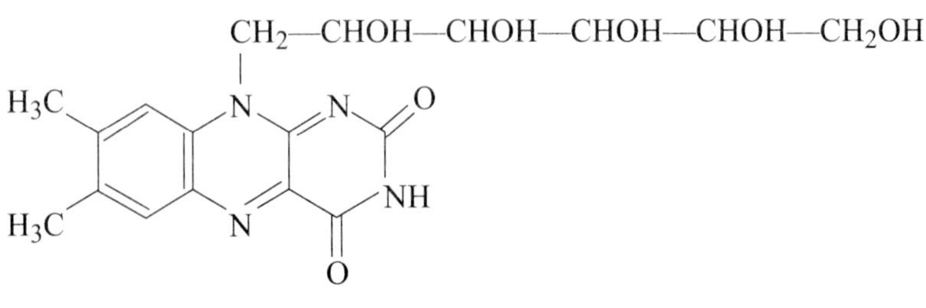

Figure 14.6 Riboflavin (Kutsal and Özbas, 1989).

Riboflavin can be synthesized by many microorganisms like: bacteria, fungi, algae and protozoa; among fungi by species: *Saccharomyces fragilis, Candida tropicalis, C. flareri, C. arborea, C. lactis, C. ghosi, Aspergillus niger, A. flavus, Penicillium chrysogenum, Trichoderma viride, Mycelium radicis atrovirens, Mortierella nana* and species of genus *Fusarium* (Strzelczyk, 1987).

Among these riboflavin producing fungi, two closely related ascomycetes, *Eremothecium ashbyii* and *Ashbya gossypii* are capable of synthesizing very large amounts of this vitamin (Ainsworth and Sussman, 1965). These fungi are assuming to possess the most important positions in the riboflavin biotechnology.

There are many reports on factors influencing riboflavin biosynthesis. Purines stimulate riboflavin overproduction in all of the organisms studied. Glycine, purine precursor, stimulates overproduction in *Candida* species and *A. gossypii*. The stimulatory effect of purines and glycine is not due to an effect on growth, but is specific for riboflavin overproduction (Demain, 1972).

Table 14.3 Effects of various substrates on specific growth (μ) and riboflavin production (v) rates and riboflavin yields ($Y_{P/SCo}$) for *E. ashybyii* and *A. gossypii* (30°C, stirring rate: 100 rpm, initial pH=6,5) (Kutsal and Özbas, 1989).

Substrates	*Initial substrate concentration (g/liter)*	E ashybyii μ^a (h^{-1})	E ashybyii $v \times 10^3$ *(gR/gmol per h)*	E ashybyii $Y_{P/SCo}$	A gossypii μ^a (h^{-1})	A gossypii $v \times 10^3$ *(gR/gmol per h)*	A gossypii $Y_{P/SCo}$
Glucose	20.0	0.1854	3.57	3.6	0.1538	1.25	3.1
Glycerol	20.0	0.1733	2.27	3.1	0.1315	0.89	1.3
Sunflower oil	20.0	0.1669	1.45	3.0	0.2245	1.00	2.9
Glucose + glycerol	5.0 + 15.0	0.2044	4.16	4.5	0.1979	1.32	1.7
Glucose + glycerol	7.5 + 12.5	0.2006	4.10	3.7	0.2153	1.5	1.9
Glucose + sunflower oil	5.0 + 15.0	0.2438	5.56	9.2	0.2323	1.49	5.2
Glucose + sunflower oil	10.0 + 10.0	0.2358	4.75	6.8	0.2635	4.05	8.8
Whey	12.5	0.1379	1.48	6.5	-	-	-
Whey	30.0	-	-	-	0.1529	0.24	2.2

[a] calculated in the region of exponential growth (Monod kinetic model)

Amino acids provide carbon source and not nitrogen in biosynthesis of riboflavin. L-treonine, L-serine, L-tyrosine stimulated riboflavin synthesis, whereas L-glutamate, L-aspartate and L-asparagine stimulated both growth and riboflavin synthesis by *E. ashbyii* (Goodwin, 1963a, b; Robinson, 1966).

Pyridoxine (Vitamin B_6)

The generic term vitamin B_6 refers to all 3 hydroxy-2-methylpyridine derivates that exhibit the biological activity of pyridoxine.

Pyridoxine (3 hydroxy-4,5-bis-(hydroxymethyl)-2-methyl pyridine) should not be used as generic term synonymous with vitamin B_6. Two other forms of vitamin $B_{6,}$ pyridoxal (PL) and pyridoxamina (PM) differ from pyridoxine (PN) in the respective location of an aldehyde and amine group at the 4 position of the pyridine ring structure.

The pyridoxal-5′-phosphate (PLP) is the metabolically active form of vitamin B_6 (Ubbink, 2000).

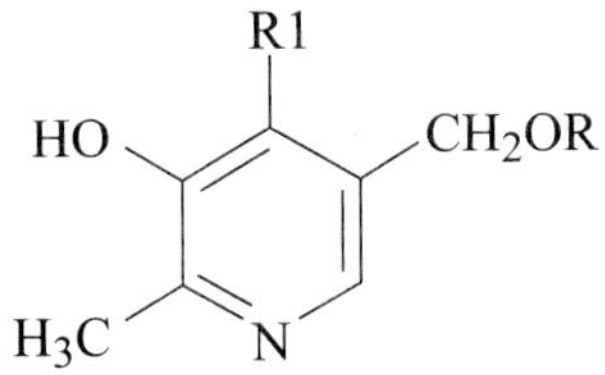

Pyridoxine: $R_1 = CH_2OH$, $R_2 = H$
Pyridoxal: $R_1 = CHO$, $R_2 = H$
Pyridoxamine: $R_1 = CH_2NH_2$, $R_2 = H$
Pyridoxine 5'-phosphate: $R_1 = CHO$, $R_2 = PO_3H_2$
Pyridoxal 5'-phosphate: $R_1 = CH_2OH$, $R_2 = PO_3H_2$
Pyridoxamine 5'-phosphate: $R_1 = CH_2NH_2$, $R_2 = PO_3H_2$

Figure 14.7 Chemical structure of vitamin B_6 (Tani, 1989).

Pyridoxal or pyridoxamine is a coenzyme for various enzymes associated with the metabolism of amino acids, amines, saccharides, fatty acids, etc. (Tani, 1989). It controls the absorption, metabolism and conversion of amino acids into neurotransmitters, antibodies, digestive enzymes, muscles and tissues in the body. The forming of histamine, serotonin, dopamine and adrenaline are dependent on vitamin B_6 (Bernstein, 1990).

Extensive screening for vitamin B_6 producers among microorganisms has been performed. Almost all microorganisms studied by Tani *et al.* (1972) and Tani (1989) synthesized vitamin B_6 intracellularly in amounts of 0.05–0.3 mg/g of dry cells and excreted it in amounts of 0.05–0.5 mg/litre of culture medium. However, such productivity was insufficient to establish an industrial microbiological process.

Pichia guilliermondii (yeast) synthesize the highest amounts of extracellular vitamin B_6 (25 mg L^{-1}). Vitamin B_6 was also found in culture filtrates of *Candida albicans* (0–6 mg L^{-1}) and *Saccharomyces marxianus* (0–2 mg L^{-1}).

Biotin

Biotin, hexahydro-2-oxo-1H thieno (3,4 d) imidazole-4-pentanoic acid is required by all living cells but only biosynthesized by plants, bacteria and fungi. Biotin is produced by the bacteria in the intestinal tract. It is a sulfur-containing vitamin that work with a number of enzymes responsible for physiological reactions in the body.

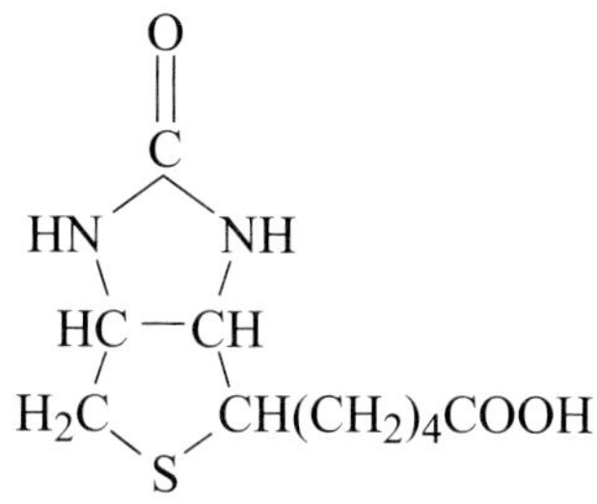

Figure 14.8 Chemical structure of (+)-biotin (Izumi and Yamada, 1989).

Biotin is a coenzyme that uses fatty acids and amino acids. It is essential to form purines, which are important parts of DNA and RNA. The biotin-dependent enzyme group includes three classes of enzymes: carboxylases, transcarboxylases and decarboxylases (Moss and Lane, 1971). In all the reactions catalyzed by these enzymes, biotin acts as a CO_2 carrier via N_1-carboxybiotin.

However, the informations about the production of biotin by fungi are scarce. Dahm *et al.* (1989) detected this vitamin in the culture medium of soil-borne fungus of *Macrophomina phaseolina* (3.5 μg L^{-1}). Mycorrhizal fungi studied by Strzelczyk *et al.* (1991) did not produce biotin.

Table 14.4 Biotin-vitamin producing abilities[a] and biotin biosynthetic enzyme activities of various yeasts (Izumi and Yamada, 1989).

Strain	*True biotin*[b]	*Total biotin*	*Specific activity*[c] *(units/mg protein)*			
	(μg/ml)	*(μg/ml)*	*1*	*2*	*3*	*4*
Saccharomyces kloeckerianus IFIO 0016	tr[d]	0.71	tr	tr	0.06	0.21
Lipomyces starkeyi IFO 0678	0.015	0.02	tr	2.90	tr	0.07
Sporobolomyces salmonicolor IFO 0374	0.085	0.49	0.23	1.30	tr	0.06
Sporobolomyces salmonicolor IFO 1038	0.068	0.30	0.57	1.08	tr	0.03
Sporobolomyces coprophilus IFO 1442	0.100	0.16	0.14	0.31	tr	0.01
Rhodotorula glutinis IFO 0415	0.066	0.12	tr	0.10	tr	0.03

[a] The amounts of biotin-vitamers produced by organism grown with pimelic aid

[b] The amount of biotin-vitamer given by the bioassay using Lactobacillus plantarum, which includes biotin and biotin sulfoxide

[c] 1,pimelyl-CoA synthetase; KAPA synthetase; 3,DAPA aminotransferase; 4,DTB synthetase

[d] Trace

Commercial production of biotin depends on the chemical process.

Pantothenic Acid (Vitamin B_5)

Pantothenic acid, (R-or D – (+) – (2,4 dihydroxy-3,3dimethyl – 1-oxobutyl) –β-alanine could be obtained from a variety of plants and animal tissues (pantothenic acid, meaning "from everywhere"). Pantothenic acid occurs in all types of living organisms in free form and in conjugated forms such as coenzyme A, pantetheine and phosphopantetheine. Vitamin B_5 is essential for the function of brain and adrenal gland that controls the hormones that handles physical and mental stress.

Microorganisms (and plants) are able to produce pantothenic acid from the precursors pantoic acid and β-alanine. Animal, plants and microorganisms can convert pantothenic acid to 4′-phosphopantetheine and coenzyme A, the metabolically active forms of the vitamin (Shimizu and Yamada, 1989). Coenzyme A can form high energy bonds with acetic acid via its sulfhydryl group to yield acetyl-coenzyme A (activated acetate).

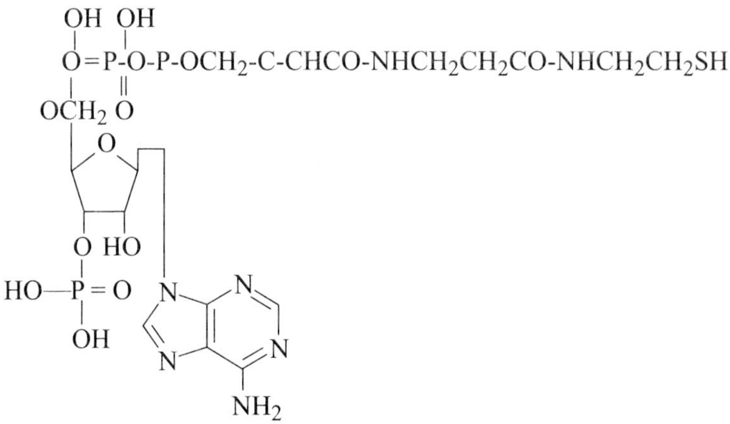

Figure 14.9 The structure of coenzyme A (Shimizu and Yamada, 1989).

Among B-group vitamins produced by soil-borne fungi, pantothenic acid was detected in the culture of all fungi studied and in large quantities, even 830 μg L^{-1} *(Suillus bovinus)* (Dahm *et al.*, 1989; Strzelczyk *et al.*,1991).

The commercial production process of pantothenic acid involves reaction yielding racemic pantolactone from isobutyraldehyde, formaldehyde and cyanide to D-pantolactone and condensation of D-pantolactone with β-alanine.

Recently, in this process, an efficient enzymatic method using the fungal enzyme lactonohydrolase as the catalyst has been introduced. Several filamentous fungi of the genera *Fusarium, Gibberella, Cylindrocarpon* show high activity of this enzyme (Shimizu, 2001).

Nicotinic Acid and Nicotinamide (Vitamin PP)

Pellagra (term consisting of a synthesis of the Italian word from *pelle* (skin) and *agre* (rough) was introduced by Frappoli in 1771 (Shibata and Taguchi, 2000).

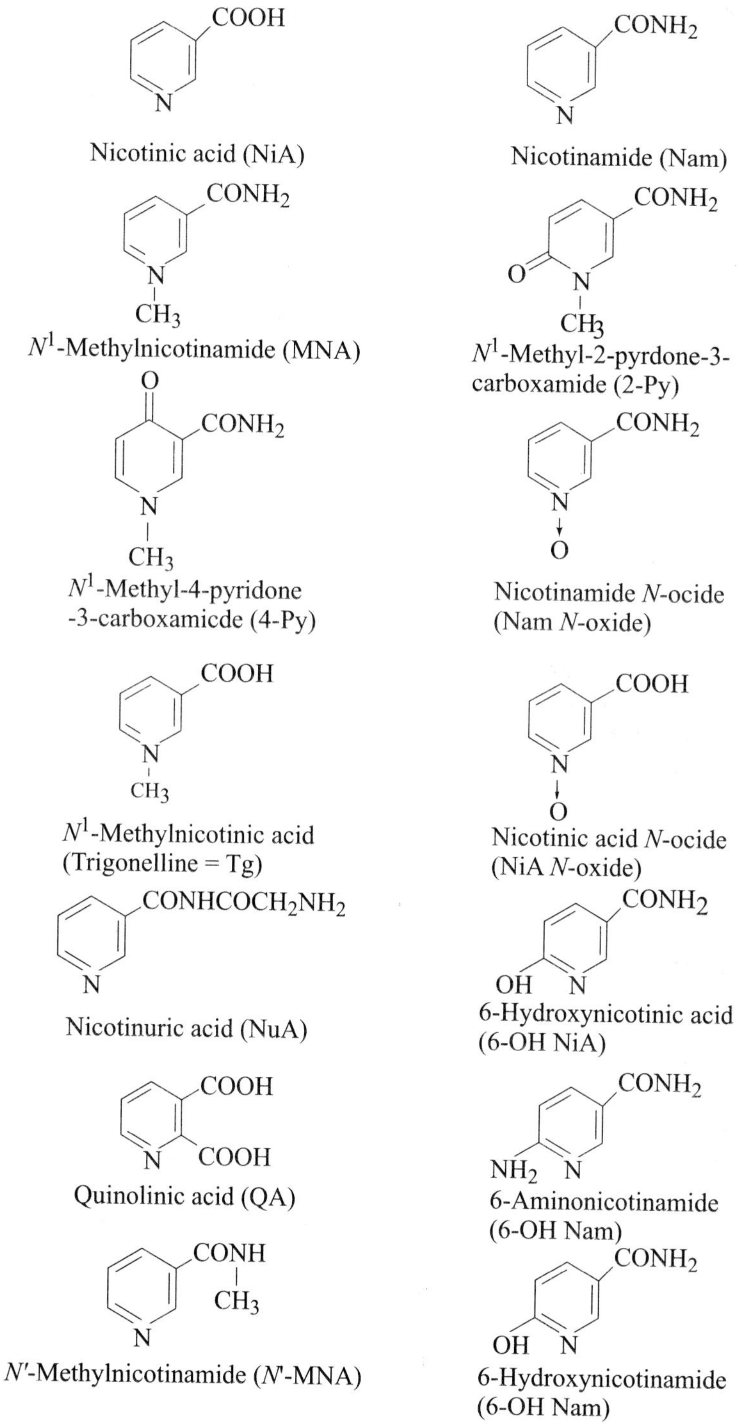

Figure 14.10 Structures of nicotinamide, nicotinic acid, and related compounds (Shibata and Taguchi, 2000).

Now, it is accepted that pellagra is caused by a nicotinic acid or nicotinamide deficiency. Pellagra is a typical hypovitaminosis disorder. This disorder is characterized by three symptoms: dermatitis, diarrhea and dementia.

Nicotinic acid is contained mainly in plant food such as coffee, cereals and seeds, but not in animal foods.

Pyridine-3-carboxylic acid (nicotinic acid) was first discovered as an oxidation product of nicotine by Huber in 1867. Nicotinic acid was first isolated from rice bran by Suzuki in 1912 and from yeasts by Funk in 1913 (Shibata and Taguchi, 2000). Subsequently, nicotinamide was found to be an integral part of NAD^+ (von Euler *et al.*, 1936) and $NADP^+$ (Warburg and Christian, 1935). The biochemical function of nicotinic acid and nicotinamide actually comes down to that of NAD^+ and $NADP^+$ as coenzymes of dehydrogenases. Over 350 enzymes require NAD and NADP as coenzymes.

Extraction from yeast cells and enzymatic synthesis are performed for the commercial production of NAD. *Saccharomyces cerevisiae, S. carlsbergensis* and *S. uvarum* have been shown to be sources of NAD.

In a growth medium containing adenine and nicotinamide 12 mg/g dry cells and 42 mg/g dry cells *S. cerevisiae and S. carlsbergensis*, respectively, were noted (Sakai *et al.*, 1973a, b). Strzelczyk *et al.* (1991) detected nicotinic acid in the culture of mycorrhizal fungus *Suillus bovinus*, mainly in the neutral (pH 7) media. Amounts of nicotinic acid were high:1.2 mg/g dry weight of mycelium. Considerably lower amounts of nicotinic acid were detected in the culture of the soil borne filamentous fungi: *Trichoderma viride* (18.2 μg/g dry mass), *Mycelium radicis atrovirens* (4.1 μg/g dry mass) and *Mortierella nana* (21.4 μg/g dry mass (Strzelczyk, 1987).

Various detailed methods of the industrial production of vitamins were reviewed by Vandamme (1989) and Rehm and Reed (2001).

Methods for analysis of vitamins

Various methods have been designed to vitamins analysis: enzymatic, microbiological assay, fluorescence spectrophotometry, immunoassays, chromatographic methods, masspectrometry. Assay methods are based on use of the defined vitamin requiring microorganisms such as *Lactobacillus* species (*L. plantarum, L. casei, L. arabinosus), Saccharomyces cerevisiae, S. uvarum, S. carlsbergensis, Staphylococcus aureus, Escherichia coli, Phycomycetes blakesleeanus.*

Assay methods for detection of thiamine is based on the use of thiamine requiring fungus *Phycomyces blakesleeanus* and strains of bacteria such as *Staphylococcus aureus, Escherichia coli,* and *Lactobacillus* species (Thomas, 1966; Strzelczyk, 1987; Dahm *et al.*, 1989; Strzelczyk *et al.*, 1991). However, chemical methods are usually preferred to microbiological methods since such analyses can be performed rapidly.

For the differential determination of thiamine and its phosphate esters electrophoresis, thin-layer chromatography and ion-exchange chromatography can be used. In recent years, high performance liquid chromatography (HPLC) has been developed for determination of thiamine.

The intensity of the fluorescent products is measured with a spectrophotometer. The minimum detection by these methods has been reported to be 0.1 pmol (Sanemori *et al.*, 1980). Thiamine has also been estimated manometrically by measuring the evolution of CO_2 from pyruvate in the presence of the apoenzyme of yeast pyruvate decarboxylase and spectrophotometrically by measuring acetylaldehyde–depend reduction of $NADH_2$ using alcohol dehydrogenase (Kajiro, 1957; Kawasaki and Sanemori, 1985). *Riboflavin* can be assayed by physico-chemical or microbiological methods. The most commonly used physico-chemical method involves measurement of the characteristic yellow-green fluorescence, which has a maximum at 565 nm at pH 6. Microbiological assays of riboflavin is based on the growth of some riboflavin-requiring microorganisms. In classic microbiological assay for riboflavin with *Lactobacillus casei* or more sensitive *Leuconostoc mesenteroides* which is measured by titrating the lactic acid (Strzelczyk, 1987; Kutsal and Özbas, 1989).

Test microorganisms used for the microbiological assay of *pantothenic acid* are *Lactobacillus plantarum (L. arabinosus), L. casei and Saccharomyces uvarum (S. carlsbergensis). Lactobacillus plantarum* is suitable for determination of unconjugated pantothenate in samples. Chemical and physical methods in determination of pantothenic acid have also been used, but they have low sensitivity (Strzelczyk and Leniarska, 1985; Airas,1986; Strzelczyk, 1987; Dahm *et al.*, 1989; Shimizu and Yamada, 1989; Strzelczyk *et al.*, 1991). A number of microbiological procedures are established for the estimation of *pyridoxine.* The method with *Saccharomyces cerevisiae, S. carlsbergensis, S. uvarum* is widely used for bioassay of this vitamin in non-esterified form. An assay employs *Lactobacillus casei,* which responds to pyridoxal, and *Streptococcus faecalis,* which responds to pyridoxal and pyridoxamine. Chemical methods based on reactions with 2,6 dichloroquinone chlorimide and diazotized p-aminoacetophenone are less sensitive and non-specific. The best method is high performance liquid chromatography, applied for quantitative and simultaneous detection of all vitamin B_6 compounds by determining UV absorption and fluorescence (Tani, 1989). The microbiological assay has been most commonly used among three methods (microbiological, chemical and enzymatic) for the determination of *biotin.* Microbiological assay uses the microorganisms *Lactobacillus plantarum, Saccharomyces cerevisiae, Bacillus subtilis* and biotin-requiring mutant of *Escherichia coli* (Scheiner, 1985; Strzelczyk, 1987; Dahm *et al.*, 1989; Strzelczyk *et al.*, 1991). Bioassay as a method was not frequently used for *ascorbic acid* determination. Most estimations are based on colorimetric methods using blue dye 2,6-dichloroindophenol (Hughes, 1983). In the presence of L-ascorbic acid, dye is reduced, resulting in colourless or pink colour solution. Titration of L-ascorbic acid with 2,6-dichloroindophenol is the most frequently used oxido-reduction applicable method. Gas-liquid chromatography was also adapted for quantitative determination of L-ascorbic acid (Schlack, 1974).

Conclusion

Fungi are the first organisms that mankind has used in biotechnological processes for food production (beer, honey, wine bread, silage etc.).

Many fungal metabolites are well known to possess some biological activities (mycotoxins, antibiotics, vitamins). Vitamins are used for human nutrition and therapy and as an animal feed additive.

The world production of vitamins is estimated to be 10 t^{a-1} Vitamin B_{12}; 25 t^{a-1} Biotin; 2,500 t^{a-1} Vitamin B_2, Vitamin B_6 and Vitamin A; 22,000 t^{a-1} Vitamin E; 70,000 t^{a-1} Vitamin C.

Many vitamins are produced on industrial scale by biotechnological processes, although chemical synthesis is still dominant method for several vitamins. Biotechnological processes involved in fermentation and microbial transformation have many valuable properties for the practical purposes.

Most reactions involve only one compound that is structurally related and proceed by distinguishing almost completely between steroisomers or regioisomers.

Microorganisms performing unique biochemical reactions have been discovered through extensive screening among a wide variety of microorganisms.

Some fungi show specific environmental requirements for growth. The vitamin production by fungi may be increased through optimization of culture conditions. However, the culture requirements are traditionally optimized by the one factor at a time method, i.e. varying one factor while keeping all others at certain levels. The selection of high vitamin-producing mutants cause the most important increases in vitamins productivity.

References

Abico, Y. (1975). Metabolism of coenzyme A. In: Metabolic pathways. 3rd edn. Vol. 7. Metabolism of sulfur compounds. Greenberg D.M. Ed. Academic Press, New York: pp 1–25.

Akagi, M., and Kumaoka, H. (1963). Effect of sulfur-containing amino acids on the production of thiamine by *Escherichia coli*. *J. Vitaminol.* 9: 183–187.

Ainsworth, G.C. and Sussman, A.S. (1965). The fungi. Academic Press, New York: pp 37, 492.

Airas, R.K. (1986). Pantothenase-based assay of pantothenic acid. *Meth. Enzymol.* 122: 33–35.

Andrewes, A.G. and Starr, M.P. (1976). (3*R*, 3'*R*)-Astaxanthin from the yeast *Phaffia rhodozyma*. *Phytochem.* 15: 1009–1011.

Avalos, J. and Cerdá-Olmedo, E. (1987). Carotenoid mutants of *Giberella fujikuroi*. *Curr. Genet.* 11: 505–511.

Avalos, J., Youssar, L., Arrach, N. and Schmidhauser, T.J. (2001). New mutants of the carotenoid pathway of *Neurospora crassa*. XXI Fungal Genetics Conference, Asilomar, California. *Appl. Env. Microbiol.* 65: 2558–2564.

Barua, A.B., Olson, J.A., Furr H.C. and van Breemen R.B. (2000). Vitamin A and carotenoids. In: Modern Chromatographic analysis of vitamins. De Leenheer A.P., Lambert W.E., Van Bocxlaer J.F. Eds. Marcel Dekker, Inc., New York, Basel: pp 1–74.

Bergman, K., Eslava, A.P. and Cerdá-Olmedo, E. (1973). Mutants of *Phycomyces* with abnormal phototropism. *Molec. Gen. Genet.* 123: 1–16.

Bernstein, A.L. (1990). Vitamin B_6 in clinical neurology. Ann. N.Y. Acad. Scie. 585: 250–260.

Bertrand, G. (1896). Préparation biochimique du sorbose. C.R. Hebd. Séanc. Acad. Sci., Paris 122: 900–903.

Bills, C.E., Massengale, O.N. and Prickett, P.S. (1930). Factors determining the ergosterol content of yeasts. I. Species. *J. Biol. Chem.* 87: 259–64.

Bindl, E., Lang, W. and Rau, W. (1970). Untersugungen uber die lichtabhangige Carotinoidsynthese. VI. Zeitcher Verlauf der Synthese der einzelnen Carotenoide bei *Fusarium aquaeductuum* under verschiedenen Induktionsbedingungen. *Planta* 94: 156–174.

Cerdá-Olmedo, E. (1987). Carotene. In: Phycomyces. Cerda-Olmedo E. and Lipson E.D. Eds. Cold Spring Harbor Laboratory, Cold Spring Harbor, NY: pp 199–222.

Cerdá-Olmedo, E. (1989). Production of carotenoids with fungi. In: Biotechnology of vitamins, pigments and growth factors. Vandamme E.J. Ed. Elsevier Applied Science, London: pp 27–42

Cerdá-Olmedo, E. and Hüttermann, A. (1986). Forderung and hemmung der carotinsynthese bei *Phycomyces* durch aromaten. *Angew. Botanik* 60: 59–70.

Combs, G. F.(1992). The vitamins. Fundamental aspects in nutrition and health. Academic Press. San Diego: pp 179–203.

Conly, J.M., Stein, K., Worobetz, L. and Rutledge-Harding, S. (1994). The contribution of vitamin K2 (menaquinones) produced by the intestinal microflora to human nutritional requirements for vitamin K. *Am. J. Gastroenterol* 89(6): 311–316.

Dam, H. (1935). The antihaemorrhagic vitamin of the chick (letter). *Nature* 135: 652–653.

Dahm, H., Reddy, M.N. and Strzelczyk, E. (1989). Production of B-group vitamins by some soil borne fungi pathogenic to groundnut (*Arachis hypogea* L.). *Bull. Polish Acad. Sci., Biol. Sci.* 37(1-3): 13–17.

Delić, V., Šunić, D. and Vlašić, D. (1989). Microbial reactions for the synthesis of vitamin C (L-ascorbic acid). In: Biotechnology of vitamins, pigments and growth factors. Vandamme E.J. Ed. Elsevier Applied Science, London: pp 299–334.

Demain, A.L. (1972). Riboflavin oversynthesis. In: Annual Reviews in Microbiology Vol. 26. Clyton C.E., Raffel S., Starr M.P. Eds. George Benta Inc., USA: pp 369–388.

Diplock, A.T. (Ed.) (1985). Fat soluble vitamins. Their biochemistry and applications. Heinman, London.

Diplock, A.T., Green, J., Edwin, E.E. and Bunyan, J. (1961). Tocopherol, ubiquinones and ubichromenols in yeasts and mushrooms. *Nature,* 189: 749–750.

Dulaney, E.L., Stapley, E.O. and Simpf, K. (1954). Studies on ergosterol production by yeasts. *Appl. Microbiol.* 2: 371–379.

Eder, L. and Dunant, Y. (1980). Thiamine and cholinergic transmission in the electric organ of *Torpedo*. I. Cellular localization and functional changes of thiamine and thiamine phosphate esters. *J. Neurochem.* 35: 1278–1286.

Egi, Y., Koyama, S., Shikata, H., Yamada, K. and Kawasaki, T. (1986). Content of thiamin phosphate esters in mammalian tissues – an extremely high concentration of thiamin triphosphate in pig skeletal muscle. *Biochem Int.* 12: 385–390.

El-Jack, M., Mackenzie, A. and Bramley, P.M. (1988). The photoregulation of carotenoid biosynthesis in *Aspergillus giganteus* mut. *alba*. *Planta* 174: 59–66.

Estramareix, B. and Therisod, M. (1984). Biosynthesis of thiamine. 5-Aminoimidazole ribotide as the precursor of all the carbon atoms of the pyrimidine moiety. *J. Am. Chem. Soc.* 106: 3857–3860.

Fauler, G., Muntean, W. and Leis, H.J. (2000). Vitamin K. In: Modern chromatographic analysis of vitamins. De Leenheer A.P., Lambert W.E., Van Bocxlaer J.F. Eds. Marcel Dekker, Inc., New York, Basel: pp 229–270.

Funk 1913

Gaby, S.K., Bendich, A., Singh, V.N. and Machlin, L.J. (1991). Vitamin intake in health, a scientific review. Marcel Dekker, New York: pp 103–161.

Gal'tsowa, R.D., Novichkova, A.T. and Vakina, I.P. (1959). Effect of glucose on ergosterol synthesis by yeasts. *Mikrobiologiya* 28: 502–506.

Geiger-Huber, M. and Galli, H. (1945). Über den Nachweis der L-Ascorbinsäure ais Stoffwechsel-produkt von *Aspergillus niger*. *Helv. Chim. Acta* 28: 248–250.

Goodwin, T.W. (1963a). Vitamins. In: Biochemistry of industrial micro-organisms. Rainbow C. and Rose A.H. Eds. Academic Press, London: pp 151–158.

Goodwin, T.W. (1963b). Riboflavin and related compounds. In: The biosynthesis of vitamins and related compounds. Goodwin T.W. Ed. Academic Press, London: pp 24–35.

Grue-Sørensen, G., White, R.L. and Spenser, I.D. (1986). Thiamin biosynthesis in *Saccharomyces cerevisiae*. Origin of the pyrimidine unit. *J. Am. Chem. Soc.* 108: 146–158.

Haj-Ahmad, Y., Bilinski, C.A., Russel, I. and Stewart, G.G. (1992). Thiamin secretion in yeasts. *Can. J. Microbiol.* 38: 1156–1161.

Han, Y.F., Liang, Z.Q. and Liu, A.Y. (2004). Fat-soluble ingredients from *Paecilomyces* spp. Journal of Huazhong Agricultural University 23(1): 171–172.

Harding, R.W., Huang, P.C. and Mitchell, H.K. (1969). Photochemical studies of the carotenoid biosynthetic pathway in *Neurospora crassa*. *Archs. Biochem. Biophys.* 129: 696–707.

Hodges, S.J., Bejui, J., Leclercq, M. and Delmas, P.D. (1993). Detection and measurement of vitamins K_1 and K_2 in human cortical and trabecular bone. *J. Bone Miner. Res.* 8(8): 1005–1008.

Hughes, D.E. (1983). Titrimetric determination of ascorbic acid with 2,6-dichlorophenol indophenol in commercial liquid diets. *J. Pharm. Sci.* 72(2): 126–129.

Hughes, P.E. and Tove, S.B. (1982). Occurence of α-tocopherolquinone and α-tocopherolquinol in microorganisms. *J. Bacteriol.* 151: 1397–1402.

Iwashima, A. (1980). Absorption and membrane transport of thiamine. In: Vitaminology, Vol. II. Vitamin Society of Japan Ed. Tokyokagakudojin, Tokyo: pp 30–39.

Izumi, Y. and Yamada, H. (1989). Microbial production of biotin. In: Biotechnology of vitamins, pigments and growth factors. Vandamme E.J. Ed. Elsevier Applied Science, London: pp 231–256.

Jones, G. and Makin, H.L.J. (2000). Vitamin Ds. In: Modern chromatographic analysis of vitamins. De Leenheer A.P., Lambert W.E., Van Bocxlaer J.F. Eds. Marcel Dekker, Inc., New York, Basel: pp 75–141.

Kajiro, Y. (1957). Quantitative determination of thiamine pyrophosphate using apocarboxylase and alcohol dehydrogenase. *J. Biochem.* 44: 827–838.

Kawasaki, T. and Egi, Y. (2000). Thiamine. In: Modern chromatographic analysis of vitamins. De Leenheer A.P., Lambert W.E., Van Bocxlaer J.F. Eds. Marcel Dekker, Inc., New York, Basel: pp 365–389.

Kawasaki, T. and Sanemori, H. (1985). Vitamin B_1: thiamines. In: Modern chromatographic analysis of the vitamins. De Ruyter M.G.M. Ed. Marcel Dekker, New York: pp 385–411.

Klein, H.P. (1955). Synthesis of lipids in resting cells of *Saccharomyces cerevisiae*. *J. Bacteriol.* 69: 620–627.

Krampitz, L.O. (1969). Catalytic functions of thiamine diphosphate. *Annu. Rev. Biochem.* 38: 213–240.

Kransinski, S.D. and Russell, K.M. (1985). The prevalence of vitamin K deficiency in chronic gastrointestinal disordes. *Ann. J. Cl. Nutr.* 41:639

Kumaoka, H. and Brown, G.M. (1967). Biosynthesis of thiamine. VI. Incorporation of formate into carbon atom two of the pyrimidine moiety of thiamine. *Archs. Biochem. Biophys.* 122: 378–284.

Kutsal, T. and Özbas, M.T. (1989). Microbial production of vitamin B_2 (riboflavin). In: Biotechnology of vitamins, pigments and growth factors. Vandamme E.J. Ed. Elsevier Applied Science, London: pp 149–166.

Lilly, V.G., Barnett, H.L. and Krause, R.F. (1960). The production of carotene by *Phycomyces blakesleeanus*. West Virginia Univ. *Agr. Exper. Sta.*, Bulletin 441T, Morgantown.

Linder, M.C. (1991). Nutrition and metabolism of vitamins. In: Nutritional biochemistry and metabolism with clinical applications. Linder M.C. Ed. Elsevier Science, New York: pp 111–189.

Margalith, P. (1989). Vitamin D: the biotechnology of ergosterol. In: Biotechnology of vitamins, pigments and growth factors. Vandamme E.J. Ed. Elsevier Applied Science, London: pp 81–93.

Moss, J. and Lane, D. (1971). The biotin-dependent enzymes. *Adv. Enzymol.* 35: 321–442.

Nelis, H.J., D'Haese, E. and Vermis, K. (2000). Vitamin E. In: Modern chromatographic analysis of vitamins. De Leenheer A.P., Lambert W.E., Van Bocxlaer J.F. Eds. Marcel Dekker, Inc., New York, Basel: pp143–228.

Nielsen, P. (2000). Flavins. In: Modern chromatographic analysis of vitamins. De Leenheer A.P., Lambert W.E., Van Bocxlaer J.F. Eds. Marcel Dekker, Inc., New York, Basel: pp 391–433.

Nyyssõnen, K., Salonen, J.T. and Parviainen, M.T. (2000). Ascorbic acid. In: Modern Chromatographic analysis of vitamins. De Leenheer A.P., Lambert W.E., Van Bocxlaer J.F. Eds. Marcel Dekker, Inc., New York, Basel: pp 271–300.

Obata, Y., Nara, K., Tarui, K., Mochizuki, K. and Isono, M. (1975). L-ascorbic acid. Japan Patent No. 75 22113.

Okazaki, H., Kanzaki, T., Doi, M., Nara, K. and Motizuki, K. (1968). 2-keto-L-gulonic acid. II. Identification of metabolic products from sorbitol. *Agric. Biol. Chem.* 32: 1250–1255.

Okazaki, H., Kanzaki, T., Sasayama, K. and Taketa, Y. (1969). 2-keto-L-gulonic acid. III. Evaluation of the pathway of sorbitol metabolism in *Gluconobacter melanogenus*. *Agric. Biol. Chem.* 33: 207–211.

Packer, L. and Fuchs, J. (1993). Vitamin E in health and disease. Marcel Dekker, New York: pp 977-982.

Ramakrishnan, C.V. and Desai, P.J. (1956). Effect of addition of iron, cobalt and ascorbic acid to the medium on the synthesis of ascorbic acid in molds. *Current Sci.* 25: 189–190.

Rehm. H.-J. and Reed, G. (2001). Biotechnology. Vol. 10. Weinheim: Wiley-VCH.

Robinson, F.A. (1966). The vitamin co-factors of enzyme systems. Pergamon Press, Braunschweig: pp 160–163.

Roland, J.F., Cayle, T., Dinwoodie, R.C. and Mehnert, D.W. (1985). Bioconversion production of ascorbic acid. WO 85/01745.

Roland, J.F., Cayle, T., Dinwoodie, R.C. and Mehnert, D.W. (1986). Fermentation production of ascorbic acid from L-galactonic substrate. US Patent No. 4 565 659.

Rose, R.C. and Bode, A.M. (1992). Tissue-mediated regeneration of ascorbic acid: is the process enzymatic? *Enzyme* 46: 196–203.

Sakai, T., Uchida, T. and Chibata, I. (1973a). Production of nicotinamide adenine dinucleotide by *Saccharomyces carlsbergensis*. *Agric. Biol. Chem.* 37: 1041–1048.

Sakai, T., Uchida T. and Chibata, I. (1973b). Accumulation of nicotinamide adenine dinucleotide in baker's yeasts by secondary culture. *Agric. Biol. Chem.* 37: 1049–1056.

Sanemori, H. and Kawasaki, T. (1980). Purification and properties of thiamine pyrophosphokinase in *Paracoccus denitrificans*. *J. Biochem.* 88: 223–230.

Sanemori, H., Ueki, H. and Kawasaki, T. (1980). Reversed-phase high-performance liquid chromatographic analysis of thiamine phosphate esters at subpicomole levels. *Analyt. Biochem.* 107(2): 451–455.

Sastry, K.S. and Sarma, P.S. (1957). Glucuronic acid, a precursor of ascorbic acid in *Aspergillus niger*. *Nature, Lond.* 179: 44–45.

Schechter, I., Sweat F.W. and Bloch K. (1970). Comparative properties of 2,3 oxidosqualene-lanosterol cyclase from yeast and liver. *Biochim. Biophys. Acta.* 220: 463–468.

Scheiner, J. (1985). Biotin. In: Methods of vitamin assay. 4th edn. Augustin J., Klein B.P., Becker D.A., Venugopal P.B. Eds. John Wiley & Sons, New York: pp 535–53.

Schlack, J.E. (1974). Quantitative determination of L-ascorbic acid by gas-liquid chromatography. *J. Assoc. Off Anal. Chem.* 57(6): 1346–1348.

Shibata, K. and Taguchi, H. (2000). Nicotinic acid and nicotinamide. In: Modern Chromatographic analysis of vitamins. De Leenheer A.P., Lambert W.E., Van Bocxlaer J.F. Eds. Marcel Dekker, Inc., New York, Basel: pp 325–364.

Shimizu, S. (2001). 11 Vitamins and related compounds: microbial production. In: Biotechnology. Vol. 10. Rehm H.-J. and Reed G. Eds. Weinheim: Wiley-VCH: pp 320–340.

Shimizu, S. and Yamada, H. (1989). Pantothenic acid (vitamin B_5), coenzyme A and related compounds. In: Biotechnology of vitamins, pigments and growth factors. Vandamme E.J. Ed. Elsevier Applied Science, London: pp 199–219.

Silhankova, L. (1985). Yeasts mutants excreting vitamin B_1 and their use in the production of thiamine rich beers. *J. Inst. Brew.* 91: 78–81.

Strzelczyk, E. (1987). Production of B-group vitamins by non-mycorrhizal fungi associated with the roots of forest trees. *Acta Microbiologica Polonica* 36(4): 337–344

Strzelczyk, E. and Leniarska, U. (1985). Production of B-group vitamins by mycorrhizal fungi and actinomycetes isolated from the root zone of pine (*Pinus sylvestris* L.). *Plant and Soil* 86: 387–394.

Strzelczyk, E., Dahm, H. and Pachlewski, R. (1991). B-group vitamins production by mycorrhizal fungi in response to pH (*in vitro* studies). *Plant and Soil* 137: 237–241

Sutter, R.P. (1970). Mutations affecting sexual development in *Phycomyces blakesleeanus*. *Proc. Natl. Acad. Sci. USA* 72 (1): 127–130.

Suzuki 1912

Tada, M. and Shiroishi, M. (1982). Mechanism of photoregulated carotenogenesis in *Rhodotorula minuta*. I. Photocontrol of carotenoid production. *Plant. Cell. Physiol.* 23: 541–547.

Tani, Y. (1989). Microbial production of vitamin B_6 and derivatives. In: Biotechnology of vitamins, pigments and growth factors. Vandamme E.J. Ed. Elsevier Applied Science, London: pp 221–230.

Tani ,Y., Nakamatsu, T., Izumi, Y. and Ogata, K. (1972). Studies on vitamin B_6 metabolism in microorganisms. Part XI. Extracellular formation of vitamin B_6 by marine and terrestrial microorganisms and its control. *Agric. Biol. Chem.* 41: 1749–1754.

Tani, Y., Nishise, H., Morita H. and Yamada, H. (1984). Vitamin B_6 biosynthesis and glycolate reductase in *Flavobacterium* sp. 238-7. *J. Nutr. Sci. Vitaminol.* 30: 415–420.

Thomas, M.H (1966). Microbiological assay techniques. In: Methods of vitamin assay. Association of Vitamin Chemists Ed. Interscience Publishers, New York: pp 37–62.

Ubbink, J.B. (2000). Vitamin B_6. In: Modern chromatographic analysis of vitamins. De Leenheer A.P., Lambert W.E., Van Bocxlaer J.F. Eds. Marcel Dekker, Inc., New York, Basel: pp 435–477.

Usdin, V.R. and Burell, R.C. (1952). Preliminary study of the lipids of *Rhodotorula gracilis*. *Archs. Biochem. Biophys.* 36: 172–177.

Valadon, L.R.G. and Mummery, R.S. (1973). Effect of certain inhibitors of carotenogenesis in *Verticillium agaricum*. *Microbios* 7: 173–80.

Vandamme, E.J. (1989). Biotechnology of vitamins, pigments and growth factors. Elsevier Applied Science, London: pp 1–11.

Vaskivnyuk, V.Y. and Getman, E.I. (1984). Biosynthesis of carotenoids in light in continuous cultures of the yeast *Rhodosporidium diobovatum* (in Russian). *Priklad. Biokhim. Mikrobiol.* 20: 480–483.

Vermeer, C., Jie, K.S. and Knapen, M.H. (1995). Role of vitamin K in bone metabolism. *Annu. Rev. Nutr.* 15: 1–22.

von Euler, H., Ahlers, H. and Schlenk, F. (1936). Chemical investigations on highly purified cozymase. *Z. Physiol. Chem.* 124: 113–126.

Warburg, O. and Christian, W. (1935). Coenzyme problem. *Biochem.* Z. 275: 464.

Yamada, K., Nakahara, T. and Fukui, S. (1971). Petroleum microbiology and vitamin production. In: Biochemical and industrial aspects of fermentation. Sakaguchi K., Uemara T., Kinoshita S. Eds. Kodnasha, Tokyo: pp 61–90.

Yin, G.L. *et al.* (1980). Studies on the production of vitamin C precursor 2-keto-L-gulonic acid from L-sorbose by fermentation. I. Isolation, screening and identification of 2-keto-L-gulonic acid producing bacteria. *Wei. Sheng. Wu. Hsueh. Pao.* 20: 246–251.

15

Fungi from Indoor Environments in USA: An Overview

Tara Dubey

MACS LAB INC. 3137 Diablo Ave., Hayward, CA 94544, USA
E-mail: tdubey@hotmail.com and tara@kellcomacsb.com

INTRODUCTION

In Western countries like Canada and USA, approximately 95% of time is spent inside the air-conditioned buildings. One of the major health concerns for the residents is asthma, allergy and pulmonary diseases. This has been observed more frequently in day care centers, schools and elderly care centers. Mold species like *Stachybotrys chartarum* (atra) growing inside the buildings due to increased humidity, sufficient nutrition and favorable temperature, are found to be responsible for such health problems as they produce chemical substances called mycotoxins. This paper presents an overview of fungi pathogenic to human and methods currently used to analyze different fungal species from problem and non-problem areas.

In West people mostly spend their time in closed buildings, which are regulated by HVAC system. Many of these buildings have excessive moisture problems due to seepage or running water. Prolonged accumulation of humidity results into the growth of molds and other microbes responsible for allergic problems like asthma, hay fever, running nose and other pulmonary diseases. Overgrowing molds cause damage of walls, ceilings, wooden doors and windows and carpeted area. Majority of states in USA are governed by mold regulation law, which requires a regular inspection of buildings for indoor air quality (IAQ) assurance. Building owners may have to face a lawsuit if an expert reports any unacceptable degree of growth of any mould genera. Standards for low, moderate or heavy mould growth may vary from very few spores to a few hundred spores depending on the type of fungus and type of residents. Very small amount of mold growth may be harmful for children, elderly people, for those persons having incompetent immunity level, and for the patients of AIDS, cancer and other diseases. This causes an extra responsibility for the analyst who is analyzing the quality and quantity of fungal species.

Most recent concern is the toxic effect of *Stachybotrys chartarum* (atra), which is found to be human pathogenic due to the toxic substances like Trichothecen and Satratoxin. This species grows very fast in humid conditions with sufficient nutrient supply and is reported as poisonous in living as well as in dead conditions. This paper explains the standard methods used for sampling and analyzing indoor air quality of buildings and to understand the levels of distribution of common indoor fungal taxa when compared to outdoor environment.

Major Sources

Indoor dusts collected by vacuuming, carpets, gypsum wall-boards, sheet-rocks, house plants, ventilation screens, wooden closets, shoe corners, bathroom tiles and tubs with wet surfaces are major sources if ventilation is poor.

Fungi Pathogenic to Human

The impact of indoor fungal contamination on human health has received ever-increasing attention since the early 1970s mainly due to declining IAQ. In North America this decline can be correlated with the following factors

- More white collar jobs and less industrial out door jobs.
- Increasing economic dependence on fossil fuels and oil crisis causing poor ventilation.
- The movement to design and construct energy efficient buildings.
- Prolonged indoor activities than outdoors like TV and computer involvement.

Very little is known about inhalation exposure to mycotoxins in human. The allerginicity of some common taxa like *Alternaria, Cladosporium, Penicillium* and *Aspergillus* species have been studied but lack of standard fungal extracts for skin test and serum antibody testing has made assessment of fungal allergy difficult (Gallup and Valesco, 1999). The response of fungal antigens varies with the age of individual, dose and length of exposure. It can be serious issue for immune comprised, AIDS, Organ Transplant, Infants and Elderly. It may not be a big issue for persons with competent immune system. Dead spores of some fungi can also produce harmful toxic compounds.

Allergic reactions are classified into four types and allergy may be caused by a combination of more than one type.

Type I: Atopic and Anaphylactic Hypersensitivity cause asthma and hayfever and may have acute reaction to genetically susceptible individuals.

Type II: Antibody dependent cytotoxic hypersensitivity involves human immunoglobulin antibodies and complement reaction with antigen on target cells and cause cell destruction. This is common in Rh blood group incompatibility, drug reactions and transfusion reactions.

Type III: Immune complex mediated hypersensitivity. Involves antigen and antibody and cause serum sickness, hypersensitive pneumonitis like "Farmer's Lung" due to inhalation of dust containing high number of microbial spores.

Type IV: Cell mediated hypersensitivity. Involves delayed allergic reaction to bacteria, virus and fungi. Examples, Dermatitis, Rejection of organ transplantation.

Table 15.1 Selected hypersensitivity pneumonitides with probable microbial etiologies (Pope *et al.*,1993; Park *et al.*, 1994; Hunninghake and Richardson, 1998).

Disease	*Source*	*Probable Allergen*
Bagassosis	Moldy bagasse (sugar cane)	Thermophilic actinomycetes
Cephalosporium HP	Basement sewage contamination	*Cephalosporium* spp. (=*Acremonium*)
Cheese washer's lung	Moldy cheese	*Penicillium casei* (=*P. roquefortii*)
Compost lung	Compost	*Aspergillus* spp.
Farmer's lung	Moldy hay, grain or silage	*Aspergillus fumigatus* and thermophilic actinomycetes
Hot tub lung	Mold on ceiling	*Cladosporium* spp.
Housewife's lung	Moldy wooden flooring	*Penicillium expansum* and other moulds
Humidifier/Air-conditioner lung	Contaminated water or coils in humidifiers and air-conditioners	*Aureobasidium pullulans*, *Cephalosporium* spp., *Penicillium* spp. and thermophilic actinomycetes
Japanese summer house HP	Bird droppings, house dust	*Trichosporon cutaneum*
Lycoperdonosis	Puffballs	*Lycoperdon* spp.
Malt worker's lung	Moldy barley	*Aspergillus fumigatus* or A. clavatus
Maple bark disease	Maple bark	*Cryptostroma corticale*
Mushroom worker's lung	Mushroom compost	Thermophilic actinomycetes and other microorganisms
Potato riddler's lung	Moldy hay around potatoes	*Aspergillus* spp. and thermophilic actinomycetes
Sauna taker's lung	Contaminated sauna water	*Cladosporium* spp. and others
Thatched roof disease	Dried grasses and other leaves	*Sacchoromonospora viridis*
Tobacco worker's disease	Moldy tobacco	*Aspergillus* spp.
Winegrower's lung	Moldy grapes	*Botrytis cinerea*
Wood trimmer's disease	Contaminated wood trimmings	*Rhizopus* spp. And *Mucor* spp.
Woodman's disease	Oak and maple trees	*Penicillium* spp.
Woodworker's lung	Oak, cedar and mahogany dusts, pine and spruce pulp	*Alternaria* spp. and wood dust

Table 15.2 Major mycotoxins produced by indoor fungi (Rodricks *et al.* 1977; Samson *et al.*, 1996; Burge and Ammann 1999).

Mycotoxin	*Primary health effect*	*Fungal producers*
Aflatoxins	Carcinogens, hepatotoxins	*Aspergillus flavus* *A. parasiticus*
Citrinin	Nephrotoxin	*Penicillium citrinum* *P. verrucosum*
Cyclosporin	Immunosuppressant	*Tolypocladium inflatum*
Fumonisins	Carcinogens, neurotoxins	*Fusarium moniliforme* (=*F. verticillioides*) *F. proliferatum*
Ochratoxin A	Carcinogen	*As. ochraceus* *P. verrucosum*
Patulin	Protein synthesis inhibitor, nephrotoxin	*A. terreus* *Paecilomyces variotii* *P. expansum* *P. griseofulvum* *P. roquefortii*
Sterigmatocystin	Carcinogen, hepatotoxin	*A. nidulans* *A. versicolor* *Chaetomium* spp.
Trichothecenes, macrocyclic		
Satratoxins	Protein synthesis inhibitors	*Stachybotrys chartarum* *Myrothecium* spp.
Trichothecenes, non-macrocyclic		
Deoxynivalenol (vomitoxin)	Emetic	*F. cerealis* *F. culmorum* *F. graminearum*
T-2 toxin	Hemorrhagic, emetic, carcinogen	*F. sporotrichioides*
Verrucosidin	Neurotoxin	*P. aurantiogriseum* group
Xanthomegnin	Hepatotoxin, nephrotoxin	*A.. ochraceus* *P. aurantiogriseum* group
Zeralenone	Estrogenic	*Fusarium* spp.

Toxicity of *Stachybotrys chartarum (atra)*

Recently, this fungus has become a notorious as a mycotoxin producer that can cause animal and human mycotoxicosis. It also is responsible for "Sick building syndrome". This fungus is found in extra moist conditions like flooded homes, water damage from broken pipes, floor leaks, condensation, etc. and many times it has forced school children to evacuate the building due to possible health problems. Growth of this fungus has resulted in multimillion-dollar litigation and caused serious problems for homeowners and building managers who must deal with human issues and remediation. The fungus produces a variety of macro cyclic trichothecenes and related trichoverroids: roridinE and L-2, Satratoxins F, G and H. These have potential toxicity to inhibit protein synthesis. Cyclosporin, Phenylspirodrimanes are other immunosuppressive agents causing pulmonary hemorrhage in infants. Some of these compounds are also responsible for causing memory loss, hearing problems and depression among adults.

Principles of Air Quality Sampling

Different methods involved in mold growth assessment from problem and non-problem buildings are based on following assumptions :

1. All fungi are not molds but all molds are fungi.
2. Generally indoor fungal spores come from outside environment.
3. Mostly raw counts of spores of a fungal species from indoor air sample is much lesser than the raw counts of spores of same fungus from outdoor air samples.
4. Spores of some genera grow rapidly if indoor environment is favorable for growth and may cause damage of building as well as toxic effect to the human.
5. The judgment that "presence of specific number of spores from indoor samples is a health problem" varies depending on the potential toxicity of fungal species. Very few spores of *Stachybotrys chartarum* (atra) may be a great concern for residents, on the other hand presence of a hundred spores without growing structures of *Cladosporium* species or *Penicillium/ Aspergillus* species or basidiospores of unidentified genera sampled on a rainy day may not be very alarming since these spores are found in abundance in outdoor atmosphere.
6. Comparison between indoor and outdoor growth levels of individual fungal species is essential.
7. It is beneficial to compare the results by using different sampling techniques.

Commonly used Techniques for Sampling Non-viable and Viable Culturable Fungi

Non-viable Air or Spore Trap Method

These are non-culturable air samples collected by accelerating air containing mold spores through a narrow opening and impacting suspended particles onto a slide inside a Zefon Air-o-cell cassette (horizontal trace-1400 micron), or Allergenco MK-3, Burkard (vertical trace-1450 micron) or Cyclix-D or Micro-5 (circular type) covered with an optically clear sticky material for 15 or more minutes. Trace collected on slide is fixed with Lacto-cotton blue stain and examined under the microscope. Total number of spores for each genus present on trace are counted and interpreted as raw count/cubic meter of air filtered. 80–90 % of particles in size of mold spores and pollen are trapped by this method. Spore counted/cubic meter of air are converted according to amount of air filtered (15-75 liter ranging from 15 to 60 minute duration).

Viable Air or Fungal Culture Method

Andersen's Air Sampler technique (Andersen, 1958) is used by passing air over a culture plate with suitable culture medium (MEA, CYA, Cellulose Agar, DG-18) at sampling location and transporting sealed plates to lab for analysis. Plate is marked with impactions (200 or 400). These plates are incubated at room temperature (25°–30°C) for 5 days and number of colonies growing are identified and counted as CFU/cubic meter of air sampled. These samplers may vary from single stage to two stages or four stages.

Dust samples

These samples are analyzed for colony forming units of each species present per gram of sample by using standard soil dilution plate method.

Non-viable Bulk or Surface Sampling by Direct Microscopic Examination Method

Object of this method is to see if mould growth is seen on the sampled surface. This may vary from samples collected as tape lifts from different locations inside a building, swabs collected from surface, bulk samples like sheet rock, dry wall paper, fiber glass, insulation material, piece of a carpet, wood chips, textile or any other suspicious material brought to lab as a bulk sample to be examined by making tape lifts, dust samples collected by vacuum cleaner. These samples are examined on a microscopic slide for any fungi, showing symptoms of growing hyphae and spores. These taxa are estimated to a degree of poor, moderate or heavy growth. Sometimes these analysis are required in exact quantity which needs counting of spores of each taxa found in per unit field of a tape sample. Presence of substantial amounts of pollen, mites, insect parts, bacteria, algae or yeast growth is mentioned as special comments which may be helpful to find out the source of allergy.

Molecular Method Includes PCR Technique

Polymerase Chain Reaction by REAL TIME PCR

Real-time polymerase chain reaction (PCR) based typing is most recent and preferred technique for performing DNA typing of mold. 5′ flurogenic nuclease assay is performed by using the ABI Prism® 7000 Sequence Detection System.Very little sample is required to perform the analysis, as the technique is extremely sensitive and precise. This is quick and accurate method of analysis, which can be done in 2–3 hours duration. This avoids waiting time to grow fungi on different media and compare morphology.

Polymerase Chain Reaction (PCR)

The method amplifies a mold's DNA base sequence into millions of copies using a heat-stable polymerase. By attaching a fluorescent probe to a unique region of a mold's DNA, one is able to qualitatively identify and quantitatively measure a species of mold from the sample. Real Time PCR machine is used for a quick detection of fungal species by using specific primers set for Polymerase Chain Reactions for different species of *Aspergillus, Penicillium, Stachybotrys* or any other genera. This specific device eliminates the longer process of gel electrophoresis and other steps and gets the accurate results quickly in case of rush or emergency. This provides results that are more accurate when litigation is involved.

Reaction mixture contains a heat stable polymerase that extends oligonucleotides in the 5′ to 3′ direction along the length of the template, isolated sample DNA, Primer for annelation in forward and reverse direction. Oligo probe with a reporter and quencher dye attached to recognize a special sequence by fluorescent signal and Deoxynucleotides (dATP, dTTP, dCTP, dGTP) to provide energy and nucleotides for the synthesis of amplifying DNA strands.

Major Steps Involved in This Method

1. Samples are collected on special Polycarbonate filters in a sealed cassette similar to non-viable air.
2. Filter is processed to isolate DNA of fungi present on the filter by using buffer, and bead beating technique followed by centrifugation for one minute and separation of DNA from supernate. This process takes maximum of 5 minutes.
3. Isolated DNA is subjected to amplification by adding Taqman® (7.5 μl), Universal PCR Master Mix (Applied biosystems Int; P/N 4304437) primers and probes (2.1 μl) for each fungal species e.g., AfumiF1 and AfumiR1(*A. fumigatus*) and sterile water (ddi 2.4μl). A 96 well reaction plate (MicroAmp® Applied biosystems;4316567) is used to mix the reaction mixture with sample DNA. After centrifugation, this well plate is ready to go inside the PCR machine for 2 hours to complete 40 cycles of Polymerization. Each well of this plate is labeled for specific fungal species to be detected by using **ABI Prism® 7000 Sequence Detection System with Notebook Computer (Applied Biosystems;403080; universal voltage, 100–240 V, 50–60 Hz.).**
4. After completion of 40 cycles, results can be seen on computer as CT Values. Results are compared with standards for particular species found in the sample and number of spores/samples is calculated with the help of standard values. For all analysis one endogenous positive control (DNA of Known fungal species e.g., *Geotrichum candidum*) and one negative control (no DNA) are used.
5. Samples are qualitatively and quantitatively analyzed against a predetermined standard and endogenous control to confirm identifies and provides cells per ml quantification.
6. Quantification: Obtained CT value for each well sample contained in reported result. Obtain a ΔCT calibrator value for each sample by

substracting mean reference sequence CT value (CT, reference) from mean standard sequence CT value (CT, standard).

7. Substract ΔCT calibrator from ΔCT unknown to get ΔCT.
8. Calculate ratio of standard sequences in unknown and calibrator samples using equation equation: ae –ΔCT, ae = amplification efficiency of standard TaqMan assay. . Multiply ratio of standard sequences in the unknown and calibrator samples to obtain number of standard cells in the unknown samples.
9. Validate all standards and endogenous control in triplets. Unknown CT values are compared to the validated calibrator using the ΔCT equation. Since only 3 μl (3%) sample DNA is used; to calculate total number of spores in 100 μl of sample the value retained in above equation is multiplied by a factor of 33.33 to give a cell/ml quantitation.

Data Collection and Analysis

Detail information is required to reach any conclusion by person who interprets these results and for making remedial decisions. Given is an example of dataset for distribution of indoor fungi from Daly City of San Francisco, California. Samples were collected from different locations of a single residence by using Spore trap and Anderson's viable culture techniques. It is helpful to estimate the seriousness of problem if data from different methods are compared.

Table 15.3 Sample of data showing raw counts fungal spores collected by Spore trap method.

Sampling Date – 01/10/02

Location – Daly City, CA

Temperature: 25°C

Humidity: 90%

Room type	***Outdoor***	***Basement - Complaint area 1***	***Bedroom – Non-complaint area***	***Hallway to bathroom – Complaint area***
Background debris (1-4+)	2+	2+	2+	2+
Counts	Raw ct. Spores/m^3	Raw ct. Spores/m^3	Raw ct. Spores/m^3	Raw ct. Spores/m^3
Alternaria sp.	3 40			1 13
Arthrinium sp.				
Ascospores	36 480			
Aureobasidium sp.				
Basidiospores	144 1920	36 180		3 40
Bipolaris/ Drechslera group				

(Contd.)

Table: 15.3 *(Cont.)*

Room type	***Outdoor***	***Basement - Complaint area 1***	***Bedroom – Non complaint area***	***Hallway to bathroom – Complaint area***
Botrytis sp.	1 13			
Chaetomium sp.				8 107
Cladosporium sp.	68 907	264 3520	4 53	20 267
Curvularia sp.	2 27			
Epicoccum sp.	1 13			
Fusarium sp.				
Myrothecium sp.				
Nigrospora sp.				
Penicillium/ Aspergillus types	8 107	144 1920	8 107	68 907
Pithomyces sp.				
Rusts				
Smut, Periconia, Myxomycetes	12 160	2 27	1 13	1 13
Stachybotrys chartarum (atra)		4 53		4 53
Stemphylium sp.				
Torula sp.				1 13
Ulocladium sp.				
Unidentified brown	1 13		1 13	
Unidentified colorless				
Total spores/M3	3680	**5700**	372	1613
Sample volume (liters)	75	75	75	75
Comments for clumps: Indication that fungus is growing inside and may need an action.		40 of the raw count *Pen/ Aspergillus* type of spores were present as a **single growing clump**.		Total raw count is lesser but presence of *Chaetomium and Stachybotrys* may cause a problem.

- Most of these spores are not seen with culturable methods, although some may appear as non-sporulating fungi. Most of the basidiospores are "mushroom" spores while rusts and smuts are plant pathogens.
- The spores of *Aspergillus* and *Penicillium* (and *Acremonium, Paecilomyces*) are small and round with little differences to identify on spore traps.

- Background debris is an indication of the amount of non-biological particulate matter present on slide (dust in air) are graded as 1 + to 4 + as highest amount. This also indicates the visibility for the analyst to read the slide. High debris of 4 + or > 4 + may obscure the small spores like Pen/Asp group and should be regarded as minimal counts and may be higher than reported.

Table 15.4 Sample data showing raw counts of culturable air fungi collected by Anderson Air Sampler on MEA plates after 5 days of incubation at room temperature

Date of sampling – 01/10/02

Location- Daly City, CA

Room type	*Outdoors*		*Basement*		*Bedroom*		*Hallway to Bathroom*	
Sample volume (liters)	85.0		85.0		85.0		85.0	
Colony counts	Raw ct = cfu/m^3		Raw ct = cfu/m^3		Raw ct = cfu/m^3		Raw ct = cfu/m^3	
Acremonium sp.								
Alternaria sp.	2	24						
Aspergillus flavus								
Aspergillus fumigatus								
Aspergillus nidulans								
Aspergillus niger								
Aspergillus ochraceus								
Aspergillus versicolor	2	24	68	883			27	330
Aureobasidium sp.	1	12						
Basidiomycetes								
Bipolaris/Drechslera								
Botrytis sp.								
Chaetomium sp.							6	71
Cladosporium sp.	120	1684	20	247	6	71	4	47
Curvularia sp.								
Epicoccum sp.								
Fusarium sp.								
Mucor/ Rhizopus sp.								
Non-sporulating sps.	1	12	1	12				
Paecilomyces sp.								
Penicillium sp.	3	35	80	1048	20	247		
Phoma sp.								
Stachybotrys chartarum (atra)							1	12
Ulocladium sp.								
Yeasts								
Positive Hole	400		400		400		400	
Total Colony forming units/M^3	1791		2190		318		460	

Notes: Positive hole correction chart is used for all calculations.

Analyst is responsible for reporting of fungal genera but interpretation of these data is left to the person who does field work. Variation is an inherent part of biological sampling. Presence or absence of a few genera in small numbers should not be considered abnormal.

Normal Spore Levels

Indoor spore levels usually average 30–80 % of outdoor spore level at the time of sampling with same general distribution of spore type. Indoor air may be 5–15% of outdoor air at the time of sampling. It also depends on amount of exchanged air between outdoor and indoor areas. In a residence with open doors, windows or heavy foot traffic it may be 95% while high-rise office buildings with little air exchange may average 2%.

Problem Interior

A substantial increase of one or two spore types, which are inconsistent with and non-reflective of the outside distribution of spore types, is indicative of an indoor reservoir of mold growth. Presence of higher number of *Aspergillus versicolor* in hallway to bathroom may become a problem for whole interior if not cured immediately as spores can multiply rapidly. Similarly, number of spores of *Cladosporium* species may be lesser than outdoors but it may cause a problem if allowed to grow.

Table 15.5 Showing different levels of fungal growth in different rooms of same building.

Description of Background debris/ bulk sample	*Miscellaneous spores*	*Mold growth showing mycelium or sporulating structures*	*Special comments*	*General Impression*
Scant	Very few	From kitchen platform None	None	Normal trapping
Light	Few	From bathroom wall 1+ *Cladosporium* species 2+ *Stachybotrys chartarum* (atra)	Very few mites present	Mold growth
Heavy	Variety	From garage 2+ *Cladosporium* species 1+ *Chaetomium* species	None	Mold growth

(Cont.)

(Cont.)

Moderate	Wide variety	Swab from living room carpet <1+ *Chaetomium* species	None	Minimal mold growth
Wet wall paper as bulk sample	Few	West wall layer 3+ *Aspergillus species* 1+ *Acremonium* species		Mold growth
Wooden frame as bulk sample	Very few	2+ *Ceratocystis/ Ophiostoma* 3+ basidiomycete 1+ *Aureobasidium* species	Growth of a wood decay basidiomycete which could cause wood deterioration	Mold growth and growth of a wood decay basidiomycete.

Direct Microscopic Examination (Wet mount)

Direct examination also indicates the possibility of growth of specific fungi on wood, cellulose and other wet substratum.

General Pattern of Fungal Distribution

- Normal distribution: when spores count is more from outside samples than from inside samples.
- Problem: when inside spore counts are higher than outside spore counts.
- Higher counts of basidiospores are seen when it rains as mushrooms discharge spores in high moisture conditions.
- Higher spore counts are recorded during growing season when pathogenic fungi are abundant in outdoor samples.
- Higher spore counts are seen in winter from indoor samples due to good temperature and condensation, and temperature is very low outside in case of snow.
- Viable air and bulk analysis is the best reliable method method to know how many spores are alive.
- Twelve most common mold types in the air are: *Alternaria, Aspergillus, Cladosporium, Coprinus, Epicoccum, Ganoderma, Leptosphaeria like, Stachybotrys, Penicillium, Pithomyces,* hyaline basidiospores and other ascospores.

Species Diversity

Seasonal distribution of indoor fungi varies with temperature, humidity, substratum and source from outdoors.

Many species found growing on tape lift are not culturable on growth media like MEA, CYA or DG-18, when they are sampled by viable air methods on Andersons plates to count colony forming units (CFU)/cubic centimeters of air.

Species of *Ascotricha, Ceratocystis/Ophiostoma, Graphium, Gonatobotryium, Myxotrichum and Gymnoascus* are common on wooden surface tape lifts but not able to grow on culture media limited time.

Table 15.6 Given is an example of frequency of fungal occurrence in Northern California during year 2002 as seen by viable air culture method.

Genera	*Per cent of total number of samples collected on Anderson plates*
Emericella nidulans	10
Eurotium amstelodami	10
Mortierella species	10
Sepedonium species	10
Verticillium species	10
Chrysoniella species	10–15
Beauvaria species	10
Bipolaris/Dreschlera	10–20
Penicillium simplicessimum	10–20
Fusarium species	20–30
Alternaria species	25–30
A.ustus	25–30
Aspergillus flavus	25–30
A.fumigatus	25–30
A.glaucus	25–30
Botrytis species	25–30
Curvularia species	25–30
Epicoccum nigrum	25–30
Mucor species	25–30
Rhodotorula specie	25–30
Phoma species	25–40
Aspergillus niger	30–40
A.sydowii	30–40
Chaetomium globosum	30–40
Chrysosporium species	30–40
Penicillium brevicompactum	30–40
Rhizopus stolonifer	30–40
Stachybotrys chartarum (atra)	30–40
Acremonium strictum	40–50
Paecilomyces species	40–50
Ulocladium species	40–50
Aureobasidium pollutants	50–60
Yeasts	50–60
C. sphaerospermum	50–60
Cladosporium herbarum	70–75
A. versicolor	75–80
Penicillium chrysogenum	75–80

Interpretation of Results

General interpretation for mold growth must consider the facts like

1. Have more than one sample from outside
2. Inside samples should be in low quantity than outside
3. Same fungal type as in outside samples
4. Same basic distribution of spores
5. Look for indicator fungal species
6. Consider magnitude of fungal species present
7. Consider exchange rate of indoor air with outdoor
8. Consider weather (in a rainy day you can see more basidiospores than normal days)

Analyst is advised to limit the interpretation upto the questions

Is there any growth in building?
Are people exposed to fungi?
Is there any evidence of building defect?

We can not interpret any possible health concern based on these numbers because there is no Threshhold Limit Values (**TLV**) for bioaerosols like bacteria and fungi. There are no assayable biological contaminants (endotoxins, mycotoxins, etc.). However, dose response relationship for some bioaerosols have been observed in experimental studies. So far, these values are known for other biological substances like grain, dusts, nicotine and sucrose.

Remediation

Mostly, it is the responsibility of owners or building managers to make a remedial decision, which is done by certified companies, and methods vary from chemical spray to tearing, burning or rebuilding of property. All these conditions indicate a primary responsibility of an analyst who reads the samples and provides data on mould growth to a level of normal or problem conditions for a sampled area. In genera, infectious using other methods like awareness of potential for infection controls diseases, controlled work practice to minimize exposure and personal protective equipment.

What is the significance of mold analysis for developing countries?

Sampling of residences for mold growth has been proved to be of great help to Physicians in tracing the root cause of chronic respiratory and other diseases (Nelson, 2001). Home and building owners have millions of questions on molds and need straightforward advise to correct a problem. In Western countries, these questions are answered by agencies like EPA, PHD, Agricultural Extension services and many private microbiological companies. In developed countries, increased number of reliable resources with practical experience of dealing with health problems due to microbes is highly essential. Scientific conclusions from laboratories need to be linked with real world to be applicable for solving any problem in daily life caused due to microbes.

In tropical countries indoor air quality (IAQ) may not be a serious health problem for residences with good air circulation but with an increase in air-conditioned atmosphere inside hospitals, computer centers, child-care centers, offices, luxurious hotel and residences; there is always a chance of increased humidity which may cause a heavy mold growth. Most of the child-care centers need to be tested for allergic molds since children are at highest risk of allergic symptoms. It is the case for hospitals and other health centers where patients are with less tolerance and poor immunity and are likely to be subjected to type II, III allergies and I. Introducing new inspection rules for sensitive buildings, Operation rooms, ICU, Pediatric nurseries for regular assessment of mold growth and proper remediation on daily basis can be very helpful in avoiding major complications during treatments. Checking the quality of wood used in house construction, furnishing along with the quality of carpets being used may eliminate the chances of fungal growth on these sources.

Role of Public health departments to educate general public and make laws for regular inspection of buildings.

Regular education and inspection by public health departments may help to create a healthy indoor environment. Awareness for fungal genera responsible for serious respiratory or other diseases like cancer and knowledge of proper use of chemical (10% Clorox bleach spray is most common method used in Western households for protection from microbial growth) without any side effect can be of much help. Regulation act for inspection and mold analysis of hospitals and operation rooms for patient's safety can prevent many post surgery problems.

Future Prospects of IAQ

We are still in need of scientifically approved standards for spore levels for different environments. Reliable data on dose response studies are in great need. Special attention should be paid to those toxins produced by commonly distributed human pathogenic species including *Aspergillus fumigates, A. vesicular, A. niger, A. ustus, Acremonium sps., Beauveria sps., Cladosporium sps., Fusarium sps., Paecilomyces sps., Penicillium marneffei, Rhizopus sps., Stachybotrys chartarum (atra)* and *Trichoderma sps.* Identification of remaining unknown genera and their volatile compounds is always an open field for researchers. Ecological aspect of succession of indoor fungal species on various substrata in natural and controlled conditions can add to the information on carpet, paper, wood, textile and leather deterioration process and prevention measures for such industries.

References

Andersen, A.A. (1958). New sampler for the collection, sizing and enumeration of viable airborne particles, *Journal of Bacteriology* 76:471–484.

Nelson, B.D. (2001). "*Stachybotrys chartarum*: The Toxic Indoor Mold", APS Feature Story, http://www.apsnet.org/online/feature/stachybotrys/.

Burge, H.A., (1995). "Bioaerosol", Lewis publishers, Boca Raton.

Burge, H.A. and Ammann, H.M. (1999). Fungal toxins and beta-(1-3)-D-Glucans" In Bioaerosols: Assessment and control.J. Macher (Ed.) Cincinnati: American conference of Governmental Industrial Hygienist pp 24.1–24.13

Gallup, J. and M. Valesco (1999). "Characteristics of some commonly encountered fungal genera" EMLAB Inc

DEHS (2002). "Indoor fungi resources", http://www.dehs.umn.edu/iaq/fungus/references.html#26

Haugland, R.A., and J.L. Heckman (1998). "Identification of putative sequence specific pcr primers for detection of the toxicogenic fungal species *Stachybotrys chartarum*. Molecular and Cellular probes. 12: 387–96.

Haugland, R., Heckman, J.L., Wymer L.J. (1999). "Evaluation of different methods for the extraction of DNA from fungal conidia by quantitative competitive PCR analysis Journal of Microbiological methods 37: 164 –167.

Hunninghake, G.W. and Richardson, H.B. (1998). Hypersensitivity pneumonitis and eosinophilic pneumonias. In Harrson's Principles of Internal medicine 14th ed.' Chap.253.pp 1426-1429. NewYork Mc Graw-Hill.

Park, H.S., Jung, K.S., Kim, S.O. and Kim S.J. (1994). Hypersensitivity pneumonitis induced by *Penicillium expansum* in a home environment. Clin. Exp. Allergy:24:383–385.

Pope, A.M., Patterson, R., Burge, H. (Eds.). (1993). Indoor Allergens: Assessing and controlling Adverse health effects of Indoor Allergens Division of Health promotion and Prevention. Institute of medicine. Washington D.C. : National Academy Press. 308 pp.

Rodricks, J., Heseltine, C.W. and Mehlmn, M.A. (1977). Mycotoxin in human and Animal Health. Park Forest South, Illinois Pathotox Publishers 807 pp.

Roe, J.D., Haugland, RA, Vesper, J, Wymer, L.J. (2001). Quantification of *Stachybotrys chartarum* conodia in indoor dust using real time, fluorescent probe-based detection of PCR products. Journal of Exposure Analysis and Environmental Epidemiology. 11: 1–9.

Samson, R.A. , Hoekstra, E.S., Frisvad, J.C. and Filtenburg, O. (1996). Introduction to Food born fungi, 3rd ed. Central Bureau Voor Schimmelcultures, Baarn.

Scott, J.A. (2001). "Studies on indoor fungi", a thesis submitted in conformity with the requirements for the degree of Doctor of Philosophy in Mycology, Graduate Department of Botany in the University of Toronto, Canada.

16

Recent Methods for Molecular Study of Fungal Community Structure with Special Reference to DGGE and TGGE

A. Arif Khan, Preeti Bhatnagar, Sudhir K. Jain□ S.K. Deshmukh* and M.K. Rai**

Department of Microbiology, College of Life Sciences, Cancer Hospital and Research Institute, Gwalior (M.P.)

□S.O.S. in Microbiology, Vikram University, Ujjain (M.P.)

*Department of Natural Products, Nichoals Piramal Research Centre, Goregaon, Mumbai, M.S.

**Department of Biotechnology, SGB Amravati University, Amravati–444 602, M.S.

□*E-mail: sudhirkjain1@rediffmail.com.*

INTRODUCTION

From depth of the ocean to the apex of the mountains, fungi are found in almost every environment. Fungal ecology deals with the fungal species, their prevalence, and the process they perform in the environment. So, identification and evaluation of prevalence of various fungi is of prime importance during the study of fungal community structure. But, at present our knowledge about role of fungi in the ecosystem is very poor, such as, in soil ecosystem fungi play an important role in the decomposition of plant residue, releasing nutrients that stimulate plants growth. Some fungi possess antagonistic activity towards plant pathogen (Lumsden, 1981). Some fungi cause various diseases in humans (Jain *et al.*, 2002, 2003 and 2004). Most of the fungal biologists are focusing on the fungal kingdom that is cultivable, can be visualized by the microscope and identified by the morphological and biochemical tests. In order to isolate diverse novel fungi from different environments, we are following few model ideal cultivation techniques. Everyday mycologists are doing enormous efforts to isolate the fungi from diverse environments and reporting newly isolated species, but all these species belong to the cultivable strains. It is now well recognized among mycologist that only a small fractions of all fungi have been isolated and characterized. Although, cultivation methods are cornerstone in the identification of fungal identification, but these methods can cause some problems in identification. These methods rely on experienced, skilled laboratory staff, the ability of the organism to be cultured. In order to assess the very wide fungal community these methods are totally impractical, because of time-consuming nature of these methods that are prone to contamination and error.

It has been estimated that global fungal diversity ranges from 0.5 to 9.9 million fungal species (Hawksworth, 2001). Such as the occurrence of fungi specially saprophytic and keratinophiles in soil have been reported by many workers (Jain and Agrawal, 1977; Deshmukh and Agrawal, 1982, 1983; Deshmukh, 1985). Mycologists throughout the world have currently identified and classified around 74,000 fungal species, representing only 5% of estimated fungal diversity (Kennedy *et al.*, 2003). Most of fungal species are yet to be identified; the reason is lack of global exploration and efforts and low cultivability of most fungi. Most of the environments are oligotrophic, so it presents very different nutrient conditions then that of standard (rich) mycological media. So most of the fungi growing on this media will be fast growing species adapted to new nutritional condition, but these organisms possibly will not represent dominant organism of the ecosystem.

So, to obtain a better understanding of the role of fungal diversity in the maintenance of ecosystem, other approaches, that complement the traditional microbiological procedures, are needed. Fungal ecologists deal with a variety of microorganisms, in some instances the fungi are not the dominant member of the microbial community. In such situation the mycologist should have a knowledge of some chemical difference in the fungi from other member of the microbial community. Some scientists use fungal cell wall components like chitin and ergosterol as an indicator. But these constituents can also be present in some other group of organisms such as crustacean, arthropods and microalgae (Bermingham *et al*, 1995), so phospholipids fatty acids (PLFAs) can be used as another marker regarding distribution, diversity and activity of fungal mycelia in soil (Zelles, 1999). But presently, very few numbers of fungal specific chemical markers are available, and these markers provide only information about the total fungal biomass present in the soil.

After the pioneering work of Woese (1987), it has been clear that all organisms can be classified by comparative sequence analysis of their small subunit ribosomal RNA. Fungi consist of 80S ribosome, which consist of two subunits, one large subunit called (60S) and another small (40S) subunit, each subunit is made up of characteristic rRNA associated with proteins. Large subunit consist of 28S, 5.8S and 5S rRNA while small subunit consist of 18S rRNA. Gene for 18S rRNA is highly conserved because this molecule is necessary for the formation of initiation complex during translation, so any mutation in the sequence of 18S rRNA gene led to lethal mutation. The genes for each subunit of rRNA is separated by the internally transcribed spacer (ITS), which are highly variable both in length and sequence composition. These genes for rRNA with ITS constitutes single cluster of ribosomal gene. On a chromosome various cluster of ribosomal genes are present and separated from each other by intergenic spacer (IGS). By exploiting the rRNA sequence or the sequence of ITS/IGS, one can identify the fungi. With the advent of PCR technology, it has become feasible to directly amplify the rDNA gene or reverse transcribing the rRNA to DNA and then amplifying it. Universal primers are available for fungi that amplify ITS region of virtually all fungi (White *et al.*, 1990). These primers are highly important for fungal community study, because for the community study, one has to deal with various unidentified fungal species. Cloning of PCR amplified 18S rRNA and sequencing can give clear idea about the fungal species composition present in the environment. For the study of fungal ecology, we have to study the interaction among the microorganisms and between the microorganisms and their environment as well as, we will have to study success ional population changes in a microbial community with time. However, because cloning is a time-consuming and laborious process, this approach is not suited for the study of large number of fungi with their interactions and prevalence, which is a important part of fungal ecology (Marshall, 1993).

Thus, to study fungal community, it is necessary to have a rapid and efficient technique, which can analyze a large number of sample in a short time period.

Molecular Fingerprinting Techniques

For the study of fungal community structure, various methods are available. These methods are divided into two types. One is called direct method while another is called indirect method. In direct method, nucleic acid is extracted from organism and analyzed directly. But in case of indirect methods, nucleic acid is extracted form organism and first amplified by the PCR and then analyzed (Muyzer, 1999).

Direct method for genetic identification includes Low Molecular Weight (LMW) RNA fingerprinting. In this technique, total RNA is isolated from an organism and then separated by high resolution polyacrylamide gel electrophoresis (PAGE). Due to the high resolution of PAGE, large size molecule of RNA can not enter inside the gel and we get the separation profile on the basis of small size RNA (5S RNA, tRNA). This profile will be different for every type of organism and can be used for the fungal community analysis. But this technique can cause some errors like RNA molecule rapidly degrade in environment, and this can lead to erroneous results by formation of additional bands as well as these small molecules do not provide the accurate information about fungal community composition (Hofle, 1998).

Other methods of molecular fingerprinting are categorized as indirect method, because here an additional method is used to characterize the organism in community. Indirect methods for community analysis include SSCP (Single Stranded Conformation Polymorphism). As the name suggest, in this method characterization of organism is done on the basis of differences in the folded conformation of single stranded DNA. Here, the DNA specific to the small subunit RNA is amplified by the primers and denatured to become single stranded, this single stranded DNA is directly electrophorsed on gel. The DNA molecule different in sequence acquire different folded conformation and can be separated by electrophoresis. But the use of only very short fragments of DNA limit the application of this technique for fungal community analysis (Lee *et al.*, 1996).

However, RAPD (Randomly amplified Polymorphic DNA) method is also used as an indirect fingerprinting method for analysis of fungal community, but experimental conditions like, quantity of template DNA and the conditions of PCR experiment greatly affect reproducibility of the results. Thus, this method is not routinely used for fungal community analysis (Hadrys, 1992).

Another most widely used method for the molecular identification of fungi is the Restriction fragment length polymorphism (RFLP) of PCR amplified 18S ribosomal DNA fragments. This method is also known as ARDRA (Amplified Ribosomal DNA Restriction Analysis). In this method, ITS region of 18S rDNA gene is amplified by the use of fungi specific primer to amplify virtually all known and unknown fungal species present in the ecosystem. This amplified population of 18S rDNA genes is digested by combination of restriction enzymes. Because these restriction enzymes are sequence specific, so it generates different sized fragments of 18S rDNA that depends on the sequence of gene. So, this pattern is also used for the identification of fungi. But in this method single DNA molecule generate various fragments, and during the analysis of large number of organisms (as in fungal community study), this method create very complicated situation. Thus, this method also have its limitation to be used as a tool for the study of fungal community structure (Viaud *et al.*, 2000).

However, above described problem has been slightly circumvented by the use of another indirect method of molecular fingerprinting. This method is a variant of RFLP and known as T-RFLP (Terminal Restriction Fragment Length Polymorphism). T-RFLP method uses fluorescent labelled primers to label the terminal end of newly synthesized DNA strand.

When these terminally labeled DNA are cleaved by restriction enzyme they generate various size fragments of different species. It depends on the distance of first restriction site from terminally labelled end. These labeled fragments of different size can be detected on the basis of fluorescence. This method will select only terminal fragments for analysis while omitting all others. So this method slightly overcomes the problem of complex results obtained by ARDRA during analysis of a large number of organisms (Kozdroj and van Elsas, 2001). This method is very advantageous in assessing soil fungal diversity during ecological studies. Another technique similar to T-RFLP is known as ARISA (Amplified Ribosomal Intergenic spacer Analysis), this technique is similar to T-RFLP except whole amplicons are analyzed rather then terminal fragments produced by restriction digestion (Anderson and Cairney, 2004).

However, all these the methods are well suited for the fungal identification with some limitations. With the help of sequencing and similarity searching of the isolated DNA segments in BLAST (Basic Local Alignment Search Tool), it is possible to accurately identify the fungi in short time duration. But these methods should be cross checked with traditional morphological and biochemical methods to get hundred percent assurances of the results. But when we discuss about, fungal ecology we are not concerning about one or two fungi, actually we are studying about all fungi present in the ecosystem, their prevalence and interactions. So, in this situation a method is required, that can address about total fungal diversity present in an ecosystem with their differences in less laborious manner, as well as, which can solve the problem of identification. Recently, another genetic fingerprinting technique denaturing gradient gel electrophoresis (DGGE) and its related technique Temperature gradient gel electrophoresis (TGGE) has been introduced in microbial ecology. Within a short period of time this method has become a major technique used in many laboratories by environmental microbiologists to study microbial ecology.

DGGE and TGGE: Theory and Practice

DGGE as well as TGGE both are genetic fingerprinting techniques which are capable of separate individual sequence from a complex mixture of different genetic material. DNA fragments of same length but with different sequence can be efficiently separated by the use of DGGE (Fischer and Lerman, 1979, 1983; Myers *et al.*, 1987; Muyzer and Smalla, 1998) or by the use of TGGE (Rosenbaum and Reisner, 1987; Reisner *et al.*, 1991), because DNA molecules having different sequence will be having different malting temperatures. When these DNA molecules will be separated by the electrophoresis in the gel containing increasing concentration of either chemical (in DGGE) or physical (in TGGE) denaturant, their separation will be based on the decreased electrophoretic mobility of a partially melted double stranded DNA molecule in polyacrylamide gel containing a linear gradient of DNA denaturant. Actually, melting temperature of double stranded DNA molecule depends on hydrogen bonds between two complementary base pairs, however another interaction, named stacking interaction between neighboring bases on the same strand also helps in maintenance of structure of double stranded DNA molecule. As G and C bind to each other with the help of three hydrogen bonds, in comparison of two hydrogen bonds required for the binding of A and T. Thus, more energy is required to cleave G and C bonds in comparison to A and T bonds. Hence, the DNA molecules having different sequence composition of A and T and G and C, will be having different Tm (Melting temperature). Similar situation is with chemical denaturant, more concentration of chemical denaturant will be required for the cleaving G and C bond in comparison to A and T bonds. Changes in the base sequences, as small as single base may alter the stacking significantly enough to modify the Tm by over 1°C (Hope, 2004). (Fig. 16.1)

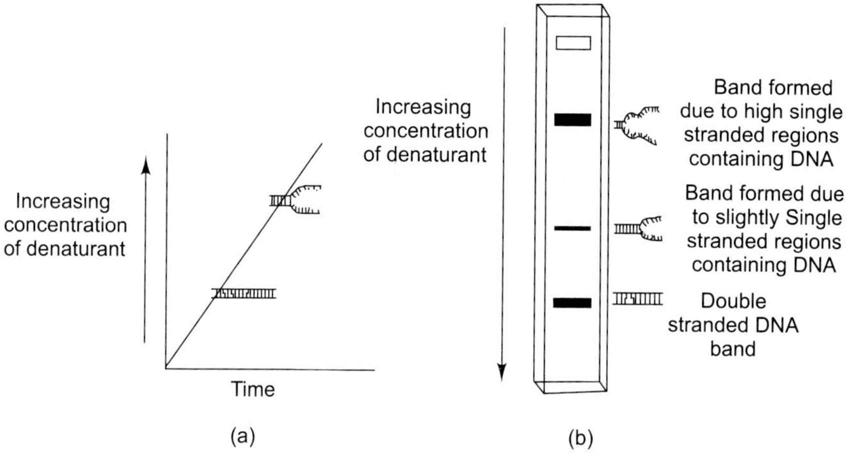

Figure 16.1 Diagram shows principle behind DGGE and TGGE. (a) As the concentration of denaturant (either temperature or chemical like Urea/Formamide) increases, the dsDNA converts to ssDNA form. Conversion of dsDNA to ssDNA starts first at the region containing higher concentration of AT base pairs (See text for details). (b) DNA molecules containing dsDNA form can move fastest in the gel, while DNA molecules with more ssDNA regions will be retarded in gel. So, different DNA molecules will form bands in denaturing gel on the basis of relative sequence differences.

In DGGE, increasing concentration of urea or formamide is used to create denaturing gradient, while in case of TGGE this concentration gradient is formed by the use of increasing temperature throughout the gel. So, once a domain with the lowest melting temperature reaches its melting temperature at particular position in denaturing gradient gel, it will be converted into partially double stranded structure and its further movement will be retarded. The DNA molecules with different sequences will retard at different positions in denaturing gel and will form different bands (Muyzer and Smalla, 1998). But complete denaturation of DNA molecules can be a problem during this type of electrophoresis. So, this denaturation of DNA molecule is usually prevented by incorporation of GC clamp (GC rich sequence of 30–50 nucleotides) at one side of DNA molecule. Because this GC clamp has high melting temperature, it prevents complete denaturation of DNA molecule during electrophoresis (Myers *et al.*, 1985; Sheffield *et al.*, 1989; Muyzer *et al.*, 1997). For incorporation of GC clamp a short sequence of GC nucleotides is added to the 5′ end of PCR primer (required for amplification of population of target DNA) and coamplified, Simultaneous amplification of GC clamp will take place with target DNA and it will be incorporated at the side of every amplified DNA fragments (Sheffield *et al.*, 1989; Sheffield *et al.*, 1992). However, use of GC clamp can lead to development of artifacts during PCR amplification, so now another strategy to use DNA region, itself consist of high GC region has been devised. In this method, scientist utilized high melting temperature region of the tRNA gene located in the middle of intergenic spacer region, it has been verified that this DNA fragment can be more efficiently amplified by the PCR in comparison to the region amplified by the GC clamp attached primer (Yasuda and Shiaris, 2005).

DNA bands obtained by the DGGE and TGGE can be easily visualized by staining with ethidium bromide. However, now another stain SYBR is used as an alternative to ethidium bromide. The advantage of SYBR is the lack of background staining, so it shows very clear bands in which less dominant DNA fragments can be visualized (Muyzer *et al.*, 1997). Beside it some researcher also used silver staining as a detection system; however, the use of silver staining prevents further analysis of bands by sequencing or hybridization (Falske *et al.*, 1996; Heuer and Smalla, 1997). For practical applications of DGGE and TGGE various equipments are available from commercial companies. Anybody, who wants to perform this procedure can avail these equipments form market. These equipments consist of special system for making chemical gradient (in DGGE) and temperature gradient (in TGGE).

Advantages of DGGE and TGGE

As we have discussed that most commonly used marker for profiling microbes is small subunit RNA due to its conserved nature. For the study of microbial ecology, within a short period of time DGGE and TGGE has become an exceptional tool to study species diversity and community dynamics. DGGE of PCR amplified 16S rRNA gene was first used to profile community structure of microbial mat and bacterial biofilms (Muyzer et al., 1993) and thereafter it has been used to analyze microbial community structure in numerous cases evident by number of publications. DGGE and TGGE show various advantages over other molecular fingerprinting techniques in order to study microbial community structure.

Denaturing gel analysis of PCR amplified DNA molecule offer an attractive way to study community complexity. Fungal diversity present in various environment has been assessed by the use of DGGE and TGGE. Smit *et al.* (1999) have studied fungal diversity present in the wheat rhizosphere by temperature gradient gel electrophoresis (TGGE), during this analysis they used 18S rDNA fragments for TGGE analysis and they found results are good reproducible with little variation (Smit *et al.*, 1999). However, by blotting of DGGE/TGGE profile on nitrocellulose membrane and its further hybridization with radioactive probes can give more detailed picture about microbial diversity present in environment (Amann *et al.*, 1992). DGGE and TGGE profile can be directly analyzed by the presence and absence of bands, as well as intensity of the bands also gives a semi-quantitative measure of relative abundance of sequence present in sample (Hope, 2004). Beside this, by performing RT-PCR analysis of SSU-RNA followed by denaturing gel analysis, one can determine expression level and activity of more prominent microbes in environment (Zoetendal et al., 1998) (Fig. 16.2).

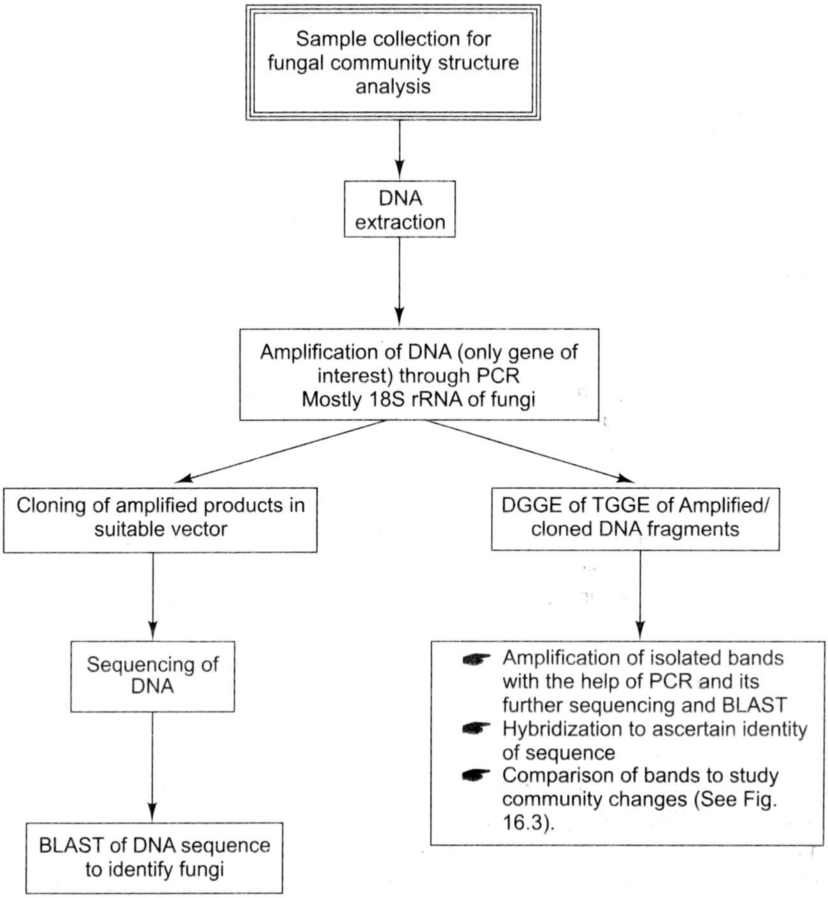

Figure 16.2 Steps in Denaturing gradient gel electrphoresis (DGGE) and Temperature gradient gel electrophoresis (TGGE) for study of fungal community and diversity.

For the ecological study of fungi, it is necessary to study environmental changes with time over along period. If we take cloning technique in our consideration, then we will find it is not feasible to manage large number of organisms present in the environment by cloning. As well as repeated cloning for the same sample at different time interval will be very laborious. In this situation, DGGE/TGGE can give an alternative for accurate study of community changes with time. By the use of DGGE/TGGE many samples taken at different time interval can be analyzed simultaneously. And we can interpret that which species number is increasing and which species number is decreasing with time (Muyzer and Smalla, 1998) (Fig. 16.3).

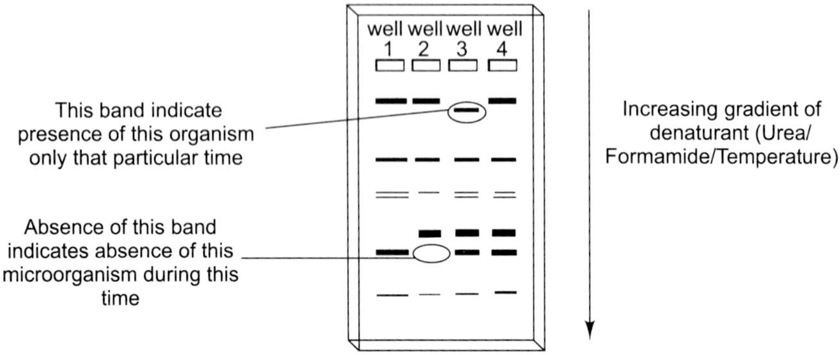

Figure 16.3 Diagrammatic illustration for use of DGGE and TGGE to study changes in soil microbial community structure with time. Soil samples collected from same place after every three months interval for one year. DNA isolated from all these soil samples are amplified by PCR and loaded on same denaturing gel in different wells. Same bands are representing presence of same organism, while new bands indicate presence of new microorganism in particular time and absence of any band indicate the absence of that microorganism in that particular time.

Besides these advantages for the study of fungal community, DGGE and TGGE also offer an attractive way to compare efficacy and reproducibility of different DNA extraction protocols. Various studies have been performed on bacteria. Lie sack *et al.* (1997) compared two different types of lysis method for isolation of DNA. They performed bead mill homogenization versus combination of freeze thawing, lysozyme and SDS treatment (Liesack *et al.*, 1997).

DGGE and TGGE can also be used for the screening of clone libraries. Usually RFLP and colony hybridization is used to screen clone libraries. But by cloning of desired gene in vector and its amplification by nested PCR followed by analysis of PCR products through DGGE/TGGE can greatly estimate abundance of particular clone in library. TGGE (Muyzer and Smalla, 1998) and DGGE (Kowalschuk *et al.*, 1997; Muyzer and Smalla, 1998) have been used to screen clone libraries of 16S rDNA inserts obtained from natural environment.

Limitations of DGGE and TGGE

As with all other molecular techniques, DGGE and TGGE are not exceptional to be free from limitations. There are few limitations of these techniques that should be overcome to make this technique an ideal. One of the limitations of DGGE and TGGE is that it can separate fragments only less then 500 bp. Due to this limitation it is not possible to make definitive species identification and to infer phylogenetic relationship (Myers *et al.*,1985).

The technique is based on relative sequence composition of nucleotide in the DNA. Hence, this cannot separate sequences having different sequence arrangement but same sequence composition. However, in some instances, comigration of non-related sequence can also lead to problem during interpretation of results as well as retrieving clean sequence from bands (Muyzer and Smalla, 1998).

Although, the use DGGE and TGGE in fungal ecology is in infancy and most of the experiments are being performed on bacteria. But due to its simplicity and capacity for simultaneous analysis of multiple samples, their future perspectives are promising in fungal ecology. PCR amplification of SSU-RNA or rDNA gene followed by DGGE and TGGE, can give a comparative results of number of total fungi present in environment, their relative abundance, as well as relative activity of fungi present in environment. These techniques can complement laborious procedure of cloning for community analysis. In this respect, these techniques are exceptionally suited for analysis of community changes with time, required for study of fungal ecology. This aspect is most important reason for the popularity of this technique in microbial ecological study. Relative easiness, reliability, reproducibility and fast speed of DGGE/TGGE is also popularizing these techniques among microbial ecologist. Beside this, less technological expertise is required, in comparison to cloning for DGGE and TGGE.

In conclusion, DGGE and TGGE have become exceptional tools for microbial ecological study. This technique is well suited to study species diversity and community complexity. However, an integrated approach for molecular and cultivation technique can simultaneously give an exact picture of fungal ecology. Within a short period of time, this technique has become an important tool and in near future, no doubt this technique will become core facility for fungal ecological studies.

References

Amann, R.I., Stromley, J., Devereux, R., Key, R. and Stahl, D.A. (1992). Molecular and microscopic identification of sulphate reducing bacteria in multispecies biofilms. *Appl. Environ. Microbiol.* 58: 614–623.

Anderson, I.C., and Cairney, J.W.G. (2004). Diversity and ecology of soil fungal communities: increased understanding through the application of molecular techniques. *Environ Microbiol.* 6: 769–779.

Bermingham, S., Maltby, L., and Cooke, R.C. (1995). A critical assessment of the validity of ergosterol as an indicator of fungal biomass. *Mycological Res.* 99: 479–484.

Deshmukh, S.K. (1985). Isolation of dermatophytes and other keratinophilic fungi in soils of Madhya Pradesh (India). *Mykosen.* 26(11): 574–577.

Deshmukh, S.K. and Agrawal, S.C. (1982). *Malbranchea sulfurea* (Miche) Sigler and Charmichael: A new record from India. *Current Science.* 51(16): 785–786.

Deshmukh, S.K. and Agrawal, S.C. (1983). Prevalence of dermatophytes and other keratinophilic fungi from soil of Mussoorie (India). *Mykosen.* 28(2): 98–101.

Falske, A., Engelen, B., Nubel U., and Backhaus H. (1996). Direct ribosomal isolation from soil to extract bacterial rRNA for community analysis. *Appl. Environ. Microbiol.* 62: 4162–4167.

Fischer, C.R. and Lerman, L.S. (1979). Length independent separation of DNA restriction fragments in two-dimensional gel electrophoresis. *Cell* 16: 191–200.

Fischer, C.R., and Lerman, L.S. (1983). DNA fragments differing by single base pair substitutions are separated in denaturing gradient gels: correspondence with melting theory. *Proc. Natl. Acad. Sci.* USA 80: 1579–1583.

Hadrys, H., Balick, M., and Schierwater, B. (1992). Application of Random amplified polymorphic DNA (RAPD) in molecular ecology. *Mol. Ecol.* 1: 55–63.

Hawksworth, D.L. (2001). The magnitude of fungal diversity: the 1.5 million species estimate revisited. *Mycological Research* 105**:** 1422–1432**.**

Heuer, H. and Small K. (1997). Application of denaturing gradient gel electrophoresis (DGGE) and temperature gradient gel electrophoresis (TGGE) for studying soil microbial communities. *In: Modern soil microbiology*. (Eds.) Van Elsas J.D., Trevors, J.T. and Wellington, E.M.H. Marcel Dekker, New York. pp. 353–373.

Hofle, M.G. (1998). Genotyping of bacterial isolates from the environment using low molecular weight RNA fingerprints. *Mol. Microb. Ecol. Manual* 3.3.7:1–23.

Hope, M. (2004). Density Gradient Gel Electrophoresis. *Microbiologist.* **Jun**: 38–39.

Jain, P.C. and Agrawal, S.C. (1977). Keratinophilic fungi from the soils of Mount Abu (India).*Geobios*. 4: 136–138

Jain, S.K., Agrawal S.C. and Jain, P.C. (2003). Subcutaneous Phaeohyphomycosis on face caused by *Cladopheophora bantiana. Mycoses*. 46(5 and 6):137–138.

Jain, S.K., Singh, V. and Gupta, S.K. (2002) Immunological study of onychomycosis caused by *Candida albicans. J. of Immunology and Immunopathology*. 4 (1 and 2): 1-3.

Jain, S.K., Shrivastava, A. and Agrawal S.C. (2004). Incidence of *Candidiasis* in Leukoplekia cases in cancer suspected patients. *In: Microbiology and Biotechnology for Sustainable Development* (Ed.) Jain, P.C., CBS Publishers and distributors, New Delhi, India, pp.354–357.

Kennedy, N. and Clipson, N. (2003). Fingerprinting of the fungal community. *Mycologist* 17: 158–164.

Kowalchuk, G.A., Stephen, J.R., de Boer, W., Prosser, J.I., Embley, T.M., and Woldendorp, J.W. (1997). Analysis of ammonia oxidizing bacteria of the ß subdivision of the class Proteobacteria in coastal sand dunes by denaturing gradient gel electrophoresis and sequencing of PCR amplified 16S ribosomal DNA fragments. *Appl. Environ. Microbiol.* 63**:** 1489–1497.

Kozrodj, J., and van Elsas, J.D. (2001). Structural diversity of microorganisms in chemically perturbed soil assessed by molecular and cytochemical approach. *J Microbiol Methods* 43: 197–212.

Lee, D.H., Zo, Y.G., and Kim, S.J. (1996). Non-radioactive method to study genetic profiles of natural bacterial communities by single stranded conformation polymorphism. *Appl._Environ.Microbiol.* 62: 3112–3120.

Liesack, W., Janssen, P.H., Rainey, F.A., Ward-Rainey, N.L., and Stackebrandt, E. (1997). Microbial diversity in soil: The need for a combined approach using molecular and cultivation techniques. *In: Modern soil Microbiology.* (Eds.) Van Elsas, J.D., Trevors, J.T., Wellington, E.M.H, Marcel Dekker, New York. pp. 375–439.

Lumsden, R.D. (1981). Ecology of mycoparasitism. *In: The fungal community* (Ed.) D.T. Wicklow and G.C. Carroll Marcel Dekker, Inc., New York, N.Y., p. 295–328

Marshall, K.C. (1993). Microbial ecology: whither goes thou. *In: Trends in Microbial Ecology* (Eds) Guerrerro R., Pedros- Alio C., Spanish Society for Microbiology, Barcelona. pp. 5–8.

Muyzer, G. (1999). Genetic fingerprinting of microbial community-present status and future perspectives. *In: Methods of Microbial Community Analysis:* Microbial Biosystem: New Frontiers (Ed) Bell C.R., Brylinsky M., Johnson-Green P. Atlanta Canada Society for Microbial Ecology, Halifax, Canada.

Muyzer, G., Brinkhoff, T., Nubel, U., Santegoeds, C., Schafer H., and Wawer, C. (1997). Denaturing gradient gel electrophoresis (DGGE) in microbial ecology. *In : Molecular Microbial Ecology Manual* . (Eds.) Akkermans ADL, van Elsas J.D., de Bruijin, F.J. Kluwar Academic Publishers, Dordrecht, the Netherland. 3.4.4: pp. 1–27

Muyzer, G., de Wall, E.C., and Uitterlinden, A.G. (1993). Profiling of complex microbial populations by denaturing gradient gel electrophoresis analysis of polymerase chain reaction amplified genes coding for 16S rRNA. *Appl. Environ. Microbiol.* 59: 695–700.

Muyzer, G. and Smalla, K. (1998). Application of denaturing gradient gel electrophoresis (DGGE) and temperature gradient gel electrophoresis (TGGE) in microbial ecology. *Antonie Von Leeuwenhoek.* 73: 127–141.

Myers, R.M., Fischer, S.G., Lerman, L.S., and Maniatis, T. (1985). Nearly all single base substitution in DNA fragments joined to a GC clamp can be detected by the denaturing gradient gel electrophoresis. *Nucleic Acid Res.* 13: 3131–3145.

Myers, R.S., Maniatis, T., and Lerman, L.S. (1987). Detection and localization of single base changes by denaturing gradient gel electrophoresis. *Methods Enzymol.* 155: 501–527.

Reisner, D., Henko, K., and Steger, G. (1991). Temperature gradient gel electrophoresis: a method for the analysis of conformational transitions and mutations in nucleic acids and proteins. *In: Advances in Electrophoresis* 4: 169–250.

Rosenbaum, V., and Reisner, D. (1987). Temperature gradient gel electrophoresis; thermodynamic analysis of nucleic acid and protein in purified form and in cellular extracts. *Biophys. Chem.* 26: 235–246.

Sheffield, V.C., Beck, J.S., Stone, E.M. and Myers, R.M. (1992). A simple and efficient method for attachment of a 40-base pair, GC-rich sequence to PCR amplified DNA. *BioTechniques* 12**:** 386–387.

Sheffield, V.C., Cox, D.R., and Myers, R.M. (1989). Attachment of a 40-bp G + C rich sequence (GC clamp) to genomic DNA fragments by polymerase chain reaction results in improved detection of single base changes. *Proc Natl. Acad. Sci.* USA. 86: 232–236.

Smit, E., Leeflang, P., Glandorf, B., and van Elsas, J.D. (1999). Analysis of fungal diversity in the wheat rhizosphere by sequencing of cloned PCR-amplified genes encoding 18S rRNA and Temperature gradient gel electrophoresis. *Appl. Environ.Microbiol.* 65: 2614–2621.

Viaud, M., Pasquier, A., and Brygoo, Y. (2000). Diversity of soil fungi studied by PCR-RFLP of ITS. *Mycol Res* 104: 1027–1032.

White, T.J., Brun T., Lee S., and Taylor J. (1990) Amplification and direct sequencing of fungal ribosomal RNA genes for phylogenetics, *In: PCR protocols. A guide to methods and applications* (Eds.) M.A. Innis, D.H. Gelfand, J.J. Sninsky, and T.J. White Academic Press, San Diago, USA:315–320.

Woese, C.R. (1987). Bacterial Evolution. *Microbiol. Rev.* 51: 221–271.

Yasuda, M., and Shiaris, M.P.(2005). Differentiation of bacterial strains by thermal gradient gel electrophoresis using non-GC-Clamped PCR primers for the 16S–23S rDNA intergenic spacer region. *FEMS Microbiol. Letters*. 243: 235–242.

Zelles, L. (1999). Fatty acid pattern of phospholipids and lipopolysaccharides in the characterization of microbial community in soil: A Review. *Biol. Fertil. Soils* 29: 111–129.

Zoetendal, E.G., Akkermans, A.D., and de Vos, W.M. (1998). Temperature gradient gel electrophoresis analysis of 16S rRNA from human fecal samples reveals stable and host specific communities of active bacteria. *Appl. Environ. Microbiol.* 64:3854–3859.

17

Tinea Capitis: Causal Organisms, Prognosis and Therapy

Sonali Wankhede and Mahendra Rai

Department of Biotechnology, SGB Amravati University, Amravati-444 602, Maharashtra;
E-mail: mkrai123@rediffmail.com

INTRODUCTION

Mycotic infection of the scalp is worldwide disease which is creating a havoc nowadays (Ali-Shtaych and Arda, 1986; Gruseck *et al.*, 1996). Such problematic infections are caused by dermatophytes which require keratin for growth, and so they are restricted to hair, nails and superficial skin (Aly, 1994). Thus, these fungi do not infect mucosal surfaces. Dermatophytes are referred to as "tinea" infections.

Dermatophytes include several distinct clinical entities, depending on the anatomic site and etiologic agents involved. Clinically the conditions include tinea capitis, tinea favosa (favus resulting from infection by *Trichophyton schoenleinii*), tinea corporis (ringworm of glabrous skin), tinea imbricata (ringworm resulting from infection by *Trichophyton concentricum*), tinea cruris (ringworm of groin), tinea unguinum or onychomycosis (ringworm of nail), tinea pedis (ringworm of feet), tinea barbae (ringworm of beard), and tinea manuum (ringworm of hand).

Tinea capitis is a common dermatophyte infection of the scalp in children (Lobato *et al.*, 1997). Tinea capitis is caused predominantly by *Trichophyton* or *Microsporum* species (Bocobo *et al.*, 1965). Clinical presentations are seborrheic-like scale, 'black dot' pattern, inflammatory tinea capitis with kerion and tiny pustules in the scalp (Martinez-Roig *et al.,* 1982; Sperlling, 1991; Pakilha-Goncalves, 1992; Ghorpade and Ramanan,1994; Aste *et al.*, 1996; Gargoom *et al.*, 2000). The clinical diagnosis should be confirmed by mycological examination (Head *et al.,* 1984; Borchers, 1985; Hebert, 1988; Hubbard and de Triquet, 1992; Rai 1992, Rai and Shrivastava, 1992; Rai and Thakre, 1996; Rai et al., 2004).

Tinea infections can be spread by anthropophilic, geophilic or zoophilic organisms (Lunder, 1992: Korstnje and Staats, 1994). Fomites are important vectors of tinea capitis. (Mackenzie, 1963; Davis, 1980; Hebert, 1988). Asymptomatic carriers can be significant reservoirs of infection (Ive, 1966; Polonelli *et al.*, 1982).

To get rid of tinea capitis, the main treatment that will be considered are the traditional agents like, griseofulvin (Davis, 1980; Gan *et al.*, 1987; Elewski and Hay, 1994; Caceres-Rios *et al.* , 2000) and the oral antimycotics like terbinafine (Goodfield, 1992; McClellan, 1999; Caceres-Rios *et al.*, 2000; Fuller, 2001), itraconazole (Lopez-Gomez *et al.*, 1994; Elewski, 1994; Leeming and Elliot (1995) and fluconazole (Lazar and Hilligoss,1990; Lyman and Walsh, 1992; Bisson, 1999). Carriers may be beneficial by applying selenium sulphide shampoo (Allen *et al.*, 1982; Givens *et al.*, 1995).

The main goal of this chapter is to discuss mode of infection, clinical symptoms, prognosis and treatment of tinea capitis.

History

The term originally indicated larvae of insects that fed on cloths and books. Subsequently, it meant parasitic infestation of the skin. By the mid 16th century, the term was used to describe disease of hairy scalp. The term ringworm referred to skin disease that assumed a ring form, including tinea.

At the beginning of the 19th century, Alibert (1806), used the term favus to describe the honey like exudates in some scalp infections. The parasitic nature of favus was reported in the first half of the 19th century (Hakendorf *et al.*, 1965). Gruby (1841) demonstrated that the fungus could be cultured and transmitted to other humans. Between 1841 and 1844, Gruby described the *in vivo* characteristics and clinical manifestations of *Trichophyton schoenleinii.* He also described *Microsporum audouinii.* Remak (1842,1845) named the organism responsible for favus as *Achorion schoenlenii,* which was later renamed *T. schoenleinii.* Charles Robin (1853), described several types of dermatophytoses in the book, *Histoire naturelle des vegetaux parasites* (Kwon-Chung, 1992). Ramond Sabouraud (1910), published the book *Les Teignes* which classified dermatophytes into four genera, *Trichophyton, Epidermophyton, Microsporum* and *Achorion.* Despite the establishment of taxonomic criteria for dermatophytes by Sabouraud, diseases continued to be described based on the presentation of clinical disease or minor morphological differences amongst the fungi. In 1934, Emmons used spore morphology and accessory structures of fungal organisms to define the present day taxonomic system for dermatophytes, *Trichophyton, Microsporum* and *Epidermophyton* species. In the few years following 1959 sexual states of dermatophytes were recognized (Dawson *et al.*,1959; Griffin 1960; Stockdale, 1961, 1967 and Weitzman, 1964).

The early treatments of tinea capitis were difficult to apply, often painful to the patient and largely unsuccessful (Rosenthal, 1960). In 1641, in Lyon, France Guy de Chauliac discussed the use of epilaton of hair using mechanical measures (Rosenthal, 1960). In the 19th century in france, the pitch helmet *Capellus piceus,* or *la calotte* was used to treat tinea capitis. The calotte consisted of an adhesive, resinous plaster placed over the entire scalp, left in place for 2 to 3 days, and then forcibly removed so that both diseased and normal hair were traumatically epilated (Rosenthal, 1960). In some instances the application of irritating solutions and ointments had systemic adverse effects, some of which were serious or even fatal (Rosenthal, 1960). Topical measures to try and achieve cure were sometimes carried out over months to years. From the beginning of the 19th century, epilation of the scalp in patients with ringworm was performed using fingernails and later forceps (Rosenthal, 1960).

The use of X-rays to treat scalp ringworm was first suggested by Freund (1897). This involved administering two or three fractionated doses that resulted defluvium; however, in many instances this was associated with permanent alopecia resulting from the use of excess radiation (Rosenthal, 1960; Freund *et al.*,1897). Studies have demonstrated an increase in tumors of the brain and nervous system in some children following radiotherapy for tinea capitis (Modan *et al.*, 1974; Ron *et al.*, 1998).

The first major breakthrough in the effective treatment of tinea capitis came in 1958 when Gentles reported that griseofulvin was effective in treating guinea pigs that had been experimentally infected with fungal organisms (Gentles, 1958; Oxford *et al.,* 1939). A few months later, the effectiveness of griseofulvin was demonstrated in human dermatophytosis (Williams *et al.*,1958; Blank and Roth, 1959; Russel *et al.*, 1960; Maibach *et al* ., 1960; Gupta *et al.* , 1993,1994).

Background

Tinea capitis is a disease caused by superficial fungal infection of the skin of the scalp, eyebrows, and eyelashes, with a propensity for attacking hair shafts and follicles (Assaf and Weil, 1994). This infection is predominantly found in prepubertal, school going children (Rippon, 1982; Badillet, 1991), where males are more infected than females (Malhotra *et al.*, 1979; Ravits and Himmerstein, 1983; Pereiro mingllens *et al.,* 1991; Khorsravi, 1994). This may be due to shortness of hair, which facilitates easy reach of fungal spore to the scalp. Adults are rarely infected (Malhotra *et al.*, 1979; Leeming, 1995), this is believed to be due to the higher fatty acid of the scalp, which have dermatophyte inhibiting property.

Clinical presentation of tinea capitis varies from a scaly non-inflamed dermatosis resembling seborrheic dermatitis to an inflammatory disease with scaly erythematous lesions and hair loss or alopecia that may progress to severely inflamed deep abscesses termed kerion (Sperlling, 1991).

Mode of Infection

Tinea capitis infections are common in temperate countries (Aly, 1994; Schwartz and Janniger, 1995). Some dermatophytes are spread directly from one person to another (anthropophilic), some are transmitted to human from soil (geophilic), and still others spread to humans from animal hosts (zoophilic) (Lunder, 1992; Korstnje and Staats, 1994). Transmission of dermatophytes can also occur indirectly from fomites e.g., upholstery, hairbrushes (Mackenzie ,1963), hats (Davis, 1980; Hebert, 1988). Minor trauma is an important predisposing factor for seeding of the fungi on the scalp to cause infection.

Causal Agents

Tinea capitis is caused by *Trichophyton* and *Microsporum* (Table 17.1). Tinea capitis is the most common pediatric dermatophyte infection worldwide. The age predilection is believed to result from the presence of *Pityrosporum orbiculare (Malassezia furfur)*, which is the part of normal flora, and from the fungistatic properties of fatty acids of short and medium chains in postpubertal sebum (Rippon, 1988).

Table 17.1 Species of *Microsporum* and *Trichophyton* causing tinea capitis

Pathogen	*Type of infection*	*Author and year*
Microsporum audounii	The most common strain to cause *T.capitis*.	Rothman *et al.* (1947), Kligman (1952), Bocobo, *et al.* (1965), Elewski (1996)
M. canis	Infection comes from animals, can cause highly inflammatory lesions.	Kligman (1952), Marples (1956), Alteras *et al.* (1986), Lunder (1992), Pereiro *et al.* (1997)
Trichophyton mentagrophytes	Cause highly inflammatory tinea capitis.	Rook *et al.* (1982), Couatarmanac'h *et al.* (1989), Sberna *et al.* (1993)
T. tonsurans	Causes an exceedingly chronic infection. The fungal infections often are familiar.	Rudolph (1979), Malhotra *et al.* (1979), Babel & Baughman (1989), Elewski (1994), Leeming and Elliot (1995)
T. violaceum	Causes the clinical lesions called "black dot" ringworm.	Ghorpade and Ramanan (1994), Aste *et al.* (1996),
T. schoennleini	Causes the clinical type of T.capitis known as favus.	Gargoom *et al.* (2000) Elewski and Hay (1994)

Causative agents of tinea capitis include keratinophilic fungi and dermatophytes. These molds are usually present in non-living cornified layers of skin, its appendages and sometimes are capable of invading the outermost layer of skin, stratum corneum, or other keratinized skin appendages derived from epidermis, such as hair and nails (Rook and Dawber, 1982).

Mortality

Rippon (1988) stated that classification and severity of tinea capitis depends on the site of formation of their arthroconidia. Ectoyhrix infection is defined as fragmentation of the mycelium into conidia around the hair shaft or just beneath the cuticle of the hair, with destruction of the cuticle. Some mild ringworm or prepubertal tinea capitis infections are of the ectothrix type, also termed the gray-patch type. Depending on the extent of associated inflammation, lesions may heal with scarring.

Endothrix infections are noted in which arthrospores are present within the hair shaft. These infections tend to progress, become chronic, and may last into adult life. Lessions can be eradicated by systemic antifungal treatment.

Incidence of Tinea capitis may vary by sex, depending on the causative fungal organism. In *Microsporum audounii,* related tinea capitis, boys are affected much more commonly (Malhotra *et al.*, 1979; Ravits and Himmerstein, 1983; Pereiro *et al.*, 1991; Khorsravi *et al.*, 1994). The infection rate has been reported to be up to 5 times higher in boys than in girls. However, the reverse is true after puberty. Girls and boys are affected equally by *Trichophyton* infections of the scalp both adult women are infected more frequently than man (Vidimos *et al.*, 1991; Gianni *et al.*, 1995; Aste *et al.*, 1996; Cremer *et al.*, 1997) By *Microsporum canis* male children are usually infected (Alteras *et al.*, 1986; Pereiro *et al.*, 1997).

Tinea capitis occurs primarily in children and occasionally in other age groups. It is seen most commonly in children younger than 10 years. Peak age range is in patients aged 3–7 years.

Clinical Symptoms

1. It initiates often in a subtle way and in the beginning resembles dandruff.
2. Erythematous papule grows around a hair shaft on the scalp, eyebrows, or eyelashes.
3. Within a few days, the red papule becomes paler and scaly, and the hairs appear discolored, lusterless and brittle. The break off a few millimeters above the scalp skin surface.
4. The lesion spreads, forming numerous papules in a typical ring form. Ring formed lesions may coalesce with other infected areas.
5. Pruritus usually is minimal but may be intense at times
6. Alopecia is common in infected areas
7. Inflammation may be mild or severe. Deep boggy red areas characterized by a severe acute inflammatory infiltrate with pustule formation are termed “kerions” or “kerion celsi” (Martinez-Roig *et al.*, 1982; Sperlling 1991; Pakilha-Goncalves,1992).
8. Favus is a severe form of tinea capitis, it may give the scalp a waxy or doughy appearance (Elewski and Hay, 1994).

Physical Symptoms

Physical picture of tinea capitis infection is limited to the skin of scalp, eyebrows or eyelashes.

1. Lesion begins as red papules with progression to grayish ring formed patches containing perifollicular papules.
2. Pustules with inflamed crusts, exudate, matted infected hair, and debris may be seen.
3. Black dot tinea capitis refers to an infection with fracture of the hair, leaving the infected dark stubs visible in the follicular orifices (Hebert,1988; Elewski, 1996).

4. Kerion celsi may progress to a patchy or diffuse distribution and to severe hair loss with scarring alopecia (Pomeranz and Fairley, 1994; Howard and Frieden, 1995).

Prognosis

Ringworm can usually be cured, but recurrence is common. Chronic infection develops in one patient in five. It can take six to 12 months for new hair to cover bald patches, and three to 12 months to cure infected fingernails. Toenail infections do not always respond to treatment.

Continuous shedding of fungal spores may last several months despite active treatment; therefore, keeping patients with tinea capitis out of school is impractical.

Topical and Systemic Treatment

Choice of treatment is determined by the species of fungus concerned, the degree of inflammation, and in some cases, by the immunologic and nutritional status of the patient.

Systemic administration of griseofulvin provided the 1st oral therapy of tinea capitis. Gupta *et al.* (1994) reported newer alternative antifungal medications, such as ketoconazole, itraconazole, terbinafine, and fluconazole, which have been stated as effective alternative therapeutic agents for tinea capitis (Bickers, 1994; Drake *et al.*, 1996). Among these agents, itraconazole and terbinafine are used most commonly (Gupta *et al.*, 1999).

Selenium sulfide shampoo may reduce the risk of spreading the infection early in course of therapy by the number of viable spores that are shed (Allen *et al.*, 1982; Givens *et al.*, 1995).

Therapy

Systemic treatment

In general, systemic therapy is indicated for treatment of tinea capitis (Smith, 1993; Noble *et al.*, 1998; Table 17.2).

Table 17.2 Antifungal agents used in tinea capitis.

Antifungal	*Infection*
Tolnaflate	It is a narrow-spectrum antifungal (Barrelt-bee, 1986). It is effective in most dermatophytoses (Weinstein *et al.*, 1964).
Haloprogin	Introduced after tolnaftate, haloprogin combines equivalent efficacy with a broadened fungal spectrum (Hermann, 1972). It is associated with higher rates of irritant dermatitis.
Ciclopirox	It is broad-spectrum antimycotic, effective against dermatophytes, yeasts, and some bacteria (Aly *et al.*, 1989; Gupta *et al.*, 1998).

Griseofulvin (Fulvicin)

Antibiotic derived from a *Penicillium* species that is deposited in the keratin precursor cells that are replaced gradually by non-infected tissue. As a result, new keratin becomes highly resistant to fungal invasions. Active against dermatophytes but not against yeasts or bacteria. Absorption is enhanced when fatty food is taken simultaneously. Accumulates in keratin of the stratum corneum, hair, and nails. Has a long record of safety, but newer regimens may prove more cost effective (Davis, 1980; Gan *et al.*, 1987; Elewski and Hay, 1994: Caceres-Rios *et al.*, 2000).

Contraindications: Documented hypersensitivity, Porphyra, Hepatocellular failure.

Interaction: May decrease hypoprothrombinemic activity of warfarin (adjust dose); co-administration decreases contraceptive effects, resulting in breakthrough bleeding, amenorrhea, or unintended pregnancy; may reduce effects of cyclosporine and salicylates; barbiturates.

Pregnancy: Safety for use during pregnancy has not been established.

Precautions: Oral form found to be embryotoxic and teratogenic to pregnant rats; therefore, do not prescribe for women contemplating pregnancy; patients are encouraged to take medication after fatty meal to increase absorption from the alimentary tract.

Adverse effect: Headache, nausea, fatigue, abdominal discomfort, or transient rash; less common adverse reactions include urticaria, diarrhoea and photosensitivity; may precipitate occur intermittent porphyra.

Alternative Medication

Gupta *et al.* (1998) has been reported new alternative drugs to compensate the time spending cure rate of older ones.

Azoles

This class includes many over-the-counter and prescription preparations. Members of this class include clotrimazole, econazole (Spectazole), ketoconazole (Nizoral), miconazole (Micatin), oxiconazole (Oxistat), and sulconazole (Exelderm). These agents have broad-spectrum activity. Ketoconazole (Lester, 1995), sulconazole and oxiconazole (Polak, 1984) require only once-daily application because of their long durability in the superficial layers of the skin. Clotrimazole, miconazole, and econazole require twice-daily application.

Itraconazole (sporanox)

It is azole with activity against many dermatophytes, *Malassezia furfur* and some moulds. It has long life in skin It is effective for tinea capitis caused by both

Trichophyton and *Microsporum* (Lopez-Gomez *et al.,* 1994; Elewski, 1994; Leeming and Elliot, 1995). Results of several clinical trials indicated that itraconazole is a safe and effective alternative to griseofulvin-failed cases. Itraconazole has slightly higher cure rate in children with tinea capitis infection caused by *T. violaceum* (Leggendre and Esola-Macre, 1990; Elewski, 1994; Greer, 1996).

Contraindications: Documented hypersensitivity; concomitant administration with HMG-CoA reductase inhibitors (e.g., lovastatin, simvastatin), astemizole, cisapride, midazolam, triazolam, or terfenadine are contraindicated.

Interactions: Antacids may reduce absorption of itraconazole; edema may occur with coadministration of calcium channel blockers (e.g., amlodipine, nifedipine); hypoglycemia may occur with sulfonylureas; may increase tacrolimus and cyclosporine plasma concentrations when high doses are used; rhabdomyolysis may occur with coadministration of HMG-CoA reductase inhibitors (lovastatin or simvastatin); coadministration with cisapride can cause cardiac rhythm abnormalities and death.

Pregnancy: Safety for use during pregnancy has not been established.

Precautions: Caution in hepatic insufficiencies, since it is poorly water soluble; should be taken with fatty meal to improve absorption.

Adverse effect: Gastrointestinal disturbances, headache, rarely afoliative dermatitis.

Ketoconazole (Nizoral)

It is a broad-spectrum synthetic antifungal compound of the azole group. When orally administered, it is active against anthropophilic dermatophytes. It is hydrophilic and high concentration of the drug develops within skin, making it potentially beneficial for treating superficial dermatophytosis. Delivery of this drug to the skin is accomplished through normal blood circulation and sweat. Some excretion occurs into sebum and epidermal basal layer. In the presence of normal gastric acidity, it is well absorbed (Gan *et al.*, 1987; Vargo *et al.*, 1993).

Contraindications: Since hepatic toxicity is noted in adult patients treated with ketoconazole, is not considered appropriate therapy for pediatric tinea capitis; coadministration of ketoconazole with terfenadine, astemizole, cisapride, or triazolam is contraindicated.

Interactions: Coadministration of ketoconazole and drugs primarily metabolized by enzyme system may result in increased plasma concentrations of drugs (e.g., cyclosporine, warfarin, terfenadine, astemizole); rifampin and phenytoin may decrease ketoconazole levels.

Pregnancy: Safety for use during pregnancy has not been established.

Precautions: Significant incidence of idiosyncratic hepatic toxicity (primarily hepatocellular type) in 1 in 15,000; rare fatalities have occurred from severe hepatic necrosis; reversible transient mild elevations of liver enzymes can occur.

Adverse effect: Headache and nausea, treatment have to stopped if ALT or SGPT rise two to three fold, inhibit androgen biosynthesis.

Fluconazole (Diflucan)

A triazole compound that is relatively water-soluble and well absorbed upon ingestion. Peak plasma concentration is achieved within 1–2 h after oral administration. Drug is distributed widely to body tissues, and fluids free without binding to plasma proteins. Drug has a long half-life of 22–30 h in adults, and steady-state levels are reached within 6–10 d after initiation of treatment. Most of the drug is excreted unchanged in urine with little hepatic metabolism. Eliminated slower from skin than from plasma, which contributes therapeutic benefit against superficial dermatophytosis, even after dosage has been discontinued.

Contraindications: Documented hypersensitivity.

Interactions: Levels may increase with hydrochlorothiazides; fluconazole levels may decrease with chronic coadministration of rifampin; coadministration of fluconazole may decrease phenytoin clearance; effects of anticoagulants may increase with fluconazole coadministration; increases in cyclosporine concentrations may occur when administered concurrently.

Pregnancy: Safety for use during pregnancy has not been established.

Precautions: Adverse reaction is noticed in patient who received more than 7 days of medication; reaction is mild and reversible when drug is discontinued.

Adverse effect: Nausea, vomiting, abnormal liver function tests.

Allylamines

According to Shuster and Ryder (1990), Naftifine (Naftin) and terbinafine (Lamisil) are applied once daily and remain active in the skin for up to one week after application. Both the agents have fungicidal activity (Goodfield, 1992; Mc Clellan, 1999; Fuller, 2001).

Terbinafine (Lamisil)

It is an allylamine with antifungal properties. Well absorbed upon oral administration (Goodfield, 1992; McClellan, 1999; Caceres-Rios, 2000; Fuller, 2001).

Peak plasma concentration is reached in approximately 2 h. Drug has strong plasma protein binding. Preferential uptake into fat resulted in relatively high concentration in the skin. Concentration within the stratum corneum reaches 75 times that of plasma concentration during first 2 weeks of therapy. Antifungal activity remains in the skin for 2 months after plasma concentration has depleted, following cessation of medication. High cure rates of fungal infections in children are reported (Meinicke, 1984; Hill *et al.*, 1992; Rademaker and Havill, 1998).

Contraindications: Documented hypersensitivity.

Interactions: May decrease cyclosporine effects; toxicity of terbinafine may increase with rifampin and cimetidine

Pregnancy: Usually safe but benefits must outweigh the risks.

Precautions: Discontinue if chemical irritation or signs of hepatobiliary dysfunction develop.

Adverse effect: Slightly low cure rate, nausea, headache, vomiting.

Topical Treatment

Along with the oral drugs topical treatment is often used adjunctively to decrease the shedding of viable organisms. Topical agents may be useful in preventing the relapse of chronic tinea capitis. Concomitant use of topical therapy with the newer oral antifungal agents may result in a more rapid cure, a possibility that will be evaluated in clinical trials. To check the rate of spreading of dermatophytes topical agents play important role (Weinstein *et al.*, 1964; Barrelt-bee *et al.*, 1986).

Topical selenium sulfide has been shown to improve the cure rate when it is used in combination with oral griseofulvin and thus hasten response and prevent spread of disease (Allen *et al.*, 1982; Givens *et al.* , 1995).

Another topical agents also plays vital activity to restrict the growth of harmful dermatophytes, these topical agents are Nystanin; the imidazoles such as Clotrimazole, Miconazole, Econazole, Ketoconazole, and Bifonazole; Terbinafine and Tolnaflate (Weinstein *et al.*, 1964; Meinicke *et al.*, 1984; Barrelt-bee *et al.* 1986; Hill *et al.*, 1992). The drugs in this therapeutic class are chemically distinct from each other. All drugs when applied topically have a good margin of safety, and most show a high degree of effectiveness.

Parallel to oral suspension Loporox cream is applied to the scalp and selenium shampoo is used to wash hair daily (Allen *et al.* , 1982; Givens *et al.*, 1995).

Disadvantages

Although topical drugs can provide immediate reductions in infectivity, they are free of systemic adverse effects and are relatively inexpensive.

There are major problems with compliance as the patient finds it difficult to continue treatment or to know where to apply the cream, once the inflammatory signs have settled. In practice, most topical drugs would need to be continue for some time after disappearance of the symptoms of visible signs. Topical drugs may be difficult to use in certain areas e.g., on the hair and nails and in some more sensitive areas.

Along with this some adverse reactions are most common such as increase in normal hair loss and irritation occurs, abnormal hair textures, scalp pustules, mild dryness of the skin and itching, etc.

Future Prospects

Due to various side effects (Smith,1993), long duration of treatment and high cost of drugs, treatments have not been successful in some cases. Now, this is the time to select an appropriate therapy. Further, the proper diagnosis should be carried out and the effects of essential oils of various herbs are to be studied in the categorical manner viz. sex, age group, natural inhabitants, geographical conditions. Also the category wise study of the pathogenicity of dermatophytes is needed. These factors should be considered simultaneously and causes of infections, remedies and effectiveness of drugs to be studied.

In recent years, there has been a gradual revival of interest in the use of medicinal plants in developed as well as developing countries, because herbal medicines have been reported to be safe and without any adverse effect. Thus, searches for new drugs with better and cheaper substitutes from plant resources are natural choice. Recently, some products of higher plant origin have shown to be an effective source of chemotherapeutic agents and with strong fungicidal activity (Bahuguna and Kushwaha, 1993; Bickers, 1994; Lyss *et al.*, 1997).

These products promote us to explore other plant products, which could be exploited as effective antifungal, especially against tinea capitis.

Conclusion

After decades of frustration and disappointment with this stubborn tinea infection, dermatologists and other clinicians now have access to drugs with high cure rates and excellent safety profiles. Moreover, short treatment times increase patient compliance, reduce treatment costs, and allow patients to feel hopeful that their unsightly infections will be ended.

Perhaps the most important task of the clinician is accurate diagnosis of the causal agent. Selection of an optimal antifungal drug whose spectrum of activity encompasses the infecting microorganism can proceed only with accurate diagnosis.

In the last decade, there have been significant advances in the development of effective and safe drugs for tinea capitis, but unfortunately their adverse reactions, dearness and longer time to cure hindered the fulfillment of expectations. Against this trend are the new antifungal drugs itraconazole, ketoconazole, terbinafine and fluconazole. Along with the oral suspension topical treatment is necessary. It is widely accepted that griseofulvin with selenium sulfide shampoo helps in eradicating the infection fast. Advantages of these drugs are overshadowed by the fact that they are having a lot of side effects, such as nausea, headache, vomiting, while less common adverse reactions are abdominal discomfort, transient rash, urticaria, diarrhea and photosensitivity, etc. But perhaps the single, area most deserving of our attention in the near future is that of improving diagnostic methods. Diagnostic methodology and fungal susceptibility testing lag behind therapeutic advances. We should turn our attention to these problems by seeking the solutions through the essential oils of medicinal plants.

References

Alibert, J.L. (1806).Description des malaides de la peau observees a l'hopital Saint-Louis et exposition des meilleures methodes suivies pour leur traitement, barrios, paris 129.

Ali-Shtaych, M.S. and Arda, H.M. (1986). A study of tinea capitis in Jorden (west bank). *J. Trop. Med. Hygiene.,* 19: 137–141.

Allen, H.B., Honing, P.J., Leyden, J.J. (1982). Selenium sulphide: adjunctive therapy for tinea capitis. *Pediatrics.* 69: 81–3.

Alteras, L., Feuerman, E.J. David, M. and Segal, R. (1986). The increasing role of *Microsporum canis* in the variety of dermatophytic manifestations reported from Israel. *Mycopathology.* 94: 105–107.

Aly, R. (1994). Ecology and epidemiology of dermatophyte infections. *J. Am. Acad. Dermatol.* 31: S21–S25.

Aly, R., Maibach, H.I., Bagatell, F.K., Dittmar, W., Hanel, H. and Falenga, V. (1989). Ciclopirox olamine lotion 1%: Bioequivalence to ciclopirox olamine cream 1% and clinical efficiency in tinea pedis *Clin. Ther.* 11: 298–303.

Assaf, R.R. and Weil, M.L. (1994). The superficial mycoses. *Dermatol Clin.* 14(1): 57-67

Aste, N., Pau, M. and Biggio, P. (1996). Tinea capitis in adults. *Mycoses* 39: 299–301.

Babel, D.E. and Baughman, S.A. (1989). Evaluation of the adult carrier state in juvenile tinea capitis caused by *trichophyton tonsurans. J.Am.Acad. Dermatol.* 21: 1209–12.

Badillet, G. (1991). Dermatophytecs et Dermatophytes. In; Badillet, G. (ed.) *Atlas Clinique et Biologigue, 3rd edn. Paris; Edition Varia, pp.* 10–35.

Bahuguna, S. and Kushwaha, R.K.S. (1993). *Int. Journal of cosmetic science.* 15 (1): 1–5.

Barrelt-bee, K.J., Lane, A.C., Turner, R.W. (1986). The mode of antifungal action of tolnaflate. *J. Med. Vet. Mycol.* 24: 155–60.

Bickers, D.R. (1994). Antifungal therapy: Potential interactions with other classes of drugs *J. Am. Acad. Dermatol*. 31: S87–90.

Bisson, R. (1999). Fluconazole. In: Bisson, R. ed. Compendium of pharmaceuticals and specialties, 34th edn. Toronto webcom Ltd. 510–513.

Blank, H. Roth, F.J. (1959). The treatment of dermatomycoses with orally administered griseofulvin. *Arch Dermatol.* 79: 259–266.

Bocobo, F.C., Eadie, G.A., Miedler, L.J.(1965). Epidemiologic study of Tinea capitis caused by *T. tonsurance* and *M. audouinii. Pub Health Rep.* 80:285–287.

Borchers, S.W. (1985).Moistened gauze technic to aid in diagnosis of tinea capitis. *J Am Acad Dermatol.* 13: 672–673.

Caceres-Rios, H., Rueda, M., Ballcona, R., Bustamante, B. (2000). Comparison of terbinafine and griseofulvin in the treatment of tinea capitis. *J Am Acad Dermatol.* 42: 80–84.

Charles, R. (1853). *Histoire naturelle des vegetaux parasites.*

Couatarmanac'h, A., Mignard, M.H., Guigen, C., Aznar, C., Chevrant-Breton, J. (1989). Teigne a *Trichophyton mentagrophytes* d' evolution prolonge sous griseofulvine et ketoconazole, guerie par l'itraconazole. *Bull. Soc. Fr. Mycol. Med.* 18:191–194.

Cremer, G., Bournerias, I., Vandemeleubroucke, E., Houin, R. and Revuz, J. (1997). Tinea capitis in adults; misdiagnosis or reappearance? *Dermatology.* 194, 8–11.

Davis,R.R.(1980). Griseofulvin In D.C.E. Speller(ed.), Antifungal Chemotherapy. John Wiley and sons Ltd., Chichester, England. 149–182.

Dawson ,C.O., Gentles, J.C.(1959). Perfect state of *Keratinomyces ajelloi. Nature* 183: 1345–1346.

Drake, L.A., Dinehart, S.M., Farmer, E.R. (1996). Guidelines of care for superficial mycotic infections of the skin: tinea infections. *J. Am. Acad Dermatol.* 34 (2 pt1): 282–6.

Elewski, B. (1996). Tinea capitis. *Dermatol Clin.* 14(1): 23–31.

Elewski, B.E. (1994). Tinea capitis; Itraconazole in *Trichophyton tonsurans* infection. *J. Am. Acad. Dermatol.* 31: 65–7.

Elewski, B.E. and Hay, R.J. (1994). International summit on cutaneous antifungal therapy, focus on tinea capitis, Boston, Massachusetts, November 11-15, *Pediatr. Dermatol.* 1996; 13 (1); 69–77.

Emmon, CW. (1934). Dermatophytes; natural groupings based on the form of spores and accessory organs. *Arch Dermatol Syphilol* 30: 337–362.

Fuller, L.C., Smith, C.H., cerio, R., (2001). A randomized comparison of 4 weeks of terbinafine verses 8 weeks of griseofulvin in the treatment of tinea capitis. *J. Dermatol.* 144: 321–327.

Gan, V.N., Petruska, M., Ginsburg, C.M. (1987). Epidemiology and treatment of tinea capitis: Ketoconazole Vs griseofulvin. *Pedtr. Infect. Dis. J.* 6: 46–9.

Gargoom, A.M., Elyazachi, M.B., Al.-Ani, S.M., Duweb, G.A. (2000). Tinea capitis in Benghazi, Libya. *Int. J. Dermatol* 39: 263–5.

Gentles, J.C.(1958). Experimental ringworm in guinea pigs: oral treatment with griseofulvin. *Nature* 182: 476–477.

Ghorpade,A.,Ramanan, C. (1994). Tinea capitis and corporis due to *Trichophyton violaceum* in six-day-old infant. *Int. J. Dermatol.* 33:219–220.

Gianni, C., Betti, R., Perotta, E. and Crosti, C. (1995) Tinea capitis in adults. *Mycoses* 38: 329–331.

Givens, T.G., Murray, M.M., Baker, R.C.(1995). comparison of 1% and 2.5% selenium sulphide in the treatment of tinea capitis. *Arch. Pediatr. Adolesc. Med.* 149: 808–811.

Goodfield, M. J. D.(1992). Short duration therapy with terbinafine for dermatophyte onychomycosis: a multicentre trial. *Br. J. Dermatol,* 126 (suppl. 39): 33–35 [medline]

Greer, D.L. (1996). Treatment of tinea capitis with itraconazole *J. Am. Acad. Dermatol.* 35: 637–8.

Griffin, D.N. (1960). The rediscovery of *Gymnoacus gypseum*, the perfect state of *Microsporum gypseum*; and a note on *Trychophyton terrestre*. Trans Br Mycol Soc 43: 637–641.

Gruby, D. (1841). Sur Les Micodermes qui constituent la teigne faveuse. *CR Acad Sci. (Paris)* 13: 309–312.

Gruseck, E., Splanemann, V., Bleck, O., Ring, J. and Abeck D. (1996). Oral terbinafine in tinea capitis in children *mycoses* 39: 237–240.

Gupta, A.K., Einarson, T.R., Summerbell, R.C., Shear, N.H. (1998). An overview of topical antifungal therapy in dermatomycoses. *A North American Perspective. Drug* 55: 645–75.

Gupta, A.K., Hofstarder S.L., Adam P., Summerbell, R.C.(1999 May-June) Tinea capitis, An overview with emphasis on management. *Pediatr. Dermatol* 16(3): 171–189. [medline]

Gupta, A.K., Sauder, D.N., Shear, N.H. (1993). New Antifungal Agents. *Curr Opin Dermatol.* 2: 200–206.

Gupta, A.K., Sauder, D.N., Shear, N.H. (1994). Antifungal agents: An overview. Part I. *J Am Acad Dermatol.*30: 677–698.

Hakendorf, J.A., Donald, G.F., Linn, H.W.(1965). Favus. *Australian J Dermatol* 8: 22–30.

Head, E.S., Henry, J.C., Macdonald, E.M. (1984). The cotton-swab technic for the culture of dermatophyte infections: its efficacy and merit. *J Am Acad Dermatol.*11:797–801.

Hebert, A.A. (1988). Tinea capitis. *Curr. Concepts Arch. Dermatol.* 124: 1554–1557.

Hermann, H.W. (1972). Clinical efficacy studies of haloprogin, topical antimicrobial agent. *Arch. Dermatol.* 106: 839–42.

Hill, S., Thomas R., Smith, S.G., Finlay, A.Y. (1992). An investigation of the pharmacokinetics of topical terbinafine 1% cream. *Br. J. Dermatol.* 127: 396–400.

Howard, R. and Frieden, I.J. (1995). Tinea capitis: new perspectives on an old disease. *Semin Dermatol* 14: 2–8.

Hubbard, T.W., de Triquet, J.M. (1992). Brush-culture method for diagnosis of tinea capitis. Pediatrics. 90: 416–418.

Ive, F.A. (1966). The carrier stage of tinea capitis in Nigeria. *Br J Dermatol.* 78: 219–221.

Khorsravi, A.R. Aghamirian, M.R. and Mahmoudi, M. (1994). Dermatophytosis in Iran. *Mycoses.* 37: 43–48.

Kligman, A.M.(1952). The pathogenesis of Tinea capitis due to *Microsporum audouinii* and *Microsporum canis. J. Invest Dermatol.* 18:231–246.

Korstnje, M.J. and Staats, C.G. (1994). Tinea capitis in northwestern Europe 1963–93;etiologic agents and their changing prevalence. *Int. J. Dermatol.* 33: 548–549.

Kwon-Chung, K.J., Bennett, J.E.(1992). Dermatophytoses. In; Kwon- Chung KJ, Bennett JE, eds. *Medical mycology*. Philaldelphia: Lea and Febiger, 107–170.

Lazar, J.D., Hilligoss, D.M. (1990). The clinical pharmacology of fluconazole. *Semin Oncol.* 17: 14–18.

Leeming, J.G. and Elliot, T.S. (1995). The emergence of *Trichophyton tonsurans* tinea capitis in Birmingham, U.K. *Br. J. Dermatol* 133: 929–931.

Leggendre R. and Esola-Macre J. (1990). Itraconazole in the treatment of tinea capitis. *J. Am. Acad. Dermatol.* 23: 559– 60.

Lester, M. (1995). Ketoconazole 2% cream in the treatment of tinea infection. *Cutis* 55: 181–183.

Lobato, M.N., Vugia, D.J. Frieden, I.J. (1997). Tinea capitis in California children: A population-based study of a growing epidemic. *Pediatrics.* 99: 551–554.

Lopez-Gomez, S., Del Palacic, A., van custem, J., Soledad Cucteara, M., Iglesias, C., Rodrigues-Noriga, A. (1994). Itraconazole verses griseofulvin in the treatment of tinea capitis; A double blind randomized study in children. *Int Dermatol.* 33: 743–7.

Lunder, M. and Lunder, M. (1992). Is microsporum canis infection about to become a serious dermatological problem? *Dermatology* 184: 87–89.

Lyman, C.A. Walsh, T.J. (1992). Systemically administered antifungal agents. A review of their clinical pharmacology and therapeutic applications. *Drugs.* 44: 9–35.

Lyss, G., Schmidt, T.J., Merfort, I., Pahi, H.L., (1997) Helenalin. *Biological chemistry* 378(9): 951–961.

Mackenzie, D.W.R. (1963). 'Hairbrush diagnosis' in detection and eradication of non-fluorescent scalp ringworm. *BMJ.* 2: 363–365.

Maibach, H.I., Kligman, A.M. (1960). Short-term treatment of onychomycosis with griseofulvin. *Arch Dermatol,* 81: 733–734.

Malhotra, Y.K., Garg, M.P., Kanrvar, A.J. and Nagrajan, S. (1979). A study of tinea capitis in Libya (Benghazi). *Sabouraudia* 17: 181–183.

Marples, M.J. (1956). The ecology of *microsporum canis* bodin in New Zealand. *J Hygiene.*54:378–387.

Martinez-Roig, A., Llorens-Terol, J., Torres, J.M. (1982). Erythema nodosum and kerion of the scalp. *Am J Dis Child.* 136: 440–442.

Mc. Clellan, K.J., Wiseman, C.R., Markham. (1999). A Terbinafine. An update of its use in superficial mycoses. *Drugs* 58: 179–202 [medline]

Meinicke, K., Striegel, C., Weidinger, G. (1984). Treatment of dermatomycoses with naftin. Therapeutic effectiveness following once and twice daily administration. *Mycosen.* 27: 608–14.

Modan, B., Baidatz, D., Mart H., Steinitz, R., Levin, S.G.(1974). Radiation-induced head and neck tumors. *Lancet* 1: 277–279.

Noble, S.L., Forbes, R.C., Stamm, P.L.(1998). Diagnosis and management of common Tinea Infections. *Am Fam Physician* 58(1): 163–78.

Oxford, A.E., Raistrick, H., Simonart, P.(1939). Studies in Thebiochemistry of micro-organisms; griseofulvin, C17H17O6Cl, a metabolic product of penicllium gresio-fulvum. *Biochem J.* 33:240–248.

Pakilha-Goncalves A. (1992). Inflammatory tinea capitis (kerion) mimicking dissecting folliculitis. *Int. J. Dermatol* 1992: 31: 66.

Pereiro mingllens, M., Pereiro, M. and Pereiro M. Jr. (1991). Review of dermatophytosis in Galicia from 1951 to 1987, and comparison with other areas of Spain. *Mycopathalogia* 113: 65–78.

Pereiro, M. Jr., Rodriguez, B. and Toribio J.(1997) Unusual findings in two cases of dermatophytosis due to *Microsporum canis*. J. *Med. Mycol.*7: 162–165.

Polak, A. (1984).Antifungal activity of four antifungal drugs in the cutaneous retention time test. *Sabouradia.* 22: 501–3.

Polonelli, L., Garcovich, A., Morace, G. (1982). Dermatophytes carriers among school children. *Mykosen.*25: 254–257.

Pomeranz, A.J. and Fairley, J.A. (1994). Management errors leading to unnecessary hospitalization for kerion. *Pediatrics.* 93: 986–988.

Rademaker, M. and Havill, S. (1998). Griseofulvin and terbinafine in the treatment of tinea capitis in children NZ *Med. J.*111: 55–57.

Rai M.K. : Tinea capitis due to *Trichophyton rubrum* in an adult woman, *J of Dermatol Venereol and Leprol*, 58: 213–214, 1992.

Rai M.K. and K.K. Shrivastava: On the incidence of Raktaja Krimis (dermatophytes) in Chhindwara, M.P., *Ancient Science of Life* 12 (1 & 2) : 221–226, 1992.

Rai, M.K. and R.P. Thakre : Dermatophytes: Retrospects and Prospects, *Nagpur Univ. Jour.*, 7: 36–43, 1996.

Rai, Mahendra, Nandkishore Chikhale, Manjusha Choukhande and Milind Dudhane (2004). *Raktaja Krimis* (Dermatophytes), In: Scientific basis of Ayurveda, CRC press, USA, pp 479–489.

Ravits, M.S. and Himmerstein, R. (1983). Tinea capitis in the New York City area. *Arch. Dermatol.* 119: 532–533.

Remak, R. (1845). Diagnostische und pathogenetische Unterschungen in der klinik des Geh. Raths Dr. Schonlein auf dessn Veranlassung angestellt und mit Benutzung anderweitiger Beobaschtungen veroffentlicht. Berlin, A. Hirschwald , 242.

Remak, R.(1842). Gelungere Impfung des favus. *Med Atg* 11: 137.

Rippon, J.W. (1982) Dermatophytosis and Dermatomycosis. In; Rippon, J.W. (ed.) *Medical mycology: The Pathogenic fungi and the Pathogenic actinomycetes.*

Rippon, J.W. (1988). Tinea capitis in dermatophytosis and dermatomycosis. In; Rippon J.W. *medical mycology; the pathogenic fungi and the Pathogenic Actinomycetes.* Philadelphia: WB. Saunders pp.: 186–96.

Ron, E., Modan, B., Boice, J.D.(1998). Tumors of the brain and nervous system after radiotherapy in childhood. *New Eng Med J.* 319 : 1033–1039.

Rook, A., Dawber, R. (1982). Ringworm of the scalp. In; Infestations: Diseases of the Hair and Scalp. London: Blackwell Scientific Pub: 367–85.

Rosenthal, T. (1960). Perspectives in ringworm of the scalp. *Arch Dermatol.* 82: 851–856.

Rothman, S., Smiljanic, A.L., Weitkamp, A.W. (1947). The spontaneous cure of Tinea capitis in puberty. *J Invest dermatol.* 8:81–98.

Rudolph, A.H. (1979). The clinical recognition of tinea capitis from *Trichophyton tonsurans. JAMA.* 242: 1170.

Russel, B., Frain-Bell, W., Stevenson, C.J. (1960). Chronic ringworm infection of the skin and nails treated with griseofulvin. *Lancet.*1: 1140–1147.

Sabouraud, R.(1910). Les Teignes. Paris, Masson.

Sberna, F., Farella, V., Geti, V. (1993). Epidemiology of the dematophytoses in the Florence area of Italy: 1985–1990. *Trychophyton mentagrophyte*s, Epidermophyton floccosum and *Microsporum gypsum* infections. *Mycopathologia.*122: 153–162.

Schwartz, R.A. and Janniger, C.K. (1995) Tinea capitis. *Cutis* 55: 29–33.

Shuster, I. and Ryder, N.S. (1990). Allylamines mode of selectivity of action compared to azoles antifungals and biological fate in mammalian organisms. *J. Dermatol Treat.* 1 (suppl. 2): 7–9

Smith, E.B. (1993).Topical Antifungal Drugs in the treatment of Tinea Infections. *J Am Acad Dermatol* 28(5 pt 1): S24–8.

Sperlling, L.C. (1991). Inflammatory tinea capitis (kerion) mimicking dissectiong cellutities: occurance in two adolescents. *Int. J. Dermatol.* 30: 190–2.

Stockdale, P.M. (1967). *Nannizzia persicolor.* Sp. nov., the perfect state of *Trichophyton persicolor. Sabouraudia.* 5: 355–359.

Stockdale, P.N. (1961). *Nannizzia incurvata* gen. nov., sp. nov., a perfect state of *microsporum* (Bodin) Gujrat et Grigorakis. *Sabouraudia.* 1: 41–48.

Vargo, Karen, Cohen, Bernad, A. (Jul 1993). *The American Academy of Pediaytrics.* Volume 92(1) : 155–157.

Vidimos, A.T., Camisa, C. and Tomecki, K.J. (1991). Tinea capitis in three adults *Int. J. Dermatol.* 30: 206–207.

Weinstein, M.J., Oden, E.M., Moss, E. (1964). Antifungal properties of tolnaflate *in vitro* and *in vivo. Antimicrob. Agents Chemother.* 3: 595–601.

Weitzman, I. (1964). Variation in *Microsoprum gypseum*. A genetic study of pleomorphism. *Sabouraudia* 3: 195–204.

Williams, D.I., Marten, R.H., Sarkany, I. (1958). Treatment of ringworm with griseofulvin. *Lancet.*, 2: 1212–1213.

Zitterl, E.K., Sosa, S. J.J., Schubert, Z.M., Della, L.R., Tubaro, A., Bertoldi, M., Franz, C. (1997). *Journal of Ethnopharmacology.* 57(2): 137–144.

18

The Health Implications of Antifungal Drugs and Mycotoxin Interactions in Chronic Toxigenic Mold Exposures

Ebere C. Anyanwu[1] and Ijeoma Kanu[2]

[1]Cahers Neurosciences Research, 41 MCH, 190 College Road, Birmingham B8 3TG, UK
E-mail: ebereanyanwu@msn.com
[2]Department of Microbiology, Abia State University, Uturu, Nigeria.
Ijay2001kay@yahoo.com

INTRODUCTION

Water-damaged buildings are known to be susceptible to contamination by toxigenic molds to which the exposed occupants suffer from various illnesses associated with mycotoxins released into the indoor environment by the toxigenic molds (Etzel *et al.*, 1998). Consequently, the issue of water-damaged buildings has attracted much attention in recent times because of the escalating health and property insurance costs nationwide. In the last twenty years or so, increasing number of associated illnesses with potentially fatal outcomes in children and older adults have been reported. The most common mycotoxin-producing molds that have been implicated in these illnesses include *Fusarium, Trichoderma, Stachybotrys Cladosporium, Penicillium, Aspergillus*, and *Alternaria* (Miller, 1992; Gravesen *et al.*, 1994). They have the potential to produce extremely potent secondary metabolites some of which are beneficial (antibiotics) and some are toxic (mycotoxins) to humans and animals. They are natural contaminants of cereals and other edible commodities throughout the world. The main medium of human toxigenic mold infection is through inhalation of the spores, and the most common route being ingestion, but it may also involve dermal, respiratory, and parenteral routes, the last being associated with drug abuse (Solomon, 1974, Kozak *et al.*, 1980; Lebowitz, 1983; Fergusson *et al.*, 1984) Although, some efforts have been made by different government environmental health agencies to regulate the presence of mycotoxins in foods, there is still a huge problem concerning the specific effects of different mycotoxins released by several toxigenic molds that are identified with one another (ecologically compatible), each affecting the structural, biochemical and functional activity of the other (Paraica and Domijan, 2001). Understanding the human effects of mycotoxins and antifungal drugs interactions is therefore, the first step forward and of great value in preventing and treating various illnesses associated with chronic toxigenic mold exposures (Castell *et al.*, 1997; Wolf and Henderson, 1998). Furthermore, because there are no specific clinical tools for assessing the dose-effect of mycotoxins, it is the aim of this chapter to review the structure and functions of mycotoxins, their relationships and the implications in the treatment of fugal infections. It is hoped that these would form the basis for further research work.

Mycotoxins

Mycotoxins are toxic chemicals released by indoor toxigenic molds. Under favorable physiological conditions, they undergo metabolic processes to produce yet other toxic chemicals compounds known as volatile organic compounds (VOCs), which are responsible for the moldy odors often perceived in contaminated indoor environments. There are already several mycotoxins produced by toxigenic molds (Table 18.1) some of which have been studied extensively. Mycotoxins are lipid-soluble and are readily absorbed by the intestinal lining, airways, and skin. Their toxic effects are thought to be associated with exposure to toxins on the surface of the mold spores, not with the growth of the mold in the body. Some toxigenic molds are capable of producing one or more mycotoxins that are limited to particular genera (specific), while others can produce the same mycotoxins that are not specific to any genera ("common mycotoxins").

Table 18.1 A list of some of the known mycotoxins.

Aflatoxins	Fusaric Acid	Penicillic Acid
Alternariol	Fusariocin	Penitrem
Citreoviridin	Fusarins	Phomopsin
Citrinin	Islanditoxin	Roridins
Cyclopiazonic Acid	Luteoskyrin	Rubratoxin
Deoxynivalenol	Moniliformin	Rubratoxin
Diacetoxyscirpenol	Monoacetoxyscirpenol	Slaframine
Dicoumarol	Neosolaniol	Sporidesmin
Ergotamine	Ochratoxins	Stachbotryotoxins
Ergo Toxins	Oosporein	Sterigmatocystin
Fumitremorgen	Paspalitrems	T-2 Toxin, Tremorgens
Fumonisins	Patulin	Zearalenone

The genera-specific and genera-common mycotoxins are presented in Table 18.2. In the genera-common mycotoxins, *Aspergillus* species produce aflatoxins and Sterigmatocystin. Aflatoxins are perhaps the best-known and most researched mycotoxins in the world. *Aspergillus* and *Penicillium* produce 4 genera-common mycotoxins including: citrinin (nephrotoxic), ochratoxin (carcinogenic), patulin, and gliotoxin (cytotoxic and immunosuppressive effects). *Penicillium* alone produces Cladosporin and emodin. Emodin has antibiotic properties as well as mycotoxic effects. *Penicillium* and *Memnoniella echinata* commonly produce griseofulvin. Grieseofulvin exhibits tumorgenic, carcinogenic, tetratogenic, and hepatotoxic reactions. Cyclopiazonic acid is produced by *Penicillium* and *Aspergillus* (Motta *et al.*, 2001). *Tolypocladium inflatum* specifically produce cyclosporin A. Cyclosporin A is reported to be immunosuppressive and a potential carcinogen, and used in medicine after organ transplants for the prevention of organ rejection (Qian *et al.*, 2002).

Table 18.2. A list of toxigenic molds and specific and common mycotoxins.

Type of mold	*Mycotoxin*	*Extent of production*
1. *Aspergillus flavus* & other A. species	Aflatoxins Sterigmatocystin	Specific
2. Aspergillus + Penicillium	Citrinin Ochratoxin Patulin Gliotoxin	Common
3. Penicillium + Cladiosporin	Emodin	Common
4. Penicillium + Memnoniella	Griseofulvin	Common
5. *Tolypocladium inflatum*	Cyclosporin	Specific
6. Fusarium species	Deoxynivalenol Or vomitoxin Fumonisin B1 T-2 toxin	Specific
7. *Stachybotrys chartarum*	Satratoxin H	Specific
8. Myrothecium + Stachybotrys	Roridin A + 3 Isomers	Common
9. Myrothecium	Verrucarin A	Specific

Fusarium produces high concentrations of trichothecenes, ochratoxin A and zearalenone, deoxynivalenol, also known as DON or vomitoxin, fumonisin B1 and T-2 Toxin (Ono *et al.*, 2001; Eskola *et al.*, 2001; Pietri *et al.*, 2001). Deoxynivalenol has been found to cause reduced weight gain, suppressed feeding, and at higher concentration levels there can be an emetic response in animals and humans (Eskola *et al.*, 2001; Pietri *et al.*, 2001) T-2 Toxin (a trichothecene) induces hemorrhagic, immunosuppressive reactions, nausea and vomiting, DNA damage and cell death. *Stachybotrys* and *Myrothecium* produce roridin *A*. Roridin *A* is a trichothecene mycotoxin that has a cosmopolitan distribution. *Myrothecium* specifically produces verrucarin *A* mycotoxin. Verrucarin A is a macrocyclic trichothecene triesters of the sesquiterpene alcohol verrucarol, and is synonymous with muconomycin *A*. They are both closely related to roridin mycotoxin.

Specifically, some strains of *Stachybotrys chatarum* produce satratoxin H that is implicated in very high cytotoxicity and several environmental allergic reactions (Anyanwu *et al.*, 2006).

Structural Properties of Mycotoxins

Mycotoxins are complex organic chemical compounds with molecular weights generally greater than 200 mass units produced when the mycelium is subjected to nutrient limitations during growth (Anyanwu *et al.*, 2006). Most of these mycotoxins are structurally related to one another (Fakhoury and Woloshuk, 2001). A single mold is capable of producing several mycotoxins that structurally have the propensity to cause harm to humans. With the aid of a combination of X-ray crystallographic analysis, NMR spectroscopy, and chemical transformations, it has been shown that mycotoxins undergo the process of isomerism. *Myrothecium verrucaria,* for example, produces all four roridin E isomers, diastereomeric at the C-6' and C-13' centers, the exceptions being those congeners having a C-4' ketone group in the macrolide ring (Jarvis and Wang, 1999). A representative structure of mycotoxins is shown in Fig. 18.1. Generally, they are a complex macrocyclic group some of which consist of a series of sesquiterpene polyalcohols, esters, and several ergosterol epoxides (Jarvis, 1992; Jarvis and Wang, 1996; Laurent *et al.*, 2000).

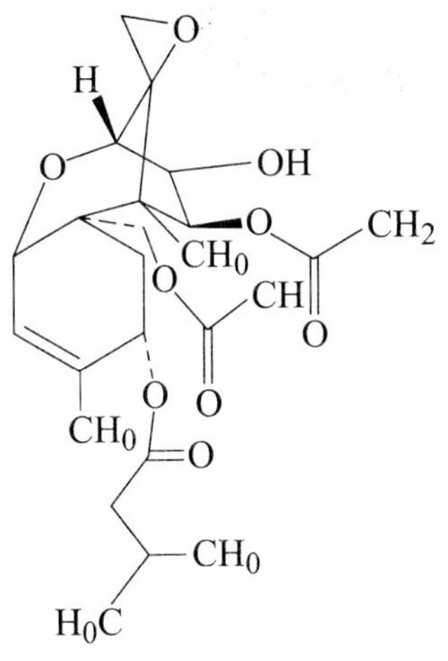

Figure 18.1 Representative Molecular Structure of T-2 Toxin.

Physiological Properties

In general, mycotoxins exert various biological activities, including insecticidal, antifungal, antibacterial and antiviral. They are also known to exhibit phytotoxic and cytotoxic qualities and inhibit a high degree of cytostaticity; the latter property makes them attractive antitumor candidates. Mycotoxins have also been associated with a wide variety of human and animal intoxications throughout the world (Ellenhorn and Bareeloux, 1988; Jarvis, 1992; Jarvis *et al.*, 1996; Loukaci *et al.*, 2000;).

Electrophilic Properties

Although, the electrophilic properties of mycotoxins have not been fully documented however, there are some indications that mycotoxins in general have electrophilic properties. Such indications are found in the published works of Fliege and Metzler on the electrophilic properties of patulin (Fliege and Metzler, 2000). In these reports, the electrophilic properties and non-enzymatic reaction of patulin with thiol-containing tripeptide glutathione and metabolic degradation product N-acetyl-L-cysteine (NAC) were demonstrated. The major products formed in the reaction of PAT with glutathione (GSH) were of the same structural type as obtained with NAC. Mycotoxin patulin was also found to induce intra- and intermolecular protein cross-links *in vitro* involving cysteine, lysine, and histidine side chains, and alpha-amino groups (Fliege and Metzler, 2000).

Antibiotic Properties of Mycotoxins

Some mycotoxins (e.g., blasticidin A and citrinin) have antibiotic properties that show strong inhibitory activity toward other mycotoxins production by significantly reducing the expression of genes encoding their biosynthetic enzymes and regulatory genes (Fliege and Metzler, 1999; Ammar *et al.*, 2000). In the studies of the antibiotic and toxigenic properties of 106 mold strains (Zaichenko *et al.*, 1992; Zaichenko *et al.*, 1997), which were the potential producers of macrocyclic trichothecenes, it was shown that 40% had high antibiotic activity. Others, such as *Laetisaria fuciformis* molds produce diffusible compounds that are inhibitory to a range of Gram-positive and -negative bacteria but not to the mould species tested (Tiwari *et al.*, 1986; Zaichenko *et al.*, 1992; Di menna *et al.*, 1996; Stormer and Hoiby, 1996).

Dose-effects of Mycotoxins—how safe are they?

There is no evidence in the literature to demonstrate the dose levels of mycotoxin (s) that could be safe to humans. Whatever regulation available now is purely precautionary and arbitrary. The determination as to whether or not a given concentration of mycotoxin is safe will depend on a number of factors including chemical class and chemical structure of the mycotoxins. The length of time of exposure to the mycotoxin (s) may vary from a few days to several years. It is assumed from animal experimental derivations that the risks of mycotoxigenesis increase as exposure time increases. In humans however, no observable effect levels (NOEL) for mycotoxins exists. The adult LD50 (lethal dose to 50% of exposed individuals) of trichothecenes is estimated to be only 35 milligrams (El-Makawy *et al.*, 2001). As the research interest in mycotoxicity evolves, more and more mycotoxins are identified and the unsolved problems of dose-effect continuously complicate our systematic understanding of the pathogenesis of mycotoxins. The complication stems, at least in part, from the fact that mycotoxins have functional groups that exert different health effects on both humans and animals; their metabolites are more frequently reported as potent carcinogenic substance known, and if one chemical bond is changed in the structure of the molecule, its toxicity can be reduced dramatically.

Does Mycotoxic Effect Depend on age and dose dependent?

Another important question is, whether there is any tolerable dose and whether mycotoxic effects are age and dose dependent. In this case, there are no specific answers because of the multiplicity of mycotoxins released by a single mold and the fact that many molds release similar mycotoxins. From the available literatures that pertain to "fungal drug" use, it appears as though age and dose inferences are mostly based on animal models of mycotoxicity, and little of humans. Thuvander *et al.* (2001) estimated the intake of some mycotoxins from food in Sweden and found that mycotoxin levels in the food samples were generally below the European/national maximum limits. However, the intake of ochratoxin A, patulin and trichothecenes was found to be lower than the temporary daily intake values (tTDI) proposed for these toxins by international expert groups. Although, the intake of trichothecenes (expressed as T-2 toxin equivalents) in children with a high consumption of cereals was close to the tTDI for T-2 toxin, such a value would be seriously unreliable (Thuvander *et al.*, 2001; Galvano *et al.*, 2001; Skaug *et al.*, 2001).

Pathophysiological Implications

The symptoms of mycotoxin poisoning are largely dependent on the type of mycotoxin, the dose, route of exposure, and the sensitivity of the person exposed. Here are just a few of the many short-term symptoms that have been documented in the scientific and medical literature. The effects acute exposures to toxigenic molds could occur within hours or days of exposure and may lead to headaches, coughing, skin rashes, fever, nausea, sore throat, nosebleeds, congestion, fatigue, depression, and diarrhea. However, chronic exposures may lead to kidney and respiratory disorders, liver cancer and lung cancer (glotoxin, aflatoxin), brain disease (citroviridin, satratoxin), immune suppression, skin disease, respiratory disorders, birth defects, and organ damage (trichothecenes, zearalenone) (Nielson *et al.*, 1999; Gravesen *et al.*, 1999; Ochmanski and Barabaz, 2000). In general, the severity of the pathophysiological effects of mycotoxins probably depend at least in part, on the concentration of toxic molds in the environment, the type of the metabolites produced, the dose-effect of mycotoxins in the exposed individuals, and the individual constitutional variability. It has been observed in our center that the dominance of a particular species of toxigenic molds such as *Penicillium* may influence the pathophysiological effects of other fungal mycotoxins in the patients' serum. Hence, whenever there is one or no *Penicillium* species present, for example, the concentrations of all other toxigenic molds are affected. In other words, it could be that the mycotoxins from one dominant toxigenic species could suppress the activity of the mycotoxins from other molds by masking the reactive regions of such mycotoxins (;Yamada *et al.*, 2000; El-Makawy *et al.*, 2001). Besides health effects such as headaches, fatigue, and respiratory diseases, mycotoxins especially, mycotoxins have been shown to be carcinogenic in laboratory animals (Paraica and Domijan, 2001; Gelderblom *et al.*, 2001). The mechanisms for cancer induction by mycotoxins include the possible role of oxidative damage during initiation and the disruption of lipid metabolism, integrity of cellular membranes, and altered growth-regulatory responses as important events during promotion (Gelderblom *et al.*, 2001). Lipid peroxidation is one of the main manifestations of oxidative damage and has been found to play an important role in the toxicity and carcinogenesis of many carcinogens such as mycotoxins (Rastogi *et al.*, 2001).

Problems of "Undesired" Toxicological Effects

One of the problems in elucidating the mechanism of mycotoxicity is being able to determine the minimal tolerable blood level in humans. The mycotoxin dose levels, the ligand-receptor, and their complexes are not yet known. However, it is assumed that if the ligand is endogenous, the effect will have a physiologically undesired toxicological effect. In addition, if the reaction is not endogenous, the effects are expected to be pharmacological (Fig. 18.2). For a mycotoxicity to have toxicologically undesired effects, three major manifestations must be observed in both adult and children, either at short- or long-term endpoints:

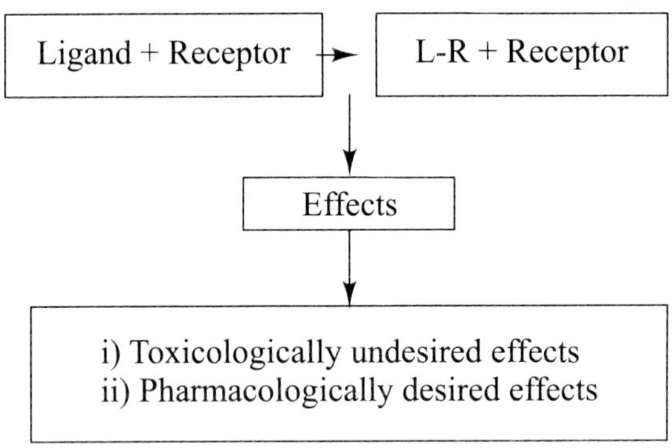

Figure 18.2. Proposed basic mechanism of mycotoxic effects.

Behavioral functions (adults) such as impaired memory and learning ability must be affected. The problem here is that the concentration to which behavioral changes are effected is not yet determined. Furthermore, changes in individual constitution and genetic make-up may affect the levels of behavioral changes.

Peripheral nerve conduction functions in adults must be affected i.e., there should be a decreased nerve conduction velocity. The question here is: at what concentration is nerve conduction velocity affected ? Is the effect transient ?

What neurological signs and symptoms are manifested in children? If these signs and symptoms are at high levels of concentration, can the possibility of encephalopathy be ruled out ?

Neurodevelopmental and Pharmacologically Desired Effects

Neurological effects of mycotoxins are those effects on the structure or functioning of the central nervous system or peripheral nervous system that result from exposures to toxigenic molds. They can cause morphological changes that lead to generalize damage to nerve cells (neuropathy), injury to axons (axonopathy), or destruction of the myelin sheath (myelinopathy), or neurobehavioral dysfunction (IPCS, 1986; Needleman, 1995; Norton, 1996). If mycotoxins affect development; it then implies that the anatomical and functional "hard wiring" and differentiation in the CNS may be affected. There could be an overexpression of the proteolipid protein (PLP) gene, endosomal/lysosomal accumulation of cholesterol and this accumulation is shown to perturb the process of myelination, and impair the viability of oligodendrocytes (Office of Tech. Assess, 1990). Impairment of oligodendrocytes and myelin, could lead to the molecular cascade leading to white matter damage (Simons *et al.*, 2002). If that happens, the adverse effects may occur during synaptogenesis. That means that cell: cell adhesion may be affected too (Kadhim *et al.*, 2002). On the other hand, if a mycotoxin has a pharmacological effect, the cellular homeostasis and the ionic mechanism of neurotransmission may be affected. These effects may either lead to the inhibition or facilitation of transmitter release and modulation of ionic conductance. In children, for example, mycotoxins affect memory and attention, and in adults mycotoxins produce peripheral neurotoxicity, while in the CNS they affect mood and affects (Calabresi, 2002). However, as far as developmental and pharmacological effects are concerned, further questions are asked:

- Are these changes related to each other and how?
- Which molecular actions of mycotoxins are related to one or more of these effects?
- Are neurotransmitter release and transmitter-gated ion channels only at high concentration in the processes of mycotoxicity?
- How do mycotoxins enter neurons and affect intracellular sites such as second messenger and ion-storage system?
- What are the mycotoxin binding proteins in brain that may help to determine their toxicity?
- What cellular event are most sensitive to mycotoxin activation of protein kinase C, *in vitro* or *in vivo* at what concentration?
- Can the mitochondrial release of Ca2+ sensitive nM range of mycotoxin be measured?
- Does DA-sensitive adenylate cyclase sensitive to micromolar range of mycotoxic; is Na/ATPase also sensitive to mycotoxins, probably to very low concentration?

- Are there any phosphorylation reactions associated with learning and do these mechanisms connect with behavior/learning?

Until these questions are fully answered in full through compelling evidence-based research outcomes, the understanding of the cause-effects of mycotoxins and their interactions with antifungal drugs will remain a serious health problem.

Treatment of Illnesses due to Toxigenic Mold Exposures

Consequently, treatment of diseases due to toxigenic mold exposures is a complex endeavor that requires proper knowledge and extensive experience of the nature and behavior of the molds. Most of the antifungal drugs are derived from the mycotoxins, for example, the antifungal polyenes (amphotericin B, fungizone), griseofulvin, flucytosine, azoles (imidazoles and triazoles), allylamines and others. Most of them have similar chemical and physical properties to mycotoxins. Itraconazole, the azole derivative that is active against *Aspergillus*, may be administered only in mild cases of aspergillus infections in immunocompromised patients. Although, intravenous amphotexicin B (AMB) and oral itraconazole remain the standard of care for invasive aspergillosis, their effectiveness against other mycotoxins is not fully established. The overall response rate was 39%, in other words, there are variable results in treating patients with amphotericin B, itraconazole or both (Kretschmer *et al.*, 1968). Therefore, in treating fungal infections, the structural and preferential effectiveness must be considered.

Structural Consideration

A number of antifungal drugs have the ability to degrade and inactivate certain mycotoxins. They do so by mechanistically "masking" the functional groups or degrading mycotoxins to their non-active state through the processes of ammoniation and sulfitization (Heathcote and Hibbert, 1978; Wyllie and Morchause, 1978; Just-Nbling and Stille, 1991). For example, is that those mycotoxins that do not contain amino and sulfite functional groups (trichothecenes) can easily respond to antifungal drugs that contain amino and sulfite groups (acid-base reaction). The mechanism for this action appears to involve hydrolysis of the lactone ring and subsequently, converts the parent mycotoxin compounds to numerous products that exhibit greatly decreased toxicity (Heathcote and Hibbert, 1978; Wyllie and Morchause, 1978; Just-Nbling and Stille, 1991; Eaton and Groopman, 1994).

Preferential Effectiveness Consideration

Certain toxigenic molds dominate and exert inhibitory effects on the activities of other molds. Since most antifungal drugs are derived from mycotoxins, it then implies that their effectiveness will probably depend on the predominant mold mycotoxins present in the patients' serum. The antifungal drug would preferentially, interact with the mycotoxins it has inhibitory affinity for. For example, if a patient is presenting with infections due to mycotoxins with azole structure, and is treated with the antifungal azole drugs, the clinical response may not be a 100% effective. Although the patient's condition may apparently improve slightly, the azole mycotoxin will relapse with time, as we have observed in our center. Another example is the case of fluconazole and clotrimazole that are the most commonly prescribed treatments for oral candidiasis, one would find that the two drugs are functionally similar, and the difference between their effectiveness monotherapeutically and polytherapeutically is not much. Hence, the clinical response from two groups of patients (Miller *et al.*, 1988) treated with both drugs was statistically equivalent in both groups (98% of fluconazole recipients and 94% of clotrimazole recipients). However, the efficacy of elimination of *C. albicans* with fluconazole was 65%, while clotrimazole was 48%, and when two antifungal drugs with different functional groups are compared, significant differences were possible (Marriott *et al.*, 1993; Pons *et al.*, 1993; Millon *et al.*, 1994). In addition, separate treatment of patients with itraconazole 200 mg daily and ketoconazole 200 mg daily showed no significant difference in time to clearing infection between the two treatments. The differences in these cases may be due to the types of unrelated mycotoxins present in the patient's serum, and not necessarily on the superiority or efficacy of the individual drugs (Smith *et al.*, 1991; De Repentigny *et al.*, 1992; Pons *et al.*, 1995).

Conclusion

Mycotoxins, antifungal drugs interactions and the implications in the treatment of illnesses due to indoor chronic toxigenic mold exposures have been reviewed. It is very important that the structural and functional relationships of the mycotoxins with one another and with antifungal drugs be fully understood so that a proper treatment strategy could be carried out with a higher level of confidence. It is appropriately suggested that if a patient is sensitive to those mycotoxins that can be produced for example, by *Penicillium* species then it would not be advisable to prescribe the patients with penicillin medication of any type otherwise, the patient would be at risks of adverse reaction to the penicillin. It is important also that the identities, tolerable doses, long-term and short-term adverse effects of the mycotoxins are determined to avert or minimized the scenario of adverse antifungal drug reactions.

References

Ammar, H., Michaelis, G., Lisowsky, T. (2000). A screen of yeast respiratory mutants for sensitivity against the mycotoxin citrinin identifies the vacuolar ATPase as an essential factor for the toxicity mechanism. *Curr. Genet.* 37(5): 277–284.

Anyanwu, EC., Kanu I., Nwachukwu, N.C., Saleh M.A. (2005). Chronic Environmental exposure to *Alternaria tenuis* may manifest symptoms of neuropsychological illnesses: A study of 12 cases. *Journal of Applied Sciences and Environmantal Management* 9(3): 45-51.

Calabresi, L., Gomaraschi, M., Villa, B., Omoboni, L., Dmitrieff, C., Franceschini, G. (2002). Elevated Soluble Cellular Adhesion Molecules in Subjects With Low HDL-Cholesterol. Arterioscler. *Thromb. Vasc. Biol.* 22(4): 656–661Kretschmer, R., Janeway, C.A., Rosen, F.S. (1968). Immunologic amnesia. Study of an 11-year-old girl with recurrent severe infections associated with dysgammaglobulinemia, lymphopenia and lymphocytotoxic antibody, resulting in loss of immunologic memory. *Pediatr. Res.* 2 (1): 7–16

Castell, J.V., Gomez-Lechon, M.J., Ponsoda, X., Bort, R. (1997). The use of cultured hepatocytes to investigate the mechanisms of drug hepatotoxicity. *Cell Biol Toxicol.*13(4-5):331–338

de Repentigny. L., Phaneuf, M., Mathieu, L.G. (1992). Itraconazole vs. ketaconazole in HIV-positive patients with oropharyngeal and/or esophageal candidiasis. 32nd ICAAC, Abstract 1117.

di Menna, M.E., Burch, G., Smith, B.L., Garthwaite, I. (1996). Toxic and antibiotic properties of red thread, *Laetisaria fuciformis. Mycopathologia* 135(3): 187–190.

Eaton, D.L., Groopman, J.D. (1994). The Toxicology of Aflatoxins. Academic Press, New York; pp.383–426.

Ellenhorn, M.J., Barceloux, D.G. (1988). In Medical Toxicology: Diagnosis and Treatment of Human Poisoning. Merck Index, 12th Ed., Elsevier, New York; pp.1312–1314."

El-Makawy. A., Hassanane, M.S., Abd Alla, E.S. (2001). Genotoxic evaluation for the estrogenic mycotoxin zearalenone. Reprod. *Nutr. Dev.* 41(1): 79–89.

Eskola, M., Parikka, P., Rizzo, A. (2001) Trichothecenes, ochratoxin A and zearalenone contamination and fusarium infection in Finnish cereal samples in 1998. *Food Addit Contam.* 18(8): 707–718.

Etzel, R.A,. Balk, S.J., Beare, C.F., Miller, M.D., Shannon, M.W., Shea, C.M. (1998). Toxic Effects of Indoor Molds (RE9736) *Pediatrics* 101(4): 712–714.

Fakhoury, A.M., Woloshuk, C.P. (2001). Inhibition of growth of *Aspergillus flavus* and fungal alpha-amylases by a lectin-like protein from *Lablab purpureus. Mol. Plant Microbe Interact.*14(8):955–961.

Fergusson, R.J., Milne, L.J., Crompton, G.K. (1984). Penicillium allergic alveolitis: faulty installation of central heating. *Thorax* 39:294–298.

Fliege, R., Metzler, M. (1999). The mycotoxin patulin induces intra- and intermolecular protein cross-links *in vitro* involving cysteine, lysine, and histidine side chains, and alpha-amino groups. *Chem Biol Interact.* 123(2): 85–103.

Fliege, R., Metzler, M. (2000). Electrophilic properties of patulin. Adduct structures and reaction pathways with 4-bromothiophenol and other model nucleophiles. Chem *Res Toxicol.* 13(5): 363–72.

Fliege, R., Metzler, M. (2000). Electrophilic properties of patulin. N-acetylcysteine and glutathione adducts. *Chem Res Toxicol.* 13(5): 373–381.

Galvano, F., Galofaro, V., Ritieni, A., Bognanno, M., De Angelis, A., Galvano, G. (2001). Survey of the occurrence of aflatoxin M1 in dairy products marketed in Italy: second year of observation. *Food Addit. Contam.* 18(7): 644-646.

Gelderblom, W.C., Abel, S., Smurs, C.M., Marnewick, J., Marasas, W.F., Lemmer, E.R., Ramljak, D. (2001). Fumonisin-induced hepatocarcinogenesis: mechanisms related to cancer initiation and promotion. *Environ Health Perspect.* 109 (Suppl 2): 291–300.

Gravesen, S., Frisvad, J.C., Samson, R.A.(1994). Microfungi. Copenhagen, Denmark: Munksgaard Publishingprovide 11436–1.

Gravesen, S., Nielsen, P.A., Iversen, R., Nielsen, K.F. (1999). Microfungal contamination of damp buildings—examples of risk constructions and risk materials. Environ. Health Perspect.; 107 (Suppl. 3): 505–508.

Health and Welfare Canada Working Group on Fungi and Indoor Air: Significance of fungi in indoor air: Report of a Working Group (1987). *Can J Public Health*, 78 (2): S1–S14.

Heathcote, J.G., Hibbert, J.R. (1978) Aflatoxins: Chemical and biological aspect. Elsevier, New York, pp.173–186. International Programme on Chemical Safety (IPCS). Principles and Methods for the Assessment of Neurotoxicity Associated with Exposure to Chemicals. Environmental Health Criteria, no. 60. Geneva: World Health Organization, 1986.

Jarvis, B.B. (1992). Macrocyclic and other novel trichothecenes: their structure, synthesis, and biological significance. *Acc. Chem. Res.* 15: 388–395.

Jarvis, B.B., Wang, S. (1999). Stereochemistry of the roridins. Diastereomers of roridin *E. J. Nat. Prod.* 62(9):1284–1289.

Jarvis, B.B., Wang, S., Ammon, H.L. (1996). Trichoverroid stereoisomers. *J. Nat. Prod.* 59(3): 254–261.

Just-Nubling, G., Stille, W. (1991). Therapy of systemic mycoses in immunodeficiency. *Immun. Infekt.* 19(4): 116–120.

Kadhim, H., Tabarki, B., De Prez, C., Rona, A.M., Sebire, G. (2002). Interleukin-2 in the pathogenesis of perinatal white matter damage. *Neurology* 58(7): 1125–1128.

Kozak, P.P., Gallup, J., Cummins, L.H. (1980). Currently available methods for home mould surveys: II. Examples of problem homes studied. *Ann Allergy* 45:167–176.

Laurent, D., Guella, G., Roquebert, M.F., Farinole, F., Mancini, I., Pietra, F. (2000). Cytotoxins, mycotoxins and drugs from a new deuteromycete, *Acremonium neo-caledoniae*, from the southwestern lagoon of New Caledonia. *Planta Med.* 66(1): 63–66.

Lebowitz, M.D. (1983). Health effects of indoor pollutants. *Annu Rev Public Health* 4:203–221.

Loukaci, A., Kayser, O., Bindseil, K., Siems, K., Frevert, J., Abreu, P.M. (2000). New trichothecenes isolated from *Holarrhena floribunda*. *J. Nat. Prod.*63(1): 52–56.

Marriott, D.J, Jones, P.D., Hoy, J.F., Speed, B.R., Harkness, J.L. (1993). Fluconazole once a week as a secondary prophylaxis against oropharyngeal candidiasis in HIV-infected patients. Med. J. *Australia 158*: 312–316.

Miller, J.D. (1992). Fungi as contaminants in indoor air. *Atmospheric Environ.*, 26:2163–2172.

Miller, J.D., Laflamme, A.M., Sobol, Y. (1988) Fungi and fungal products in some Canadian houses. *Int. Biodeterioration* 24:103–120.

Millon, L., Manteaux, A., Reboux, G., Drobacheff, C., Monod, M., Barale, T., Michel-Briand, Y. (1994). Fluconazole-resistant recurrent oral candidiasis in human immunodeficiency virus-positive patients: persistence of *Candida albicans* strains with the same genotype. *J. Clin. Micro.* 32(4): 1115–1118.

Motta, S.D., Valente-Soares, L.M. (2001). Survey of Brazilian tomato products for alternariol, alternariol monomethyl ether, tenuazonic acid and cyclopiazonic acid.*Food Addit Contam.* 18(7): 630–634.

Needleman, H.L. (1995). Behavioral Toxicology. Environmental Health Perspectives. 103 (Suppl. 6): 77–79.

Nielsen, K.F., Gravesen, S., Nielsen, P.A., Andersen, B., Thrane, U., Frisvad, J.C. (1999). Production of mycotoxins on artificially and naturally infested building materials. *Mycopathologia*145(1): 43–56.

Norton, S. (1996). Toxic Responses of the Central Nervous System. Chapter 13 in Casarett and Doull's Toxicology, (Eds. C. Klaassen, M. Amdur, and J. Doull). New York: Pergamon Press, pp.141–75.

Ochmanski, W., Barabaz, W. (2000). Microbiological threat from buildings and rooms and its influence on human health (sick building syndrome). Przegl. Lek.; 57 (7-8): 419–423.

Office of Technology Assessment. Neurotoxicity: Identifying and Controlling Poisons of the Nervous System (1990). Washington, D.C.: Government Printing Office, pp 235–249.

Ono, E.Y., Ono, M.A., Funo, F.Y., Medinal, A.E., Oliveira, T.C., Kawamura, O., Ueno, Y., Hirooka, E.Y. (2001). Evaluation of fumonisin-aflatoxin co-occurrence in Brazilian corn hybrids by ELISA. *Food Addit Contam.* 18(8): 719–729.

Paraica, M., Domijan, A.M. (2001). Contamination of food with mycotoxins and human health. *Arh Hig Rada Toksikol.*52(1): 23–35.

Pietri, A., Bertuzzi, T., Pallaroni, L., Piva, G. (2001). Occurrence of ochratoxin A in Italian wines. Food Addit Contam. 18(7): 647–654.

Pons, V., Greenspan, D., Debruin, M. (1993). Therapy for oropharyngeal candidiasis in HIV-infected patients: a randomized, prospective multicenter study of oral fluconazole versus clotrimazole troches. *JAIDS* 6:1311–1316.

Pons, V., Greenspan, D., Lozada-Nur, F., McPhail, L., Gallant, J.E., Tunkel, A., Johnson, C.C., McCarty, J., Panzer, H., Levenstein, M., Barranco, A., Green S. (1995). Comparative clinical study of oral suspension fluconazole versus topical liquid nystatin in the treatment of oropharyngeal candidiasis in AIDS. 35th ICAAC, Abstract 1221 A, San Francisco.

Qian, Y., Chen, G., Huang, J. (2002). Treatment of organ transplant rejection with low doses of CsA and CTLA4-Ig. *Zhonghua Wai Ke Za Zhi.*40(2): 150–152.

Rastogi, R., Srivastava, A.K., Rastogi, A.K. (2001). Long term effect of aflatoxin B (1) on lipid peroxidation in rat liver and kidney: effect of picroliv and silymarin. *Phytother. Res.* 15(4): 307–310.

Sakuda, S., Ikeda, H., Nakamura, T., Kawachi, R., Kondo, T., Ono, M., Sakurada, M., Inagaki, H., Ito, R., Nagasawa, H. (2000). Blasticidin A derivatives with highly specific inhibitory activity toward aflatoxin production in *Aspergillus parasiticus*. *J Antibiot (Tokyo)* 53(12): 1378–1384.

Simons, M., Kramer, E.M., Macchi, P., Rathke-Hartlieb, S., Trotter, J., Nave, K.A., Schulz. J.B. (2002). Overexpression of the myelin proteolipid protein leads to accumulation of cholesterol and proteolipid protein in endosomes/lysosomes: implications for Pelizaeus-Merzbacher disease. *J. Cell Biol.* 157(2): 327–336.

Skaug, M.A., Helland, I., Solvoll, K., Saugstad, O.D. (2001). Presence of ochratoxin A in human milk in relation to dietary intake. *Food Addit. Contam.* 18(4): 321–327.

Smith, D.E., Midgley, J., Allan, M., Connolly, G.M., Gazzard, B.G. (1991). Itraconazole versus ketoconazole in the treatment of oral and esophageal candidiasis in patients infected with HIV. *AIDS* 5: 1367–1371.

Solomon, W.R. (1974). Fungus aerosols arising from cold-mist vaporizers. *J Allergy*; 54:222–228

Stormer, F.C., Hoiby, E.A. (1996). Citrinin, ochratoxin A and iron. Possible implications for their biological function and induction of nephropathy. *Mycopathologia* 134(2): 103–7.

Thuvander, A., Moller, T., Barbieri, H.E., Jansson, A., Salomonsson, A.C., Olsen, M. (2001). Dietary intake of some important mycotoxins by the Swedish population. *Food Addit. Contam.* 18(8): 696–706.

Tiwari, R.P., Singh, G., Vadehra, D.V. (1986). Drug resistance patterns and susceptibility to aflatoxin B1 of strains of *Escherichia coli* and *Staphylococcus aureus*. *J Med. Microbiol.* 22(2): 115–118.

Wolf, C.R., Henderson, C.J. (1998). Use of transgenic animals in understanding molecular mechanisms of toxicity. *J Pharm Pharmacol.*50(6):567–574.

Wyllie, T.D., Morchause, L.G. (1978). Mycotoxin Fungi, Mycotoxins, Mycotoxicoses-An Encyclopedic Handbook.Vols.1, 2, and 3. Marcel Dekker, Inc. New york. pp.392–394.

Yamada, A., Kataoka, T., Nagai, K. (2000). The fungal metabolite gliotoxin: immunosuppressive activity on CTL-mediated cytotoxicity. *Immunol. Lett.* 71(1): 27–32.

Zaichenko, A.M., Nagornaia, S.S., Kirillova, L.M., Ganova, L.A., Kriukova, O.V. (1992). The antibiotic properties of macrocyclic trichothecene mycotoxins. *Mikrobiol. Zh.* 54(3): 34–40.

Zaichenko, O.M., Kyrylova, L.M., Rubezhniak, I.H., Andriienko, O.V. (1997). Comparative characteristics of some cultural and toxigenic properties of the *Dendrodochium* Bonorden and *Myrothecium* Tode ex Fr. representatives. *Mikrobiol. Z.* 59 (3): 33–41.

19

The Impact of the Internet on Mycology

Thomas J. Volk

Department of Biology, 3024 Cowley Hall, University of Wisconsin-La Crosse, La Crosse WI 54601
E-mail:volk.thom@uwlax.edu, web page [http://TomVolkFungi.net]

INTRODUCTION

The Internet has changed the way we do mycology. It has promoted fast and easy communication between researchers through electronic mail and newsgroups. One of the newest tools for reaching a broad range of people and educating them about mycology is the World Wide Web, which can include video and still images as well as text. Many researchers, teachers and amateur mycologists, especially in developing countries, do not have access to books or journals in the field of mycology. Increasing numbers of web sites are available that provide up-to-date information and illustrations from nearly every aspect of mycology. Keys and databases are increasingly available, including such valuable resources as herbarium and nomenclature databases. One problem is the lack of peer review on the internet; anyone with a computer can put up a web site, leading to much misinformation that can be interpreted as fact. This paper will discuss some advantages, problems and the future of the internet as a source of information. Tom Volk's Fungi [http://TomVolkFungi.net], The Mycological Society of America's web page [http://MSAfungi.org], Mushroom Observer [http://mushroom observer. org], and Kathie Hodge's "Mycological Resources on the Internet" [http://mycology. cornell.edu/] will be used as a basis for discussion.

I have had an online internet presence since 1994, when Mike Clayton of the Botany Department at the University of Wisconsin-Madison [http://botit. botany.wisc.edu] convinced me to start scanning in my large collection of fungi slides and put them online. Electronic education technology was quite primitive then, with very few people even having an Internet connection or even a web browser on their computer. I put several hundred pictures online on a Gopher server, and almost no one found them. There were no good search engines at the time, except for the very simple and not-very-effective Archie and its companion

Jughead from the University of Minnesota, so it was only by word of mouth and through email mailing lists that anyone could find it. In 1995, browsers began to become more common, and it became apparent that the Internet was going to be "the next big thing." So I bought myself a book on writing HTML (HyperText Markup Language) and in one afternoon constructed a primitive web page that was not much more than a couple small pictures and a link to my Gopher images. In the first month November 1995 when just 158 hits (viewers) were recorded for that main page, probably mostly from my logging on to admire my work (and to make sure it was still there!). Since then about 1.1 million hits have been recorded on that slowly evolving main page, with dozens of other pages being gradually added as I learned more about HTML and the internet became more sophisticated. The most hits I have had on that main page in one month is 21,000, which is pretty good for a site without any pictures of naked people. I have received email from more than 110 countries on all seven continents where my web page has been viewed.

In those early days without good search engines, many web pages consisted entirely of links to other people's pages, many of which contained only links to other people's pages. Sometimes it was hard to find pages that actually contained any information. Now, we have one important links page in Mycology, Kathie Hodge's Mycological Resources on the Internet [http://mycology.cornell.edu/]. One other problem was that there seemed to be many Internet sites that someone put up and then just forgot about, never updating with new information or new pictures or anything. This still seems to be a problem today. The major advantage of the internet is that it allows for frequent updating, so that information on any page should theoretically be current.

Another important thing I learned early on is that if I wanted people to return to my page, there had to be something new to make them want to return. In January 1997, I started my "Fungus of the Month" pages (FOTM), in which I would choose an appropriate fungus each month to describe and let readers know why it was an interesting fungus. This would force me to update my page once a month. So far there are more than 12 years' worth of FOTM's, with a new one coming every month. I try to relate the FOTM to a holiday (e.g., Christmas, Thanksgiving, US Independence Day, St. Patrick's Day, etc.) or a fungus that is particularly prevalent during that month. I try to make readers aware that each fungus has something interesting about it, some kind of story or human use or place in history or just beauty.

People have asked me, "Why do you put some much time and energy into producing these web pages for free?" My goal and impetus for putting so much work into my web pages is that I really love fungi, and I want everyone else to love them too. There are so many interesting fungi with interesting stories about them. I hope that by making my web pages fun, free and easily accessible, people from all walks of life can learn more about why fungi are important in their lives. My goal is for all biology teachers all over the world to increase their teaching about these important organisms, or in some cases to even just mention the "F" word (Fungi) in their classes. My department, dean and chancellor also recognize the importance of my web pages in promoting mycology and promoting our department and school. I hope that I have been able to make an impact on Mycology from a relatively small school (9000 students) in the middle of nowhere. Outside sources seem to be noticing. Among other awards, in 1999 Science (AAAS) magazine honored my website in its "NetWatch" column. I have also been awarded the lifetime award for "Contibutions to amateur Mycology" from the North American Mycological Association.

Designing Web Pages

Please see some example of web pages in mycology. There are certainly many other great examples.

- Tom Volk's Fungi [http://TomVolkFungi.net]
- The Mycological Society of America's (MSA) web page [http://MSAfungi.org]
- Mushroom observer [http://mushroom observer.org]
- Kathie Hodge's Mycological Resources on the Internet [http:// mycology.cornell.edu]

When designing a web page, think of who your target audience will be. Why are you writing this page? What kind of information do you want to convey? Here are a few reasons I can think of for writing web pages. You may have reasons of your own.

- Education: college, high school, grade school—see [http://botit.botany.wisc.edu/toms_fungi/teachers.html]
- Using web pages in the classroom. Computer projection systems or self-study and enrichment
- Provide information for amateur mycologists (see [http://www.namyco.org] for the website of the North American Mycological Association, NAMA)
- Keys to your favorite groups of fungi
- Questions from readers, get feedback from experts.
- Customized text development
- Research: Literature search for students, researchers
- Mushroom cultivation information
- Promote organization or business
- Provide links to other resources
- Provide images for classroom/research use
- Adjuncts to or substitutes for paper publication; Scientific papers can be put online directly.
- Get people working on the same thing in touch with one another. e.g., Mushroom observer [http://Mushroom Observer.org]. "The purpose of this site is to record observations about mushroom, help people identify mushrooms they aren't familiar with, and expand the community around the scientific exploration of mushroom." See also Jon Palmers, Chestnut Fungi [http://Chestnut Fungi.com], a database of chesnut my corrhizae, including pictures and BLAST search capability for DNA sequences.

Databases

Online databases ensure wide availability, free up room on your hard disk, and can be easily kept up-to-date. User inputs and error corrections are facilitated. Hard copies of databases are usually out of date as soon as they are printed. Moreover, printed copies are not easily searchable.

1. "Literature: library searches, and individual researchers' web pages.
2. Herbarium and culture collections databases e.g., Systematic Botany and Mycology, National Fungus Collection, Beltsville, MD Fungal Databases [http://nT.ars-grin.gov/fungaldatabases/] CABI Biosciences database [http://www.speciesfungorum.org/], or see an entire list of databases at [http://mycology.cornell.edu/fcollect.html].
3. Mushroom observer.org is attempting to do this on a world wide basis.
4. Mycoflora (or mycota) of an area. These inventories are important for cataloguing biodiversity.
5. Images of fungi. Image databases allow internet-wide searches for illustrations, e.g., MSA slide collection compiled by Don Ruch at [http://www.bsu.edu/classes/ruch/msa/description.html]
6. Keys and identification resources
   a. Synoptic keys are easily implemented using database interfaces.
   b. Dichotomous keys can be easily constructed (e.g., see this key to North, American *Armillaria* species at [http://botit.botany.wisc.edu/toms_fungi/armkey.html] and the more sophisticated site by Yves Lemoreux and Jean Desprès, *Tricholoma* of Québec at [http://www.mycomontreal.qc.ca/tricho/home.htm]).

Web Page Development

Once you've decided what you would like to write about, you'll have to create your web pages. When I first started with my web pages, the only way to write them was to do it manually by writing your own code in HTML (HyperText Markup Language). It is really not very complicated, but it can be difficult to get the pages to look exactly as you would like them to look all the time on all sorts of browsers. To view the HTML code for any web page select "View Page Source" or "View Source" in the pull-down menu of your browser. Things have become much easier now that there are a number of WYSIWYG (pronounced Wissy-wig— What You See Is What You Get) editors for producing web pages. There are too many to mention here, and I have no favorites, since I continue to write my own HTML code from scratch. I find that it gives me much more flexibility, and I can do whatever I want. Animated GIF files can add some splash to your pages with minimal tinkering. MapEdit can add "clickability" to your pictures for site navigation. To make more complex pages, with all the bells and whistles and popups, you'll have to learn more complicated programs such as JAVA, JAVA script and Flash. However, making a simple HTML page with WYSIWYG editors is very easy nowadays. My students make their own web pages at [http://Multipleorganisms.net]

As a favor to your audience, think about how they will be accessing your pages. Bottlenecks in data transfer can be caused by your Internet connection, your computer, the server computer, and the number of people accessing the server. Web pages heavy on graphics and special effects are slow to load. Who'd like to read a book if it takes 3 minutes to turn the page? Keep your user in mind when you design your pages. Many web designers with high speed Ethernet, T1 or DSL connections forget about their end users and place large files (images, music, or java) on their web pages that may take several minutes to download over phone line connections. Remember, much of the world is still operating on slow modems. Internet users in some countries actually have to pay a certain charge per byte that comes across the line. Many internet users are charged per time online by their Internet Service Provider (ISP). Also remember that not all WWW surfers are using the same web browser software. Did you know that there are more than 500 other browsers besides MS Internet Explorer and Netscape? At a minimum, your web pages should work on these two major browser types, but you should test them on as many others as possible.

Choose a Domain Name

In the olden days, way back in the late 1990s, it was rather expensive to have your own domain, the part of the address (URL, or Uniform Resource Listing) that someone types in to find and view your page. These domain names are short, ending in .com, .net, .edu, or .gov in the United States, various others in other parts of the world. Since it was so expensive to set up your own domain name (up to USD$50 per year), everyone had these long, complicated URLs in other domains that no one wanted to repeat, much less type in. When more companies began offering redirection services to websites and new domain names, the price has decreased quite a bit. You can now reserve and keep your domain name for around USD$2-10 per year, or even less. It is much easier for potential readers to type in TomVolkFungi.net than to type in [http://botit.botany.wisc.edu/toms_fungi/]. It was even longer previously! Note that you must still have a computer or computer space as a host server for your pages. Most US universities (and probably some government and industries) offer their employees, faculty and staff a reasonable amount of space for web pages. In many parts of the world server computers have greatly decreased in price in the past few years, so it can be practical to buy your own computer to use as a server. The server I use at [http://botit.botany.wisc.edu] is a reasonably priced Macintosh computer. You can read about how to configure this relatively inexpensive computer to be used as a server at [http://botit.botany.wisc.edu/server_setup/].

One Caution

Once you have a domain name, be sure to protect it. Pay close attention to when your registration will expire, and pay that bill to your DNS (Domain Name Server) corporation right away. There are other "companies" that specialize in taking over domain names once they've expired (sometimes just minutes after they expire), then blackmailing the former domain owners into paying big bucks for their domain names back by putting up a porno site at that domain. This has happened to the old American Mushroom Institute domain name, as well as some friends who run a mushroom company. I have heard of this from plenty of other sites. When I realized this could happen, I immediately registered my domain for an additional 5 years. You can renew for up to 10 years.

Critical Issues

Information quality

Users must be alert to the potential unreliability of Internet information sources. Internet sites are seldom subject to peer review. Anyone with a computer can construct and publish a website on anything! Mushroom photographs may be misidentified. Unscrupulous webmasters could (and do) provide misleading or false information to promote a product or service. There are examples of sites that are put up because the author could not get his (flawed) information published in a peer-reviewed journal. You should always question the source for the information. Is it someone you trust, or from an institution you trust to provide accurate information? I do not claim that my site is peer reviewed, but in fact I receive corrective email as soon as I publish the smallest mistake, one time (correctly) from a 10-year-old home-schooled boy!

There has been some call and petition to review websites. One such site for all of Biology is MERLOT, Multimedia Educational Resource for Learning and Online Teaching, at [http://www.merlot.org/]. According to their website blurb, "MERLOT is a free and open resource designed primarily for faculty and students of higher education. Links to online learning materials are collected here along with annotations such as peer reviews and assignments. MERLOT is also a community of people who are involved in education. Community members help MERLOT grow by contributing materials and adding assignments and comments. Many community members make their professional information available in MERLOT's member directory." My website has been favorably reviewed at MERLOT by their committee. They also invite reviews from interested internet users. There is no specific sanctioning or review body for Mycology websites. Maybe it's time to come up with one?

Finding Information

By its very nature, the Internet has no "center." So there's no one directory to all the information that exists out there. The web is growing so fast that no one search engine can possibly include all internet sites. Everyone's favorite at this point is [http://google.com], but you might want to also try multiple search engines, or let someone else narrow down the search by using a links library, such as Kathie Hodge's WWW Virtual Library of Mycology at [http://mycology. cornell.edu].

Computer Viruses

Everyone should know from personal experience to be diligent in scanning for viruses in every file you download. You should periodically and frequently update your antiviral software—new viruses are constantly being devised and discovered. Viruses can only be transmitted in an executable file and are only harmful if the file is executed. Be especially careful of Word Processing Macro viruses. Macros are executable files! Better yet, ask your email server gurus to check and filter your email for viruses. The email server at my university screens each incoming (and outgoing?) email for viruses. If they find one, it is sequestered and rendered harmless. The rationale for providing this filtering service is that it saves our computer staff LOTS of time reconfiguring our computers and trying to save our corrupted files and drives. I am still very careful with files I have received from someone I don't know, and I NEVER open an executable file from ANYONE. Occasional glitches in our email filtering system have been known to happen, but they fortunately have been very rare. So far no one has been able to put a virus into a java or javascript web page—and Macintosh based viruses are virtually unknown. But, I suppose those are coming someday. Internet terrorists will always find a way around whatever protection we can devise.

E-mail and Mycology

Remember when e-mail was the entire Internet? Times change, but e-mail is still the most useful of Internet protocols. Lightning fast, relatively secure, widespread, inexpensive or free for most academic users, and intermediate in formality between a phone call and a letter, e-mail has found a place in our daily routines. I spend the first 20 minute to a half hour each day going through my e-mail, seeing if there's anything interesting there. I might spend an hour or two answering e-mail on a particularly hectic day. Today, with the right software, we can participate in e-mail discussion groups, automatically filter incoming e-mail into topic-specific mailboxes, send formatted text, and send files and applications as e-mail attachments. The downside? Spam- unsolicited junk mail, often in large quantities. See below for some spam coping mechanisms.

Many scientific papers and grants have now been written together by people who have never met in person. It's so easy to transfer files that it is possible to collaborate with a person halfway around the world. People can even collaborate from a distance on their daily scientific activities. The internet has begun to even out the playing field for researchers from relatively poor areas. It is now possible to achieve worldwide dispersal of your scientific data and ideas with just a few mouse clicks.

If e-mail is not quick enough for you, there are also electronic chat boards, where you can talk live with other people interested in the same topics. There is also a wide range of Instant Messenger software to chat live with someone. Many of these include an audio portion as well, so you don't even have to exercise your fingers to chat. Webcams add the special advantage of being able to see the person as you talk with them. It's certainly an improvement to be able to see someone's face to be able to gauge their reaction what you've just told him or her, or to understand what someone means by what she's already telling you. The picture and voice together make it almost as if you're together in the same room. Watch out, telephone long distance companies!

Quotes from Outrageous E-mail Requests

A major and fun benefit of putting up a web page is the great deal of email you receive from all over the world. One minor downside of having a web page is the great deal of e-mail you receive from all over the world— asking for help. I answer an average of 100–200 requests per month for information from all over the world, from grade school, high school, college and graduate students, researchers, amateur mycologists and mushroom hunters, gardeners, doctors, interested internet surfers, and other categories too strange to classify. Here are a few of the more interesting requests:

******

To whoever it may concern,

I need help with my Science Project. If you have any information on mycology please send it to me. Plus do you have any pictures of mycology please send it to me.

****

Aloha,

I recently had a fungus known as Fusarium growing on my armpit hair. This was determined by a culture. My doctor seems to think it is harmless. If you might know of this fungus could you let me know what I might do to stop it from growing on my hair. My doctor recommended just shaving it off. I did a number of times but it just keeps coming back.

****

Hey Tom, Thanks for the info. In putting the cart before the horse fashion, I did eat some of the false morels already last Sunday. And as you may have guessed without the boiling procedure you mention. Good news is I am here to talk about it I guess. If there were no immediate affects, do I fear some yet to come? The person I found them with said he had eaten them before so I thought "what the hell.

****

From an apparently disgruntled high school student (not one of mine):

"i REALLY DON'T LIKE FUNGUS.

i REALLY DON'T LIKE FUNGUS.

i REALLY DON'T LIKE FUNGUS."

(this was the entire text of the email)

****

From a high school student:

"I need to know What is Homeostasis? Could you include the folowing: Plasma membrane, Cell Transport, diffusion vs. Osmosis, Active vs. Passive Transport, and ATP and the cell? Also how do cells grow and reproduce? This is the last one What is the difference between a plant and an animal cell? I would apreciate it if you could get this to me tomorrow morning? If you can't could you get it to me by tomorrow night?"

*****

I am writing on behalf of Tom B***, keeper of the wolves, large cats, etc. at the Phoenix Zoo. Mr. B*** expressed concern over large fruiting bodies that are colonizing or appearing on wolf feces. These fungi also appear on sites of hosed down runoff from the meat-cutting lab (meat for animal consumption). The appearance of this fungus may be cause for concern because of the danger of exposure to human wolf feces and the large number of pathogens that are present and harmful to humans. The colonization is, of course, confined and zoo patrons are not directly exposed (i.e., eating the feces), however, it would be helpful to know the identity of the fungus. Can you assist Mr. B***? I have a sample in my lab that I can send you....

****

I was recently disassembling a helicopter transmission from Guatemala that had been sitting for 20 years. I came across what appeared to be a fungus on some of the internal parts. This "fungus" only attacked steel parts that were plated with cadmium, completely stripping all traces of the cadmium. It was grey in colour and for lack of a better description, fuzzy. Any information that you have to my query would be greatly appreciated.

*****

Dr. Volk: Thank you so much!! Wow! there's no one I'd like to talk to more (except maybe Paul McCartney)!

Sorry for keeping you waiting so long, I've been having some trouble with the Internet...

I live in Phoenix, AZ and, sadly, do not see many mushrooms here. Occasionally several will crop up in the park after it rains, but my resident city is not noted for its humidity. Thus, I do most of my mushroom sighting in textbooks and field guides. However, I keep a very optimistic view of mushrooms, hoping that someday I will be able to see live specimens with more regularity.

Again, thank you very much! It may be a little while before I can send you my questions, because I have to find out what things I don't know...

*****

I am an ER physician in an Air Force Hospital in Japan. Last night I admitted a lady to the hospital who had become ill after ingesting a mushroom picked on our base that was identified by a local "expert" as being moderately toxic. The name they gave was "Tengudake". I found a list of Japanese names for various mushrooms along with their scientific names but was unable to match this one. I have two pictures attached, and I was hoping you could help me identify it and obtain some information about its toxicology.

****

We found your website while beginning to investigate ways of identifying some fungi here at the Philadelphia Zoo, and were hoping you might be able to help us. Here is our situation: We have two large dead red oaks (lying down, not upright) in our outdoor gorilla exhibit, which the gorillas use for climbing, lying under, etc. The oaks have been dead for about two years and have had an increasing density and variety of fungi growing on them. We've had a wet summer, and have seen a big increase in the past few weeks. We are concerned that the gorillas might consume a toxic species.

I've attached photos of the various types we could distinguish. Most are growing in full sun or close to full sun conditions, if that is of any help in IDing them. One of the attached photos (10038) shows the "habitat". If you have time to look at these photos, I would greatly appreciate it. I guess we're not so much interested in specific ID's (although that would be great) as much as knowing whether any of these are or resemble toxic species likely to be found in this area and under these conditions.

****

I live in Dallas and have planted a bed of purple verbena—4"inch plants. Friday morning I was going to work and saw one plant completely covered in bright –almost neon–yellow stuff. Looking closer, it appeared as if someone had sprayed that canned Styrofoam-filler all over a single plant—only a few tips were visible. Some of it was kind of falling off one of the small branches and there was translucent slime underneath, across some of the now-exposed leaves (I considered that maybe a dog had vomited on the plant as a possibility because it looked so mucus-y). When I arrived home, the yellow had turned darker - golden brown - and when I stuck it with a stick, there was soil-like filler inside. I wish I had taken the time to take a photograph the stages.

Can you tell me what this is? It has killed the original plant and is sporadically moving through my garden—one plant at a time.

****

I read the answer to your question about killing mushrooms in your yard. I just spent $300 to take my 10-week old beagle puppy to the emergency animal clinic because she ate a mushroom growing in our yard. Is there some way to kill the mushrooms? Until we got the puppy, they were fine. But I like the puppy a lot more than the mushrooms.

*****

My name is P***, (of Stockholm, Sweden (for your map)). My compliments on your fungus site! I went on the net just a few moments ago, and through yahoo I ended up here. Currently doing a special project for school (all Swedish seniors in high school have to do one) on ulcerative mycosis in the platypus (*Ornithorhynchus anatinus*) caused by *Mucor amphibiorum*, I wanted to find out more about the basics of mycology to be more confident in my discussion of this fungus and its modus operandi in causing disease in the platypus. I read your Introduction to the Kingdom Fungi, and though most is not relevant to my particular fungus, I still found it quite useful. The format provides a very good introduction for people who know little about fungi (and even biology in general).

*****

Hi! I am a doctor in a big hospital in Athens, Greece, affiliated to the Medical School of Athens and my field is Infectious Diseases with a special interest in fungal infections. Right now I am trying to find some help on the web, so maybe you could help me. We have an AIDS patient with cryptococcal meningitis. He has received amphotericin B and 5FC and then fluconazole. However, the India ink stains of his CSF still show Cryptococcus although his cultures are negative . His clinical condition is good and we have also given him six intrathecal infusions of amphotericin. We have been treating him for 3 months but the infection is still there.

Could you perhaps have some useful suggestions (maybe another link)? Thank you,

*****

Tom,

Is it real? Or is it a University 'decoration' of concrete or the like? or webbed up phony?

I am the parent of a 12-year old boy who showed me your web page. He now believes mushrooms can grow just like the one you are posing next to on your home page. Is it real? or is it concrete?

I'd appreciate a response. But I'll understand if you fail to reply. Just write 'No it is fake' or 'Yes it is real'

*****

THIS ORANGE COLORED LICHEN IS THE NUCLEAR WASTE COLLECTOR FROM THE SPLEEN OF MOTHER EARTH. THESE ORGANISMS ARE THE DETECTIVES OF MOTHER-EARTH'S POLICE FORCE. THEY CAN BE LIKENED TO THE WHITE CELLS WITHIN THE PHYSICAL EMBODIMENT THAT INITIATE A DEFENSIVE SEARCH & RESCUE SIGNAL FOR THE DISEASEMENTS THERIN.

HELP, I AM NOT A MYCOLOGIST NOR A BOTANIST. I AM A PSYCHIC! IN ONE OF OUR REMOTE VIEWING SESSIONS, THIS INFORMA-

TION WAS GIVEN. I HADN'T THE SLIGHTEST IDEA AS TO WHAT TO DO WITH THIS INFO, NOR DID I KNOW ANYTHING ABOUT "TAMARINDS", AS THIS LICHEN WAS ANCIENTLY KNOWN.

I SIMPLY TYPED UP THE INFO AND SENT IT OUT TO THE 800 PEOPLE ON OUR INTERNATIONAL NETWORK IN APRIL OF 1997. I HOPED THAT SOMEONE WITH SUCH KNOWLEDGE WOULD RESEARCH AND EXPAND UPON IT.

THIS MORNING I RECEIVED AN E-MAIL LETTER FROM THAILAND, INVITING ME TO ATTEND A BIO-TECH MYCOLOGY CONFERENCE.

NOT ME! I AM MERELY A FACILITATOR FOR MY HIGHER GUIDES TO EXPAND UPON THEIR MISSION. SO NOW, WHAT AM I SUPPOSED TO DO?

IF YOU WILL E-MAIL ME YOUR FAX NUMBER, I WILL FORWARD THE CONTENT OF THIS ORIGINAL THESIS. PLEASE DO NOT DISBELIEVE THAT WHICH YOU PERSONALLY CANNOT DISPROVE. ALL THINGS ARE POSSIBLE AND IN THIS INSTANCE MY PERCEPTIONS HAVE INSPIRED ME TO BELIEVE IN THE PROBABILITIES.

SHARING LIGHT, L**

****

Dealing with Spam

Spam is unsolicited commercial e-mail. Because spammers don't like to listen to us complain, they usually disguise their return address and use a subject line designed to pique our attention. Spammers use robots to pick up e-mail addresses from USENET newsgroups, web pages, and other spammers. Spam can sometimes accumulate massively in your e-mail inbox. How can you cope with this intrusion?

1. Protect your e-mail address. Disguise it when posting to newsgroups; be careful about posting it to the web.
2. Filter spam. Devoted spam-despisers provide tips on constructing filters that automatically remove spam from your inbox. This function allows you to filter out keywords in subject lines (such as free, mortgage, printer cartridges, insurance, smut, sex, etc.), e-mail from a particular person or even from an entire domain. However, many spammers have found ways to forge their "headers," so that what you read on the subject line and what you read in the domain is not true and accurate, and thus unfilterable—so far.
3. Don't respond! NEVER!! It doesn't do any good to ask to be taken off the mailing list. Your reply merely indicates that you exist, and is an open invitation for more spam. Some companies exist to merely to sell e-mail addresses to spammers.
4. Report the spammer to their Internet Service Provider (ISP). Usually this does no good, but certain ISP's, (especially large ones) will act on preventing their subscribers from sending spam.
5. Chase the spammer. OK, this is difficult and ineffective, but potentially satisfying. There ARE ways of tracking them down.

The Future of Mycology on the Internet

Electronic publishing—the future is now

Who would have thought it would happen this fast? Articles in electronic journals can be in color, and include pictures, graphs, and lots of accessory data. They can be published quickly and inexpensively. Already some authors refer to a web page for an illustration of new species, or provide online versions of their articles, like this Mycotaxon paper on *Bridgeoporus nobilissimus* by Burdsall, Volk, and Ammirati [http://botit.botany.wisc.edu/toms_fungi/bridgeop.html] and this paper on *Armillaria nabsnona* by Volk, Burdsall, and Banik at [http://botit.botany.wisc.edu/toms_fungi/nabsnona.html]. However, the International Code of Botanical Nomenclature does not (yet?) consider internet publication valid for nomenclatural purposes!

An electronic journal would be more easily available, and can be searched and indexed much more effectively than paper publications. Articles would come out much more quickly after review, and overall costs would be lower. Problem areas include copyright and password protection—why subscribe to a journal if you can get it free online? Why should a library subscribe? There are many considerations that have yet to be worked out.

PDF format (Portable Document File, available from [http://www.Adobe.com], is flexible, indexable, and readable on any computer platform. A picture of the document, not the actual text of the document is online. As a good example of a large collection of PDF documents, all Forest Products Laboratory (Madison, Wisconsin) publications are available online at [http://www.fpl.fs.fed.us/]. Many mycological researchers have made their research papers available online as PDF articles. These are especially important for those of us whose libraries may not be able to afford to carry all of the journals that contain all of these papers. It saves time (on both ends) in asking for paper interlibrary loan or reprints from others.

Mycologia, the official journal of the Mycological Society of America, is now available online [http://www.mycologia.org/]. At first anyone will be able to read it. Once electronic publishing becomes more accepted and people get used to reading journal articles online, this journal will convert to a password protected, members- and subscribers-only site. Mycologia has chosen to use a true HTML format, along with PDF. This allows you to click within the article to bring you to another point in the article. This is especially useful for references in the text, which, upon clicking, bring you to the full citation. Eventually, those full citations would be clickable to lead you to the actually article being referenced. Some of these are already linked to the full article or at least to an abstract on PubMed. These online articles also allow for full color pictures and for enlargement of tables and charts to a very large even-more-readable size. Remember all those cladistic trees you had trouble reading? Now you can enlarge them to any size you want, and even zoom in on them. Links to GenBank for sequences can easily be included. The possibilities are endless. Inoculum, the newsletter of the MSA has been available online for the past several years, first as a text file, then as an HTML file, and now as a PDF file.

One Mycology journal, Myconet, began entirely online at [http://www.fiddmuseum.org/myconet]. According to their website, "MYCONET is an international mycological journal intended for the documentation of selected mycosystematic files published on the Internet. New files accepted for publication will simultaneously be published on the Internet and as hard copies in MYCONET. New issues will also be printed when files have been extensively updated on the Internet. New taxa are not validly published on the Internet, but are when published in the printed version.... One rationale for publishing this new journal on the Internet is that papers can be updated continuously and published as hard copies at intervals. Another one is that mailing costs have increased considerably in recent years. Anyone who needs the papers in MYCONET can easily print hard copies on a laser printer, and they are free of charge."

Another similar journal "North American Mycology" has been online for several years, beginning as the "Pacific North west Fungi" and now expanded to include the whole continent. "A feer reviewed online journal for information on fungal natural history in North America, including taxonomy, nomenclature, ecology and biogeography."

My Recommendations for the future

- Design user-friendly pages to describe your work or for teaching about your favorite subject
- Use and develop internet keys and identification guides
- Participate in collaborative online efforts like AnaTeleo, the Anamorph/Holomorph Database; [http://www.cbs.knaw.nl/cbs_home/cbs_home_html?-http://www.cbs.knaw.nl./databases/anateleo.htm~main]
- Interact in online discussion groups like mushroom observer or chat groups or whatever you can find.

Most importantly, don't let your computer gather dust! Get Online!

All references are from the World Wide Web, and links are provided with the text above. This paper will likely be online at [http://TomVolkFungi.net], with all the hyperlinks intact and clickable.

Index